Ergebnisse der Mathematik und ihrer Grenzgebiete — Volume 47

3. Folge

A Series of Modern Surveys in Mathematics

Editorial Board
M. Gromov, Bures-sur-Yvette J. Jost, Leipzig J. Kollár, Princeton
H.W. Lenstra, Jr., Leiden J. Tits, Paris D. B. Zagier, Bonn/Paris
G. M. Ziegler, Berlin

Managing Editor R. Remmert, Münster

Mark Adler
Pierre van Moerbeke
Pol Vanhaecke

Algebraic Integrability, Painlevé Geometry and Lie Algebras

Springer

Mark Adler
Pierre van Moerbeke
Department of Mathematics
Brandeis University
Waltham, MA 02254
USA
adler@brandeis.edu
vanmoerbeke@brandeis.edu

Pierre van Moerbeke
Department of Mathematics
University of Louvain
1348 Louvain-la-Neuve
Belgium
vanmoerbeke@math.ucl.ac.be

Pol Vanhaecke
Université de Poitiers
Laboratoire de Mathématiques
et Applications
UMR 6086 du C.N.R.S.
SP2MI Téléport 2
Boulevard Pierre et Marie Curie
BP 30179
86962 Futuroscope
France
pol.vanhaecke@math.univ-poitiers.fr

Library of Congress Control Number: 2004110298

Mathematics Subject Classification (2000) : 70Gxx, 37Jxx, 14Kxx, 17Bxx

ISSN 0071-1136
ISBN 3-540-22470-x Springer Berlin Heidelberg New York

Springer Berlin Heidelberg New York
is a part of Springer Science+Business Media

Printed in Germany

Cover design: *Hans Kirchner, Heidelberg*
Typesetting: Computer to film by author´s data
Drawings by the authors

Printed on acid-free paper 41/3142XT 5 4 3 2 1 0

To

Edith, Bernadette and Lieve

Their indomitable spirits
encouraged us to persevere

Preface

In the early 70's and 80's the field of integrable systems was in its prime youth: results and ideas were mushrooming all over the world. It was during the roaring 70's and 80's that a first version of the book was born, based on our research and on lectures which each of us had given. We owe many ideas to our colleagues Teruhisa Matsusaka and David Mumford, and to our inspiring graduate students (Constantin Bechlivanidis, Luc Haine, Ahmed Lesfari, Andrew McDaniel, Luis Piovan and Pol Vanhaecke).

As it stood, our first version lacked rigor and precision, was rough, disconnected and incomplete... In the early 90's new problems appeared on the horizon and the project came to a complete standstill, ultimately confined to a floppy. A few years ago, under the impulse of Pol Vanhaecke, the project was revived and gained real momentum due to his insight, vision and determination. The leap from the old to the new version is gigantic.

The book is designed as a teaching textbook and is aimed at a wide readership of mathematicians and physicists, graduate students and professionals. The main thrust of the book is to show how algebraic geometry, Lie theory and Painlevé analysis can be used to explicitly solve integrable differential equations and to construct the algebraic tori on which they linearize; at the same time, it is a play ground for the student in applying algebraic geometry and Lie theory. The book is meant to be reasonably self-contained and presents numerous examples. The latter appear throughout the text to illustrate the ideas and make up the core of the last part of the book, acting as a final movement to unite the various themes of this volume. The book contains the basic tools from Lie groups, algebraic and differential geometry needed to understand its main scope. We do not claim to cover the whole subject: we have developed a certain point of view, which we hope is sufficiently rich, and passed in silence over many other fascinating aspects of integrable geometry.

Our thanks and long time appreciation go to Suzanne D'Addato who, at an early stage, transformed a messy handwritten manuscript into the first version of the book. We also thank Antony Phan for kindly providing pictures for the Dynkin diagrams.

Mark Adler and Pierre van Moerbeke

Table of Contents

Part II Algebraic Completely Integrable Systems

Part III Examples

1 Introduction

In the late 60's and 70's a remarkable renaissance occurred around an equation, discovered in 1895 by Korteweg and de Vries, describing the evolution over time of a shallow water wave. This equation has its roots in Scott Russel's horseback journey along the Edinburgh to Glasgow canal; he followed a wave created by the prow of a boat, which ***stubbornly*** refused to change its shape over miles. This revival in the 60's was driven by a discovery of Kruskal and coworkers: the scattering data for the one-dimensional Schrödinger operator, with potential given by the solution of the KdV equation, moves in a remarkably simple way over time, while the spectrum is ***stubbornly*** preserved in time. This led to a Lax pair representation involving a fractional power of the Schrödinger operator; it ties in with later developments around coadjoint orbits in the algebra of pseudo-differential operators. Very soon it was realized that this isolated example of a "soliton equation" had many striking properties, leading to an explosion of ideas, following each other at a rapid pace.

The KdV equation is a Hamiltonian system with regard to a "symplectic structure", but is also Hamiltonian with regard to another "compatible symplectic structure", turning KdV into a bi-Hamiltonian system, which has an infinite number of constants of motion, all in involution. That is to say the KdV equation is part of an infinite hierarchy of commuting non-linear PDE's.

Besides the soliton and scattering solutions, other important solutions of KdV emerged, namely rational and algebro-geometrical solutions. The solutions in terms of theta functions established the fundamental link with curve theory. This was the royal road to the infinite-dimensional Grassmannian description of the KdV-solutions, leading to the fundamental concept of τ-function, which enjoys Plücker relations and bilinear identities. The τ-function is a far reaching generalization of classical theta functions and is a unifying theme in mathematics: representation theory, curve theory, symmetric function theory, random matrix theory, the theory of orthogonal polynomials and Painlevé theory all live under the same hat!

Very soon vast generalizations of the KdV equation appeared on the horizon, namely the KP and AKNS hierarchies, to name a few, which enjoyed many of the same features. These facts led to an age of enlightenment, where universal principles began to emerge beyond the special cases and overshadowed many of the previous investigations.

The ideas, reaped for infinite-dimensional systems, began to shed some new light on finite-dimensional systems, old and new. The old include ancient systems initiated by the giants of the 18th and 19th centuries, like Euler, Jacobi, Lagrange and Kowalewski; more recent ones include the Toda lattices. Classically, it was stated as a general principle by Noether that symmetries of mechanical systems produce constants of motion. Much later it was discovered that the splitting of a Lie algebra gives rise to functions with sufficient invariance properties and which are in involution with regard to a symplectic structure associated with coadjoint orbits. The corresponding Hamiltonian systems have Lax pair representations and are solved by a group factorization procedure; its analytic implementation often involves Riemann-Hilbert methods.

It was realized that the Lax pairs involve matrices depending on an extra parameter, pointing the way to Kac-Moody Lie algebras. In their quest for Lax pairs, researchers found more and more intricate Kac-Moody Lie algebras, some twisted and some related to the exceptional Lie algebras. The extra parameter in the Lax matrix has the advantage that its characteristic polynomial is an algebraic curve. Then comes the remarkable fact: the equations can be linearized on the Jacobian of that curve, at least under mild and practically verifiable conditions on the Lax pair.

All this knowledge, unknown classically, shed further light on a celebrated paper of Kowalewski, which stated that under a reasonable working hypothesis the only integrable rigid body motions are (*i*) the free motion of Euler about a fixed point, without gravity, (*ii*) the Lagrange top and (*iii*) a somewhat esoteric "top", discovered by Kowalewski. Kowalewski's top was integrable for no obvious reason: the configurational symmetry was not enough to do the job using Noether's Theorem! The symmetry resides at a higher level than the configuration space and was later explained by the splitting theorem; however, more to the point, the working hypothesis employed by Kowalewski is that if a system is completely integrable, the generic solutions to the system must at some moment blow up in time and instantly be finite again. "Generic" means that these solutions depend on a number of free parameters, besides time, equal to the dimension of the phase space minus one.

This can be spelled out as follows: a system, integrable in the sense of Liouville, has many constants of motion in involution with regard to the Poisson structure; this implies many commuting flows, depending on *real* times $t_1, t_2, \ldots$ and leading, in a compact situation, to real tori.

In addition, making the times t_i *complex* leads, in favorable cases, to complex tori, on which the phase variables are meromorphic functions.

In other terms, the blowup locus of the solutions gives rise to divisors which, glued onto the natural complexified phase space, turn the invariant manifolds into complex algebraic tori. These complex tori are identical to or closely related to the Jacobians obtained by means of the Lax pairs.

The Laurent solutions which are easily computed from the differential equations enable one to compute the explicit functions which embed the complex tori into projective space; thus these tori are Abelian varieties. This technology provides a powerful tool to explore their algebraic features, such as finding their periods and answering questions on how they relate to algebraic curves. Are they Jacobians, Prym varieties, or other natural objects?

Meanwhile the field has reached maturity; by now it has invaded nearly all of mathematics and has had a definite impact on whole new fields, like topological field theory and matrix models, combinatorics and number theory, etc... This general field goes under the somewhat bizarre name of "integrable mathematics".

The book is divided into three parts. Part I deals with Liouville integrability, in a real or complex setting, and with basic differential geometry, including the theory of Poisson manifolds, Lie groups and Lie algebras, finite and infinite-dimensional. Besides classical integrability theory, it contains contemporary integrability theorems, which have been discovered in the last few decades.

Part II gives the theoretical foundations of algebraic integrability. This part contains an outline of algebraic tools for non-specialists which will be employed throughout the book, like Abelian, Jacobi and Prym varieties. Exploring algebraic integrability requires tools like Laurent expansions, in order to understand the embedding of the complexified invariant manifolds in projective space and their algebraic nature.

To show that a Hamiltonian system linearizes on an Abelian variety, one may *either* construct a Lax representation of the differential equation depending on an extra-parameter and linearize on the Jacobian of the curve specified by its characteristic equation, *or* one may complete the complexified invariant manifolds by using the Laurent solutions of the differential equations. The latter method allows us in addition to identify the nature of the invariant manifolds and of the solutions of the system: in most examples the isospectral manifolds and the invariant manifolds are *different.*

Part III deals with three sets of examples which are analyzed using the tools developed in parts I and II. Among a class of left-invariant metrics on $\mathbf{SO}(4)$, our search for algebraic integrable systems leads us to three metrics.

Except for the third metric, which is new, the two others appeared previously in the context of rigid body motions in fluids, as investigated by Clebsch and Lyapounov-Steklov. For each of these metrics there is an extensive discussion of their moduli, the algebraic geometry of the invariant tori and their connection with Kac-Moody Lie algebras.

A second set of examples addresses the classification of integrable lattices among a natural class of Toda-like lattices; this turns out to be equivalent to the classification of twisted affine Lie algebras. The latter amounts to finding all the "outer" automorphisms of semi-simple Lie algebras. The blow-up locus for this system consists of several codimension one subvarieties (divisors on the algebraic tori), each associated with one point of the Dynkin diagram of the twisted Lie algebra. In fact, their intersection patterns are totally governed by the associated Dynkin subdiagrams.

A last set of examples deals with the classification of rigid body motions about a fixed point under the influence of gravity; to wit, the Euler, Lagrange and Kowalewski top. For some special initial conditions, other tops have arisen, whose solutions linearize on Abelian varieties, such as the Goryachev-Chaplygin top, which turns out to be related to a Toda system.

Mark Adler, Pierre van Moerbeke and Pol Vanhaecke

Part I

Liouville Integrable Systems

2 Lie Algebras

In this chapter we fix our conventions and terminology and we provide a quick review of the notions in differential geometry and in Lie theory that will be used. Since algebraic geometry, mainly the geometry of Abelian varieties, will only show up later and since we will need to do in that case a little more than just a review, we defer that subject to Part II of the book.

2.1 Structures on Manifolds

Our manifolds will always be either real smooth or complex holomorphic. In both cases the algebra of functions on such a manifold M will be denoted by $\mathcal{F}(M)$. Thus $\mathcal{F}(M)$ is the algebra of smooth functions on M when M is a real manifold while $\mathcal{F}(M)$ is the algebra of holomorphic functions on M when M is a complex manifold; when $M = \mathbf{C}^n$ or a smooth affine variety we will often restrict ourselves to the polynomial functions on M (usually called *regular functions* on M). Since many of the basic definitions and constructions that are given below are algebraic they apply to complex (algebraic) manifolds as well as real manifolds, and we will just write "Let M be a manifold" when our definition or construction applies to the real as well as to the complex case. Similarly, the word "map" will stand for "smooth map" (resp. "holomorphic map" or "regular map") in the case of smooth manifolds (resp. holomorphic manifolds or (non-singular) algebraic varieties).

2.1.1 Vector Fields and 1-Forms

For a manifold M and a point $m \in M$ the (real or holomorphic) *tangent space* to M at m is denoted by T_mM and its dual space, the *cotangent space* to M at m, is denoted by T^*_mM. The tangent and cotangent spaces to M form the fibers of the *tangent bundle* TM, resp. the *cotangent bundle* T^*M. A *vector field* $\mathcal{V}$ is a section of the tangent bundle while a *1-form* ω is a section of the cotangent bundle; the values of $\mathcal{V}$ and ω at $m \in M$ are simply denoted by $\mathcal{V}(m)$ and $\omega(m)$, where $\mathcal{V}(m) \in T_mM$ and $\omega(m) \in T^*_mM$. The $\mathcal{F}(M)$-modules of vector fields and 1-forms on M will be denoted by $\mathfrak{X}(M)$ and $\Omega(M)$. We will find it convenient to denote the pairing between a vector space and its dual, such as T_mM and T^*_mM, by $\langle\cdot\,,\cdot\rangle$.

For example, if $\mathcal{V} \in \mathfrak{X}(M)$ and $\omega \in \Omega(M)$ then we may define a function $\omega(\mathcal{V}) \in \mathcal{F}(M)$ by setting

$$\omega(\mathcal{V})(m) := \langle \omega(m), \mathcal{V}(m) \rangle \tag{2.1}$$

for all $m \in M$. To a function $F \in \mathcal{F}(M)$ we may associate its differential $\mathrm{d}F \in \Omega(M)$, which is a 1-form, hence can be applied to vector fields on M. This is used to associate to every vector field $\mathcal{V}$ on M a *derivation* on $\mathcal{F}(M)$: for $F \in \mathcal{F}(M)$ we define $\mathcal{V}[F] \in \mathcal{F}(M)$ by

$$\mathcal{V}[F] := \mathrm{d}F(\mathcal{V}), \tag{2.2}$$

which means in view of (2.1) that

$$\mathcal{V}[F](m) = \langle \mathrm{d}F(m), \mathcal{V}(m) \rangle \tag{2.3}$$

for $m \in M$. Saying that $\mathcal{V}$ is a derivation on $\mathcal{F}(M)$ means that if $F, H \in \mathcal{F}(M)$ then

$$\mathcal{V}[FH] = \mathcal{V}[F]H + F\mathcal{V}[H],$$

an easy consequence of (2.2) and the Leibniz rule for differentials. It follows from (2.3) that $\mathcal{V}[F](m)$ depends on $\mathcal{V}(m)$ (and F) only; it is the derivative of F at m in the direction of $\mathcal{V}(m)$, hence it is legitimate to write it as $\mathcal{V}(m)[F]$. At $m \in M$ the derivation property then reads

$$\mathcal{V}(m)[FH] = (\mathcal{V}(m)[F])H(m) + F(m)(\mathcal{V}(m)[H]);$$

one says that $\mathcal{V}(m)$ defines a *derivation* on $\mathcal{F}(M)$ *at m*.

It is a fundamental fact that, conversely, every derivation on M corresponds to a unique vector field on M and that every derivation at m corresponds to a unique tangent vector at m. As a corollary, since the commutator of two derivations is a derivation we may define the *Lie bracket* $[\mathcal{V}_1, \mathcal{V}_2]$ of $\mathcal{V}_1$, $\mathcal{V}_2 \in \mathfrak{X}(M)$ as the vector field that corresponds to the derivation $\mathcal{V}_1 \circ \mathcal{V}_2 - \mathcal{V}_2 \circ \mathcal{V}_1$. This way, $\mathfrak{X}(M)$ becomes an infinite-dimensional Lie algebra. For $F \in \mathcal{F}(M)$ and for $\mathcal{V}_1, \mathcal{V}_2 \in \mathfrak{X}(M)$ one has

$$[F\mathcal{V}_1, \mathcal{V}_2] = F[\mathcal{V}_1, \mathcal{V}_2] - \mathcal{V}_2[F]\mathcal{V}_1. \tag{2.4}$$

Notice also that if $U \subseteq M$ is a coordinate neighborhood then a derivation on $\mathcal{F}(\mathcal{U})$ is completely determined once its effect on all elements x_i of a coordinate system $(x_1, \dots, x_n)$ on $\mathcal{U}$ is known, where $n := \dim M$. Indeed, since in terms of these coordinates

$$\mathrm{d}F = \sum_{i=1}^{n} \frac{\partial F}{\partial x_i} \mathrm{d}x_i$$

we have in view of (2.2) that

$$\mathcal{V}[F] = \sum_{i=1}^{n} \frac{\partial F}{\partial x_i} \mathcal{V}[x_i]. \tag{2.5}$$

When we are dealing with a fixed vector field $\mathcal{V}$ we often write $\dot{F}$ for $\mathcal{V}[F]$, where $F \in \mathcal{F}(M)$. In this notation, the coordinate expression (2.5) takes the form

$$\dot{F} = \sum_{i=1}^{n} \frac{\partial F}{\partial x_i} \dot{x}_i.$$

There is a one-to-one correspondence between vector fields on the coordinate neighborhood U and differential equations on U of the form

$$\begin{aligned} \frac{\mathrm{d}x_1}{\mathrm{d}t} &= f_1(x_1, \ldots, x_n), \\ &\vdots \\ \frac{\mathrm{d}x_n}{\mathrm{d}t} &= f_n(x_1, \ldots, x_n), \end{aligned} \tag{2.6}$$

where $f_i \in \mathcal{F}(U)$, for $i = 1, \ldots, n$. Indeed, given a vector field $\mathcal{V}$, define the functions f_i by $f_i := \mathcal{V}[x_i]$; given the functions f_i, define $\mathcal{V}[x_i] := f_i$ and extend $\mathcal{V}$ to a derivation on $\mathcal{F}(U)$ by using (2.5). Solutions to (2.6) are easily interpreted as parametrized curves in U, whose tangent vector at each point coincides with the value of $\mathcal{V}$ at that point; we will usually consider solutions that are defined on an open ball B_ϵ around ϵ, where $B_\epsilon := \{t \in \mathbf{C} \mid |t| < \epsilon\}$ in the holomorphic case and $B_\epsilon := \{t \in \mathbf{R} \mid |t| < \epsilon\}$ in the smooth real case. For that reason a solution $x(t) = (x_1(t), \ldots, x_n(t))$ to (2.6), defined on B_ϵ, and such that $x(0) = m$ is often called an *integral curve* of $\mathcal{V}$, starting at m. The well-known uniqueness and existence theorem for differential equations can (in the holomorphic case) be formulated in terms of vector fields and integral curves as follows.

Theorem 2.1 (Picard Theorem for ODE's). *Let $\mathcal{V}$ be a holomorphic vector field on an open subset U of $\mathbf{C}^n$ and let $m \in U$. There exists an integral curve of $\mathcal{V}$, starting at m; this integral curve $x(t; m)$ is unique in the sense that any two integral curves of $\mathcal{V}$ that start at m coincide on the intersection of their domains. Moreover, $x(t; m)$ depends in a holomorphic way on m.*

The analogous theorem for smooth vector fields on open subsets of $\mathbf{R}^n$ of course also holds. The theorem and its smooth analog imply that given a vector field $\mathcal{V}$ on an n-dimensional manifold M we can find for any $m \in M$ a coordinate neighborhood U of m, with coordinates $(x_1, \ldots, x_n)$, an open subset $U' \subseteq U$ and an $\epsilon > 0$, such that the solution $x(t; m)$ is defined for $(t, m) \in (B_\epsilon \times U')$. The map

$$\begin{aligned} \Phi : B_\epsilon \times U' &\to U \\ (t, m) &\mapsto \Phi_t(m) := x(t; m) \end{aligned}$$

is called the *flow* of $\mathcal{V}$.

For a fixed $t \in B_\epsilon$ the map $\Phi_t : U' \to U$ is a biholomorphism (diffeomorphism, in the smooth case) from U' to $\Phi_t(U')$. It is customary to pretend that for small $|t|$ the local biholomorphism (diffeomorphism) Φ_t is global if we are in the case of a vector field on a manifold M and to write $\Phi_t : M \to M$, but of course — unless M is compact — the globalness needs not be true. For example, one writes the fundamental property that links directional derivatives to flows in the form

$$\mathcal{V}[F] = \frac{\mathsf{d}}{\mathsf{d}t}_{|t=0} \Phi_t^* F,$$

where $F \in \mathcal{F}(M)$ and $\Phi_t^* F = F \circ \Phi_t$. Theorem 2.1 and its smooth analog lead to the following theorem, that we will use often.

Theorem 2.2 (Straightening Theorem). *Let $\mathcal{V}$ be a vector field on a manifold M of dimension n and suppose that $\mathcal{V}(m) \neq 0$, where $m \in M$. Then there exist coordinates $x_1, \ldots, x_n$ on a neighborhood U of m such that the restriction of $\mathcal{V}$ to U is the first coordinate vector field, i.e., $\mathcal{V}[F] = \partial F / \partial x_1$.*

In the same spirit we can, intuitively speaking, parameterize a neighborhood of an analytic hypersurface by local coordinates on the hypersurface on the one hand, and by the parameter which is going with any fixed vector field, on the other hand, assuming that the vector field is transversal to the divisor. Precisely, the following theorem holds (see Figure 2.1).

Theorem 2.3. *Let M be a complex manifold of dimension n and let $\mathcal{V}$ be a holomorphic vector field on M. Suppose that $\mathcal{D}$ is an analytic hypersurface of M and let m_0 be a smooth point of $\mathcal{D}$. If $\mathcal{V}$ is transversal to $\mathcal{D}$ at m_0, then there exist neighborhoods U and V of m_0 in $\mathcal{D}$, resp. in M, and there exists $\epsilon > 0$, such that the restriction of Φ to $B_\epsilon \times U$ is a biholomorphism onto V. In addition, if U is a coordinate neighborhood of m_0 in $\mathcal{D}$, with coordinates $x_2, \ldots, x_n$ then V is a coordinate neighborhood of m_0 in M, with holomorphic coordinates $(t, x_2, \ldots, x_n)$, where $\mathcal{V} = \frac{\partial}{\partial t}$ (on V). In the latter case,*

$$\mathcal{D} \cap V = \{m \in V \mid t(m) = 0\}.$$

It follows that, under the above transversality assumption, we can write any holomorphic function F on V locally as a series

$$F(t) = t^p (f^{(0)} + f^{(1)} t + \cdots), \tag{2.7}$$

where the coefficients $f^{(0)}, f^{(1)}, \ldots$ of the series $F(t)$ are holomorphic functions on a neighborhood of m_0 in $\mathcal{D}$. By analyticity, the series $F(t)$ is actually convergent on an open neighborhood in M of an open dense subset of the irreducible component $\mathcal{D}'$ of $\mathcal{D}$ that contains m_0, and it coincides on this neighborhood with the function F. We call the series (2.7) the *Taylor series* of F with respect to $\mathcal{V}$, starting at $\mathcal{D}'$, and we denote it by $F(t; \mathcal{D}')$.

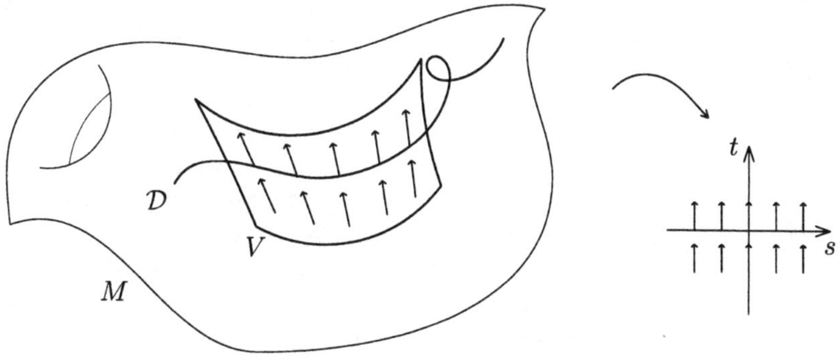

Fig. 2.1. When a holomorphic vector field $\mathcal{V}$ on a complex manifold M is transversal to a divisor $\mathcal{D}$ at $m_0 \in \mathcal{D}$ then a neighborhood V of m_0 in M admits holomorphic coordinates that come from coordinates $s = (s_1, \ldots, s_{n-1})$ on $\mathcal{D}$, plus the time coordinate t that goes with $\mathcal{V}$.

Since $\mathcal{V}$ is transversal to $\mathcal{D}'$ at m_0, the integer p in (2.7) is equal to $\text{ord}_{\mathcal{D}'}(F)$, the order of vanishing of F along $\mathcal{D}'$, if $f^{(0)}$ is not identically zero on $\mathcal{D}'$. The restriction of F to $\mathcal{D}'$ is given by substituting 0 for t in this Taylor series, i.e., by the first coefficient of its Taylor series.

The same can be done for meromorphic functions on U: writing such a function F as the ratio G/H of two holomorphic functions we define the *Laurent series* of F with respect to $\mathcal{V}$, starting at $\mathcal{D}'$, denoted $F(t; \mathcal{D}')$, to be the quotient $G(t; \mathcal{D}')/H(t; \mathcal{D}')$. In this case the series converges for small, non-zero $|t|$. The Laurent series of F is still of the form (2.7), where

$$p = \text{ord}_{\mathcal{D}'}(F) = \text{ord}_{\mathcal{D}'}(G) - \text{ord}_{\mathcal{D}'}(H)$$

is now any integer. Under the assumption that $f^{(0)}$ is not identically zero on $\mathcal{D}'$, it is still true that p is the order of vanishing of F along $\mathcal{D}'$, which in the case of negative p means that F has a pole of order $-p$ along $\mathcal{D}'$.

2.1.2 Distributions and the Frobenius Theorem

Instead of having a vector at every point of a manifold M, as is the case of a vector field on M, one may have a one-dimensional subspace of the tangent space to M, at every point of M. This is what is called a 1-dimensional *distribution* on M; a k-dimensional distribution Δ on M is then the datum of a k-dimensional subspace $\Delta(m)$ of T_mM for every $m \in M$. One says that Δ is smooth (or holomorphic) if there exist for every $m \in M$ smooth (or holomorphic) vector fields $\mathcal{V}_1, \ldots, \mathcal{V}_k$, on a neighborhood U of m, such that

$$\Delta(m) = \text{span}\{\mathcal{V}_1(m), \ldots, \mathcal{V}_k(m)\}, \qquad \text{for any } m \in U.$$

The notion of an integral curve is easily adapted to the case of a k-dimensional distribution Δ: a k-dimensional connected immersed submanifold M' of M is called an *integral manifold* of Δ if $T_mM' = \Delta(m)$ for any $m \in M'$. In contrast to the case of integral curves, integral manifolds need not exist in general, even locally. One obstruction comes from the following fact: if $\mathcal{V}_1$ and $\mathcal{V}_2$ are two vector fields on M which are tangent to some submanifold M' (such as the candidate integral manifold) then their Lie bracket $[\mathcal{V}_1, \mathcal{V}_2]$ is also tangent to M'. In order to rephrase this in the language of distributions, let us say that a vector field $\mathcal{V}$ on $U \subseteq M$ is adapted to Δ on U if $\mathcal{V}(m) \in \Delta(m)$ for every $m \in U$. In these terms the obstruction reads: if $\mathcal{V}_1$ and $\mathcal{V}_2$ are adapted to Δ on some open subset U then $[\mathcal{V}_1, \mathcal{V}_2]$ is also adapted to Δ on U. One says that Δ is an *integrable distribution* if for any $\mathcal{V}_1$ and $\mathcal{V}_2$ that are adapted to Δ on an open subset U, their commutator $[\mathcal{V}_1, \mathcal{V}_2]$ is also adapted to Δ on U. The Frobenius Theorem says that the above obstruction to the existence of integral manifolds is the only one.

Theorem 2.4 (Frobenius). *Suppose that Δ is a (smooth or holomorphic) k-dimensional distribution on M. If Δ is integrable then there exists through any point $m_0 \in M$ a unique maximal integral manifold for Δ.*

The above version of the Frobenius Theorem is the analogue of Theorem 2.1 for distributions. For a short and elementary proof, which is immediately adapted to the holomorphic case, we refer to [42] or [108]. Another version of the Frobenius Theorem is the following.

Theorem 2.5 (Frobenius). *Under the conditions of Theorem 2.4 coordinates $x_1, \dots, x_n$ can be chosen in a neighborhood U of any point $m_0 \in M$ such that*

$$\Delta(m) = \operatorname{span}\left\{\frac{\partial}{\partial x_1}(m), \dots, \frac{\partial}{\partial x_k}(m)\right\}, \qquad \textit{for any } m \in U.$$

In terms of these coordinates the integral manifold of $\Delta_{|U}$ through m_0 is given by the connected component of

$$\{m \in U \mid x_i(m) = x_i(m_0) \textit{ for } i = k+1, \dots, m\}$$

that contains m_0.

It is clear that the latter version of the Frobenius Theorem generalizes the Straightening Theorem (Theorem 2.2). It implies that the maximal integral manifolds of an integrable distribution on M form the leaves of a foliation on M.

In applications it is sometimes necessary to consider the more general concept of a *singular distribution*, in which the dimension of the subspace of T_mM may vary with m. This happens for example when one considers the singular distribution associated with the Hamiltonian vector fields on a Poisson manifold, as we will see in Section 3.4. For a good account on singular distributions we refer to [107, Appendix 3].

2.1.3 Differential Forms and Polyvector Fields

We will make frequent use of k-forms and k-vector fields on manifolds and of the operations on and between them. For $k \in \mathbf{N}$ we denote the $\mathcal{F}(M)$-module of k-forms on a manifold M by $\Omega^k(M)$; an element of $\Omega^k(M)$ is by definition a section of $\bigwedge^k T^*M$, in particular $\Omega^0(M) = \mathcal{F}(M)$ and $\Omega^1(M) = \Omega(M)$. We let

$$\Omega^*(M) := \bigoplus_{k=0}^{n} \Omega^k(M),$$

where $n := \dim(M)$. An element of $\Omega^*(M)$ will be called a *differential form.* There is an $\mathcal{F}(M)$-bilinear map

$$\wedge : \Omega^*(M) \times \Omega^*(M) \to \Omega^*(M),$$

which associates to two differential forms ω, ω' their *wedge product* $\omega \wedge \omega'$. This operation makes $\Omega^*(M)$ into a graded associative algebra over $\mathcal{F}(M)$, called the *Grassmann algebra* of M. It is graded commutative, which means that for $\omega \in \Omega^k(M)$ and $\omega' \in \Omega^l(M)$ we have

$$\omega \wedge \omega' = (-1)^{kl} \omega' \wedge \omega.$$

The fact that a 1-form can be evaluated on a vector field to produce an element of $\mathcal{F}(M)$ generalizes in two ways to a k-form ω, where $k \geqslant 1$. We can evaluate ω on k vector fields $\mathcal{V}_1, \dots, \mathcal{V}_k$, giving $\omega(\mathcal{V}_1, \dots, \mathcal{V}_k) \in \mathcal{F}(M)$; from this point of view a k-form is an $\mathcal{F}(M)$-k-linear map on $\mathfrak{X}(M)$ with values in $\mathcal{F}(M)$. Or we can insert one vector field $\mathcal{V}$ as the first argument to ω, yielding a $(k-1)$-form, which is denoted by $\imath_\mathcal{V}\omega$; from this point of view a k-form is, for $k \geqslant 1$, an $\mathcal{F}(M)$-linear map $\mathfrak{X}(M) \to \Omega^{k-1}(M)$. Notice that it is from the former point of view natural to define a k-form by prescribing its value on all k-tuples of vector fields on M; however, one still needs to check besides skew-symmetry that the k-form is indeed $\mathcal{F}(M)$-k-linear. It is convenient to extend the above definition of $\imath_\mathcal{V}$ to all of $\Omega^*(M)$ by defining $\imath_\mathcal{V}\omega = 0$, for all 0-forms, i.e. functions, ω on M.

The differential is a linear map $\mathsf{d} : \Omega^*(M) \to \Omega^*(M)$ which maps k-forms to $(k+1)$-forms according to the following formula:

$$\begin{aligned} \mathsf{d}\omega\,(\mathcal{V}_0, \dots, \mathcal{V}_k) &= \sum_{i=0}^{k} (-1)^i \mathcal{V}_i \left[\omega(\mathcal{V}_0, \dots, \widehat{\mathcal{V}_i}, \dots, \mathcal{V}_k)\right] \\ &\quad + \sum_{i<j} (-1)^{i+j} \omega\left([\mathcal{V}_i, \mathcal{V}_j], \mathcal{V}_0, \dots, \widehat{\mathcal{V}_i}, \dots, \widehat{\mathcal{V}_j}, \dots, \mathcal{V}_k\right). \end{aligned} \tag{2.8}$$

As we just pointed out one has to verify that the right hand side of this formula is $\mathcal{F}(M)$-$(k+1)$-linear, but that is an easy consequence of (2.4).

We have that $\mathsf{d}\circ\mathsf{d}=0$, which implies that each *exact differential form* (an element of $\Omega^*(M)$ that is in the image of d) is a *closed differential form* (an element $\omega\in\Omega^*(M)$ for which $\mathsf{d}\omega=0$). On a coordinate neighborhood every closed differential form is exact, but this is false for general open subsets of manifolds. The differential of a wedge satisfies the graded Leibniz rule

$$\mathsf{d}(\omega\wedge\omega')=\mathsf{d}\omega\wedge\omega'+(-1)^k\omega\wedge\mathsf{d}\omega',$$

where ω is a k-form and ω' an l-form. The differential is not an $\mathcal{F}(M)$-linear map, as is seen from the following formula: if $F\in\mathcal{F}(M)$ and $\omega\in\Omega^l(M)$ then

$$\mathsf{d}(F\omega)=\mathsf{d}F\wedge\omega+F\mathsf{d}\omega,\qquad F\in\mathcal{F}(M),\ \omega\in\Omega^k(M).$$

We will also consider k-vector fields, mainly in the case $k=2,3$, in which cases we speak of a *bivector field* or a *trivector field.* A k-vector field is by definition a section of $\bigwedge^k TM$, hence we can evaluate any k-form ω on any k-vector field P: for P a k-vector field of the form $P=\mathcal{V}_1\wedge\ldots\wedge\mathcal{V}_k$, we let $\omega(P)=\omega(\mathcal{V}_1\wedge\ldots\wedge\mathcal{V}_k):=\omega(\mathcal{V}_1,\ldots,\mathcal{V}_k)$. Since vector fields correspond to derivations we have that a 2-vector field corresponds to a skew-symmetric *biderivation*, a 3-vector field corresponds to a skew-symmetric *triderivation*, and so on. Namely, if P is a k-vector field the value of the corresponding skew-symmetric k-derivation on k functions $F_1,\ldots,F_k\in\mathcal{F}(M)$ is denoted[1], resp. defined by

$$P[F_1\wedge\cdots\wedge F_k]:=(\mathsf{d}F_1\wedge\cdots\wedge\mathsf{d}F_k)(P) \tag{2.9}$$

and we have that P is completely specified on a coordinate neighborhood U once it is known on all k-tuples $(x_{i_1},\ldots,x_{i_k})$, with $1\leqslant i_1<i_2\cdots<i_k\leqslant n=\dim M$, where the k-tuples are taken from any chosen system of coordinates $(x_1,\ldots,x_n)$ on U. Explicitly, (2.5) admits the following generalization to arbitrary k-vector fields,

$$P[F_1\wedge\ldots\wedge F_k]=\sum_{i_1,\ldots,i_k=1}^{n}\frac{\partial F_1}{\partial x_{i_1}}\cdots\frac{\partial F_k}{\partial x_{i_k}}P[x_{i_1}\wedge\ldots\wedge x_{i_k}].$$

We denote the $\mathcal{F}(M)$-module of k-vector fields by $\mathfrak{X}^k(M)$, in particular $\mathfrak{X}^0(M)=\mathcal{F}(M)$ and $\mathfrak{X}^1(M)=\mathfrak{X}(M)$, and we let

$$\mathfrak{X}^*(M):=\bigoplus_{k=0}^{n}\mathfrak{X}^k(M),$$

where $n:=\dim(M)$. An element of $\mathfrak{X}^*(M)$ is called a *polyvector field.* We can define, as in the case of differential forms, a wedge product

$$\wedge:\mathfrak{X}^*(M)\times\mathfrak{X}^*(M)\to\mathfrak{X}^*(M),$$

which makes $\mathfrak{X}^*(M)$ into a graded associative algebra which is the covariant analogue of the Grassmann algebra.

[1] We use $P[F_1\wedge\ldots\wedge F_k]$ instead of $P[F_1,\ldots,F_k]$ to avoid confusion with the notation for Lie brackets. Moreover, this notation makes sense in view of (2.9).

2.1.4 Lie Derivatives

The most important operation on k-forms and on k-vector fields is the Lie derivative. For $\mathcal{V} \in \mathfrak{X}(M)$ and $\omega \in \Omega^k(M)$ we denote the *Lie derivative* of ω in the direction of $\mathcal{V}$ by $L_{\mathcal{V}}\omega$. For $F \in \mathcal{F}(M)$ and $\mathcal{V} \in \mathfrak{X}(M)$ the Lie derivative is given by $L_{\mathcal{V}}F := \mathcal{V}[F] = \mathsf{d}F(\mathcal{V})$, while for an arbitrary k-form ω (with $k > 0$) its Lie derivative $L_{\mathcal{V}}\omega$ is the k-form whose value on $\mathcal{V}_1 \dots, \mathcal{V}_k \in \mathfrak{X}(M)$ is given by

$$L_{\mathcal{V}}\omega(\mathcal{V}_1, \dots, \mathcal{V}_k) := \mathcal{V}[\omega(\mathcal{V}_1, \dots, \mathcal{V}_k)] - \sum_{i=1}^{k} \omega(\mathcal{V}_1, \dots, [\mathcal{V}, \mathcal{V}_i], \dots, \mathcal{V}_k); \quad (2.10)$$

again one checks that the right hand side of this formula is $\mathcal{F}(M)$-k-linear, so that $L_{\mathcal{V}}\omega$ is indeed a k-form.

The Lie derivative $L_{\mathcal{V}}\omega$ measures how ω changes in the direction of $\mathcal{V}$, hence $L_{\mathcal{V}}\omega = 0$ if and only if ω is constant on the integral curves of $\mathcal{V}$. This follows immediately from the following alternative (geometric!) definition,

$$L_{\mathcal{V}}\omega = \frac{\mathsf{d}}{\mathsf{d}t}_{|t=0} \Phi_t^* \omega,$$

where Φ denotes the (local) flow of $\mathcal{V}$ on M. The most useful expression for $L_{\mathcal{V}}\omega$ is given by the following formula, known as *Cartan's Formula*

$$L_{\mathcal{V}}\omega = \mathsf{d}\imath_{\mathcal{V}}\omega + \imath_{\mathcal{V}}\mathsf{d}\omega; \quad (2.11)$$

for example, Cartan's formula implies at once that $L_{\mathcal{V}}\omega$ is $\mathcal{F}(M)$-k-linear, i.e., that it is a k-form. We will also need the formula

$$\imath_{[\mathcal{V}_1, \mathcal{V}_2]}\omega = L_{\mathcal{V}_1}\imath_{\mathcal{V}_2}\omega - \imath_{\mathcal{V}_2}L_{\mathcal{V}_1}\omega, \quad (2.12)$$

which is valid for $\mathcal{V}_1$, $\mathcal{V}_2 \in \mathfrak{X}(M)$ and $\omega \in \Omega^*(M)$. Let us prove (2.12) when ω is a two-form (this is the only case which will be used), in which case both sides of (2.12) are one-forms. For any vector field $\mathcal{V}$ we have, in view of (2.10) applied to the one-form $\imath_{\mathcal{V}_2}\omega$,

$$L_{\mathcal{V}_1}\imath_{\mathcal{V}_2}\omega(\mathcal{V}) = \mathcal{V}_1[\imath_{\mathcal{V}_2}\omega(\mathcal{V})] - \imath_{\mathcal{V}_2}\omega([\mathcal{V}_1, \mathcal{V}]) = \mathcal{V}_1[\omega(\mathcal{V}_2, \mathcal{V})] - \omega(\mathcal{V}_2, [\mathcal{V}_1, \mathcal{V}]).$$

Applying (2.10) again, but now to the two-form ω we get

$$\imath_{\mathcal{V}_2}L_{\mathcal{V}_1}\omega(\mathcal{V}) = L_{\mathcal{V}_1}\omega(\mathcal{V}_2, \mathcal{V}) = \mathcal{V}_1[\omega(\mathcal{V}_2, \mathcal{V})] - \omega([\mathcal{V}_1, \mathcal{V}_2], \mathcal{V}) - \omega(\mathcal{V}_2, [\mathcal{V}_1, \mathcal{V}]).$$

It follows that for any vector field $\mathcal{V}$

$$L_{\mathcal{V}_1}\imath_{\mathcal{V}_2}\omega(\mathcal{V}) - \imath_{\mathcal{V}_2}L_{\mathcal{V}_1}\omega(\mathcal{V}) = \omega([\mathcal{V}_1, \mathcal{V}_2], \mathcal{V}) = \imath_{[\mathcal{V}_1, \mathcal{V}_2]}\omega(\mathcal{V}),$$

showing (2.12).

The Lie derivative $L_{\mathcal{V}}\mathcal{W}$ of a vector field $\mathcal{W}$ is given by $L_{\mathcal{V}}\mathcal{W} := [\mathcal{V}, \mathcal{W}]$, while the Lie derivative of an arbitrary k-vector field P is the k-vector field $L_{\mathcal{V}}P$, defined by

$$L_{\mathcal{V}}P[F_1 \wedge \ldots \wedge F_k] := \mathcal{V}[P[F_1 \wedge \ldots \wedge F_k]] - \sum_{i=1}^{k} P[F_1 \wedge \ldots \wedge \mathcal{V}[F_i] \wedge \ldots \wedge F_k], \tag{2.13}$$

where $F_1, \ldots, F_k \in \mathcal{F}(M)$. As in the case of the Lie derivative of a differential form, the Lie derivative $L_{\mathcal{V}}P$ of a k-vector field P on M also measures how P changes in the direction of $\mathcal{V}$, hence $L_{\mathcal{V}}P = 0$ if and only if P is constant on the integral curves of the vector field $\mathcal{V}$.

Example 2.6. In order to get familiar with the notations, let us verify that $L_{\mathcal{V}}\mathcal{W} = [\mathcal{V}, \mathcal{W}]$, for any $\mathcal{V}, \mathcal{W} \in \mathfrak{X}(M)$. In fact, if $F \in \mathcal{F}(M)$ then (2.13) implies

$$(L_{\mathcal{V}}\mathcal{W})[F] = \mathcal{V}[\mathcal{W}[F]] - \mathcal{W}[\mathcal{V}[F]] = (\mathcal{V} \circ \mathcal{W} - \mathcal{W} \circ \mathcal{V})[F] = [\mathcal{V}, \mathcal{W}][F],$$

proving our claim.

2.2 Lie Groups and Lie Algebras

Unless otherwise stated, all Lie groups and Lie algebras will be defined over $\mathbf{C}$. We use the standard convention that Lie groups are denoted by boldface capital letters ($\mathbf{G}, \mathbf{H}, \ldots$) and their Lie algebras by the corresponding gothic letters ($\mathfrak{g}, \mathfrak{h}, \ldots$). The main examples of Lie groups include linear groups, i.e., Lie subgroups of $\mathbf{GL}(n) = \mathbf{GL}(\mathbf{C}^n)$, the non-commutative group of all invertible $n \times n$ matrices (with coefficients in $\mathbf{C}$), where the group operation is given by the usual product of matrices. Similarly, the main examples of Lie algebras include matrix Lie algebras, i.e., Lie subalgebras of $\mathfrak{gl}(n) = \mathfrak{gl}(\mathbf{C}^n)$, the Lie algebra of all $n \times n$ matrices (with coefficients in $\mathbf{C}$), where the *Lie bracket* is given by the commutator of matrices. In fact, according to *Ado's Theorem*, every finite-dimensional Lie algebra is isomorphic to a matrix Lie algebra, but the corresponding theorem does not hold true for (finite-dimensional) Lie groups. $\mathfrak{gl}(n)$ is the Lie algebra of $\mathbf{GL}(n)$, and the Lie algebra of a given linear group can easily be realized as a matrix Lie algebra. This is done by using the exponential map exp, which is a natural local biholomorphism $\exp : U \subseteq \mathfrak{g} \to \mathbf{G}$ between an open neighborhood U of the origin 0 of any finite-dimensional Lie algebra $\mathfrak{g}$ and an open neighborhood of the unit element $e \in \mathbf{G}$. For $X \in \mathfrak{g} = T_e\mathbf{G}$ close to 0, $\exp X$ is the Lie group element $\Phi_1(e)$ where Φ denotes the flow of the left invariant vector field on $\mathbf{G}$, defined by X. For example, taking $\mathfrak{g} = \mathfrak{gl}(n)$ and $\mathbf{G} = \mathbf{GL}(n)$ the map exp is the usual exponential of matrices and the conditions that define a subgroup of $\mathbf{GL}(n)$ are easily translated in the conditions that define its Lie algebra.

Example 2.7. Consider the linear group $\mathbf{G}$ of all orthogonal $n \times n$ matrices. For $X \in \mathfrak{gl}(n)$ and for small $|t|$, consider the invertible matrix $\exp(tX) = \mathrm{Id}_n + tX + t^2X^2 + O(t^3)$. Orthogonality of $\exp(tX)$ yields

$$\begin{aligned}\mathrm{Id}_n &= \exp(tX)\exp(tX)^\top \\ &= (\mathrm{Id}_n + tX + t^2X^2 + O(t^3))\left(\mathrm{Id}_n + tX^\top + t^2(X^2)^\top + O(t^3)\right) \\ &= \mathrm{Id}_n + t(X + X^\top) + O(t^2),\end{aligned}$$

hence the Lie algebra $\mathfrak{g}$ of $\mathbf{G}$ is given by the matrices X for which $X + X^\top = 0$, i.e., $\mathfrak{g}$ is the Lie algebra of all skew-symmetric $n \times n$ matrices.

More generally, Lie subalgebras of a finite-dimensional Lie algebra $\mathfrak{g}$ are in one-to-one correspondence with connected Lie subgroups of $\mathbf{G}$; notice however that a Lie subgroup needs not be closed in its ambient Lie group (consider the subgroup generated by a generic element of the complex torus $\mathbf{C}^2/\mathbf{Z}^2$).

The tangent space $T_g\mathbf{G}$ to $\mathbf{G}$ at an element $g \in \mathbf{G}$ is naturally identified with $\mathfrak{g}$: the left translation map $L_{g^{-1}} : \mathbf{G} \to \mathbf{G}$ maps g to e, and its differential maps $T_g\mathbf{G}$ to $\mathfrak{g}$. Similarly we can identify the cotangent spaces to $\mathbf{G}$ with $\mathfrak{g}^*$ (the dual vector space to $\mathfrak{g}$) and so on.

Example 2.8. Suppose that $\mathbf{G}$ is a linear group which is closed (as a topological subspace of $\mathbf{GL}(n)$). Elements of $T\mathbf{G}$ are then naturally represented by pairs of matrices (g, X), where a vector (g, X) acts by definition on $F \in \mathcal{F}(\mathbf{G})$ by

$$(g, X)[F] := \lim_{t\to 0} \frac{\overline{F}(g + tX) - F(g)}{t},$$

where $\overline{F}$ is any (holomorphic) extension of F to a small open neighborhood of g in its ambient space, the space of all $n \times n$ matrices (this can be done because $\mathbf{G}$ is closed). Then $\mathrm{d}L_{g^{-1}}(g, X) = (e, g^{-1}X)$, so that (g, X) gets naturally identified with the matrix $g^{-1}X$. The same result holds true when $\mathbf{G}$ is not closed, because a small neighborhood in $\mathbf{G}$ of any element g of $\mathbf{G}$ is closed in a small neighborhood of g in $\mathbf{GL}(n)$.

Since $\mathfrak{g}$ (resp. $\mathfrak{g}^*$) is a vector space, its tangent spaces are also naturally identified with $\mathfrak{g}$ (resp. $\mathfrak{g}^*$). These identification will be (ab)used in the sequel, often without further mention. A particular instance of this that will be used throughout the text is the following: if $F \in \mathcal{F}(\mathfrak{g}^*)$ and $\xi \subset \mathfrak{g}^*$ then the differential of F at ξ is a linear map

$$\mathrm{d}F(\xi) : T_\xi\mathfrak{g}^* \to \mathbf{C}$$

which under the above identifications gets naturally identified with an element of $\mathfrak{g}$. Conversely, an element $X \in \mathfrak{g}$ will often be viewed (without changing notations) as a linear map $\mathfrak{g}^* \to \mathbf{C}$, a change of perspective that is made transparent by the notations, which simply read $\langle X, \xi\rangle = \langle \xi, X\rangle$ for $X \in \mathfrak{g}$ and $\xi \in \mathfrak{g}^*$.

Lie groups appear most often through their action (always taken to be a left action) on manifolds. By assumption, if

$$\chi : \mathbf{G} \times M \to M$$

is an action, then for each $g \in \mathbf{G}$ the map $\chi_g : M \to M$, defined by $m \mapsto \chi(g, m)$ is a biholomorphism, with inverse $m \mapsto \chi(g^{-1}, m)$. We will usually write $g{\cdot}m$ or gm for $\chi(g, m)$. The action allows us to associate to each element $X \in \mathfrak{g}$ a vector field $\underline{X}$, whose value at m, denoted $\underline{X}(m)$, is the derivation on $\mathcal{F}(M)$ at m given by

$$\underline{X}(m)[F] := \frac{\mathsf{d}}{\mathsf{d}t}_{|t=0} F((\exp tX) \cdot m),$$

for all $F \in \mathcal{F}(M)$. The vector field $\underline{X}$ is called the *fundamental vector field* corresponding to $X \in \mathfrak{g}$. Its flow is given by the action of the one-parameter group $\exp tX$. The fundamental vector fields describe infinitesimally the action of $\mathbf{G}$ on M and they span the tangent space to the orbits of $\mathbf{G}$ at every point of M.

The simplest action of a Lie group $\mathbf{G}$ on a vector space S is a *linear action*, which means that for each $g \in \mathbf{G}$ one has that $\chi_g \in \mathbf{GL}(S)$. Then one can view χ as a homomorphism $\chi : \mathbf{G} \to \mathbf{GL}(S)$ and one says that χ is a *representation* of $\mathbf{G}$ on S. A subspace $T \subseteq S$ is called an *invariant subspace* if χ_g leaves T stable, i.e. $\chi_g(T) \subseteq T$, for all $g \in \mathbf{G}$. Then χ induces a representation $\mathbf{G} \to \mathbf{GL}(T)$ which is called a *subrepresentation.* A representation χ of $\mathbf{G}$ on S is called an *irreducible representation* if $\dim S > 0$ and if χ does not admit a non-trivial (i.e., with T different from $\{0\}$ and T) subrepresentation. In general an invariant subspace T may or may not have a complement T' in S which is also an invariant subspace. One says that a representation χ is a *completely reducible representation* if any invariant subspace admits a complementary subspace which is also invariant. In that case one can describe χ as a direct sum of irreducible representations.

The above terminology applies equally to the case of Lie algebra representations, with the understanding that a *representation* of $\mathfrak{g}$ on S is a Lie algebra homomorphism $\mathfrak{g} \to \operatorname{End}(S)$, the Lie bracket in $\operatorname{End}(S)$ being the commutator of endomorphisms. One also says that S is a $\mathfrak{g}$-*module.* Using the fact that $\operatorname{End}(S)$ is the Lie algebra of $\mathbf{GL}(S)$ and using our convention that we identify all tangent spaces to $\mathbf{G}$ with $\mathfrak{g}$, every representation of $\mathbf{G}$ on S leads to a representation of $\mathfrak{g}$ on S by mapping $X \in \mathfrak{g} \mapsto \underline{X} \in \operatorname{End}(S)$. The fact that this gives indeed a representation follows from the formula

$$[\underline{X}, \underline{Y}] = \underline{[X, Y]} \qquad X, Y \in \mathfrak{g}.$$

The two most important examples are the adjoint and the coadjoint action (representation) of a Lie group $\mathbf{G}$ on its Lie algebra $\mathfrak{g}$, resp. on the dual $\mathfrak{g}^*$ of its Lie algebra.

For $g \in \mathbf{G}$, we define Ad_g to be the endomorphism of $\mathfrak{g}$ which is the derivative of the conjugation map $C_g : \mathbf{G} \to \mathbf{G} : h \mapsto ghg^{-1}$ at the identity, $\operatorname{Ad}_g := \mathsf{d}C_g(e)$. The *adjoint action* or *adjoint representation* of $\mathbf{G}$ on $\mathfrak{g}$ is then given by

$$\operatorname{Ad} : \mathbf{G} \to \mathbf{GL}(\mathfrak{g}) : g \mapsto \operatorname{Ad}_g .$$

For example, if $\mathbf{G}$ is a linear group and $\mathfrak{g}$ its (matrix) Lie algebra, then it follows, as in Example 2.8, that

$$\operatorname{Ad}_g X = \mathsf{d}C_g(e)(X) = gXg^{-1},$$

where $g \in \mathbf{G}$ and $X \in \mathfrak{g}$.

The representation of $\mathfrak{g}$ on itself which corresponds to the adjoint action is called the *adjoint representation* of the Lie algebra $\mathfrak{g}$ on itself and is denoted by ad; the image of $X \in \mathfrak{g}$ under ad will, for readability, be written as ad_X. By the above definition, ad_X is the fundamental vector field $\underline{X}$ on $\mathfrak{g}$ that corresponds to the adjoint action, viewed as an endomorphism of $\mathfrak{g}$ (by identifying all tangent spaces of $\mathfrak{g}$ to $\mathfrak{g}$). Explicitly, $\operatorname{ad}_X Y = [X, Y]$, for $Y \in \mathfrak{g}$.

We now turn to the coadjoint action. For $g \in \mathbf{G}$ we define Ad_g^* by duality:

$$\left\langle \operatorname{Ad}_g^* \xi, X \right\rangle = \left\langle \xi, \operatorname{Ad}_{g^{-1}} X \right\rangle,$$

where $\xi \in \mathfrak{g}^*$ and $X \in \mathfrak{g}$. The resulting map

$$\operatorname{Ad}^* : \mathbf{G} \to \mathbf{GL}(\mathfrak{g}^*) : g \mapsto \operatorname{Ad}_g^*$$

is called the *coadjoint action* or the *coadjoint representation* of $\mathbf{G}$ on $\mathfrak{g}^*$. Its orbits are called *coadjoint orbits* and they play an important role in what follows. The representation of $\mathfrak{g}$ on $\mathfrak{g}^*$ that corresponds to the coadjoint representation is denoted by ad^* and is called the *coadjoint representation* of $\mathfrak{g}$ on $\mathfrak{g}^*$. The relation between ad and ad^* is consequently given by

$$\langle \operatorname{ad}_X^* \xi, Y \rangle = \langle \xi, -\operatorname{ad}_X Y \rangle = \langle \xi, [Y, X] \rangle, \tag{2.14}$$

where $\xi \in \mathfrak{g}^*$ and $X, Y \in \mathfrak{g}$.

A function $H \in \mathcal{F}(\mathfrak{g})$ (resp. $H \in \mathcal{F}(\mathfrak{g}^*)$) is called Ad-*invariant* (resp. Ad^*-*invariant*) if $H(\operatorname{Ad}_g X) = H(X)$ for all $g \in \mathbf{G}$ and $X \in \mathfrak{g}$ (resp. $H\left(\operatorname{Ad}_g^* \xi\right) = H(\xi)$ for all $g \in \mathbf{G}$ and $\xi \in \mathfrak{g}^*$). The algebra of Ad-invariant functions on $\mathfrak{g}$ is denoted by $\mathcal{F}(\mathfrak{g})^{\mathbf{G}}$, while the algebra of Ad^*-invariant functions on $\mathfrak{g}^*$ is denoted by $\mathcal{F}(\mathfrak{g}^*)^{\mathbf{G}}$.

In the following lemma we describe two properties of Ad^*-invariant functions that we will use. The Ad-invariant functions have similar properties, that are easily written down, and are proven in the same way, but these properties will not be used explicitly used here.

Lemma 2.9. *Let $H \in \mathcal{F}(\mathfrak{g}^*)^{\mathbf{G}}$. For any $\xi \in \mathfrak{g}^*$ and for any $X \in \mathfrak{g}$ we have that $\langle \xi, [\mathrm{d}H(\xi), X] \rangle = 0$, i.e., one has for any $\xi \in \mathfrak{g}^*$ that*

$$\mathrm{ad}^*_{\mathrm{d}H(\xi)}\, \xi = 0.$$

Moreover, for any $g \in \mathbf{G}$ and $\xi \in \mathfrak{g}^*$ the following diagram is commutative.

$$\begin{array}{ccc} \mathfrak{g}^* & \xrightarrow{\ \mathrm{d}H\ } & \mathfrak{g} \\ {\scriptstyle \mathrm{Ad}^*_g}\big\downarrow & & \big\downarrow{\scriptstyle \mathrm{Ad}_g} \\ \mathfrak{g}^* & \xrightarrow[\ \mathrm{d}H\ ]{} & \mathfrak{g} \end{array}$$

Proof. Let $X \in \mathfrak{g}$ and $\xi \in \mathfrak{g}^*$. If $H \in \mathcal{F}(\mathfrak{g}^*)^{\mathbf{G}}$ then $H(\xi) = H\left(\mathrm{Ad}^*_g \xi\right)$ for all $\xi \in \mathfrak{g}^*$ and $g \in \mathbf{G}$. Taking any $X \in \mathfrak{g}$ we therefore have that

$$\begin{aligned} \left\langle \mathrm{ad}^*_{\mathrm{d}H(\xi)}\, \xi, X \right\rangle &= -\left\langle \mathrm{ad}^*_X\, \xi, \mathrm{d}H(\xi) \right\rangle = -\,\mathrm{ad}^*_X\, \xi\, [H] \\ &= -\frac{\mathrm{d}}{\mathrm{d}t}_{|t=0} H\left(\mathrm{Ad}^*_{\exp tX}\, \xi\right) = -\frac{\mathrm{d}}{\mathrm{d}t}_{|t=0} H(\xi) = 0, \end{aligned}$$

showing the first property. In order to prove that the above diagram is commutative, differentiate for a fixed $g \in \mathbf{G}$ the identity $H = H \circ \mathrm{Ad}^*_g$ at $\xi \in \mathfrak{g}^*$. It gives

$$\mathrm{d}H(\xi) = \mathrm{d}H(\mathrm{Ad}^*_g\, \xi) \circ (\mathrm{d}\,\mathrm{Ad}^*_g)(\xi) = \mathrm{d}H(\mathrm{Ad}^*_g\, \xi) \circ \mathrm{Ad}^*_g,$$

because, for fixed g, the map Ad^*_g is a linear map. Thus, for $\eta \in \mathfrak{g}^*$, we have

$$\langle \mathrm{d}H(\xi), \eta \rangle = \left\langle \mathrm{d}H(\mathrm{Ad}^*_g\, \xi), \mathrm{Ad}^*_g\, \eta \right\rangle = \left\langle \mathrm{Ad}_{g^{-1}}\left(\mathrm{d}H(\mathrm{Ad}^*_g\, \xi)\right), \eta \right\rangle.$$

It follows that $\mathrm{d}H(\mathrm{Ad}^*_g\, \xi) = \mathrm{Ad}_g(\mathrm{d}H(\xi))$, which proves that the diagram is commutative. □

Lie algebras often come equipped with a non-degenerate symmetric bilinear form

$$\langle \cdot \,|\, \cdot \rangle : \mathfrak{g} \times \mathfrak{g} \to \mathbf{C}.$$

Such a form allows us to identify $\mathfrak{g}$ with $\mathfrak{g}^*$, simply by assigning to $X \in \mathfrak{g}$ the linear form $\hat{X}$ which maps $Y \in \mathfrak{g}$ to $\langle X \,|\, Y \rangle$, i.e.,

$$\langle \hat{X}, Y \rangle = \langle X \,|\, Y \rangle$$

for all $X, Y \in \mathfrak{g}$. Its inverse is the linear map $\mathfrak{g}^* \to \mathfrak{g} : \xi \mapsto X$, where X is the unique element of $\mathfrak{g}$ which satisfies $\langle X \,|\, Y \rangle = \langle \xi, Y \rangle$, for all $Y \in \mathfrak{g}$. A symmetric bilinear form $\langle \cdot \,|\, \cdot \rangle$ on $\mathfrak{g}$ will be called Ad-*invariant* when for any $g \in \mathbf{G}$ and for any $X, Y \in \mathfrak{g}$ one has

$$\langle \mathrm{Ad}_g X \,|\, \mathrm{Ad}_g Y \rangle = \langle X \,|\, Y \rangle.$$

Ad-invariance of $\langle\cdot\,|\,\cdot\rangle$ implies the following associativity-like rule: for any $X, Y, Z \in \mathfrak{g}$ one has

$$\langle \mathrm{ad}_Y X \,|\, Z\rangle = -\langle X \,|\, \mathrm{ad}_Y Z\rangle\,, \tag{2.15}$$

so that ad_Y is skew-symmetric with respect to $\langle\cdot\,|\,\cdot\rangle$. Ad-invariance of $\langle\cdot\,|\,\cdot\rangle$ also implies that for any $g \in \mathbf{G}$ the following diagram is commutative.

$$\begin{array}{ccc} \mathfrak{g} & \xrightarrow{\ \wedge\ } & \mathfrak{g}^* \\ {\scriptstyle \mathrm{Ad}_g}\downarrow & & \downarrow{\scriptstyle \mathrm{Ad}^*_g} \\ \mathfrak{g} & \xrightarrow[\ \wedge\ ]{} & \mathfrak{g}^* \end{array} \tag{2.16}$$

Indeed, for any $X, Y \in \mathfrak{g}$ it follows from the definitions and from Ad-invariance that

$$\begin{aligned} \left\langle \mathrm{Ad}^*_g \hat{X}, Y\right\rangle &= \left\langle \hat{X}, \mathrm{Ad}_{g^{-1}} Y\right\rangle = \langle X \,|\, \mathrm{Ad}_{g^{-1}} Y\rangle \\ &= \langle \mathrm{Ad}_g X \,|\, Y\rangle = \left\langle \widehat{\mathrm{Ad}_g X}, Y\right\rangle. \end{aligned}$$

In words: upon identifying a Lie algebra with its dual (using an non-degenerate Ad-invariant symmetric bilinear form $\langle\cdot\,|\,\cdot\rangle$), the adjoint and coadjoint actions get identified. The datum of a bilinear form on $\mathfrak{g}$ leads to a notion of orthogonality: for any subset $A \subseteq \mathfrak{g}$ the $\langle\cdot\,|\,\cdot\rangle$-*orthogonal* of A is the subspace of $\mathfrak{g}$, defined by

$$A^\perp := \{Y \in \mathfrak{g} \mid \langle X \,|\, Y\rangle = 0 \text{ for all } X \in A\}\,.$$

Example 2.10. Consider the subalgebras $\mathfrak{a}$ and $\mathfrak{b}$ of $\mathfrak{gl}(n)$ which consist respectively of the skew-symmetric and the upper triangular matrices. Obviously, $\mathfrak{gl}(n) = \mathfrak{a} \oplus \mathfrak{b}$ (direct sum of vector spaces). Consider on $\mathfrak{gl}(n)$ the non-degenerate symmetric bilinear form defined by $\langle X \,|\, Y\rangle := \mathrm{Trace}(XY)$. Then $\mathfrak{a}^\perp$ is the subspace of $\mathfrak{gl}(n)$ consisting of all symmetric matrices, while $\mathfrak{b}^\perp$ consists of all strictly upper triangular matrices.

The main example of an Ad-invariant symmetric bilinear form is the *Killing form* of $\mathfrak{g}$, which is defined by

$$\langle X \,|\, Y\rangle := \mathrm{Trace}(\mathrm{ad}_X \circ \mathrm{ad}_Y). \tag{2.17}$$

The Killing form of $\mathfrak{g}$ is non-degenerate if and only if $\mathfrak{g}$ is a semi-simple Lie algebra. Semi-simple Lie algebras and simple Lie algebras will be defined in the next section. Together with their infinite-dimensional analogues, the (twisted) affine Lie algebras, they will be the main types of Lie algebras encountered in this book.

2.3 Simple Lie Algebras

A non-empty subset $\mathfrak{h}$ of a Lie algebra $\mathfrak{g}$ is called an *ideal* when $[\mathfrak{g}, \mathfrak{h}] \subseteq \mathfrak{h}$. If $\mathfrak{g}$ contains no other ideals than 0 and itself and $\dim \mathfrak{g} > 1$ then $\mathfrak{g}$ is called a *simple Lie algebra.* A Lie algebra that is isomorphic to the direct sum of simple Lie algebras is called a *semi-simple Lie algebra.* Such a Lie algebra is characterized by the fact that its Killing form $\langle \cdot \mid \cdot \rangle$ (see (2.17)) is non-degenerate. Moreover, for any simple Lie algebra $\mathfrak{g}$ the Killing form is, up to a constant, the unique Ad-invariant symmetric bilinear form on $\mathfrak{g}$ which is non-degenerate.

2.3.1 The Classification

We describe in this paragraph the elements that appear in the classification and in the representation theory of simple Lie algebras, since we will need them in what follows. For proofs and details we refer to [87].

In this paragraph we assume that $\mathfrak{g}$ is a simple Lie algebra (over $\mathbf{C}$). Let $\mathfrak{h} \subseteq \mathfrak{g}$ be a *Cartan subalgebra* of $\mathfrak{g}$, i.e., $\mathfrak{h}$ is Abelian ($[\mathfrak{h}, \mathfrak{h}] = 0$) and self-normalizing ($x \in \mathfrak{g}$ and $[x, \mathfrak{h}] \subseteq \mathfrak{h}$ implies $x \in \mathfrak{h}$). The dimension of $\mathfrak{h}$ is called the *rank* of $\mathfrak{g}$, denoted $\operatorname{Rk} \mathfrak{g}$. It does not depend on the choice of $\mathfrak{h}$ because one shows that $\mathfrak{h}$ is unique up to an automorphism of $\mathfrak{g}$. For $X \in \mathfrak{h}$ the endomorphism $\operatorname{ad}_X : \mathfrak{g} \to \mathfrak{g}$ is diagonalizable and commutativity of $\mathfrak{h}$ implies that $\operatorname{ad}_\mathfrak{h}$ is a family of simultaneously diagonalizable endomorphisms of $\mathfrak{g}$, leading to a direct sum decomposition of $\mathfrak{g}$ into eigenspaces of $\operatorname{ad}_\mathfrak{h}$,

$$\mathfrak{g} = \mathfrak{h} \oplus \sum_{\alpha \in \Phi} \mathfrak{g}^\alpha, \tag{2.18}$$

where each subspace $\mathfrak{g}^\alpha$ can be shown to be one-dimensional. An element α of Φ is called a *root,* $\Phi \subseteq \mathfrak{h}^*$ is called a *root system* and the decomposition (2.18) is called the *root space decomposition.* A root $\alpha \in \Phi$ is a collection of eigenvalues of $\operatorname{ad}_\mathfrak{h}$ in the sense that if $E_\alpha \in \mathfrak{g}^\alpha$ then

$$[H, E_\alpha] = \langle \alpha, H \rangle E_\alpha \qquad \text{for all } H \in \mathfrak{h}.$$

Notice that 0 is not a root because we assumed that $\mathfrak{h}$ is self-normalizing. It can be shown that the root system Φ spans $\mathfrak{h}^*$ and that a (non-unique) basis Π for $\mathfrak{h}^*$ can be extracted from Φ, with the following property: any root $\alpha \in \Phi$ is a linear combination of elements of Π with coefficients in $\mathbf{Z}$ which are either all positive or all negative. Thus, $\Phi = \Phi^+ \cup \Phi^-$, where $\Phi^- = -\Phi^+$, and any element of Φ^+ is a linear combination of elements of Π with coefficients in $\mathbf{N}$. In particular, $\Pi \subseteq \Phi^+$ and the roots all belong to the lattice[2], generated by the simple roots, called the *root lattice.*

[2] This lattice is independent of the choice of simple roots Π since it is the smallest lattice in $\mathfrak{h}^*$ that contains all the roots.

We call Π a set of *simple roots* and we set $\Pi = (\alpha_1, \ldots, \alpha_l)$, where $l :=$ Rk $\mathfrak{g}$. For $\alpha \in \Phi$ we define its *height* $|\alpha|$ by

$$|\alpha| = \sum_{i=1}^{l} a_i, \quad \text{where} \quad \alpha = \sum_{i=1}^{l} a_i \alpha_i .$$

It leads to a grading of $\mathfrak{g}$,

$$\mathfrak{g} = \bigoplus_k \mathfrak{g}_k, \qquad [\mathfrak{g}_k, \mathfrak{g}_l] \subseteq [\mathfrak{g}_{k+l}], \tag{2.19}$$

where $\mathfrak{g}_k$ is, for $k \neq 0$, the span of the eigenvectors of all E_α, with $|\alpha| = k$, and $\mathfrak{g}_0 := \mathfrak{h}$. One proves that the Killing form $\langle \cdot | \cdot \rangle$ restricts to a bilinear form on $\mathfrak{h}$, which is also non-degenerate, hence the isomorphism $\mathfrak{g} \to \mathfrak{g}^*$, induced by $\langle \cdot | \cdot \rangle$, leads to an isomorphism $\mathfrak{h} \to \mathfrak{h}^*$. We mainly use its inverse, $\mathfrak{h}^* \to \mathfrak{h}$, which maps $\lambda \in \mathfrak{h}^*$ to $h_\lambda \in \mathfrak{h}$, where

$$\langle \lambda, \cdot \rangle = \langle h_\lambda \,|\, \cdot \rangle .$$

We will use, for $i = 1, \ldots, l$, the abbreviation h_i for h_{α_i}. The l-tuple $(h_1, \ldots, h_l)$ forms a basis of $\mathfrak{h}$. For $\alpha \in \Phi$ we define the following useful normalization[3] of h_α:

$$H_\alpha := 2 \frac{h_\alpha}{\langle h_\alpha \,|\, h_\alpha \rangle} . \tag{2.20}$$

Each H_α is called a *coroot* and for $i = 1, \ldots, l$ the coroot which corresponds to α_i is denoted by H_i. The coroots appear in the following refinement of the root space decomposition.

Theorem 2.11 (Chevalley). *Let $\mathfrak{g}$ be a simple Lie algebra of rank l, let $\mathfrak{h}$ be a Cartan subalgebra with root system Φ, and let $\Pi = (\alpha_1, \ldots, \alpha_l)$ denote a system of simple roots with respect to $\mathfrak{h}$. Then there exists for every $\alpha \in \Phi$ a non-zero vector $E_\alpha \in \mathfrak{g}^\alpha$ (see (2.18)) such that for any $H, H' \in \mathfrak{h}$ and $\alpha, \beta \in \Phi$*

$$\begin{aligned} [H, H'] &= 0, \\ [H, E_\alpha] &= \langle \alpha, H \rangle E_\alpha, \\ [E_\alpha, E_\beta] &= \begin{cases} H_\alpha & \text{if } \alpha + \beta = 0, \\ 0 & \text{if } \alpha + \beta \notin \Phi \cup \{0\}, \\ N_{\alpha\beta} E_{\alpha+\beta} & \text{if } \alpha + \beta \in \Phi. \end{cases} \end{aligned}$$

Here, $N_{\alpha\beta} = \pm(p+1)$, where

$$p := \max \{ n \mid \beta - n\alpha \in \Phi \} .$$

The basis $(H_1, \ldots, H_l) \cup (E_\alpha)_{\alpha \in \Phi}$ of $\mathfrak{g}$ is called a Chevalley basis *of $\mathfrak{g}$, and each E_α is called a* root vector.

[3] It is a normalization in the sense that if we replace $\langle \cdot | \cdot \rangle$ by a non-zero multiple of itself, then H_α do not change, while the h_α get divided by that factor.

For computational purposes it is useful to know that for given $\alpha, \beta \in \Phi$ the set $\{n \mid \alpha - n\beta \in \Phi\}$, consists of (a finite number of) *consecutive* integers. The choice of sign for $N_{\alpha\beta}$ for all $\alpha, \beta \in \Phi$ is non-trivial, since the signs that correspond to the different values of α and β need to satisfy several non-trivial coherence conditions, but they can be determined algorithmically (see [67]). Notice that Chevalley's Theorem implies that, in terms of a Chevalley basis, all structure constants of $\mathfrak{g}$ are integers.

The Killing form $\langle \cdot \mid \cdot \rangle$ allows us to measure angles and lengths of roots in $\Phi \subseteq \mathfrak{h}^*$. To do this, let $\mathfrak{h}^*_{\mathbf{R}}$ denote the real vector space which is spanned by Φ and define $\langle \cdot \mid \cdot \rangle_{\mathfrak{h}^*}$ to be the bilinear form on $\mathfrak{h}^*_{\mathbf{R}}$ which corresponds to the Killing form via the isomorphism $h \mapsto h_\lambda$. Thus, for $\lambda, \mu \in \mathfrak{h}^*$ we have that $\langle \lambda \mid \mu \rangle_{\mathfrak{h}^*} = \langle h_\lambda \mid h_\mu \rangle$. It turns out that $\langle \cdot \mid \cdot \rangle_{\mathfrak{h}^*}$ is positive definite, making $\mathfrak{h}^*_{\mathbf{R}}$ into a genuine Euclidean space. For $\alpha \in \Phi$, let $s_\alpha : \mathfrak{h}^*_{\mathbf{R}} \to \mathfrak{h}^*_{\mathbf{R}}$ be the linear map defined by

$$s_\alpha(\lambda) = \lambda - 2\frac{\langle h_\alpha \mid h_\lambda \rangle}{\langle h_\alpha \mid h_\alpha \rangle}\alpha = \lambda - \langle \lambda, H_\alpha \rangle \alpha, \tag{2.21}$$

where $\lambda \in \mathfrak{h}^*_{\mathbf{R}}$. This linear map is the reflection in the hyperplane orthogonal to α, since it fixes all roots which are orthogonal to α and since $s_\alpha(\alpha) = -\alpha$.

The *Weyl group* $\mathbf{W}$ is the group generated by $\{s_\alpha \mid \alpha \in \Phi\}$. One shows the following properties of the Weyl group. Every non-trivial element of $\mathbf{W}$ permutes at least two elements of Φ hence $\mathbf{W}$ is finite. Moreover, $\mathbf{W}$ is generated by the l reflections that correspond to the elements $\alpha_1, \ldots, \alpha_l$ of Π. The root system Φ consists either of one $\mathbf{W}$-orbit, in which case all roots have the same length, or it consists of two $\mathbf{W}$-orbits, where roots from one $\mathbf{W}$-orbit have a length which is different from the length of the vectors in the other $\mathbf{W}$-orbit. The two $\mathbf{W}$-orbits are then distinguished by calling its elements *short roots* or *long roots*, according to their lengths. Among the long roots there is precisely one that has maximal height. It is called the *highest long root* (with respect to $\mathfrak{h}$ and Π). Similarly, there is among the short roots precisely one that has maximal height, the *highest short root*. When all roots have the same length then the highest long root and the highest short root of course coincide.

For $\alpha, \beta \in \Pi$ the fact that

$$s_\alpha(\beta) = \beta - 2\frac{\langle h_\alpha \mid h_\beta \rangle}{\langle h_\alpha \mid h_\alpha \rangle}\alpha$$

is a root, implies for $\alpha \neq \beta$ that $s_\alpha(\beta) \in \Phi^+$, hence that

$$a_{ij} := 2\frac{\langle h_i \mid h_j \rangle}{\langle h_j \mid h_j \rangle} = \langle h_i \mid H_j \rangle = \langle \alpha_i, H_j \rangle \tag{2.22}$$

is a non-positive integer for $i \neq j$, and equals 2 for $i = j$.

The numbers a_{ij} are called the *Cartan integers* and the matrix $A = (a_{ij})$ is called the *Cartan matrix* of $\mathfrak{g}$ (with respect to $\mathfrak{h}$ and Π). It is a fundamental result that there is a bijection between the triples $(\mathfrak{g}, \mathfrak{h}, \Pi)$, modulo conjugation in $\mathfrak{g}$ and their Cartan matrices, modulo conjugation by a permutation matrix. In fact, $\mathfrak{g}$ can be reconstructed from its Cartan matrix by a set of generators and relations (see [156, Chapter VI]).

If we denote by θ_{ij} the angle between α_i and α_j then

$$\cos\theta_{ij} = \frac{\langle \alpha_i \,|\, \alpha_j \rangle_{\mathfrak{h}^*}}{\sqrt{\langle \alpha_i \,|\, \alpha_i \rangle_{\mathfrak{h}^*}}\sqrt{\langle \alpha_j \,|\, \alpha_j \rangle_{\mathfrak{h}^*}}} = \frac{\langle h_i \,|\, h_j \rangle}{\sqrt{\langle h_i \,|\, h_i \rangle}\sqrt{\langle h_j \,|\, h_j \rangle}}$$

so that (2.22) implies that $4\cos^2\theta_{ij} = a_{ij}a_{ji}$. Letting $n_{ij} := a_{ij}a_{ji}$ we have that if $i \neq j$ then $0, 1, 2, 3$ are the only possible values for n_{ij}, since a_{ij} is a non-positive integer when $i \neq j$. The *Dynkin diagram* of $\mathfrak{g}$ is the graph with l nodes labeled by $1, \dots, l$ such that the nodes i and j are joined with n_{ij} bonds. Notice that the integers n_{ij} do not contain enough information to determine the a_{ij}, i.e., to reconstruct the Cartan matrix: when $n_{ij} = 0$ then $a_{ij} = a_{ji} = 0$ and when $n_{ij} = 1$ then $a_{ij} = a_{ji} = -1$, but when $n_{ij} \in \{2, 3\}$ then there are two possibilities to assign the values -1 and $-n_{ij}$ to a_{ij} and a_{ji}. To resolve this ambiguity one adds an arrow to the double and the triple bonds in the Dynkin diagram which points to the shorter root ((2.22) shows that the two roots cannot have the same length). This way the Cartan matrix, and hence the whole structure of the simple Lie algebra, can be encoded in its Dynkin diagram. Analyzing the properties that root systems which come from a simple Lie algebra have and constructing all possible Dynkin diagrams that bear the corresponding properties one arrives at the well-known list of Dynkin diagrams, given in Table 2.1 (the labeling of the roots in the Dynkin diagram is the one that is used in most classical books on Lie algebras, in particular [37], [79] and [87]).

The coroots H_α, which were defined in (2.20), satisfy the axioms of a root system as well as the roots α, the *dual root system*, for which a system of simple roots can be chosen as $(H_1, \dots, H_l)$. It leads to a natural duality on the set of simple Lie algebras, which at the level of the Cartan matrix amounts to $A \leftrightarrow A^\top$. As it turns out, this duality is trivial except that it permutes the Lie algebras $\mathfrak{b}_l$ and $\mathfrak{c}_l$.

If $\lambda \in \mathfrak{h}^*_{\mathbf{R}}$ has the property that $\langle \lambda, H_\alpha \rangle \in \mathbf{Z}$ for all $\alpha \in \Phi$ then λ is called a *weight* and the set of all weight vectors is a lattice in $\mathfrak{h}^*_{\mathbf{R}}$ which is denoted by Λ and which is called the *weight lattice*. Clearly $\Phi \subseteq \Lambda$. A basis for the lattice Λ can be constructed as follows: for $i = 1, \dots, l$ let $\lambda_i \in \mathfrak{h}^*$ be such that $\langle \lambda_i, H_j \rangle = \delta_{ij}$, where $j = 1, \dots, l$. Each of the basis vectors λ_i is called a *fundamental dominant weight*, or a *weight* for short. Since $\Phi \subseteq \Lambda$, (2.22) implies that

$$\alpha_i = \sum_{k=1}^{l} a_{ik}\lambda_k, \qquad i = 1, \dots, l. \tag{2.23}$$

Table 2.1. Some data on simple Lie algebras. For each simple Lie algebra we list its rank, the order of its Weyl group $\mathbf{W}$, the determinant of its Cartan matrix A, the coefficients of highest long/short root in terms of the simple roots (only one is given if they are the same) and its Dynkin diagram. A label i in the Dynkin diagram refers to the root α_i. The Cartan matrix A is immediately written down from the Dynkin diagram.

$\mathfrak{g}$	Rank	$\#\mathbf{W}$	$\lvert A \rvert$	Highest long/short root	Dynkin diagram
$\mathfrak{a}_l$	$l \geqslant 1$	$(l+1)!$	$l+1$	$(1,1,\ldots,1)$	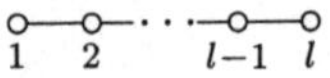
$\mathfrak{b}_l$	$l \geqslant 2$	$2^l l!$	2	$(1,2,\ldots,2)/(1,1,\ldots,1)$	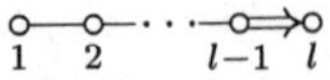
$\mathfrak{c}_l$	$l \geqslant 3$	$2^l l!$	2	$(2,\ldots,2,1)/(1,2,\ldots,2,1)$	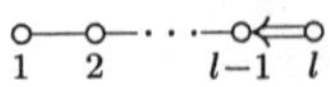
$\mathfrak{d}_l$	$l \geqslant 4$	$2^{l-1} l!$	4	$(1,2,\ldots,2,1,1)$	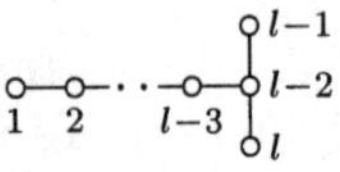
$\mathfrak{e}_6$	6	$2^7 3^4 5$	3	$(1,2,2,3,2,1)$	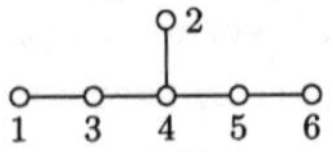
$\mathfrak{e}_7$	7	$2^{10} 3^4 5\, 7$	2	$(2,2,3,4,3,2,1)$	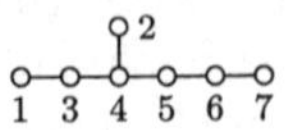
$\mathfrak{e}_8$	8	$2^{14} 3^5 5^2 7$	1	$(2,3,4,6,5,4,3,2)$	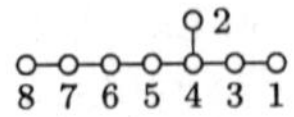
$\mathfrak{f}_4$	4	$2^7 3^2$	1	$(2,3,4,2)/(1,2,3,2)$	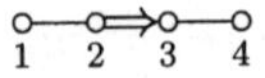
$\mathfrak{g}_2$	2	$2^2 3$	1	$(3,2)/(2,1)$	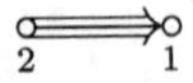

It follows from this relation that the Cartan matrix describes the change of basis from the simple roots to the fundamental dominant weights, a property that will play a fundamental rôle in our study of the periodic Toda lattice (see Chapter 9).

In the four examples that follow we give a concrete representation of the classical Lie algebras, whose root systems are $\mathfrak{a}_l$, $\mathfrak{b}_l$, $\mathfrak{c}_l$ and $\mathfrak{d}_l$, together with a choice of root vectors which, supplemented with a basis of $\mathfrak{h}$, form a Chevalley basis. We only give a choice for root vectors corresponding to the roots E_{α_i}, $E_{-\alpha_i}$, where $i = 1, \dots, l$, and to plus and minus the highest long/short root, because the other root vectors will not be needed. The choices that we make are the most appropriate for our approach to the periodic Toda lattices (Chapter 9), and are taken from [36], where one also finds explicit expressions for the other root vectors. We denote by $\mathcal{E}_{ij}$ the square matrix (of the appropriate size) which has a 1 at position (i,j) and zeros elsewhere and Δ is the $l \times l$ matrix with 1's on the anti-diagonal and zeros elsewhere, $\Delta := \sum_{i=1}^{l} \mathcal{E}_{i,l-i+1}$. Notice that the condition $A\Delta = \Delta A^\top$ (resp. $A\Delta + \Delta A^\top = 0$) means that A is symmetric (resp. skew-symmetric) with respect to its anti-diagonal.

Example 2.12. $\mathfrak{a}_l$ is the root system of the semi-simple Lie algebra $\mathfrak{sl}(l+1)$ of all traceless matrices of size $l+1$. For E_{α_i} one chooses $\mathcal{E}_{i,i+1}$ and for the root vector corresponding to the highest (long = short) root α_0 one takes $E_{\alpha_0} := \mathcal{E}_{1,l+1}$. Then $E_{-\alpha_i} := E_{\alpha_i}^\top$ for $i = 0, \dots, l$.

Example 2.13. $\mathfrak{b}_l$ is the root system of the semi-simple Lie algebra of all block matrices of size $2l+1$ of the form

$$\begin{pmatrix} A & 2\Delta w & B \\ v^\top & 0 & w^\top \\ C & 2\Delta v & D \end{pmatrix}, \qquad \text{where} \qquad \begin{aligned} A\Delta + \Delta D^\top &= 0, \\ B\Delta + \Delta B^\top &= 0, \\ C\Delta + \Delta C^\top &= 0, \end{aligned}$$

and where $A, \dots, D$ are square matrices of size l, while v and w are column vectors on length l. For the root vectors of height ± 1 we choose

$$\begin{aligned} E_{\alpha_i} &:= \mathcal{E}_{i,i+1} - \mathcal{E}_{2l-i+1,2l-i+2}, && i = 1, \dots, l-1, \\ E_{-\alpha_i} &:= \mathcal{E}_{i+1,i} - \mathcal{E}_{2l-i+2,2l-i+1}, && i = 1, \dots, l-1, \\ E_{\alpha_l} &:= 2\mathcal{E}_{l,l+1} + \mathcal{E}_{l+1,l+2}, \\ E_{-\alpha_l} &:= \mathcal{E}_{l+1,l} + 2\mathcal{E}_{l+2,l+1}, \end{aligned}$$

while for the root vectors corresponding to the highest long/short roots we choose

$$\begin{aligned} E_{\alpha} &:= \mathcal{E}_{1,2l} - \mathcal{E}_{2,2l+1}, && \alpha \text{ highest long root}, \\ E_{-\alpha} &:= \mathcal{E}_{2l,1} - \mathcal{E}_{2l+1,2}, && \alpha \text{ highest long root}, \\ E_{\alpha} &:= 2\mathcal{E}_{1,l+1} + \mathcal{E}_{l+1,2l+1}, && \alpha \text{ highest short root}, \\ E_{-\alpha} &:= \mathcal{E}_{l+1,1} + 2\mathcal{E}_{2l+1,l+1}, && \alpha \text{ highest short root}. \end{aligned}$$

Example 2.14. $\mathfrak{c}_l$ is the root system of the semi-simple Lie algebra of all block matrices of size $2l$ of the form

$$\begin{pmatrix} A & B \\ C & D \end{pmatrix}, \qquad \text{where} \qquad \begin{aligned} A\Delta + \Delta D^\top &= 0, \\ B\Delta &= \Delta B^\top, \\ C\Delta &= \Delta C^\top, \end{aligned}$$

and where $A, \ldots, D$ are square matrices of size l. For the root vectors of height ± 1 we choose

$$\begin{aligned} E_{\alpha_i} &:= \mathcal{E}_{i,i+1} - \mathcal{E}_{2l-i,2l-i+1}, && i = 1, \ldots, l-1, \\ E_{-\alpha_i} &:= \mathcal{E}_{i+1,i} - \mathcal{E}_{2l-i+1,2l-i}, && i = 1, \ldots, l-1, \\ E_{\alpha_l} &:= \mathcal{E}_{l,l+1}, \\ E_{-\alpha_l} &:= \mathcal{E}_{l+1,l}, \end{aligned}$$

while for the root vectors corresponding to the highest long/short roots we choose

$$\begin{aligned} E_\alpha &:= \mathcal{E}_{1,2l}, && \alpha \text{ highest long root}, \\ E_{-\alpha} &:= \mathcal{E}_{2l,1}, && \alpha \text{ highest long root}, \\ E_\alpha &:= \mathcal{E}_{1,2l-1} + \mathcal{E}_{2,2l}, && \alpha \text{ highest short root}, \\ E_{-\alpha} &:= \mathcal{E}_{2l-1,1} + \mathcal{E}_{2l,2}, && \alpha \text{ highest short root}. \end{aligned}$$

Example 2.15. $\mathfrak{d}_l$ is the root system of the semi-simple Lie algebra of all block matrices of size $2l$ of the form

$$\begin{pmatrix} A & B \\ C & D \end{pmatrix}, \qquad \text{where} \qquad \begin{aligned} A\Delta + \Delta D^\top &= 0, \\ B\Delta + \Delta B^\top &= 0, \\ C\Delta + \Delta C^\top &= 0, \end{aligned}$$

and where $A, \ldots, D$ are square matrices of size l. For the root vectors of height ± 1 we choose

$$\begin{aligned} E_{\alpha_i} &:= \mathcal{E}_{i,i+1} - \mathcal{E}_{2l-i,2l-i+1}, && i = 1, \ldots, l-1, \\ E_{-\alpha_i} &:= \mathcal{E}_{i+1,i} - \mathcal{E}_{2l-i+1,2l-i}, && i = 1, \ldots, l-1, \\ E_{\alpha_l} &:= \mathcal{E}_{l-1,l+1} - \mathcal{E}_{l,l+2}, \\ E_{-\alpha_l} &:= \mathcal{E}_{l+1,l-1} - \mathcal{E}_{l+2,l}, \end{aligned}$$

while for the root vectors corresponding to the highest (long = short) root α we choose

$$\begin{aligned} E_\alpha &:= \mathcal{E}_{1,2l-1} - \mathcal{E}_{2,2l}, \\ E_{-\alpha} &:= \mathcal{E}_{2l-1,1} - \mathcal{E}_{2l,2}. \end{aligned}$$

2.3.2 Invariant Functions and Exponents

We have defined in Section 2.2 the algebra of Ad-invariant functions on $\mathfrak{g}$ and the algebra of Ad^*-invariant functions on $\mathfrak{g}^*$. If $\mathfrak{g}$ is simple then these are isomorphic algebras since they correspond to each other under the isomorphism $\mathfrak{g} \to \mathfrak{g}^*$, defined by the Killing form. Moreover, they are isomorphic to a polynomial algebra which is generated by l homogeneous elements, since one proves that

$$\mathcal{F}(\mathfrak{g}^*)^{\mathbf{G}} \cong \mathcal{F}(\mathfrak{g})^{\mathbf{G}} \cong \mathbf{C}[I_1, \dots, I_l].$$

We define the *exponents* of $\mathfrak{g}$ to be the l integers $(m_1, \dots, m_l)$, where $m_i := (\deg I_i) - 1$. One has that $m_i + m_{l-i+1}$ is independent of i, and is equal to the so-called *Coxeter number* of $\mathfrak{g}$. It is a fundamental fact that the order of the Weyl group is given by

$$\#\mathbf{W} = \prod_{i=1}^{l} (m_i + 1).$$

For future use, we also give an alternative formula for the latter, namely let $\mathfrak{h}$ by a Cartan subalgebra of $\mathfrak{g}$ and let Π be a system of simple roots with respect to $\mathfrak{h}$, with Cartan matrix A. Then

$$\#\mathbf{W} = l! \prod_{i=1}^{l} \eta_i \det A, \tag{2.24}$$

where $\eta_1, \dots, \eta_l$ are the coefficients of the highest long root with respect to Π. For a proof of this (non-trivial) fact, see [37, Chapter VI no 2.4].

Below we give the list of exponents for all simple Lie algebras. Notice that a Lie algebra and its dual have the same exponents (i.e., for $\mathfrak{b}_l$ and $\mathfrak{c}_l$ they are the same; the other Lie algebras coincide with their duals, hence for those the statement is trivial).

The following proposition will play an important role in the study of Toda lattices (see Section 9.2).

Proposition 2.16. *Let* $N := \mathrm{diag}(n_1, \dots, n_l)$*, where the integers* n_i *are defined by* $\sum_{\alpha \subset \Phi_+} H_\alpha = \sum_{i=1}^{l} n_i H_i$ *and consider the linear operator*

$$\Psi : \mathfrak{h} \to \mathfrak{h}$$
$$X \mapsto \sum_{i=1}^{l} n_i H_i \langle h_i \mid X \rangle,$$

whose matrix is NA*, in the basis* $(H_1, \dots, H_l)$*. The spectrum of* Ψ*, and hence of* NA*, is expressible in terms of the exponents of* $\mathfrak{g}$ *as follows:*

$$\mathrm{Spec}(\Psi) = \{m_1(m_1+1), \dots, m_l(m_l+1)\}.$$

Table 2.2. More data on simple Lie algebras: for each type we give the l exponents and the Coxeter number, which is the sum of the i-th and $(l+1-i)$-th exponents. The dual Weyl integers n_i are defined in Proposition 2.16.

$\mathfrak{g}$	Exponents	Coxeter	n_i or $(n_1,\ldots,n_l)$
$\mathfrak{a}_l$	$1,2,\ldots,l$	$l+1$	$i(l-i+1)$
$\mathfrak{b}_l$	$1,3,5,\ldots,2l-1$	$2l$	$i(2l-i+1)-\delta_{il}\binom{l+1}{2}$
$\mathfrak{c}_l$	$1,3,5,\ldots,2l-1$	$2l$	$i(2l-i)$
$\mathfrak{d}_l$	$1,3,\ldots,2l-3,l-1$	$2l-2$	$i(2l-i-1)-(\delta_{il}+\delta_{i,l-1})\binom{l}{2}$
$\mathfrak{e}_6$	$1,4,5,7,8,11$	12	$(16,22,30,42,30,16)$
$\mathfrak{e}_7$	$1,5,7,9,11,13,17$	18	$(34,49,66,96,75,52,27)$
$\mathfrak{e}_8$	$1,7,11,13,17,19,23,29$	30	$2(46,68,91,135,110,84,57,29)$
$\mathfrak{f}_4$	$1,5,7,11$	12	$(16,30,42,22)$
$\mathfrak{g}_2$	$1,5$	6	$(10,6)$

Proof. The proposition can be checked case by case by going through the list of simple Lie algebras (see Example 2.17 below). We give a representation theory proof, which was provided to us by Eric Sommers. Let (e,f,h) be an *S-triplet* for $\mathfrak{g}$, i.e., they are non-zero elements of $\mathfrak{g}$ which satisfy the standard $\mathfrak{sl}(2)$ commutation relations

$$[h,e]=2e, \quad [h,f]=-2f, \quad [e,f]=h.$$

Let us denote the corresponding adjoint representation of $\mathfrak{sl}(2)$ on $\mathfrak{g}$ by χ. We will suppose that the S-triplet is a *principal S-triplet* which means that $\mathfrak{g}^e$, the *centralizer* of e (the subspace of all elements that commute with e), satisfies

$$\dim \mathfrak{g}^e = \min\{\mathfrak{g}^x \mid x \in \mathfrak{g}\}. \tag{2.25}$$

Then the representation χ decomposes in precisely $l = \mathrm{Rk}\,\mathfrak{g}$ irreducible subrepresentations, $\chi_i : \mathfrak{sl}(2) \to \mathrm{End}(S_i)$ $(i=1,\ldots,l)$, where $\dim S_i = 2m_i+1$. Decomposing S_i further into eigenspaces of $[h,\cdot]$ we can write for a fixed $1 \leqslant i \leqslant l$

$$S_i = \mathbf{C}v_{i,-m_i} \oplus \mathbf{C}v_{i,1-m_i} \oplus \cdots \oplus \mathbf{C}v_{i,m_i}$$

where the action of e, f and h is described by

$$\begin{aligned}
[e, v_{i,j}] &= (m_i + j + 1)v_{i,j+1}, \\
[h, v_{i,j}] &= 2j v_{i,j}, \qquad\qquad j = -m_i, \ldots, m_i, \\
[f, v_{i,j}] &= (m_i - j + 1)v_{i,j-1},
\end{aligned}$$

where $v_{i,-m_i-1} = v_{i,m_i+1} = 0$. Notice that

$$\mathfrak{h} = \bigoplus_{i=1}^{l} \mathbf{C} v_{i,0}$$

so that $\mathrm{ad}_f \circ \mathrm{ad}_e$ restricts to an endomorphism ψ of $\mathfrak{h}$, which is given by $\psi(v_{i,0}) = m_i(m_i + 1)v_{i,0}$ for $i = 1, \ldots, l$. Thus, ψ has as eigenvalues the integers $m_i(m_i + 1)$, for $i = 1, \ldots, l$.

We wish to relate ψ to Ψ. To do this we pick a particular principal S-triplet (all principal S-triplets are conjugate to each other). Choose a Chevalley basis $(H_1, \ldots, H_l) \cup (E_\alpha)_{\alpha \in \Phi}$ for $\mathfrak{g}$ and let

$$h := \sum_{\alpha \in \Phi_+} H_\alpha = \sum_{i=1}^{l} n_i H_i \tag{2.26}$$

where the latter equality is a definition of the positive integers $n_1, \ldots, n_l$. The element h, defined by (2.26), is called the *dual Weyl element* and the integers $n_1, \ldots, n_l$ are called the *dual Weyl integers.* We define for $\alpha \in \Phi$ a reflection on $\mathfrak{h}$ in analogy with the reflection s_α on $\mathfrak{h}^*$, which was defined in (2.21). For $X \in \mathfrak{h}$ let

$$\sigma_\alpha(X) := X - \langle \alpha, X \rangle H_\alpha \tag{2.27}$$

It is the reflection with respect to the hyperplane orthogonal to the coroot H_α. Indeed, if X is orthogonal to H_α then $\langle \alpha, X \rangle = \langle h_\alpha \mid X \rangle = 0$, hence σ_α fixes X, while $\sigma_\alpha(H_\alpha) = -H_\alpha$ because $\langle \alpha, H_\alpha \rangle = 2$. For $1 \leqslant i \leqslant l$ the reflection σ_{α_i} permutes all coroots H_α, with $\alpha \in \Phi^+ \setminus \{\alpha_i\}$, so that

$$\sigma_{\alpha_i}\left(\sum_{\alpha \in \Phi^+} H_\alpha \right) = \sum_{\alpha \in \Phi^+} H_\alpha - 2H_i. \tag{2.28}$$

Combining (2.27) and (2.28) we find that

$$\langle \alpha_i, h \rangle = \left\langle \alpha_i, \sum_{\alpha \in \Phi^+} H_\alpha \right\rangle = 2, \qquad \text{for } i = 1, \ldots, l, \tag{2.29}$$

which characterizes the dual Weyl element. Furthermore, let e and f be defined by

$$e := \sum_{i=1}^{l} E_{\alpha_i}, \qquad f := \sum_{i=1}^{l} n_i E_{-\alpha_i}.$$

e satisfies (2.25) and e, f, h satisfy the $\mathfrak{sl}(2)$ commutation relations, as follows from Chevalley's Theorem (Theorem 2.11) and (2.29). Thus, (e, f, h) is a principal S-triplet. For $k = 1, \dots, l$ we have that

$$\begin{aligned}\psi(H_k) &= \mathrm{ad}_f \circ \mathrm{ad}_e\, H_k \\ &= \left[\sum_{i=1}^{l} n_i E_{-\alpha_i}, \left[\sum_{j=1}^{l} E_{\alpha_j}, H_k\right]\right] \\ &= -\sum_{i,j=1}^{l} \left[n_i E_{-\alpha_i}, \langle \alpha_j, H_k\rangle E_{\alpha_j}\right] \\ &= \sum_{i=1}^{l} n_i H_i \langle \alpha_i, H_k\rangle \\ &= \sum_{i=1}^{l} n_i H_i \langle h_i \mid H_k\rangle \\ &= \Psi(H_k),\end{aligned}$$

showing that $\psi = \Psi$. Since we have shown that the eigenvalues of ψ are the integers $m_i(m_i + 1)$, where $i = 1, \dots, l$, this yields the announced eigenvalues for Ψ, and hence for the matrix NA. □

Example 2.17. Let us verify Proposition 2.16 by direct computation for one of the simple Lie algebras, say for $\mathfrak{f}_4$. We find from the last columns of Tables 2.1 and 2.2 that

$$A = \begin{pmatrix} 2 & -1 & 0 & 0 \\ -1 & 2 & -2 & 0 \\ 0 & -1 & 2 & -1 \\ 0 & 0 & -1 & 2 \end{pmatrix} \qquad N = \begin{pmatrix} 16 & 0 & 0 & 0 \\ 0 & 30 & 0 & 0 \\ 0 & 0 & 42 & 0 \\ 0 & 0 & 0 & 22 \end{pmatrix}.$$

The exponents of $\mathfrak{f}_4$ are $(1, 5, 7, 11)$, as can be read of from the second column of Table 2.2, while the eigenvalues of

$$NA = \begin{pmatrix} 32 & -16 & 0 & 0 \\ -30 & 60 & -60 & 0 \\ 0 & -42 & 84 & -42 \\ 0 & 0 & -22 & 44 \end{pmatrix}$$

are given by $2 = 1.2$, $30 = 5.6$, $56 = 7.8$, $132 = 11.12$, as follows from a direct computation. Clearly, this corresponds to the eigenvalues, predicted by Proposition 2.16.

2.4 Twisted Affine Lie Algebras

For any Lie algebra $\mathfrak{g}$ and for any element $g \in \mathbf{G}$, the linear map $\mathrm{Ad}_g : \mathfrak{g} \to \mathfrak{g}$ is an automorphism of $\mathfrak{g}$, which is called an *inner automorphism.* The group of outer automorphisms $\Gamma(\mathfrak{g})$ is by definition the group of all automorphisms, modulo the inner automorphisms. If $\mathfrak{g}$ is simple then any element of $\Gamma(\mathfrak{g})$ is represented by a (unique) automorphism of $\mathfrak{g}$ which is induced by an automorphism of the Dynkin diagram of $\mathfrak{g}$. Therefore, $\Gamma(\mathfrak{g})$ can be identified naturally with the group of automorphisms of the Dynkin diagram of $\mathfrak{g}$. By inspecting Table 2.1 one finds that only a few Dynkin diagrams admit a non-trivial automorphisms; those are given in Table 2.2.

Table 2.3. We list the simple Lie algebras $\mathfrak{g}$ which admit a non-trivial group $\Gamma(\mathfrak{g})$ of outer automorphisms. We give the possible values for the order of its elements.

$\mathfrak{g}$	Rank	$\Gamma(\mathfrak{g})$	order(ν)
$\mathfrak{a}_l$	$l > 1$	$\mathbf{Z}/2\mathbf{Z}$	2
$\mathfrak{d}_4$	$l = 4$	S_3	2, 3
$\mathfrak{d}_l$	$l > 4$	$\mathbf{Z}/2\mathbf{Z}$	2
$\mathfrak{e}_6$	6	$\mathbf{Z}/2\mathbf{Z}$	2

Let ν be an automorphism of $\mathfrak{g}$ which is induced by a diagram automorphism, and let us denote its order by m. Since $\nu^m = \mathrm{Id}_{\mathfrak{g}}$ each eigenvalue of ν has the form ε^i, where ε is a primitive m^{th} root of unity and $0 \leqslant i \leqslant m-1$. The eigenspace of ν which corresponds to this eigenvalue ε^i is denoted by $\mathfrak{g}_i$. Then the algebra $\mathfrak{g}$ admits the following finite grading:

$$\mathfrak{g} = \bigoplus_{i \in \mathbf{Z}_m} \mathfrak{g}_i \quad \text{and} \quad [\mathfrak{g}_i, \mathfrak{g}_j] \subseteq \mathfrak{g}_{i+j}.$$

We now define the (infinite-dimensional) *twisted affine Lie algebra* of $(\mathfrak{g}, \nu)$.

$$L(\mathfrak{g}, \nu) := \left\{ \sum_{j-M}^{N} \mathfrak{h}^j X_j \mid M, N \in \mathbf{Z} \text{ and } X_j \in \mathfrak{g}_{j \bmod m} \text{ for } M \leqslant j \leqslant N \right\}.$$

Notice that if we extend ν in the obvious way to elements of the form $X(\mathfrak{h}) = \sum_{j=M}^{N} \mathfrak{h}^j X_j$, then elements of $L(\mathfrak{g}, \nu)$ are characterized by the property $X(\varepsilon^p \mathfrak{h}) = \nu^p X(\mathfrak{h})$, for $p = 1, \ldots, m-1$. When $\nu = \mathrm{Id}_{\mathfrak{g}}$ then

$$L(\mathfrak{g}) := L(\mathfrak{g}, \mathrm{Id}_{\mathfrak{g}}) = \mathfrak{g} \otimes \mathbf{C}\left[\mathfrak{h}, \mathfrak{h}^{-1}\right]$$

the *affine Lie algebra* of $\mathfrak{g}$. The term *loop algebra* is also used. A natural Lie bracket on $L(\mathfrak{g}, \nu)$ is given by

$$\left[\sum_{i\leqslant N} \mathfrak{h}^i X_i, \sum_{j\leqslant M} \mathfrak{h}^j Y_j\right] = \sum_{k\leqslant M+N} \mathfrak{h}^k \left(\sum_{i+j=k} [X_i, Y_j]\right)$$

and the Killing form $\langle\cdot\,|\,\cdot\rangle$ on $\mathfrak{g}$ leads for every $k \in \mathbf{Z}$ to a non-degenerate symmetric form on $L(\mathfrak{g}, \nu)$, denoted by $\langle\cdot\,|\,\cdot\rangle_k$ which is defined by

$$\left\langle \sum_{i\leqslant N} \mathfrak{h}^i X_i \,\Big|\, \sum_{j\leqslant M} \mathfrak{h}^j Y_j \right\rangle_k := \sum_{i+j+k=0} \langle X_i \,|\, Y_j \rangle . \tag{2.30}$$

It is easy to see that each of the bilinear forms $\langle\cdot\,|\,\cdot\rangle_k$ on $L(\mathfrak{g}, \nu)$ is Ad-invariant. We will refer to $\langle\cdot\,|\,\cdot\rangle_0$ as the *Killing form* of $L(\mathfrak{g}, \nu)$.

Example 2.18. Consider the direct sum decomposition

$$L(\mathfrak{g}) = L(\mathfrak{g})_+ \oplus L(\mathfrak{g})_-,$$

where $L(\mathfrak{g})_+$ consists of those elements of $L(\mathfrak{g})$ which are polynomial in $\mathfrak{h}$, while $L(\mathfrak{g})_-$ consists of all elements of $L(\mathfrak{g})$ that are polynomial in $\mathfrak{h}^{-1}$, but without constant term. In terms of the orthogonality that is induced by the Killing form $\langle\cdot\,|\,\cdot\rangle_0$ we have that $L(\mathfrak{g})_+^\perp$ consists of those elements of $L(\mathfrak{g})$ which are polynomial in $\mathfrak{h}$, but without constant term, while $L(\mathfrak{g})_-^\perp$ consists of all elements of $L(\mathfrak{g})$ that are polynomial in $\mathfrak{h}^{-1}$.

It is possible to develop a theory of roots for twisted affine Lie algebras, which is analogous to the one for simple Lie algebras. We will first start with the easier case of (untwisted) affine Lie algebras. Let $\mathfrak{g}$ be a simple Lie algebra, let $\mathfrak{h}$ be a Cartan subalgebra with root system Φ and let $\Pi = \{\alpha_1, \ldots, \alpha_l\}$ be a system of simple roots. By definition, a *root* of $L(\mathfrak{g})$ is a pair $(\alpha, i) \neq (0,0)$, where $\alpha \in \Phi \cup \{0\}$ and $i \in \mathbf{Z}$; such pairs are added in the obvious way: $(\alpha, i) + (\beta, j) = (\alpha+\beta, i+j)$. We denote the set of all roots of $L(\mathfrak{g})$ by $\bar{\Phi}$ and we call $\bar{\Phi}$ the *root system* of $L(\mathfrak{g})$. Let α_0 denote minus the highest long root of $\mathfrak{g}$, and notice that α_0 is the unique root of $L(\mathfrak{g})$ which has the property that no decomposition of the form $(\alpha_0, 1) = (\alpha, 1) + (\beta, 0)$, with $\alpha \in \Phi$ and $\beta \in \Pi$ is possible. One calls $\bar{\alpha}_0 := (\alpha_0, 1)$ the *lowest root* of $L(\mathfrak{g})$. Define

$$\bar{\Pi} = \{\bar{\alpha}_0 = (\alpha_0, 1), \bar{\alpha}_1 = (\alpha_1, 0), \ldots, \bar{\alpha}_l = (\alpha_l, 0)\} .$$

Using the fact that $-\alpha_0$ is the highest long root of $\mathfrak{g}$, it is easy to show that every root of $L(\mathfrak{g})$ can be written uniquely as a linear combination of the elements of $\bar{\Phi}$, where all coefficients belong to $\mathbf{Z}_+$ or they all belong to $\mathbf{Z}_-$.

Thus, $\bar{\Pi}$ is the natural analogue of Π, so we will call it a *system of simple roots* for $L(\mathfrak{g})$.

To a root $(\alpha, j) \in \bar{\Phi}$ we associate the vector $E_{(\alpha,j)} := E_\alpha \mathfrak{h}^j$, where E_α is the root vector that corresponds to α (see Theorem 2.11). We will call $E_{(\alpha,j)}$ the *root vector* which corresponds to (α, j). It follows easily from Theorem 2.11 that these root vectors satisfy the following relations: for any $\bar{\alpha} = (\alpha, j) \in \bar{\Phi}$,

$$\begin{aligned} [H, E_{\bar{\alpha}}] &= \langle \alpha, H \rangle E_{\bar{\alpha}}, \qquad H \in \mathfrak{h}, \\ [E_{\bar{\alpha}}, E_{-\bar{\alpha}}] &= H_\alpha. \end{aligned} \tag{2.31}$$

The *Cartan matrix* A of $L(\mathfrak{g})$ is constructed from the system of simple roots as before, namely

$$a_{ij} := \langle \alpha_i, H_j \rangle ,$$

except that the indices i, j can now also take the value 0, besides the values $1, \dots, l$ (the numbering of the rows and columns of these bigger Cartan matrices starts from 0). The Cartan matrix of $L(\mathfrak{g})$ and its Dynkin diagram are easily computed from the one of $\mathfrak{g}$ and the coefficients $\xi_1, \dots, \xi_l$ of the highest long root (these coefficients are listed in Table 2.1). Indeed, one only needs to compute the first row and the first column of A, since the remaining block is precisely the Cartan matrix of $\mathfrak{g}$. In order to compute the first row of A, whose first element a_{00} is 2, it suffices to express that $\xi = (\xi_0 = 1, \xi_1, \dots, \xi_l)^\top$ is a (normalized) null-vector of A, which follows from the fact that

$$\alpha_0 = \sum_{i=1}^{l} \xi_i \alpha_i,$$

upon taking inner products with H_j, $0 \leqslant j \leqslant l$. Note that once again we have a duality between the system of roots $\{\bar{\alpha} \mid \alpha \in \bar{\Pi}\}$ and coroots $\{H_{\bar{\alpha}} \mid \alpha \in \bar{\Pi}\}$, which amounts to $A \leftrightarrow A^\top$, inducing a duality between the $L(\mathfrak{g}, \nu)$.

By a direct computation for each of the simple Lie algebras we find that there is in each case one non-zero entry in the first row, besides the leading 2. Therefore the same is true for the first column of A. That non-zero entry is then computed by expressing that the first element of $\xi^\top A$ is zero. The resulting matrices are given for each of the affine Lie algebras in Table 2.4; in this table the case $\mathfrak{a}_1^{(1)}$ and $\mathfrak{b}_2^{(1)}$ should be interpreted properly: the Cartan matrix of $\mathfrak{a}_1^{(1)}$ is $\begin{pmatrix} 2 & -2 \\ -2 & 2 \end{pmatrix}$, as follows from the fact that $\xi^\top = (1, 1)$.

We now turn to the case of twisted affine Lie algebras. Suppose that $\mathfrak{g}$ is a simple Lie algebra and that ν is an automorphism which corresponds to a non-trivial diagram automorphism of the Dynkin diagram of $\mathfrak{g}$. This means that $\mathfrak{g}$ is $\mathfrak{a}_l$ or $\mathfrak{d}_l$ or $\mathfrak{e}_6$ and the order of the automorphism ν is two, except in case $\mathfrak{d}_4$, for which we can also consider an automorphism of order 3.

Table 2.4. For each of the affine Lie algebras we give its Dynkin diagram, its Cartan matrix A, the normalized null-vectors ξ and $\hat{\xi}$ of $A^\top$ resp. of A (only one is given when they are the same) and the vector η that contains the coefficients of the highest weight vector.

$\mathfrak{g}$	Dynkin diagram	Cartan matrix	ξ, $\hat{\xi}$	η
$\mathfrak{a}_l^{(1)}$ $(l \geqslant 1)$	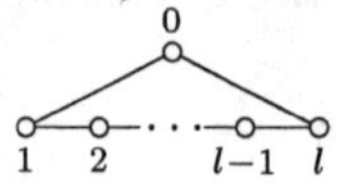	$\begin{pmatrix} 2 & -1 & & & -1 \\ -1 & 2 & & & \\ & & \ddots & & \\ & & & \ddots & -1 \\ -1 & & & -1 & 2 \end{pmatrix}$	$\begin{pmatrix} 1 \\ 1 \\ \vdots \\ 1 \end{pmatrix}$	$\begin{pmatrix} 1 \\ 1 \\ \vdots \\ 1 \end{pmatrix}$
$\mathfrak{b}_l^{(1)}$ $(l > 2)$	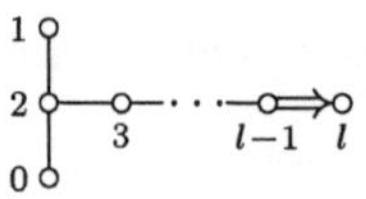	$\begin{pmatrix} 2 & 0 & -1 & 0 & & & \\ 0 & 2 & -1 & 0 & & & \\ -1 & -1 & 2 & -1 & & & \\ 0 & 0 & -1 & 2 & & & \\ & & & & \ddots & & \\ & & & & 2 & -1 & 0 \\ & & & & -1 & 2 & -2 \\ & & & & 0 & -1 & 2 \end{pmatrix}$	$\begin{pmatrix} 1 \\ 1 \\ 2 \\ \vdots \\ 2 \end{pmatrix}, \begin{pmatrix} 1 \\ 1 \\ 2 \\ \vdots \\ 2 \\ 1 \end{pmatrix}$	$\begin{pmatrix} 1 \\ 2 \\ \vdots \\ 2 \end{pmatrix}$
$\mathfrak{c}_l^{(1)}$ $(l > 1)$	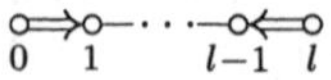	$\begin{pmatrix} 2 & -2 & 0 & 0 & & & \\ -1 & 2 & -1 & 0 & & & \\ 0 & -1 & 2 & -1 & & & \\ 0 & 0 & -1 & 2 & & & \\ & & & & \ddots & & \\ & & & & 2 & -1 & 0 \\ & & & & -1 & 2 & -1 \\ & & & & 0 & -2 & 2 \end{pmatrix}$	$\begin{pmatrix} 1 \\ 2 \\ \vdots \\ 2 \\ 1 \end{pmatrix}, \begin{pmatrix} 1 \\ 1 \\ \vdots \\ 1 \end{pmatrix}$	$\begin{pmatrix} 2 \\ \vdots \\ 2 \\ 1 \end{pmatrix}$
$\mathfrak{d}_l^{(1)}$ $(l > 3)$	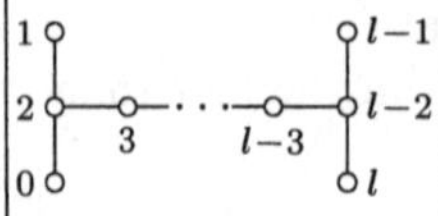	$\begin{pmatrix} 2 & 0 & -1 & 0 & & & \\ 0 & 2 & -1 & 0 & & & \\ -1 & -1 & 2 & -1 & & & \\ 0 & 0 & -1 & 2 & & & \\ & & & & \ddots & & \\ & & & & 2 & -1 & -1 \\ & & & & -1 & 2 & 0 \\ & & & & -1 & 0 & 2 \end{pmatrix}$	$\begin{pmatrix} 1 \\ 1 \\ 2 \\ \vdots \\ 2 \\ 1 \\ 1 \end{pmatrix}$	$\begin{pmatrix} 1 \\ 2 \\ \vdots \\ 2 \\ 1 \\ 1 \end{pmatrix}$
$\mathfrak{e}_6^{(1)}$	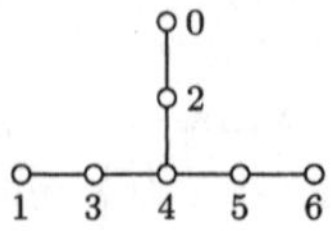	$\begin{pmatrix} 2 & 0 & -1 & 0 & 0 & 0 & 0 \\ 0 & 2 & 0 & -1 & 0 & 0 & 0 \\ -1 & 0 & 2 & 0 & -1 & 0 & 0 \\ 0 & -1 & 0 & 2 & -1 & 0 & 0 \\ 0 & 0 & -1 & -1 & 2 & -1 & 0 \\ 0 & 0 & 0 & 0 & -1 & 2 & -1 \\ 0 & 0 & 0 & 0 & 0 & -1 & 2 \end{pmatrix}$	$\begin{pmatrix} 1 \\ 1 \\ 2 \\ 2 \\ 3 \\ 2 \\ 1 \end{pmatrix}$	$\begin{pmatrix} 1 \\ 2 \\ 2 \\ 3 \\ 2 \\ 1 \end{pmatrix}$
$\mathfrak{e}_7^{(1)}$	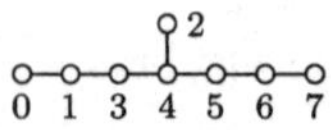	$\begin{pmatrix} 2 & -1 & & & & & & \\ -1 & 2 & 0 & -1 & & & & \\ & 0 & 2 & 0 & -1 & & & \\ & -1 & 0 & 2 & -1 & & & \\ & & -1 & -1 & 2 & -1 & & \\ & & & & -1 & 2 & -1 & \\ & & & & & -1 & 2 & -1 \\ & & & & & & -1 & 2 \end{pmatrix}$	$\begin{pmatrix} 1 \\ 2 \\ 2 \\ 3 \\ 4 \\ 3 \\ 2 \\ 1 \end{pmatrix}$	$\begin{pmatrix} 2 \\ 2 \\ 3 \\ 4 \\ 3 \\ 2 \\ 1 \end{pmatrix}$
$\mathfrak{e}_8^{(1)}$	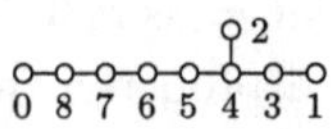	$\begin{pmatrix} 2 & 0 & & & & & & & -1 \\ 0 & 2 & 0 & -1 & & & & & \\ & 0 & 2 & 0 & -1 & & & & \\ & -1 & 0 & 2 & -1 & & & & \\ & & -1 & -1 & 2 & -1 & & & \\ & & & & -1 & 2 & -1 & & \\ & & & & & -1 & 2 & -1 & \\ & & & & & & -1 & 2 & -1 \\ -1 & & & & & & & -1 & 2 \end{pmatrix}$	$\begin{pmatrix} 1 \\ 2 \\ 3 \\ 4 \\ 6 \\ 5 \\ 4 \\ 3 \\ 2 \end{pmatrix}$	$\begin{pmatrix} 2 \\ 3 \\ 4 \\ 6 \\ 5 \\ 4 \\ 3 \\ 2 \end{pmatrix}$
$\mathfrak{f}_4^{(1)}$	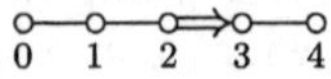	$\begin{pmatrix} 2 & -1 & 0 & 0 & 0 \\ -1 & 2 & -1 & 0 & 0 \\ 0 & -1 & 2 & -2 & 0 \\ 0 & 0 & -1 & 2 & -1 \\ 0 & 0 & 0 & -1 & 2 \end{pmatrix}$	$\begin{pmatrix} 1 \\ 2 \\ 3 \\ 4 \\ 2 \end{pmatrix}, \begin{pmatrix} 1 \\ 2 \\ 3 \\ 2 \\ 1 \end{pmatrix}$	$\begin{pmatrix} 2 \\ 3 \\ 4 \\ 2 \end{pmatrix}$
$\mathfrak{g}_2^{(1)}$	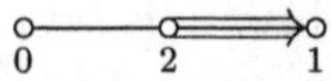	$\begin{pmatrix} 2 & 0 & -1 \\ 0 & 2 & -1 \\ -1 & -3 & 2 \end{pmatrix}$	$\begin{pmatrix} 1 \\ 3 \\ 2 \end{pmatrix}, \begin{pmatrix} 1 \\ 1 \\ 2 \end{pmatrix}$	$\begin{pmatrix} 3 \\ 2 \end{pmatrix}$

By definition a *root* of $L(\mathfrak{g},\nu)$ is a pair $(\alpha, j) \neq (0,0)$, with $\alpha \in \mathfrak{h}^*$ and $j \in \mathbf{Z}$, such that the joint eigenspace

$$\{X \in \mathfrak{g}_j \mid [H,X] = \langle \alpha, H\rangle\, X \text{ for any } H \in \mathfrak{h}\}$$

is non-trivial (in the untwisted case this definition is equivalent to the one that we have given). We denote the set of all roots of $L(\mathfrak{g},\nu)$ by $\bar{\Phi}$ and call it the *root system* of $L(\mathfrak{g},\nu)$. There are two main differences with the case of (untwisted) affine Lie algebras. First, the Lie algebra $\mathfrak{g}_0$ is different from $\mathfrak{g}$, but it is still one of the simple Lie algebras; for each $(\mathfrak{g},\nu)$ with $\nu \neq \mathrm{Id}_{\mathfrak{g}}$ the corresponding simple Lie algebra $\mathfrak{g}_0$ is given in Table 2.5. Second, the root α_0 which is used to define the lowest root $\bar{\alpha}_0 = (\alpha_0, 1)$ of $L(\mathfrak{g})$, takes now the following form

$$\begin{aligned} \alpha_0 &= -2(\text{highest short root of } \mathfrak{g}_0) && \text{if } \mathfrak{g} = \mathfrak{a}_{2l}, \\ \alpha_0 &= -(\text{highest short root of } \mathfrak{g}_0) && \text{otherwise.} \end{aligned} \tag{2.32}$$

The computation of the Cartan matrices is the same as in the case of the (untwisted) affine Lie algebras. The results are displayed in Table 2.5; in this table the case $\mathfrak{a}_2^{(2)}$ should be interpreted properly: its Cartan matrix is $\begin{pmatrix} 2 & -4 \\ -1 & 2 \end{pmatrix}$, as follows from the fact that $\xi^\top = (1,2)$. A system of simple roots $\bar{\Phi}$ can be constructed also in the twisted case, where each element now belongs to $\mathfrak{g}_0$ (see [79, pp. 505–507] for explicit formulas). It leads, as before to the same formulas (2.31) for the simple roots $\bar{\alpha}$.

The Cartan matrices that we see in Tables 2.4 and 2.5 are characterized by a few of their properties, just as in the case of the Cartan matrix of a simple Lie algebra, yielding a different approach to affine Lie algebras. Start with a collection Π of $n+1$ non-zero vectors $\alpha_0, \dots, \alpha_n$ in $\mathbf{R}^n$ and let $\langle \cdot \mid \cdot \rangle$ be an inner product on $\mathbf{R}^n$. We will say that Π is an *indecomposable system of vectors* if Π cannot be split in two sets Π_1 and Π_2 such that $\langle \Pi_1 \mid \Pi_2 \rangle = 0$. The *Cartan matrix* of Π is by definition the $(n+1)\times(n+1)$ matrix A, which is defined by

$$a_{ij} := 2\frac{\langle \alpha_i \mid \alpha_j \rangle}{\langle \alpha_j \mid \alpha_j \rangle}.$$

Then one has the following proposition.

Proposition 2.19. *Let Π be a collection of $n+1$ non-zero vectors $\alpha_0, \dots, \alpha_n$ in $(\mathbf{R}^n, \langle \cdot \mid \cdot \rangle)$ and denote its Cartan matrix by A. Suppose that Π and A satisfy the following three properties:*

(1) Π is an indecomposable system of vectors;
(2) Π spans $\mathbf{R}^n$;
(3) $a_{ij} \in \mathbf{Z}_-$ for $0 \leqslant i < j \leqslant n$.

Then A is the Cartan matrix of a (twisted) affine Lie algebra.

Table 2.5. For each of the twisted affine Lie algebras we give the type of $\mathfrak{g}_0$, its Dynkin diagram, its Cartan matrix A, the normalized null-vectors ξ and $\hat{\xi}$ of $A^\top$ resp. of A and the vector η that contains the coefficients of the highest weight vector of $\mathfrak{g}_0$.

$\mathfrak{g}$	$\mathfrak{g}_0$	Dynkin diagram	Cartan matrix	ξ, $\hat{\xi}$	η
$\mathfrak{a}_{2l}^{(2)}$ $(l\geqslant 1)$	$\mathfrak{b}_l$	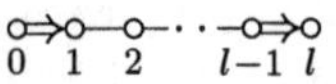	$\begin{pmatrix} 2&-2&0&0&&&\\ -1&2&-1&0&&&\\ 0&-1&2&-1&&&\\ 0&0&-1&2&&&\\ &&&&\ddots&&\\ &&&&2&-1&0\\ &&&&-1&2&-2\\ &&&&0&-1&2 \end{pmatrix}$	$\begin{pmatrix}1\\2\\ \vdots\\2\end{pmatrix}, \begin{pmatrix}1\\ \vdots\\1\\1/2\end{pmatrix}$	$\begin{pmatrix}1\\2\\ \vdots\\2\end{pmatrix}$
$\mathfrak{a}_{2l-1}^{(2)}$ $(l>2)$	$\mathfrak{c}_l$	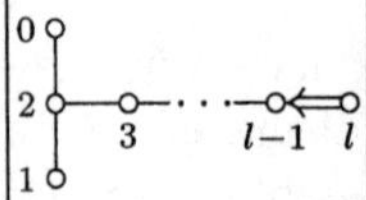	$\begin{pmatrix} 2&0&-1&0&&&\\ 0&2&-1&0&&&\\ -1&-1&2&-1&&&\\ 0&0&-1&2&&&\\ &&&&\ddots&&\\ &&&&2&-1&0\\ &&&&-1&2&-1\\ &&&&0&-2&2 \end{pmatrix}$	$\begin{pmatrix}1\\1\\2\\ \vdots\\2\\1\end{pmatrix}, \begin{pmatrix}1\\1\\2\\ \vdots\\2\end{pmatrix}$	$\begin{pmatrix}2\\ \vdots\\2\\1\end{pmatrix}$
$\mathfrak{d}_{l+1}^{(2)}$ $(l>1)$	$\mathfrak{b}_l$	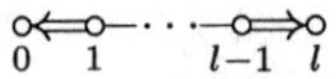	$\begin{pmatrix} 2&-1&0&0&&&\\ -2&2&-1&0&&&\\ 0&-1&2&-1&&&\\ 0&0&-1&2&&&\\ &&&&\ddots&&\\ &&&&2&-1&0\\ &&&&-1&2&-2\\ &&&&0&-1&2 \end{pmatrix}$	$\begin{pmatrix}1\\1\\ \vdots\\1\end{pmatrix}, \begin{pmatrix}1\\2\\ \vdots\\2\\1\end{pmatrix}$	$\begin{pmatrix}1\\2\\ \vdots\\2\end{pmatrix}$
$\mathfrak{e}_6^{(2)}$	$\mathfrak{f}_4$	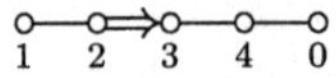	$\begin{pmatrix} 2&0&0&0&-1\\ 0&2&-1&0&0\\ 0&-1&2&-2&0\\ 0&0&-1&2&-1\\ -1&0&0&-1&2 \end{pmatrix}$	$\begin{pmatrix}1\\1\\2\\3\\2\end{pmatrix}, \begin{pmatrix}1\\2\\4\\3\\2\end{pmatrix}$	$\begin{pmatrix}2\\3\\4\\2\end{pmatrix}$
$\mathfrak{d}_4^{(3)}$	$\mathfrak{g}_2$	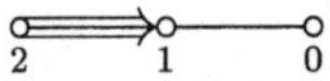	$\begin{pmatrix} 2&-1&0\\ -1&2&-1\\ 0&-3&2 \end{pmatrix}$	$\begin{pmatrix}1\\2\\1\end{pmatrix}, \begin{pmatrix}1\\2\\3\end{pmatrix}$	$\begin{pmatrix}3\\2\end{pmatrix}$

The reader can verify easily by merely looking at the tables that, conversely, each of the Cartan matrices of the (twisted) affine Lie algebras satisfies the above three properties.

We end this section with some definitions, valid for a twisted or untwisted affine Lie algebras $L(\mathfrak{g},\nu)$. Let $\bar{\Pi} = \{\bar{\alpha}_0, \bar{\alpha}_1, \ldots, \bar{\alpha}_l\}$ denote a system of simple roots of $L(\mathfrak{g},\nu)$. The *height* $|\bar{\alpha}|$ of a root $\bar{\alpha} \in \bar{\Phi}$ is defined by

$$|\bar{\alpha}| = \sum_{i=0}^{l} m_i, \quad \text{where} \quad \bar{\alpha} = \sum_{i=0}^{l} m_i \bar{\alpha}_i.$$

This induces a grading of $L(\mathfrak{g},\nu)$, similar to the grading (2.19) of $\mathfrak{g}$, namely,

$$L(\mathfrak{g},\nu) = \bigoplus_{k\in\mathbf{Z}} L_k, \quad [L_k, L_l] \subseteq L_{k+l},$$

where L_k is the span of all $E_{\bar{\alpha}}$, with $|\bar{\alpha}| = k$. The grading also leads to a natural operation of *transpose*:

$$\left(H + \sum_{\bar{\alpha}\in\bar{\Phi}} a_{\bar{\alpha}} E_{\bar{\alpha}}\right)^{\top} = H + \sum_{\bar{\alpha}\in\bar{\Phi}} a_{\bar{\alpha}} E_{-\bar{\alpha}}, \quad H \in \mathfrak{g}_0.$$

3 Poisson Manifolds

In 1809 Poisson (see [144]) introduced a bracket on smooth functions, defined on $\mathbf{R}^{2n}$, by the formula

$$\{F,G\} := \sum_{i=1}^{n} \left(\frac{\partial F}{\partial q_i} \frac{\partial G}{\partial p_i} - \frac{\partial G}{\partial q_i} \frac{\partial F}{\partial p_i} \right). \tag{3.1}$$

In this formula, F and G are arbitrary smooth functions on $\mathbf{R}^{2n}$ and $(q_1, \dots, q_n, p_1, \dots, p_n)$ are linear coordinates on $\mathbf{R}^{2n}$. He observed that if F and G are two first integrals of a mechanical system (defined on $\mathbf{R}^{2n}$) then their *Poisson bracket* $\{F, G\}$ is also a first integral. Notice that the Poisson bracket also allows one to describe the equations of motion in their most symmetric form

$$\begin{aligned} \dot{q}_i &= \{q_i, H\} \\ \dot{p}_i &= \{p_i, H\} \end{aligned} \qquad i = 1, \dots, n, \tag{3.2}$$

where $H : \mathbf{R}^{2n} \to \mathbf{R}$ is the Hamiltonian (the energy of the mechanical system, expressed in terms of position and momentum). Thirty years later, Jacobi *explained* (in [90]) Poisson's observation by showing that the bracket (3.1) satisfies the identity

$$\{\{F,G\},H\} + \{\{G,H\},F\} + \{\{H,F\},G\} = 0 \tag{3.3}$$

for all smooth functions F, G, H defined on $\mathbf{R}^{2n}$. The above identity is now known as the *Jacobi identity*. To see how Poisson's Theorem follows from the Jacobi identity it suffices to remark that K is a constant of the motion (3.2), precisely if K Poisson-commutes with H, i.e., $\{K, H\} = 0$, since

$$\dot{K} = \sum_{i=1}^{n} \frac{\partial K}{\partial q_i} \{q_i, H\} + \sum_{i=1}^{n} \frac{\partial K}{\partial p_i} \{p_i, H\} = \{K, H\}.$$

By the Jacobi identity (3.3), if F and G are constants of motion then

$$\{F,G\}^{\cdot} = \left\{\dot{F},G\right\} + \left\{F,\dot{G}\right\} = 0,$$

showing that $\{F, G\}$ is a constant of motion.

Formalizing the properties of the Poisson bracket will lead us to the notion of a Poisson manifold and we will give several basic examples of Poisson manifolds (see Section 3.1).

The Poisson bracket allows one to associate a *Hamiltonian vector field* $\mathcal{X}_H$ to any function H, defined on a Poisson manifold $(M, \{\cdot\,,\cdot\})$ by setting

$$\mathcal{X}_H := \{\cdot\,, H\},$$

and the Jacobi identity is equivalent to the fundamental formula

$$[\mathcal{X}_F, \mathcal{X}_G] = -\mathcal{X}_{\{F,G\}}.$$

In local coordinates the Hamiltonian vector field $\mathcal{X}_H$ is given by

$$\mathcal{X}_H = \sum_{j=1}^{n} \{x_j, H\} \frac{\partial}{\partial x_j}.$$

The Hamiltonian vector fields preserve the Poisson structure and they lead to a notion of rank at each point of M (see Section 3.2). Bi-Hamiltonian manifolds and bi-Hamiltonian vector fields will shortly be discussed in Section 3.3; their relevance for integrable systems will show up later.

A prime example of a Poisson manifold is that of a symplectic manifold (M, ω), i.e., ω is a closed non-degenerate two-form on M (see Example 3.4 below). Such a manifold carries a Poisson structure, which is defined for smooth functions F, G by

$$\{F, G\} := \omega(\mathcal{X}_F, \mathcal{X}_G),$$

where for $H \in \mathcal{F}(M)$ the Hamiltonian vector field $\mathcal{X}_H$ is now defined by

$$\omega(\mathcal{X}_H, \cdot) = \mathrm{d}H.$$

The following string of useful formulas shows that $\mathcal{X}_H = \{\cdot\,, H\}$ so that the above notations are coherent:

$$\{F, H\} = \omega(\mathcal{X}_F, \mathcal{X}_H) = \mathrm{d}F(\mathcal{X}_H) = \mathcal{X}_H[F].$$

We will show that, conversely, every Poisson manifold which has constant maximal rank is a symplectic manifold and that for any Poisson manifold the Hamiltonian vector fields define a generalized distribution whose leaves inherit a natural symplectic structure (see Section 3.4).

Another prime example is that of the canonical Lie-Poisson structure on the dual $\mathfrak{g}^*$ of a (finite-dimensional) Lie algebra $\mathfrak{g}$. For smooth functions F, G on $\mathfrak{g}^*$ the Poisson bracket is defined by

$$\{F, G\}(\xi) := \langle \xi, [\mathrm{d}F(\xi), \mathrm{d}G(\xi)] \rangle,$$

where $\mathrm{d}F(\xi)$ and $\mathrm{d}G(\xi)$ are interpreted as elements of $\mathfrak{g}$ when computing the bracket. The symplectic leaves of $\{\cdot\,,\cdot\}$ are in this case the coadjoint orbits. The Lie-Poisson structure of $\mathfrak{g}^*$ will be studied in detail in Section 3.5. Most of the Poisson structures in this book will be Lie-Poisson structures, or will be closely related to Lie-Poisson structures.

3.1 Basic Definitions

We start with the basic definitions. As we said, all manifolds considered here can be chosen real smooth or complex holomorphic.

Definition 3.1. Let M be a manifold. A *Poisson bracket* or *Poisson structure* on M is a Lie algebra structure $\{\cdot\,,\cdot\}$ on $\mathcal{F}(M)$, which is a biderivation on $\mathcal{F}(M)$, i.e., for any $H \in \mathcal{F}(M)$ the linear map

$$\begin{aligned} \mathcal{X}_H : \mathcal{F}(M) &\to \mathcal{F}(M) \\ F \quad &\mapsto \{F, H\} \end{aligned} \tag{3.4}$$

is a derivation on $\mathcal{F}(M)$, i.e., a vector field. The pair $(M, \{\cdot\,,\cdot\})$ is called a *Poisson manifold.* If we want to stress that M is real or complex we use the terms *real Poisson manifold* and *complex Poisson manifold.*

In many cases we will deal with Poisson structures on the affine space $M = \mathbf{R}^n$ or $M = \mathbf{C}^n$. In this case, the standard coordinates $x_1 \dots, x_n$ on M lead to *structure functions* $x_{ij} \in \mathcal{F}(M)$, which are defined by $x_{ij} := \{x_i, x_j\}$ $(1 \leqslant i, j \leqslant n)$. They satisfy

$$x_{ij} = -x_{ij}, \tag{3.5}$$

$$\sum_{l=1}^{n} \left(\frac{\partial x_{ij}}{\partial x_l} x_{lk} + \frac{\partial x_{jk}}{\partial x_l} x_{li} + \frac{\partial x_{ki}}{\partial x_l} x_{lj} \right) = 0, \tag{3.6}$$

for all $1 \leqslant i, j, k \leqslant n$. Formula (3.6) is obtained by writing the Jacobi identity for the triple (x_i, x_j, x_k) and by writing

$$\{x_{ij}, x_k\} = \sum_{l=1}^{n} \frac{\partial x_{ij}}{\partial x_l} \{x_l, x_k\},$$

a consequence of (2.5) for $\mathcal{V} = \mathcal{X}_{x_k} = \{\cdot\,, x_k\}$. Conversely, every set of functions $x_{ij} \in \mathcal{F}(M)$ (where $M = \mathbf{R}^n$ or $M = \mathbf{C}^n$, as before) satisfying the above two conditions defines a Poisson bracket on M, merely by setting

$$\{F, H\} := \sum_{i<j} x_{ij} \left(\frac{\partial F}{\partial x_i} \frac{\partial H}{\partial x_j} - \frac{\partial H}{\partial x_i} \frac{\partial F}{\partial x_j} \right) \tag{3.7}$$

for all $F, H \in \mathcal{F}(M)$, as can easily be verified. If we view the functions x_{ij} as the elements of a (skew-symmetric) matrix X then the latter formula may also be expressed in a compact form by

$$\{F, H\} = [\mathrm{d}F]^\top X\, [\mathrm{d}H], \tag{3.8}$$

where $[\mathrm{d}F]$ is the column vector which represents $\mathrm{d}F$ in the natural basis $(\mathrm{d}x_1, \dots, \mathrm{d}x_n)$, i.e., the i-th component of $[\mathrm{d}F]$ is $\frac{\partial F}{\partial x_i}$. More generally, the above formulas may also be used for calculations on coordinate neighborhoods on arbitrary manifolds, by taking the x_i to be local coordinates.

Since the Poisson bracket is a biderivation, it vanishes whenever one of its arguments is constant, so that we can associate to a Poisson bracket $\{\cdot\,,\cdot\}$ an $\mathcal{F}(M)$-bilinear map

$$\begin{aligned} P : \Omega(M) \times \Omega(M) &\to \mathcal{F}(M) \\ (\mathrm{d}F, \mathrm{d}H) &\mapsto \{F, H\}. \end{aligned}$$

P is called the *Poisson tensor* associated to $\{\cdot\,,\cdot\}$, and $\{\cdot\,,\cdot\}$ can be reconstructed from P. Recall that when we view P as a bivector field we also write $P[F \wedge H]$ for $P(\mathrm{d}F, \mathrm{d}H) = P(\mathrm{d}F \wedge \mathrm{d}H) = \{F, H\}$. In terms of coordinates, P can be written as

$$P = \sum_{i,j=1}^{n} x_{ij} \frac{\partial}{\partial x_i} \wedge \frac{\partial}{\partial x_j},$$

as follows from (3.7). We derive from P a map

$$\tilde{P} : \Omega(M) \to \mathfrak{X}(M),$$

simply by putting $\tilde{P}(\mathrm{d}H)[F] := P[F \wedge H] = \{F, H\}$. Thus, we have that

$$\tilde{P}(\mathrm{d}H) = \{\cdot\,, H\} = \mathcal{X}_H, \tag{3.9}$$

for any $H \in \mathcal{F}(M)$. The bundle map $T^*M \to TM$ that corresponds to $\tilde{P}$ will be denoted by the same letter $\tilde{P}$.

A necessary and sufficient condition for a bivector field $\{\cdot\,,\cdot\}$ to define a Poisson structure is that $[\{\cdot\,,\cdot\}, \{\cdot\,,\cdot\}]_S = 0$, where $[\cdot\,,\cdot]_S$ denotes the *Schouten-Nijenhuis bracket.* Indeed, if $\{\cdot\,,\cdot\}$ and $\{\cdot\,,\cdot\}'$ are two bivector fields on M then their Schouten-Nijenhuis bracket[1] is the trivector field, given by

$$\begin{aligned} &[\{\cdot\,,\cdot\}, \{\cdot\,,\cdot\}']_S (F, G, H) \\ &\qquad := \{\{F,G\}, H\}' + \{\{G,H\}, F\}' + \{\{H,F\}, G\}' + \\ &\qquad\quad\ \{\{F,G\}', H\} + \{\{G,H\}', F\} + \{\{H,F\}', G\} \end{aligned}$$

so that

$$[\{\cdot\,,\cdot\}, \{\cdot\,,\cdot\}]_S (F, G, H) = 2\{\{F,G\}, H\} + \mathrm{cycl}(F, G, H),$$

where $F, G, H \in \mathcal{F}(M)$ are arbitrary. It follows that $[\{\cdot\,,\cdot\}, \{\cdot\,,\cdot\}]_S = 0$ if and only if $\{\cdot\,,\cdot\}$ satisfies the Jacobi identity .

We next introduce the notion of a morphism for Poisson manifolds. Recall that if M_1 and M_2 are manifolds then a map $\phi : M_1 \to M_2$ is a morphism (i.e., a smooth, holomorphic or regular map) if and only if $\phi^*(F) \in \mathcal{F}(M_1)$ for every $F \in \mathcal{F}(M_2)$, where $\phi^*(F) := F \circ \phi$.

[1] For the formula for the Schouten-Nijenhuis bracket on arbitrary multi-vector fields, see [164, Chapter 1.2].

Definition 3.2. Let $(M_1, \{\cdot\,,\cdot\}_1)$ and $(M_2, \{\cdot\,,\cdot\}_2)$ be two Poisson manifolds. A map $\phi : M_1 \to M_2$ is called a *Poisson morphism* or a *morphism of Poisson manifolds* if

(1) ϕ is a morphism, $\phi^*(\mathcal{F}(M_2)) \subseteq \mathcal{F}(M_1)$;
(2) For all $F, G \in \mathcal{F}(M_2)$, $\phi^* \{F, G\}_2 = \{\phi^* F, \phi^* G\}_1$.

An (immersed) submanifold M' of M is called an *(immersed) Poisson submanifold* if it admits a Poisson structure for which the inclusion map $\imath : M' \hookrightarrow M$ is a Poisson morphism. Clearly such a Poisson structure on M', if it exists, is unique.

Notice that the inverse of a Poisson morphism, if it exists, is also a Poisson morphism, hence it is a *Poisson isomorphism*.

We next consider the basic examples of Poisson manifolds. These examples are important for the sequel because essentially all examples that will be encountered in this text are particular cases of them.

Example 3.3. Any constant skew-symmetric $n \times n$ matrix is the matrix of a Poisson structure on $\mathbf{C}^n$, in terms of its standard coordinates, as follows from (3.6). We refer to such a Poisson structure as a *constant Poisson structure.* By the classification theorem for skew-symmetric bilinear forms there exists a linear system of coordinates $(x_1, \ldots, x_n)$ of $\mathbf{C}^n$ with respect to which the Poisson matrix takes the form

$$X = \begin{pmatrix} 0 & \mathrm{Id}_r & 0 \\ -\mathrm{Id}_r & 0 & 0 \\ 0 & 0 & 0 \end{pmatrix}.$$

The integer $2r$ is called the *rank* of the Poisson structure; for general Poisson structures the rank will be defined in Section 3.2. This structure is often called the *canonical Poisson structure* of rank $2r$ on $\mathbf{C}^n$.

Example 3.4. Recall that a *symplectic manifold* (M, ω) is a (real or complex) manifold equipped with a (smooth or holomorphic) closed two-form ω (a *symplectic two-form*) which is non-degenerate (as a bilinear form on each tangent space). A vector field $\mathcal{X}_H$ is associated to any function $H \in C^\infty(M)$ by

$$\omega(\mathcal{X}_H, \cdot) = \mathsf{d}H(\cdot). \tag{3.10}$$

Rewriting this definition as $\imath_{\mathcal{X}_H}\omega = \mathsf{d}H$, it is clear that $\imath_{\mathcal{X}_H}\omega$ is closed (even exact) for any vector field $\mathcal{X}_H$. Combining (2.11) and (2.12) we find that for any $\mathcal{V}_1, \mathcal{V}_2 \in \mathfrak{X}(M)$,

$$\imath_{[\mathcal{V}_1,\mathcal{V}_2]} = \mathsf{d}\imath_{\mathcal{V}_1}\imath_{\mathcal{V}_2} + \imath_{\mathcal{V}_1}\mathsf{d}\imath_{\mathcal{V}_2} - \imath_{\mathcal{V}_2}\mathsf{d}\imath_{\mathcal{V}_1} - \imath_{\mathcal{V}_2}\imath_{\mathcal{V}_1}\mathsf{d}.$$

Since ω is closed, one has for vector fields $\mathcal{X}_F$ and $\mathcal{X}_G$,

$$\imath_{[\mathcal{X}_F,\mathcal{X}_G]}\omega = \mathrm{d}\imath_{\mathcal{X}_F}\imath_{\mathcal{X}_G}\omega = -\mathrm{d}\left(\omega(\mathcal{X}_F,\mathcal{X}_G)\right). \tag{3.11}$$

We define a skew-symmetric biderivation on $\mathcal{F}(M)$ by

$$\{F,G\} := \omega\left(\mathcal{X}_F,\mathcal{X}_G\right) = \mathrm{d}F(\mathcal{X}_G) = \mathcal{X}_G[F]; \tag{3.12}$$

notice that the above equalities show that our definitions (3.4) and (3.10) of $\mathcal{X}_F$ are coherent. In terms of $\{\cdot\,,\cdot\}$, (3.11) can be written as

$$\imath_{[\mathcal{X}_F,\mathcal{X}_G]}\omega = -\mathrm{d}\left\{F,G\right\} = -\imath_{\mathcal{X}_{\{F,G\}}}\omega.$$

Since ω is non-degenerate this shows that $[\mathcal{X}_F,\mathcal{X}_G] = -\mathcal{X}_{\{F,G\}}$. Applying this equation to an arbitrary $H \in \mathcal{F}(M)$ we find

$$\mathcal{X}_F\mathcal{X}_G[H] - \mathcal{X}_G\mathcal{X}_F[H] + \mathcal{X}_{\{F,G\}}[H] = 0,$$

which is precisely the Jacobi identity, as follows from (3.12). Thus, every symplectic manifold is in a natural way a Poisson manifold. Picking local coordinates $(x_1,\ldots,x_n)$ we can introduce a skew-symmetric $n\times n$ matrix Ω by

$$\Omega_{ij} := \omega\left(\frac{\partial}{\partial x_i},\frac{\partial}{\partial x_j}\right).$$

We claim that the Poisson matrix X of $\{\cdot\,,\cdot\}$ with respect to these coordinates is given by $X = -\Omega^{-1}$. To show this, let us denote by $[\mathcal{X}_F]$ the column matrix whose elements are the coefficients of $\mathcal{X}_F$ with respect to the basis $\left(\frac{\partial}{\partial x_1},\ldots,\frac{\partial}{\partial x_n}\right)$ and let us recall that we denote by $[\mathrm{d}F]$ the column matrix whose elements are the coefficients of $\mathrm{d}F$ with respect to the basis $(\mathrm{d}x_1,\ldots,\mathrm{d}x_n)$. Then (3.10) says that $\Omega\,[\mathcal{X}_F] = -[\mathrm{d}F]$ and $\omega\left(\mathcal{X}_F,\mathcal{X}_G\right) = [\mathcal{X}_F]^\top\,\Omega\,[\mathcal{X}_G]$, so that

$$\{F,G\} = [\mathcal{X}_F]^\top\Omega\,[\mathcal{X}_G] = \left(\Omega^{-1}[\mathrm{d}F]\right)^\top\,\Omega\,\Omega^{-1}\,[\mathrm{d}G] = -[\mathrm{d}F]^\top\,\Omega^{-1}\,[\mathrm{d}G].$$

Comparing this to Formula (3.8) we see that $X = -\Omega^{-1}$.

A final comment about symplectic manifolds. If (M,ω) is a symplectic manifold then the fact that the two-form ω is non-degenerate implies that the dimension n of M is even. Then $\omega^{n/2}$ is a top-form which is nowhere vanishing, i.e., a volume form. In particular, every symplectic manifold is orientable.

Example 3.5. Let $\mathfrak{g}$ be any finite-dimensional (complex) Lie algebra, with Lie bracket $[\cdot\,,\cdot]$. Then a Poisson bracket is defined on $\mathfrak{g}^*$ by putting, for $F,G \in \mathcal{F}\left(\mathfrak{g}^*\right)$

$$\{F,G\}\,(\xi) := \langle \xi, [\mathrm{d}F(\xi),\mathrm{d}G(\xi)]\rangle\,,$$

where $\mathrm{d}F(\xi)$ and $\mathrm{d}G(\xi)$ are interpreted as elements of $\mathfrak{g}$ when computing the bracket (as explained in Section 2.1). Because of its importance for what follows, this example will be dealt with in more detail in Section 3.5.

Example 3.6. It is easy to describe *all* Poisson structures on $\mathbf{C}^2$ since in this case the Jacobi identity is satisfied for any skew-symmetric biderivation on $\mathcal{F}(\mathbf{C}^2)$. Letting x and y denote the standard coordinates on $\mathbf{C}^2$ this means that every Poisson bracket on $\mathbf{C}^2$ is of the form

$$\{F, G\} = \varphi\left(\frac{\partial F}{\partial x}\frac{\partial G}{\partial y} - \frac{\partial G}{\partial x}\frac{\partial F}{\partial y}\right) \tag{3.13}$$

for some $\varphi \in \mathcal{F}(\mathbf{C}^2)$, where $F, G \in \mathcal{F}(\mathbf{C}^2)$. In fact, $\varphi = \{x, y\}$ and for any $\varphi \in \mathcal{F}(\mathbf{C}^2)$ the above formula (3.13) defines a Poisson structure on $\mathbf{C}^2$.

Example 3.7. Last and least (!) every manifold has a Poisson structure: just define $\{F, G\} := 0$ for any $F, G \in \mathcal{F}(M)$. We refer to it as the *trivial Poisson structure* on M.

3.2 Hamiltonian Mechanics

The Poisson bracket allows us to define Hamiltonian mechanics on spaces that are more general than $\mathbf{R}^{2n}$, equipped with its standard symplectic structure (see Examples 3.3 and 3.4). For $H \in \mathcal{F}(M)$, where $(M, \{\cdot\,,\cdot\})$ is a Poisson manifold, we have by definition that $\mathcal{X}_H$, defined by $\mathcal{X}_H[F] := \{F, H\}$ for all $F \in \mathcal{F}(M)$, is a derivation on $\mathcal{F}(M)$, hence a vector field on M, which can be written in terms of local coordinates $(x_1, \ldots, x_n)$ by

$$\mathcal{X}_H = \sum_{i=1}^{n} \{x_i, H\} \frac{\partial}{\partial x_i}, \tag{3.14}$$

as follows from (2.5), or also as $\dot{x}_i = \{x_i, H\}$, $i = 1, \ldots, n$, where $\dot{x}_i$ is a convenient abbreviation of $\mathcal{X}_H[x_i]$, when the function H has been fixed.

Example 3.8. Consider on $M := \mathbf{R}^{2n}$ with coordinates $(q_1, \ldots, q_n, p_1, \ldots, p_n)$ the symplectic two-form

$$\omega := \mathrm{d}q_1 \wedge \mathrm{d}p_1 + \cdots + \mathrm{d}q_n \wedge \mathrm{d}p_n.$$

For $1 \leqslant i \leqslant n$ the Hamiltonian vector field $\mathcal{X}_{q_i}$, as defined in (3.10), is given by

$$\mathrm{d}q_i = \left(\sum_{j=1}^{n} \mathrm{d}q_j \wedge \mathrm{d}p_j\right)(\mathcal{X}_{q_i}, \cdot) = \sum_{j=1}^{n} \left(\mathcal{X}_{q_i}[q_j]\mathrm{d}p_j - \mathcal{X}_{q_i}[p_j]\mathrm{d}q_j\right),$$

so that $\mathcal{X}_{q_i} = -\partial/\partial p_i$. Similarly, $\mathcal{X}_{p_i} = \partial/\partial q_i$ for $i = 1, \ldots, n$. It follows that for any $1 \leqslant i, j \leqslant n$,

$$\{q_i, p_j\} = \omega\left(\mathcal{X}_{q_i}, \mathcal{X}_{p_j}\right) = \left(\sum_{k=1}^{n} \mathrm{d}q_k \wedge \mathrm{d}p_k\right)\left(-\frac{\partial}{\partial p_i}, \frac{\partial}{\partial q_j}\right) = \delta_{ij}.$$

Similarly, $\{q_i, q_j\} := \{p_i, p_j\} := 0$ for any $1 \leqslant i, j \leqslant n$. As a consequence, for $H \in \mathcal{F}(M)$ the Hamiltonian vector field $\mathcal{X}_H$ is explicitly given by

$$\dot{q}_i = \frac{\partial H}{\partial p_i}, \qquad \dot{p}_i = -\frac{\partial H}{\partial q_i}, \tag{3.15}$$

and we recognize *Hamilton's equations.* To specialize this example further, suppose that H is of the standard form "Kinetic energy + Potential energy",

$$H = \frac{1}{2}\left(p_1^2 + \cdots + p_n^2\right) + V(q_1, \ldots, q_n), \tag{3.16}$$

then we recover *Newton's equations*

$$\ddot{q}_i = -\frac{\partial V}{\partial q_i}$$

for the motion of a particle of mass 1 in a potential field V on $\mathbf{R}^n$.

Definition 3.9. Let $(M, \{\cdot\,,\cdot\})$ be a Poisson manifold and let $H \in \mathcal{F}(M)$. The vector field $\mathcal{X}_H = \{\cdot\,, H\}$ is called the *Hamiltonian vector field* associated to the *Hamiltonian* H and we write

$$\operatorname{Ham}(M, \{\cdot\,,\cdot\}) := \{\mathcal{X}_H \mid H \in \mathcal{F}(M)\}$$

for the vector space of Hamiltonian vector fields. A function $H \in \mathcal{F}(M)$ whose Hamiltonian vector field is zero, $\mathcal{X}_H = 0$, is called a *Casimir function* or a *Casimir* and we denote

$$\operatorname{Cas}(M, \{\cdot\,,\cdot\}) := \{H \in \mathcal{F}(M) \mid \mathcal{X}_H = 0\}$$

for the (vector) space of Casimirs. When no confusion can arise, either argument in $\operatorname{Ham}(M, \{\cdot\,,\cdot\})$ and $\operatorname{Cas}(M, \{\cdot\,,\cdot\})$ is omitted. When studying Hamiltonian vector fields on a Poisson manifold $(M, \{\cdot\,,\cdot\})$ the manifold M is often referred to as *phase space.*

The Leibniz property for $\{\cdot\,,\cdot\}$ implies on the one hand that $\operatorname{Cas}(M)$ is a subalgebra of $\mathcal{F}(M)$ (for the ordinary multiplication of functions) and on the other hand that $\operatorname{Ham}(M)$ is a $\operatorname{Cas}(M)$-module (notice that $\operatorname{Ham}(M)$ is not an $\mathcal{F}(M)$-module, except in trivial cases). The latter implies that the map

$$\mathcal{F}(M) \to \operatorname{Ham}(M) : F \mapsto \mathcal{X}_F$$

is $\operatorname{Cas}(M)$-linear, while the Jacobi identity is equivalent to the fact that this map is a Lie algebra anti-homomorphism,

$$[\mathcal{X}_F, \mathcal{X}_G] = -\mathcal{X}_{\{F,G\}} \tag{3.17}$$

for all $F, G \in \mathcal{F}(M)$. In particular, $\operatorname{Ham}(M)$ is a subalgebra of the Lie algebra $\mathfrak{X}(M)$.

Example 3.10. In the case of Example 3.3 the algebra of Casimirs is generated by the linear functions $x_{2r+1}, \ldots, x_n$. On a connected symplectic manifold (Example 3.4) the constant functions are the only Casimirs. However, this property does not characterize the symplectic manifolds within the class of connected Poisson manifolds, see Example 3.27 below. In the case of the canonical Poisson structure on $\mathfrak{g}^*$ (Example 3.5) the algebra of Casimirs can be identified with the center of the universal enveloping algebra $\mathcal{U}\mathfrak{g}$ of $\mathfrak{g}$. For the trivial Poisson structure on M (Example 3.7) one obviously has that $\operatorname{Cas}(M) = \mathcal{F}(M)$.

If we specialize the above construction of Hamiltonian vector fields to a point m of M we find a linear space $\operatorname{Ham}_m(M) \subseteq T_m M$, whose dimension varies in general with m, thereby defining a generalized distribution on M (see Paragraph 2.1.2). For future use, notice that

$$\operatorname{Ham}_m(M) = \operatorname{span}\{\mathcal{X}_{x_1}(m), \cdots, \mathcal{X}_{x_n}(m)\}, \tag{3.18}$$

where $(x_1, \ldots, x_n)$ is any system of coordinates, defined on a neighborhood of m.

Definition 3.11. For $m \in M$, the dimension of $\operatorname{Ham}_m(M)$ is called the *rank* of $\{\cdot\,,\cdot\}$ at m, denoted $\operatorname{Rk}_m\{\cdot\,,\cdot\}$ and $\max\{\operatorname{Rk}_m\{\cdot\,,\cdot\} \mid m \in M\}$ is called the rank of $(M, \{\cdot\,,\cdot\})$, denoted $\operatorname{Rk}\{\cdot\,,\cdot\}$. We say that $\{\cdot\,,\cdot\}$ has *maximal rank* at m when $\operatorname{Rk}_m\{\cdot\,,\cdot\} = \dim M$, and that $\{\cdot\,,\cdot\}$ has *maximal rank* on a subset M' of M when $\{\cdot\,,\cdot\}$ has maximal rank at each point of M'. $(M, \{\cdot\,,\cdot\})$ is called a *regular Poisson manifold* when $\operatorname{Rk}_m\{\cdot\,,\cdot\} = \operatorname{Rk}\{\cdot\,,\cdot\}$, independently of $m \in M$.

If we choose local coordinates $(x_1, \ldots, x_n)$ on a neighborhood of m then we see that $\operatorname{Rk}_m\{\cdot\,,\cdot\}$ equals the number of independent columns of the Poisson matrix $X := (\{x_i, x_j\})_{1 \leqslant i,j \leqslant n}$ of $\{\cdot\,,\cdot\}$ with respect to $(x_1, \ldots, x_n)$ at the point m, hence it is the rank of X at m. It is also the rank at m of the bundle map $\tilde{P} : T^*M \to TM$ that corresponds to $\{\cdot\,,\cdot\}$, as follows from

$$\mathcal{X}_{x_j} = \sum_{i=1}^{n} \{x_i, x_j\} \frac{\partial}{\partial x_i},$$

see (3.14) and (3.18). Skew-symmetry of X implies that the rank of a Poisson structure at a point is always even.

Example 3.12. The Poisson structures in Examples 3.3 and 3.7 are regular (of respective rank $2r$ and 0). The same is true for a symplectic manifold (Example 3.4) because if M is a symplectic manifold then $\operatorname{Ham}_m(M) = T_m M$ for any $m \in M$; hence the (even) rank of a symplectic manifold equals its dimension. The canonical Poisson structure on the dual $\mathfrak{g}^*$ of a Lie algebra $\mathfrak{g}$ is never regular, unless $\mathfrak{g}$ is Abelian: the rank at $0 \in \mathfrak{g}$ is always zero. In the case of Example 3.6, if φ is a polynomial then the Poisson structure is regular if and only if φ is constant; otherwise the rank is two except at the plane algebraic curve Γ_φ defined by $\varphi = 0$.

Proposition 3.13. *Let $(M, \{\cdot\,,\cdot\})$ be a Poisson manifold and let $s \in \mathbf{N}$. The subset $M_{(s)}$ of M, defined by*

$$M_{(s)} := \{m \in M \mid \mathrm{Rk}_m \{\cdot\,,\cdot\} \geqslant 2s\} \tag{3.19}$$

is open.

Proof. Since $M_{(s)} = \mathrm{Rk}^{-1}\{2t \mid t \geqslant s\}$, it is sufficient to show that the map $\mathrm{Rk} : M \to \mathbf{Z}$ is lower semi-continuous. Let s be such that $M_{(s)}$ is non-empty, i.e., $2s \leqslant \mathrm{Rk}\{\cdot\,,\cdot\}$, let $m \in M_{(s)}$ and let $(x_1, \dots, x_n)$ be local coordinates on a neighborhood U of m. The rank of $\{\cdot\,,\cdot\}$ at $p \in U$ is the rank of $\{\cdot\,,\cdot\}(p)$, the Poisson matrix of $\{\cdot\,,\cdot\}$ with respect to $(x_1, \dots, x_n)$, evaluated at p. Hence, the restriction of Rk to U is the composition of the map $U \to \mathfrak{gl}(n)$, defined by $m \mapsto (\{x_i, x_j\}(m))_{1 \leqslant i,j \leqslant n}$ and the lower semi-continuous map $\mathfrak{gl}(n) \to \mathbf{Z}$ which assigns to a matrix its rank. □

In particular we have that, if we denote the rank of $\{\cdot\,,\cdot\}$ by $2r$, then $M_{(r)}$ is a non-empty open subset of M and the restriction of $\{\cdot\,,\cdot\}$ to $M_{(r)}$ is regular (of rank $2r$).

Example 3.14. The rank needs not attain its maximum on a *dense* open subset. Let φ be a function on $\mathbf{R}^2$ which is positive on the interior of the unit disc and which is zero elsewhere. Then the rank of the Poisson bracket on $\mathbf{R}^2$, which is defined by φ as in Example 3.6, is maximal only on the inside of the disk.

The case in which the Poisson bracket has maximal rank on all of M is described in the following proposition.

Proposition 3.15. *Let $(M, \{\cdot\,,\cdot\})$ be a Poisson manifold and suppose that $\{\cdot\,,\cdot\}$ has maximal rank on M. Then M admits a symplectic structure ω for which $\{\cdot\,,\cdot\}$ is the corresponding Poisson structure.*

Proof. Since the bundle map $\tilde{P} : T^*M \to TM$ has maximal rank in every point it is invertible; we define ω to be the two-form which corresponds to its inverse $\tilde{P}^{-1}$. Since $\tilde{P}(\mathrm{d}H) = \mathcal{X}_H$, for any $H \in \mathcal{F}(M)$, see (3.9), this means that we are defining

$$\omega(\mathcal{X}_G, \mathcal{X}_H) := \tilde{P}^{-1}(\mathcal{X}_G)(\mathcal{X}_H) = \mathrm{d}G(\mathcal{X}_H) = \{G, H\}, \tag{3.20}$$

for $G, H \in \mathcal{F}(M)$. Obviously, ω is non-degenerate. Applying Formula (2.8) to *Hamiltonian* vector fields we get

$$\begin{aligned} \mathrm{d}\omega(\mathcal{X}_F, \mathcal{X}_G, \mathcal{X}_H) &= \mathcal{X}_F[\omega(\mathcal{X}_G, \mathcal{X}_H)] + \omega(\mathcal{X}_F, [\mathcal{X}_G, \mathcal{X}_H]) + \mathrm{cycl}(\mathcal{X}_F, \mathcal{X}_G, \mathcal{X}_H) \\ &= 2(\{\{F, G\}, H\} + \{\{G, H\}, F\} + \{\{H, F\}, G\}) \\ &= 0. \end{aligned}$$

We have used that

$$\mathcal{X}_F\left[\omega\left(\mathcal{X}_G,\mathcal{X}_H\right)\right]=\left\{\{G,H\},F\right\}=\omega\left(\mathcal{X}_F,[\mathcal{X}_G,\mathcal{X}_H]\right),$$

as follows from (3.20) and (3.17). Since $T_mM=\mathrm{Ham}_m(\{\cdot\,,\cdot\})$ for any $m\in M$ this shows that ω is closed, hence that ω is a symplectic structure. □

Explicitly, for $m\in M$ and $v,w\in T_mM$, let $F,G\in\mathcal{F}(M)$ be such that $v=\mathcal{X}_F(m)$ and $w=\mathcal{X}_G(m)$. Then the above construction of ω amounts to defining $\omega_m(v,w):=\{F,G\}(m)$, independently of the choices made for F and G.

Proposition 3.16. *Let* $(M_i,\{\cdot\,,\cdot\}_i)$, $(i=1,2)$ *be two Poisson manifolds, let* $m\in M_1$ *and let* $\phi:M_1\to M_2$ *be a Poisson morphism. Then* $\mathrm{Rk}_m\{\cdot\,,\cdot\}_1\geqslant\mathrm{Rk}_{\phi(m)}\{\cdot\,,\cdot\}_2$.

Proof. Consider the linear map $\mathsf{d}\phi(m):T_mM_1\to T_{\phi(m)}M_2$. We claim that $\mathrm{Ham}_{\phi(m)}(M_2)$ is contained in $\mathsf{d}\phi(\mathrm{Ham}_m(M_1))$. Then

$$\mathrm{Rk}_m\{\cdot\,,\cdot\}_1=\dim\mathrm{Ham}_m(M_1)\geqslant\dim\mathrm{Ham}_{\phi(m)}(M_2)=\mathrm{Rk}_{\phi(m)}\{\cdot\,,\cdot\}_2\,.$$

To show the above claim, write a given $v\in\mathrm{Ham}_{\phi(m)}(M_2)$ as $\mathcal{X}_F(\phi(m))$, where $F\in\mathcal{F}(M_2)$. Then

$$\mathsf{d}\phi\left(\mathcal{X}_{F\circ\phi}(m)\right)[G]=\{G\circ\phi,F\circ\phi\}(m)=\{G,F\}(\phi(m))=\mathcal{X}_F(\phi(m))[G],$$

for any $G\in\mathcal{F}(M_2)$, so that $v=\mathsf{d}\phi\left(\mathcal{X}_{F\circ\phi}(m)\right)\in\mathsf{d}\phi\left(\mathrm{Ham}_m(M_1)\right)$. □

The proposition implies that a point m of a Poisson manifold is only a Poisson submanifold when the rank of the Poisson structure vanishes at m. More generally, the inequality

$$\dim\mathrm{Ham}_m\left(M,\{\cdot\,,\cdot\}\right)\leqslant\dim M$$

implies in view of the proposition that a necessary condition for a submanifold M' of M to be a Poisson submanifold is that $\dim M'\geqslant\mathrm{Rk}_m\{\cdot\,,\cdot\}$ for all $m\in M'$. A refinement of this statement will be given in Section 3.4.

The Jacobi identity has another interesting consequence, given in the following proposition.

Proposition 3.17. *Let* $(M,\{\cdot\,,\cdot\})$, *be a Poisson manifold and let* $H\in\mathcal{F}(M)$. *The Hamiltonian vector field* $\mathcal{X}_H$ *leaves* $\{\cdot\,,\cdot\}$ *invariant.*

Proof. Let P denote the bivector field that corresponds to $\{\cdot\,,\cdot\}$, so that $\{F,G\}=P[F\wedge G]$ for $F,G\in\mathcal{F}(M)$. According to Section 2.1 we need to show that $L_{\mathcal{X}_H}P=0$. Using (2.13) we find for any $F,G\in\mathcal{F}(M)$ that

$$\begin{aligned}L_{\mathcal{X}_H}P[F\wedge G]&=\mathcal{X}_H[P[F\wedge G]]-P[\mathcal{X}_H[F]\wedge G]-P[F\wedge\mathcal{X}_H[G]],\\&-\{\{F,G\},H\}-\{\{F,H\},G\}-\{F,\{G,H\}\},\end{aligned}\tag{3.21}$$

which evaluates to zero in view of the Jacobi identity. Notice that another way to state the vanishing of (3.21) is that all Hamiltonian vector fields are derivations of the Poisson bracket. □

It follows for example that for any integer s the open subset $M_{(s)}$, defined in (3.19) is invariant for all Hamiltonian flows.

Definition 3.18. Let $(M, \{\cdot\,,\cdot\})$, be a Poisson manifold and let $\mathcal{V} \in \mathfrak{X}(M)$. Then $\mathcal{V}$ is called a *Poisson vector field* if $\mathcal{V}$ leaves $\{\cdot\,,\cdot\}$ invariant, i.e., $L_{\mathcal{V}}P = 0$, where P denotes the bivector field that corresponds to $\{\cdot\,,\cdot\}$.

Proposition 3.17 states that all Hamiltonian vector fields are Poisson vector fields; however the converse needs not be true. In terms of Poisson cohomology (see [169, Chapter 2.1]) one has that Poisson vector fields are 1-cocycles while Hamiltonian vector fields are 1-coboundaries. Poisson vector fields appear naturally in the context of group actions, as follows from the following proposition.

Proposition 3.19. *Let* $\mathbf{G}$ *be a Lie group which acts on a Poisson manifold* $(M, \{\cdot\,,\cdot\})$. *Denoting the action by* $\chi : \mathbf{G} \times M \to M$, *let* $X \in \mathfrak{g}$ *and suppose that for any* $|t|$ *sufficiently small one has that* $\chi_{\exp(tX)} : M \to M$ *is a morphism of Poisson manifolds. Then* $\underline{X}$, *the fundamental vector field corresponding to* X, *is a Poisson vector field. As a consequence, if for every* $g \in \mathbf{G}$ *the map* $\chi_g : M \to M$ *is a Poisson morphism then all fundamental vector fields of the action are Poisson vector fields.*

Proof. In order to show that $L_{\underline{X}}P = 0$ we need to check, according to (2.13), that the Leibniz property

$$\underline{X}\left[\{F, G\}\right] = \{\underline{X}[F], G\} + \{F, \underline{X}[G]\},$$

holds for any $F, G \in \mathcal{F}(M)$. The fact that $\chi_{\exp(tX)}$ is Poisson means that

$$\{F, G\} \circ \chi_{\exp(tX)} = \left\{F \circ \chi_{\exp(tX)}, G \circ \chi_{\exp(tX)}\right\}.$$

Therefore,

$$\begin{aligned}
\underline{X}\left[\{F, G\}\right] &= \frac{\mathsf{d}}{\mathsf{d}t}_{|t=0} \{F, G\} \circ \chi_{\exp(tX)} \\
&= \frac{\mathsf{d}}{\mathsf{d}t}_{|t=0} \left\{F \circ \chi_{\exp(tX)}, G \circ \chi_{\exp(tX)}\right\} \\
&= \left\{\frac{\mathsf{d}}{\mathsf{d}t}_{|t=0} F \circ \chi_{\exp(tX)}, G\right\} + \left\{F, \frac{\mathsf{d}}{\mathsf{d}t}_{|t=0} G \circ \chi_{\exp(tX)}\right\} \\
&= \{\underline{X}[F], G\} + \{F, \underline{X}[G]\},
\end{aligned}$$

as we needed to show. □

Given a group action $\chi : \mathbf{G} \times M \to M$ such that for any $g \in \mathbf{G}$ the map $\chi_g : M \to M$ is Poisson (w.r.t. some given Poisson structure on M) it is in view of Propositions 3.17 and 3.19 a natural question to ask if the fundamental vector fields of the action are Hamiltonian. If this is so, then the action is called a *Hamiltonian action* and one can build a map $\mu : M \to \mathfrak{g}^*$ by defining for any $m \in M$ and for any $X \in \mathfrak{g}$,

$$\langle \mu(m), X \rangle = \mathcal{J}_X(m), \tag{3.22}$$

where $\mathcal{J}_X$ is a Hamiltonian for X (the Hamiltonian is only determined up to a Casimir; one picks a particular Hamiltonian $\mathcal{J}_X$ for all X in a basis of $\mathfrak{g}$ and then one extends by linearity). The map μ is called the *momentum map* with values in $\mathfrak{g}^*$, and its dual map $\mathcal{J}$ is called the *co-momentum map.*

Hamiltonian actions and their momentum maps have been studied mainly in the context of symplectic manifolds, and the theory of Hamiltonian actions, such as torus actions, on symplectic manifolds has become a subject of its own. See e.g., [19] and [73] and the references in these books. More recent developments include generalizations to Poisson maps (see Definition 3.41 below), and momentum maps that take values in a Lie group (see [64] and [16]).

3.3 Bi-Hamiltonian Manifolds and Vector Fields

Most of the Poisson manifolds that we will consider carry another Poisson structure, revelant for the Hamiltonian vector fields that we study. We give the basic definitions and refer to later chapters for the relation to integrability.

Definition 3.20. Let M be a manifold and let $\{\cdot\,,\cdot\}_1, \ldots, \{\cdot\,,\cdot\}_s$ be s Poisson structures such that any linear combination of them is also a Poisson structure (i.e., any linear combination of them satisfies the Jacobi identity). Then these Poisson brackets are called *compatible Poisson structures*, and M, equipped with these Poisson structures is called a *multi-Hamiltonian manifold.* In the cases $s = 2, 3$ one usually speaks of a *bi-Hamiltonian*, respectively a *tri-Hamiltonian manifold.*

Let P_1 and P_2 denote the Poisson bivectors that correspond to two Poisson structures $\{\cdot\,,\cdot\}_1$ and $\{\cdot\,,\cdot\}_2$ on M. For non-zero scalars λ_1 and λ_2 we have that $\lambda_1 \{\cdot\,,\cdot\}_1 + \lambda_2 \{\cdot\,,\cdot\}_2$ is a Poisson bracket if and only if

$$[\lambda_1 P_1 + \lambda_2 P_2, \lambda_1 P_1 + \lambda_2 P_2]_S = 0,$$

which is equivalent to $[P_1, P_2]_S = 0$, since $[P_1, P_1] = [P_2, P_2] = 0$. Therefore, two Poisson structures $\{\cdot\,,\cdot\}_1$ and $\{\cdot\,,\cdot\}_2$ are compatible if and only if their sum satisfies the Jacobi identity. Similarly, s Poisson structures $\{\cdot\,,\cdot\}_1, \ldots, \{\cdot\,,\cdot\}_s$ are compatible if and only if for any $1 \leqslant i < j \leqslant s$ one has that $\{\cdot\,,\cdot\}_i + \{\cdot\,,\cdot\}_j$ satisfies the Jacobi identity, i.e., if and only if these s Poisson structures are pairwise compatible.

Proposition 3.21. *Let $(M, \{\cdot\,,\cdot\}_1)$ be a Poisson manifold, let $\mathcal{V}$ be any vector field on M and denote $\{\cdot\,,\cdot\}_2 := L_{\mathcal{V}} \{\cdot\,,\cdot\}_1$. If $\{\cdot\,,\cdot\}_2$ satisfies the Jacobi identity, then $\{\cdot\,,\cdot\}_1$ and $\{\cdot\,,\cdot\}_2$ are compatible Poisson structures.*

Proof. By the above remark we have that $\{\cdot,\cdot\}_1$ and $\{\cdot,\cdot\}_2$ are compatible Poisson structures if and only if their sum $\{\cdot,\cdot\}_1 + \{\cdot,\cdot\}_2$ satisfies the Jacobi identity. Since $\{\cdot,\cdot\}_1$ and $\{\cdot,\cdot\}_2$ satisfy the Jacobi identity this means that we need to check that

$$\{\{F,G\}_1, H\}_2 + \{\{F,G\}_2, H\}_1 + \mathrm{cycl}(F,G,H) = 0. \tag{3.23}$$

Using Formula (2.13), $\{F,G\}_2 = \mathcal{V}[\{F,G\}_1] - \{\mathcal{V}[F],G\}_1 - \{F,\mathcal{V}[G]\}_1$, so we can express all brackets $\{\cdot,\cdot\}_2$ that appear in (3.23) in terms of $\{\cdot,\cdot\}_1$, yielding the equality in (3.23) in view of the Jacobi identity for $\{\cdot,\cdot\}_1$, applied to 4 cycles. □

Example 3.22. Any two constant Poisson structures on $\mathbf{R}^n$ or $\mathbf{C}^n$ are compatible. Poisson cohomology arguments show that if one of them is of maximal rank, then the other one is a Lie derivative of it.

Definition 3.23. Suppose that $(M, \{\cdot,\cdot\}_1, \{\cdot,\cdot\}_2)$ is a bi-Hamiltonian manifold, where $\{\cdot,\cdot\}_1$ and $\{\cdot,\cdot\}_2$ are essentially different in the sense that $\{\cdot,\cdot\}_2$ is not a scalar multiple of $\{\cdot,\cdot\}_1$. A vector field $\mathcal{V}$ is called a *bi-Hamiltonian vector field* if it is Hamiltonian with respect to both Poisson structures.

Example 3.24. Let $F : \mathbf{C}^2 \to \mathbf{C}$ be a holomorphic function. We decompose F in its real and complex parts, $F = G + \sqrt{-1}H$, where we view G and H as smooth functions $\mathbf{R}^4 \to \mathbf{R}$. The standard coordinates z_1 and z_2 on $\mathbf{C}^2$ will accordingly be decomposed as $z_1 = x_1 + \sqrt{-1}y_1$ and $z_2 = x_2 + \sqrt{-1}y_2$. We consider on $\mathbf{C}^2$ the holomorphic symplectic two-form $\Omega := \mathrm{d}z_1 \wedge \mathrm{d}z_2$. We write $\Omega = \omega_0 + \sqrt{-1}\omega_1$, where ω_0 and ω_1 are the (real) symplectic two-forms on $\mathbf{R}^4$, given by

$$\begin{aligned}\omega_0 &= \mathrm{d}x_1 \wedge \mathrm{d}x_2 - \mathrm{d}y_1 \wedge \mathrm{d}y_2,\\ \omega_1 &= \mathrm{d}x_1 \wedge \mathrm{d}y_2 - \mathrm{d}x_2 \wedge \mathrm{d}y_1.\end{aligned}$$

The corresponding Poisson structures $\{\cdot,\cdot\}_0$ and $\{\cdot,\cdot\}_1$ are compatible since they are both constant. The Cauchy-Riemann equations for F,

$$\frac{\partial G}{\partial x_i} = \frac{\partial H}{\partial y_i} \quad \text{and} \quad \frac{\partial G}{\partial y_i} = -\frac{\partial H}{\partial x_i} \qquad (i = 1, 2) \tag{3.24}$$

imply that $\{\cdot, G\}_0 = \{\cdot, H\}_1$. To check this, compute that

$$\{x_1, G\}_0 = \frac{\partial G}{\partial x_2} = \frac{\partial H}{\partial y_2} = \{x_1, H\}_1,$$

and similarly for the other coordinate functions x_2, y_1, y_2. It follows that the vector field $\{\cdot, G\}_0$ is bi-Hamiltonian. One computes in the same way that $\{\cdot, H\}_0 = -\{\cdot, G\}_1$, so that this vector field is also bi-Hamiltonian. In fact, each of the equations $\{\cdot, G\}_0 = \{\cdot, H\}_1$ and $\{\cdot, H\}_0 = -\{\cdot, G\}_1$ is equivalent to the Cauchy-Riemann equations (3.24) for $G + \sqrt{-1}H$.

Another fundamental example is given in Section 3.5.

3.4 Local and Global Structure

In this section we describe the Poisson bracket in the neighborhood of any point of a Poisson manifold and deduce from it a (global) decomposition of the Poisson manifold into symplectic leaves (of varying dimensions). The key is the *Splitting Theorem*, due to A. Weinstein (see [173] or [39]), which generalizes the *classical Darboux Theorem*, which corresponds to the case of a symplectic manifold (or a regular Poisson manifold).

Theorem 3.25 (Splitting Theorem). *Suppose that $(M, \{\cdot\,,\cdot\})$ is a Poisson manifold of dimension n, let $m \in M$ be arbitrary and denote the rank of $\{\cdot\,,\cdot\}$ at m by $2r$. There exists a coordinate neighborhood U of m with coordinates $(q_1, \dots, q_r, p_1, \dots, p_r, z_1 \dots, z_s)$ centered at m, such that, on U,*

$$\{\cdot\,,\cdot\} = \sum_{i=1}^{r} \frac{\partial}{\partial q_i} \wedge \frac{\partial}{\partial p_i} + \frac{1}{2} \sum_{k,l=1}^{s} \phi_{kl}(z) \frac{\partial}{\partial z_k} \wedge \frac{\partial}{\partial z_l}, \tag{3.25}$$

where the functions ϕ_{kl} are smooth (or holomorphic) functions, which depend on $z_1, \dots, z_s$ only, and which vanish at m.

Proof. We suppose that $r > 0$ and we show the existence of coordinates $(q_1, p_1, z_1 \dots, z_{n-2})$ on a neighborhood U of m, for which (3.25) holds, where the functions ϕ_{kl} are smooth (or holomorphic) and depend on $z_1, \dots, z_{n-2}$ only. The proof then follows by induction on r.

Since $r > 0$ we may find a function p_1 such that $\mathcal{X}_{p_1}(m) \neq 0$. By the Straightening Theorem (Theorem 2.2) there exists a neighborhood V of m and a function q_1 on it, such that $\mathcal{X}_{p_1} = \frac{\partial}{\partial q_1}$ on V. Notice that $\{q_1, p_1\} = \mathcal{X}_{p_1}[q_1] = \frac{\partial q_1}{\partial q_1} = 1$. It follows that $\mathcal{X}_{p_1}$ and $\mathcal{X}_{q_1}$ define an integrable distribution of rank 2 on a neighborhood W of m: the vector fields $\mathcal{X}_{p_1}$ and $\mathcal{X}_{q_1}$ are independent on a neighborhood of m because they are independent at m, and the vector space spanned by these vector fields forms a Lie subalgebra of $\mathfrak{X}(W)$ since

$$[\mathcal{X}_{p_1}, \mathcal{X}_{q_1}] = \mathcal{X}_{\{q_1, p_1\}} = \mathcal{X}_1 = 0.$$

By the Frobenius Theorem (Theorem 2.4) they define a foliation near m with two-dimensional leaves. On a neighborhood U of m we can then find functions $z_1, \dots, z_{n-2}$ such that $\mathrm{d}z_1, \dots, \mathrm{d}z_{n-2}$ are independent on U and such that $\mathcal{X}_{p_1}[z_i] = \mathcal{X}_{q_1}[z_i] = 0$, by the description of the leaves of the foliation. It follows that $p_1, q_1, z_1, \dots, z_{n-2}$ are coordinates on U and that, on U,

$$\{\cdot\,,\cdot\} = \frac{\partial}{\partial q_1} \wedge \frac{\partial}{\partial p_1} + \frac{1}{2} \sum_{k<l} \{z_k, z_l\} \frac{\partial}{\partial z_k} \wedge \frac{\partial}{\partial z_l}.$$

In order to show that $\{z_k, z_l\}$ is independent of q_1 and p_1 it suffices to show that $\{\{z_k, z_l\}, p_1\} = 0 = \{\{z_k, z_l\}, q_1\}$, an easy consequence of the Jacobi identity. □

The rank of $\{\cdot\,,\cdot\}$ at m is $2r$ but is not necessarily constant on a neighborhood of m. When the rank is constant and equal to $2r$ on a neighborhood of m, the neighborhood U of m can be chosen such that, on U, the functions ϕ_{kl} vanish, yielding the following *canonical brackets* for the above coordinates:

$$\{q_i, q_j\} = \{p_i, p_j\} = \{q_i, z_k\} = \{p_i, z_k\} = \{z_k, z_l\} = 0, \qquad \{q_i, p_j\} = \delta_{ij},$$

where $1 \leqslant i, j \leqslant r$ and $1 \leqslant k, l \leqslant s$. In this form Weinstein's Splitting Theorem is usually referred to as the *Darboux Theorem* and the above local coordinates are called *Darboux coordinates* or *canonical coordinates.*

Moreover, when the Poisson structure is of constant rank $2r$ in a neighborhood U of m, then the Hamiltonian vector fields define a distribution of rank $2r$ on U and this distribution is integrable because $[\mathcal{X}_F, \mathcal{X}_G] = \mathcal{X}_{\{G,F\}}$. Thus, we have a (regular) foliation of U, where each leaf has dimension $2r$. The leaves do not necessarily carry the induced topology, (see Example 3.27 below), but they inherit a Poisson structure from $\{\cdot\,,\cdot\}$; since the rank of this Poisson structure is $2r$ each leaf carries a symplectic form and U admits a natural (set-theoretical) decomposition into symplectic leaves.

Surprisingly, such a decomposition exists in the neighborhood of any point m of a Poisson manifold (i.e., the rank needs not be constant in a neighborhood of m), but the dimension of the leaves will not be constant in general. This is the content of the following theorem.

Theorem 3.26. *Let $(M, \{\cdot\,,\cdot\})$ be a Poisson manifold. The (singular) distribution on M defined by the Hamiltonian vector fields is integrable in the sense that every $m \in M$ has a coordinate neighborhood U which is, in a unique way, a disjoint union of symplectic manifolds U_i which are Poisson submanifolds of U. The resulting (singular) foliation is called the* symplectic foliation *and each of its leaves is called a* symplectic leaf.

Proof. For $m \in M$, let U denote a coordinate neighborhood of m with coordinates $(q_1, \ldots, q_r, p_1, \ldots, p_r, z_1, \ldots, z_s)$ in terms of which $\{\cdot\,,\cdot\}$ takes the form (3.25), as obtained in the proof of Theorem 3.25. The equations $z_1 = \cdots = z_s = 0$ define a $2r$-dimensional submanifold M' of U which passes through m and to which the Poisson structure restricts, giving

$$\{\cdot\,,\cdot\}_{|M'} = \sum_{i=1}^{r} \frac{\partial}{\partial q_i} \wedge \frac{\partial}{\partial p_i}.$$

M' is tangent to the distribution in a neighborhood of m, and $\dim M' = \mathrm{Rk}_p \{\cdot\,,\cdot\}$ for p in a neighborhood of m in M', hence M' coincides with *the*[2] integral manifold of the distribution, passing through m. Since $\{\cdot\,,\cdot\}_{M'}$ is of maximal rank on M' it comes from a symplectic structure (see Proposition 3.15). □

[2] For a singular distribution the integral manifold passing through any point is unique, just as in the case of a regular distribution.

Another useful characterization of the leaf passing through m is that its restriction to U is the smallest arcwise connected Poisson submanifold of U which contains m.

Example 3.27. All Casimir functions are constant on each symplectic leaf. In good cases most or all of the symplectic leaves are level sets of the Casimir functions, but this is not true in general. Consider for example a constant bivector field P on $\mathbf{R}^3$. It descends to any torus $\mathbf{R}^3/\Lambda$, where Λ is a lattice in $\mathbf{R}^3$. Unless there is a special relation between P and Λ, all symplectic leaves are dense on the torus, hence cannot be the level sets of a (Casimir) function.

Example 3.28. In the previous example there still exist local Casimirs around any point, which single out the symplectic leaves (locally). In the present example we show that on a Poisson manifold which is not a symplectic manifold, non-trivial Casimirs may even not exist locally. Take on $\mathbf{R}^3$ with respect to the coordinates (x, y, z) the following Poisson matrix,

$$\begin{pmatrix} 0 & 0 & x \\ 0 & 0 & y \\ -x & -y & 0 \end{pmatrix}.$$

The symplectic leaves of dimension 0 are the points on the Z-axis, while the symplectic leaves of dimension 2 are the half-planes which are the connected components of $(\alpha x + \beta y = 0) \setminus (x = y = 0)$ $(\alpha, \beta \in \mathbf{R})$. This symplectic foliation is called the *open book foliation*. Taking any point on the Z-axis it is clear that on no neighborhood of it the symplectic foliation is given as the level sets of one or several functions.

Another interesting example, the symplectic foliation of the Lie-Poisson structure on $\mathfrak{g}^*$ will be discussed in Section 3.5.

3.5 The Lie-Poisson Structure of $\mathfrak{g}^*$

In this section we show that the dual of a finite-dimensional Lie algebra $\mathfrak{g}$ carries a natural Poisson structure. We will also explain how the construction is modified in the case of certain infinite-dimensional Lie algebras.

Suppose that $\mathfrak{g}$ is a finite-dimensional (complex) Lie algebra, with Lie bracket $[\cdot\,,\cdot]$. As we already indicated in Section 2.2, we can view an element $X \in \mathfrak{g}$ as a linear function on $\mathfrak{g}^*$. For the clarity of the exposition we will violate in this section several times our convention, adapted in that section, that we do not distinguish X notationally from the linear function on $\mathfrak{g}^*$ that it defines.

Thus, to $X \in \mathfrak{g}$ we associate a linear function X_* on the dual vector space $\mathfrak{g}^*$, which is defined by

$$X_* : \mathfrak{g}^* \to \mathbf{C} : \xi \mapsto \langle \xi, X \rangle .$$

The vector space of linear functions on $\mathfrak{g}^*$ forms a Lie algebra, isomorphic to $\mathfrak{g}$, by setting $[X_*, Y_*]_{\mathfrak{g}^*} := [X, Y]_*$. It follows that $\mathfrak{g}^*$ admits a Poisson structure $\{\cdot\,,\cdot\}$ whose structure functions are linear functions with the structure constants of $\mathfrak{g}$ as coefficients: $\{\cdot\,,\cdot\}$ is the unique skew-symmetric biderivation on $\mathcal{F}(\mathfrak{g}^*)$ such that

$$\{X_*, Y_*\} = [X_*, Y_*]_{\mathfrak{g}^*} = [X, Y]_* \tag{3.26}$$

for any $X, Y \in \mathfrak{g}$. Explicitly, for $F, G \in \mathcal{F}(\mathfrak{g}^*)$ and for $\xi \in \mathfrak{g}^*$ the bracket is given by

$$\{F, G\}(\xi) = [\mathrm{d}F(\xi), \mathrm{d}G(\xi)]_{\mathfrak{g}^*}(\xi). \tag{3.27}$$

To prove this it suffices to notice that it is valid for linear functions on $\mathfrak{g}^*$, because of (3.26), and that for fixed G and ξ both sides of (3.27) are a derivation (in F) at ξ. Formula (3.27) is, according to our conventions from Section 2.2, written in the form

$$\{F, G\}(\xi) = \langle \xi, [\mathrm{d}F(\xi), \mathrm{d}G(\xi)] \rangle . \tag{3.28}$$

This Poisson structure on $\mathfrak{g}^*$ is known as the *Lie-Poisson structure on* $\mathfrak{g}^*$ or the *canonical Poisson structure on* $\mathfrak{g}^*$. Notice that every Poisson structure on $\mathbf{C}^n$ which is linear in the sense that the Poisson bracket of any two linear functions is linear, is obtained in this way. Rewriting (3.28) for $H = G$ as

$$\mathcal{X}_H(\xi)[F] = \langle \xi, -\operatorname{ad}_{\mathrm{d}H(\xi)} \mathrm{d}F(\xi) \rangle = \left\langle \operatorname{ad}^*_{\mathrm{d}H(\xi)} \xi, \mathrm{d}F(\xi) \right\rangle$$

we find that the Hamiltonian vector field $\mathcal{X}_H$ is given, at $\xi \in \mathfrak{g}^*$ by

$$\mathcal{X}_H(\xi) = \operatorname{ad}^*_{\mathrm{d}H(\xi)} \xi. \tag{3.29}$$

By identifying $\mathfrak{g}$ with its dual we obtain a linear Poisson structure on $\mathfrak{g}$. Let us analyze the (most natural) case in which the identification is done by means of an Ad-invariant non-degenerate bilinear form on $\mathfrak{g}$,

$$\langle \cdot \,|\, \cdot \rangle : \mathfrak{g} \times \mathfrak{g} \to \mathbf{C},$$

such as the Killing form in the case of a semi-simple Lie algebra. Recall from (2.15) that Ad-invariance implies that

$$\langle X \,|\, [Y, Z] \rangle = \langle [X, Y] \,|\, Z \rangle \tag{3.30}$$

for all $X, Y, Z \in \mathfrak{g}$. Non-degeneracy of $\langle \cdot \,|\, \cdot \rangle$ means that the map $\chi : \mathfrak{g} \to \mathfrak{g}^*$, defined by $X \mapsto \langle X | \cdot \rangle$, is an isomorphism; let us denote its inverse by $\psi : \mathfrak{g}^* \to \mathfrak{g}$. These isomorphisms allow us to associate to a function on $\mathfrak{g}^*$ a function on $\mathfrak{g}$ and vice versa, simply by composing with χ or with ψ. Clearly, $\mathfrak{g}$ then has a unique Poisson structure $\{\cdot\,,\cdot\}_{\mathfrak{g}}$ with respect to which χ is an isomorphism of Poisson manifolds.

We call $\{\cdot\,,\cdot\}_{\mathfrak{g}}$ the *Lie-Poisson structure on* $\mathfrak{g}$ with respect to $\langle\cdot\,|\,\cdot\rangle$. Explicitly, let $F, G \in \mathcal{F}(\mathfrak{g})$ and compute their Lie-Poisson bracket $\{F,G\}_{\mathfrak{g}}$ (with respect to $\langle\cdot\,|\,\cdot\rangle$) at $X \in \mathfrak{g}$ by using (3.28) as follows.

$$\begin{aligned}\{F,G\}_{\mathfrak{g}}(X) &= \{\psi^*F, \psi^*G\}(\chi(X)) \\ &= \langle \chi(X), [\mathsf{d}(\psi^*F)(\chi(X)), \mathsf{d}(\psi^*G)(\chi(X))]\rangle \\ &= \langle X \,|\, [\psi(\mathsf{d}F(X)), \psi(\mathsf{d}G(X))]\rangle\,.\end{aligned} \tag{3.31}$$

We have used that $\mathsf{d}(\psi^*F)(\chi(X))$, viewed as an element of $\mathfrak{g}$ is precisely $\psi(\mathsf{d}F(X))$. To check the latter, first use the chain rule and the fact that ψ is a linear map to find that

$$\mathsf{d}(\psi^*F)(\chi(X)) = \mathsf{d}(F\circ\psi)(\chi(X)) = \mathsf{d}F(X)\circ\psi,$$

so that, for any $\xi \in \mathfrak{g}^*$,

$$\langle \mathsf{d}(\psi^*F)(\chi(X)), \xi\rangle = \langle \mathsf{d}F(X), \psi(\xi)\rangle = \langle \psi(\mathsf{d}F(X)) \,|\, \psi(\xi)\rangle = \langle \xi, \psi(\mathsf{d}F(X))\rangle\,.$$

Formula (3.31) is usually written in the following form,

$$\{F,G\}(X) = \langle X \,|\, [\nabla F(X), \nabla G(X)]\rangle\,, \tag{3.32}$$

where $\nabla F(X)$, the *gradient* of F at X (with respect to $\langle\cdot\,|\,\cdot\rangle$) is defined, for $F \in \mathcal{F}(\mathfrak{g})$ and $X \in \mathfrak{g}$ by

$$\langle \nabla F(X) \,|\, Y\rangle = \langle \mathsf{d}F(X), Y\rangle\,,$$

for all $Y \in \mathfrak{g}$, which is equivalent to saying that $\nabla F(X) = \psi(\mathsf{d}F(X))$. Since F is a function on a vector space, the latter definition can also be written in the following form,

$$\langle \nabla F(X) \,|\, Y\rangle = \frac{d}{dt}_{|t=0} F(X+tY), \tag{3.33}$$

which is the most useful form for explicit computation. The Hamiltonian vector fields $\mathcal{X}_H$ on $(\mathfrak{g}, \{\cdot\,,\cdot\})$ take a particularly simple form. In fact, (3.30) implies that (3.32), for $H = G$, can be written as

$$\{F,H\}(X) = \langle [\nabla H(X), X] \,|\, \nabla F(X)\rangle = \langle \mathsf{d}F(X), [\nabla H(X), X]\rangle\,;$$

if we compare this to

$$\{F,H\}(X) = \chi_H(X)[F] = \langle \mathsf{d}F(X), \chi_H(X)\rangle\,,$$

then we find that $\dot{X} := \chi_H(X)$ is given by the *Lax equation*

$$\dot{X} = [\nabla H(X), X]\,. \tag{3.34}$$

This follows also immediately from (3.29) as ad and ad* get identified by χ (see (2.16)). Equations of form (3.34) will come up repeatedly in this book, see Sections 4.4, 4.5, 6.4 and the examples (part III).

Ad-invariance of $\langle \cdot | \cdot \rangle$ implies, in view of (2.16), also that χ establishes a one-to-one correspondence between Ad^*-invariant functions on $\mathfrak{g}^*$ and Ad-invariant functions on $\mathfrak{g}$. Since the Casimirs of the Lie-Poisson structure on $\mathfrak{g}^*$ are the Ad^*-invariant functions on $\mathfrak{g}^*$, as we show in the following proposition, the Casimirs of the Lie-Poisson structure on $\mathfrak{g}$ are the Ad-invariant functions on $\mathfrak{g}$ (when $\mathfrak{g}$ and $\mathfrak{g}^*$ are identified by using an Ad-invariant non-degenerate bilinear form on $\mathfrak{g}$).

Proposition 3.29. *Let $\mathfrak{g}$ be a finite-dimensional Lie algebra and let $\mathbf{G}$ be the simply connected Lie group whose Lie algebra is $\mathfrak{g}$. The symplectic leaves of the Lie-Poisson structure on $\mathfrak{g}^*$ are the coadjoint orbits of $\mathbf{G}$.*

Proof. Recall that the fundamental vector fields of a group action span all tangent spaces to the orbits of the action. Hence, the vector fields ad_X^* span the tangent space to the coadjoint orbits at any point of $\mathfrak{g}^*$, so that for any $\xi \in \mathfrak{g}^*$

$$T_\xi \mathcal{O}_\xi = \{\mathrm{ad}_X^* \xi \mid X \in \mathfrak{g}\},$$

where $\mathcal{O}_\xi = \left\{\mathrm{Ad}_g^*(\xi) \mid g \in \mathbf{G}\right\}$ is the coadjoint orbit passing through ξ. We have from (3.29) that

$$\mathrm{Ham}_\xi(\mathfrak{g}^*) = \left\{\mathrm{ad}_{\mathrm{d}G(\xi)}^* \xi \mid G \in \mathcal{F}(\mathfrak{g}^*)\right\} = \{\mathrm{ad}_X^* \xi \mid X \in \mathfrak{g}\} = T_\xi \mathcal{O}_\xi,$$

where we used that, given $X \in \mathfrak{g}$, the linear function G on $\mathfrak{g}^*$, defined by $G(\xi) = \langle X, \xi \rangle$ realizes $\mathrm{d}G(\xi) = X$. This proves the proposition. □

Example 3.30. For any manifold M the (infinite-dimensional) vector space $\mathfrak{X}(M)$ of vector fields on M forms a Lie algebra. It is natural to wonder if this leads to a Poisson structure on a (finite-dimensional) manifold. The answer is yes, namely it leads to a Poisson structure on the cotangent bundle T^*M of M. To show this, we associate to any vector field $\mathcal{V}$ on M a function $\tilde{\mathcal{V}}$ on T^*M by defining $\tilde{\mathcal{V}}(\omega_m) := \langle \omega_m, \mathcal{V}(m) \rangle$, where ω_m is any covector in the fiber over $m \in M$. Similarly, every function $F \in \mathcal{F}(M)$ leads to a function $\tilde{F} := \pi^* F$ on T^*M, where $\pi : T^*M \to M$ is the natural projection map. The Poisson bracket of two functions $\tilde{\mathcal{V}_1}$ and $\tilde{\mathcal{V}_2}$ is then defined, as in the case of the Lie-Poisson structure on the dual of a Lie algebra, by

$$\left\{\tilde{\mathcal{V}_1}, \tilde{\mathcal{V}_2}\right\} := \widetilde{[\mathcal{V}_1, \mathcal{V}_2]}. \tag{3.35}$$

For such functions the Jacobi identity follows from the Jacobi identity for the Lie bracket of vector fields. We wish to extend this bracket to a biderivation on $\mathcal{F}(T^*M)$. Using first (3.35) and then (2.4) we have that

$$\left\{\widetilde{F\mathcal{V}_1}, \widetilde{G\mathcal{V}_2}\right\} = \widetilde{[F\mathcal{V}_1, G\mathcal{V}_2]} = \widetilde{FG}\left\{\tilde{\mathcal{V}_1}, \tilde{\mathcal{V}_2}\right\} - \widetilde{G\mathcal{V}_1}\, \widetilde{\mathcal{V}_2[F]} + \widetilde{F\mathcal{V}_2}\, \widetilde{\mathcal{V}_1[G]}.$$

If the Leibniz property for $\{\cdot,\cdot\}$ holds then

$$\left\{\tilde{F}\tilde{\mathcal{V}}_1, \tilde{G}\tilde{\mathcal{V}}_2\right\} = \widetilde{FG}\left\{\tilde{\mathcal{V}}_1, \tilde{\mathcal{V}}_2\right\} + \widetilde{G\mathcal{V}_1}\left\{\tilde{F}, \tilde{\mathcal{V}}_2\right\} + \widetilde{F\mathcal{V}_2}\left\{\tilde{\mathcal{V}}_1, \tilde{G}\right\} + \widetilde{\mathcal{V}_1\mathcal{V}_2}\left\{\tilde{F}, \tilde{G}\right\}.$$

If we compare the two expressions for $\left\{\widetilde{F\mathcal{V}_1}, \widetilde{G\mathcal{V}_2}\right\} = \left\{\tilde{F}\tilde{\mathcal{V}}_1, \tilde{G}\tilde{\mathcal{V}}_2\right\}$ that we have obtained we see that, if we want to extend this bracket to a biderivation on $\mathcal{F}(T^*M)$ then we must define

$$\left\{\tilde{F}, \tilde{G}\right\} := 0, \qquad \left\{\tilde{\mathcal{V}}, \tilde{F}\right\} := \widetilde{\mathcal{V}[F]},$$

for all $F, G \in \mathcal{F}(M)$ and $\mathcal{V} \in \mathfrak{X}(M)$. These definitions extend uniquely to a biderivation on $\mathcal{F}(T^*M)$, which automatically satisfies the Jacobi identity. If we take local coordinates $x_1, \dots, x_n$ on a coordinate neighborhood U of M then $\widetilde{x_1}, \dots, \widetilde{x_n}, \widetilde{\partial/\partial x_1}, \dots, \widetilde{\partial/\partial x_n}$ are Darboux coordinates on T^*U, since

$$\{\widetilde{x_i}, \widetilde{x_j}\} = \left\{\widetilde{\partial/\partial x_i}, \widetilde{\partial/\partial x_j}\right\} = 0, \qquad \left\{\widetilde{\partial/\partial x_i}, \widetilde{x_j}\right\} = \delta_{ij},$$

where $1 \leqslant i, j \leqslant n$. In particular, $\{\cdot,\cdot\}$ has maximal rank on T^*M, so that T^*M carries a (natural[3]) symplectic structure.

Example 3.31. Another example is the dual of an affine Lie algebra $L(\mathfrak{g}, \nu)$. Recalling that such a Lie algebra has a Killing form $\langle\cdot\,|\,\cdot\rangle$ we define the (restricted) dual $L(\mathfrak{g}, \nu)^*$ to be the vector space of all linear forms on $L(\mathfrak{g}, \nu)$ of the form $\langle X(\mathfrak{h})\,|\,\cdot\rangle$, where $X(\mathfrak{h}) \in L(\mathfrak{g}, \nu)$. Any element $Y(\mathfrak{h}) \in L(\mathfrak{g}, \nu)$ then defines a (linear) function $Y(\mathfrak{h})_*$ on $L(\mathfrak{g}, \nu)^*$ via $\langle X(\mathfrak{h})\,|\,\cdot\rangle \mapsto \langle X(\mathfrak{h})\,|\,Y(\mathfrak{h})\rangle$. By definition the algebra of functions on the restricted dual, denoted $\mathcal{F}(L(\mathfrak{g}, \nu)^*)$, is generated by the functions $Y(\mathfrak{h})_*$, where $Y(\mathfrak{h}) \in L(\mathfrak{g}, \nu)$ and a Poisson bracket is defined on on $\mathcal{F}(L(\mathfrak{g}, \nu)^*)$ by

$$\{X(\mathfrak{h})_*, Y(\mathfrak{h})_*\} := \left[X(\mathfrak{h}), Y(\mathfrak{h})\right]_*.$$

The reason why we include this infinite-dimensional example in a book on finite-dimensional integrable systems is that the construction of the latter often involves a (twisted) affine Lie algebra, which then gets further restricted to a finite-dimensional subspace of it. In each of those examples one could start out with an appropriate finite-dimensional truncation of the loop algebra, but this would in fact complicate the construction and it would obscure the (often very simple) nature of the Poisson structure.

In many cases of interest one deals with Poisson brackets on a vector space which are *almost* linear in the sense that the Poisson bracket of two linear functions is of degree at most one (i.e., it may also contain constants). This case is considered in the following proposition.

[3] To be ultra-precise, the same Poisson structure with opposite sign, as is obtained by putting a minus sign in (3.35), is just as natural; the two corresponding symplectic structures lead to opposite orientations of T^*M when $\dim M$ is odd.

Proposition 3.32. *Let $\mathfrak{g}$ be a Lie algebra and let $\{\cdot\,,\cdot\}_0$ be a constant Poisson structure on $\mathfrak{g}^*$. The Lie-Poisson structure $\{\cdot\,,\cdot\}$ on $\mathfrak{g}^*$ and the constant Poisson structure $\{\cdot\,,\cdot\}_0$ on $\mathfrak{g}^*$ are compatible if and only if the bilinear map C, defined by*

$$C : \mathfrak{g} \wedge \mathfrak{g} \to \mathbf{C} : (X,Y) \mapsto \{X_*, Y_*\}_0\,, \tag{3.36}$$

is a 2-cocycle in the cohomology of $\mathfrak{g}$ associated with the trivial representation of $\mathfrak{g}$ on $\mathbf{C}$. In this case the Poisson structure $\{\cdot\,,\cdot\}_0 + \{\cdot\,,\cdot\}$ is called a modified Lie-Poisson structure.

Proof. We have, as in (3.23), that $\{\cdot\,,\cdot\}_0$ and $\{\cdot\,,\cdot\}$ are compatible if and only if

$$\{\{X_*, Y_*\}, Z_*\}_0 + \{\{Y_*, Z_*\}, X_*\}_0 + \{\{Z_*, X_*\}, Y_*\}_0 = 0 \tag{3.37}$$

for any $X, Y, Z \in \mathfrak{g}$; indeed $\{X_*, Y_*\}_0$ is constant so that $\{\{X_*, Y_*\}_0\,, Z_*\} = 0$ and all the terms in $\{\{X_*, Y_*\}_0\,, Z_*\} + \mathrm{cycl}(X,Y,Z)$ vanish. In terms of the linear map C, defined by (3.36) we have in view of (3.26) that

$$\{\{X_*, Y_*\}, Z_*\}_0 = C([X,Y], Z).$$

Therefore (3.37) becomes

$$C\,([X,Y],Z) + C\,([Y,Z],X) + C\,([Z,X],Y) = 0,$$

which means precisely that C is a 2-cocycle in the cohomology of $\mathfrak{g}$ associated with the trivial representation of $\mathfrak{g}$ on $\mathbf{C}$ (see [107, Appendix 5]). □

As an application, suppose that $H^2(\mathfrak{g}) = 0$, which is e.g. the case when $\mathfrak{g}$ is semi-simple (see [107, Appendix 5]). For any Poisson structure $\{\cdot\,,\cdot\}_0$ which is compatible with the Lie-Poisson structure $\{\cdot\,,\cdot\}$ on $\mathfrak{g}$ we have that $(\mathfrak{g}^*, \{\cdot\,,\cdot\} + \{\cdot\,,\cdot\}_0)$ and $(\mathfrak{g}^*, \{\cdot\,,\cdot\})$ are isomorphic, the isomorphism (of Poisson manifolds!) being an *affine* transformation $\mathfrak{g}^* \to \mathfrak{g}^*$.

3.6 Constructing New Poisson Manifolds from Old Ones

In this section we describe some basic methods to construct new Poisson manifolds from given ones. We already noticed (in Section 3.2) that a submanifold of a Poisson manifold is, in general, not a Poisson submanifold. We will give precise conditions for this to happen in the proposition that follows.

Proposition 3.33. *Let $(M, \{\cdot\,,\cdot\})$ be a Poisson manifold and let M' be a (possibly immersed) submanifold. There exists a Poisson structure $\{\cdot\,,\cdot\}_{M'}$ for which M' is an (immersed) Poisson submanifold of M if and only if the restriction of every Hamiltonian vector field on M to M' is tangent to M'.*

Proof. The restriction of an immersion $\imath : M' \to M$ to a small neighborhood, in M', of any point in M', is an embedding. Therefore, it suffices to prove the proposition in the case in which M' is a submanifold of M, i.e., M' is a subset of M and the inclusion map $\imath : M' \hookrightarrow M$ is an embedding.

For any open subset U of M let us denote by $\mathcal{I}_{M'}(U)$ the ideal of all functions on U that vanish on $M' \cap U$. Thus, if $F \in \mathcal{F}(U)$ then $F \in \mathcal{I}_{M'}(U)$ if and only if $\imath^* F = 0$. We first show that a vector field $\mathcal{V}$ on M is tangent to M' at all points of M' if and only if for any open subset U of M, the ideal $\mathcal{I}_{M'}(U)$ has the property that $\mathcal{V}[\mathcal{I}_{M'}(U)] \subseteq \mathcal{I}_{M'}(U)$, i.e., $\mathcal{V}[F] \in \mathcal{I}_{M'}(U)$ for any $F \in \mathcal{I}_{M'}(U)$. Since M' is a submanifold, say of dimension p, there exists for any $m \in M'$ a coordinate neighborhood U containing m with coordinates $(x_1, \ldots, x_n)$, such that $M' \cap U$ is given by the equations $x_{p+1} = \cdots = x_n = 0$. In terms of these coordinates the ideal $\mathcal{I}_{M'}(U)$ is generated by $x_{p+1}, \ldots, x_n$ and the condition $\mathcal{V}[\mathcal{I}_{M'}(U)] \subseteq \mathcal{I}_{M'}(U)$ takes the form $\mathcal{V}[x_i] = \sum_{j=p+1}^{n} h_{ij} x_j$, for $i = p+1, \ldots, n$. By uniqueness, the integral curves $x(t)$ of $\mathcal{V}$ in U that start at points of M' ($x_{p+1}(0) = \cdots = x_n(0) = 0$) will have $x_{p+1}(t) = \cdots = x_n(t) = 0$ ($|t|$ small), so they stay in M' and $\mathcal{V}$ is tangent to M'.

Suppose now that $\{\cdot\,,\cdot\}_{M'}$ is a Poisson structure on M' for which the inclusion map $\imath : M' \to M$ is a Poisson map. Let U be any open subset of M and let $F \in \mathcal{I}_{M'}(U)$, so that $\imath^* F = 0$. For any $G \in \mathcal{F}(U)$ it follows that

$$\imath^*(\mathcal{X}_G[F]) = \imath^* \{F, G\} = \{\imath^* F, \imath^* G\}_{M'} = 0,$$

showing that $\mathcal{X}_G[F] \in \mathcal{I}_{M'}(U)$. By the above characterization, all Hamiltonian vector fields $\mathcal{X}_G$ on M are tangent to M' at points of M'.

Suppose, on the other hand, that all Hamiltonian vector fields on M are tangent to M' at points of M'. Since M' is a submanifold of M we may consider any open subset V of M' which is small enough so that every function on V extends to an open subset U of M, with $V = U \cap M'$. We may then define, for $f, g \in \mathcal{F}(V)$ their bracket $\{f, g\}_{M'}$ by $\{f, g\}_{M'} := \imath^* \{F, G\}$, where $F, G \in \mathcal{F}(U)$ are any functions for which $f = \imath^* F$ and $g = \imath^* G$. To see that this is well-defined, notice that the difference between two extensions of f (or g) belongs to the ideal $\mathcal{I}_{M'}(U)$ and take $H \in \mathcal{I}_{M'}(U)$. Then, $\imath^* \{H, G\} = \imath^*(\mathcal{X}_G[H]) = 0$, for any function $G \in \mathcal{F}(U)$, since $\mathcal{X}_G[H] \in \mathcal{I}_{M'}(U)$, again by the above characterization of tangent vector fields. By construction, $\imath$ is a Poisson map, implying also that

$$\imath^* \{\{F, G\}, H\} = \{\{\imath^* F, \imath^* G\}_{M'}, \imath^* H\}_{M'}$$

for $F, G, H \in \mathcal{F}(M)$, leading to the Jacobi identity for $\{\cdot\,,\cdot\}_{M'}$. □

Example 3.34. The symplectic leaves of a Poisson manifold are immersed submanifolds to which all Hamiltonian vector fields are tangent. Therefore, Proposition 3.33 yields another proof that these leaves carry a Poisson structure. More generally, the proposition implies that any (immersed) submanifold M' of M of that is the union of symplectic leaves carries a Poisson structure, making it into an (immersed) Poisson submanifold.

Example 3.35. Suppose that $F_1, \ldots, F_s$ are Casimirs and that $\mathbf{c} = (c_1 \ldots, c_s)$ are constants such that

$$\mathcal{A}_{\mathbf{c}} := \bigcap_{i=1}^{s} \{m \in M \mid F_i(m) = c_i\}$$

is a (non-empty) submanifold of M. Since $\mathcal{X}_F[F_i] = \{F_i, F\} = 0$ for all $F \in \mathcal{F}(M)$ and $1 \leqslant i \leqslant s$, all Hamiltonian vector fields $\mathcal{X}_F$ are tangent to $\mathcal{A}_{\mathbf{c}}$ and $\mathcal{A}_{\mathbf{c}}$ is a Poisson submanifold of M.

Example 3.36. Let $m \in M$ and suppose that M' is a Poisson submanifold, passing through m. Then M' contains, at least in a neighborhood of m, the symplectic leaf of M which passes through m, so M' is locally a union of symplectic leaves.

A second standard construction is the product of Poisson manifolds.

Proposition 3.37. *Let $(M_1, \{\cdot\,,\cdot\}_1)$ and $(M_2, \{\cdot\,,\cdot\}_2)$ be two Poisson manifolds. The product $M_1 \times M_2$ has a natural Poisson bracket such that the two projection maps $\pi_i : M_1 \times M_2 \to M_i$ are Poisson morphisms.*

Proof. In order for π_1 and π_2 to be Poisson morphisms it is necessary and sufficient to define $\{\pi_1^* F_1, \pi_1^* G_1\} := \pi_1^* \{F_1, G_1\}_1$ and $\{\pi_2^* F_2, \pi_2^* G_2\} := \pi_2^* \{F_2, G_2\}_2$ for any $F_1, G_1 \in \mathcal{F}(M_1)$ and $F_2, G_2 \in \mathcal{F}(M_2)$. We define in addition $\{\pi_1^* F_1, \pi_2^* F_2\} := 0$ for any $F_1 \in \mathcal{F}(M_1)$ and $F_2 \in \mathcal{F}(M_2)$. These definitions extend uniquely to a skew-symmetric biderivation on $\mathcal{F}(M_1 \times M_2)$, which we denote by $\{\cdot\,,\cdot\}$. Notice that the Poisson matrix of $\{\cdot\,,\cdot\}$ with respect to the system of local coordinates coming from local coordinates on M_1 and on M_2 has a block form, where each block is (the pull-back under π_i^* of) the Poisson matrix with respect to those local coordinates on M_1 and on M_2. Therefore, the Jacobi identity is satisfied. □

Notice that the fact that the matrix of $\{\cdot\,,\cdot\}$ has block form also shows that the rank at $(m_1, m_2) \in M_1 \times M_2$ of $\{\cdot\,,\cdot\}$ is given by the sum $\mathrm{Rk}_{m_1} \{\cdot\,,\cdot\}_1 + \mathrm{Rk}_{m_2} \{\cdot\,,\cdot\}_2$. In particular, $\mathrm{Rk}\{\cdot\,,\cdot\} = \mathrm{Rk}\{\cdot\,,\cdot\}_1 + \mathrm{Rk}\{\cdot\,,\cdot\}_2$.

Definition 3.38. The Poisson bracket on $M_1 \times M_2$ given by Proposition 3.37 is called the *product bracket.*

The above construction of the product of two Poisson manifolds is easily generalized to the product of several Poisson manifolds.

Example 3.39. Let $\mathbf{G}$ be a Lie group with multiplication $\chi : \mathbf{G} \times \mathbf{G} \to \mathbf{G}$. If $\{\cdot\,,\cdot\}$ is a Poisson structure on $\mathbf{G}$ such that χ is a Poisson morphism, the Poisson bracket on $\mathbf{G} \times \mathbf{G}$ being the product bracket, then $(\mathbf{G}, \{\cdot\,,\cdot\})$ is called a *Lie-Poisson group.*

Example 3.40. Theorem 3.25 can be restated as follows: every point m in a Poisson manifold $(M, \{\cdot\,,\cdot\})$ has a coordinate neighborhood which is the product of a symplectic manifold of dimension $\mathrm{Rk}_m \{\cdot\,,\cdot\}$ and a Poisson manifold which has rank 0 at the point that corresponds to m.

A final construction that we give here consists of the most simple form of *Poisson reduction:* we consider a group acting in a Poisson way on a Poisson manifold.

Definition 3.41. Let $(M, \{\cdot\,,\cdot\})$ be a Poisson manifold and let $(\mathbf{G}, \{\cdot\,,\cdot\}_{\mathbf{G}})$ be a Lie group, equipped with a Poisson structure, acting (say on the left) on M. The action is called a *Poisson action* if the map $\mathbf{G} \times M \to M : (g, m) \mapsto gm$ is a Poisson morphism, where one considers the product bracket on $\mathbf{G} \times M$.

An interesting example is the diagonal adjoint action of a Lie-Poisson group $(\mathbf{G}, \{\cdot\,,\cdot\})$ on $\left(\mathbf{G}^d, \{\cdot\,,\cdot\}^d\right)$, where $\mathbf{G}^d = \mathbf{G} \times \cdots \mathbf{G}$ (d factors) and where $\{\cdot\,,\cdot\}^d$ denotes the product bracket on $\mathbf{G}^d$.

Notice that we do not demand, in the above definition, that $(\mathbf{G}, \{\cdot\,,\cdot\}_{\mathbf{G}})$ be a Lie-Poisson group. We neither demand this in the following proposition.

Proposition 3.42. *Let $(M, \{\cdot\,,\cdot\})$ be a Poisson manifold and let $\mathbf{G}$ be a Lie group acting on M. We assume that $\mathbf{G}$ comes also equipped with a Poisson structure $\{\cdot\,,\cdot\}_{\mathbf{G}}$. If the action is a Poisson action, then the algebra $\mathcal{F}(M)^{\mathbf{G}}$ of $\mathbf{G}$-invariant functions is a* Poisson subalgebra *of $\mathcal{F}(M)$, i.e., it is closed under $\{\cdot\,,\cdot\}$.*

Proof. Let us denote the action $\mathbf{G} \times M \to M$ by χ and the projection $\mathbf{G} \times M \to M$ by π_2. Then $F \in \mathcal{F}(M)$ is $\mathbf{G}$-invariant if and only if $F \circ \chi = F \circ \pi_2$. Thus, if $F, G \in \mathcal{F}(M)^{\mathbf{G}}$ and χ is Poisson then

$$\chi^* \{F, G\}_M = \{\chi^* F, \chi^* G\}_{\mathbf{G} \times M} = \{\pi_2^* F, \pi_2^* G\}_{\mathbf{G} \times M} = \pi_2^* \{F, G\}_M$$

and we see that the bracket of any two $\mathbf{G}$-invariant functions is $\mathbf{G}$-invariant. Therefore the subalgebra $\mathcal{F}(M)^{\mathbf{G}}$ of $\mathcal{F}(M)$ is, in addition, a Lie subalgebra of $(\mathcal{F}(M), \{\cdot\,,\cdot\})$, making it a Poisson subalgebra. □

If the quotient $M/\mathbf{G}$ is a smooth manifold then we may identify $\mathcal{F}(M/\mathbf{G})$ with $\mathcal{F}(M)^{\mathbf{G}}$ and the above proposition states that $M/\mathbf{G}$ carries a Poisson structure for which the quotient map $M \to M/\mathbf{G}$ is a morphism of Poisson manifolds. In general there is at least an open subset of M which is smooth and on it we get, by the above construction, a Poisson structure, the *reduced Poisson structure.* The quotient structure usually has Casimirs, so that we get Poisson or symplectic structures on the level sets of them. The case in which symplectic structures are found by the above procedure corresponds to what is classically known as *symplectic reduction.*

Example 3.43. Let $(M, \{\cdot\,,\cdot\})$ be a Poisson manifold and let $\left(M^d, \{\cdot\,,\cdot\}^d\right)$ denote the product $M \times \cdots \times M$ (d factors), equipped with the product bracket. The group S_d of permutations of $\{1, \ldots, d\}$ acts on M^d by permuting the factors; since this group is discrete it only carries the trivial Poisson structure. The action is Poisson, hence the algebra of symmetric functions in d variables on M carries a natural Poisson structure, coming from the Poisson structure on M.

For a few other constructions, that will not be used in this text, we refer to [169].

4 Integrable Systems on Poisson Manifolds

In this chapter we give the basic definitions of Liouville integrable systems on Poisson manifolds, we prove some key propositions and we give simple examples to illustrate the theory. While the definition of Liouville integrability is given on a general Poisson manifold, we will restrict ourselves to *real* Poisson manifolds in Section 4.3, where we will discuss the classical Liouville Theorem and the Action-Angle Theorem, which are, as such, only valid in the real case. For a complex version of the Liouville Theorem, we refer to Section 6.3. Lax equations, which often represent a vector field of an integrable system, are the subject of Sections 4.4 and 4.5.

4.1 Functions in Involution

Definition 4.1. Let $(M, \{\cdot\,,\cdot\})$ be a Poisson manifold and let $F, G \in \mathcal{F}(M)$. We say that F and G are in *involution* if $\{F, G\} = 0$. For a subset $\mathbf{F}$ of $\mathcal{F}(M)$ we say that $\mathbf{F}$ is *involutive* if any two elements of $\mathbf{F}$ are in involution.

Example 4.2. Let X be a bi-Hamiltonian vector field on a bi-Hamiltonian manifold $(M, \{\cdot\,,\cdot\}_1, \{\cdot\,,\cdot\}_2)$. The functions F and G for which $X = \{\cdot\,, F\}_1 = \{\cdot\,, G\}_2$ are in involution with respect to both brackets. Indeed, $\{F, G\}_2 = \{F, F\}_1 = 0$ and $\{G, F\}_1 = \{G, G\}_2 = 0$. More generally, suppose that we have a *bi-Hamiltonian hierarchy*, i.e., a sequence of functions $\mathbf{F} = \{F_i \mid i \in \mathbf{Z}\}$ such that

$$\{\cdot\,, F_i\}_2 = \{\cdot\,, F_{i+1}\}_1\,, \qquad (i \in \mathbf{Z})\,.$$

In this case one has for any $i < j \in \mathbf{Z}$

$$\begin{aligned}
\{F_i, F_j\}_1 &= \{F_i, F_{j-1}\}_2 \\
&= \{F_{i+1}, F_{j-1}\}_1 \\
&= \quad \dots \\
&= \{F_j, F_i\}_1\,,
\end{aligned}$$

so that $\{F_i, F_j\}_1 = 0$ by skew-symmetry. It follows that $\mathbf{F}$ is involutive with respect to $\{\cdot\,,\cdot\}_1$. Notice that $\mathbf{F}$ is also involutive with respect to $\{\cdot\,,\cdot\}_2$, since $\{F_i, F_j\}_2 = \{F_i, F_{j+1}\}_1$.

Example 4.3. Let $M \subseteq \mathfrak{gl}(n)$ be an affine subspace of the Lie algebra of $(n \times n)$-matrices and let $H \in \mathcal{F}(M)$. Suppose that there exists a Poisson structure on M such that the Hamiltonian vector field $\mathcal{X}_H$ takes the form

$$\dot{X} = [X, Y] \tag{4.1}$$

where $X \in M$ and $Y : M \to \mathfrak{gl}(n)$ is any map, and $\dot{X}$ is a shorthand for $\mathcal{X}_H(X)$, viewed as an element of $\mathfrak{gl}(n)$. As we have seen in Section 3.5, if a vector field on a Lie algebra $\mathfrak{g}$ is Hamiltonian with respect to the Lie-Poisson bracket on $\mathfrak{g}$ (identified with its dual using an Ad-invariant non-degenerate bilinear form) then it is of this form. Notice that in order for (4.1) to make sense, the function Y and the affine subspace M of $\mathfrak{gl}(n)$ must be chosen such that for any $X_0 \in M$, with corresponding value of Y denoted by Y_0, the commutator $[X_0, Y_0]$ belongs to the vector subspace of $\mathfrak{gl}(n)$ that corresponds to M. We claim that all coefficients of the characteristic polynomial $|\mu \operatorname{Id}_n - X|$, which we view as functions on M, are in involution with H. To see this, let $X(t)$ be an integral curve of (4.1), defined for t in a neighborhood of 0, and denote by $Y(t)$ the corresponding value of Y at $X(t) \in M$. Take any $i > 0$ and use Trace(AB) = Trace(BA) (twice!) to compute

$$\begin{aligned}
\left\{\operatorname{Trace} X^i, H\right\}(X(t)) &= \frac{\mathrm{d}}{\mathrm{d}t} \operatorname{Trace} X^i(t) \\
&= i \operatorname{Trace} X^{i-1}(t) \frac{\mathrm{d}X}{\mathrm{d}t}(t) \\
&= i \operatorname{Trace}\left(X^i(t)Y(t) - X^{i-1}(t)Y(t)X(t)\right) = 0.
\end{aligned}$$

Thus, the functions $H_i : X \mapsto \operatorname{Trace} X^i$ are in involution with H. The same is true for each coefficient of the characteristic polynomial of X because each such coefficient is a polynomial in the functions H_i. An equation (i.e., vector field) on M of the form (4.1) is called a *Lax equation* (with values in $\mathfrak{gl}(n)$). Lax equations will be discussed in more detail in Sections 4.4, 4.5 and 6.4, and we will give a Lax equation for each of our main examples (Part III).

If $\mathbf{F} = (F_1, \dots, F_s)$ is an s-tuple, where $F_i \in \mathcal{F}(M)$ for $1 \leqslant i \leqslant s$, then $\mathbf{F}$ defines a map to $\mathbf{C}^s$ (or $\mathbf{R}^s$), which we will denote by the same letter. A *fiber* of $\mathbf{F}$ is a fiber of $\mathbf{F}$ as a map: it is a common level set of the functions F_i. The fiber of $\mathbf{F}$ that passes through $m \in M$ will be denoted by $\mathbf{F}_m$,

$$\mathbf{F}_m := \{p \in M \mid F_i(p) = F_i(m) \text{ for } i = 1, \dots, s\};$$

for $\mathbf{c} \in \mathbf{C}^s$ we will also use the notation $\mathbf{F}_\mathbf{c}$ for the fiber $\mathbf{F}^{-1}(\mathbf{c})$ over $\mathbf{c}$, so that for any $m \in M$, one has that $\mathbf{F}_m = \mathbf{F}^{-1}(\mathbf{F}(m)) = \mathbf{F}_{\mathbf{F}(m)}$. By Sard's Theorem[1] (see [80, Chapter 3.1.]), the set of regular values of $\mathbf{F}$ is a residual subset (hence a dense subset) of $\mathbf{C}^s$ (resp. $\mathbf{R}^s$). By the inverse function theorem, the fiber $\mathbf{F}_\mathbf{c}$ over each regular value $\mathbf{c}$ that lies in the image of $\mathbf{F}$ is non-singular.

[1] For algebraic varieties, Sard's Theorem can be considerably strengthened, namely the set of regular values of $\mathbf{F}$ contains a Zariski open subset, see [134, Chapter 3].

Since $\{F, G\} = \mathcal{X}_G F$ and since $[\mathcal{X}_F, \mathcal{X}_G] = -\mathcal{X}_{\{F,G\}}$ for any $F, G \in \mathcal{F}(M)$ the proof of the following proposition is immediate.

Proposition 4.4. *Let $(M, \{\cdot\,,\cdot\})$ be a Poisson manifold and assume that $\mathbf{F} = (F_1, \dots, F_s)$ is involutive. The Hamiltonian vector fields $\mathcal{X}_{F_i}$, $1 \leqslant i \leqslant s$, commute and for any $m \in M$ they are tangent to the non-singular part of $\mathbf{F}_m$.*
□

The proof of the following proposition is also immediate, in view of the Leibniz rule for Poisson brackets.

Proposition 4.5. *Let $(M, \{\cdot\,,\cdot\})$ be a Poisson manifold and assume that $\mathbf{F} = (F_1, \dots, F_s)$ is involutive. The subalgebra of $\mathcal{F}(M)$, generated by the functions F_i is also involutive.* □

A third proposition, classically known as the *Poisson Theorem*, is an immediate consequence of the Jacobi identity for $\{\cdot\,,\cdot\}$.

Proposition 4.6 (Poisson). *Let $(M, \{\cdot\,,\cdot\})$ be a Poisson manifold and let $F, G, H \in \mathcal{F}(M)$. If $\{F, H\} = 0$ and $\{G, H\} = 0$ then $\{\{F, G\}, H\} = 0$.*
□

Example 4.7. Returning to Example 3.8, suppose that $\mathbf{F} = (F_1, \dots, F_s)$, where each of the functions F_i is in involution with H ($\mathbf{F}$ needs not be involutive). $\mathcal{X}_H$ is tangent to each of the hypersurfaces $F_i =$ constant, hence all functions F_i are constant on the trajectories of the vector field $\mathcal{X}_H$. For this reason, functions in involution with H are classically called *constants of motion.* Finding (independent) constants of motion being very useful for the explicit integration of the equations of motion (3.15), as we will see in the next paragraph, the constants of motion are also referred to as *first integrals.* Stated in this language, Poisson's Theorem says that the Poisson bracket of two first integrals is a first integral. Compare this statement to the less appealing, but equivalent, statement that the centralizer of any function $H \in \mathcal{F}(M)$ is a Lie subalgebra of $\mathcal{F}(M)$.

Another classical theorem that leads to constants of motion is *Noether's Theorem*, which we give in a Hamiltonian form (the classical version is in a Lagrangian form). Recall from Section 3.2 that if the action of a Lie group $\mathbf{G}$ on a manifold M is a Hamiltonian action then we can construct a co-momentum map, which is a linear map $\mathcal{J} : \mathfrak{g} \to \mathcal{F}(M)$ having the property that for any $X \in \mathfrak{g}$ the function $\mathcal{J}_X$ is a Hamiltonian for the fundamental vector field $\underline{X}$, i.e.,

$$\underline{X} = \mathcal{X}_{\mathcal{J}_X} = \{\cdot\,, \mathcal{J}_X\}\,.$$

The Noether Theorem, in its Hamiltonian form, then states that the co-momentum map yields constants of motion for any G-invariant Hamiltonian.

Theorem 4.8 (Noether). *Let* $\mathbf{G}$ *be a group which acts on a Poisson manifold* $(M, \{\cdot,\cdot\})$ *and assume that the action is Hamiltonian, with co-momentum map* $\mathcal{J}$. *If* H *is a* $\mathbf{G}$*-invariant function then for any* $X \in \mathfrak{g}$ *the function* $\mathcal{J}_X$ *is a constant of motion for* $\mathcal{X}_H$.

Proof. Let $X \in \mathfrak{g}$ and $m \in M$ be arbitrary. We denote for $g \in \mathbf{G}$ by $\chi_g : M \to M$ the diffeomorphism that comes from the action. By $\mathbf{G}$-invariance of H, i.e., $H \circ \chi_g = H$ for any $g \in \mathbf{G}$, we have that

$$
\begin{aligned}
-\mathcal{X}_H(m)[\mathcal{J}_X] &= \{H, \mathcal{J}_X\}(m) = \mathcal{X}_{\mathcal{J}_X}(m)[H] = \underline{X}(m)[H] = \\
&= \frac{\mathsf{d}}{\mathsf{d}t}_{|t=0} H\left(\chi_{\exp tX}(m)\right) = \frac{\mathsf{d}}{\mathsf{d}t}_{|t=0} H(m) = 0.
\end{aligned}
$$

Since $m \in M$ is arbitrary this shows that $\mathcal{J}_X$ is a constant of motion of χ_H.
□

Remark 4.9. The point of view that we wish to emphasize is that Hamiltonian vector fields that are generated by group actions a priori lead to Lie algebra induced constants of the motion for Hamiltonian vector fields where the Hamiltonian itself is invariant under the group action. The Adler-Kostant-Symes Theorem that we will give later in this chapter will be a theorem in this historic tradition.

Example 4.10. As an application of Noether's Theorem we consider the classical two-body problem, which is a conservative system of two point masses m_1 and m_2 in $\mathbf{R}^3$, where the interactive potential $U(q_1, q_2)$ depends only on the vector that joins the particles, $U(q_1, q_2) = U(Q)$, where $Q = q_1 - q_2 \in \mathbf{R}^3$. Denoting the standard Euclidean norm on $\mathbf{R}^3$ by $||\cdot||$ the total energy is given by

$$
H = \frac{1}{2}\left(\frac{||p_1||^2}{m_1} + \frac{||p_2||^2}{m_2}\right) + U(Q),
$$

which we view as a function on $\mathbf{R}^{12} \cong T^*\mathbf{R}^6$, equipped with the standard symplectic structure. Clearly, H is invariant with respect to the natural translation action of $\mathbf{R}^3$, given by $V \cdot (q_1, p_1, q_2, p_2) = (q_1 + V, p_1, q_2 + V, p_2)$, where $V \in \mathbf{G} = \mathbf{R}^3$. The fundamental vector fields of this action are given by $\dot{q}_1 = \dot{q}_2 = X$, $\dot{p}_1 = \dot{p}_2 = 0$, where $X \in \mathfrak{g} = \mathbf{R}^3$. It follows that the linear map

$$
\begin{aligned}
\mathcal{J} : \mathfrak{g} = \mathbf{R}^3 &\to \mathcal{F}(\mathbf{R}^3) \\
X &\mapsto \langle X \,|\, p_1 + p_2 \rangle
\end{aligned}
$$

is a co-momentum map for the action, where $\langle \cdot \,|\, \cdot \rangle$ denotes the standard inner product on $\mathbf{R}^3$. Indeed, $\mathcal{X}_{\langle X \,|\, p_1 + p_2 \rangle} q_i = X$ and $\mathcal{X}_{\langle X \,|\, p_1 + p_2 \rangle} p_i = 0$, for $i = 1, 2$. By Noether's Theorem the three components of the total linear momentum $P := p_1 + p_2$ are constants of motion.

We have moreover that $H = H_{cm} + H_e$ completely separates:

$$H_{cm} = \frac{||P||^2}{2(m_1+m_2)}, \qquad H_e = \frac{m}{2}\left|\left|\frac{p_1}{m_1} - \frac{p_2}{m_2}\right|\right|^2 + U(Q),$$

where m is the reduced mass,

$$\frac{1}{m} = \frac{1}{m_1} + \frac{1}{m_2}.$$

As P is constant the center of mass of the two particles moves in a linear way, as dictated by H_{cm}; let us therefore ignore this part of the total energy and focus on H_e, where we assume now that $U(Q)$ depends only on the distance of the particles, $U(Q) = U(\rho)$, where $\rho = ||Q||$. Since H has then spherical (i.e., $\mathbf{SO(3)}$) symmetry we have again by Noether's Theorem that all three components of its angular momentum are preserved, and hence the reduced particle evolves in a plane $\mathbf{R}^2$, with coordinates (ρ, θ) and the conserved momentum in that plane, $\ell = m\rho^2\theta$ may be used to eliminate θ, leaving us with a Hamiltonian of the form (3.16) with

$$V(\rho) = U(\rho) + \frac{\ell^2}{2m\rho^2}.$$

For more information on the two-body problem, see [65, pp. 331-332].

Definition 4.11. Let $(M, \{\cdot\,,\cdot\})$ be a Poisson manifold and suppose that $\mathbf{F} = (F_1, \dots, F_s)$, where $F_i \in \mathcal{F}(M)$ for $1 \leqslant i \leqslant s$. We say that $\mathbf{F}$ is *independent* when the open subset on which the differentials $\mathrm{d}F_1, \dots, \mathrm{d}F_s$ are independent is dense in M.

Thus, $\mathbf{F} = (F_1, \dots, F_s)$ is independent if and only if the set

$$\mathcal{U}_{\mathbf{F}} := \{m \in M \mid \mathrm{d}F_1(m) \wedge \cdots \wedge \mathrm{d}F_s(m) \neq 0\} \tag{4.2}$$

is a dense open subset of M. Taking local coordinates $x_1, \dots, x_n$ on a neighborhood of $m \in M$ we have that $m \in \mathcal{U}_{\mathbf{F}}$ if and only if

$$\operatorname{Rk}\left(\frac{\partial F_i}{\partial x_j}(m)\right)_{\substack{1\leqslant i\leqslant s\\ 1\leqslant j\leqslant n}} = s.$$

One obviously has that $s \leqslant \dim M$ when $(F_1, \dots, F_s)$ is independent. For functions that are in involution, and in particular for Casimirs, one has stronger restrictions, as given in the following proposition. Recall from Chapter 3 that for any integer s we denote by $M_{(s)}$ the open subset of points of M where the rank is at least $2s$. Also, recall that we have defined the rank of a Poisson manifold $(M, \{\cdot\,,\cdot\})$ as the maximal rank attained by $\{\cdot\,,\cdot\}$ at points of M.

Proposition 4.12. *Let $(M, \{\cdot,\cdot\})$ be a Poisson manifold of rank $2r$ and suppose that $(F_1, \dots, F_s)$ is independent.*

(1) If $F_1, \dots, F_s$ are Casimirs then $s \leqslant \dim M - 2r$;
(2) If $(F_1, \dots, F_s)$ is involutive then $s \leqslant \dim M - r$;
(3) If $(F_1, \dots, F_s)$ is involutive with $s = \dim M - r$ then

$$\dim \operatorname{span} \{\mathcal{X}_{F_1}(m), \dots, \mathcal{X}_{F_s}(m)\} \leqslant r$$

for any $m \in \mathcal{U}_{\mathbf{F}}$, with equality if $m \in \mathcal{U}_{\mathbf{F}} \cap M_{(r)}$.

Proof. For $m \in M$ let $\tilde{P}_m$ denote the linear map $T_m^*M \to T_mM$, which corresponds to the Poisson structure. Explicitly, for $F \in \mathcal{F}(M)$,

$$\tilde{P}_m(\mathrm{d}F(m)) = \mathcal{X}_F(m) = \{\cdot, F\}(m).$$

The rank of $\tilde{P}_m$ is $\mathrm{Rk}_m \{\cdot,\cdot\}$ and for every $F \in \mathrm{Cas}(M)$ the covector $\mathrm{d}F(m)$ belongs to $\operatorname{Ker} \tilde{P}_m$, whose dimension is $\dim M - \mathrm{Rk}_m \{\cdot,\cdot\}$. Let $\mathbf{F} = (F_1, \dots, F_s)$ be independent and let m be an element of the non-empty (open) set $\mathcal{U}_{\mathbf{F}} \cap M_{(r)}$. Suppose first that each element of $\mathbf{F}$ is a Casimir. Since $\mathrm{d}F_1(m), \dots, \mathrm{d}F_s(m)$ are independent we have that

$$s \leqslant \dim \operatorname{Ker} \tilde{P}_m = \dim M - 2r$$

and *(1)* follows. Next, suppose that $\mathbf{F}$ is involutive and consider the fiber $\mathbf{F}_m$, where m is still taken from $\mathcal{U}_{\mathbf{F}} \cap M_{(r)}$, so that the restriction of $\mathbf{F}_m$ to a neighborhood U of m is a submanifold of dimension $\dim M - s$ of U, passing through m. This dimension is an upper bound for the dimension d_m of $\operatorname{span} \{\mathcal{X}_{F_1}(m), \dots, \mathcal{X}_{F_s}(m)\}$, because these s vectors are tangent to that fiber at m. Moreover, $d_m \geqslant s - \dim \operatorname{Ker} \tilde{P}_m = s + 2r - \dim M$, because the differentials $\mathrm{d}F_1, \dots, \mathrm{d}F_s$ are independent at m. Combining the two inequalities for d_m (still assuming that $m \in \mathcal{U}_{\mathbf{F}} \cap M_{(r)}$), we get

$$s + 2r - \dim M = s - \dim \operatorname{Ker} \tilde{P}_m \leqslant d_m \leqslant \dim M - s, \tag{4.3}$$

leading to *(2)*. Third, suppose that $s = \dim M - r$. For $m \in \mathcal{U}_{\mathbf{F}} \cap M_{(r)}$ we deduce from (4.3) that

$$r = s - \dim \operatorname{Ker} \tilde{P}_m \leqslant d_m \leqslant r,$$

so that $\dim \operatorname{span} \{\mathcal{X}_{F_1}(m), \dots, \mathcal{X}_{F_s}(m)\} = d_m = r$. □

When the Poisson structure on M is algebraic or analytic, as will be the case in all our examples, the proposition implies that when $(F_1, \dots, F_s)$ is involutive, with $s = \dim M - r$, then the vector fields $\mathcal{X}_{F_1}, \dots, \mathcal{X}_{F_s}$ are independent on a dense open subset of M.

4.2 Liouville Integrability

We now give the definition of (Liouville) integrability, we give the simplest examples and we show that integrable systems can be solved by quadratures.

Definition 4.13. Let $(M, \{\cdot\,,\cdot\})$ be a Poisson manifold of rank $2r$ and let $\mathbf{F} = (F_1, \dots, F_s)$ be involutive and independent, with $s = \dim M - r$. Then we say that $\mathbf{F}$ is *completely integrable* and that $(M, \{\cdot\,,\cdot\}, \mathbf{F})$ is an *integrable system* or a *completely integrable system*. The vector fields $\mathcal{X}_{F_i}$ are then called *integrable vector fields* and the map $\mathbf{F}$ is called the *momentum map*. We say that the integer r is the number of *degrees of freedom* of the integrable system and we call $2r$ its *rank*. In order to distinguish this notion of integrability from other notions of integrability one also says that $(M, \{\cdot\,,\cdot\}, \mathbf{F})$ is *Liouville integrable*.

We speak of a *real integrable system* and of a *complex integrable system* when we need to be precise about the real or complex nature of its phase space M.

Notice that $2r$ is the dimension of the symplectic leaves of maximal dimension (typically the generic leaf) and that r is the number of independent commuting Hamiltonian vector fields on such a leaf.

Example 4.14. On $M := \mathbf{R}^{2n}$ consider the Poisson structure coming from the standard symplectic structure $\mathrm{d}q_1 \wedge \mathrm{d}p_1 + \cdots + \mathrm{d}q_n \wedge \mathrm{d}p_n$, where $p_1, \dots, q_n$ are linear coordinates on M. The potential energy $V := \frac{1}{2}\sum_i \nu_i q_i^2$, where each ν_i is positive, leads to the Hamiltonian

$$H := \frac{1}{2}\sum_{i=1}^n p_i^2 + \frac{1}{2}\sum_{i=1}^n \nu_i q_i^2,$$

(see Example 3.8), which in physical terms is the Hamiltonian of the *n-dimensional harmonic oscillator*. It is easy to check that the n functions $F_i := (p_i^2 + \nu_i q_i^2)/2$ $(1 \leqslant i \leqslant n)$, are independent and that they are in involution. Therefore, $\mathbf{F} := (F_1, \dots, F_n)$ is completely integrable. The fibers of the momentum map $\mathbf{F}$ over $(c_1, \dots, c_n) \in (\mathbf{R}_{>0})^n$ are products of circles $p_i^2 + \nu_i q_i^2 = c_i$, hence they are n-dimensional tori. Notice that $H = \sum_{i=1}^n F_i$, so that we may replace e.g. F_1 by H to find that $(H, F_2, \dots, F_n)$ is completely integrable. When all ν_i are equal one speaks of an *isotropic oscillator*. Notice that in this case each of the functions $q_i p_j - q_j p_i$ is a constant of motion, but these functions are not all in involution.

Example 4.15. The previous example is a special case of a general construction which, given integrable systems $(M_i, \{\cdot\,,\cdot\}_i, \mathbf{F}_i)$, for $1 \leqslant i \leqslant s$, allows one to construct an integrable system $(M, \{\cdot\,,\cdot\}, \mathbf{F})$, where $M := M_1 \times \cdots \times M_s$. For $\{\cdot\,,\cdot\}$ one takes the product of the Poisson structures $\{\cdot\,,\cdot\}_i$ (see Proposition 3.37) and one takes $\mathbf{F} := \mathbf{F}_1 \times \cdots \times \mathbf{F}_s$. The (easy) verification that this defines an integrable system is left to the reader.

Example 4.16. For the canonical Poisson structure $\{\cdot\,,\cdot\}$ of rank $2r$ on $\mathbf{C}^n$ (see Example 3.3) let $\mathbf{F} := (x_{r+1}, \dots, x_n)$. Then $\mathbf{F}$ is completely integrable and $(\mathbf{C}^n, \{\cdot\,,\cdot\}, \mathbf{F})$ is a (complex) integrable system. Each fiber of $\mathbf{F}$ is a complex r-dimensional plane. All symplectic leaves are $2r$-dimensional and are of the form $\{x \in \mathbf{C}^n \mid x_i = c_i,\ i \geqslant 2r+1\}$, where the constants c_i are arbitrary. The functions $x_{r+1}, \dots, x_{2r}$ yield independent Hamiltonians on such a leaf. Notice that in this case $(x_1, \dots, x_r, x_{r+2}, \dots x_n)$ is independent and that each of these $n-1$ functions is in involution with x_1; this does not contradict Proposition 4.12 however, because $(x_1, \dots, x_n)$ is not involutive.

Example 4.17. Picking up Example 3.24 again, let $F : \mathbf{C}^2 \to \mathbf{C}$ be a holomorphic function and let ω_0 and ω_1 be the real and imaginary parts of $\Omega = \mathrm{d}z_1 \wedge \mathrm{d}z_2$. We know from Example 4.2 that the functions G and H which are defined by $F = G + \sqrt{-1}H$ are in involution with respect to ω_1 and ω_2. We claim that if F is not constant then G and H are independent (in the sense of Definition 4.11). Indeed, if $\mathrm{d}G(m)$ and $\mathrm{d}H(m)$ are dependent then

$$\mathrm{Rk}\begin{pmatrix} \frac{\partial G}{\partial x_1}(m) & \frac{\partial G}{\partial y_1}(m) & \frac{\partial G}{\partial x_2}(m) & \frac{\partial G}{\partial y_2}(m) \\ \frac{\partial H}{\partial x_1}(m) & \frac{\partial H}{\partial y_1}(m) & \frac{\partial H}{\partial x_2}(m) & \frac{\partial H}{\partial y_2}(m) \end{pmatrix} < 2.$$

Combined with the Cauchy-Riemann equations (3.24), this condition implies that $\mathrm{d}G(m) = \mathrm{d}H(m) = 0$. Therefore m is a common zero of the holomorphic functions $\partial F/\partial z_1$ and $\partial F/\partial z_2$ so that m is contained in a non-trivial (because F is not constant) analytic subvariety of $\mathbf{C}^2$. It follows that $\mathrm{d}G$ and $\mathrm{d}H$ are independent on a dense subset of $\mathbf{R}^4$. This shows that $\mathbf{F} := (G, H)$ is completely integrable and that each of $\left(\mathbf{R}^4, \{\cdot\,,\cdot\}_0, \mathbf{F}\right)$ and $\left(\mathbf{R}^4, \{\cdot\,,\cdot\}_1, \mathbf{F}\right)$ is a (real) integrable system.

Example 4.18. Let r and s be fixed integers such that $s \geqslant r \geqslant 1$ and define $t := s-r$. Let U be a non-empty open subset of $\mathbf{R}^s$ on which we denote the standard coordinates by $p_1, \dots, p_r, y_1, \dots, y_t$. We also consider $\mathbf{T}^r := (\mathbf{R}/\mathbf{Z})^r$ with local coordinates $q_1 \dots, q_r$ that come from the standard coordinates on the universal covering space $\mathbf{R}^r$ of $\mathbf{T}^r$. We describe an elementary integrable system on $M := \mathbf{T}^r \times U$ which will serve later as a local model, on a neighborhood of a generic compact invariant manifold (if any) of the momentum map of a real integrable system (Theorem 4.32 below). We denote the coordinates on M by the same letters $q_1, \dots, p_r, y_1, \dots, y_t$ as the coordinates on each of the factors of M. We consider the Poisson structure $\{\cdot\,,\cdot\}$ for which all functions y_k are Casimirs and for which $\{q_i, p_j\} = \delta_{ij}$, where $1 \leqslant i, j \leqslant r$. This Poisson structure can be seen as coming from the standard structure of rank $2r$ on $\mathbf{R}^{2r+t}$ by using a Poisson reduction (see Proposition 3.42). Letting $\mathbf{F} := (p_1, \dots, p_r, y_1, \dots, y_t)$ we have that $\mathbf{F}$ is independent, hence completely integrable and each fiber of $\mathbf{F}$ is an r-dimensional torus $\mathbf{T}^r$.

More generally, let $F'_1, \dots, F'_s$ be independent functions on U, which we consider as functions on $\mathbf{T}^r \times U$. Then $\mathbf{F}' := (F'_1, \dots, F'_s)$ is completely integrable. Each fiber of $\mathbf{F}'$ is a disjoint union of tori; indeed, $\mathbf{F}'$, viewed as a map $U \to \mathbf{R}^s$ needs not be injective. For $1 \leqslant k \leqslant s$ the integrable vector field $\mathcal{X}_{F'_k}$ is given by

$$\dot{q}_i = \frac{\partial F'_k}{\partial p_i}, \qquad \dot{p}_i = 0, \qquad \dot{y}_j = 0, \tag{4.4}$$

where $1 \leqslant i \leqslant r$ and $1 \leqslant j \leqslant t$. The integration of (4.4) is trivial.

Example 4.19. In one degree of freedom integrability is trivial in the following sense. Suppose that the rank of $(M, \{\cdot\,,\cdot\})$ is two and that there exist $n-2$ independent Casimirs $F_1, \dots, F_{n-2}$, where $n = \dim M$. For a generic function F one has that $\mathbf{F} := (F_1, \dots, F_{n-2}, F)$ is independent, hence completely integrable. Notice that the fibers of the momentum map are in this case 1-dimensional.

Remark 4.20. Integrable systems often depend on parameters in the sense that one does not consider one particular Hamiltonian, but a whole class of them, parametrized by a one or several parameters. When the integrable systems come from physics these parameters usually have a physical meaning, such as moment of inertia, mass, spring constant, and so on. We still use the singular term "integrable system" in this case. It is then understood that all claims, such as independence of the functions, integrability, and so on, are valid for generic values of the parameters, i.e., when the parameters are fixed to generic values (say, taken in some non-empty Zariski open subset, when the parameter space is an algebraic variety). Given an integrable system that depends on parameters one may also think of it as a single integrable system by replacing phase space by the product of phase space with the parameter space. The Poisson structure is then extended to this larger phase space by declaring the parameters, that have now become phase variables, to be Casimirs. Similarly one enlarges the algebra of functions in involution by adding the parameters as extra functions. This point of view is of course consistent with the above convention that all claims are valid for generic values of the parameters. See [169] for formal details on constructions of this type.

In order to explain the terminology *integrable* we show that the system of differential equations representing an integrable vector field is solvable by quadratures, a classical notion that will be defined below. We start with an example with one degree of freedom, where we show how the equations of motion can be integrated explicitly, by using only algebraic operations, the process of taking inverse functions and integration.

Example 4.21. We consider an integrable system on $\mathbf{R}^2$, as in Example 3.8 with $n = 1$. Let us write p for p_1 and q for q_1. Then the Hamiltonian is given by

$$H := \frac{p^2}{2} + V(q) \tag{4.5}$$

and it is, of course, a constant of motion. We fix a point $(q_0, p_0) \in \mathbf{R}^2$ for which $\mathrm{d}H(q_0, p_0) \neq 0$, and we denote the value of H at (q_0, p_0) by h_0. Using $\dot{q} = p$ (4.5) implies that the integral curve which starts at (q_0, p_0) satisfies the differential equation

$$\mathrm{d}t = \frac{\mathrm{d}q}{\sqrt{2\left(h_0 - V(q)\right)}}.$$

The above denominator does not vanish in a neighborhood of q_0, except maybe at q_0, because $\mathrm{d}H(q_0, p_0) \neq 0$. Integrating both sides we find

$$t = \int_{q_0}^{q} \frac{\mathrm{d}q}{\sqrt{2\left(h_0 - V(q)\right)}}$$

which defines q (and hence also p) implicitly as a function of q. The obtained functions $(q(t), p(t))$ define, for $|t|$ small, the integral curve of $\mathcal{X}_H$, starting at (q_0, p_0), hence they integrate the equations of motion for the initial condition $(q(0), p(0)) = (q_0, p_0)$. Notice that q was obtained by using only algebraic operations, inverting a function (the inverse function theorem) and integration.

We now show how the explicit integration of the equations of motion is done in the case of an arbitrary integrable system. To do this we suppose that $(M, \{\cdot\,,\cdot\}, \mathbf{F})$ is an integrable system of rank $2r$ and we write $\mathbf{F} = (F_1, \dots, F_s)$. We choose an arbitrary $m_0 \in \mathcal{U}_{\mathbf{F}} \cap M_{(r)}$ (this corresponds exactly to the condition $\mathrm{d}H(q_0, p_0) \neq 0$ in Example 4.21) and we show how the integral curve, starting at m_0, of each of the Hamiltonian vector fields $\mathcal{X}_{F_i}$ can be obtained locally by using only algebraic operations, the inverse function theorem and integration. This is what is meant precisely when saying that the system is *solvable by quadratures.*

In view of item *(3)* of Proposition 4.12 we may suppose that the elements of $\mathbf{F}$ are ordered in such a way that the Hamiltonian vector fields $\mathcal{X}_{F_1}, \dots, \mathcal{X}_{F_r}$ are independent at m_0. Then these vector fields are independent on an open neighborhood of m_0; we may suppose that this neighborhood is contained in $\mathcal{U}_{\mathbf{F}} \cap M_{(r)}$ since the latter is open (and contains m_0). Let us denote by U the intersection of this neighborhood with the fiber of $\mathbf{F}$ that contains m_0, which is, by the cited item, r-dimensional. By shrinking U further, if necessary, we may assume that U is a coordinate neighborhood of m_0 in the fiber of $\mathbf{F}$ through m_0. Since the vector fields $\mathcal{X}_{F_1}, \dots, \mathcal{X}_{F_r}$ are independent at every point of U there exist unique 1-forms $\omega_1, \dots, \omega_r$ on U such that $\omega_i\left(\mathcal{X}_{F_j}\right) =$

δ_{ij}, for $1 \leqslant i, j \leqslant r$. These 1-forms can be computed using linear algebra only. Indeed, for $F \in \mathcal{F}(M)$ we have on U that

$$\mathsf{d}F = \sum_{i=1}^{r} \{F, F_i\} \, \omega_i, \tag{4.6}$$

(for a proof, evaluate both sides on the vector fields $\mathcal{X}_{F_1}, \ldots, \mathcal{X}_{F_r}$, which span the tangent space to U at any point of U). Choose now r functions $\phi_1 \ldots, \phi_r$ on U whose differentials are independent at every point $m \in U$, i.e., choose a system of coordinates on U (these functions may for example be chosen as r of the elements of a system of coordinates of M around m_0). Then

$$\begin{pmatrix} \mathsf{d}\phi_1 \\ \vdots \\ \mathsf{d}\phi_r \end{pmatrix} = \begin{pmatrix} \{\phi_1, F_1\} & \cdots & \{\phi_1, F_r\} \\ \vdots & & \vdots \\ \{\phi_r, F_1\} & \cdots & \{\phi_r, F_r\} \end{pmatrix} \begin{pmatrix} \omega_1 \\ \vdots \\ \omega_r \end{pmatrix}$$

and the $r \times r$ matrix in this expression is invertible at any $m \in U$ because the vectors $\mathcal{X}_{F_i}(m)$ span T_mU at every $m \in U$ and because $(\phi_1, \ldots, \phi_r)$ is a system of coordinates on U. For $1 \leqslant i \leqslant r$ the 1-form ω_i is closed: the vector fields $\mathcal{X}_{F_1}, \ldots \mathcal{X}_{F_r}$ span the tangent space to U at every point of U and (see (2.8))

$$\mathsf{d}\omega_i \left(\mathcal{X}_{F_j}, \mathcal{X}_{F_k}\right) = \mathcal{X}_{F_j}[\omega_i\left(\mathcal{X}_{F_k}\right)] - \mathcal{X}_{F_k}[\omega_i\left(\mathcal{X}_{F_j}\right)] - \omega_i\left(\left[\mathcal{X}_{F_j}, \mathcal{X}_{F_k}\right]\right)$$

for any $1 \leqslant j, k \leqslant r$, which evaluates to zero because $\omega_i\left(\mathcal{X}_{F_j}\right)$ is constant and because the vector fields $\mathcal{X}_{F_j}$ commute. Since U is a coordinate neighborhood these closed forms are exact and we may integrate each of the 1-forms $\omega_1, \ldots, \omega_r$, to obtain r functions $t_1, \ldots, t_r$; we choose the constants in these functions such that m_0 corresponds to $t_1 = \cdots = t_r = 0$. Notice that these functions provide a system of coordinates on U because $\mathsf{d}t_1 \wedge \ldots \wedge \mathsf{d}t_r \neq 0$ on U, hence we can, by the inverse function theorem, write the coordinates $(\phi_1, \ldots, \phi_r)$ locally, around m_0, in terms of $(t_1 \ldots, t_r)$. By construction, $\frac{\partial}{\partial t_i} = \mathcal{X}_{F_i}$ on U. Therefore the resulting functions $\phi_i\left(t_1, \ldots, t_r\right)$ provide the integral curve of $\mathcal{X}_{F_i}$ that passes through m_0 (as contained in U) by putting $t_1, \ldots, t_{i-1}, t_{i+1}, \ldots, t_r$ equal to zero. Using the equations $F_i = c_i$ we get the corresponding integral curve of $\mathcal{X}_{F_i}$ as a curve in M, via the implicit function theorem. One similarly determines the integral curve that corresponds to a linear combination of the vector fields $\mathcal{X}_{F_i}$.

Remark 4.22. Notice that the above proof works in the real as well as in the complex case; indeed, U being a coordinate neighborhood there is no ambiguity for the paths used in computing the complex integrals.

Remark 4.23. It is easy to show that the integrable vector fields $\mathcal{X}_{F_1}, \ldots, \mathcal{X}_{F_r}$ define an integrable r-dimensional distribution on a neighborhood of m_0 (see Proposition 4.24 below). The Frobenius Theorem (Theorem 2.5) implies that there exist coordinates $(t_1, \ldots, t_r)$ in a neighborhood of m_0, such that $\mathcal{X}_{F_i} = \frac{\partial}{\partial t_i}$ for $i = 1, \ldots, r$. The point of the above proof is to show that these t_i can be constructed by using only algebraic operations, the implicit function theorem and integration. The fact that this can be done in the case of a distribution defined by commuting vectors can be generalized to general integrable distributions, giving more general conditions under which a Hamiltonian system can be solved by quadratures (see e.g. [59, Theorem 3.6]).

4.3 The Liouville Theorem and the Action-Angle Theorem

The *Liouville Theorem* is often considered as the motivation of the above Definition 4.13 of a completely integrable system. This theorem describes the generic manifolds, traced out by the flows of the integrable vector fields of a *real* integrable system, assuming compactness of these manifolds or completeness of these flows on them: these manifolds are tori or cylinders and the flow on them is linear. The modern version of this theorem is due to Arnold and for this reason the theorem is sometimes (see e.g. [1]) referred to as the *Arnold-Liouville Theorem*, while often the latter name is reserved for the more elaborate theorem which proves the existence of action-angle coordinates (Theorem 4.32). For a complex version of the Liouville Theorem, see Section 6.3.

We first define the submanifolds that will be described in the real case by the Liouville Theorem and that will lie at the basis of one of the definitions of an a.c.i. system in the complex case.

Proposition 4.24. *Let $(M, \{\cdot, \cdot\}, \mathbf{F})$ be an integrable system of rank $2r$, with $\mathbf{F} = (F_1, \ldots, F_s)$. The open subset $\mathcal{U}_{\mathbf{F}} \cap M_{(r)}$ is preserved by the flows of the integrable vector fields $\mathcal{X}_{F_i}$, $i = 1, \ldots, s$, which define a distribution $\mathcal{D}$ of rank r on $\mathcal{U}_{\mathbf{F}} \cap M_{(r)}$, integrable in the sense of Frobenius.*

Proof. Let Φ denote the flow of one of the vector fields $\mathcal{X}_{F_i}$. We have (for small $|t|$) that $\Phi_t^* F_j = F_j$ for $j = 1, \ldots, s$ because $\mathcal{X}_{F_i}$ is tangent to the fibers of $\mathbf{F}$. Therefore,

$$\Phi_t^* \left(\mathrm{d}F_1 \wedge \ldots \wedge \mathrm{d}F_s\right) = \mathrm{d}\Phi_t^* F_1 \wedge \ldots \wedge \mathrm{d}\Phi_t^* F_s = \mathrm{d}F_1 \wedge \ldots \wedge \mathrm{d}F_s,$$

so that the s-form $\mathrm{d}F_1 \wedge \ldots \wedge \mathrm{d}F_s$ is preserved by the flow of $\mathcal{X}_{F_i}$. It follows that $\mathcal{U}_{\mathbf{F}}$, as defined by (4.2), is preserved by the flow of $\mathcal{X}_{F_i}$. But Proposition 3.17 implies that the open subset $M_{(r)}$ is also preserved by these flows. Hence their intersection also. The integrable vector fields $\mathcal{X}_{F_i}$ define on $\mathcal{U}_{\mathbf{F}} \cap M_{(r)}$ a distribution of rank r, as follows from item *(3)* in Proposition 4.12. This distribution is integrable because the vector fields $\mathcal{X}_{F_i}$ commute. □

Definition 4.25. For $m \in \mathcal{U}_{\mathbf{F}} \cap M_{(r)}$ the maximal integral manifold of $\mathcal{D}$, passing through m, is called the *invariant manifold* (of $\mathbf{F}$) passing through m, and is denoted by $\mathbf{F}'_m$.

Thus, the invariant manifold of $\mathbf{F}$ that passes through m is by definition the immersed submanifold which is traced out by the flow of the integrable vector fields, starting at m. In the following proposition we give an alternative description of the invariant manifold $\mathbf{F}'_m$, which on the one hand explains the notation $\mathbf{F}'_m$ (recall that $\mathbf{F}_m$ denotes the fiber of $\mathbf{F}$ which passes through m), and on the other hand shows that $\mathbf{F}'_m$ is actually an (embedded) submanifold of M.

Remark 4.26. The integrable vector fields $\mathcal{X}_{F_i}$ define on M a generalized distribution, which can also be shown to be integrable, hence we can define $\mathbf{F}'_m$ for any $m \in M$. We will in this book only restrict our attention to the invariant manifolds $\mathbf{F}'_m$, where $m \in \mathcal{U}_{\mathbf{F}} \cap M_{(r)}$, although the other ones also deserve to be studied; see Section 6.1, starting from Definition 6.2.

Proposition 4.27. *Let $(M, \{\cdot\,,\cdot\}, \mathbf{F})$ be an integrable system and suppose that $m \in \mathcal{U}_{\mathbf{F}} \cap M_{(r)}$. The invariant manifold $\mathbf{F}'_m$ of $\mathbf{F}$ is the connected component of $\mathbf{F}_m \cap \mathcal{U}_{\mathbf{F}} \cap M_{(r)}$ which contains m. In particular, $\mathbf{F}'_m$ is an (embedded) submanifold of M.*

Proof. Let us denote by $\mathbf{F}^0_m$ the connected component of $\mathbf{F}_m \cap \mathcal{U}_{\mathbf{F}} \cap M_{(r)}$ which contains m. Since $\mathbf{F}^0_m$ is a connected submanifold of dimension r, whose tangent space at each of its points m coincides with $\mathcal{D}(m)$, we have that $\mathbf{F}^0_m \subseteq \mathbf{F}'_m$. On the other hand, $\mathbf{F}'_m \subset \mathcal{U}_{\mathbf{F}} \cap M_{(r)}$, by definition, while $\mathbf{F}'_m \subseteq \mathbf{F}_m$ since $\mathcal{D}$ is tangent to $\mathbf{F}_m$. Since $\mathbf{F}'_m$ contains m and is connected it follows that $\mathbf{F}'_m \subseteq \mathbf{F}^0_m$. □

We now come to the Liouville Theorem for *real* integrable systems.

Theorem 4.28 (Liouville Theorem). *Let $(M, \{\cdot\,,\cdot\}, \mathbf{F})$ be a real integrable system of rank $2r$, where $\mathbf{F} = (F_1, \dots, F_s)$. For $m \in \mathcal{U}_{\mathbf{F}} \cap M_{(r)}$, let $\mathbf{F}'_m$ denote the invariant manifold of $\mathbf{F}$ that passes through m.*

(1) If $\mathbf{F}'_m$ is compact then there exists a diffeomorphism from $\mathbf{F}'_m$ to the torus $\mathbf{T}^r = (\mathbf{R}/\mathbf{Z})^r$, under which the vector fields $\mathcal{X}_{F_1}, \dots, \mathcal{X}_{F_s}$ are mapped to linear (i.e., translation-invariant) vector fields.
(2) If $\mathbf{F}'_m$ is not compact, but the flow of each of the vector fields $\mathcal{X}_{F_i}$ (where $i = 1, \dots, s$) is complete on $\mathbf{F}'_m$ then there exists a diffeomorphism from $\mathbf{F}'_m$ to a cylinder $\mathbf{R}^{r-q} \times \mathbf{T}^q$ $(0 \leqslant q < r)$, under which the vector fields $\mathcal{X}_{F_i}$ are mapped to linear vector fields.

Proof. We suppose that the flow $\Phi^{(i)}$ of each of the integrable vector fields $\mathcal{X}_{F_i}$ is complete on $\mathbf{F}'_m$ and that the functions F_i are ordered such that $\mathcal{X}_{F_1}, \dots, \mathcal{X}_{F_r}$ give the r independent vector fields. Then their completeness and commutativity imply that we can define an action $\mathbf{R}^r \times \mathbf{F}'_m \to \mathbf{F}'_m$ by

$$((t_1, \dots, t_r), m) \mapsto \Phi^{(1)}_{t_1} \circ \Phi^{(2)}_{t_2} \circ \dots \circ \Phi^{(r)}_{t_r}(m).$$

Since $\mathbf{F}'_m$ is the integral manifold through m of the distribution defined by the integrable vector fields, the action is transitive on $\mathbf{F}'_m$ and $\mathbf{F}'_m$ becomes a homogeneous space. The action is also locally free because the vector fields $\mathcal{X}_{F_i}$ are independent at any point of $\mathbf{F}'_m$. Therefore the stabilizer is a discrete subgroup H of $\mathbf{R}^r$ and $\mathbf{F}'_m$ is diffeomorphic to $\mathbf{R}^r/H$. If $\mathbf{F}'_m$ is compact then H must be a lattice, so $\mathbf{R}^r/H$ is a torus, smoothly embedded into M. Otherwise H is a discrete subgroup whose rank q is at most $r-1$ and $\mathbf{R}^r/H$ is isomorphic to $\mathbf{R}^{r-q} \times \mathbf{T}^q$. By construction, the vector fields $\mathcal{X}_{F_i}$ are mapped to linear vector fields in both cases. □

The tori that appear in Liouville's Theorem are often called *Liouville tori.*

Remark 4.29. For most integrable systems that come from classical mechanics (spinning tops, systems of oscillators, penduli, ...) the energy function is proper, so that *all* fibers $\mathbf{F}_m$ of the momentum map are compact. This does however not mean that each invariant manifold $\mathbf{F}'_m$, with $m \in \mathcal{U}_{\mathbf{F}} \cap M_{(r)}$ is compact, but it does imply that the flow of the integrable vector fields is complete on each such $\mathbf{F}'_m$. Therefore, each $\mathbf{F}'_m$ is a torus or a cylinder (of dimension r). In order to see how a cylinder can appear in this setting, think of a pinched torus which appears as a singular fiber of the momentum map. The integrable vector fields necessarily vanish at the singular point, and for m different from that point, $\mathbf{F}'_m$ is the pinched torus, minus its singular point, which is a cylinder. Many integrable systems of interest do not satisfy either assumption of the Liouville Theorem and the topology of the fibers of the momentum map has to be determined in a different way (see [20], [169]).

Example 4.30. In the case of Example 3.24 we find that the fibers of the momentum map are given by $F = c$, where c is any complex constant. Taking for F any polynomial in two variables, the fiber becomes an (affine) plane algebraic curve and every such curve appears in this way. The fiber will be smooth for generic values of c and it will be a topological surface of genus g, with a few points removed. Thus, the topological types that appear as the fibers of the momentum map of an integrable system are in the non-compact case much more general than the ones that appear in the Liouville Theorem.

Definition 4.31. Let $(M, \{\cdot\,,\cdot\}, \mathbf{F})$ be a real integrable system of dimension n and rank $2r$. We say that $(M, \{\cdot\,,\cdot\}, \mathbf{F})$ admits *action-angle coordinates* around $m \in M$ if there exists

(1) an open neighborhood U_m of m in M;
(2) an open subset U of $\mathbf{R}^{n-r}$;
(3) a diffeomorphism $\phi : U_m \to \mathbf{T}^r \times U$;
(4) an independent $(n-r)$-tuple $\mathbf{E}$ of functions, defined on U;

such that, when $\mathbf{T}^r \times U$ is equipped with the Poisson structure from Example 4.18, then

(1) ϕ is a Poisson map;
(2) $\phi^*\mathbf{E} = \mathbf{F}$ on U_m.

In the latter formula we view the elements of $\mathbf{E}$ as functions on $\mathbf{T}^r \times U$, so strictly speaking, (2) should be written as $\phi^*\pi^*\mathbf{E} = \mathbf{F}$, where $\pi : \mathbf{T}^r \times U \to U$ is the projection on the second component.

The following diagram may be helpful for visualizing the spaces and maps involved.

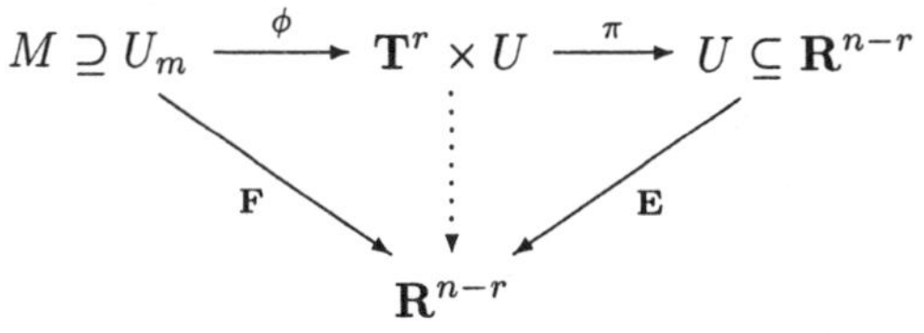

The coordinates on U_m which come from the natural coordinates on $\mathbf{T}^r \times U$ are called *action-angle coordinates.* In plain language, after a change of coordinates which respects the Poisson structure, U_m looks like the canonical Example 4.18.

The following theorem is called the *Action-Angle Theorem* and is in combination with the Liouville Theorem sometimes called the *Arnold-Liouville Theorem.* Simply put it says that neighborhoods of compact invariant manifolds $\mathbf{F}'_m$ look like the canonical Example 4.18.

Theorem 4.32 (Action-Angle Theorem). *Let $(M, \{\cdot\,,\cdot\}, \mathbf{F})$ be a real integrable system and suppose that $m \in \mathcal{U}_\mathbf{F} \cap M_{(r)}$ is such that $\mathbf{F}'_m$ is compact. Then $(M, \{\cdot\,,\cdot\}, \mathbf{F})$ admits action-angle coordinates around m.*

For a complete proof of the above action-angle theorem in case M is symplectic we refer to [107]. The adaption to the case of a general real Poisson manifold is left to the reader.

Example 4.33. There is at least one case in which action-angle coordinates can easily be computed: the case of an isotropic oscillator (see Example 4.14). Recall that the functions in involution for a harmonic oscillator are the functions $F_i = \left(p_i^2 + \nu_i q_i^2\right)/2$. For an isotropic oscillator all ν_i are equal; we take them here equal to 1. Thus, $\mathbf{F} = (F_1, \dots, F_r)$, with $F_i = (p_i^2 + q_i^2)/2$. In the notation of Definition 4.31 we take $U := (\mathbf{R}_{>0})^r$ and $U_m := \left(\mathbf{R}^2 \setminus \{0,0\}\right)^r$ and we take for $\mathbf{E} = (E_1, \dots, E_r)$ the standard coordinates on $(\mathbf{R}_{>0})^r$. We define the map $\psi = \phi^{-1}$ by

$$\begin{aligned} \psi : \qquad \mathbf{T}^r \times (\mathbf{R}_{>0})^r &\to \left(\mathbf{R}^2 \setminus \{0,0\}\right)^r \\ (\theta_1, \dots, \theta_r, E_1, \dots, E_r) &\mapsto (\sqrt{2E_i}\cos\theta_i, \sqrt{2E_i}\sin\theta_i)_{i=1,\dots,r}. \end{aligned}$$

It is clear from the formulas that ψ and its inverse ϕ are diffeomorphisms. We need to verify that ϕ is a Poisson map and that $\phi^*\mathbf{E} = \mathbf{F}$ on U_m. In terms of ψ this means that we need to verify that ψ is a Poisson map and that $\psi^*\mathbf{F} = \mathbf{E}$ on $\mathbf{T}^r \times U$. But $\psi^* F_i = \frac{1}{2}\left(2E_i\cos^2\theta_i + 2E_i\sin^2\theta_i\right) = E_i$ for $i = 1, \dots, r$. Thus, ψ preserves indeed the constants of motion.

Since $\{E_i, \theta_j\} = \delta_{ij}$, while the other brackets are zero we have that

$$\begin{aligned}\{\psi^* q_i, \psi^* p_j\} &= \left\{\sqrt{2E_i}\cos\theta_i, \sqrt{2E_j}\sin\theta_j\right\} \\ &= \left(\frac{\cos\theta_i\cos\theta_j\sqrt{2E_j}}{\sqrt{2E_i}} + \frac{\sin\theta_i\sin\theta_j\sqrt{2E_i}}{\sqrt{2E_j}}\right)\delta_{ij} \\ &= \delta_{ij} \\ &= \psi^*\{q_i, p_j\},\end{aligned}$$

for $1 \leqslant i, j \leqslant r$. Similarly, $\{\psi^* q_i, \psi^* q_j\} = 0 = \psi^*\{q_i, q_j\}$ and $\{\psi^* p_i, \psi^* p_j\} = 0 = \psi^*\{p_i, p_j\}$. Thus, ψ is indeed a Poisson map.

In general it is difficult to compute explicit action-angle variables for a given integrable system. It is already non-trivial to find whether or not some of the fibers of the momentum map of the integrable system are compact and to localize them; furthermore the fibration by Liouville tori is in general not trivial and global action-angle variables do not exist due to monodromy. For this, see the original article [48] and also [21] and [29].

4.4 The Adler-Kostant-Symes Theorem(s)

We have seen in Example 4.3 that a Hamiltonian vector field that can be written in the Lax form $\dot{X} = [X, Y]$, admits each coefficient of the characteristic polynomial of X as a constant of motion. A *Lax equation* with values in $\mathfrak{g}$, where $\mathfrak{g}$ is any Lie algebra, is a vector field on a (finite-dimensional) affine subspace M of $\mathfrak{g}$, which has the form $\dot{X} = [X, Y]$, where $X \in M$ and Y is a map from M to $\mathfrak{g}$. In our case, the vector field that is given by a Lax equation is always a Hamiltonian vector field with respect to some Poisson structure. The Lie algebras that we consider are either finite-dimensional Lie algebras or (twisted) affine Lie algebras. In the latter case, we also use the terminology Lax equation with a spectral parameter, see Paragraph 6.4. In part III we give Lax equations the integrable systems that we consider.

The purpose of this section is to present a method that allows one to construct Lax equations from a Lie algebra splitting. We show in addition that the obtained constants of motion are in involution and that, while they may not necessarily form a large enough set to make the system integrable, the Lax equation can be integrated explicitly upon using the underlying Lie group. This collection of fundamental results is called the *Adler-Kostant-Symes Theorem.* We will present two versions of this theorem. The first one is on the dual $\mathfrak{g}^*$ of a Lie algebra and has the advantage of admitting a more transparent proof, but the Hamiltonian vector fields are not in the standard Lax form (because they are defined on $\mathfrak{g}^*$). For the second one it is assumed that $\mathfrak{g}$ comes equipped with a non-degenerate pairing $\langle\cdot\,|\,\cdot\rangle$, which is Ad-invariant. The latter form yields Lax equations and is most suited towards applications.

4.4.1 Lie Algebra Splitting

Let $\mathfrak{g}$ be a Lie algebra and recall from Section 3.5 that the Lie-Poisson bracket on $\mathfrak{g}^*$ (the dual of $\mathfrak{g}$ when $\mathfrak{g}$ is finite-dimensional; the restricted dual of $\mathfrak{g}$ when $\mathfrak{g}$ is a (twisted) affine Lie algebra) is given by

$$\{F, G\}(\xi) = \langle \xi, [\mathrm{d}F(\xi), \mathrm{d}G(\xi)] \rangle,$$

where $F, G \in \mathcal{F}(\mathfrak{g}^*)$ and $\xi \in \mathfrak{g}^*$, and where $\mathrm{d}F(\xi)$ and $\mathrm{d}G(\xi)$ are interpreted as elements of $\mathfrak{g}$ when computing the Lie bracket. Recall also that if we view $X \in \mathfrak{g}$ as a linear function on $\mathfrak{g}^*$ we still denote it by the same letter X.

We suppose that we are given a *Lie algebra splitting* of $\mathfrak{g}$, i.e., that $\mathfrak{g}$ is a direct sum (as a vector space) of two Lie subalgebras, $\mathfrak{g} = \mathfrak{g}_+ \oplus \mathfrak{g}_-$. We denote the corresponding projection operators by

$$P_+ : \mathfrak{g} \to \mathfrak{g}_+ \qquad \text{and} \qquad P_- : \mathfrak{g} \to \mathfrak{g}_-.$$

For $X \in \mathfrak{g}$ we define $X_+ := P_+(X)$ and $X_- := P_-(X)$. The splitting leads to a direct sum decomposition of $\mathfrak{g}^*$. Namely,

$$\mathfrak{g}^* = \mathrm{Ann}(\mathfrak{g}_-) \oplus \mathrm{Ann}(\mathfrak{g}_+) \cong \mathfrak{g}_+^* \oplus \mathfrak{g}_-^*, \tag{4.7}$$

where for a subset $A \subseteq \mathfrak{g}$, its annihilator $\mathrm{Ann}(A)$ is defined as the subspace of $\mathfrak{g}^*$, given by

$$\mathrm{Ann}(A) := \{\xi \in \mathfrak{g}^* \mid \langle \xi, X \rangle = 0 \text{ for all } X \in A\}.$$

Indeed, dualizing the projection maps P_+ and P_- we find two injective linear maps

$$P_+^* = \imath_+ : \mathfrak{g}_+^* \to \mathfrak{g}^* \qquad \text{and} \qquad P_-^* = \imath_- : \mathfrak{g}_-^* \to \mathfrak{g}^*,$$

which are explicitly given by $\langle \imath_\pm(\phi), X \rangle = \langle \phi, X_\pm \rangle$, for $\phi \in \mathfrak{g}_\pm^*$ and $X \in \mathfrak{g}$. Thus, $\imath_\pm(\phi)$ is a natural extension of $\phi \in \mathfrak{g}_\pm^*$ to a linear function on $\mathfrak{g}$. We have that $\mathfrak{g}_\pm^* \cong \imath_\pm(\mathfrak{g}_\pm^*) = \mathrm{Ann}(\mathfrak{g}_\mp)$, which yields the isomorphism in (4.7).

Let R denote the endomorphism of $\mathfrak{g}$ which is given by $R := P_+ - P_-$. For $X, Y \in \mathfrak{g}$, let

$$[X, Y]_R := \frac{1}{2}\left([RX, Y] + [X, RY]\right) = [X_+, Y_+] - [X_-, Y_-]. \tag{4.8}$$

Lemma 4.34. *Let $\mathfrak{g} = \mathfrak{g}_+ \oplus \mathfrak{g}_-$ be a Lie algebra splitting and let $[\cdot\,,\cdot]_R$ denote the bracket, defined by (4.8).*

*(1) $[\cdot\,,\cdot]_R$ defines a Lie bracket on $\mathfrak{g}$, called an R-*bracket. *Its Lie-Poisson bracket on $\mathfrak{g}^*$ is called a* Lie-Poisson R-bracket *and is denoted by $\{\cdot\,,\cdot\}_R$.*

(2) Denoting by $\{\cdot\,,\cdot\}_+$ the Lie-Poisson structure on the dual $\mathfrak{g}_+^*$ of the Lie algebra $\mathfrak{g}_+$, the linear map

$$\imath_+ : (\mathfrak{g}_+^*, \{\cdot\,,\cdot\}_+) \to (\mathfrak{g}^*, \{\cdot\,,\cdot\}_R)$$

is a Poisson map.

Proof. $\mathfrak{g}_+$ and $\mathfrak{g}_-$ being subalgebras of $\mathfrak{g}$ we have that $P_+([X,Y]_R) = [X_+, Y_+]$, so that

$$P_+([[X,Y]_R, Z]_R) = [[X_+, Y_+], Z_+].$$

Similarly $P_-([[X,Y]_R, Z]_R) = [[X_-, Y_-], Z_-]$. It follows that $[\cdot,\cdot]_R$ satisfies the Jacobi identity and defines another Lie algebra structure on $\mathfrak{g}$. This proves *(1)*. Since $\mathfrak{g}^*$ admits a system of linear coordinates, taken from $\mathfrak{g}$, where elements of $\mathfrak{g}$ are viewed as linear functions on $\mathfrak{g}^*$, it follows that in order to prove *(2)*, it suffices to prove that if X and Y belong to $\mathfrak{g}$ then

$$\imath_+^* \{X,Y\}_R = \{\imath_+^* X, \imath_+^* Y\}_+ . \tag{4.9}$$

For $\phi \in \mathfrak{g}_+^*$ and $X \in \mathfrak{g}$ we have that $\langle \imath_+^* X, \phi\rangle = \langle \phi, X_+\rangle$, since $\imath_+^* = P_+$. On the one hand it follows that if $\phi \in \mathfrak{g}_+^*$ and $X, Y \in \mathfrak{g}$ and then, as functions on $\mathfrak{g}^*$, they satisfy

$$\{\imath_+^* X, \imath_+^* Y\}_+ (\phi) = \langle \phi, [X_+, Y_+]\rangle .$$

On the other hand, since $P_+([X,Y]_R) = [X_+, Y_+]$ we have that

$$\imath_+^* \{X,Y\}_R (\phi) = \{X,Y\}_R (\imath_+(\phi)) = \langle \imath_+(\phi), [X,Y]_R\rangle = \langle \phi, [X_+, Y_+]\rangle ,$$

Since ϕ is arbitrary this shows (4.9) for any $X, Y \in \mathfrak{g}$ and completes the proof of *(2)*. □

Remark 4.35. The linear map $\imath_- : (\mathfrak{g}_-^*, \{\cdot,\cdot\}_-) \to (\mathfrak{g}^*, \{\cdot,\cdot\}_R)$, where $\{\cdot,\cdot\}_-$ denotes the Lie-Poisson structure on $\mathfrak{g}_-^*$ is an *anti-Poisson morphism*, which means that

$$\imath_-^* \{X,Y\}_R = -\{\imath_-^* X, \imath_-^* Y\}_- ,$$

for any $X, Y \in \mathfrak{g}$. The reason for this asymmetry between $\mathfrak{g}_+$ and $\mathfrak{g}_-$ is that $P_+[X,Y]_R = [X,Y]_+$ while $P_-[X,Y]_R = -[X,Y]_-$.

4.4.2 The AKS Theorem on $\mathfrak{g}^*$

In order to formulate and prove the Adler-Kostant-Symes Theorem on $\mathfrak{g}^*$, we need some more notation. For $\epsilon \in \mathfrak{g}^*$ and for $H \in \mathcal{F}(\mathfrak{g}^*)$ we introduce two functions H_ϵ and $\tilde{H}_\epsilon$ by

$$\begin{aligned} H_\epsilon &: \mathfrak{g}_+^* \to \mathbf{C} : \phi \mapsto H(\epsilon + \phi), \\ \tilde{H}_\epsilon &: \mathfrak{g}^* \to \mathbf{C} : \xi \mapsto H(\epsilon + \xi). \end{aligned}$$

Notice that H_ϵ is just the restriction of $\tilde{H}_\epsilon$ to $\mathfrak{g}_+^*$, i.e., $\imath_+^* \tilde{H}_\epsilon = H_\epsilon$ (we identify ϕ with $\imath_+(\phi)$ to make the notation lighter). We also need the group analogue of the splitting of $\mathfrak{g}$. Therefore, we denote by $\mathbf{G}$ the (connected, simply-connected) Lie group whose Lie algebra is $\mathfrak{g}$ and we denote the subgroups of

$\mathbf{G}$ that correspond to $\mathfrak{g}_+$ and $\mathfrak{g}_-$ by $\mathbf{G}_+$ and $\mathbf{G}_-$. The subset $\mathbf{G}_+\mathbf{G}_-$ of $\mathbf{G}$ contains an open neighborhood of the origin in $\mathbf{G}$, but $\mathbf{G}_+\mathbf{G}_-$ is in general neither dense nor open in $\mathbf{G}$. We recall from Section 2.2 that $\mathcal{F}(\mathfrak{g}^*)^{\mathbf{G}}$ denotes the algebra of Ad^*-invariant functions.

Theorem 4.36 (AKS Theorem on $\mathfrak{g}^*$). *Suppose that $\mathfrak{g} = \mathfrak{g}_+ \oplus \mathfrak{g}_-$ is a Lie algebra splitting, let $F, H \in \mathcal{F}(\mathfrak{g}^*)^{\mathbf{G}}$ and suppose that $\epsilon \in \mathfrak{g}^*$ satisfies*

$$\langle \epsilon, [\mathfrak{g}_+, \mathfrak{g}_+] \rangle = 0 = \langle \epsilon, [\mathfrak{g}_-, \mathfrak{g}_-] \rangle . \tag{4.10}$$

Then

(1) $\{F, H\}_R = 0$ and $\{F_\epsilon, H_\epsilon\}_+ = 0$;

(2) The Hamiltonian vector fields $\mathcal{X}_H := \{\cdot\,, H\}_R$ and $\mathcal{X}_{H_\epsilon} := \{\cdot\,, H_\epsilon\}_+$ are given by

$$\mathcal{X}_H(\xi) = \frac{1}{2}\,\mathrm{ad}^*_{R(\mathrm{d}H(\xi))}\,\xi = \pm\,\mathrm{ad}^*_{(\mathrm{d}H(\xi))_\pm}\,\xi, \tag{4.11}$$

$$\mathcal{X}_{H_\epsilon}(\phi) = \frac{1}{2}\,\mathrm{ad}^*_{R(\mathrm{d}H(\xi))}\,\xi = \pm\,\mathrm{ad}^*_{(\mathrm{d}H(\xi))_\pm}\,\xi, \tag{4.12}$$

where, on the second line, $\xi := \phi + \epsilon$;

(3) For $\xi_0 \in \mathfrak{g}^*$ and for $|t|$ small, let $g_+(t)$ and $g_-(t)$ denote the smooth curves in $\mathbf{G}_+$ resp. $\mathbf{G}_-$ which solve the factorization problem

$$\exp(-t\mathrm{d}H(\xi_0)) = g_+(t)^{-1} g_-(t), \qquad g_\pm(0) = e. \tag{4.13}$$

Then the integral curve of $\mathcal{X}_H$ which starts at ξ_0 is given for $|t|$ small by

$$\xi(t) = \mathrm{Ad}^*_{g_+(t)}\,\xi_0 = \mathrm{Ad}^*_{g_-(t)}\,\xi_0. \tag{4.14}$$

The same formulas provide the integral curves of $\mathcal{X}_{H_\epsilon}$, by replacing $\xi(t)$ by $\phi(t) + \epsilon$ and ξ_0 by $\phi_0 + \epsilon$.

Proof. If $F, H \in \mathcal{F}(\mathfrak{g}^*)^{\mathbf{G}}$ then it follows from the first equality in Lemma 2.9 that

$$\{F, H\}_R(\xi) = \frac{1}{2}\langle \xi, [R(\mathrm{d}F(\xi)), \mathrm{d}H(\xi)] \rangle + \frac{1}{2}\langle \xi, [\mathrm{d}F(\xi), R(\mathrm{d}H(\xi))] \rangle = 0.$$

This shows the first part of *(1)*. In view of item *(2)* in Lemma 4.34, it is for the second part of *(1)* sufficient to prove that $\left\{\tilde{F}_\epsilon, \tilde{H}_\epsilon\right\}_R(\xi) = 0$ for any $\xi \in \mathfrak{g}^*$ that vanishes on $\mathfrak{g}_-$. For such a ξ, we find by using (4.8) and (4.10) and by applying the first equality in Lemma 2.9 to $\tilde{F}_\epsilon$ as well as to $\tilde{H}_\epsilon$,

$$\begin{aligned}
\left\{\tilde{F}_\epsilon, \tilde{H}_\epsilon\right\}_R(\xi) &= \left\langle \xi, \left[(\mathrm{d}\tilde{F}_\epsilon(\xi))_+, (\mathrm{d}\tilde{H}_\epsilon(\xi))_+\right]\right\rangle \\
&= \left\langle \xi + \epsilon, \left[(\mathrm{d}\tilde{F}_\epsilon(\xi))_+, (\mathrm{d}\tilde{H}_\epsilon(\xi))_+\right]\right\rangle \\
&= -\left\langle \xi + \epsilon, \left[(\mathrm{d}\tilde{F}_\epsilon(\xi))_-, (\mathrm{d}\tilde{H}_\epsilon(\xi))_+\right]\right\rangle \\
&= \left\langle \xi + \epsilon, \left[(\mathrm{d}\tilde{F}_\epsilon(\xi))_-, (\mathrm{d}\tilde{H}_\epsilon(\xi))_-\right]\right\rangle \\
&= 0.
\end{aligned}$$

This proves that they are in involution with respect to the Lie-Poisson R-bracket. In order to write down the Hamiltonian vector field $\mathcal{X}_H$, for $H \in \mathcal{F}(\mathfrak{g}^*)^{\mathbf{G}}$, take any $F \in \mathcal{F}(\mathfrak{g}^*)$ and let $\xi \in \mathfrak{g}^*$. By definition of $\mathcal{X}_H$ we have that

$$\langle \mathrm{d}F(\xi), \mathcal{X}_H(\xi)\rangle = \{F, H\}_R(\xi). \tag{4.15}$$

But Lemma 2.9, applied to H, gives

$$\{F, H\}_R(\xi) = \frac{1}{2}\langle \xi, [\mathrm{d}F(\xi), R(\mathrm{d}H(\xi))]\rangle = \frac{1}{2}\left\langle \mathrm{ad}^*_{R(\mathrm{d}H(\xi))}\,\xi, \mathrm{d}F(\xi)\right\rangle,$$

leading to the first expression for $\mathcal{X}_H$ in (4.11). The other expressions follow at once from the obvious formulas

$$\begin{aligned} R(\mathrm{d}H(\xi)) &= (\mathrm{d}H(\xi))_+ - (\mathrm{d}H(\xi))_- \\ &= 2(\mathrm{d}H(\xi))_+ - \mathrm{d}H(\xi) = \mathrm{d}H(\xi) - 2(\mathrm{d}H(\xi))_- \end{aligned}$$

and Lemma 2.9. The computation that leads to $\mathcal{X}_{H_\epsilon}$ is more subtle, so we also give it. For $F \in \mathcal{F}(\mathfrak{g}^*)$ and $\phi \in \mathfrak{g}_+^*$ we have as in (4.15) that

$$\langle \mathrm{d}F_\epsilon(\phi), \mathcal{X}_{H_\epsilon}(\phi)\rangle = \{F_\epsilon, H_\epsilon\}_+(\phi), \tag{4.16}$$

and note that the vectors $\mathrm{d}F_\epsilon(\phi)$ span $\mathfrak{g}_+^{**} \cong \mathfrak{g}_+$ as F runs through $\mathcal{F}(\mathfrak{g}^*)$. Formula (4.16) can also be written as

$$\left\langle \mathrm{d}\tilde{F}_\epsilon(\imath_+(\phi)), \imath_+(\mathcal{X}_{H_\epsilon}(\phi))\right\rangle = \{F_\epsilon, H_\epsilon\}_+(\phi), \tag{4.17}$$

because $\mathrm{d}F_\epsilon = \mathrm{d}\imath_+^*\tilde{F}_\epsilon = \imath_+^*\mathrm{d}\tilde{F}_\epsilon$. Since $\imath_+$ is a Poisson map (see Lemma 4.34) we have that

$$\{F_\epsilon, H_\epsilon\}_+(\phi) = \left\{\imath_+^*\tilde{F}_\epsilon, \imath_+^*\tilde{H}_\epsilon\right\}_+(\phi) = \left\{\tilde{F}_\epsilon, \tilde{H}_\epsilon\right\}_R(\imath_+(\phi)).$$

Since the latter expression is of the form $\langle \imath_+(\phi), \star\rangle$, where $\star$ belongs to $[\mathfrak{g}_+, \mathfrak{g}_+] + [\mathfrak{g}_-, \mathfrak{g}_-]$, the conditions (4.10) imply that it equals $\langle \imath_+(\phi) + \epsilon, \star\rangle$. We can now use the invariance of H again to obtain

$$\begin{aligned} \{F_\epsilon, H_\epsilon\}_+(\phi) &= \left\{\tilde{F}_\epsilon, \tilde{H}_\epsilon\right\}_R(\imath_+(\phi) + \epsilon) \\ &= \frac{1}{2}\left\langle \imath_+(\phi) + \epsilon, \left[\mathrm{d}\tilde{F}_\epsilon(\imath_+(\phi)), R(\mathrm{d}\tilde{H}_\epsilon(\imath_+(\phi)))\right]\right\rangle \\ &= \frac{1}{2}\left\langle \mathrm{ad}^*_{R(\mathrm{d}\tilde{H}_\epsilon(\imath_+(\phi)))}\,\imath_+(\phi) + \epsilon, \mathrm{d}\tilde{F}_\epsilon(\imath_+(\phi))\right\rangle. \end{aligned}$$

Comparing the last expression to (4.17) leads to the announced expression for $\mathcal{X}_{H_\epsilon}$, upon suppressing $\imath_+$ through our identifications.

We now turn to the integral curves of $\mathcal{X}_H$. We first show that $\mathrm{Ad}^*_{g_+(t)}\,\xi_0 = \mathrm{Ad}^*_{g_-(t)}\,\xi_0$ (the second equality in (4.14)). Since Ad^* is a group homomorphism, the factorization (4.13) implies that

$$\mathrm{Ad}^*_{\exp(-t\mathrm{d}H(\xi_0))}\,\xi_0 = \mathrm{Ad}^*_{g_+(t)^{-1}}\,\mathrm{Ad}^*_{g_-(t)}\,\xi_0,$$

for any $\xi_0 \in \mathfrak{g}^*$. We have for any $X \in \mathfrak{g}$ that

$$\begin{aligned}\left\langle \mathrm{Ad}^*_{\exp(-t\mathrm{d}H(\xi_0))}\,\xi_0, X\right\rangle &= \left\langle \xi_0, \mathrm{Ad}_{\exp(t\mathrm{d}H(\xi_0))}\,X\right\rangle \\ &= \left\langle \xi_0, \exp\left(t\,\mathrm{ad}_{\mathrm{d}H(\xi_0)}\right) X\right\rangle \\ &= \langle \xi_0, X\rangle + t\,\langle \xi_0, [\mathrm{d}H(\xi_0), \star]\rangle \\ &= \langle \xi_0, X\rangle,\end{aligned}$$

where the value of $\star$ (which depends on t) is irrelevant, by Lemma 2.9. This shows that $\mathrm{Ad}^*_{\exp(-t\mathrm{d}H(\xi_0))}\,\xi_0 = \xi_0$, and hence that $\mathrm{Ad}^*_{g_+(t)}\,\xi_0 = \mathrm{Ad}^*_{g_-(t)}\,\xi_0$.

We now show that $\xi(t) := \mathrm{Ad}^*_{g_+(t)}\,\xi_0$ is (for $|t|$ small) a solution to (4.11), which amounts to proving that

$$\frac{\mathrm{d}}{\mathrm{d}t}\,\mathrm{Ad}^*_{g_+(t)}\,\xi_0 = \mathrm{ad}^*_{(\mathrm{d}H(\xi(t)))_+}\,\xi(t). \tag{4.18}$$

In order to simplify the proof (mainly the notation), we will assume that **G** is a linear group, so that Ad and Ad^* are just given by conjugation. Then the left hand side of (4.18) is given by $\mathrm{ad}^*_{\dot{g}_+(t)g_+(t)^{-1}}\,\xi(t)$, so it suffices to show that

$$\dot{g}_+(t)g_+(t)^{-1} = (\mathrm{d}H(\xi(t)))_+. \tag{4.19}$$

If we differentiate the identity $g_+(t)\exp(-t\mathrm{d}H(\xi_0)) = g_-(t)$ with respect to t and multiply both sides of the result by $g_-(t)^{-1}$ we get

$$\dot{g}_+(t)g_+(t)^{-1} - g_-(t)\mathrm{d}H(\xi_0)g_-(t)^{-1} = \dot{g}_-(t)g_-(t)^{-1}.$$

The Ad^*-invariance of H implies, by the commutativity of the diagram in Lemma 2.9, that

$$\mathrm{d}H(\xi(t)) = \mathrm{d}H\left(\mathrm{Ad}^*_{g_-(t)}\,\xi_0\right) = \mathrm{Ad}_{g_-(t)}\,\mathrm{d}H(\xi_0) = g_-(t)\mathrm{d}H(\xi_0)g_-(t)^{-1},$$

so that

$$\dot{g}_+(t)g_+(t)^{-1} - \mathrm{d}H(\xi(t)) = \dot{g}_-(t)g_-(t)^{-1}.$$

If we take the + part of both sides of this equation then we find (4.19), as was to be shown. The proof in the case of X_{H_ϵ} is similar. □

4.4.3 *R*-Brackets and Double Lie Algebras

It is easily seen from the above proof that the first part of *(1)* and the first part of *(2)* in the Theorem 4.36 hold true for any endomorphism $R : \mathfrak{g} \to \mathfrak{g}$ for which

$$[X, Y]_R = \frac{1}{2}\left([RX, Y] + [X, RY]\right)$$

defines a Lie bracket on $\mathfrak{g}$, which is then called an *R-bracket* and $\mathfrak{g}$, equipped with the two Lie algebra structures $[\cdot\,,\cdot]$ and $[\cdot\,,\cdot]_R$ is called a *double Lie algebra.* Letting B_R denote the bilinear map $\mathfrak{g} \times \mathfrak{g} \to \mathfrak{g}$, defined by

$$B_R(X, Y) = [RX, RY] - R\left([RX, Y] + [X, RY]\right),$$

the condition that $[\cdot\,,\cdot]_R$ defines a Lie bracket (satisfies the Jacobi identity) is easily rewritten as

$$[B_R(X, Y), Z] + \operatorname{cycl}(X, Y, Z) = 0. \tag{4.20}$$

Solutions R to the *Yang-Baxter equation* $B_R = 0$, i.e., solutions R to

$$\forall X, Y \in \mathfrak{g} \;:\; [RX, RY] = R\left([RX, Y] + [X, RY]\right),$$

yield particular solutions to (4.20), and the same is true for the solutions to the more general *modified Yang-Baxter equation*

$$B_R(X, Y) = -c[X, Y],$$

where $c \in \mathbf{C}$ is a constant. The endomorphism $R = P_+ - P_-$ that we have considered corresponds to $c = 1$. Notice that when $c \neq 0$ we may rescale[2] R so as to get $c = 1$. But this case presents nothing new: if we define $\mathfrak{g}_+$ and $\mathfrak{g}_-$ by

$$\mathfrak{g}_\pm := \{X \pm RX \mid X \in \mathfrak{g}\}$$

then $\mathfrak{g}_+$ and $\mathfrak{g}_-$ are subalgebras of $\mathfrak{g}$. Indeed, if we define

$$Z := [RX, Y] + [X, RY]$$

then

$$\begin{aligned}[X \pm RX, Y \pm RY] &= [X, Y] + [RX, RY] \pm \left([X, RY] + [RX, Y]\right) \\ &= \pm(Z \pm RZ),\end{aligned}$$

a consequence of the modified Yang-Baxter equation $B_R(X, Y) = -[X, Y]$. Thus, we have a Lie algebra splitting $\mathfrak{g} = \mathfrak{g}_+ \oplus \mathfrak{g}_-$, and RX is nothing but the difference between the projection operators $P_+ : \mathfrak{g} \to \mathfrak{g}_+$ and $P_- : \mathfrak{g} \to \mathfrak{g}_-$, as in the AKS Theorem.

[2] When $\mathfrak{g}$ is a real Lie algebra the rescaling yields $c = \pm 1$.

The case $c = 0$ (or $c \leqslant 0$ when $\mathfrak{g}$ is a real Lie algebra) might lead to something new, but in that case the method that we presented to produce the solutions (integral curves), i.e., item *(3)* of Theorem 4.36, fails and no general integration method for that case is known. Finally, it is only in the case in which R comes from a Lie algebra splitting that one gets also an ϵ-version of the AKS Theorem, as can be seen from our proof. We stress that for the majority of examples it is the ϵ-version of the theorem that needs to be used.

4.4.4 The AKS Theorem on $\mathfrak{g}$

We will now transcribe the AKS Theorem to $\mathfrak{g}$, assuming that we are given an Ad-invariant non-degenerate bilinear form on $\mathfrak{g}$,

$$\langle \cdot \,|\, \cdot \rangle : \mathfrak{g} \times \mathfrak{g} \to \mathbf{C}.$$

Recall from Section 3.5 that the Lie-Poisson structure on $\mathfrak{g}$ with respect to $\langle \cdot \,|\, \cdot \rangle$ is given by

$$\{F, H\}(X) = \langle X \,|\, [\nabla F(X), \nabla H(X)] \rangle ,$$

where $\nabla F(X)$ is the gradient of F at X (with respect to $\langle \cdot \,|\, \cdot \rangle$). We also recall that the isomorphism $\mathfrak{g} \to \mathfrak{g}^*$, which is induced by $\langle \cdot \,|\, \cdot \rangle$, establishes a one-to-one correspondence between Ad^*-invariant functions on $\mathfrak{g}^*$ and Ad-invariant functions on $\mathfrak{g}$, and that the Hamiltonian vector fields $\mathcal{X}_H$ on $(\mathfrak{g}, \{\cdot\,,\cdot\})$ take the simple form

$$\dot{X} = [\nabla H(X), X] .$$

See Section 3.5 for details.

Non-degeneracy of $\langle \cdot \,|\, \cdot \rangle$ implies that $\mathfrak{g}$ admits another (vector space) direct sum decomposition, namely as $\mathfrak{g} = \mathfrak{g}_+^\perp \oplus \mathfrak{g}_-^\perp$ where $\mathfrak{g}_+^\perp$ (resp. $\mathfrak{g}_-^\perp$) is the orthogonal complement (with respect to $\langle \cdot \,|\, \cdot \rangle$) of $\mathfrak{g}_+$ (resp. $\mathfrak{g}_-$). In view of the isomorphisms

$$\mathfrak{g}_+^\perp \cong \mathfrak{g}_-^* \qquad \text{and} \qquad \mathfrak{g}_-^\perp \cong \mathfrak{g}_+^*,$$

which are induced by $\langle \cdot \,|\, \cdot \rangle$, the two subspaces $\mathfrak{g}_+^\perp$ and $\mathfrak{g}_-^\perp$ of $\mathfrak{g}$ (which are not Lie subalgebras) carry a linear Poisson structure, which we will denote by $\{\cdot\,,\cdot\}_{\mathfrak{g}_+^\perp}$, resp. $\{\cdot\,,\cdot\}_{\mathfrak{g}_-^\perp}$. The element of $\mathfrak{g}$ that corresponds to ϵ will still be denoted by ϵ, and for $H \in \mathcal{F}(\mathfrak{g})$ we get now a function $\tilde{H}_\epsilon \in \mathcal{F}(\mathfrak{g})$ and a function $H_\epsilon \in \mathcal{F}(\mathfrak{g}_-^\perp)$, defined as before. In the formulas in the theorem that follows we will leave once again the inclusion $\imath_+ : \mathfrak{g}_-^\perp \to \mathfrak{g}$ out in favor of the readability of the formulas.

We are now ready to formulate the Adler-Kostant-Symes on $\mathfrak{g}$. The proof follows by transcribing Theorem 4.36 to $\mathfrak{g}$.

Theorem 4.37 (AKS Theorem on $\mathfrak{g}$). *Suppose that $\mathfrak{g} = \mathfrak{g}_+ \oplus \mathfrak{g}_-$ is a Lie algebra splitting and that $\langle \cdot \,|\, \cdot \rangle$ is an* Ad*-invariant non-degenerate bilinear form on $\mathfrak{g}$, leading to a vector space splitting*

$$\mathfrak{g} = \mathfrak{g}_+^\perp \oplus \mathfrak{g}_-^\perp \cong \mathfrak{g}_-^* \oplus \mathfrak{g}_+^*.$$

Let $F, H \in \mathcal{F}(\mathfrak{g})^{\mathbf{G}}$ and suppose that $\epsilon \in \mathfrak{g}$ satisfies

$$[\epsilon, \mathfrak{g}_+] \in \mathfrak{g}_+^\perp, \qquad [\epsilon, \mathfrak{g}_-] \in \mathfrak{g}_-^\perp.$$

Then

(1) $\{F, H\}_R = 0$ *and* $\{F_\epsilon, H_\epsilon\}_{\mathfrak{g}_-^\perp} = 0$;

(2) The Hamiltonian vector fields $\mathcal{X}_H := \{\cdot\,, H\}_R$ and $\mathcal{X}_{H_\epsilon} := \{\cdot\,, H_\epsilon\}_{\mathfrak{g}_-^\perp}$ are respectively given by

$$\mathcal{X}_H(X) = -\frac{1}{2}[X, R(\nabla H(X))] = \pm\,[X, (\nabla H(X))_\mp] \tag{4.21}$$

and

$$\mathcal{X}_{H_\epsilon}(Y) = -\frac{1}{2}[Y, R(\nabla H(Y))] = \pm[Y, (\nabla H(Y))_\mp], \tag{4.22}$$

where $Y \in \mathfrak{g}_-^\perp + \epsilon$;

(3) For $X_0 \in \mathfrak{g}$ and for $|t|$ small, let $g_+(t)$ and $g_-(t)$ denote the smooth curves in $\mathbf{G}_+$ resp. $\mathbf{G}_-$ which solve the factorization problem

$$\exp(-t\nabla H(X_0)) = g_+(t)^{-1} g_-(t), \qquad g_\pm(0) = e.$$

Then the integral curve of $\mathcal{X}_H$ which starts at X_0 is given for $|t|$ small by

$$X(t) = \operatorname{Ad}_{g_+(t)} X_0 = \operatorname{Ad}_{g_-(t)} X_0.$$

We illustrate the AKS Theorem (on $\mathfrak{g}$) by the following simple example.

Example 4.38. Consider the Lie algebra splitting

$$\mathfrak{g} = \mathfrak{gl}(n+1) = \Delta_{n+1}^{\leqslant} \oplus \Delta_{n+1}^{>} = \mathfrak{g}_+ \oplus \mathfrak{g}_-,$$

where $\Delta_{n+1}^{\leqslant}$ (resp. $\Delta_{n+1}^{>}$) denotes the Lie algebra of lower triangular matrices (resp. of strictly upper triangular matrices). Letting $\langle A \,|\, B \rangle = \operatorname{Trace}(AB)$ we have that $\mathfrak{g}_-^\perp = \left(\Delta_{n+1}^{>}\right)^\perp = \Delta_{n+1}^{\geqslant}$, the vector space of all upper triangular matrices in $\mathfrak{gl}(n+1)$. If F is a linear function on $\mathfrak{gl}(n+1)$ then $\nabla F(X) = \nabla F$ is independent of X and it follows from (3.33) that it is given by $\langle \nabla F \,|\, Y \rangle = F(Y)$, where $Y \in \mathfrak{gl}(n+1)$. Denoting by x_{ij} the linear function that picks the element (i,j) of a matrix, it follows that $\nabla x_{ij} = \mathcal{E}_{ji}$, the matrix whose entry at position (k,l) is $\delta_{jk}\delta_{il}$. The Poisson bracket on $\Delta_{n+1}^{\geqslant} \cong \mathfrak{g}_+^*$ is therefore given by

$$\{x_{ij}, x_{kl}\}(X) = \langle X \,|\, [\mathcal{E}_{ji}, \mathcal{E}_{lk}] \rangle,$$

where $j \geqslant i$ and $l \geqslant k$. Let us denote by B_p the set of all p-band matrices in $\Delta_{n+1}^{\geqslant}$. Thus

$$B_p := \left\{ X \in \Delta_{n+1}^{\geqslant} \mid x_{ij}(X) = 0 \text{ if } j - i > p \right\}.$$

All Hamiltonian vector fields on $\Delta_{n+1}^{\geqslant}$ are tangent to B_p (at points of B_p): letting $j - i > p$ and $l \geqslant k$ we have that

$$\mathcal{X}_{x_{kl}}[x_{ij}](X) = \{x_{ij}, x_{kl}\}(X) = \operatorname{Trace}(X[\mathcal{E}_{ji}, \mathcal{E}_{lk}]) = X_{kj}\delta_{il} - X_{il}\delta_{jk} = 0,$$

for any $X \in B_p$, since $j - k > p$ and $l - i > p$. Thus, all Hamiltonian vector fields annihilate the ideal defining B_p, which implies, in view of Proposition 3.33 that they are tangent to B_p and that B_p is a Poisson subspace of $\Delta_{n+1}^{>}$. We take $p := 1$ and

$$\epsilon := \begin{pmatrix} 0 & & & 0 \\ 1 & 0 & & \\ & \ddots & \ddots & \\ 0 & & 1 & 0 \end{pmatrix}.$$

It is easy to verify that $[\epsilon, \mathfrak{g}_-] \subseteq \mathfrak{g}_-^{\perp}$ and that $[\epsilon, \mathfrak{g}_+] \subseteq \mathfrak{g}_+^{\perp}$. Taking $H : Y \mapsto -\frac{1}{2} \operatorname{Trace} Y^2$, so that $\nabla H(Y) = -Y$ we find from (4.22) the Hamiltonian vector field $\mathcal{X}_{H_\epsilon}$, given by the Lax equation

$$\dot{Y} = [Y, Y_+] = -[Y, Y_-],$$

where

$$Y := \begin{pmatrix} b_1 & a_1 & & & 0 \\ 1 & b_2 & a_2 & & \\ & \ddots & \ddots & \ddots & \\ & & 1 & b_n & a_n \\ 0 & & & 1 & b_{n+1} \end{pmatrix}$$

and where

$$Y_- = \begin{pmatrix} 0 & a_1 & & & 0 \\ 0 & 0 & a_2 & & \\ & \ddots & \ddots & \ddots & \\ & & 0 & 0 & a_n \\ 0 & & & 0 & 0 \end{pmatrix}, \qquad Y_+ = \begin{pmatrix} b_1 & 0 & & & 0 \\ 1 & b_2 & 0 & & \\ & \ddots & \ddots & \ddots & \\ & & 1 & b_n & 0 \\ 0 & & & 1 & b_{n+1} \end{pmatrix}.$$

For future reference, we also wish to write down the vector field that goes with $K := -\frac{1}{3}\operatorname{Trace} Y^3$. Since $\nabla K(Y) = -Y^2$ it is given by

$$\dot{Y} = -\left[Y, (Y^2)_-\right],$$

where Y is as above and

$$(Y^2)_- = \begin{pmatrix} 0 & a_1(b_1+b_2) & a_1a_2 & & & 0 \\ & 0 & a_2(b_2+b_3) & a_2a_3 & & \\ & & \ddots & \ddots & \ddots & \\ & & & 0 & a_{n-1}(b_{n-1}+b_n) & a_{n-1}a_n \\ 0 & & & & 0 & a_n(b_n+b_{n-1}) \end{pmatrix}$$

Example 4.39. We now give a class of examples where the Lie algebra is a finite-dimensional subspace of the affine Lie algebra

$$L(\mathfrak{g}) = \left\{ X(\mathfrak{h}) = \sum_{j=M}^{N} \mathfrak{h}^j X_j \mid M,\, N \in \mathbf{Z} \text{ and } X_j \in \mathfrak{g} \text{ for } M \leqslant j \leqslant N \right\},$$

where $\mathfrak{g}$ is a Lie subalgebra of $\mathfrak{gl}(n)$, that is assumed to carry a non-degenerate Ad-invariant symmetric bilinear form $\langle \cdot \,|\, \cdot \rangle$. Recall from Section 2.4 that there is a natural Lie bracket on $L(\mathfrak{g})$, which is given by

$$\left[\sum_{i\leqslant N} \mathfrak{h}^i X_i, \sum_{j\leqslant M} \mathfrak{h}^j Y_j\right] = \sum_{k\leqslant M+N} \mathfrak{h}^k \left(\sum_{i+j=k} [X_i, Y_j]\right)$$

and that $\langle \cdot \,|\, \cdot \rangle$ leads to a sequence of Ad-invariant symmetric bilinear forms $\langle \cdot \,|\, \cdot \rangle_k$ on $L(\mathfrak{g})$ upon defining

$$\left\langle \sum_{i\leqslant N} \mathfrak{h}^i X_i \,\Big|\, \sum_{j\leqslant M} \mathfrak{h}^j Y_j \right\rangle_k := \sum_{i+j+k=0} \langle X_i \,|\, Y_j \rangle .$$

Moreover, the non-degeneracy of $\langle \cdot \,|\, \cdot \rangle$ implies that $\langle \cdot \,|\, \cdot \rangle_k$ is non-degenerate, for any k. In the sequel we take $k = 1$ and we write

$$L(\mathfrak{g}) = L(\mathfrak{g})_+ \oplus L(\mathfrak{g})_-$$

where

$$L(\mathfrak{g})_+ = L(\mathfrak{g})_+^\perp = \left\{ X(\mathfrak{h}) = \sum_{j=0}^{N} \mathfrak{h}^j X_j \mid N \in \mathbf{N} \right\} \subseteq L(\mathfrak{g}),$$

$$L(\mathfrak{g})_- = L(\mathfrak{g})_-^\perp = \left\{ X(\mathfrak{h}) = \sum_{j=-M}^{-1} \mathfrak{h}^j X_j \mid M \in \mathbf{N}^* \right\} \subseteq L(\mathfrak{g}),$$

the subalgebra of $L(\mathfrak{g})$ consisting of polynomial elements (in $\mathfrak{h}$), resp. the subalgebra of $L(\mathfrak{g})$ consisting of polynomial elements in $\mathfrak{h}^{-1}$, without constant term. Notice that with respect to this splitting

$$\pi_+\left(\sum_i X_i\mathfrak{h}^i\right) = \sum_{i\geqslant 0} X_i\mathfrak{h}^i, \quad \text{and} \quad \pi_-\left(\sum_i X_i\mathfrak{h}^i\right) = \sum_{i<0} X_i\mathfrak{h}^i,$$

i.e., $\pi_+(X) = X_+$ is the polynomial part of $X = X(\mathfrak{h})$ while $\pi_-(X) = X_-$ is its Laurent part. Let α and γ be two fixed diagonal matrices, $\alpha = \operatorname{diag}(\alpha_1,\dots,\alpha_n)$ and $\gamma = \operatorname{diag}(\gamma_1,\dots,\gamma_n)$, where we assume that all α_i are different. We consider the affine subspace of $L(\mathfrak{g})_+$, defined by

$$\Gamma_m(\alpha,\gamma) = \alpha\mathfrak{h}^m + \gamma\mathfrak{h}^{m-1} + \mathcal{A}_{m-1},$$

where

$$\mathcal{A}_{m-1} := \left\{\sum_{j=0}^{m-1} \mathfrak{h}^j X_j \mid \operatorname{diag} X_{m-1} = 0\right\} \subset L(\mathfrak{g})_+.$$

We claim that $\Gamma_m(\alpha,\gamma)$ is a Poisson subspace of $(L(\mathfrak{g})_+, \{\cdot\,,\cdot\}_-)$, where $\{\cdot\,,\cdot\}_-$ is, in analogy[3] with the notation using in Lemma 4.34, the Lie-Poisson structure on $L(\mathfrak{g})^*_- \cong L(\mathfrak{g})^\perp_+ = L(\mathfrak{g})_+$. Since the symplectic leaves of the Lie-Poisson structure are the coadjoint orbits (Proposition 3.29) and since we have identified $L(\mathfrak{g})$ with its dual by using an Ad-invariant form this amounts to showing that for any $X \in L(\mathfrak{g})_-$ the fundamental vector field ad^*_X of the coadjoint action is tangent to $\Gamma_m(\alpha,\gamma)$. The coadjoint action is given, under the isomorphism $L(\mathfrak{g})_+ \cong L(\mathfrak{g})^*_-$, by $\operatorname{ad}^*_X Y = [X,Y]_+$, where $X \in L(\mathfrak{g})_-$ and $Y \in L(\mathfrak{g})_+ \cong L(\mathfrak{g})^*_-$; indeed, for any $Z \in L(\mathfrak{g})_-$,

$$\begin{aligned}\langle \operatorname{ad}^*_X Y, Z\rangle &= -\langle Y,[X,Z]\rangle = -\langle Y \mid [X,Z]\rangle_1\\ &= \langle [Y,X] \mid Z\rangle_1 = \langle [X,Y]_+ \mid Z\rangle_1 = \langle [X,Y]_+, Z\rangle.\end{aligned}$$

Notice that the Lie group of $L(\mathfrak{g})_-$ is infinite-dimensional, but this is inessential because only a finite-dimensional quotient of this group has a non-trivial action on $\Gamma_m(\alpha,\gamma)$. To check tangency, let $X = \sum_{i\leqslant -1} X_i\mathfrak{h}^i \in L(\mathfrak{g})_-$ to find

$$\begin{aligned}\operatorname{ad}^*_X(\alpha\mathfrak{h}^m + \gamma\mathfrak{h}^{m-1} + \mathcal{A}_{m-1}) &= \left[\sum_{i\leqslant -1} X_i\mathfrak{h}^i, \alpha\mathfrak{h}^m + \gamma\mathfrak{h}^{m-1} + \mathcal{A}_{m-1}\right]_+\\ &= [X_{-1},\alpha]\mathfrak{h}^{m-1} + O(\mathfrak{h}^{m-2}) \in \mathcal{A}_{m-1},\end{aligned}$$

where the vector space $\mathcal{A}_{m-1}$ is naturally identified with the tangent space of the affine space $\Gamma_m(\alpha,\gamma)$ at any of its points. Notice that all elements on the diagonal of $[X_{-1},\alpha]$ are indeed zero, as is obvious from the fact that α is a diagonal matrix. Thus, $\Gamma_m(\alpha,\gamma)$ is a (finite-dimensional) Poisson submanifold of $(L(\mathfrak{g})_+, \{\cdot\,,\cdot\}_-)$.

[3] Notice that since we will work with $L(\mathfrak{g})^*_-$, rather than with $L(\mathfrak{g})^*_+$ (as we did in the AKS Theorem) we will have an extra sign in the Hamiltonian vector fields.

We now specialize to $\mathfrak{g} = \mathfrak{gl}(n)$ and we take $\langle X \mid Y \rangle = \mathrm{Trace}(XY)$, for $X, Y \in \mathfrak{gl}(n)$, so that

$$\langle X(\mathfrak{h}) \mid Y(\mathfrak{h}) \rangle_1 = \mathrm{Res}_{\mathfrak{h}=0} \, \mathrm{Trace}\, X(\mathfrak{h}) Y(\mathfrak{h}).$$

We take a function f of one variable (typically, think of a polynomial or a logarithm) and we define a function H_f on $L(\mathfrak{g})_+$ by

$$H_f(X(\mathfrak{h})) := \left\langle f\left(X(\mathfrak{h})\mathfrak{h}^{-j}\right) \mid \mathfrak{h}^k \right\rangle_1 .$$

Then H is Ad-invariant. Indeed, let $g = g(\mathfrak{h})$ be an element of the loop group of $L(\mathfrak{g})$ and let $X = X(\mathfrak{h})$ be an arbitrary element of $L(\mathfrak{g})_+$. Then

$$\begin{aligned} H_f(gXg^{-1}) &= \left\langle f\left(gXg^{-1}\mathfrak{h}^{-j}\right) \mid \mathfrak{h}^k \right\rangle_1 = \left\langle g f\left(X\mathfrak{h}^{-j}\right) g^{-1} \mid \mathfrak{h}^k \right\rangle_1 \\ &= \left\langle f\left(X\mathfrak{h}^{-j}\right) \mid g^{-1}\mathfrak{h}^k g \right\rangle_1 = \left\langle f\left(X\mathfrak{h}^{-j}\right) \mid \mathfrak{h}^k \right\rangle_1 = H_f(X) \end{aligned}$$

We have that

$$\nabla H_f(X) = f'\left(X\mathfrak{h}^{-j}\right)\mathfrak{h}^{k-j}, \tag{4.23}$$

where $f'(z)$ it the ordinary derivative of $f(z)$. Indeed, assuming first that $f(z) = z^i$ we compute by using (3.33) for $Y = Y(\mathfrak{h}) \in L(\mathfrak{g})$

$$\begin{aligned} \langle \nabla H_f(X) \mid Y \rangle_1 &= \frac{\mathsf{d}}{\mathsf{d}t}_{|t=0} H_f(X + tY) \\ &= \frac{\mathsf{d}}{\mathsf{d}t}_{|t=0} \left\langle (X + tY)^i \mathfrak{h}^{-ij} \mid \mathfrak{h}^k \right\rangle_1 \\ &= \text{the linear term in } t \text{ in } \left\langle (X + tY)^i \mathfrak{h}^{-ij} \mid \mathfrak{h}^k \right\rangle_1 \\ &= \left\langle (X^{i-1}Y + X^{i-2}YX + \cdots + YX^{i-1})\mathfrak{h}^{-ij} \mid \mathfrak{h}^k \right\rangle_1 \\ &= \left\langle i(X\mathfrak{h}^{-j})^{i-1}\mathfrak{h}^{-j} \mid Y\mathfrak{h}^k \right\rangle_1 \\ &= \left\langle f'(X\mathfrak{h}^{-j})\mathfrak{h}^{k-j} \mid Y \right\rangle_1 . \end{aligned}$$

This shows (4.23) when f is of the form $f(z) = z^i$. Since (4.23) is linear in f the formula remains correct for f when f is a polynomial and, more generally, when f is an analytic function. It follows from (4.21) that the Hamiltonian vector field on $(L(\mathfrak{g})_+, \{\cdot,\cdot\}_-)$ that is associated to H_f is given by (see the previous footnote for the extra minus sign)

$$\dot{X} = \left[X, \left(f'\left(X\mathfrak{h}^{-j}\right)\mathfrak{h}^{k-j}\right)_+\right],$$

Table 4.1. We display four examples where the Hamiltonian H is given by (4.24), with Hamiltonian vector field, given by the Lax equation (4.25). In the table, $T_{xx} = x \otimes x$ while $S_{xy} = x \otimes y + y \otimes x$. The last column gives a Hamiltonian $\tilde{H}$ which yields the same Hamiltonian vector field $\mathcal{X}_H$, but with respect to the standard Poisson structure $\{x_i, y_j\} = \delta_{ij}$. They yield, in that order, (a) the Arnold-Euler equations, (b) geodesic flow on an ellipsoid, (c) the Neumann problem of n harmonic oscillators, confined to the sphere and (d) the central force problem on an ellipsoid.

$X = X(\mathfrak{h})$	$f(z)$	$\tilde{H}$
$\alpha\mathfrak{h} + T_{xy}$	$\frac{2}{3}z^{3/2}$	$\frac{1}{2}\sum\sqrt{\alpha_i}H_i$
$\alpha\mathfrak{h}^2 + T_{xy}\mathfrak{h} - T_{xx}$	$\ln(z)$	$\frac{1}{2}\sum\alpha_i^{-1}F_i$
$\alpha\mathfrak{h}^2 + T_{xy}\mathfrak{h} - T_{xx}$	$z^2/2$	$\frac{1}{2}\sum\alpha_i F_i$
$\alpha\mathfrak{h}^2 + T_{xy}\mathfrak{h} + S_{xy} - \alpha$	$\ln(z)$	$\frac{1}{2}\sum\alpha_i^{-1}G_i$

where $X = X(\mathfrak{h}) \in \alpha\mathfrak{h}^m + \gamma\mathfrak{h}^{m-1} + \sum_{i=0}^{m-1} X_i\mathfrak{h}^i$, with $\operatorname{diag}(X_{m-1}) = 0$. If we specialize this further to the case $j = m$, $k = m+1$ then $H(X) = \left\langle f\left(X\mathfrak{h}^{-m}\right) \mid \mathfrak{h}^{m+1}\right\rangle_1$ and the vector field becomes

$$\dot{X} = [X, \beta\mathfrak{h} + E],$$

where $\beta = f'(\alpha)$ and $E = \operatorname{ad}_\beta \operatorname{ad}_\alpha^{-1} X_{m-1} + f''(\alpha)\gamma$, i.e., for $1 \leqslant i, j \leqslant n$,

$$E_{ij} = (1 - \delta_{ij})\left(\frac{f'(\alpha_i) - f'(\alpha_j)}{\alpha_i - \alpha_j}\right)(X_{m-1})_{ij} + \delta_{ij} f''(\alpha_i)\gamma_i,$$

the above being easily checked for the case $f(z) = z^k$ and then the general case follows by linearity. A few relevant special cases are given in Table 4.1; see [7] and [8] for more information on these examples and for other important examples. In the given examples $\Gamma_m(\alpha)$ does not depend on γ, the Hamiltonian is always of the form

$$H(X) = \left\langle f\left(X\mathfrak{h}^{-m}\right) \mid \mathfrak{h}^{m+1}\right\rangle_1, \tag{4.24}$$

and the Hamiltonian vector field $\mathcal{X}_H$ on $\Gamma_m(\alpha)$ is always of the form

$$\dot{X} = \left[X, \beta\mathfrak{h} + \operatorname{ad}_\beta \operatorname{ad}_\alpha^{-1} T_{xy}\right], \tag{4.25}$$

where $T_{xy} := x \otimes y - y \otimes x$, for $x, y \in \mathbf{C}^n$, and $\beta := f'(\alpha)$. We also denote

$$F_i := x_i^2 + H_i, \quad G_i := -2x_iy_i + H_i, \quad H_i := \sum_{j \neq i} \frac{(x_iy_j - x_jy_i)^2}{\alpha_i - \alpha_j},$$

where $1 \leqslant i, j \leqslant n$.

4.5 Lax Operators and r-matrices

In the previous section we have seen how a Lie algebra splitting $\mathfrak{g} = \mathfrak{g}_+ \oplus \mathfrak{g}_-$ leads to a collection of commuting functions, and hence to a collection of commuting vector fields on the dual $\mathfrak{g}^*$ of $\mathfrak{g}$. In order to make these into commuting vector fields on $\mathfrak{g}$, which can be written as Lax equations, we used an Ad-invariant non-degenerate symmetric bilinear form on $\mathfrak{g}$ (such as the Killing form when $\mathfrak{g}$ is semi-simple). In the present section we describe a different approach to Lax equations, where the essential ingredient is a Lie algebra structure on $\mathfrak{g}^*$ (which yields a linear Poisson structure on $\mathfrak{g}$). In our exposition we will not describe the theory in its utmost generality, restricting ourselves to the most important class of such Lie algebra structures. See [152] or [99] for more details.

We start with a Lie algebra $\mathfrak{g}$ and we equip $\mathfrak{g} \otimes \mathfrak{g}$ with the structure of a $\mathfrak{g}$-module by letting

$$\begin{aligned} \mathfrak{g} \times (\mathfrak{g} \otimes \mathfrak{g}) &\to \mathfrak{g} \otimes \mathfrak{g} \\ (X, r) &\mapsto [X \otimes 1 + 1 \otimes X, r], \end{aligned} \tag{4.26}$$

where $[X \otimes 1, r]$ and $[1 \otimes X, r]$ are, for r of the form $r = Y \otimes Z$, explicitly given by

$$\begin{aligned} [X \otimes 1, Y \otimes Z] &= [X, Y] \otimes Z, \\ [1 \otimes X, Y \otimes Z] &= Y \otimes [X, Z]. \end{aligned} \tag{4.27}$$

For a fixed $r \in \mathfrak{g} \otimes \mathfrak{g}$ the action (4.26) leads to a linear map

$$\begin{aligned} \delta r : \mathfrak{g} &\to \mathfrak{g} \otimes \mathfrak{g} \\ X &\mapsto [X \otimes 1 + 1 \otimes X, r]. \end{aligned}$$

Let us suppose that the dual of δr defines a Lie algebra structure

$$[\cdot\,,\cdot]_r : \mathfrak{g}^* \times \mathfrak{g}^* \to \mathfrak{g}^*$$

on $\mathfrak{g}^*$ (rather general sufficient conditions for this to happen will be given shortly); for $\xi, \eta \in \mathfrak{g}^*$ and $X \in \mathfrak{g}$, this means by definition that

$$\langle [\xi, \eta]_r, X \rangle = \langle \xi \wedge \eta, \delta r(X) \rangle = \langle \xi \wedge \eta, [X \otimes 1 + 1 \otimes X, r] \rangle \,. \tag{4.28}$$

Then $\mathfrak{g}$ inherits a linear Poisson structure $\{\cdot\,,\cdot\}_r$ from $[\cdot\,,\cdot]_r$, since $\mathfrak{g} \cong \mathfrak{g}^{**}$ is then the dual of a Lie algebra. Explicitly, if $F, G \in \mathcal{F}(\mathfrak{g})$ then

$$\begin{aligned} \{F, G\}_r (X) &= \langle [\mathsf{d}F(X), \mathsf{d}G(X)]_r, X \rangle \\ &= \langle \mathsf{d}F(X) \wedge \mathsf{d}G(X), [X \otimes 1 + 1 \otimes X, r] \rangle \,. \end{aligned}$$

where $\mathsf{d}F(X)$ and $\mathsf{d}G(X)$ are interpreted as elements of $\mathfrak{g}^*$. In particular, if we take for F and G elements ξ and η of $\mathfrak{g}^*$ then this formula becomes

$$\{\xi,\eta\}_r(X) = \langle \xi\wedge\eta, [X\otimes 1 + 1\otimes X, r]\rangle . \tag{4.29}$$

In order to get sufficient conditions for $[\cdot\,,\cdot]_r$ to define a Lie algebra structure on $\mathfrak{g}^*$, let us denote by r^+ (resp. r^-) the symmetric (resp. skew-symmetric) part of r. Notice that $\delta r^+(X)$ (resp. $\delta r^-(X)$) is a symmetric (resp. skew-symmetric) element of $\mathfrak{g}\otimes\mathfrak{g}$, for any $X\in\mathfrak{g}$. Indeed, for a symmetric/skew-symmetric element $s_\pm = Y\otimes Z \pm Z\otimes Y$ of $\mathfrak{g}\otimes\mathfrak{g}$ and $X\in\mathfrak{g}$ we have that

$$\begin{aligned}\delta s_\pm(X) &= [X\otimes 1 + 1\otimes X, Y\otimes Z \pm Z\otimes Y]\\ &= [X,Y]\otimes Z \pm Z\otimes [X,Y] + Y\otimes [X,Z] \pm [X,Z]\otimes Y,\end{aligned}$$

showing our claim. It follows that skew-symmetry of $[\cdot\,,\cdot]_r$ is guaranteed if we assume r to be skew-symmetric, $r = r^-$, which we will do from now on. In order to analyze the Jacobi identity for $[\cdot\,,\cdot]_r$ we introduce the linear map $\underline{r}:\mathfrak{g}^*\to\mathfrak{g}$, which is defined for $\xi\in\mathfrak{g}^*$ by

$$\langle \eta, \underline{r}(\xi)\rangle = \langle \xi\wedge\eta, r\rangle . \tag{4.30}$$

In terms of $\underline{r}$ we get the following explicit formula for $[\cdot\,,\cdot]_r$.

Lemma 4.40. *Let $r\in\mathfrak{g}\otimes\mathfrak{g}$ be skew-symmetric. For any $\xi,\eta\in\mathfrak{g}^*$,*

$$[\xi,\eta]_r = \operatorname{ad}^*_{\underline{r}(\xi)}\eta - \operatorname{ad}^*_{\underline{r}(\eta)}\xi. \tag{4.31}$$

Proof. Since the formula is linear in r it suffices to show it for r of the form $r = Y\wedge Z$, with $Y,Z\in\mathfrak{g}$. We do this in two steps. First we show that

$$[\xi,\eta]_{Y\wedge Z} = \langle\xi,Y\rangle \operatorname{ad}^*_Z\eta - \langle\xi,Z\rangle\operatorname{ad}^*_Y\eta - (\xi\leftrightarrow\eta). \tag{4.32}$$

To do this, take any $X\in\mathfrak{g}$ and use (4.28), (4.27) and (2.14) to get

$$\begin{aligned}\langle [\xi,\eta]_{Y\wedge Z}, X\rangle &= \langle \xi\wedge\eta, [X\otimes 1 + 1\otimes X, Y\wedge Z]\rangle\\ &= \langle \xi\wedge\eta, [X,Y]\wedge Z + Y\wedge [X,Z]\rangle\\ &= \langle\xi,Y\rangle\langle\eta,[X,Z]\rangle - \langle\xi,Z\rangle\langle\eta,[X,Y]\rangle - (\xi\leftrightarrow\eta)\\ &= \langle\xi,Y\rangle\langle\operatorname{ad}^*_Z\eta, X\rangle - \langle\xi,Z\rangle\langle\operatorname{ad}^*_Y\eta,X\rangle - (\xi\leftrightarrow\eta).\end{aligned}$$

This shows (4.32). In order to show (4.31) it suffices now to show that

$$\operatorname{ad}^*_{\underline{Y\wedge Z}(\xi)}\eta = \langle\xi,Y\rangle\operatorname{ad}^*_Z\eta - \langle\xi,Z\rangle\operatorname{ad}^*_Y\eta. \tag{4.33}$$

Taking again any $X\in\mathfrak{g}$, use now (2.14) and (4.30) to compute

$$\begin{aligned}\left\langle \operatorname{ad}^*_{\underline{Y\wedge Z}(\xi)}\eta, X\right\rangle &= -\langle \operatorname{ad}^*_X\eta, \underline{Y\wedge Z}(\xi)\rangle\\ &= -\langle \xi\wedge\operatorname{ad}^*_X\eta, Y\wedge Z\rangle\\ &= -\langle\xi,Y\rangle\langle\operatorname{ad}^*_X\eta, Z\rangle + \langle\xi,Z\rangle\langle\operatorname{ad}^*_X\eta,Y\rangle\\ &= \langle\xi,Y\rangle\langle\operatorname{ad}^*_Z\eta,X\rangle - \langle\xi,Z\rangle\langle\operatorname{ad}^*_Y\eta,X\rangle .\end{aligned}$$

This establishes (4.33) and hence also (4.31). □

Proposition 4.41. *If $r \in \mathfrak{g} \otimes \mathfrak{g}$ is skew-symmetric and satisfies*

$$\langle \zeta, [\underline{r}(\xi), \underline{r}(\eta)] \rangle + \langle \xi, [\underline{r}(\eta), \underline{r}(\zeta)] \rangle + \langle \eta, [\underline{r}(\zeta), \underline{r}(\xi)] \rangle = 0, \tag{4.34}$$

for all $\zeta, \xi, \xi \in \mathfrak{g}^*$, then $[\cdot\,,\cdot]_r$ is a Lie bracket on $\mathfrak{g}^*$.

Proof. We have already shown that if r is skew-symmetric then $[\cdot\,,\cdot]_r$ is skew-symmetric. Thus, it remains to be shown that if r satisfies, in addition, (4.34) then $[\cdot\,,\cdot]_r$ satisfies the Jacobi identity

$$[[\xi,\eta]_r\,,\zeta]_r + [[\eta,\zeta]_r\,,\xi]_r + [[\zeta,\xi]_r\,,\eta]_r = 0. \tag{4.35}$$

We prove that

$$[[\xi,\eta]_r\,,\zeta]_r = \mathrm{ad}^*_{\underline{r}(\xi)}\,\mathrm{ad}^*_{\underline{r}(\eta)}\,\zeta - \mathrm{ad}^*_{\underline{r}(\zeta)}\,\mathrm{ad}^*_{\underline{r}(\xi)}\,\eta - (\xi \leftrightarrow \eta); \tag{4.36}$$

(4.35) follows from it trivially. Using Lemma 4.40 (twice) we have that

$$\begin{aligned} [[\xi,\eta]_r\,,\zeta]_r &= \left[\mathrm{ad}^*_{\underline{r}(\xi)}\,\eta,\zeta\right]_r - (\xi \leftrightarrow \eta) \\ &= \mathrm{ad}^*_{\underline{r}\left(\mathrm{ad}^*_{\underline{r}(\xi)}\,\eta\right)}\,\zeta - \mathrm{ad}^*_{\underline{r}(\zeta)}\,\mathrm{ad}^*_{\underline{r}(\xi)}\,\eta - (\xi \leftrightarrow \eta), \end{aligned}$$

so (4.36) is proven when we show that

$$\mathrm{ad}^*_{\underline{r}\left(\mathrm{ad}^*_{\underline{r}(\xi)}\,\eta\right)}\,\zeta - \mathrm{ad}^*_{\underline{r}(\xi)}\,\mathrm{ad}^*_{\underline{r}(\eta)}\,\zeta - (\xi \leftrightarrow \eta) = 0. \tag{4.37}$$

To show the latter, let $X \in \mathfrak{g}$ and use (2.14) and the skew-symmetry of r to find that

$$\begin{aligned} \left\langle \mathrm{ad}^*_{\underline{r}\left(\mathrm{ad}^*_{\underline{r}(\xi)}\,\eta\right)}\,\zeta, X \right\rangle &= -\left\langle \mathrm{ad}^*_X\,\zeta, \underline{r}\left(\mathrm{ad}^*_{\underline{r}(\xi)}\,\eta\right)\right\rangle \\ &= \left\langle \underline{r}\left(\mathrm{ad}^*_X\,\zeta\right), \mathrm{ad}^*_{\underline{r}(\xi)}\,\eta\right\rangle \\ &= \langle \eta, [\underline{r}\left(\mathrm{ad}^*_X\,\zeta\right), \underline{r}(\xi)]\rangle\,. \end{aligned}$$

If we substitute this in (4.37), evaluated at X, and we use that ad^* is a representation, $[\mathrm{ad}^*_Y, \mathrm{ad}^*_Z] = \mathrm{ad}^*_{[Y,Z]}$, for any $Y, Z \in \mathfrak{g}$, then we find

$$\begin{aligned} &\langle \eta, [\underline{r}\left(\mathrm{ad}^*_X\,\zeta\right), \underline{r}(\xi)]\rangle - \langle \xi, [\underline{r}\left(\mathrm{ad}^*_X\,\zeta\right), \underline{r}(\eta)]\rangle - \left\langle [\mathrm{ad}^*_{\underline{r}(\xi)}, \mathrm{ad}^*_{\underline{r}(\eta)}]\zeta, X\right\rangle \\ &= \langle \eta, [\underline{r}\left(\mathrm{ad}^*_X\,\zeta\right), \underline{r}(\xi)]\rangle + \langle \xi, [\underline{r}(\eta), \underline{r}\left(\mathrm{ad}^*_X\,\zeta\right)]\rangle - \left\langle \mathrm{ad}^*_{[\underline{r}(\xi),\underline{r}(\eta)]}\,\zeta, X\right\rangle \\ &= \langle \eta, [\underline{r}\left(\mathrm{ad}^*_X\,\zeta\right), \underline{r}(\xi)]\rangle + \langle \xi, [\underline{r}(\eta), \underline{r}\left(\mathrm{ad}^*_X\,\zeta\right)]\rangle + \langle \mathrm{ad}^*_X\,\zeta, [\underline{r}(\xi), \underline{r}(\eta)]\rangle \end{aligned}$$

which is zero in view of (4.34), applied to the triple $(\xi, \eta, \mathrm{ad}^*_X\,\zeta)$. □

If $r \in \mathfrak{g} \otimes \mathfrak{g}$ is skew-symmetric and satisfies (4.34) then we will say that r is an *r-matrix*, although some authors reserve this terms for the particular case in which $\mathfrak{g}$ is a Lie algebra of matrices.

We now wish to apply this formalism to Lax operators. For a manifold M and a Lie algebra $\mathfrak{g}$ we denote by $\mathfrak{g}^M$ the space of functions on M with values in $\mathfrak{g}$. Elements of $\mathfrak{g}^M$ may be viewed as matrices over the ring $\mathcal{F}(M)$, hence the Lie bracket $[\cdot\,,\cdot]$ on $\mathfrak{g}$ leads to a Lie bracket on $\mathfrak{g}^M$, which is also denoted by $[\cdot\,,\cdot]$.

Definition 4.42. Let $(M, \{\cdot\,,\cdot\})$ be a Poisson manifold and let $\mathfrak{g}$ be a Lie algebra, equipped with a Poisson bracket $\{\cdot\,,\cdot\}_{\mathfrak{g}}$. An element $L \in \mathfrak{g}^M$ is called a *Lax operator* in $(\mathfrak{g}, \{\cdot\,,\cdot\}_{\mathfrak{g}})$ if

$$L : (M, \{\cdot\,,\cdot\}) \to (\mathfrak{g}, \{\cdot\,,\cdot\}_{\mathfrak{g}})$$

is a Poisson map. When $\{\cdot\,,\cdot\}_{\mathfrak{g}}$ is the Poisson structure that comes from an r-bracket r, then we also say that L is a Lax operator with r-matrix r.

An important particular case for what follows is obtained by taking $M = \mathfrak{g}$ and $L = \mathrm{Id}_{\mathfrak{g}}$.

For $K, L \in \mathfrak{g}^M$ we define an element $\{K \overset{\otimes}{,} L\} \in (\mathfrak{g} \otimes \mathfrak{g})^M$. To do this, we pick any basis $E_1, \dots, E_N$ of $\mathfrak{g}$, with dual basis $\xi_1, \dots, \xi_N$, and we define

$$\{K \overset{\otimes}{,} L\} := \sum_{i,j=1}^{N} \{K^*\xi_i, L^*\xi_j\}\, E_i \otimes E_j, \tag{4.38}$$

which is easily seen to be independent of the chosen basis for $\mathfrak{g}$. If L is a Lax operator in $\mathfrak{g}$ with r-matrix r, as defined above, then (4.38) implies

$$\{L \overset{\otimes}{,} L\} := \sum_{i,j=1}^{N} L^* \{\xi_i, \xi_j\}_r\, E_i \otimes E_j,$$

so we find, by (4.29), that for any $m \in M$,

$$\begin{aligned} \{L \overset{\otimes}{,} L\}(m) &= \sum_{i,j=1}^{N} \langle \{\xi_i, \xi_j\}_r\,, L(m)\rangle\, E_i \otimes E_j \\ &= \sum_{i,j=1}^{N} \langle \xi_i \wedge \xi_j, [L(m) \otimes 1 + 1 \otimes L(m), r]\rangle\, E_i \otimes E_j \\ &= [L(m) \otimes 1 + 1 \otimes L(m), r]. \end{aligned}$$

This yields a proof of the following result.

Proposition 4.43. *Let $(M, \{\cdot\,,\cdot\})$ be a Poisson manifold and let $\mathfrak{g}$ be a Lie algebra. If $L \in \mathfrak{g}^M$ is a Lax operator in $\mathfrak{g}$ with r-matrix r, then*

$$\{L \overset{\otimes}{,} L\} = [L \otimes 1 + 1 \otimes L, r].$$

□

We now specialize the above construction to the case in which $\mathfrak{g} = \mathfrak{gl}(E)$, where E is a finite-dimensional vector space, whose dimension will be denoted by n. The extra feature that we will use is the fact that we can multiply (compose) elements of $\mathfrak{gl}(E)$. Thus, we are also able to multiply elements of $\mathfrak{g}^M$. Another extra feature, which is very useful for explicit computations, is that upon choosing a basis $\mathcal{E} = (e_1, \dots, e_n)$ for E, one can easily write down explicitly a $n^2 \times n^2$ matrix for $\left\{K \overset{\otimes}{,} L\right\}$ from the matrices for K and L with respect to $\mathcal{E}$. In fact, if we write the (i,j)-th entry of the matrices for K and L by K_{ij} and L_{ij} then the matrix of $\left\{K \overset{\otimes}{,} L\right\}$ with respect to the basis $e_1 \otimes e_1, e_1 \otimes e_2, \dots, e_n \otimes e_n$ is given by (recall that K_{ij} and L_{ij} are functions (on M))

$$\left\{K \overset{\otimes}{,} L\right\} = \begin{pmatrix} \{K_{11}, L_{11}\} & \dots & \{K_{11}, L_{1n}\} & \dots & \{K_{1n}, L_{11}\} & \dots & \{K_{1n}, L_{1n}\} \\ \vdots & & \vdots & & \vdots & & \vdots \\ \{K_{11}, L_{n1}\} & \dots & \{K_{11}, L_{nn}\} & \dots & \{K_{1n}, L_{n1}\} & \dots & \{K_{1n}, L_{nn}\} \\ \{K_{21}, L_{11}\} & \dots & \{K_{21}, L_{1n}\} & \dots & \{K_{2n}, L_{11}\} & \dots & \{K_{2n}, L_{1n}\} \\ \vdots & & \vdots & & \vdots & & \vdots \\ \{K_{n1}, L_{n1}\} & \dots & \{K_{n1}, L_{nn}\} & \dots & \{K_{nn}, L_{n1}\} & \dots & \{K_{nn}, L_{nn}\} \end{pmatrix} \tag{4.39}$$

The relevance of r-matrices and their Poisson brackets for the theory of integrable systems is given by the following theorem.

Theorem 4.44. *Let $(M, \{\cdot\,,\cdot\})$ be a Poisson manifold and let $\mathfrak{g} = \mathfrak{gl}(E)$, where E is a finite-dimensional vector space. If $L \in \mathfrak{g}^M$ is a Lax operator in $\mathfrak{g}$ with r-matrix r, then the functions $H_i \in \mathcal{F}(M)$, which are defined by $H_i(m) := \operatorname{Trace} L^i(m)$, for $m \in M$, are in involution with respect to $\{\cdot\,,\cdot\}$, where $i \in \mathbf{N}^*$. Moreover, the Hamiltonian vector field associated to H_i has the Lax form*

$$\mathcal{X}_{H_i} L = \left[i \operatorname{Trace}_2(1 \otimes L^{i-1} r), L\right].$$

Proof. We first establish a few formulas[4] for computing with $\left\{\cdot \overset{\otimes}{,} \cdot\right\}$. It is seen at once from (4.39) that the Leibniz property for $\{\cdot\,,\cdot\}$ implies the following Leibniz properties for $\left\{\cdot \overset{\otimes}{,} \cdot\right\}$,

$$\begin{aligned} \left\{K_1 K_2 \overset{\otimes}{,} L\right\} &= K_1 \otimes 1 \left\{K_2 \overset{\otimes}{,} L\right\} + \left\{K_1 \overset{\otimes}{,} L\right\} K_2 \otimes 1, \\ \left\{L \overset{\otimes}{,} K_1 K_2\right\} &= 1 \otimes K_1 \left\{L \overset{\otimes}{,} K_2\right\} + \left\{L \overset{\otimes}{,} K_1\right\} 1 \otimes K_2, \end{aligned} \tag{4.40}$$

where $K_1, K_2, L \in \mathfrak{g}^M$.

[4] We minimize the use of brackets in the formulas, but they remain unambiguous: for example, if $r = a \otimes b$ then $1 \otimes L r = a \otimes (Lb)$.

Also, notice that for any $\alpha \in \mathfrak{g} \otimes \mathfrak{g}$ and for any integer k we have that

$$\begin{aligned} L^p \otimes 1[\alpha, L \otimes 1 + 1 \otimes L] &= [L^p \otimes 1\, \alpha, L \otimes 1 + 1 \otimes L], \\ [\alpha, L \otimes 1 + 1 \otimes L] L^p \otimes 1 &= [\alpha\, L^p \otimes 1\,, L \otimes 1 + 1 \otimes L]. \end{aligned} \tag{4.41}$$

The proof follows trivially from the fact that $L^p \otimes 1$ commutes with $L \otimes 1$ as well as with $1 \otimes L$.

We use these formulas to show that if $L \in \mathfrak{g}^M$ is a Lax operator in $\mathfrak{g}$ with r-matrix r, then

$$\left\{L^p \overset{\otimes}{,} L^q\right\} = \sum_{i=0}^{p-1} \sum_{j=0}^{q-1} \left[L^{p-i-1} \otimes L^{q-j-1} r L^i \otimes L^j, L \otimes 1 + 1 \otimes L\right]. \tag{4.42}$$

The proof goes by induction on p and q; let us just show that if the formula is true for (p, q) and for $(1, q)$ then it is true for $(p+1, q)$. Indeed, using (4.40) and (4.41),

$$\begin{aligned} \left\{L^{p+1} \overset{\otimes}{,} L^q\right\} &= L \otimes 1 \left\{L^p \overset{\otimes}{,} L^q\right\} + \left\{L \overset{\otimes}{,} L^q\right\} L^p \otimes 1 \\ &= \sum_{i=0}^{p-1} \sum_{j=0}^{q-1} \left[L^{p-i} \otimes L^{q-j-1} r L^i \otimes L^j, L \otimes 1 + 1 \otimes L\right] \\ &\qquad + \sum_{j=0}^{q-1} \left[1 \otimes L^{q-j-1} r L^p \otimes L^j, L \otimes 1 + 1 \otimes L\right] \\ &= \sum_{i=0}^{p} \sum_{j=0}^{q-1} \left[L^{p-i} \otimes L^{q-j-1} r L^i \otimes L^j, L \otimes 1 + 1 \otimes L\right]. \end{aligned}$$

As a corollary, notice that each term in $\left\{L^p \overset{\otimes}{,} L^q\right\}$ is a commutator, hence the trace of $\left\{L^p \overset{\otimes}{,} L^q\right\} = 0$ is zero, for any $p, q \geqslant 1$. But

$$\operatorname{Trace}\left\{K \overset{\otimes}{,} L\right\} = \{\operatorname{Trace} K, \operatorname{Trace} L\},$$

for any $K, L \in \mathfrak{g}^M$, as is most easily seen from (4.39). It follows that for any $p, q \geqslant 1$,

$$\{\operatorname{Trace} L^p, \operatorname{Trace} L^q\} = \operatorname{Trace}\left\{L^p \overset{\otimes}{,} L^q\right\} = 0.$$

We use the above formula (4.42) to show that the Hamiltonian vector fields $\mathcal{X}_{H_i}$ can be written in Lax form. To do this, we define a linear map $\operatorname{Trace}_2 : \mathfrak{g} \otimes \mathfrak{g} \to \mathfrak{g}$, *trace in the second component*, which is given for $a \otimes b$ by $\operatorname{Trace}_2(a \otimes b) = a \operatorname{Trace}(b)$. For example, if $X, Y \in \mathfrak{g}$ then

$$\operatorname{Trace}_2[X \otimes Y, 1 \otimes L] = X \operatorname{Trace}[Y, L] = 0.$$

Now $\mathcal{X}_{H_i} L$ is a matrix whose entry at position (j, k) is given by

$$\mathcal{X}_{H_i} L_{jk} = \left\{L_{jk}, \operatorname{Trace}(L^i)\right\} = \left(\operatorname{Trace}_2 \left\{L \overset{\otimes}{,} L^i\right\}\right)_{jk}.$$

It follows, using (4.42), that

$$\begin{aligned}\mathcal{X}_{H_i} L &= \mathrm{Trace}_2 \left\{ L \overset{\otimes}{,} L^i \right\} \\ &= \sum_{j=0}^{i-1} \mathrm{Trace}_2 \left[1 \otimes L^{i-j-1}\, r\, 1 \otimes L^j, L \otimes 1 \right] \\ &= \left[i\, \mathrm{Trace}_2 (1 \otimes L^{i-1} r), L \right].\end{aligned}$$

In order to prove the last equality, the reader may first assume that r is a decomposable tensor, i.e., that r is of the form $X \otimes Y$, with $X, Y \in \mathfrak{g}$ and then write the bracket as a commutator; the proof for general r then follows from the fact that any r is a finite sum of decomposable tensors. □

The fact that the traces of the powers of L are in involution can also be restated by saying that the coefficients of the characteristic polynomial of L, which are elements of $\mathcal{F}(M)$, are in involution or by saying that the eigenvalues of L are in involution (the latter may, strictly speaking, be only smooth or holomorphic on a dense open subset of M).

Suppose that $(M, \{\cdot\,,\cdot\})$ is a Poisson manifold and that $L \in \mathfrak{g}^M$, where $\mathfrak{g} = \mathfrak{gl}(E)$ as above. If there exist functions $A, B \in (\mathfrak{g} \otimes \mathfrak{g})^M$ such that

$$\left\{ L \overset{\otimes}{,} L \right\} = [1 \otimes L, A] + [L \otimes 1, B] \tag{4.43}$$

then the traces of the powers of L are also in involution. The proof is essentially the same as the proof of the above theorem. In the theorem $A = B = r$; moreover, A and B are elements of $\mathfrak{g} \otimes \mathfrak{g}$, i.e., they are constant functions $M \to \mathfrak{g} \otimes \mathfrak{g}$. We show now, following [25] that, conversely, if the traces of L are in involution then — under some genericity assumption of L — there exist functions A, B, defined on a dense open subset of M, such that (4.43) holds.

Theorem 4.45 (Babelon-Viallet). *Let $(M, \{\cdot\,,\cdot\})$ is a Poisson manifold and let $L \in \mathfrak{g}^M$, where $\mathfrak{g} = \mathfrak{gl}(E)$ and E is a finite-dimensional vector space. Suppose that*

(1) There exists an open dense subset U of M such that $L(m)$ is diagonalizable for all $m \in U$;
(2) The coefficients of the characteristic polynomial of L are in involution.

Then there exists an open dense subset $V \subseteq U$ of M, and smooth functions $A, B : V \to \mathfrak{g} \otimes \mathfrak{g}$, such that

$$\left\{ L \overset{\otimes}{,} L \right\} = [1 \otimes L, A] + [L \otimes 1, B].$$

Proof. By assumption, we can diagonalize L on U. By passing to a smaller subset V, which can still be chosen open and dense, this implies that we have a smooth map (the diagonalizing map) $S : V \to \mathbf{GL}(E)$, such that $L = S^{-1} \Lambda S$, where Λ is a diagonal matrix.

Now $\left\{L \overset{\otimes}{,} L\right\} = \left\{S^{-1}\Lambda S \overset{\otimes}{,} S^{-1}\Lambda S\right\}$, and the latter is computed by using the derivation property (4.40). This gives *only* 8 terms, since $\left\{\Lambda \overset{\otimes}{,} \Lambda\right\} = 0$. Each of them contains either $\left\{\Lambda \overset{\otimes}{,} S\right\}$ or $\left\{S \overset{\otimes}{,} \Lambda\right\}$ or $\left\{S \overset{\otimes}{,} S\right\}$. We write each of them in the form $[1 \otimes L, \star]$ or $[L \otimes 1, \star]$. First, for the (two) terms that contain $\left\{\Lambda \overset{\otimes}{,} S\right\}$ we have explicitly

$$\begin{aligned} S^{-1} \otimes (S^{-1}\Lambda)\left\{\Lambda \overset{\otimes}{,} S\right\} S \otimes 1 - S^{-1} \otimes S^{-1}\left\{\Lambda \overset{\otimes}{,} S\right\} S \otimes (S^{-1}\Lambda S)& \\ = (1 \otimes (S^{-1}\Lambda S))(S^{-1} \otimes S^{-1})\left\{\Lambda \overset{\otimes}{,} S\right\} S \otimes 1& \\ -(S^{-1} \otimes S^{-1})\left\{\Lambda \overset{\otimes}{,} S\right\}(S \otimes 1)(1 \otimes (S^{-1}\Lambda S))& \\ = \left[1 \otimes (S^{-1}\Lambda S), S^{-1} \otimes S^{-1}\left\{\Lambda \overset{\otimes}{,} S\right\} S \otimes 1\right]& \\ = \left[1 \otimes L, S^{-1} \otimes S^{-1}\left\{\Lambda \overset{\otimes}{,} S\right\} S \otimes 1\right].& \end{aligned}$$

Similarly, the (two) terms that contain $\left\{S \overset{\otimes}{,} \Lambda\right\}$ are given by

$$\left[L \otimes 1, S^{-1} \otimes S^{-1}\left\{S \overset{\otimes}{,} \Lambda\right\} 1 \otimes S\right].$$

The four remaining terms, which contain $\left\{S \overset{\otimes}{,} S\right\}$, are given by

$$\begin{aligned} &(S^{-1}\Lambda) \otimes (S^{-1}\Lambda)\left\{S \overset{\otimes}{,} S\right\} + S^{-1} \otimes S^{-1}\left\{S \overset{\otimes}{,} S\right\}(S^{-1}\Lambda S) \otimes (S^{-1}\Lambda S) - \\ &(S^{-1}\Lambda) \otimes S^{-1}\left\{S \overset{\otimes}{,} S\right\} 1 \otimes (S^{-1}\Lambda S) - S^{-1} \otimes (S^{-1}\Lambda)\left\{S \overset{\otimes}{,} S\right\}(S^{-1}\Lambda S) \otimes 1 \end{aligned}$$

which can be rewritten in the following way:

$$\begin{aligned} &[(S^{-1}\Lambda S) \otimes 1, S^{-1} \otimes (S^{-1}\Lambda)\left\{S \overset{\otimes}{,} S\right\}] - \\ &\quad [(S^{-1}\Lambda S) \otimes 1, S^{-1} \otimes S^{-1}\left\{S \overset{\otimes}{,} S\right\} 1 \otimes (S^{-1}\Lambda S)] \\ &= \left[L \otimes 1, S^{-1} \otimes (LS^{-1})\left\{S \overset{\otimes}{,} S\right\} - S^{-1} \otimes S^{-1}\left\{S \overset{\otimes}{,} S\right\} 1 \otimes L\right] \\ &= \left[L \otimes 1, \left[1 \otimes L, S^{-1} \otimes S^{-1}\left\{S \overset{\otimes}{,} S\right\}\right]\right]. \end{aligned}$$

Notice that, in view of the Jacobi identity and because $[L \otimes 1, 1 \otimes L] = 0$, the latter can also be written as

$$\left[1 \otimes L, \left[L \otimes 1, S^{-1} \otimes S^{-1}\left\{S \overset{\otimes}{,} S\right\}\right]\right].$$

This shows that the eight terms in $\left\{L \overset{\otimes}{,} L\right\}$ can be written in the desired commutator form. □

Example 4.46. We illustrate the Babelon-Viallet Theorem by a simple non-linear example. Consider the quadratic Poisson structure

$$\left\{L \overset{\otimes}{,} L\right\} = [L \otimes L, r] \tag{4.44}$$

where

$$r := \begin{pmatrix} 0 & 0 & 0 & 0 \\ 0 & 0 & 1 & 0 \\ 0 & -1 & 0 & 0 \\ 0 & 0 & 0 & 0 \end{pmatrix} \qquad \text{and} \qquad L := \begin{pmatrix} P & Q \\ R & S \end{pmatrix}.$$

Explicitly, (4.44) is given by

$$\begin{pmatrix} \{P,P\} & \{P,Q\} & \{Q,P\} & \{Q,Q\} \\ \{P,R\} & \{P,S\} & \{Q,R\} & \{Q,S\} \\ \{R,P\} & \{R,Q\} & \{S,P\} & \{S,Q\} \\ \{R,R\} & \{R,S\} & \{S,R\} & \{S,S\} \end{pmatrix}$$
$$= \left[\begin{pmatrix} P^2 & PQ & QP & Q^2 \\ PR & PS & QR & QS \\ RP & RQ & SP & SQ \\ R^2 & RS & SR & S^2 \end{pmatrix}, \begin{pmatrix} 0 & 0 & 0 & 0 \\ 0 & 0 & 1 & 0 \\ 0 & -1 & 0 & 0 \\ 0 & 0 & 0 & 0 \end{pmatrix} \right]$$
$$= \begin{pmatrix} 0 & -PQ & PQ & 0 \\ -PR & -2QR & 0 & -QS \\ PR & 0 & 2QR & QS \\ 0 & -RS & RS & 0 \end{pmatrix}.$$

It is easy verified that the Jacobi identity is satisfied and that the coefficients of the characteristic polynomial of L are in involution (in fact, $\det L$ is a Casimir, and the above Poisson structure is most naturally interpreted as (an extension of) a Poisson structure on $\mathbf{GL}(2)$ or on $\mathbf{SL}(2)$, which makes them into Lie-Poisson groups). According to the Babelon-Viallet Theorem there must exist matrices A and B such that

$$\{L \overset{\otimes}{,} L\} = [1 \otimes L, A] + [L \otimes 1, B]. \tag{4.45}$$

A simple example of such matrices is given by

$$A = \begin{pmatrix} 0 & 0 & 0 & 0 \\ 0 & 0 & P & 0 \\ 0 & -P & 0 & -Q \\ 0 & 0 & R & 0 \end{pmatrix} \quad \text{and} \quad B = \begin{pmatrix} P & 0 & Q & 0 \\ 0 & P & P & Q \\ R & -P & S & 0 \\ 0 & -R & 0 & P \end{pmatrix},$$

as is easily checked by direct computation, using

$$1 \otimes L = \begin{pmatrix} P & Q & 0 & 0 \\ R & S & 0 & 0 \\ 0 & 0 & P & Q \\ 0 & 0 & R & S \end{pmatrix} \qquad L \otimes 1 = \begin{pmatrix} P & 0 & Q & 0 \\ 0 & P & 0 & Q \\ R & 0 & S & 0 \\ 0 & R & 0 & S \end{pmatrix}.$$

The matrices A and B that satisfy (4.45) are, of course, not unique.

Part II

Algebraic Completely Integrable Systems

5 The Geometry of Abelian Varieties

In this chapter we give a down-to-earth introduction to the theory of Abelian varieties, with the purpose of making this book reasonably self-contained and accessible to the applied community. The basic algebraic-geometric tools that we present here are well-known and can be found scattered around in the excellent books by Fay [52], Lange-Birkenhake [105], Mumford [129], Weil [172], and especially Griffiths and Harris [69].

5.1 Algebraic Varieties versus Complex Manifolds

In this section we fix the algebraic-geometric terminology and notations that we will use and we recall the basic correspondence between divisors and line bundles, with emphasis on the Kodaira map, which embeds, under certain conditions, a complex manifold into projective space.

5.1.1 Notations and Terminology

We will assume that the reader is familiar with the basic definitions of complex manifolds and algebraic varieties. Our convention is that an *affine variety* is an irreducible closed subset of $\mathbf{C}^N$ (some authors do not assume it to be irreducible). We say that a complex manifold, which we always assume to be connected, is *projective* if it admits a *projective embedding*, i.e., a holomorphic embedding into projective space (projective embeddings will be studied in Paragraph 5.1.3). An affine (resp. projective) variety of dimension 1 is called an *affine curve* (resp. *projective curve*), while any 1-dimensional subvariety of a general algebraic variety (for example the support of a divisor in an algebraic surface) is simply called a *curve.* An *analytic curve* in a complex manifold M is by definition a holomorphic map from an open subset of $\mathbf{C}$ into M. We will sometimes remove from an affine variety $\mathcal{A}$ in $\mathbf{C}^n$ the zero locus of a polynomial $P \in \mathbf{C}[x_1, \dots, x_n]$. The resulting algebraic variety is often called a quasi-projective variety, but we will still call it an affine variety, because it can be naturally viewed as a closed subset of $\mathbf{C}^{n+1}$, by adding to the polynomial equations $P_i = 0$ for $\mathcal{A}$ (where $P_i \in \mathbf{C}[x_1, \dots, x_n]$) the equation $x_{n+1}P = 1$.

On a complex manifold M we will use, besides the constant sheaves $\mathbf{Z}, \mathbf{Q}, \mathbf{R}$ and $\mathbf{C}$, the sheaves, defined for any connected open subset U by

$$\begin{aligned}
\mathcal{O}_M(U) &= \text{the ring (algebra) of holomorphic functions on } U,\\
\mathcal{M}_M(U) &= \text{the field of meromorphic functions on } U,\\
\mathcal{O}^*_M(U) &= \text{the group of nowhere vanishing holomorphic functions on } U,\\
\mathcal{M}^*_M(U) &= \text{the group of non-zero elements in } \mathcal{M}(U),\\
\Omega^p_M(U) &= \text{the vector space of holomorphic } p\text{-forms on } U.
\end{aligned}$$

Another sheaf, introduced in Paragraph 5.1.3, is the sheaf $\mathcal{O}_M(\mathcal{D})$ of a divisor $\mathcal{D}$. For an open subset $U \subseteq M$ the group $\mathcal{O}_M(\mathcal{D})(U)$ is defined by

$$\mathcal{O}_M(\mathcal{D})(U) = \text{the } \mathcal{O}_M(U)\text{-module of holomorphic sections of } [\mathcal{D}],$$

where $[\mathcal{D}]$ denotes the line bundle associated to $\mathcal{D}$. Putting $\mathcal{L} := [\mathcal{D}]$ we also write $\mathcal{O}(\mathcal{L})$ for $\mathcal{O}(\mathcal{D})$. We also use the common notation $\Omega^p(\mathcal{L})$ for the sheaf $\Omega^p \otimes \mathcal{L}$, where $p \in \mathbf{N}$, and we write $\Omega^p(\mathcal{D})$ for $\Omega^p(\mathcal{L})$ if $\mathcal{L} = [\mathcal{D}]$. For a sheaf $\mathcal{S}_M$ on M the i-th Čech cohomology group is denoted by $H^i(M, \mathcal{S})$, since it is clear that we are talking about a sheaf on M, and we use the standard abbreviation $h^i(M, \mathcal{S})$ for $\dim H^i(M, \mathcal{S})$. For a concise introduction to sheaves and cohomology, see [69, Section 0.3].

5.1.2 Divisors and Line Bundles

We first briefly recall from [69, Section 1.1] how meromorphic functions, divisors and line bundles on a complex manifold M are related. Everything is encoded in part of the long exact Čech cohomology sequence associated to the exact sequence of sheafs on M

$$0 \longrightarrow \mathcal{O}^*_M \longrightarrow \mathcal{M}^*_M \longrightarrow \mathcal{M}^*_M/\mathcal{O}^*_M \longrightarrow 0.$$

The portion of interest is given by

$$\cdots \longrightarrow H^0(M, \mathcal{M}^*) \longrightarrow H^0(M, \mathcal{M}^*/\mathcal{O}^*) \longrightarrow H^1(M, \mathcal{O}^*) \longrightarrow \cdots \tag{5.1}$$

i.e.,

$$\cdots \longrightarrow \mathcal{M}^*(M) \longrightarrow \operatorname{Div}(M) \longrightarrow \operatorname{Pic}(M) \longrightarrow \cdots$$

First, $H^0(M, \mathcal{M}^*)$ consists of the global sections of $\mathcal{M}^*_M$, i.e., the group $\mathcal{M}^*(M) := \mathcal{M}^*_M(M)$ of meromorphic functions on M that are not identically zero. Second, $H^0(M, \mathcal{M}^*/\mathcal{O}^*)$ is naturally isomorphic to the group $\operatorname{Div}(M)$ of divisors on M: a *divisor* $\mathcal{D}$ on M is a formal, locally finite[1] sum

$$\mathcal{D} = \sum_V a_V V, \tag{5.2}$$

[1] Since we will only consider divisors on compact complex manifolds M the sum will actually be finite.

where the sum is over all irreducible hypersurfaces (codimension 1 analytic subvarieties) of M and each coefficient a_V is an integer. Given an irreducible hypersurface V of M, we can find an open cover $\{U_\alpha\}_{\alpha\in I}$ of M such that for each $\alpha \in I$ the intersection $U_\alpha \cap V$ is given as the zero locus of a holomorphic function f_α on U_α. The collection of meromorphic functions $\{f_\alpha^{a_V}\}_{\alpha\in I}$ is clearly a global section of $\mathcal{M}_M^*/\mathcal{O}_M^*$ since on the overlap $U_{\alpha\beta} := U_\alpha \cap U_\beta$ the quotient $(f_\alpha/f_\beta)^{a_V}$ is holomorphic and nowhere zero. Reversing the argument one associates to a global section of $\mathcal{M}_M^*/\mathcal{O}_M^*$ a divisor on M by taking the zero and pole locus of the meromorphic functions that represent the section locally on each U_α. Thus, $H^0(M, \mathcal{M}^*/\mathcal{O}^*)$ and $\mathrm{Div}(M)$ are naturally isomorphic and the first part of (5.1) is, under this isomorphism, given by

$$\mathcal{M}^*(M) \longrightarrow \mathrm{Div}(M)$$

and the homomorphism between these groups is the map that associates to a meromorphic function f the formal difference between its divisor of zeros and its divisor of poles, denoted (f). Recall also that a divisor $\mathcal{D} = \sum a_V V$ is called an *effective divisor* when all a_V are non-negative. This is often denoted as $\mathcal{D} \geqslant 0$, leading to a partial ordering on $\mathrm{Div}(M)$ by putting $\mathcal{D} \geqslant \mathcal{D}'$ whenever $\mathcal{D} - \mathcal{D}' \geqslant 0$. If we denote the coefficients in the expansions (5.2) of $\mathcal{D}$ and of $\mathcal{D}'$ by a_V and a'_V then the inequality $\mathcal{D} \geqslant \mathcal{D}'$ means precisely that $a_V \geqslant a'_V$ for all irreducible hypersurfaces V of M.

Next, the group $H^1(M, \mathcal{O}^*)$ is naturally isomorphic to the *Picard group* of M, denoted $\mathrm{Pic}(M)$, which is by definition the group of (isomorphism classes of) holomorphic line bundles on M. Indeed, given a line bundle $\mathcal{L}$ on M, we can find an open cover $\{U_\alpha\}_{\alpha\in I}$ of M such that for each $\alpha \in I$ there exist *local trivializations*

$$\phi_\alpha : \mathcal{L} \longrightarrow U_\alpha \times \mathbf{C}$$

and holomorphic *transition functions*

$$g_{\alpha\beta} : U_\alpha \cap U_\beta \longrightarrow \mathbf{C}^*,$$

which are defined by $\phi_\alpha\phi_\beta^{-1}(p, 1) = (p, g_{\alpha\beta}(p))$, for $p \in U_\alpha \cap U_\beta$. The collection of transition functions $g_{\alpha\beta}$ define a Čech 1-cochain, which is a Čech 1-cocycle because they satisfy

$$g_{\alpha\beta}g_{\beta\alpha} = 1, \qquad g_{\alpha\beta}g_{\beta\gamma}g_{\gamma\alpha} = 1$$

on $U_\alpha \cap U_\beta$ resp. on $U_\alpha \cap U_\beta \cap U_\gamma$. But this cocycle is only defined up to a coboundary, because we can replace each of the ϕ_α by $\phi_\alpha f_\alpha$, where each f_α is a nowhere vanishing holomorphic function on U_α. Indeed, the corresponding new transition functions will differ by the coboundary of the 0-cocycle formed by the f_α.

Reversing the argument shows that $H^1(M, \mathcal{O}^*)$ and $\text{Pic}(M)$ are naturally isomorphic. Thus, the second part of (5.1) is given by $\text{Div}(M) \longrightarrow \text{Pic}(M)$ and the (connecting) homomorphism associates to a divisor $\mathcal{D}$ on M, given by local defining functions f_α, the line bundle whose transition functions are defined by $g_{\alpha\beta} := f_\alpha / f_\beta$. We will denote the line bundle which corresponds to $\mathcal{D}$ by $[\mathcal{D}]$. We will denote the group operation in $\text{Pic}(M)$ by $\otimes$ because it corresponds to the tensor product of line bundles, and we write tensor powers as simple exponents. The fact that $[\cdot] : \text{Div}(M) \to \text{Pic}(M)$ is a homomorphism then leads to the following formulas

$$[\mathcal{D} + \mathcal{D}'] = [\mathcal{D}] \otimes [\mathcal{D}'] \qquad [k\mathcal{D}] = [\mathcal{D}]^k,$$

for $\mathcal{D}, \mathcal{D}' \in \text{Div}(M)$ and $k \in \mathbf{Z}$.

The final piece of information contained in (5.1) is exactness at the level of $H^0(M, \mathcal{M}^*/\mathcal{O}^*) \cong \text{Div}(M)$: it says that $[\mathcal{D}]$ is (isomorphic to) the trivial line bundle $M \times \mathbf{C}$ on M precisely when $\mathcal{D}$ is a *principal divisor*, i.e., when $\mathcal{D}$ is the divisor of a meromorphic function. A useful notion in this context is linear equivalence of divisors: two divisors $\mathcal{D}$ and $\mathcal{D}'$ are *linearly equivalent*, denoted $\mathcal{D} \sim \mathcal{D}'$ when $\mathcal{D} - \mathcal{D}'$ is a principal divisor. Thus, the exactness of (5.1) at the level of $H^0(M, \mathcal{M}^*/\mathcal{O}^*)$ says that divisors lead to isomorphic line bundles, precisely when they are linearly equivalent. For example, let ω_1 and ω_2 be two holomorphic n-forms on M, where $\dim M = n$ (such an n-form is also called a *top-form*). There exists a meromorphic function f on M, such that $\omega_1 = f\omega_2$, sometimes written as $f = \omega_1/\omega_2$. Then $(\omega_1) - (\omega_2) = (f)$ so that the zero locus of ω_1 and of ω_2 are linearly equivalent divisors, $(\omega_1) \sim (\omega_2)$; conversely, if ω is a top-form then any effective divisor which is linearly equivalent to (ω) is the zero locus of a top-form on M. It follows that all top-forms on M define one and the same line bundle on M. This line bundle is called the *canonical bundle* of M, denoted $\mathcal{K}_M$, while any divisor (ω), with ω a top-form on M, is called a *canonical divisor*. Holomorphic sections of $\mathcal{K}_M$ are holomorphic n-forms, so that $\mathcal{O}_M(\mathcal{K}_M) = \Omega^n_M$, which implies that we have, for any line bundle $\mathcal{L}$ on M, a natural sheaf isomorphism

$$\mathcal{O}_M(\mathcal{L}) \cong \Omega^n_M(\mathcal{L} \otimes \mathcal{K}_M^{-1}). \tag{5.3}$$

To finish this section we list some other general facts on the cohomology of line bundles and we apply it to the Poincaré Residue map. The *holomorphic Euler characteristic* $\chi(\mathcal{L})$ of a line bundle $\mathcal{L}$ over a compact complex manifold M is the integer

$$\chi(\mathcal{L}) = \sum_{p \geqslant 0} (-1)^p h^p(M, \mathcal{O}(\mathcal{L})),$$

which is a topological invariant of $\mathcal{L}$ and M. In particular the holomorphic Euler characteristic of the trivial line bundle over M is denoted by $\chi(\mathcal{O}_M)$ and is a topological invariant of M. A very useful theorem for computing dimensions of cohomology groups which are related to holomorphic differential forms is the *Kodaira-Serre Duality Theorem*.

Theorem 5.1 (Kodaira-Serre Duality). *Let M be a compact complex manifold of dimension n. For any p, q with $0 \leqslant p, q \leqslant n$ there is a natural non-degenerate pairing*

$$H^q(M, \Omega^p) \bigotimes H^{n-q}(M, \Omega^{n-p}) \longrightarrow H^n(M, \Omega^n) \cong \mathbf{C}.$$

In particular, $h^q(M, \Omega^p) = h^{n-q}(M, \Omega^{n-p})$.

To see how this pairing comes about, let us first recall the *Dolbeault Theorem*, which states that for any $p, q \in \mathbf{N}$ there is a natural isomorphism

$$H^q(M, \Omega^p) \cong H^{p,q}_{\bar{\partial}}(M),$$

where M is any complex manifold (the Dolbeault cohomology groups $H^{p,q}_{\bar{\partial}}(M)$ are the complex analogs of the de Rham cohomology groups, see [69, Chapter 0.2]). At the level of the Dolbeault cohomology groups the pairing is given by the ordinary wedge of forms. This pairing is non-degenerate (on closed forms) when M is compact, since by the Hodge Theorem, the Dolbeault cohomology groups $H^{p,q}_{\bar{\partial}}(M)$ are in turn isomorphic to the groups $\mathcal{H}^{p,q}(M)$ of harmonic (p, q)-forms on M (when M is compact); this isomorphism is not natural since it involves the choice of a hermitian metric.

The fact that $h^q(M, \Omega^p) = h^{n-q}(M, \Omega^{n-p})$ is still valid when p or q are larger than n, because in that case all cohomology groups vanish. Thus, we have on a compact complex manifold of dimension n that $h^q(M, \Omega^p) = 0$ as soon as $q > n$ or $p > n$. In the case of Kähler manifolds, a more general statement on the vanishing of cohomology groups is given by the *Kodaira-Nakano Vanishing Theorem.*

Theorem 5.2 (Kodaira-Nakano Vanishing). *Let $\mathcal{L}$ be a positive line bundle on a compact Kähler manifold M. Then $H^q(M, \Omega^p(\mathcal{L})) = 0$ whenever $p + q > n$.*

Let us explain the elements that appear in this theorem and show how they are related. If we fix a hermitian metric on a line bundle $\mathcal{L}$ then the curvature form $\Theta_{\mathcal{L}}$ of the metric connection on $\mathcal{L}$ is a closed $(1, 1)$-form. We say that $\mathcal{L}$ is a *positive line bundle*, if there exists a metric on $\mathcal{L}$ such that the $(1, 1)$-form $(i/2\pi)\Theta_{\mathcal{L}}$ is positive. In terms of complex local coordinates $z_1, \dots, z_n$ $(n = \dim M)$ this means that

$$\Theta_{\mathcal{L}} = \sum_{i,j} h_{ij}(z) \mathrm{d}z_i \wedge \mathrm{d}\bar{z}_j,$$

where the matrix $(h_{ij}(z))$ is positive Hermitian for each z. The de Rham cohomology class of $\frac{\sqrt{-1}}{2\pi}\Theta_{\mathcal{L}}$ depends on $\mathcal{L}$ only, and not on the chosen metric. Indeed, consider the exponential sheaf sequence

$$0 \longrightarrow \mathbf{Z} \xrightarrow{i} \mathcal{O}_M \xrightarrow{\exp} \mathcal{O}^*_M \longrightarrow 0.$$

We are interested in the following part of its long exact sequence,

$$\cdots \longrightarrow H^1(M,\mathcal{O}^*) \xrightarrow{c_1} H^2(M,\mathbf{Z}) \longrightarrow H^2(M,\mathcal{O}) \longrightarrow \cdots \tag{5.4}$$

The cohomology class $c_1(\mathcal{L})$ is called the *Chern class* of $\mathcal{L}$ and it is a topological invariant of $\mathcal{L}$ because the metric was not used to construct it. The line bundles $\mathcal{L}$ on M with trivial Chern class, $c_1(\mathcal{L}) = 0$ form a subgroup of the Picard group, denoted $\mathrm{Pic}^0(M)$. The Chern class is related to the curvature form by

$$c_1(\mathcal{L}) = \left[\frac{\sqrt{-1}}{2\pi}\Theta_{\mathcal{L}}\right],$$

where the latter denotes the cohomology class of $\frac{\sqrt{-1}}{2\pi}\Theta_{\mathcal{L}}$.

Any complex manifold M admits many Hermitian metrics, i.e. Hermitian inner products on the complex tangent spaces. The real part $\Re g$ of such a metric g gives a Riemannian metric on M, while $-\frac{1}{2}\Im g$ gives a positive $(1,1)$-form ω on M. The metric g is called a *Kähler metric* when ω is closed, and a *Kähler manifold* is a complex manifold, equipped with a Kähler metric. The 2-form ω then makes M into a (real) symplectic manifold.

We now recall the Poincaré Residue map and we specialize it to the case of a Kähler manifold. Let M be a compact complex manifold of dimension n and let $\mathcal{D}$ be a non-singular analytic hypersurface of M. We define a map of sheafs

$$\mathrm{PR} : \Omega^n_M(\mathcal{D}) \longrightarrow \Omega^{n-1}_{\mathcal{D}}.$$

To do this, let $U \subseteq M$ be an open subset, small enough such that $\mathcal{D}$ admits a local defining function f on U. For $\omega \in \Omega^n_M(\mathcal{D})(U)$ a meromorphic n-form on U with at worst a simple pole along $\mathcal{D}$, let $\tilde{\omega}$ be a holomorphic $(n-1)$-form on U such that

$$\omega = \frac{\mathrm{d}f}{f} \wedge \tilde{\omega}.$$

We have that $\tilde{\omega}_{|U\cap\mathcal{D}}$ is independent of the chosen $\tilde{\omega}$, since $\mathrm{d}f_{|U\cap\mathcal{D}} = 0$, and we define $\mathrm{PR}(\omega) := \tilde{\omega}_{|U\cap\mathcal{D}}$. If $(z_1,\ldots,z_n)$ is a system of local coordinates on U then ω can be written as

$$\omega = \frac{g\,\mathrm{d}z_1 \wedge \ldots \wedge \mathrm{d}z_n}{f}$$

where g is a holomorphic function on U. Then the Poincaré Residue of ω is given by

$$\mathrm{PR}(\omega) = (-1)^{i-1}\frac{g\,\mathrm{d}z_1 \wedge \ldots \wedge \widehat{\mathrm{d}z_i} \wedge \ldots \wedge \mathrm{d}z_n}{\partial f/\partial z_i}\bigg|_{f=0}, \tag{5.5}$$

where i is chosen such that $\partial f/\partial z_i \neq 0$ (besides this, the choice of i is irrelevant, as follows easily from differentiating $f = 0$).

Applied to global sections, the Poincaré Residue map associates to any meromorphic n-form on M with a simple pole along $\mathcal{D}$ at most, a holomorphic $(n-1)$-form on $\mathcal{D}$, called its *Poincaré Residue.* From the exact sheaf sequence

$$0 \longrightarrow \Omega_M^n \longrightarrow \Omega_M^n(\mathcal{D}) \stackrel{\mathrm{PR}}{\longrightarrow} \Omega_{\mathcal{D}}^{n-1} \longrightarrow 0,$$

we get a long exact sequence

$$\begin{aligned} 0 \longrightarrow H^0(M,\Omega^n) &\longrightarrow H^0(M,\Omega^n(\mathcal{D})) \longrightarrow H^0(\mathcal{D},\Omega^{n-1}) \\ \longrightarrow H^1(M,\Omega^n) &\longrightarrow H^1(M,\Omega^n(\mathcal{D})) \longrightarrow \qquad \cdots \end{aligned} \tag{5.6}$$

If we assume that M is, in addition, a Kähler manifold, and that $[\mathcal{D}]$ is positive, then $H^1(M,\Omega^n(\mathcal{D}))=0$, in view of the Kodaira-Nakano Vanishing Theorem, so that

$$h^0(\mathcal{D},\Omega^{n-1}) = h^0(M,\Omega^n(\mathcal{D})) - h^0(M,\Omega^n) + h^1(M,\Omega^n).$$

The integer $h^0(\mathcal{D},\Omega^{n-1})$ is the dimension of the space of top-forms on $\mathcal{D}$ and is called the *geometric genus* or simply the *genus* of $\mathcal{D}$, denoted $g(\mathcal{D})$. By Kodaira-Serre Duality and by the symmetry of the Hodge diamond for compact Kähler manifolds (see [69, Section 0.7]) we have

$$H^1(M,\Omega^n) \cong H^{n-1}(M,\mathcal{O}) \cong H^0(M,\Omega^{n-1}), \tag{5.7}$$

so that the genus of $\mathcal{D}$ is given by

$$g(\mathcal{D}) = h^0(M,\Omega^n(\mathcal{D})) - h^0(M,\Omega^n) + h^0(M,\Omega^{n-1}). \tag{5.8}$$

The right hand side of this equation will be computed more explicitly later in the case when M is a complex torus.

5.1.3 Projective Embeddings of Complex Manifolds

In order to describe how line bundles lead to projective embeddings we need the notion of a section of a line bundle, which we will also rephrase in terms of divisors. Let $M=\{U_\alpha\}_{\alpha\in I}$ be a complex manifold, let $\mathcal{L}$ be a (holomorphic) line bundle on M and let $\{g_{\alpha\beta}\}_{\alpha,\beta\in I}$ be transition functions for $\mathcal{L}$. A (global) *section* of $\mathcal{L}$ is given by a set of holomorphic functions s_α defined on each U_α such that $s_\alpha = g_{\alpha\beta}s_\beta$ for $\alpha,\beta\in I$. For U an open subset of M we define $\mathcal{O}_M(\mathcal{L})(U)$ to be the sections of $\mathcal{L}_{|U}$. Equipped with the natural restriction maps, $\mathcal{O}_M(\mathcal{L})$ becomes a sheaf, the *sheaf associated to* $\mathcal{L}$. When $\mathcal{L}$ is the trivial line bundle over M then $\mathcal{O}_M(\mathcal{L})\cong\mathcal{O}_M$. Also, when $\mathcal{L}=[\mathcal{D}]$ then we write $\mathcal{O}_M(\mathcal{D})$ for $\mathcal{O}_M(\mathcal{L})$.

We can express the space $H^0(M,\mathcal{O}(\mathcal{D}))$ of global sections of $[\mathcal{D}]$ in terms of divisors by using

$$L(\mathcal{D}) := \{f\in\mathcal{M}^*(M) \mid (f)\geqslant -\mathcal{D}\}.$$

Writing $\mathcal{D} = \sum_i n_i \mathcal{D}_i$, where all analytic hypersurface $\mathcal{D}_i$ are irreducible and disjoint, $L(\mathcal{D})$ is the vector space of functions holomorphic on $M \setminus \mathcal{D}$, with at worst a pole of order n_i along $\mathcal{D}_i$ when $n_i > 0$ and a zero of order at least $-n_i$ when $n_i \leqslant 0$. Let s_0 be a holomorphic section of $[\mathcal{D}]$ for which $(s_0) = \mathcal{D}$ and let $f \in L(\mathcal{D})$. Then fs_0 provides another holomorphic section of $[\mathcal{D}]$ and every section of $[\mathcal{D}]$ is obtained in this way. This means that we have an isomorphism

$$H^0(M, \mathcal{O}(\mathcal{D})) \cong L(\mathcal{D}) \tag{5.9}$$

This isomorphism is not natural since it involves the choice of s_0.

When M is compact, the vector space $H^0(M, \mathcal{O}(\mathcal{L}))$ of sections of $\mathcal{L}$ over M is finite-dimensional, for any line bundle $\mathcal{L}$. The projective space that is associated to $H^0(M, \mathcal{O}(\mathcal{L}))$ is denoted by $\mathbf{P}H^0(M, \mathcal{O}(\mathcal{L}))$. We associate to $\mathcal{L}$ a holomorphic map from (an open subset of) M into the dual projective space $(\mathbf{P}H^0(M, \mathcal{O}(\mathcal{L})))^*$.

To describe this map, let U denote the open subset of M consisting of those points $p \in U$ for which there exists at least one section of $\mathcal{L}$ which does not vanish at p (notice that since the transition functions are non-vanishing, the question whether or not a section vanishes at p makes sense, as does the ratio of two sections at p, although sections do not have a value at p). For $p \in U$, denote by H_p those global sections of $\mathcal{O}(\mathcal{L})$ that vanish at p,

$$H_p := \left\{ s \in H^0(M, \mathcal{O}(\mathcal{L})) \mid s(p) = 0 \right\}.$$

Clearly, H_p is a hyperplane of $H^0(M, \mathcal{O}(\mathcal{L}))$, hence we can associate to it a point H_p^* of the dual projective space $(\mathbf{P}H^0(M, \mathcal{O}(\mathcal{L})))^*$. If U is non-empty then we get a holomorphic map

$$\begin{aligned} \varphi_{\mathcal{L}} : U &\to \left(\mathbf{P}H^0(M, \mathcal{O}(\mathcal{L}))\right)^* \\ p &\mapsto \quad H_p^* \end{aligned}$$

which is called the *Kodaira map* associated to $\mathcal{L}$. A point $p \in M$ at which all sections of $\mathcal{L}$ vanish is called a *base point* of $\mathcal{L}$ and the set of all base points is called the *base locus* of $\mathcal{L}$. In terms of a basis $(s_0, \dots, s_N)$ of $H^0(M, \mathcal{O}(\mathcal{L}))$ the Kodaira map $\varphi_{\mathcal{L}}$ is given explicitly given by

$$\varphi_{\mathcal{L}} : p \in U \mapsto (s_0(p) : \cdots : s_N(p)) \in \mathbf{P}^N \cong \left(\mathbf{P}H^0(M, \mathcal{O}(\mathcal{L}))\right)^*.$$

If $\mathcal{L} = [\mathcal{D}]$ and if $(1, z_1, \dots, z_N)$ is a basis of $L(\mathcal{D})$ then the restriction of $\varphi_{\mathcal{D}} := \varphi_{[\mathcal{D}]}$ to $M \setminus \mathcal{D}$ can also be written as

$$\varphi_{\mathcal{D}} : p \in M \setminus \mathcal{D} \mapsto (1 : z_1(p) : \cdots : z_N(p)) \in \mathbf{P}^N.$$

If M admits a line bundle $\mathcal{L}$ which is base point free and which defines a holomorphic embedding of M into some $\mathbf{P}^N$, then one says that M is *projective*.

This is equivalent to saying that M admits a holomorphic embedding into some projective space. Indeed, suppose that $\varphi : M \to \mathbf{P}^N$ is any holomorphic embedding of M into $\mathbf{P}^N$, such that the image of M by φ is not contained in a hyperplane of $\mathbf{P}^N$. We may pull the hyperplane bundle $\mathcal{H}$ on $\mathbf{P}^N$ ($\mathcal{H}$ is the line bundle corresponding to any hyperplane of $\mathbf{P}^N$) back to M to find a base point free (since $M \not\subseteq H$) line bundle $\mathcal{L}$ on M, such that the Kodaira map associated to $\mathcal{L}$ is precisely the given embedding (in fact, let H be a generic hyperplane of $\mathbf{P}^N$ and let $\mathcal{D} := H \cap \varphi(M)$, then $\varphi = \varphi_{\mathcal{D}}$). Therefore, the above definition is equivalent to saying that M is projective when it admits a holomorphic embedding in some projective space.

The *Kodaira Embedding Theorem* gives conditions under which $\varphi_{\mathcal{L}}$ defines an embedding.

Theorem 5.3 (Kodaira Embedding). *If $\mathcal{L}$ is a positive line bundle on a compact complex manifold M, then for some positive integer the Kodaira map $\varphi_{\mathcal{L}^k}$ embeds M into $\mathbf{P}^N$. Moreover, M possesses a positive line bundle if and only if M has a closed positive $(1,1)$-form ω with cohomology class $[\omega] \in H^2(M, \mathbf{Z})$.*

Since $c_1(\mathcal{L})$ is an integral cohomology class we have that a (positive) line bundle gives rise to a closed (positive) $(1,1)$-form ω whose cohomology class is integral. The converse is however also true: if ω is a (positive) $(1,1)$-form, whose cohomology class is integral, then there exists a (positive) line bundle on M such that $c_1(\mathcal{L}) = [\omega]$. In fact, the homomorphism $\imath_*$ in (5.4), which is induced by the inclusion (of sheafs) $\imath : \mathbf{Z} \to \mathcal{O}_M$, corresponds to taking the $(0,2)$-part of ω in the Hodge decomposition, hence $\imath_*[\omega]$ vanishes, because ω is a $(1,1)$-form. By exactness of (5.4) at $H^2(M, \mathbf{Z})$ we conclude that $[\omega]$ is in the image of c_1, i.e., $[\omega] = c_1(\mathcal{L})$ for some line bundle $\mathcal{L}$.

A line bundle $\mathcal{L}$ for which the Kodaira map $\varphi_{\mathcal{L}}$ embeds M in projective space is called a *very ample line bundle*, while it is called an *ample line bundle* when some positive power of $\mathcal{L}$ embeds. A divisor $\mathcal{D}$ on M is called (very) ample when $[\mathcal{D}]$ is (very) ample. We say that an ample line bundle $\mathcal{L}$ is a *normally generated line bundle* if the canonical map

$$\operatorname{Sym}^n H^0(M, \mathcal{L}) \longrightarrow H^0(M, \mathcal{L}^n)$$

is surjective for every $n \geqslant 2$. Letting $\mathcal{L} = [\mathcal{D}]$ this can be rephrased in terms of $L(\mathcal{D})$ as the surjectivity of the map $\operatorname{Sym}^n L(\mathcal{D}) \to L(n\mathcal{D})$ which maps a monomial, consisting of an unordered collection of elements of $L(\mathcal{D})$ simply to their product, which is an element of $L(n\mathcal{D})$. In plain words, if the line bundle $[\mathcal{D}]$ is normally generated then every function on M with a pole of order n at worst along $\mathcal{D}$ is a product of n functions with a simple pole at worst along $\mathcal{D}$. It can be shown that a line bundle which is normally generated is very ample, but the converse is not true, in general. We will encounter concrete examples of this when analyzing the algebraic geometry of certain integrable systems.

A closed positive $(1,1)$-form ω whose cohomology class $[\omega]$ belongs to $H^2(M,\mathbf{Z})$ is called a *Hodge form*. In these terms the Kodaira Embedding Theorem states that a compact complex manifold M is projective if and only if M has a Hodge form. Since the $(1,1)$-form which is associated to a Kähler metric is positive and closed, the Kodaira Embedding Theorem admits the following specialization to Kähler manifolds.

Theorem 5.4. *Let M be a compact Kähler manifold. If the cohomology class of the $(1,1)$-form of its Kähler metric is rational, then M is projective.*

Conversely any holomorphic subvariety M of $\mathbf{P}^N$ inherits a Hermitian metric from the Fubini-Study metric on $\mathbf{P}^N$. The associated $(1,1)$-form ω is positive and closed, so that M is a Kähler manifold. Since the cohomology class of ω is moreover integral, Theorem 5.4 completely characterizes non-singular projective varieties. They are also characterized in terms of the size of their field of meromorphic functions, as stated in the following theorem.

Theorem 5.5 (Moishezon). *A compact Kähler manifold of dimension n is projective if and only if it admits n algebraically independent meromorphic functions.*

The adjective "projective" for complex manifolds that can be embedded in projective space $\mathbf{P}^N$ suggests that their image in $\mathbf{P}^N$, which is a priori just a complex submanifold of $\mathbf{P}^N$, is a projective variety, i.e., the zero locus of a set of homogeneous polynomials. That this is so follows from the fundamental

Theorem 5.6 (Chow). *Any closed analytic subset of $\mathbf{P}^N$ is a projective variety.*

By an analytic subset we mean any subset that is locally given as the zero locus of a collection of holomorphic functions. The Chow Theorem is part of the G.A.G.A. principle, which states that any globally defined object in or on an algebraic variety that is locally analytic is globally algebraic (see [153]). The Chow Theorem is a consequence of the *Weierstrass Preparation Theorem*, which we will also use later. Let $\mathbf{C}\{x_1,\ldots,x_n\}$ denote the ring of convergent power series, i.e., germs of analytic functions, at $0 \in \mathbf{C}^n$.

Theorem 5.7 (Weierstrass Preparation). *If $f \in \mathbf{C}\{x_1,\ldots,x_n\}$ does not vanish identically on the X_n axis, i.e., there exist $\alpha \in \mathbf{C}^*$ and $d \in \mathbf{N}$ such that*

$$f(0,\ldots,0,x_n) = \alpha x_n^d + O\left(x_n^{d+1}\right),$$

then f admits a unique factorization

$$f = u(x_n^d + a_1 x_n^{d-1} + \cdots + a_d),$$

where $u \in \mathbf{C}\{x_1,\ldots,x_n\}$ with $u(0) \neq 0$ and $a_i \in \mathbf{C}\{x_1,\ldots,x_{n-1}\}$ with $a_i(0) = 0$ for $1 \leqslant i \leqslant d$.

As a corollary, each analytic hypersurface is locally given as the zero locus of a holomorphic function $x_n^d + a_1 x_n^{d-1} + \cdots + a_d$, which is a polynomial in (at least) one of its variables.

5.1.4 Riemann Surfaces and Algebraic Curves

A complex manifold of dimension 1 is called a *Riemann surface.* As topological spaces they are orientable surfaces, hence a compact Riemann surface Γ is a sphere with g handles. The integer g is called the *topological genus* of Γ and is related to its Euler characteristic $\chi(\Gamma)$ by $\chi(\Gamma) = 2 - 2g$. The topological genus of Γ is most often computed from the *Riemann-Hurwitz formula*, which relates the Euler characteristics of two Riemann surfaces Γ_1 and Γ_2, related by a ramified cover $\pi : \Gamma_1 \to \Gamma_2$ of degree n. The formula reads

$$\chi(\Gamma_1) = n\chi(\Gamma_2) - \sum_{p\in\Gamma_1} (\nu_\pi(p) - 1),$$

where $\nu_\pi(p)$ is the *ramification index* of π at p (to compute $\nu_\pi(p)$, write π in terms of local coordinates x and u, centered at p and at $\pi(p)$ as $u = x^d$, then d is the ramification index; it equals 1 except for a finite number of points $p \in \Gamma_1$, the *ramification points*). In particular, if $\pi : \Gamma_1 \to \Gamma_2$ is a double cover, then

$$g(\Gamma_1) = 2g(\Gamma_2) - 1 + \frac{1}{2}\#\text{branch points}. \tag{5.10}$$

In particular the number of branch points of a double cover is always even.

We claim that the topological genus of Γ equals the (geometric) genus of Γ, i.e., Γ admits precisely g independent holomorphic differentials, $g = h^0(\Gamma, \Omega^1)$. Indeed, the short exact sequence of sheaves on Γ

$$0 \longrightarrow \mathbf{C} \longrightarrow \mathcal{O}_\Gamma \stackrel{\mathsf{d}}{\longrightarrow} \Omega^1_\Gamma \longrightarrow 0,$$

gives a long exact sequence in cohomology, namely

$$\begin{aligned} 0 \longrightarrow H^0(\Gamma,\mathbf{C}) &\longrightarrow H^0(\Gamma,\mathcal{O}) \longrightarrow H^0(\Gamma,\Omega^1)\\ \longrightarrow H^1(\Gamma,\mathbf{C}) &\longrightarrow H^1(\Gamma,\mathcal{O}) \longrightarrow H^1(\Gamma,\Omega^1)\\ \longrightarrow H^2(\Gamma,\mathbf{C}) &\longrightarrow 0, \end{aligned}$$

where $H^2(\Gamma, \mathcal{O}) = 0$ follows from Kodaira-Serre Duality. The latter also implies that

$$H^0(\Gamma,\mathbf{C}) \cong H^0(\Gamma,\mathcal{O}) \cong H^1(\Gamma,\Omega^1) \cong H^2(\Gamma,\mathbf{C}) \cong \mathbf{C}$$

and that

$$H^1(\Gamma,\mathcal{O}) \cong H^0(\Gamma,\Omega^1).$$

Therefore, counting dimensions in the above exact sequence, we find

$$h^0(\Gamma,\Omega^1) = \frac{1}{2}h^1(\Gamma,\mathbf{C}) = g,$$

proving the claim.

Divisors on a compact Riemann surface Γ are particularly simple, because they are finite (formal) sums of points on Γ, i.e., $\mathrm{Div}(\Gamma)$ is the free group on Γ. There is a natural homomorphism $\deg : \mathrm{Div}(\Gamma) \to \mathbf{Z}$, defined by

$$\deg\left(\sum_{i=1}^{n} a_i \mathcal{D}_i\right) = \sum_{i=1}^{n} a_i.$$

If f is a meromorphic function on Γ then $\mathrm{d}f/f$ is a meromorphic function with residue at $p \in \Gamma$ equal to the order of vanishing of f at p so that, by Stokes' theorem

$$\deg(f) = 0, \qquad \text{for any } f \in \mathcal{M}(\Gamma).$$

As a corollary, linearly equivalent divisors on Γ have the same degree. We may therefore define the degree of a line bundle $\mathcal{L}$ on Γ as the degree of any divisor $\mathcal{D}$ for which $\mathcal{L} = [\mathcal{D}]$.

With this definition $\deg(\mathcal{L})$ corresponds to the Chern class $c_1(\mathcal{L})$ under the natural isomorphism $H^2(M, \mathbf{Z}) \cong \mathbf{Z}$, where M is equipped with its natural orientation, and $\mathcal{L}$ is positive if and only if it has positive degree. For example the canonical bundle on Γ and any canonical divisor on Γ have degree $2g-2$. In this respect, notice that since $\dim \Gamma = 1$, the sheaves Ω^1_Γ and $\mathcal{O}_\Gamma(\mathcal{K}_\Gamma)$ are naturally isomorphic.

Any compact Riemann surface Γ is a projective variety, as follows from Theorem 5.4. Indeed, it is a Kähler manifold because any metric on Γ is Kähler (the differential of its associated $(1,1)$-form ω has degree three, so it is zero). In order to make the cohomology class of ω rational (or integral) it suffices to multiply the metric by an appropriate constant (one can e.g. multiply the metric with a constant such that the integral of ω over Γ equals 1). It follows that compact Riemann surfaces and non-singular projective curves are essentially the same objects, except that a specific line bundle on Γ has to be chosen to realize Γ as a projective curve. The dimension of the space of sections of a line bundle $\mathcal{L} = [\mathcal{D}]$ on Γ is given by the *Riemann-Roch Theorem.*

Theorem 5.8 (Riemann-Roch). *Let Γ be a compact Riemann surface of genus g, let $\mathcal{D}$ be a divisor on Γ. Then the following equivalent formulas hold:*

(i) $\chi([\mathcal{D}]) = \chi(\mathcal{O}_\Gamma) + \deg \mathcal{D}$,
(ii) $h^0(\Gamma, \mathcal{O}(\mathcal{D})) = h^0(\Gamma, \Omega^1(-\mathcal{D})) + \deg \mathcal{D} - g + 1$,

We show how the equivalence of *(i)* and *(ii)* follows from the Kodaira-Serre Duality Theorem (Theorem 5.1). The holomorphic Euler characteristic $\chi(\mathcal{L})$ of $\mathcal{L}$ equals

$$\chi(\mathcal{L}) = h^0(\Gamma, \mathcal{O}(\mathcal{L})) - h^1(\Gamma, \mathcal{O}(\mathcal{L})), \tag{5.11}$$

because $h^q(\Gamma, \mathcal{O}(\mathcal{L})) = 0$ when $q > \dim \Gamma = 1$.

Furthermore, we have as in Theorem 5.1 that $H^1(\Gamma, \mathcal{O}(\mathcal{L}))$ is canonically isomorphic to the dual of $H^0(\Gamma, \Omega^1(\mathcal{L}^{-1}))$ so that

$$\chi(\mathcal{L}) = h^0(\Gamma, \mathcal{O}(\mathcal{L})) - h^0(\Gamma, \Omega^1(\mathcal{L}^{-1})).$$

Taking for $\mathcal{L}$ the line bundle $[\mathcal{D}]$ and the trivial line bundle over Γ, this gives respectively

$$\chi([\mathcal{D}]) = h^0(\Gamma, \mathcal{O}(\mathcal{D})) - h^0(\Gamma, \Omega^1(-\mathcal{D})),$$
$$\chi(\mathcal{O}_\Gamma) = h^0(\Gamma, \mathcal{O}) - h^0(\Gamma, \Omega^1) = 1 - g,$$

proving the equivalence of *(i)* and *(ii)*.

As an application of the Riemann-Roch Theorem, take for $\mathcal{D}$ a canonical divisor, $[\mathcal{D}] = \mathcal{K}_\Gamma$ and use $\Omega^1_\Gamma(-\mathcal{K}_\Gamma) \cong \mathcal{O}_\Gamma$, as follows from (5.3), to find

$$h^0(\Gamma, \mathcal{O}(\mathcal{K}_\Gamma)) = h^0(\Gamma, \mathcal{O}) + \deg \mathcal{K}_\Gamma - g + 1 = 2 - g + \deg \mathcal{K}_\Gamma.$$

Therefore,

$$\deg \mathcal{K}_\Gamma = h^0(\Gamma, \mathcal{O}(\mathcal{K}_\Gamma)) + g - 2 = h^0(\Gamma, \Omega^1) + g - 2 = 2g - 2,$$

giving the announced formula for the degree of the canonical bundle.

Most compact Riemann surface of genus $g > 2$ can be embedded in projective space in a canonical way, namely by using their canonical bundle $\mathcal{K}_\Gamma$. Suppose that Γ has genus $g \geqslant 2$ and let $(\omega_1, \ldots, \omega_g)$ be a basis of $H^0(\Gamma, \Omega^1)$. In terms of a local coordinate z, we can write $\omega_i = f_i(z)\mathrm{d}z$, and the Kodaira map $\imath_{\mathcal{K}_\Gamma}$ is given by

$$\begin{aligned} \imath_{\mathcal{K}_\Gamma} : \Gamma &\to \mathbf{P}^{g-1} \\ P &\mapsto (f_1(P) : \cdots : f_g(P)). \end{aligned}$$

This map has no base points by the Riemann-Roch Theorem, hence $\imath_{\mathcal{K}_\Gamma}$ is an embedding when it is injective and immersive. A compact Riemann surface for which the above map is not an embedding is called a *hyperelliptic Riemann surface* (a compact Riemann surfaces of genus 1 being called an *elliptic Riemann surface*), while any curve whose (compact) Riemann surface is hyperelliptic is called a *hyperelliptic curve* (one speaks of an *elliptic curve* in the genus 1 case). Hyperelliptic Riemann surfaces are characterized as follows.

Theorem 5.9. *Let Γ be a compact Riemann surface of genus $g \geqslant 2$.*

(1) Γ is hyperelliptic if and only if it has a non-constant meromorphic function f for which $(f) + P + Q \geqslant 0$, for some points $P, Q \in \Gamma$;
(2) Γ is hyperelliptic if and only if Γ is the Riemann surface of an algebraic curve, given by an affine equation of the form $y^2 = f(x)$, where f is a monic polynomial of degree $2g+1$ or $2g+2$ without multiple roots;
(3) If $g = 2$ then Γ is hyperelliptic.

A point P on a compact Riemann surface of genus g is called a *Weierstrass point* if there exists a function which has a pole of order at most g at P and which is holomorphic elsewhere. A hyperelliptic Riemann surface Γ of genus g has $2g+2$ Weierstrass points which are the points for which there is a function with a double pole in one point only: if

$$y^2 = \prod_{i=1}^{2g+2} (x - x_i)$$

is an equation for an affine model $\Gamma^{(0)}$ of Γ, then the Weierstrass points are the points $P_i := (x_i, 0)$, where $i = 1, \dots, 2g+2$, i.e., they are the branch points of the double cover $(x,y) \mapsto x$. As function with a double pole at the point P_i only, one can take the function $1/(x - x(P_i)) = 1/(x - x_i)$. For $i \neq j$ the function $(x - x(P_i))/(x - x(P_j))$ has as divisor $2P_i - 2P_j$ so that $2P_i \sim 2P_j$, for any pair of Weierstrass points (P_i, P_j) on Γ. If $\Gamma^{(0)}$ is an affine model for Γ which is such that there is only one point ∞ in $\Gamma \setminus \Gamma^{(0)}$, so that $\Gamma^{(0)}$ is given by an equation

$$y^2 = \prod_{i=1}^{2g+1} (x - x_i)$$

then ∞ is a Weierstrass point, x has a double pole at ∞ and the function $x - x_i$ realizes the linear equivalence between $2P_i$ and 2∞, where $P_i = (x_i, 0)$.

It is easy to construct explicitly a basis for $H^0(\Gamma, \Omega^1)$ if Γ is hyperelliptic: writing $y^2 = f(x)$ as above, the g independent differentials

$$\omega_i := \frac{x^{i-1}\mathrm{d}x}{y}$$

are holomorphic, as is easily verified by picking a uniformizing parameter at infinity and at the Weierstrass points of Γ. Since these differentials are independent, they form a basis of the vector space of all holomorphic differentials on the Riemann surface. Thus, Kodaira map is given, on $\Gamma^{(0)}$, by

$$\begin{aligned} \imath_{\mathcal{K}_\Gamma} : \Gamma^{(0)} &\to \mathbf{P}^{g-1} \\ P &\mapsto (1 : x : \cdots : x^{g-1}) \end{aligned}$$

and we see why in this case the Kodaira map is not an embedding: the involution which is defined on $\Gamma^{(0)}$ by $(x,y) \mapsto (x, -y)$ extends to an involution σ on Γ, called its *hyperelliptic involution*; for any $P \in \Gamma$ we obviously have that P and $\sigma(P)$ are mapped to the same point by $\imath_{\mathcal{K}_\Gamma}$. The closure of $\imath_{\mathcal{K}_\Gamma}(\Gamma^{(0)})$ is isomorphic to the projective line $\mathbf{P}$ and under this isomorphism the Kodaira map corresponds to the $2:1$ cover of $\mathbf{P}$, defined by $x \mapsto (1 : x)$.

For non-hyperelliptic Riemann surfaces, one has the following theorem.

Theorem 5.10 (Noether). *Let Γ be a compact Riemann surface which is not hyperelliptic. Its canonical bundle $\mathcal{K}_\Gamma$ is normally generated.*

We finish this paragraph with a lemma on Laurent tails that will be used in our analysis of Lax equations (Section 6.4). For $\mathcal{D}$ a divisor on Γ we introduce the quotient sheaf

$$\mathcal{O}_{\mathcal{D}}(\mathcal{D}) := \mathcal{O}_\Gamma(\mathcal{D})/\mathcal{O}_\Gamma.$$

Writing $\mathcal{D} = \sum_{i=1}^l n_i p_i$ we can write a section η of $\mathcal{O}_{\mathcal{D}}(\mathcal{D})$ as $\sum_{i=1}^l \eta_i$ where each η_i is a polynomial without constant term, and of degree at most n_i in z_i^{-1}, where z_i is any local parameter, centered at p_i. One calls each η_i a *Laurent tail.* Since $L(\mathcal{D}) = H^0(\Gamma, \mathcal{O}_\Gamma(\mathcal{D}))$, any element of $L(\mathcal{D})$ leads to a collection of Laurent tails. The following lemma deals with the converse.

Lemma 5.11. *Let Γ be a compact Riemann surface and let $\mathcal{D} = \sum_{i=1}^l n_i p_i$ be an effective divisor on Γ. A section $\eta = \sum_{i=1}^l \eta_i$ of $\mathcal{O}_{\mathcal{D}}(\mathcal{D})$ comes from an element of $L(\mathcal{D})$ if and only if*

$$\sum_{i=1}^l \operatorname{Res}_{p_i}(\eta_i \omega) = 0$$

for all holomorphic differentials ω on Γ.

One direction of the proof is immediate: if η is meromorphic on the curve Γ with $(\eta) \geqslant -\mathcal{D}$, then by the residue theorem

$$\sum_{i=1}^l \operatorname{Res}_{p_i}(\eta_i \omega) = \sum_{i=1}^l \operatorname{Res}_{p_i}(\eta \omega) = 0$$

for every holomorphic differential ω. The other direction is an application of Riemann-Roch (see [17, Chapter I.2] for a proof).

5.2 Abelian Varieties

By definition, an *Abelian variety* is a complex torus $\mathbf{C}^r/\Lambda$ (Λ a lattice in $\mathbf{C}^r$) which is projective, i.e., which can be holomorphically embedded in projective space. A finite surjective group homomorphism $\phi : A \to B$ between Abelian varieties A and B is called an *isogeny* and the two Abelian varieties A and B are said to be *isogenous Abelian varieties*, denoted $A \simeq B$. The cardinality of the kernel of ϕ is called the *degree* of ϕ, denoted $\deg \phi$; it is the degree of ϕ as a holomorphic map. We often "forget" that the Abelian variety A has an origin; then we view A merely as a homogeneous space, acted upon by A itself. In this case we may still speak of an isogeny since, as we will see, a holomorphic map ϕ between Abelian varieties A and B (without origin) becomes a group homomorphism when picking an arbitrary point P as origin on A and its image $\phi(P)$ as origin on B.

We will assume that the reader is familiar with the basic facts on complex tori, such as their characterization as connected compact complex Lie groups (yes, such groups are automatically commutative!) and the fundamental rôle that their universal covering space (which is an affine space) plays in their study. For example, a complex torus has in the neighborhood of any point a system of particularly simple coordinates, namely the coordinates that come from its universal covering space. They are unique up to a linear transformation and are called *linear coordinates*; any translation in the group is, locally, described by a linear map in terms of linear coordinates. As another example, we give three different descriptions of the vector fields on complex tori (or Abelian varieties) that are fundamental in this book.

Proposition 5.12. *For a vector field $\mathcal{V}$ on a complex torus $\mathbf{T}^r$ the following conditions are equivalent.*

(i) $\mathcal{V}$ is holomorphic;
(ii) $\mathcal{V}$ is translation invariant;
(iii) $\mathcal{V}$ is constant in terms of some (hence any) system of linear coordinates;
(iv) $\mathcal{V}$ is the projection of a constant vector field on the universal covering space $\mathbf{C}^r$ of $\mathcal{V}^r$.

A vector field on $\mathcal{V}^r$ satisfying the above conditions will be called a linear vector field *on $\mathbf{T}^r$.*

Notice that as a corollary any two holomorphic vector fields on $\mathbf{T}^r$ commute, and that the dimension of the vector space of holomorphic vector fields on $\mathbf{T}^r$ is r. In fact, this vector space is spanned by r commuting holomorphic vector fields $\mathcal{V}_1, \dots, \mathcal{V}_r$ which are the projections of r independent constant vector fields on the universal covering space $\mathbf{C}^r$ of $\mathbf{T}^r$. Therefore, we may introduce linear coordinates $t_1, \dots, t_r$ on (the neighborhood of any point in) $\mathbf{T}^r$, such that $\mathcal{V}_i = \partial/\partial t_i$, i.e., such that for any holomorphic or meromorphic function f, defined on an open subset U of $\mathbf{T}^r$, one has

$$\mathcal{V}_i[f] = \frac{\partial f}{\partial t_i}.$$

5.2.1 The Riemann Conditions

We first explain what it means for a complex torus to be an Abelian variety, leading to several important notions and we refer to the Paragraphs 5.2.3 and 5.2.4 below for particular classes of Abelian varieties.

A complex torus $\mathbf{C}^r/\Lambda$, where Λ is a lattice, is a Kähler manifold: since the Euclidean metric on $\mathbf{C}^r$, which is Kähler, is translation invariant, it descends to $\mathbf{C}^r/\Lambda$. Therefore, the Moishezon Theorem (Theorem 5.5) implies

Theorem 5.13 (Moishezon). *A torus $\mathbf{T}^r = \mathbf{C}^r/\Lambda$ is an Abelian variety if and only if it admits r algebraically independent meromorphic functions.*

On the other hand, according to the Kodaira Embedding Theorem, $\mathbf{T}^r$ is an Abelian variety if and only if it has a Hodge form ω. By averaging ω over the torus we may assume that ω is constant with respect to any system of linear coordinates on $\mathbf{T}^r$. Let $(\lambda_1, \dots, \lambda_{2r})$ be a basis for Λ as a $\mathbf{Z}$-module. In terms of the real coordinates $x_1, \dots, x_{2r}$ which correspond to this basis, ω takes the form

$$\omega = \frac{1}{2} \sum_{i,j=1}^{r} q_{ij} \mathrm{d}x_i \wedge \mathrm{d}x_j,$$

where each $q_{ij} = -q_{ji}$ is a constant. Notice that such an ω is automatically closed. Integrality of the cohomology class of ω is equivalent to the fact that for all $1 \leqslant i, j \leqslant r$ the integral of ω over the rectangle formed by λ_i, λ_j is an integer, i.e., that all q_{ij} are integers. The skew-symmetric matrix Q with entries q_{ij} defines a skew-symmetric quadratic form, which can be put in the form

$$Q = \begin{pmatrix} 0 & \Delta_\delta \\ -\Delta_\delta & 0 \end{pmatrix}, \qquad \text{where} \qquad \Delta_\delta = \begin{pmatrix} \delta_1 & & 0 \\ & \ddots & \\ 0 & & \delta_r \end{pmatrix}, \tag{5.12}$$

with all δ_i positive integers such that $\delta_i \mid \delta_{i+1}$ $(1 \le i < r)$. Indeed, since ω is positive it is non-degenerate and each δ_i is different from 0. For $i = 1, \dots, r$ we define the vector $e_i := \lambda_i / \delta_i$. Then $(e_1, \dots, e_r)$ forms a *complex* basis for $\mathbf{C}^r$ and the $r \times 2r$ matrix Λ, defined by

$$\Lambda = (\lambda_{\alpha i})_{\substack{\alpha=1,\dots,r \\ i=1,\dots,2r}}, \qquad \text{where} \qquad \lambda_i = \sum_{\alpha=1}^{r} \lambda_{\alpha i} e_\alpha,$$

takes the form $\Lambda = \begin{pmatrix} \Delta_\delta & Z \end{pmatrix}$. Notice that, by a slight abuse of notation, we use the same notation Λ for the lattice and for the matrix representing a basis of Λ with respect to a basis of $\mathbf{C}^r$ (the i-th column of the matrix Λ is the coordinate vector of λ_i with respect to $(e_1, \dots, e_r)$). The matrix Λ is called the *period matrix* of the lattice Λ. Writing the Hodge form in terms of the complex coordinates $(z_1, \dots, z_r)$ which correspond to $(e_1, \dots, e_r)$ we find that ω is a $(1,1)$-form if and only if Z is symmetric and that the positivity of ω is equivalent to $\Im Z > 0$. This gives precisely the famous

Theorem 5.14 (Riemann conditions). *A complex torus $T^r = \mathbf{C}^r/\Lambda$ is an Abelian variety if and only if there exists a $\mathbf{Z}$-module basis for Λ and a complex basis for $\mathbf{C}^r$ such that*

$$\Lambda = \begin{pmatrix} \Delta_\delta & Z \end{pmatrix},$$

where Δ_δ is a matrix over $\mathbf{Z}$, of the form (5.12), and where Z is a symmetric matrix whose imaginary part is positive definite.

In terms of the real coordinates $(x_1, \dots, x_{2r})$ *which correspond to the* $\mathbf{Z}$*-basis for* Λ*, the Hodge form* ω *takes the form*

$$\omega = \sum_{\alpha=1}^{r} \delta_\alpha \mathrm{d}x_\alpha \wedge \mathrm{d}x_{\alpha+r}.$$

The cohomology class $[\omega]$ of ω is called the *polarization* of the Abelian variety $\mathbf{T}^r$ and the pair $(\mathbf{T}^r, [\omega])$ is called a *polarized Abelian variety.* The integers $\delta_1, \dots, \delta_r$ are called the *elementary divisors* of the polarization; they are invariants of the cohomology class $[\omega]$. Also, $(\delta_1, \dots, \delta_r)$ is called the *polarization type* of the Abelian variety.

A polarization $[\omega]$ is called a *principal polarization* when its type equals $(1, \dots, 1)$. A polarized Abelian variety whose polarization is a principal polarization is called a *principally polarized Abelian variety.* Every polarized Abelian variety $\mathbf{T}^r$ is related to a principally polarized Abelian variety in at least two different ways. Indeed, if we write the period matrix of $\mathbf{T}^r$ as $\Lambda_1 = (\Delta_r \ Z)$ and we define a new period matrix by $\Lambda_2 := (\mathrm{Id}_r \ Z)$ then the (lattice) inclusion $\Lambda_1 \subseteq \Lambda_2$ induces an isogeny $\mathbf{C}^r/\Lambda_1 \to \mathbf{C}^r/\Lambda_2$ of degree $\prod_{i=1}^r \delta_i$, which exhibits the Abelian variety $\mathbf{T}^r$ as an unramified cover of a principally polarized Abelian variety. Alternatively, the inclusion $\delta_r \Lambda_2 \subseteq \Lambda_1$ induces an isogeny $\mathbf{C}^r/(\delta_r \Lambda_2) \to \mathbf{C}^r/\Lambda_1$ of degree $\prod_{i=1}^r \delta_r/\delta_i$, which exhibits the Abelian variety $\mathbf{T}^r$ as being covered by an Abelian variety which is conformal to a principally polarized Abelian variety.

As we explained in the previous section, the Hodge form is the curvature form of a positive line bundle $\mathcal{L}$, which in turn comes from an ample divisor $\mathcal{D}$ on $\mathbf{T}^r$, i.e., $\omega = [(\sqrt{-1}/2\pi)\Theta_{\mathcal{L}}]$ and $\mathcal{L} = [\mathcal{D}]$. But not every (effective) divisor on an Abelian variety needs to be ample, and this is related to the reducibility of the Abelian variety, as we explain now.

An Abelian variety $\mathbf{T}^r$ will be called an *irreducible Abelian variety,* when $\mathbf{T}^r$ does not contain any Abelian subvariety, otherwise it will be called a *reducible Abelian variety.* In this respect the following theorem is fundamental.

Theorem 5.15 (Poincaré Reducibility). *Let* $\mathbf{T}^r$ *be an Abelian variety and suppose that* $\mathbf{T}^r$ *contains an Abelian subvariety* B*. Then there exists an Abelian subvariety* C *of* $\mathbf{T}^r$ *such that* $B \cap C$ *is a finite subgroup of* $\mathbf{T}^r$ *and there exists a surjective homomorphism* $h : B \oplus C \longrightarrow \mathbf{T}^r$ *with kernel* $B \cap C$*.*

In other words, the Poincaré Reducibility Theorem states that if an Abelian variety contains an Abelian subvariety, then it is isogenous to a product of this subvariety and another Abelian subvariety. In this context we also wish to point out that if $\phi : A \to B$ is a holomorphic map between two Abelian varieties A and B then ϕ is the composition of a group homomorphism $\phi_0 : A \to B$, composed with a translation $\tau : B \to B$. Thus, holomorphic maps between Abelian varieties are group homomorphisms, upon picking appropriate origins, as we asserted at the beginning of this section.

The cohomology of complex tori, in particular Abelian varieties, can be computed explicitly because the (commutative!) group structure allows one to identify all tangent and cotangent spaces with the universal covering space of the torus, objects can be averaged over the torus to yield translation invariant objects in the same cohomology class and so on. Moreover, the canonical bundle on any complex torus is trivial, because the standard volume form on $\mathbf{C}^r$ descends to $\mathbf{T}^r$, yielding a top-form that has no zeros. It follows that the natural sheaf isomorphism (5.3) simplifies in the case of an Abelian variety $\mathbf{T}^r$ to

$$\mathcal{O}_{\mathbf{T}^r}(\mathcal{L}) \cong \Omega^r_{\mathbf{T}^r}(\mathcal{L}), \tag{5.13}$$

so that for example $h^0(\mathbf{T}^r, \Omega^r) = h^0(\mathbf{T}^r, \mathcal{O}) = 1$. As an application of (5.13), let us compute the Euler characteristic $\chi(\mathcal{L})$ of a positive line bundle $\mathcal{L} = [\mathcal{D}]$ on an Abelian variety $\mathbf{T}^r$. It yields

$$\chi(\mathcal{L}) = \sum_{p=0}^{r} (-1)^p h^p(\mathbf{T}^r, \mathcal{O}(\mathcal{L})) = \sum_{p=0}^{r} (-1)^p h^p(\mathbf{T}^r, \Omega^r(\mathcal{L})),$$

which simplifies in view of the Kodaira-Nakano Vanishing Theorem to

$$\chi(\mathcal{L}) = h^0(\mathbf{T}^r, \Omega^r(\mathcal{L})) = h^0(\mathbf{T}^r, \mathcal{O}(\mathcal{L})) = \dim L(\mathcal{D}). \tag{5.14}$$

An Abelian variety $\mathbf{T}^r = \mathbf{C}^r/\Lambda$ has a natural involution $(-1)_{\mathbf{T}^r}$, induced by the reflection about the origin of $\mathbf{C}^r$. The set of fixed points of $(-1)_{\mathbf{T}^r}$ is exactly the 2-torsion subgroup $(\frac{1}{2}\Lambda)/\Lambda \subset \mathbf{T}^r$ of $\mathbf{T}^r$. Elements of this subgroup are called *half-periods* and there are 2^{2r} of them. The quotient $\mathbf{T}^r/(-1)_{\mathbf{T}^r}$ is a projective variety, called the *Kummer variety* of $\mathbf{T}^r$. It has 2^{2r} singular points, which come from the half-periods. We will come back to the Kummer variety in Paragraph 5.2.3.

5.2.2 Line Bundles on Abelian Varieties and Theta Functions

In this section we explain how positive line bundles on an Abelian variety can be explicitly described in terms of multipliers and how their sections can be described by theta functions. Let $\mathbf{T}^r = \mathbf{C}^r/\Lambda$ be a complex torus and let $\mathcal{L}$ be a line bundle on $\mathbf{T}^r$. Then $\pi^*\mathcal{L}$ is trivial ($\pi : \mathbf{C}^r \to \mathbf{T}^r$) because $\mathbf{C}^r$ is contractible. Hence, there exists a global trivialization

$$\phi : \pi^*\mathcal{L} \longrightarrow \mathbf{C}^r \times \mathbf{C}.$$

For $z \in \mathbf{C}^r$, let us denote by $(\pi^*\mathcal{L})_z$ the line of $\pi^*\mathcal{L}$ that sits over z, and let ϕ_z denote the restriction of ϕ to $(\pi^*\mathcal{L})_z$; since the first component of ϕ_z is constant (it is equal to z) we will think of ϕ_z as taking its values in $\mathbf{C}$. Since $(\pi^*\mathcal{L})_z = (\pi^*\mathcal{L})_{z+\lambda}$ we get

$$\mathbf{C} \xleftarrow{\phi_z} (\pi^*\mathcal{L})_z = (\pi^*\mathcal{L})_{z+\lambda} \xrightarrow{\phi_{z+\lambda}} \mathbf{C},$$

giving a linear automorphism $\mathbf{C} \to \mathbf{C}$, i.e., multiplication by a non-zero number $e_\lambda(z)$. The functions $\{e_\lambda \in \mathcal{O}^*(\mathbf{C}^g)\}_{\lambda \in \Lambda}$ are called the *multipliers* of $\mathcal{L}$ and they satisfy

$$e_{\lambda'}(z+\lambda)\, e_\lambda(z) = e_\lambda(z+\lambda')\, e_{\lambda'}(z) = e_{\lambda+\lambda'}(z).$$

Conversely, multipliers which satisfy these relations define a unique line bundle with these multipliers.

When the complex torus is a polarized Abelian variety $(\mathbf{T}^r, [\omega])$ then a line bundle $\mathcal{L}$ for which $c_1(\mathcal{L}) = [\omega]$ can be described by using multipliers of a simple character in the following way. Since $[\omega]$ is an integral cohomology class, a representative ω of $[\omega]$ and linear *real* coordinates $(x_1, \dots, x_{2r})$ on $\mathbf{C}^r$ can be chosen such that

$$\omega = \sum_{i=1}^{r} \delta_i \mathrm{d}x_i \wedge \mathrm{d}x_{r+i},$$

where the δ_i are positive integers. Let $(\lambda_1, \dots, \lambda_{2r})$ be the basis of Λ, defined by $\mathrm{d}x_i(\lambda_j) = \delta_{ij}$, where $1 \leqslant i, j \leqslant 2r$ and define linear *complex* coordinates $(z_1, \dots, z_r)$ on $\mathbf{C}^r$ by $\mathrm{d}z_i(\lambda_j) = \delta_j \delta_{ij}$. Then we get the following theorem.

Theorem 5.16. *The line bundle $\mathcal{L}$ on $\mathbf{T}^r$ with multipliers*

$$e_{\lambda_\alpha}(z) = 1, \qquad e_{\lambda_{r+\alpha}}(z) = e^{-2\pi i z_\alpha}, \qquad \alpha = 1\dots, r,$$

has Chern class $c_1(\mathcal{L}) = [\omega]$. Moreover, any line bundle $\mathcal{L}'$ on $\mathbf{T}^r$ for which $c_1(\mathcal{L}') = c_1(\mathcal{L})$ is a translate (in $\mathbf{T}^r$) of $\mathcal{L}$.

The fact that the line bundle is given by simple multipliers allows us to construct explicitly its holomorphic sections; they can be seen as functions on $\mathbf{C}^r$ which are periodic in r directions and "quasi-periodic" in r other directions. The number of independent holomorphic sections is given by

$$h^0(\mathbf{T}^r, \mathcal{O}(\mathcal{L})) = \prod_{i=1}^{r} \delta_i, \tag{5.15}$$

where $(\delta_1, \dots, \delta_r)$ are the elementary divisors of the polarization $c_1(\mathcal{L})$. For a line bundle defining a principal polarization, for example, there is only one section which, as a quasi-periodic function on $\mathbf{C}^r$, is given by the *Riemann theta function*

$$\vartheta(z) = \sum_{l \in \mathbf{Z}^r} e^{\pi i \langle l, Zl\rangle} e^{2\pi i \langle l, z\rangle},$$

where we have written the lattice defining $\mathbf{T}^r$ as $\Lambda = (\mathrm{Id}_r \ Z)$. Its divisor of zeros, denoted Θ, is determined uniquely by $\mathcal{L}$, hence up to a translation by $c_1(\mathcal{L})$ and is called the *Riemann theta divisor.* Also, observe that if an ample line bundle $\mathcal{L}$ defines a polarization of type $(\delta_1, \dots, \delta_r)$ on $\mathbf{T}^r$, then the line bundle $\mathcal{L}^k$ has Chern class $c_1(\mathcal{L}^k) = kc_1(\mathcal{L})$, since c_1 is a homomorphism (it is the connecting homomorphism in (5.4)). Therefore $\mathcal{L}^k$ defines a polarization of type $(k\delta_1, \dots, k\delta_r)$ and

$$h^0(\mathbf{T}^r, \mathcal{O}(\mathcal{L}^k)) = k^r \prod_{i=1}^{r} \delta_i = k^r h^0(\mathbf{T}^r, \mathcal{O}(\mathcal{L})).$$

Applied to an ample divisor $\mathcal{D}$ on $\mathbf{T}^r$ this means that

$$\dim L(k\mathcal{D}) = k^r \dim L(\mathcal{D}).$$

The group of all line bundles of degree 0 on a polarized Abelian variety $\mathbf{T}^r$ is a complex torus, called its *dual* and denoted by $\widehat{\mathbf{T}^r}$. If $\mathbf{T}^r$ corresponds to a period matrix $(\Delta_\delta\ Z)$ then a "dual" basis can be picked such that the matrix defining the lattice defining $\widehat{\mathbf{T}^r}$ is given by

$$\left(\delta_r \Delta_\delta^{-1}, \delta_r \Delta_\delta^{-1} Z \Delta_\delta^{-1}\right). \tag{5.16}$$

In this representation it is easy to check the Riemann conditions, which show that the dual is indeed an Abelian variety. For $\mathcal{L}$ a fixed positive line bundle on $\mathbf{T}^r$ one defines an isogeny between $\mathbf{T}^r$ and its dual by $v \mapsto \mathcal{L}^{-1} \otimes \tau_v^* \mathcal{L}$. Here, τ_v denotes the translation in $\mathbf{T}^r$ over v; one may think of the line bundle $\tau_v^* \mathcal{L}$ as the translate of $\mathcal{L}$ over $-v$. The degree of this isogeny is $\prod \delta_i^2$; in particular, a principally polarized Abelian variety is isomorphic to its dual. Also, the dual of an irreducible Abelian variety is irreducible.

If $\mathbf{T}^r$ is irreducible then the line bundle of any effective divisor is ample; moreover, we have the following theorem.

Theorem 5.17 (Lefschetz). *Let $\mathcal{L}$ be an ample line bundle on an Abelian variety $\mathbf{T}^r$. Then $\mathcal{L}^n$ is very ample for $n \geqslant 3$.*

For example, if $\mathcal{L}$ defines a principal polarization on an irreducible Abelian variety $\mathbf{T}^r$, then $\mathcal{L}^3$ induces a polarization of type $(3, 3, \dots, 3)$, and hence every irreducible principally polarized Abelian variety can be embedded in $\mathbf{P}H^0(\mathbf{T}^r, \mathcal{O}(\mathcal{L}^3))^*$, which is by (5.15) isomorphic to $\mathbf{P}^{3^r - 1}$. In the case of Abelian surfaces, the following theorem, which is due to Ramanan (see [145]) is an improvement of the Lefschetz Theorem in the case of Abelian surfaces.

Theorem 5.18 (Ramanan). *Let $\mathbf{T}^2$ be an irreducible Abelian surface and let $\mathcal{D}$ be an ample divisor on $\mathbf{T}^2$ defining a polarization of type (δ_1, δ_2). Then $\mathcal{D}$ is very ample if one of the following conditions is satisfied.*

(1) $\delta_1 = 1$ *and* $\delta_2 \geqslant 5$;
(2) $\delta_1 = 2$ *and* $\delta_2 \geqslant 4$;
(3) $\delta_1 \geqslant 3$.

In the following theorem we describe when a line bundle on an Abelian variety is normally generated. The Lefschetz Theorem is a consequence of it (since normally generated line bundles are always very ample).

Theorem 5.19 (Koizumi-Mumford Criterion). *Let $\mathcal{L}$ be an ample line bundle on an Abelian variety $\mathbf{T}^r$. Then $\mathcal{L}^3$ is normally generated.*

The above Ramanan Theorem fails to hold when "very ample" is replaced by "normally generated". In fact, we will encounter examples of polarized Abelian surfaces $(\mathbf{T}^2, [\omega])$ that are of polarization type $(2,4)$ but for which the line bundle that defines the polarization is not normally generated.

We now turn to symmetric line bundles and symmetric divisors on Abelian varieties (for details, see [105, Section 4.6]). A line bundle $\mathcal{L}$ on an Abelian variety $\mathbf{T}^r$ is called a *symmetric line bundle* if $(-1)^*_{\mathbf{T}^r}\mathcal{L} \cong \mathcal{L}$. Then $(-1)_{\mathbf{T}^r}$ can be lifted to an involution $(-1)_{\mathcal{L}}$ on the total space of $\mathcal{L}$. The latter involution is $\mathbf{C}$-linear on the fibers of $\mathcal{L}$ and it can be normalized so that it is the identity on the fiber over the origin of $\mathbf{T}^r$. Since the involution which $(-1)_{\mathcal{L}}$ induces on the fiber over each half period is linear it is either identity or multiplication by -1. If it is identity the corresponding half period is called *even*, otherwise it is called *odd*; in particular, according to the above normalization, the origin is always an even half period. The induced involution $s \to (-1)_{\mathcal{L}} s (-1)_{\mathbf{T}^r}$ on $H^0(\mathbf{T}^r, \mathcal{L})$ leads to a splitting of $H^0(\mathbf{T}^r, \mathcal{L})$ into $(+1)$ and (-1) spaces, whose elements are called *even sections* and *odd sections*,

$$H^0(\mathbf{T}^r, \mathcal{L}) = H^0(\mathbf{T}^r, \mathcal{L})^+ \oplus H^0(\mathbf{T}^r, \mathcal{L})^- .$$

For an ample line bundle $\mathcal{L}$ on $\mathbf{T}^r$, the dimensions of $H^0(\mathbf{T}^r, \mathcal{L})^+$ and of $H^0(\mathbf{T}^r, \mathcal{L})^-$ are given by the following proposition.

Proposition 5.20. *Let $\mathcal{L}$ be an ample line bundle on an Abelian variety $\mathbf{T}^r$. If $\mathcal{L}$ induces on $\mathbf{T}^r$ a polarization of type $(\delta_1, \dots \delta_s, \delta_{s+1}, \dots, \delta_r)$, where δ_s is odd and δ_{s+1} is even, then $h^0(\mathbf{T}^r, \mathcal{L})^+$ is equal to one of the following:*

$$h^0(\mathbf{T}^r, \mathcal{L})^+ = \frac{1}{2} h^0(\mathbf{T}^r, \mathcal{L}) + \epsilon 2^{r-s-1}, \qquad \epsilon \in \{0, 1, -1\} .$$

In either case, the formula for $h^0(\mathbf{T}^r, \mathcal{L})^-$ follows from it.

Since the symmetric line bundles $\mathcal{L}$ which we will consider come from explicitly given divisors, we will translate the notion of a symmetric line bundle and its section in the language of divisors. A divisor (or curve) $\mathcal{D}$ on $\mathbf{T}^r$ is called *symmetric* if $(-1)^*_{\mathbf{T}^r}\mathcal{D} = \mathcal{D}$, which means that the reflection of $\mathcal{D}$ with respect to the origin is $\mathcal{D}$. The line bundle of a symmetric divisor is symmetric and the even and odd sections of a symmetric line bundle are symmetric divisors; therefore, working with symmetric divisors is just as general as working with symmetric line bundles. We will call a symmetric divisor even or odd according to whether it is defined by an even or odd section. When $\mathbf{T}^r$ is an Abelian surface ($r = 2$) then it is easy to see that an even (resp. odd) divisor $\mathcal{D}$ has even (resp. odd) multiplicity at the even half periods (in particular at the origin).

5.2.3 Jacobian Varieties

Let Γ be a compact Riemann surface of genus $g \geqslant 1$ and view $H_1(\Gamma, \mathbf{Z})$ as a subgroup of $H^0(\Gamma, \Omega^1)^*$ via the natural injective homomorphism Ψ which maps $\gamma \in H_1(\Gamma, \mathbf{Z})$ to the linear map

$$\begin{aligned} \Psi(\gamma) : H^0(\Gamma, \Omega^1) &\to \mathbf{C} \\ \omega \quad &\mapsto \oint_\gamma \omega. \end{aligned} \tag{5.17}$$

The *Jacobian* or *Jacobian variety* of Γ is defined by

$$\mathrm{Jac}(\Gamma) = \frac{H^0(\Gamma, \Omega^1)^*}{H_1(\Gamma, \mathbf{Z})}. \tag{5.18}$$

$\mathrm{Jac}(\Gamma)$ is a complex torus of dimension g because $h^0(\Gamma, \Omega^1) = g$ and because $H_1(\Gamma, \mathbf{Z})$ is a lattice in $H^0(\Gamma, \Omega^1)^*$. Choose a basis $\vec{\omega} = {}^t(\omega_1, \dots, \omega_g)$ of $H^0(\Gamma, \Omega^1)$ and choose a *symplectic basis* for $H_1(\Gamma, \mathbf{Z})$, i.e., a system of $2g$ generators $(A_1, \dots, A_g, B_1, \dots, B_g)$ for $H_1(\Gamma, \mathbf{Z})$ such that $A_i \cdot A_j = B_i \cdot B_j = 0$ and $A_i \cdot B_j = \delta_{ij}$, for $1 \leqslant i, j \leqslant g$. Then the lattice $\Lambda \subset \mathbf{C}^g$ consisting of all vectors in $\mathbf{C}^g$ of the form $\oint_\gamma \vec{\omega}$, with γ running through $H_1(\Gamma, \mathbf{Z})$ is conveniently represented as the column space (over $\mathbf{Z}$) of the matrix

$$\begin{pmatrix} \oint_{A_1} \omega_1 & \cdots & \oint_{A_g} \omega_1 & \oint_{B_1} \omega_1 & \cdots & \oint_{B_g} \omega_1 \\ \vdots & & \vdots & \vdots & & \vdots \\ \oint_{A_1} \omega_g & \cdots & \oint_{A_g} \omega_g & \oint_{B_1} \omega_g & \cdots & \oint_{B_g} \omega_g \end{pmatrix}, \tag{5.19}$$

called the *period matrix* of $\mathrm{Jac}(\Gamma)$ (with respect to the chosen bases). The first $g \times g$ block of the period matrix (5.19) is called the *matrix of A-periods* and the last $g \times g$ block the *matrix of B-periods*. Finally, Λ itself is called the *period lattice* of $\mathrm{Jac}(\Gamma)$.

It can be shown that the matrix of A-periods is invertible, hence we may assume that it is equal to Id_g by switching to an appropriate basis for $H^0(\Gamma, \Omega^1)$. Writing the new period matrix as $(\mathrm{Id}_g \;\; Z)$ the matrix Z is symmetric and $\Im Z$ is positive definite. Thus

Theorem 5.21. *For any compact Riemann surface of genus g, the complex torus* $\mathrm{Jac}(\Gamma)$ *is a principally polarized Abelian variety of dimension g.*

The following converse also holds: every irreducible principally polarized Abelian variety of dimension 2 or 3 is the Jacobian of a compact Riemann surface of genus 2 or 3. In higher dimensions this is no longer true (as can be checked by an easy dimension count), and there is the famous *Schotky problem* which asks for a characterization of those matrices Z for which $(\mathrm{Id}_g \;\; Z)$ is a Jacobi variety (see [125], [134] and [157]).

On the other hand, once one knows that a principally polarized Abelian variety is the Jacobian of a compact Riemann surface, the latter is uniquely determined by its Jacobian; this is the content of the classical *Torelli Theorem.*

Since $\mathrm{Jac}(\Gamma)$ is a principally polarized Abelian variety of dimension g, the Lefschetz Theorem implies that it can be embedded in $\mathbf{P}^{3^g-1}$, by using the sections of $[3\Theta]$. However, the sections of $[2\Theta]$ never embed $\mathrm{Jac}(\Gamma)$ in projective space, but rather they embed its Kummer variety $\mathrm{Kum}(\Gamma)$ in projective space. An important particular case is that of the Kummer surface $\mathrm{Kum}(\Gamma)$, where Γ is a hyperelliptic Riemann surface of genus 2. The line bundle $[2\Theta]$ that corresponds to twice the principal polarization on $\mathrm{Jac}(\Gamma)$ has in this case 4 independent sections and the associated Kodaira map, which maps $\mathrm{Jac}(\Gamma)$ into $\mathbf{P}^3$, factors through $\mathrm{Kum}(\Gamma)$, realizing the Kummer surface as a surface in $\mathbf{P}^3$.

Being two-dimensional the image is given by a single equation; to compute the degree of this equation, we use the fact that this degree is given by $\int_{\mathrm{Kum}(\Gamma)} \Omega$, where Ω is associated $(1,1)$-form of the standard Kähler structure on $\mathbf{P}^3$. Clearly this is twice the volume of $\mathrm{Kum}(\Gamma)$, which itself is half the volume of the Jacobi surface (with the polarization of type $(2,2)$). For $\omega = 2dx_1 \wedge dx_2 + 2dx_3 \wedge dx_4$ we get $\int \omega^2 = 8$, hence the Jacobi surface has degree 8, its volume is 4, the volume of $\mathrm{Kum}(\Gamma)$ is 2 and the degree of $\mathrm{Kum}(\Gamma)$ is 4. Since the (-1) involution on an Abelian surface has 16 fixed points, its 16 half-periods, its Kummer surface has 16 singular points. In the case of the Kummer surface $\mathrm{Kum}(\Gamma)$ one has, in addition, 16 planes in $\mathbf{P}^3$ which touch $\mathrm{Kum}(\Gamma)$ along a conic. These 16 planes and the 16 singular points form a beautiful configuration on $\mathbf{P}^3$, called the 16_6 configuration (each of the 16 planes contains 6 of the points and through each of the 16 points pass 6 of the planes).

The 16_6 configuration is useful for determining an explicit equation for the Kummer surface, as a quartic in $\mathbf{P}^3$ (see [105, Sections 10.2 and 10.3]), but it also shows up naturally, together with the Kummer surface itself, in the classical theory of the quadric line complex (see [69, Chapter 6]). For a Riemann surface Γ of genus bigger than 2 the Kummer variety $\mathrm{Kum}(\Gamma)$ is also embedded in projective space by the Kodaira map associated to $[2\Theta]$, but it is only embedded as a hypersurface when Γ has genus 2.

As an algebraic variety, the Jacobian of Γ also admits a definition in terms of divisors on Γ. Consider $\mathrm{Div}(\Gamma)/\sim$, the group of all divisors on Γ modulo linear equivalence. Since linearly equivalent divisors have the same degree, deg induces a homomorphism

$$\deg\sim\colon \frac{\mathrm{Div}(\Gamma)}{\sim} \longrightarrow \mathbf{Z},$$

whose kernel is called the *algebraic Jacobian* of Γ. In view of the basic correspondence between divisors and line bundles on Γ, the algebraic Jacobian corresponds to $\mathrm{Pic}^0(\Gamma)$, the group of line bundles of degree zero on Γ.

In order to link the two definitions of the Jacobian, consider the *Abel-Jacobi map*

$$\begin{aligned} \mathrm{Ab}: \qquad \mathrm{Div}^0(\Gamma) \qquad &\to \mathbf{C}^g/\Lambda \\ \mathcal{D} = \sum_{i=1}^n (P_i - Q_i) &\mapsto \int_{\mathcal{D}} \vec{\omega} := \sum_{i=1}^n \int_{Q_i}^{P_i} \vec{\omega} \bmod \Lambda \end{aligned}$$

with $\vec{\omega}$ and Λ as defined above. In the sequel we will leave out "mod Λ" when it is clear that the *Abel sum* is taken modulo Λ. The Abel map Ab induces an injective map $\mathrm{Pic}^0(\Gamma) \to \mathrm{Jac}(\Gamma)$, as follows from the following theorem.

Theorem 5.22 (Abel). *For any divisor $\mathcal{D}$ on Γ of degree 0, $\mathrm{Ab}(\mathcal{D}) = 0$ if and only if $\mathcal{D}$ is a principal divisor, i.e., $\mathcal{D} = (f)$ for some meromorphic function f on Γ.*

Surjectivity of Ab is a consequence of the *Jacobi Inversion Theorem*, which we formulate in a stronger form.

Theorem 5.23 (Jacobi Inversion). *Let $Q_1, \dots, Q_g$ be arbitrary points on Γ. For any $v \in \mathrm{Jac}(\Gamma)$ there exist $P_1, \dots, P_g$ such that*

$$\mathrm{Ab}(P_1 + \dots + P_g - Q_1 - \dots - Q_g) = v.$$

Combining the Abel Theorem with the Jacobi Inversion Theorem we see that Ab induces a bijection between $\mathrm{Pic}^0(\Gamma)$ and $\mathrm{Jac}(\Gamma)$. In terms of the algebraic Jacobian it is easy to describe an injective map of Γ into its algebraic Jacobian, which leads via the Abel map to an embedding of Γ into $\mathrm{Jac}(\Gamma)$. Namely for any fixed $Q \in \Gamma$ we associate to any $P \in \Gamma$ the divisor class $[P - Q]$, which is injective in view of the Riemann-Roch Theorem. The resulting embedding is given by

$$\begin{aligned} \alpha_Q : \Gamma &\to \mathrm{Jac}(\Gamma) \\ P &\mapsto \mathrm{Ab}(P - Q) = \int_Q^P \vec{\omega} \end{aligned} \tag{5.20}$$

In terms of the Abel map, there is the following beautiful description of the Riemann theta divisor.

Theorem 5.24 (Riemann). *Let P be an arbitrary point on Γ. There exist points $Q_1, \dots, Q_g$ such that the theta divisor Θ of $\mathrm{Jac}(\Gamma)$ is given by*

$$\Theta = \{\mathrm{Ab}(P_1 + \dots + P_{g-1} + P - Q_1 - \dots - Q_g) \mid P_1, \dots, P_{g-1} \in \Gamma\}.$$

It follows that when $g = 2$ then the theta divisor is, up to a translation, the image of the embedding α_Q of Γ in its Jacobian. For an intrinsic description of the theta divisor, one considers the homogeneous space $\mathrm{Pic}^{g-1}(\Gamma)$ of line bundles on Γ of degree $g - 1$. Then

$$\Theta = \left\{\mathcal{L} \in \mathrm{Pic}^{g-1}(\Gamma) \mid h^0(\Gamma, \mathcal{O}(\mathcal{L})) > 0\right\}.$$

For this reason, one often calls $\mathrm{Pic}^{g-1}(\Gamma)$, rather than $\mathrm{Pic}^0(\Gamma)$, the Jacobian of Γ.

We now state the universal property of the Jacobian.

Theorem 5.25. *Suppose that* $\mathbf{T}^r$ *is an Abelian variety and that* Γ *is a complex compact Riemann surface of genus* g. *Let* $Q \in \Gamma$, *let* γ *be a holomorphic map* $\Gamma \to \mathbf{T}^r$ *and denote the translation in* $\mathbf{T}^r$ *by an element* $v \in \mathbf{T}^r$ *by* τ_v. *There exists a unique group homomorphism* $\tilde{\gamma} : \mathrm{Jac}(\Gamma) \to \mathbf{T}^r$ *such that the following diagram commutes*

$$\begin{array}{ccc} \Gamma & \xrightarrow{\ \gamma\ } & \mathbf{T}^r \\ {\scriptstyle \alpha_Q}\downarrow & & \downarrow{\scriptstyle \tau_{-\gamma(Q)}} \\ \mathrm{Jac}(\Gamma) & \xrightarrow[\tilde{\gamma}]{} & \mathbf{T}^r \end{array}$$

In view of the Jacobi Inversion Theorem it is clear that the homomorphism $\tilde{\gamma}$ admits the following description:

$$\begin{aligned} \tilde{\gamma} : \qquad \mathrm{Jac}(\Gamma) &\to \mathbf{T}^r \\ \mathrm{Ab}\left(\textstyle\sum_{i=1}^g P_i - gQ\right) &\mapsto \textstyle\sum_{i=1}^g \gamma(P_i) - g\gamma(Q), \end{aligned} \tag{5.21}$$

where the latter sum is, of course, a sum in $\mathbf{T}^r$. Intuitively speaking, $\tilde{\gamma}$ is obtained by extending γ by linearity.

Example 5.26. Suppose that Γ is an irreducible non-singular curve of genus g, contained in an irreducible Abelian variety $\mathbf{T}^r$. If we fix any point $Q \in \Gamma$ and we apply the universal property of the Jacobian we get a group homomorphism

$$\tilde{\gamma} : \mathrm{Jac}(\Gamma) \to \mathbf{T}^r,$$

which extends the embedding $\gamma : \Gamma \to \mathbf{T}^r$. Since the image of $\mathrm{Jac}(\Gamma)$ by $\tilde{\gamma}$ is a non-zero Abelian subvariety of $\mathbf{T}^r$ and since $\mathbf{T}^r$ is irreducible, the map $\tilde{\gamma}$ is surjective. The dual homomorphism

$$\hat{\gamma} : \widehat{\mathbf{T}^r} \longrightarrow \mathrm{Jac}(\Gamma)$$

has finite kernel, because $\widehat{\mathbf{T}^r}$ is irreducible (as $\mathbf{T}^r$ is) and because $\hat{\gamma}$ has a non-trivial image (it contains Γ). Therefore, $\widehat{\mathbf{T}^r}$ is isogenous to $\hat{\gamma}(\widehat{\mathbf{T}^r})$, which is an Abelian subvariety of $\mathrm{Jac}(\Gamma)$. By the Poincaré Reducibility Theorem, $\mathrm{Jac}(\Gamma)$ contains another Abelian subvariety A such that $\mathrm{Jac}(\Gamma)$ is equal to $\widehat{\mathbf{T}^r} \oplus A$, up to an isogeny. Since $\mathbf{T}^r$ and $\widehat{\mathbf{T}^r}$ are isogenous, $\mathrm{Jac}(\Gamma)$ is also equal to $\mathbf{T}^r \oplus A$, up to an isogeny, i.e.,

$$\Gamma \subset \mathbf{T}^r \quad \Longrightarrow \quad \mathrm{Jac}(\Gamma) \simeq \mathbf{T}^r \oplus A,$$

where A is an Abelian subvariety of $\mathrm{Jac}(\Gamma)$, of dimension $g - r$.

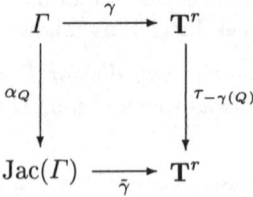

The universal property of the Jacobian says that the Jacobian of a Riemann surface can be defined by a universal property with repect to morphisms from the Riemann surface into Abelian varieties. The generalization of this universal property to higher-dimensional varieties leads to the notion of the Albanese variety (in the sense of Serre, see [154]). The definitions that we will give and the theorems that we will prove are valid in the algebraic as well as in the holomorphic category, because the target space of the morphisms that we will consider is always an Abelian variety; accordingly, the resulting universal object, the Albanese variety, will also be an Abelian variety.

As we will see, the Albanese variety allows one to reconstruct an Abelian variety from any of its affine parts. This property will prove useful when we consider the families of affine parts of Abelian varieties that appear as the generic fibers of the momentum map of a.c.i. systems, see Chapter 6).

Definition 5.27. For M a given variety, a morphism $\jmath : M \to \mathbf{T}^r$ with target an Abelian variety $\mathbf{T}^r$ is called a *universal morphism* for M, if any morphism $\imath : M \to A$, with A an Abelian variety, factorizes (uniquely) through $\mathbf{T}^r$ in the sense that there exists a commutative diagram

$$\begin{array}{ccc} M & \xrightarrow{\ \jmath\ } & \mathbf{T}^r \\ & \searrow^{\imath} & \downarrow{\scriptstyle\varphi} \\ & & A \end{array}$$

where $\varphi : \mathbf{T}^r \to A$ is an affine[2] morphism.

If $f : M \to \mathbf{T}^r$ is a universal morphism for M then $f(M)$ spans the algebraic group $\mathbf{T}^r$, but in general the dimension r of $\mathbf{T}^r$ is different from the dimension of M (see Example 5.29 below). The existence and uniqueness of a universal morphism for a given variety is stated in the following result.

Theorem 5.28. *For any algebraic variety M, a universal morphism $\jmath$: $M \to \mathbf{T}^r$ exists. Moreover, the couple $(\mathbf{T}^r, \jmath)$ is unique up to isomorphism. The Abelian variety $\mathbf{T}^r$ (or, precisely, the isomorphism class of $(\mathbf{T}^r, \jmath)$) is called the* Albanese variety *of M and is denoted by* $\mathrm{Alb}(M)$.

Similar definitions and a similar result hold for other classes of target algebraic groups such as $(\mathbf{C}^*)^r$ or extensions of an Abelian variety by an algebraic group $(\mathbf{C}^*)^r$.

Example 5.29. For a compact Riemann surface Γ and for a point $Q \in \Gamma$ the natural inclusion map $\alpha_Q : \Gamma \to \mathrm{Jac}(\Gamma)$, defined in (5.20), is a universal morphism for Γ. This follows immediately from the universal property for Jacobians (Theorem 5.25). Thus, $\mathrm{Alb}(\Gamma) \cong \mathrm{Jac}(\Gamma)$.

[2] Affine in the sense that φ commutes with translations.

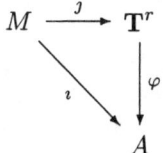

As we said, the main example that we will consider is the case in which M is an affine part of an Abelian variety, in which case the latter Abelian variety is the Albanese variety of M. This case is described by the following proposition.

Proposition 5.30. *Let $\mathbf{T}^r$ be an Abelian variety, and let $M \subseteq \mathbf{T}^r$ be an open subvariety. If $\imath : M \hookrightarrow \mathbf{T}^r$ is the inclusion, then $(\mathbf{T}^r, \imath)$ is the Albanese variety of M. As a consequence, any compactification A of M, where A is an Abelian variety, is isomorphic to $\mathbf{T}^r$.*

Proof. Let $\jmath : M \to \mathrm{Alb}(M)$ be a universal morphism for M, so that we have a commutative diagram

$$\begin{array}{ccc} M & \xrightarrow{\ \jmath\ } & \mathrm{Alb}(M) \\ & \searrow^{\imath} & \downarrow{\scriptstyle\varphi} \\ & & \mathbf{T}^r \end{array}$$

where φ is an affine morphism. We need to show that φ is an isomorphism. To do this we use the basic result which says that any rational function, defined on an open subset of a variety V, with target an Abelian variety, extends to the whole variety V. Applied to $\jmath : M \subseteq \mathbf{T}^r \to \mathrm{Alb}(M)$ we get a morphism $\psi : \mathbf{T}^r \to \mathrm{Alb}(M)$, with $\psi \circ \imath = \jmath$. Since $\jmath(M)$ and $\imath(M)$ span $\mathrm{Alb}(M)$ resp. $\mathbf{T}^r$, it is clear that ψ and φ are inverse to each other. □

Notice that in the above proof M is any (Zariski-) open subvariety of $\mathbf{T}^r$. It follows that the Albanese variety of M is a birational invariant of M.

If M is, in addition, smooth and compact then the Albanese variety of M can be described in cohomological terms, as in (5.18), namely

$$\mathrm{Alb}(M) = \frac{H^0(M, \Omega^1)^*}{H_1(M, \mathbf{Z})},$$

where $H_1(M, \mathbf{Z})$ is viewed as a subgroup of $H^0(M, \Omega^1)^*$ as in (5.17); the morphism $\jmath : M \to \mathrm{Alb}(M)$ is given, for $m \in M$ by the linear map (viewed as an element of $H^0(M, \Omega^1)^*/H_1(M, \mathbf{Z})$)

$$\begin{aligned} \jmath(m) : H^0(M, \Omega^1) &\to \mathbf{C} \\ \omega \quad &\mapsto \int_{m_0}^{m} \omega, \end{aligned}$$

where m_0 is a base point on M. When M is smooth and compact then $\mathrm{Pic}^0(M)$ and $\mathrm{Alb}(M)$ are, in a natural way, dual Abelian varieties (see [69, pp. 331–332]).

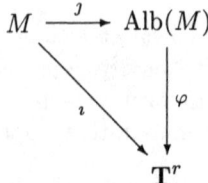

5.2.4 Prym Varieties

As Jacobi varieties did, Prym varieties provide another important class of Abelian varieties, which are in certain cases principally polarized (some authors reserve the term Prym variety to those cases). Prym varieties arise in the context of Riemann surfaces having an involution. Suppose that Γ is a compact Riemann surface, which is equipped with an involution σ. Let $\mathcal{E}$ be the compact Riemann surface, which is the quotient of Γ by σ and denote the quotient map $\Gamma \to \mathcal{E}$ by π.

Since π is a double cover, the Riemann-Hurwitz Formula (5.10) yields the genus g of Γ in terms of the genus g_0 of $\mathcal{E}$,

$$g = 2g_0 + n - 1,$$

where $2n$ denotes the number of ramification points. The involution σ extends linearly to an involution on $\mathrm{Div}(\Gamma)$, still denoted by σ: for $p_1, \dots, p_N \in \Gamma$ and $a_1, \dots, a_N \in \mathbf{Z}$,

$$\sigma\left(\sum_{i=1}^{N} a_i p_i\right) = \sum_{i=1}^{N} a_i \sigma(p_i).$$

Notice that σ respects the degree and that it sends principal divisors to principal divisors, since σ sends (f) to $(f \circ \sigma)$, where f is any meromorphic function on Γ. Therefore σ induces an involution on the Jacobian of Γ, which is still denoted by σ. We will say that a divisor $\mathcal{D}$ of degree 0 on Γ is an *odd divisor* (with respect to σ) if $\mathcal{D} + \sigma(\mathcal{D}) \sim 0$, which by the Abel Theorem means that $\mathrm{Ab}(\mathcal{D} + \sigma(\mathcal{D})) = 0 \in \mathrm{Jac}(\Gamma)$. The image under Ab of the odd divisors forms a subgroup of $\mathrm{Jac}(\Gamma)$ which is called the *Prym variety* of (Γ, σ), denoted $\mathrm{Prym}(\Gamma/\mathcal{E})$. By the Poincaré Reducibility Theorem there is an isogeny

$$\mathrm{Prym}(\Gamma/\mathcal{E}) \oplus A \longrightarrow \mathrm{Jac}(\Gamma),$$

where A is an Abelian subvariety of $\mathrm{Jac}(\Gamma)$.

The Prym variety can also be defined in the following way. The quotient map $\pi : \Gamma \to \mathcal{E}$ extends uniquely to a group homomorphism $\mathrm{Div}(\Gamma) \to \mathrm{Div}(\mathcal{E})$, still denoted by π, simply by setting $\pi(\sum a_i P_i) = \sum a_i \pi(P_i)$. If $\mathcal{D}$ is a principal divisor on Γ then $\pi(\mathcal{D})$ is a principal divisor on $\mathcal{E}$. Indeed, if $(f) = \mathcal{D}$ then the product $f(f \circ \sigma)$ of f with its conjugate (with respect to σ) $f \circ \sigma$ is σ-invariant, hence there exists a function g on $\mathcal{E}$, such that $g \circ \pi = f(f \circ \sigma)$; then $(g) = \pi(\mathcal{D})$. It follows that π induces a surjective homomorphism

$$\mathrm{Nm} : \mathrm{Jac}(\Gamma) \longrightarrow \mathrm{Jac}(\mathcal{E}),$$

which is called the *norm map*. The connected component of 0 of Ker(Nm) consists precisely of the odd divisors, i.e., it coincides with $\mathrm{Prym}(\Gamma/\mathcal{E})$ (see [131]). The dual homomorphism

$$\widehat{\mathrm{Nm}} : \mathrm{Jac}(\mathcal{E}) \longrightarrow \mathrm{Jac}(\Gamma)$$

yields, in view of the Poincaré Reducibility Theorem, the isogeny

$$\mathrm{Jac}(\Gamma) \simeq \mathrm{Prym}(\Gamma/\mathcal{E}) \oplus \mathrm{Jac}(\mathcal{E}).$$

We wish to compute rather explicitly the period matrix of $\mathrm{Prym}(\Gamma/\mathcal{E})$ and deduce from it its polarization type, more precisely the polarization type that $\mathrm{Prym}(\Gamma/\mathcal{E})$ inherits from the principal polarization on $\mathrm{Jac}(\Gamma)$. We will exclude the case $n = 0$ (no branch points), which is slightly different from the present case, because on the one hand we will not encounter it later and on the other hand because that case is presented in a very comprehensive form in [17, Appendix C].

The space $H^0(\Gamma, \Omega^1)$ of holomorphic differentials on Γ decomposes under σ as

$$H^0(\Gamma, \Omega^1) = H^0(\Gamma, \Omega^1)_+ \oplus H^0(\Gamma, \Omega^1)_-$$

where elements of $H^0(\Gamma, \Omega^1)_+$ are called *even differentials* and satisfy $\sigma^*\omega = \omega$, while elements of $H^0(\Gamma, \Omega^1)_-$ are called *odd differentials* and satisfy $\sigma^*\omega = -\omega$. Since a holomorphic differential on Γ is even precisely when it descends to $\mathcal{E}$, we have that

$$h^0(\Gamma, \Omega^1)_+ = g_0, \qquad h^0(\Gamma, \Omega^1)_- = g - g_0 = g_0 + n - 1.$$

If $\mathcal{D}$ is an odd divisor then for any even differential ω^+ on Γ one has by $\mathrm{Ab}(\mathcal{D} + \sigma(\mathcal{D})) = 0$ that

$$\int_{\mathcal{D}} \omega^+ = -\int_{\sigma(\mathcal{D})} \omega^+ = -\int_{\mathcal{D}} \sigma^*\omega^+ = -\int_{\mathcal{D}} \omega^+ = 0,$$

so that the period matrix of $\mathrm{Prym}(\Gamma/\mathcal{E})$ can be computed by integrating a basis for the odd cycles on Γ over a basis for the space of odd differentials. In order to construct a basis for the odd cycles, let A_i, B_i $(i = 1, \dots, g_0)$ be cycles on $\mathcal{E}$ whose homology classes form a symplectic basis for the first homology space of $\mathcal{E}$. We may assume that none of these cycles passes through the branch points of π. As can be seen from Figure 5.1, they can be lifted via π to $4g_0$ cycles

$$\tilde{A}_1, \dots, \tilde{A}_{g_0}, \tilde{B}_1, \dots, \tilde{B}_{g_0}, \sigma(\tilde{A}_1), \dots, \sigma(\tilde{A}_{g_0}), \sigma(\tilde{B}_1), \dots, \sigma(\tilde{B}_{g_0})$$

and this can be done in such a way that there exist additional cycles

$$C_1, \dots, C_{n-1}, D_1, \dots, D_{n-1},$$

so that the homology classes of all of these $2g = 4g_0 + 2(n-1)$ cycles on Γ form a symplectic basis for $H_1(\Gamma, \mathbf{Z})$, i.e., the only non-trivial intersection, in homology, is given by

$$\tilde{A}_i \cdot \tilde{B}_i = \sigma(\tilde{A}_i) \cdot \sigma(\tilde{B}_i) = C_i \cdot D_i = 1.$$

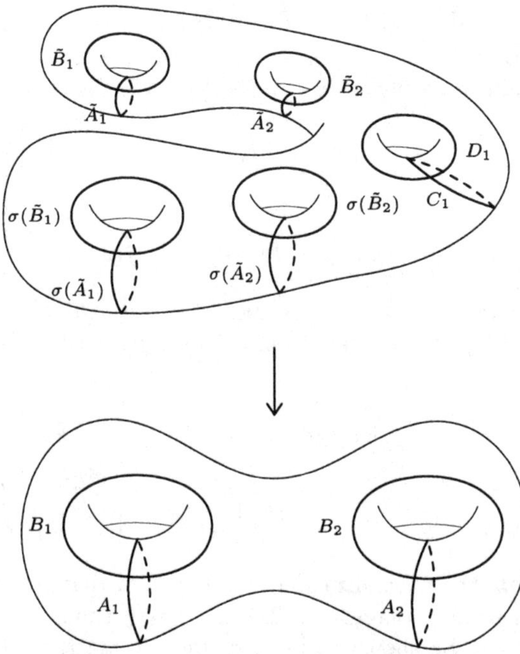

Fig. 5.1. If Γ is a double cover of $\mathcal{E}$ then a symplectic basis can be chosen for $H_1(\Gamma, \mathbf{Z})$ from a symplectic basis for $H_1(\mathcal{E}, \mathbf{Z})$. The genera of the two Riemann surfaces are related by $g(\Gamma) = 2g(\mathcal{E}) + n - 1$, where $2n$ is the number of branch points of the cover.

In order to construct a basis for the space of odd holomorphic differentials, let $\omega_1, \ldots, \omega_g$ be differentials that are normalized with respect to the above symplectic basis, where we view the cycles $\tilde{A}_i, \sigma(\tilde{A}_i)$ and C_j as the "A-cycles", i.e.,

$$\begin{aligned} &\oint_{\tilde{A}_j} \omega_i = \delta_{ij} && i,j = 1, \ldots, g_0, \\ &\oint_{\sigma(\tilde{A}_j)} \omega_{i+g_0} = \delta_{ij} && i,j = 1, \ldots, g_0, \\ &\oint_{C_j} \omega_{i+2g_0} = \delta_{ij} && i,j = 1, \ldots, n-1. \end{aligned} \tag{5.22}$$

Then a basis for the space of even resp. odd holomorphic differentials is given by

$$\begin{aligned} \omega_i^+ &:= \omega_i + \omega_{i+g_0}, && i = 1, \ldots, g_0, \\ \omega_i^- &:= \omega_i - \omega_{i+g_0}, && i = 1, \ldots, g_0, \\ \omega_{g_0+i}^- &:= \omega_{i+2g_0}, && i = 1, \ldots, n-1. \end{aligned}$$

On the other hand, a basis for the odd cycles is given by

$$A_1^-, \ldots, A_{g_0}^-, C_1^-, \ldots, C_{n-1}^-, B_1^-, \ldots, B_{g_0}^-, D_1^-, \ldots, D_{n-1}^-,$$

where $A_i^- := \tilde{A}_i - \sigma(\tilde{A}_i)$ and $B_i^- := \tilde{B}_i - \sigma(\tilde{B}_i)$, for $i = 1, \dots, g_0$, while $C_j^- := C_j$ and $D_j^- := D_j$ for $j = 1, \dots, n-1$. The period matrix of $\mathrm{Prym}(\Gamma/\mathcal{E})$ is in terms of the above basis the $(g - g_0) \times (2g - 2g_0)$ matrix

$$\begin{pmatrix} \oint_{A_i^-} \omega_1^- & \oint_{C_j^-} \omega_1^- & \oint_{B_i^-} \omega_1^- & \oint_{D_j^-} \omega_1^- \\ \vdots & \vdots & \vdots & \vdots \\ \oint_{A_i^-} \omega_{g-g_0}^- & \oint_{C_j^-} \omega_{g-g_0}^- & \oint_{B_i^-} \omega_{g-g_0}^- & \oint_{D_j^-} \omega_{g-g_0}^- \end{pmatrix}, \tag{5.23}$$

where i takes all values from 1 to g_0 and j takes all values from 1 to $n-1$. Using only the parity of the differentials ω_i^- we can rewrite this period matrix as follows:

$$\begin{pmatrix} 2\oint_{\tilde{A}_i} \omega_1^- & \oint_{C_j} \omega_1^- & 2\oint_{\tilde{B}_i} \omega_1^- & \oint_{D_j} \omega_1^- \\ \vdots & \vdots & \vdots & \vdots \\ 2\oint_{\tilde{A}_i} \omega_{g-g_0}^- & \oint_{C_j} \omega_{g-g_0}^- & 2\oint_{\tilde{B}_i} \omega_{g-g_0}^- & \oint_{D_j} \omega_{g-g_0}^- \end{pmatrix}. \tag{5.24}$$

Using, in addition, the normalization (5.22) we find that this matrix has the form $(\Delta\ Z)$, where $\Delta = \mathrm{diag}(2, \dots, 2, 1, \dots, 1)$ (g_0 times 2 and $n-1$ times 1) and where Z can be shown to be a symmetric matrix, whose imaginary part is positive definite. It follows that $\mathrm{Prym}(\Gamma/\mathcal{E})$ inherits from the principal polarization on $\mathrm{Jac}(\Gamma)$ a polarization of type $(1, \dots, 1, 2, \dots, 2)$ ($n-1$ times 1 and g_0 times 2). This polarization is twice a principal polarization when $n = 1$. A similar computation shows that this is also the case when $n = 0$ (see [17, Appendix C]).

Example 5.31. Suppose that Γ is a non-singular curve in an irreducible Abelian surface $\mathbf{T}^2$, and assume that Γ is a double (possibly ramified) cover of a non-singular curve $\mathcal{E}$. We have according to Example 5.26 that $\widehat{\mathbf{T}^2}$ is isogenous to an Abelian surface B, contained in $\mathrm{Jac}(\Gamma)$. But $\widehat{\mathbf{T}^2}$ is irreducible, hence B also. It follows that B, hence also $\mathbf{T}^2$, must be isogenous to an Abelian surface, contained in $\mathrm{Prym}(\Gamma/\mathcal{E})$ or in $\mathrm{Jac}(\mathcal{E})$. In particular, assume that $\mathcal{E}$ is an elliptic curve and that the cover has four ramification points, so that Γ has genus three. Then $\mathrm{Jac}(\mathcal{E}) = \mathcal{E}$ is 1-dimensional, while $\mathrm{Prym}(\Gamma/\mathcal{E})$ is two-dimensional, hence $\mathbf{T}^2$ is isogenous to $\mathrm{Prym}(\Gamma/\mathcal{E})$.

Example 5.32. If an irreducible Abelian surface $\mathbf{T}^2$ contains a non-singular curve $\mathcal{D}$ of genus 3 then it is isomorphic to the dual of $\mathrm{Prym}(\mathcal{D}/\mathcal{E})$, where $\mathcal{E}$ is some elliptic curve. In fact, let us denote by $\gamma : \mathcal{D} \to \mathbf{T}^2$ the inclusion map and let $\tilde{\gamma} : \mathrm{Jac}(\mathcal{D}) \to \mathbf{T}^2$ be the corresponding homomorphism, as given by the universal property of $\mathrm{Jac}(\mathcal{D})$ (Theorem 5.25). Then $\tilde{\gamma}$ is surjective since $\mathbf{T}^2$ is irreducible. The connected component of $\mathrm{Ker}\,\tilde{\gamma}$ is an elliptic curve, which is denoted by $\mathcal{E}$. It leads to an exact sequence

$$0 \longrightarrow \widehat{\mathbf{T}^2} \longrightarrow \mathrm{Jac}(\mathcal{D}) \longrightarrow \mathcal{E} \longrightarrow 0,$$

where we have used that $\mathrm{Jac}(\mathcal{D})$ and $\mathcal{E}$ are isomorphic (in a canonical way) to their duals. Now $\mathcal{D}$ injects naturally in its Jacobian and its composition with the surjective map $\rho : \mathrm{Jac}(\mathcal{D}) \to \mathcal{E}$ gives a covering map $\pi : \mathcal{D} \to \mathcal{E}$. According to the definition of the norm map we have that ρ is the norm map of π. Thus, $\widehat{\mathbf{T}^2}$ is isomorphic to the kernel of $\mathrm{Nm} : \mathrm{Jac}(\mathcal{D}) \to \mathcal{E}$. Since $\mathbf{T}^2$ is connected, $\widehat{\mathbf{T}^2}$ is isomorphic to $\mathrm{Prym}(\mathcal{D}/\mathcal{E})$. By Riemann-Hurwitz the cover $\mathcal{D} \to \mathcal{E}$ has four ramification points, so that the polarization on $\mathrm{Prym}(\mathcal{D}/\mathcal{E})$ that is induced by the principal polarization on $\mathrm{Jac}(\Gamma)$ has type $(1, 2)$.

5.2.5 Families of Abelian Varieties

We first make precise what we mean by a family of (affine parts of) Abelian varieties.

Definition 5.33. Let $\mathbf{F} : V \to S$ be a surjective morphism between connected algebraic varieties. We will say that $\mathbf{F}$ is an *algebraic family* of Abelian varieties (resp. of affine parts of Abelian varieties) if for any $\mathbf{c} \in S$ the fiber $\mathbf{F}^{-1}(\mathbf{c})$ is isomorphic to an Abelian variety (resp. to an affine part of an Abelian variety). We speak of a *smooth algebraic family* if, in addition, V and S are smooth.

Smooth algebraic families of affine parts of Abelian varieties will show up when we get to a.c.i. systems (in Chapter 6 and the chapters that follow). For any construction that applies to a particular type of algebraic variety (such as an Abelian variety) one may wonder if the same thing can be done for a family of algebraic varieties of this type. For example, considering a surjective morphism $\pi : V \to S$, between smooth algebraic varieties V and S, where all fibers of π are compact Riemann surfaces (i.e., π is a smooth algebraic family of compact Riemann surfaces) one may wonder if one can replace in this family every Riemann surface by its Jacobian, to obtain a smooth algebraic family $\overline{\pi} : \overline{V} \to S$ where

$$\overline{\pi}^{-1}(\mathbf{c}) \cong \mathrm{Jac}(\pi^{-1}(\mathbf{c})),$$

for any $\mathbf{c} \in S$. The latter family is then called the *relative Jacobian* of $\pi : V \to S$ or simply of V (one thinks of V as a Riemann surface, parametrized by the points of S), and is denoted by $\mathrm{Jac}(V/S)$. One defines similarly, for any surjective morphism $\pi : V \to S$ with smooth fibers, the *relative Picard variety*[3] $\mathrm{Pic}^0(V/S)$ and its dual, the *relative Albanese variety* $\mathrm{Alb}(V/S)$; the latter satisfies a universal property, just like the Albanese variety (see Paragraph 5.2.3), where the universal morphism takes values in a family of Abelian varieties, rather than in a single Abelian variety.

[3] Recall from Section 5.1 that the Picard variety $\mathrm{Pic}(M)$ of a holomorphic manifold M is the group of holomorphic line bundles on M and the $\mathrm{Pic}^0(M)$ is the subgroup of $\mathrm{Pic}(M)$ that consists of the line bundles with trivial Chern class.

In general these relative objects exist, but only after replacing the base S of the family (the parameter space) by a non-empty Zariski open subset $U \subseteq S$. For the case of the Picard and Albanese varieties, which are the cases that we will use, the result is given by the following theorem (see [70] and [71] for a very general and abstract treatment of the relative Picard and Albanese varieties; they are in the latter referred to as Picard and Albanese schemes).

Theorem 5.34. *Let V and S be smooth algebraic varieties and assume that $\mathbf{F}: V \to S$ is a morphism with smooth fibers. Then there exists over a non-empty Zariski open subset U of S a relative Picard variety* $\mathrm{Pic}^0(\mathbf{F}^{-1}(U)/U)$.

We now state what we mean by a partial compactification of a family of affine parts of Abelian varieties. Such families will show up later in the context of integrable systems and their compactification will be important in Section 6.2.1.

Definition 5.35. Suppose that $\mathbf{F}: V \to S$ is a smooth algebraic family of affine parts of Abelian varieties. We say that $\mathbf{F}$ admits a partial compactification over a (Zariski) open subset $U \subseteq S$ if there exists a smooth algebraic family $\overline{\mathbf{F}}: \overline{V} \to U$ of Abelian varieties, a morphism $\imath : \mathbf{F}^{-1}(U) \to \overline{V}$ and an analytic hypersurface $\mathcal{D}$ on $\overline{V}$ such that such that

(1) the following diagram commutes,

$$\begin{array}{ccc} \mathbf{F}^{-1}(U) & \xrightarrow{\ \imath\ } & \overline{V} \\ {\scriptstyle \mathbf{F}}\downarrow & \swarrow {\scriptstyle \overline{\mathbf{F}}} & \\ U & & \end{array}$$

(2) $\imath$ restricts to an isomorphism $\mathbf{F}^{-1}(U) \to \overline{V} \setminus \mathcal{D}$.

Notice that in view of Proposition 5.30, for any $\mathbf{c} \in S$ the Abelian variety $\overline{\mathbf{F}^{-1}(\mathbf{c})}$ that compactifies $\mathbf{F}^{-1}(\mathbf{c})$ is unique, up to isomorphism, and it is the Albanese variety of $\mathbf{F}^{-1}(\mathbf{c})$. This suggests that partial compactifications should be constructed via the relative Albanese variety. We do this in the following theorem by first constructing the relative Picard variety, as given by Theorem 5.34. The proof of this theorem was kindly provided to us by José Bertin.

Theorem 5.36. *Let $\mathbf{F}: V \to S$ be a smooth algebraic family of affine parts of Abelian varieties. Then there exists, for some non-empty Zariski open subset U of S a partial compactification of $\mathbf{F}$ over U. This partial compactification is the relative Albanese variety* $\mathrm{Alb}(\mathbf{F}^{-1}(U)/U)$.

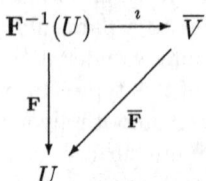

Proof. We first make a (singular) partial compactification of $\mathbf{F}$ over S; we do this by embedding $\mathbf{F}$ in a projective family. This can be done as follows. We may assume the base S is an affine set because the theorem is to be shown only for an open subset U of S. Then we way assume that the total space V is also affine, simply by taking an affine open subset of V. Then taking generators for the algebras of regular functions on S and V, we may realize V as a closed subvariety of $\mathbf{C}^n \times S$. Then taking the closure $\overline{V}$ of V in the relative projective space $\mathbf{P}^n \times S$, we get a projective compactification of our family, that is

$$V \subset \overline{V} \xrightarrow{\pi} S$$

the morphism π being projective. Note the variety $\overline{V}$ is no longer smooth. But we may desingularize it, by Hironanka's theorem.

Precisely, there is a projective morphism $W \to \overline{V}$ which is a product of blow-ups, where W is smooth. Since V is non-singular the center of each of these blow-ups can be chosen in the boundary $\overline{V} - V$; with this choice, V is naturally isomorphic to an open subset of W. In conclusion, we can choose a projective compactification $V \subset \overline{V} \xrightarrow{\pi} S$, with $\overline{V}$ smooth. Since the map will be smooth over the fibers that do not contain the centers (there are a finite number of them) we may, after replacing the base S by an open subset U', assume that the morphism $\pi : \overline{V} \to U'$ is smooth and projective.

By Theorem 5.34 the relative Picard variety exists, say over an open subset $U \subseteq U'$, and the fiber over $\mathbf{c} \in U$ is the Picard variety $\mathrm{Pic}^0(\pi^{-1}(\mathbf{c}))$. Taking its dual we get the relative Albanese variety, whose fiber over $\mathbf{c}$ is the Albanese variety of $\pi^{-1}(\mathbf{c})$ but also of $\mathbf{F}^{-1}(\mathbf{c})$, since we know the Albanese variety is a birational invariant. It follows that a partial compactification of $\mathbf{F}$ exists, at least over a Zariski open subset U of the base. □

5.3 Divisors in Abelian Varieties

In general, any ample divisor contains a lot of information about its ambient algebraic variety. The tools that will be developed in the Chapters 6 and 7 are based on this fact, in the particular case of effective divisors on irreducible Abelian varieties; recall in this respect that any effective divisor on an irreducible Abelian variety is ample. Let us first compute the restriction of the Kodaira map $\varphi_{\mathcal{D}}$ to $\mathcal{D}'$, where $\mathcal{D}'$ is an irreducible component (taken with multiplicity 1) of a very ample divisor $\mathcal{D}$ on an irreducible Abelian variety $\mathbf{T}^r$, where $\mathcal{D}$ is assumed effective. To do this, we use a holomorphic vector field $\mathcal{V}$ on $\mathbf{T}^r$. Let σ be a section of $[\mathcal{D}]$ for which $(\sigma) = \mathcal{D}$ and define for $i = 0, \dots, N$ a section of $[\mathcal{D}]$ by $\sigma_i := z_i\sigma$, where $(1 = z_0, z_1, \dots, z_N)$ is a basis for $L(\mathcal{D})$. Let us denote by $z_0(t;\mathcal{D}'), \dots, z_N(t;\mathcal{D}')$ the Laurent series of $z_0, \dots, z_N$ with respect of $\mathcal{V}$, starting at $\mathcal{D}'$ (see Paragraph 2.1.1),

$$z_i(t;\mathcal{D}') = \frac{1}{t^p}\left(z_i^{(0)} + z_i^{(1)}t + O(t^2)\right),$$

where the coefficients $z_i^{(k)}$ are meromorphic functions on $\mathcal{D}'$, with $z^{(0)} \neq 0$ (but some of the $z_i^{(0)}$ may be zero) so that $p \in \mathbf{N}^*$ is the multiplicity of $\mathcal{D}'$ in $\mathcal{D}$. The specialization of the series $z_i(t;\mathcal{D}')$ to a generic point $m \in \mathcal{D}'$ will be denoted by $z_i(t;m)$; the coefficients $z_i^{(j)}$ are then also denoted by $z_i^{(j)}(m)$. If the multiplicity of $\mathcal{D}'$ in $\mathcal{D}$ is 1, so that $p=1$, then each $z_i(t;\mathcal{D}')$ has at most a simple pole along $\mathcal{D}'$ and for a generic $m \in \mathcal{D}'$,

$$\begin{aligned}\varphi_{\mathcal{D}}(m) &= \lim_{t\to 0}(z_0(t;m):z_1(t;m):\cdots:z_N(t;m))\\ &= \lim_{t\to 0} t(z_0(t;m):z_1(t;m):\cdots:z_N(t;m))\\ &= \left(0:z_1^{(0)}(m):\cdots:z_N^{(0)}(m)\right),\end{aligned}$$

so that the residues of the functions z_i embed a Zariski open subset U of $\mathcal{D}'$, since $\mathcal{D}$ is very ample.

For points $m' \in \mathcal{D}' \setminus U$, i.e., points at which at least one of the functions $z_i^{(0)}$ has a pole, or for which all the above residues vanish, $\varphi_{\mathcal{D}}(m')$ is computed from

$$\varphi_{\mathcal{D}}(m') = \lim_{m\to m'} \varphi_{\mathcal{D}}(m) = \lim_{m\to m'} \left(0:z_1^{(0)}(m):\cdots:z_N^{(0)}(m)\right),$$

with the limit of course being taken in $\mathbf{P}^N$. If the multiplicity p of $\mathcal{D}'$ in $\mathcal{D}$ is larger than 1 then by the same argument the leading terms in t of those $z_i(t;\mathcal{D}')$ that have a pole of order p (and no less) along $\mathcal{D}'$, embed (a Zariski open subset of) $\mathcal{D}'$ in $\mathbf{P}^N$.

Next, let us show that, in terms of certain variables, any holomorphic vector field on an irreducible Abelian variety is quadratic. Let $\mathcal{D}$ be an effective divisor on an irreducible Abelian variety $\mathbf{T}^r$ and suppose that $[\mathcal{D}]$ is normally generated (in particular $\mathcal{D}$ is very ample). Also, let $\mathcal{V}$ be any holomorphic vector field on $\mathbf{T}^r$ and write $\dot{f}$ for $\mathcal{V}[f]$. Introduce for $f,g \in \mathcal{M}(\mathbf{T}^r)$ the *Wronskian* $W(f,g)$ of f and g with respect to $\mathcal{V}$ as

$$W(f,g) = \dot{f}g - f\dot{g}.$$

We claim that if $f,g \in L(\mathcal{D})$ then $W(f,g) \in L(2\mathcal{D})$. To prove this it suffices to show that if f and g have a pole of order p along some irreducible analytic hypersurface $\mathcal{D}'$ then $W(f,g)$ has a pole of order $2p$ (at worst) along $\mathcal{D}'$. But

$$\mathrm{ord}_{t=0}\, W(f,g)(t;\mathcal{D}') \leqslant \mathrm{ord}_{t=0}\, f(t;\mathcal{D}') + \mathrm{ord}_{t=0}\, g(t;\mathcal{D}'),$$

since the leading term of the series cancels out, and the result follows. Since $[\mathcal{D}]$ is normally generated the natural map $\mathrm{Sym}^2 L(\mathcal{D}) \to L(2\mathcal{D})$ is surjective, and $W(f,g)$ can be written as a quadratic polynomial in $L(\mathcal{D})$. Let us apply this to a basis $(z_0,\ldots,z_N)$ of $L(\mathcal{D})$. Fix any l such that $0 \leqslant l \leqslant N$ and define $y_i := z_i/z_l$, where $0 \leqslant i \leqslant N$.

Then

$$\dot{y}_i = \frac{W(z_i, z_l)}{z_l^2} = \frac{1}{z_l^2} \sum_{j,k=0}^{N} a_{jk}^{(i)} z_j z_k = \sum_{j,k=0}^{N} a_{jk}^{(i)} y_j y_k,$$

for some constants $a_{jk}^{(i)}$. Thus, any holomorphic vector field on $\mathbf{T}^r$ is quadratic when expressed in terms of the ratios $z_0/z_l, \ldots, z_N/z_l$, where l is fixed but arbitrary. Thus, if we are in an affine chart of $\mathbf{P}^N$ defined by $z_l \neq 0$, i.e., such that $y_0, \ldots, y_N$ define affine coordinates, then the above provides in this chart a description of the vector field as a quadratic vector field, in terms of these coordinates.

The fact that some elements of $L(2\mathcal{D})$ can be constructed by taking the Wronskian of two functions in $L(\mathcal{D})$ yields a useful tool for constructing embeddings of Abelian varieties in projective space. As was shown by Luis Piovan in [141] a basis of $L(2\mathcal{D})$ may, in certain cases, be constructed from a basis of $L(\mathcal{D})$ by adding these Wronskians to the products $z_i z_j$ of sections of $L(\mathcal{D})$.

Theorem 5.37 (Piovan). *Let $\mathcal{D}$ be a symmetric divisor on a Jacobi surface* $\mathrm{Jac}(\Gamma)$*, and assume that $[\mathcal{D}]$ induces on* $\mathrm{Jac}(\Gamma)$ *twice the principal polarization. If $z_0, \ldots, z_3$ is a basis of $L(\mathcal{D})$ and $\mathcal{V}$ is a generic vector field on* $\mathrm{Jac}(\Gamma)$ *then the products $z_i z_j$ and the Wronskians $W(z_i, z_j)$, where $0 \leqslant i \leqslant j \leqslant 3$, generate $L(2\mathcal{D})$.*

5.3.1 The Case of Non-singular Divisors

We assume in this paragraph that $\mathcal{D}$ is a non-singular divisor on an irreducible Abelian variety $\mathbf{T}^r$. We will relate its genus $g(\mathcal{D})$, which we defined as the dimension of $H^0(\mathcal{D}, \Omega^{r-1})$, on the one hand to the polarization type that it induces on $\mathbf{T}^r$, and on the other hand to its self-intersection. We will also explain how a basis for the above space $H^0(\mathcal{D}, \Omega^{r-1})$ can be computed from a basis of $L(\mathcal{D})$ and a basis for the space of holomorphic vector fields on $\mathbf{T}^r$. In the next section we will see how all this works for $\mathcal{D}$ singular, with $g(\mathcal{D})$ replaced by the arithmetic genus.

Theorem 5.38. *Let $\mathcal{D}$ be a non-singular analytic hypersurface of an irreducible Abelian variety $\mathbf{T}^r$ and let $(\delta_1, \ldots, \delta_r)$ denote the polarization type that $\mathcal{D}$ induces on $\mathbf{T}^r$. Then the following string of equalities hold.*

$$\chi([\mathcal{D}]) = \dim L(\mathcal{D}) = \delta_1 \ldots \delta_r = g(\mathcal{D}) - r + 1 - \frac{c_1([\mathcal{D}])^r}{r!} = \frac{\mathcal{D}^r}{r!}. \tag{5.25}$$

Proof. Since $\mathbf{T}^r$ is irreducible, $[\mathcal{D}]$ is a positive line bundle and the first equality follows from (5.14). The second equality follows from (5.15) by using (5.9). Since $\mathbf{T}^r$ is a compact Kähler manifold, $g(\mathcal{D})$ is given by (5.8), which specializes to

$$g(\mathcal{D}) = h^0(\mathbf{T}^r, \Omega^r(\mathcal{D})) + r - 1. \tag{5.26}$$

Indeed, $h^0(\mathbf{T}^r, \Omega^r) = h^0(\mathbf{T}^r, \mathcal{O})) = 1$, as a consequence of (5.13), and $h^0(\mathbf{T}^r, \Omega^{r-1}) = r$ since $H^0(\mathbf{T}^r, \Omega^{r-1})$ is spanned by the r forms $\mathrm{d}t_1 \wedge \ldots \wedge \widehat{\mathrm{d}t_i} \wedge \ldots \wedge \mathrm{d}t_r$, where the t_i are linear coordinates on $\mathbf{T}^r$. Using (5.14), the third equality in (5.25) follows from (5.26). Writing $c_1([\mathcal{D}]) = [\sum_\alpha \delta_\alpha \mathrm{d}x_\alpha \wedge \mathrm{d}x_{r+\alpha}]$ one finds

$$c_1([\mathcal{D}])^r = r!\delta_1 \ldots \delta_r [\mathrm{d}x_1 \wedge \ldots \wedge \mathrm{d}x_{2r}],$$

so that, when interpreted as an element of $\mathbf{Z}$, $c_1([\mathcal{D}])^r = r!\delta_1 \ldots \delta_r$. This yields the fourth equality in (5.25). Finally, $c_1([\mathcal{D}])$ is the Poincaré dual of the cycle, carried by $\mathcal{D}$ and the wedge of (cohomology classes of) two-forms corresponds, under the Poincaré Duality to the intersection of cycles. Therefore,

$$c_1([\mathcal{D}])^r = c_1([\mathcal{D}]) \wedge \ldots \wedge c_1([\mathcal{D}]) = \mathcal{D} \cdot \mathcal{D} \cdots \mathcal{D} = \mathcal{D}^r,$$

which leads to the last equality in (5.25). □

We show in the following example how part of the string of formulas in Theorem 5.38 can be obtained in the case of Abelian surfaces from the general formulas for curves on algebraic surfaces.

Example 5.39. Suppose that $\mathcal{D}$ is a non-singular curve on $\mathbf{T}^2$, which induces a polarization on $\mathbf{T}^2$ of type (δ_1, δ_2). The adjunction formula for a non-singular divisor $\mathcal{D}$ on a compact complex surface M, which expresses its canonical bundle in terms of its normal bundle and the canonical bundle on M, reads

$$\mathcal{K}_\mathcal{D} = \mathcal{K}_M \otimes \mathcal{O}_\mathcal{D}(\mathcal{D}),$$

which yields the following formula for the genus of $\mathcal{D}$,

$$g(\mathcal{D}) = \frac{\mathcal{D} \cdot \mathcal{D} + \mathcal{K}_M \cdot \mathcal{D}}{2} + 1.$$

In particular, since $\mathcal{K}_{\mathbf{T}^2} = 0$ we have that $g(\mathcal{D}) = \frac{\mathcal{D} \cdot \mathcal{D}}{2} + 1$ for $M = \mathbf{T}^2$. The Riemann-Roch theorem for line bundles on a surface tells you that

$$\chi([\mathcal{D}]) = \frac{\mathcal{D} \cdot \mathcal{D} - \mathcal{K}_M \cdot \mathcal{D}}{2} + \chi(\mathcal{O}_M);$$

where the last term is, according to the *Noether formula*, given by

$$\chi(\mathcal{O}_M) = \frac{1}{12}(\mathcal{K}_M \cdot \mathcal{K}_M + \chi(M)).$$

For an Abelian surface $\mathbf{T}^2$ the topological Euler characteristic $\chi(\mathbf{T}^2)$ is zero, so that $\chi(\mathcal{O}_{\mathbf{T}^2}) = 0$ and

$$\chi([\mathcal{D}]) = \frac{\mathcal{D} \cdot \mathcal{D}}{2} = g(\mathcal{D}) - 1,$$

which proves part of (5.25) in the case of an irreducible Abelian surface.

We now come to the construction of basis for the space of top-forms on $\mathcal{D}$.

Theorem 5.40. *Let $\mathcal{D}$ be a non-singular analytic hypersurface of an irreducible Abelian variety $\mathbf{T}^r$ and let $(t_1,\dots,t_r)$ be a system of linear coordinates on $\mathbf{T}^r$. Choose a basis $(z_0=1,z_1,\dots,z_N)$ of $L(\mathcal{D})$ and define for $i=1,\dots,r$, a top-form on $\mathcal{D}$ by*

$$\omega_i := \mathsf{d}t_1\wedge\ldots\wedge\widehat{\mathsf{d}t_i}\wedge\ldots\wedge\mathsf{d}t_r\Big|_{\mathcal{D}}. \tag{5.27}$$

Then

$$H^0(\mathcal{D},\Omega^{r-1}) = \operatorname{span}\left\{z_1^{(0)}\omega_1,\dots,z_N^{(0)}\omega_1\right\}\oplus\operatorname{span}\left\{\omega_1,\dots,\omega_r\right\}, \tag{5.28}$$

where $z^0_{(1)},\dots,z_N^{(0)}$ are the residues of the series $z_1(t_1;\mathcal{D}),\dots,z_N(t_1;\mathcal{D})$.

Proof. Consider the exact sequence (5.6), which reduces for $\mathbf{T}^r$ by the Kodaira-Nakano Vanishing Theorem and by (5.7) to

$$\begin{aligned}0 \longrightarrow H^0(\mathbf{T}^r,\Omega^r)\longrightarrow H^0(\mathbf{T}^r,\Omega^r(\mathcal{D}))\stackrel{\mathrm{PR}}{\longrightarrow} H^0(\mathcal{D},\Omega^{r-1})\\ \longrightarrow H^0(\mathbf{T}^r,\Omega^{r-1})\longrightarrow 0.\end{aligned}$$

It follows that $H^0(\mathcal{D},\Omega^{r-1})$ is the direct sum of the space of top-forms on $\mathcal{D}$ which are restrictions of holomorphic $(r-1)$-forms on $\mathbf{T}^r$, and the image of PR. Since the Poincaré Residue of a holomorphic r-form is zero,

$$H^0(\mathcal{D},\Omega^{r-1}) = \operatorname{span}\left\{\mathrm{PR}(z_1\omega),\dots,\mathrm{PR}(z_N\omega)\right\}\oplus H^0(\mathbf{T}^r,\Omega^{r-1}), \tag{5.29}$$

where ω is the top-form $\mathsf{d}t_1\wedge\ldots\wedge\mathsf{d}t_r$ on $\mathbf{T}^r$, and where we view $H^0(\mathbf{T}^r,\Omega^{r-1})$ as a subspace of $H^0(\mathcal{D},\Omega^{r-1})$, by restriction. We have that $h^0(\mathbf{T}^r,\Omega^{r-1})=r$ because a basis for the space of holomorphic $(r-1)$-forms on $\mathbf{T}^r$ is given by the forms $\mathsf{d}t_1\wedge\ldots\wedge\widehat{\mathsf{d}t_i}\wedge\ldots\wedge\mathsf{d}t_r$, $(i=1,\dots,r)$. Substituted in (5.29) this yields, combined with (5.25),

$$\dim\operatorname{span}\left\{\mathrm{PR}(z_1\omega),\dots,\mathrm{PR}(z_N\omega)\right\} = g(\mathcal{D})-r = \dim L(\mathcal{D})-1 = N,$$

which shows that the Poincaré Residues $\mathrm{PR}(z_1\omega),\dots,\mathrm{PR}(z_N\omega)$ are linearly independent. In order to compute $\mathrm{PR}(z_i\omega)$, for $i=1,\dots,N$, consider the Laurent series[4] of z_i with respect to $\partial/\partial t_1$, starting at $\mathcal{D}$,

$$z_i(t_1;\mathcal{D}) = \frac{1}{t_1}\left(z_i^{(0)}+z_i^{(1)}t_1+O(t_1^2)\right).$$

[4] We will see in Chapter 7 how the first few terms of these series can be explicitly computed for an important class of vector fields.

Then, by (5.5),

$$\begin{aligned}\mathrm{PR}(z_i\omega) &= \left.\frac{\mathrm{d}t_2 \wedge \ldots \wedge \mathrm{d}t_r}{\frac{\partial}{\partial t_1}(1/z_i)}\right|_{\mathcal{D}} \\ &= \left.\frac{\mathrm{d}t_2 \wedge \ldots \wedge \mathrm{d}t_r}{\frac{\partial}{\partial t_1}(t_1/z_i^{(0)} + O(t_1^2))}\right|_{t_1=0} = z_i^{(0)}\omega_1,\end{aligned}$$

yielding the first factor in (5.28). For the second factor in (5.28), we restrict the above basis $\mathrm{d}t_1 \wedge \ldots \wedge \widehat{\mathrm{d}t_i} \wedge \ldots \wedge \mathrm{d}t_r$, $(i = 1, \ldots, r)$, to $\mathcal{D}$, which gives, by definition, the top-forms $\omega_1, \ldots, \omega_r$ on $\mathcal{D}$. □

We will see in Paragraph 7.6.7 how the $(r-1)$-forms ω_i, defined by (5.27), can be explicitly computed when an explicit basis for the space of holomorphic vector fields on $\mathbf{T}^r$ is known.

5.3.2 The Case of Singular Divisors

The divisors that we will encounter in this book will often be singular. All formulas and notions that cease to make sense (holomorphic differentials, genus, canonical bundle, Poincaré Residue, ...) admit a natural adaption in such a way that all the formulas from the previous paragraph still hold (in particular, they still make sense). This has been worked out in detail only in the case of divisors on a complex surface, i.e., in the case of curves, which is the only case that we will need. We present a few elements of this theory here and refer for a systematic treatment of it to [28].

Let $\mathcal{D}$ be a curve in a non-singular algebraic surface M. If $\mathcal{D}$ is non-singular and connected then its (geometric) genus g is given by $g(\mathcal{D}) = h^0(\mathcal{D}, \Omega^1)$, which can in view of Kodaira-Serre Duality also be written as

$$g(\mathcal{D}) = h^0(\mathcal{D}, \Omega^1) = h^1(\mathcal{D}, \mathcal{O}). \tag{5.30}$$

Now (5.11) yields in the case of the trivial line bundle $h^1(\mathcal{D}, \mathcal{O}) = 1 - \chi(\mathcal{O}_{\mathcal{D}})$, so that (5.30) becomes

$$g(\mathcal{D}) = 1 - \chi(\mathcal{O}_{\mathcal{D}}). \tag{5.31}$$

Since $\chi(\mathcal{O}_{\mathcal{D}})$ makes also sense for singular curves, or even for arbitrary divisors on a smooth algebraic surface, the right hand side in (5.31) is defined to be the *arithmetic genus* of $\mathcal{D}$, denoted $g_a(\mathcal{D})$. Thus, $g_a(\mathcal{D}) := 1 - \chi(\mathcal{O}_{\mathcal{D}})$. The *geometric genus* of $\mathcal{D}$ is by definition the genus of the (minimal) desingularization of $\mathcal{D}$. Thus, for an analytic hypersurfaces, $g_a(\mathcal{D}) = g(\mathcal{D})$ if and only if $\mathcal{D}$ is non-singular; moreover, since $\chi(\mathcal{O}_{\mathcal{D}})$ is invariant under deformation, $g_a(\mathcal{D})$ is the genus of any smooth curve in M, which is a small deformation of $\mathcal{D}$.

Notice that since intersection multiplicities of divisors are also invariant under deformation, (5.25) implies that, for an effective divisor $\mathcal{D}$ on an Abelian variety $\mathbf{T}^2$,

$$g_a(\mathcal{D}) = \frac{\mathcal{D}^2}{2} + 1, \qquad (5.32)$$

and (5.25) still holds when $g(\mathcal{D})$ is replaced by $g_a(\mathcal{D})$.

Example 5.41. Suppose that $\mathcal{D}$ is a non-singular analytic hypersurface of an irreducible Abelian variety $\mathbf{T}^2$ and consider $k\mathcal{D}$, where $k \in \mathbf{N}^*$. Since $(k\mathcal{D})^2 = k^2\mathcal{D}^2$ by the above formula, the arithmetic genus of $k\mathcal{D}$ is given by

$$g_a(k\mathcal{D}) = k^2(g(\mathcal{D}) - 1) + 1,$$

where $g(\mathcal{D}) = g_a(\mathcal{D})$ because $\mathcal{D}$ is non-singular.

A close investigation of the blow-up $\pi : \tilde{\mathcal{D}} \to \mathcal{D}$ near the singularities of $\mathcal{D}$ yields a relation between the geometric and the arithmetic genus of a singular divisor, as given in the following proposition.

Proposition 5.42. *Let $\mathcal{D}$ be an analytic hypersurface on an irreducible Abelian surface $\mathbf{T}^2$ and let $\pi : \tilde{\mathcal{D}} \to \mathcal{D}$ be its normalization. Let $\mathcal{D}_1, \ldots, \mathcal{D}_s$ denote the irreducible components of $\mathcal{D}$ and denote the connected components of $\tilde{\mathcal{D}}$ (which are the normalization of $\mathcal{D}_1, \ldots, \mathcal{D}_s$) by $\tilde{\mathcal{D}}_1, \ldots, \tilde{\mathcal{D}}_s$. Then*

$$g_a(\mathcal{D}) = \sum_{i=1}^{s} g(\tilde{\mathcal{D}}_i) + 1 - s + \chi(Z_{\mathcal{D}}),$$

where $Z_{\mathcal{D}}$ is the quotient sheaf $\pi_*\mathcal{O}_{\tilde{\mathcal{D}}}/\mathcal{O}_{\mathcal{D}}$ on $\mathcal{D}$.

Proof. Consider the *normalization sequence* of π, which is the exact sequence of sheaves on $\mathcal{D}$, given by

$$0 \longrightarrow \mathcal{O}_{\mathcal{D}} \longrightarrow \pi_*\mathcal{O}_{\tilde{\mathcal{D}}} \longrightarrow Z_{\mathcal{D}} \longrightarrow 0, \qquad (5.33)$$

where $\pi : \tilde{\mathcal{D}} \to \mathcal{D}$. From its long exact cohomology sequence we have that

$$\chi(\mathcal{O}_{\mathcal{D}}) - \chi(\pi_*\mathcal{O}_{\tilde{\mathcal{D}}}) + \chi(Z_{\mathcal{D}}) = 0. \qquad (5.34)$$

Now

$$\chi(\mathcal{O}_{\mathcal{D}}) = 1 - g_a(\mathcal{D}),$$
$$\chi(\pi_*\mathcal{O}_{\tilde{\mathcal{D}}}) = \sum_{i=1}^{s} \chi(\mathcal{O}_{\tilde{\mathcal{D}}_i}) = s - \sum_{i=1}^{s} g(\tilde{\mathcal{D}}_i),$$

where s denotes the number of connected components of $\tilde{\mathcal{D}}$. It suffices now to substitute these expressions in (5.34). $\square$

Remark 5.43. If $\mathcal{D}_1, \ldots, \mathcal{D}_s$ are non-singular curves on an Abelian surface $\mathbf{T}^2$ and $\kappa_1, \ldots, \kappa_s$ are any positive integers, then (5.32) implies that

$$g_a\left(\sum_{i=1}^{s} \kappa_i \mathcal{D}_i\right) = \frac{1}{2}\left(\sum_{i=1}^{s} \kappa_i \mathcal{D}_i\right)^2 + 1$$
$$= \sum_{i=1}^{s} \kappa_i^2 \frac{\mathcal{D}_i^2}{2} + \sum_{1 \leqslant i < j \leqslant s} \kappa_i \kappa_k \mathcal{D}_i \cdot \mathcal{D}_j + 1,$$

so that

$$g_a\left(\sum_{i=1}^{s} \kappa_i \mathcal{D}_i\right) = \sum_{i=1}^{s} \kappa_i^2 (g(\mathcal{D}_i) - 1) + \sum_{1 \leqslant i < j \leqslant s} \kappa_i \kappa_k \chi(\mathcal{D}_i \cup \mathcal{D}_j) + 1.$$

We show how to compute $\chi(Z_{\mathcal{D}})$, which is the contribution by the singular points of a singular curve to its arithmetic genus. We will do this in the case of a planar singularity, i.e., the tangents to the s branches of $\mathcal{D}$ at m lie in a plane. Notice that the computation is completely local; if $m_1, \ldots, m_k$ denote the singular points of $\mathcal{D}$ then $\chi(Z_{\mathcal{D}}) = \sum_{i=1}^{k} \chi(Z_{m_i})$, where Z_{m_i} is the restriction of $Z_{\mathcal{D}}$ to a small neighborhood of the singular point m_i. The integer $\chi(Z_{m_i})$ is an invariant of the singularity m_i. Letting m be a singular point of $\mathcal{D}$ we have that $\chi(Z_m) = h^0(\tilde{\mathcal{D}}, Z_m)$, which is in view of (5.33) the dimension of the space of (germs of) holomorphic functions on the different branches around m taken separately, modulo the (germs of) holomorphic functions on the curve $\mathcal{D}$ near m. Explicitly, let $\mathcal{D}_i$ denote the i-th branch of $\mathcal{D}$ at m, with local defining equation $f_i(x, y) = 0$, where $i = 1, \ldots, s$, and let $\gamma_i : U \subseteq \mathbf{C} \to \mathbf{C}^2$ be a local parameterization of $\mathcal{D}_i$ at m, where U is an open neighborhood of the origin in $\mathbf{C}$. These parameterizations, taken together, lead to a map

$$\begin{aligned} \Psi : \mathbf{C}[[x,y]]/\textstyle\prod_{i=1}^{s} f_i(x,y) &\to \textstyle\prod_{i=1}^{s} \mathbf{C}[[\varsigma_i]] \\ h(x,y) &\mapsto (h \circ \gamma_1(\varsigma_1), \ldots, h \circ \gamma_s(\varsigma_s)), \end{aligned}$$

which maps a germ of holomorphic functions on $\mathcal{D}$ at m to its value at each of the branches separately. It follows that $\chi(Z_m)$ is given by

$$\chi(Z_m) = \dim \frac{\prod_{i=1}^{s} \mathbf{C}[[\varsigma_i]]}{\Psi\left(\mathbf{C}[[x,y]]/\prod_{i=1}^{s} f_i(x,y)\right)}. \tag{5.35}$$

Remark 5.44. There is an alternative way to compute $\chi(Z_m)$, which goes as follows. Suppose that m_0 is a singular point of a curve $\mathcal{D}_0$, and explicitly blow the surface M up in a neighborhood of m_0, with center m_0. Consider the proper (or strict) transform $\mathcal{D}_1$ of $\mathcal{D}_0$, which is the closure of the inverse image, under the blow-up, of $\mathcal{D}_0 \setminus \{m_0\}$. The points on $\mathcal{D}_1$ are called *infinitely near points* on $\mathcal{D}_0$ of the first order. The points over m_0 may or may not be singular points. If one of them is singular, call it m_1 and do another blow-up, now with center m_1, which yields $\mathcal{D}_2$, the proper transform of $\mathcal{D}_1$, whose points are called infinitely near points of the second order. This may again produce singular points, and we may still have singular points left from the first blow-up (these are now viewed as points on $\mathcal{D}_2$; indeed a blow-up is a biholomorphism, away from its center). In any case, we repeat the above process, calling one of the singular infinitely near points p_i, doing a blow-up with center p_i and calling the proper transform $\mathcal{D}_{i+1}$, whose points are called infinitely near points of order $i+1$. It is a fundamental fact that we get only a finite number of singular (infinitely near) points in this way. For $i = 0, 1, \dots,$ the order of m_i on $\mathcal{D}_i$ is denoted by e_i. With this notation, *Clebsch's Formula*, states that

$$\chi(Z_{m_0}) = \sum_i \frac{e_i(e_i - 1)}{2}, \tag{5.36}$$

where the sum is over as many terms as needed to get rid of all singularities, produced when blowing up recursively m_0 and all its infinitely near singular points. For more details on this method we refer to [89, Chapter 9]; the method will be illustrated in example 5.48 below.

We show in the examples below how the data of Table 5.1 were computed.

Table 5.1. For the simplest planar singularities we give the Euler characteristic χ and the number s of branches at that point.

singularity	equation	picture	χ	s
node	$y^2 = x^2$		1	2
cusp	$y^2 = x^3$		1	1
n-node	$y^n = x^n$		$\binom{n}{2}$	n
tacnode	$y^2 = x^{2n}$	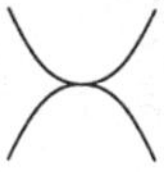	n	2
cusp	$y^2 = x^{2n+1}$		n	1

Example 5.45. The simplest example of a singular point is that of a *node*, also called an *ordinary double point* or *crunode*, which is a double point with distinct tangent lines. We may assume that the singular point m is the origin, and that the equation of the curve is given (around 0) by $\mathcal{D} : y^2 = x^2$. The two branches are then given by $\mathcal{D}_1 : y = x$ and $\mathcal{D}_2 : y = -x$. Let $h(x, y)$ be a germ of a holomorphic function on $\mathcal{D}$ at 0. Since we can replace in the series for h all terms that contain y^i with $i \geqslant 2$ by using the relation $y^2 = x^2$, we can write h, up to irrelevant terms, as

$$h(x, y) = a_0 + a_1 x + a_2 x^2 + \cdots + y(b_1 + b_2 x + \cdots).$$

A local parameterization for $\mathcal{D}_1$ and for $\mathcal{D}_2$ is given by $\gamma_1(\varsigma_1) = (\varsigma_1, \varsigma_1)$ and $\gamma_2(\varsigma_2) = (\varsigma_2, -\varsigma_2)$. This yields

$$\begin{aligned} h \circ \gamma_1(\varsigma_1) &= a_0 + (a_1 + b_1)\varsigma_1 + (a_2 + b_2)\varsigma_1^2 + \cdots, \\ h \circ \gamma_2(\varsigma_2) &= a_0 + (a_1 - b_1)\varsigma_2 + (a_2 - b_2)\varsigma_2^2 + \cdots. \end{aligned}$$

Since all the coefficients of these two series, taken together, are independent, except for the leading ones (which coincide) we find from (5.35) that $\chi(Z_m) = 1$ and we have that $s = 2$, since there are two branches at 0.

Example 5.46. The previous example is easily generalized to a curve $\mathcal{D}$ with an n-node (ordinary n-fold point), which means that at m it has at n branches with distinct tangents. We may assume in this case that $m = 0$ and that the equation of the curve is given (around 0) by $\mathcal{D} : y^n = x^n$. The n branches are then given by $\mathcal{D}_i : y = \epsilon^i x$, where ϵ is a primitive n-th root of unity and $i = 0, \ldots, n-1$. Let $h(x, y)$ be a germ of a holomorphic function on $\mathcal{D}$ at 0. As above, we can write h, up to terms that are irrelevant, as

$$\begin{aligned} h(x, y) = a_{00} + a_{10}x + a_{20}x^2 + \cdots &+ y(a_{11} + a_{21}x + \cdots) \\ &+ \cdots \\ &+ y^{n-1}(a_{n-1,n-1} + a_{n,n-1}x + \cdots). \end{aligned}$$

Using the local parameterization $\gamma_i(\varsigma_i) = (\varsigma_i, \epsilon^i \varsigma_i)$ for $\mathcal{D}_i$ we find

$$h \circ \gamma_i(\varsigma_i) = a_{00} + (a_{10} + \epsilon^i a_{11})\varsigma_i + \cdots + (a_{n-1,0} + \cdots + \epsilon^{i(n-1)} a_{n-1,n-1})\varsigma_i^{n-1} + \cdots$$

It follows that starting from the term in ς_i^{n-1} all the coefficients of these n series are independent; for the term in ς_i^j, where $0 \leqslant j < n-1$, there are only $j + 1$ independent constants that appear. Summing up for $0 \leqslant j < n-1$ we find that

$$\chi(Z_m) = \binom{n}{2} = \frac{n(n-1)}{2}.$$

Example 5.47. We now turn to a *tacnode*, which is a double point m at which two branches meet with common tangents at m. We may assume that an equation of the curve is given (around $m = 0$) by $\mathcal{D} : y^2 = x^{2n}$. The 2 branches are in this case given by $\mathcal{D}_1 : y = x^n$ and $\mathcal{D}_2 : y = -x^n$. A germ of a holomorphic function on $\mathcal{D}$ at 0 can be written as

$$h(x, y) = a_0 + a_1 x + a_2 x^2 + \cdots + y(b_1 + b_2 x + \cdots).$$

Local parameterizations γ_1 and γ_2 are given by $\gamma_1(\varsigma_1) = (\varsigma_1, \varsigma_1^n)$ and $\gamma_1(\varsigma_1) = (\varsigma_1, -\varsigma_1^n)$. It is clear that in this case the first n terms of the series $h(\gamma_1(\varsigma_1))$ and $h(\gamma_2(\varsigma_2))$ are identical, and that the other terms are independent. Therefore, $\chi(Z_m) = n$.

Example 5.48. Our final example is that of an *cusp*, which is by definition a double point with only one branch. If $\mathcal{D}$ has a cusp at 0 then, locally, $\mathcal{D}$ is given by an equation $y^2 = x^{2n+1}$, where $n \geqslant 1$. A local parameterization of $\mathcal{D}$ around 0 is given by $\gamma(\varsigma) = (\varsigma^2, \varsigma^{2n+1})$ and a germ of a holomorphic function on $\mathcal{D}$ at 0 can be written as

$$h(x, y) = a_0 + a_2 x + a_4 x^2 + \cdots + y(b_{2n+1} + b_{2n+3} x + \cdots).$$

Upon substituting we find that

$$h \circ \gamma(\varsigma) = a_0 + a_2 \varsigma^2 + a_4 \varsigma^4 + \cdots + a_{2n} \varsigma^{2n} + b_{2n+1} \varsigma^{2n+1} + a_{2n+2} \varsigma^{2n+2} \cdots,$$

and we see that there are n coefficients in the series that vanish, namely the coefficients of $\varsigma, \varsigma^3, \ldots, \varsigma^{2n-1}$. Therefore, $\chi(Z_m) = n$. For this example, let us also illustrate Clebsch's method. We do a first blow-up of a neighborhood of $(0, 0)$ by taking $x_1 := x$ and $y_1 := y/x$. The proper transform of $\mathcal{D}$ is then given by $\mathcal{D}_1 : y_1^2 = x_1^{2n-1}$, and notice that $\mathcal{D}_1$ has the origin as singular point (if $n > 1$), in fact it is a double point, just as $(0, 0)$ on the original curve $\mathcal{D}$. Repeating this procedure n times we get at $\mathcal{D}_n : y_n^2 = x$, which is a non-singular curve. Since each of the curves $\mathcal{D}, \mathcal{D}_1, \ldots, \mathcal{D}_{n-1}$ has a single double point, i.e., $e_i = 2$ for $i = 0, \ldots, n-1$ we get by Clebsch's formula (5.36) that $\chi(Z_m) = n$, as before.

For a non-singular curve $\mathcal{D}$ on a complex surface M the adjunction formula $\mathcal{K}_\mathcal{D} = \mathcal{K}_M \otimes \mathcal{O}_\mathcal{D}(\mathcal{D})$ expresses its canonical bundle in terms of its normal bundle and the canonical bundle on M. For a singular divisor $\mathcal{D}$ on a non-singular compact surface M the sheaf $\mathcal{O}_\mathcal{D}(\mathcal{D})$ still makes sense, so the canonical line bundle is formally defined as the sheaf $\mathcal{K}_\mathcal{D} := \mathcal{K}_M \otimes \mathcal{O}_\mathcal{D}(\mathcal{D})$. One calls $\mathcal{K}_\mathcal{D}$ the *dualizing sheaf* on $\mathcal{D}$. If we tensor the exact sequence

$$0 \longrightarrow \mathcal{O}_M \longrightarrow \mathcal{O}_M(\mathcal{D}) \longrightarrow \mathcal{O}_\mathcal{D}(\mathcal{D}) \longrightarrow 0$$

with $\mathcal{K}_M$ then we get the *residue sequence*

$$0 \longrightarrow \mathcal{K}_M \longrightarrow \mathcal{K}_M \otimes \mathcal{O}_M(\mathcal{D}) \xrightarrow{r} \mathcal{K}_\mathcal{D} \longrightarrow 0.$$

Assuming $\mathcal{D}$ to be an analytic hypersurface, r admits the following description, similar to the Poincaré Residue map. Let ω be a meromorphic 2-form, with at worst a single pole along $\mathcal{D}$. If $f = 0$ is a local defining equation for $\mathcal{D}$ then (on U)

$$\omega = \frac{g\,\mathrm{d}z_1 \wedge \mathrm{d}z_2}{f}$$

where g is a holomorphic function on U. Letting $\nu : \tilde{\mathcal{D}} \to \mathcal{D}$ denote the normalization map of $\mathcal{D}$ we define

$$r'(\omega) := \nu^* \left(\left. \frac{g\,\mathrm{d}z_2}{\partial f / \partial z_1} \right|_{f=0} \right),$$

a definition which is easily shown to be independent of the chosen local coordinates (z_1, z_2). It follows that r' is a globally defined morphism of $\mathcal{K}_M \otimes \mathcal{O}_M(\mathcal{D})$ into the sheaf of meromorphic 1-forms on the normalization $\tilde{\mathcal{D}}$ of $\mathcal{D}$. This means, upon identifying r' with r that the sheaf of holomorphic 1-forms on $\mathcal{D}$ is, by definition, the sheaf of meromorphic differentials on $\tilde{\mathcal{D}}$ of the form $r'(\omega)$. Notice that in the case of a torus, $M = \mathbf{T}^2$, the above definition simplifies to $\mathcal{K}_{\mathcal{D}} := \mathcal{O}_{\mathcal{D}}(\mathcal{D})$. Also, with this definition the number of independent holomorphic differentials is equal to the (arithmetic) genus of $\mathcal{D}$, and a basis of the space of holomorphic differentials on $\mathcal{D}$ is computed as before from Theorem 5.40.

Having the definition of holomorphic differentials at hand for singular curves we may now define, as before, an embedded curve $\mathcal{D}$ to be hyperelliptic if the Kodaira map, associated to its canonical bundle $\mathcal{K}_{\mathcal{D}}$ fails to be an embedding. For non-hyperelliptic embedded curves, Theorem 5.10 still holds.

Theorem 5.49 (Noether). *Let $\mathcal{D}$ be a (possibly singular) curve, embedded in a smooth algebraic surface. If $\mathcal{D}$ is not hyperelliptic then its canonical bundle $\mathcal{K}_{\mathcal{D}}$ is normally generated.*

6 A.c.i. Systems

Many integrable systems from classical mechanics admit a complexification, where phase space and time are complexified, and the geometry of the (complex) momentum map is the best possible complex analogue of the geometry that appears in the Liouville Theorem (Theorem 4.28). Namely, in many relevant examples the generic complexified fiber is an affine part of an *Abelian variety* (a compact algebraic torus, see Chapter 5) and the integrable vector fields are translation invariant, when restricted to any of these tori. Such integrable systems are the main topic of this book, and we will call them algebraic completely integrable systems, following the original definition of Adler and van Moerbeke (see [14]).

A precise definition of algebraic complete integrability, and a natural generalization of it, are given in Section 6.1, where we also present two elementary examples, whose algebraic complete integrability can be shown without reference to abstract theorems. Two necessary conditions for algebraic complete integrability will be given in Section 6.2. When we get to the examples, then we will see that these conditions can be used in a very efficient way to single out, from a given family of Hamiltonians on a Poisson manifold, those that may lead to an algebraic completely integrable system. In many examples these conditions turn out to be sufficient. However, some extra techniques are needed to prove algebraic complete integrability. We aim in that direction when we prove, in Section 6.3, two complex versions of the Liouville Theorem. These theorems are geared towards the examples, but for a full implementation of these theorems we need to wait until Chapter 7. Lax equations, with a parameter, are a very common source of a.c.i. systems. This will be explained in Section 6.4, where we will explain the geometry of Lax equations with a parameter, and where we will prove a criterion, which gives necessary and sufficient conditions for the divisor map, corresponding to the Lax equation, to linearize the equations of motion (on the Jacobian of the spectral curve, defined by the characteristic polynomial of the Lax operator).

The notion of an a.c.i. system will be further specialized in the next chapter to a particular type of a.c.i. systems in which everything can be computed explicitly.

6.1 Definitions and First Examples

The integrable systems that we will deal with in the rest of this book are complex integrable systems $(M, \{\cdot,\cdot\}, \mathbf{F})$, where M is a non-singular affine variety. Recall from Section 2.1 that the algebra of functions $\mathcal{F}(M)$ that we consider on M is the algebra of regular functions, so that $\{\cdot,\cdot\}$ is a Poisson bracket on regular functions, and each F_i in $\mathbf{F} = (F_1, \dots, F_s)$ is a regular function. In the particular case of M being an affine space this means that $\mathcal{F}(M)$ is just the algebra of polynomial functions on $\mathbf{C}^n$, i.e., $\mathcal{F}(\mathbf{C}^n) = \mathbf{C}[x_1, \dots, x_n]$.

Definition 6.1. Let $(M, \{\cdot,\cdot\}, \mathbf{F})$ be a complex integrable system, where M is a non-singular affine variety and where $\mathbf{F} = (F_1, \dots, F_s)$. We say that $(M, \{\cdot,\cdot\}, \mathbf{F})$ is an *algebraic completely integrable system* or an *a.c.i. system* if for generic[1] $\mathbf{c} \in \mathbf{C}^s$ the fiber $\mathbf{F_c}$ is an affine part of an Abelian variety and if the Hamiltonian vector fields $\mathcal{X}_{F_i}$ are translation invariant, when restricted to these fibers. In the particular case in which M is an affine space $\mathbf{C}^n$ we will call $(\mathbf{C}^n, \{\cdot,\cdot\}, \mathbf{F})$ a *polynomial a.c.i. system.* When the generic Abelian variety of the a.c.i. system is irreducible we speak of an *irreducible a.c.i. system.*

For certain examples of interest, that will not be treated in this book, the flow of the integrable vector fields induces on the generic invariant manifold the structure of a local group and the group law is algebraic, thereby corresponding precisely to Painlevé's original idea of algebraic integrability (as opposed to transcendentality, which leads to new functions, functions that are not obtained via algebraic manipulations, the inverse function theorem, and integration by quadratures). In order to cover these cases, we propose here the following definition that generalizes Definition 6.1. For the definition and a characterization of $\mathbf{F}'_m$, see Definition 4.25, Proposition 4.27 and Remark 4.26.

Definition 6.2. Let $(M, \{\cdot,\cdot\}, \mathbf{F})$ be a complex integrable system, where M is a (non-singular) affine variety, and where $\mathbf{F} = (F_1, \dots, F_s)$. We say that $(M, \{\cdot,\cdot\}, \mathbf{F})$ is a *generalized a.c.i. system* if for generic $m \in M$ the integrable vector fields $\mathcal{X}_{F_1}, \dots, \mathcal{X}_{F_s}$ define the local action of an algebraic group on $\mathbf{F}'_m$.

Notice that the algebraic group is necessarily commutative because the vector fields $\mathcal{X}_{F_i}$ commute pairwise.

When $(M, \{\cdot,\cdot\}, \mathbf{F})$ is an a.c.i. system, the generic fiber $\mathbf{F_c}$ of the momentum map will constitute an entire invariant manifold $\mathbf{F}'_m$, so that in this case the generic invariant manifold is an affine part of an (r-dimensional) Abelian variety.

[1] We use the standard terminology from algebraic geometry that "property P is true for generic $q \in N$" to denote that there exists a non-empty Zariski open subset U of the affine variety N such that property P is true for every $q \in U$.

Moreover, the integrable vector fields define the local action of $\mathbf{C}^r$ on this affine part, and this action is algebraic because it coincides, by definition, with the natural local action of the Abelian variety on an affine part of itself. Therefore Definition 6.2 is indeed a generalization of Definition 6.1.

In the two examples that follow we use elementary tools to prove their algebraic complete integrability, and we show, in addition, that the integrable vector fields define the local action of an algebraic group on $\mathbf{F}'_m$ for *all* $m \in M$. We conjecture that this is true for any generalized a.c.i. system.

Example 6.3. On $\mathbf{C}^3$ with coordinates (x, y, z) we consider the quadratic Poisson structure $\{\cdot\,,\cdot\}$ given by the following Poisson matrix.

$$X = \begin{pmatrix} 0 & -xy & xz \\ xy & 0 & -yz \\ -xz & yz & 0 \end{pmatrix}.$$

One easily checks that $K := xyz$ is a Casimir and that $M_{(1)}$ is precisely $\mathbf{C}^3$ minus the three coordinate axes (recall from Proposition 3.13 that $M_{(s)}$ denotes the open set of points where the rank of $\{\cdot\,,\cdot\}$ is at least $2s$). As Hamiltonian we choose $H := x + y + z$. Then the Hamiltonian vector field $\mathcal{X}_H$ is given by

$$\begin{aligned} \dot{x} &= x(z - y), \\ \dot{y} &= y(x - z), \\ \dot{z} &= z(y - x). \end{aligned}$$

Then $\mathbf{F} := (H, K)$ is integrable because $\mathcal{U}_{\mathbf{F}} = M_{(1)} \setminus \Delta$, is a dense subset of $\mathbf{C}^3$, where $\Delta := \{(a, a, a) \mid a \in \mathbf{C}\}$. Notice that $\mathbf{C}^3 \setminus \mathcal{U}_{\mathbf{F}}$ is precisely the locus where $\mathcal{X}_H$ vanishes. The image of the three coordinate axes under the momentum map is the line $K = 0$, while the image of Δ is the cubic curve $27K = H^3$. Let us investigate all fibers of the momentum map. We fix $m = (x_0, y_0, z_0) \in \mathbf{C}^3$ and we denote by h, k the values of the functions H and K at this point: $h := x_0 + y_0 + z_0$ and $k := x_0 y_0 z_0$, and we let $\mathbf{c} := (h, k)$. We need to distinguish 4 cases for $\mathbf{c}$, corresponding to

(1) a generic point of the HK-plane;
(2) the intersection of the cubic curve and the line (the origin);
(3) a point on the line, different from the origin;
(4) a point on the cubic curve, different from the origin.

See Figure 6.1. We will deal with each of these cases separately, in the above order. The first case is of course the generic case: the invariant manifolds and the flow can in each of the other three cases be seen as a limiting case of the invariant manifolds and of the flow for the generic case.

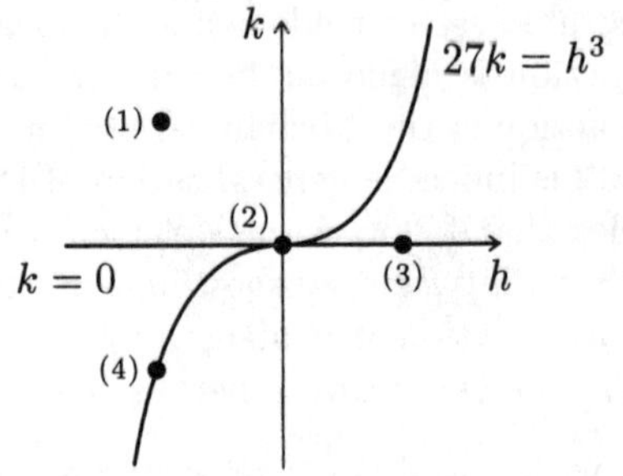

Fig. 6.1. The HK-plane, which contains the image points of the momentum map contains two special curves, the line $K = 0$ and the cubic curve $27K = H^3$. The fibers $\mathbf{F_c}$ of the momentum map over points $\mathbf{c} = (h, k)$ in the HK-plane, and the local action, induced on it by the vector field $\mathcal{X}_H$, depend very much on whether (h, k) belongs to neither of these curves, to one of them or to both. The labels $(1), \ldots, (4)$ correspond to the four different cases that we need to consider.

- $\mathbf{c} = (h, k) \in \mathbf{C}^2$, with $h^3 \neq 27k$ and $k \neq 0$. The fiber $\mathbf{F_c}$ is given by

$$\begin{aligned} x + y + z &= h, \\ xyz &= k. \end{aligned} \tag{6.1}$$

We can rewrite this curve as an isomorphic affine curve in $\mathbf{C}^2$, to wit

$$xy(h - x - y) = k.$$

Since $h^3 \neq 27k$ this cubic curve has no singular points and it is an affine part of an elliptic curve. Moreover, since this curve is irreducible, and since $\mathcal{X}_H$ does not vanish at any point of $\mathbf{F_c}$, the fiber $\mathbf{F_c}$ coincides with the invariant manifold $\mathbf{F}'_m$. Let us consider[2] the 1-form, corresponding to $\mathrm{d}t$, i.e.,

$$\omega := \frac{\mathrm{d}x}{x(z-y)} = \frac{\mathrm{d}y}{y(x-z)} = \frac{\mathrm{d}z}{z(y-x)}$$

on $\mathbf{F}_m$. The conditions on h and k guarantee that ω is holomorphic on $\mathbf{F}_m$; we check that ω extends to a holomorphic 1-form on the smooth compactification of $\mathbf{F}_m$, which is a compact Riemann surface of genus 1. To do this, notice that the points to be added correspond to the points where one or several of x, y, z tend to infinity. A closer inspection of (6.1) shows that precisely two of them must tend to infinity (in such a way that their sum remains finite) and that the other one must tend to zero. Hence there are 3 such points, which correspond under a cyclic permutation of the coordinates.

[2] In order to check by computation that these three expressions define the same 1-form, eliminate one of $\mathrm{d}x, \mathrm{d}y, \mathrm{d}z$ from the differential of the equations in (6.1); however, since $\omega = \mathrm{d}t$ it is obvious that they define the same 1-form.

It therefore suffices to check our statement for e.g. the point at infinity, which is given in terms of a local parameter ς by $(x,y,z) = (\varsigma^{-1}, -\varsigma^{-1} + h + O(\varsigma^2), -k\varsigma^2 + O(\varsigma^3))$. In terms of ς we find that ω is given by

$$\omega = -\frac{\mathrm{d}\varsigma}{1 - h\varsigma + O(\varsigma^2)},$$

which means that ω is holomorphic (and non-zero) at infinity, as we claimed. Now notice that the restriction of ω to $\mathbf{F}_m$ is the 1-form which is dual to the vector field $\mathcal{X}_H$, since

$$\langle \omega, \mathcal{X}_H \rangle = \left\langle \frac{\mathrm{d}x}{x(z-y)}, \mathcal{X}_H \right\rangle = \frac{\dot{x}}{x(z-y)} = 1.$$

It follows that $\mathcal{X}_H$ is a holomorphic, hence linear, vector field on the elliptic curve (Riemann surface). The local action is in this case the action of the elliptic curve on an affine part of itself.

- $\mathbf{c} = (h,k) = (0,0)$. In this case the fiber $\mathbf{F}_\mathbf{c}$ is given by

$$\begin{aligned} x + y + z &= 0, \\ xyz &= 0. \end{aligned}$$

$\mathbf{F}_\mathbf{c}$ consists of three straight lines, passing through the origin of $\mathbf{C}^3$, and it is the union of four invariant manifolds. Namely, if $m = (0,0,0)$ then $\mathbf{F}'_m = \{m\}$; of course, $\mathcal{X}_H$ vanishes at this point. Otherwise, say $m = (x_0, -x_0, 0)$, with $x_0 \neq 0$. The differential equation that describes $\mathcal{X}_H$ on the line $z = 0 = x + y$ (minus the origin) is given by

$$\begin{aligned} \dot{x} &= -xy, \\ \dot{y} &= xy, \\ \dot{z} &= 0, \end{aligned} \tag{6.2}$$

which is easily integrated as

$$(x(t), y(t), z(t)) = \left(\frac{x_0}{1 - x_0 t}, -\frac{x_0}{1 - x_0 t}, 0 \right). \tag{6.3}$$

Hence the flow of the vector field is given by the following (local) action of the additive group $\mathbf{C}$ on that line (minus the origin)

$$(g, (a, -a, 0)) \mapsto \left((a^{-1} - g)^{-1}, -(a^{-1} - g)^{-1}, 0 \right),$$

and this action takes you from any point on the line, different from zero, to any other point on the line, different from zero.

- $\mathbf{c} = (h,k) = (h,0)$, with $h \neq 0$. Again, $\mathbf{F_c}$ consists of three lines, but now they form a triangle. At each vertex of the triangle (the points $(h,0,0)$, $(0,h,0)$ and $(0,0,h)$) the vector field $\mathcal{X}_H$ vanishes. Let us choose another point on one of the three lines, say $m = (x_0, y_0, 0)$, with $x_0 y_0 \neq 0$. Let $h := x_0 + y_0$ and consider the line

$$
\begin{aligned}
x + y &= h,\\
z &= 0.
\end{aligned}
$$

The differential equation that describes $\mathcal{X}_H$ on this line is the same as in (6.2) and is now integrated as

$$
(x(t), y(t), z(t)) = \left(\frac{x_0 h e^{-ht}}{y_0 + x_0 e^{-ht}}, \frac{y_0 h}{y_0 + x_0 e^{-ht}}, 0\right). \tag{6.4}
$$

So $\mathbf{F_c}$ consists in this case, besides the vertices of the triangle, of three invariant manifolds, each of which is a line, minus two points, and on each of which $\mathcal{X}_H$ induces the following local action of $\mathbf{C}^*$,

$$
(g, (x_0, y_0, 0)) \mapsto ((g(x_0^{-1} - h^{-1}) + h^{-1})^{-1}, (g^{-1}(y_0^{-1} - h^{-1}) + h^{-1})^{-1}, 0).
$$

Notice that the algebraic group depends on the constants of motion only, as it should be, and that (6.3) is a limit of (6.4) as $h \to 0$ (implying that $y_0 \to -x_0$).

- $\mathbf{c} = (h,k) = (3c, c^3)$, where $c \neq 0$. The fiber $\mathbf{F_c}$ is now given by

$$
\begin{aligned}
x + y + z &= 3c,\\
xyz &= c^3.
\end{aligned}
$$

This affine curve has precisely one singular point, the point (c,c,c), as one verifies immediately by rewriting the curve as an isomorphic affine curve in $\mathbf{C}^2$, to wit

$$
xy(3c - x - y) = c^3.
$$

Notice that the vector field vanishes at the singular point. Letting $u := (y - c)/(x - c)$ we find the following rational parameterization of $\mathbf{F_c}$,

$$
(x, y, z) = \left(-c\frac{1}{u(1+u)}, -c\frac{u^2}{1+u}, c\frac{(1+u)^2}{u}\right). \tag{6.5}
$$

It follows that $\mathbf{F_c}$ is a rational curve. Rewriting any of the differential equations for $\mathcal{X}_H$ in terms of u we find $\dot{u} = -c(1+u+u^2)$ which is easily integrated as

$$
u(t) = -\frac{1}{2} - \frac{\sqrt{-3}}{2}\frac{e^{\alpha(t)} - e^{-\alpha(t)}}{e^{\alpha(t)} + e^{-\alpha(t)}}, \tag{6.6}
$$

where $\alpha(t) = \frac{\sqrt{-3}}{2}(k - tc)$ and k is the integration constant, which is related to $m = (x_0, y_0, z_0) \neq (c,c,c)$ by

$$
k = \frac{2}{\sqrt{3}} \arctan\left(\frac{2}{\sqrt{3}}\left(\frac{y_0 - c}{x_0 - c}\right) + \frac{1}{\sqrt{3}}\right).
$$

If we substitute the expression (6.6) for u in (6.5) we find, after some simplification,

$$\begin{aligned} x(t) &= c\frac{\left(e^{2\alpha(t)}+1\right)^3}{e^{6\alpha(t)}+1} \\ y(t) &= c - \frac{6ce^{2\alpha(t)}}{(1-\sqrt{-3})e^{4\alpha(t)}+2e^{2\alpha(t)}+1+\sqrt{-3}} \\ z(t) &= c - \frac{6ce^{2\alpha(t)}}{(1+\sqrt{-3})e^{4\alpha(t)}+2e^{2\alpha(t)}+1-\sqrt{-3}} \end{aligned}$$

Thus, we find that if $m \neq (c, c, c)$ then on $\mathbf{F}'_m$ the flow of the vector fields $\mathcal{X}_H$ corresponds to an action of $\mathbf{C}^*$, which finally amounts to $e^{-c\sqrt{-3}(t_1+t_2)} = e^{-c\sqrt{-3}t_1}e^{-c\sqrt{-3}t_2}$, with c constant. The algebraic group depends again on the constants of motion, i.e. on c, only.

Example 6.4. For the second example, let λ_1, λ_2 and λ_3 be three arbitrary complex numbers, which will play the rôle of parameters in the system that we will describe. On $M = \mathbf{C}^3$ we take x, y and z as coordinates and we consider the quadratic vector field $\mathcal{V}$ on $\mathbf{C}^3$, given by

$$\begin{aligned} \dot{x} &= (\lambda_3 - \lambda_2)yz, \\ \dot{y} &= (\lambda_1 - \lambda_3)zx, \\ \dot{z} &= (\lambda_2 - \lambda_1)xy. \end{aligned} \tag{6.7}$$

They are Euler's equations, which describe the free rotation of a rigid body, usually referred to as the *Euler top* or *Euler-Poinsot top* (see Chapter 10 for more on rigid bodies). We introduce the following abbreviation $\lambda_{ij} := \lambda_i - \lambda_j$ and we suppose in the sequel that all λ_{ij}, with $i \neq j$, are non-zero because otherwise $\mathcal{V}$ is tangent to one of the coordinate planes, in which case the equations (6.7) are linear differential equations and their integration is immediate.

The vector field $\mathcal{V}$ admits then the following two independent constants of motion:

$$\begin{aligned} K &:= \frac{1}{2}(x^2 + y^2 + z^2), \\ H &:= \frac{1}{2}(\lambda_1 x^2 + \lambda_2 y^2 + \lambda_3 z^2). \end{aligned}$$

Let $\{\cdot\,,\cdot\}_1$ and $\{\cdot\,,\cdot\}_2$ denote the linear Poisson structures defined by the following two matrices

$$\begin{pmatrix} 0 & \lambda_3 z & -\lambda_2 y \\ -\lambda_3 z & 0 & \lambda_1 x \\ \lambda_2 y & -\lambda_1 x & 0 \end{pmatrix} \qquad \text{resp.} \qquad \begin{pmatrix} 0 & -z & y \\ z & 0 & -x \\ -y & x & 0 \end{pmatrix}.$$

Then we have that $\mathcal{V} = \{\cdot\,, K\}_1 = \{\cdot\,, H\}_2$, while $\{\cdot\,, K\}_2 = \{\cdot\,, H\}_1 = 0$, so that $\mathcal{V}$ is a bi-Hamiltonian vector field. Since

$$\mathrm{d}K \wedge \mathrm{d}H = \lambda_{32} yz \,\mathrm{d}y \wedge \mathrm{d}z + \lambda_{13} zx \,\mathrm{d}z \wedge \mathrm{d}x + \lambda_{21} xy \,\mathrm{d}x \wedge \mathrm{d}y$$

the subset $\mathcal{U}_\mathbf{F}$ (where $\mathrm{d}K \wedge \mathrm{d}H = 0$) consists of $\mathbf{C}^3$ minus the three coordinate axes. Notice that $M_{(1)} \supseteq \mathcal{U}_\mathbf{F}$, irrespective of which of the Poisson structures $\{\cdot,\cdot\}_1$ or $\{\cdot,\cdot\}_2$ is considered, so that $\mathcal{U}_\mathbf{F} \cap M_{(1)} = \mathcal{U}_\mathbf{F}$. The image of these three axes by $\mathbf{F}$ consists of the three lines l_i, given by $h = \lambda_i k$, where $1 \leqslant i \leqslant 3$. Let k and h denote the values of K and H at m, where $m \in \mathbf{C}^3$, and denote $\mathbf{c} := (k, h)$. We need to consider three cases (see Figure 6.2):

(1) (k, h) does not belong to one of the lines l_i;
(2) (k, h) belongs to precisely one of the lines l_i;
(3) $(k, h) = (0, 0)$.

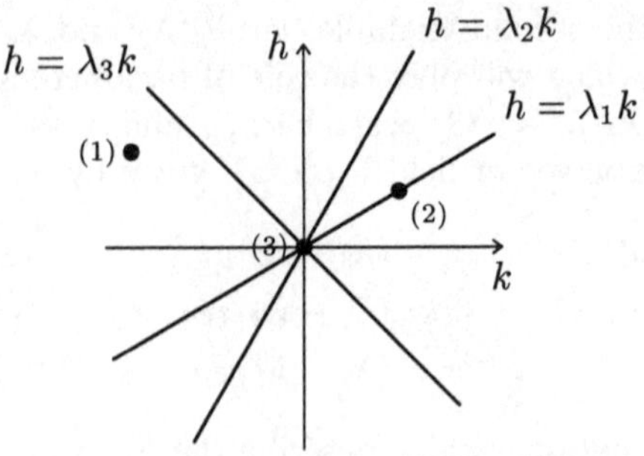

Fig. 6.2. In this case the KH-plane contains three special curves, which are straight lines through O; cfr. Figure 6.1.

- $\mathbf{c} = (k, k) \in \mathbf{C}^2 \setminus (l_1 \cup l_2 \cup l_3)$. For any point m in the fiber $\mathbf{F_c}$ the invariant manifold $\mathbf{F}'_m$ coincides with $\mathbf{F_c}$, and is given by

$$\begin{aligned} x^2 + y^2 + z^2 &= 2k, \\ \lambda_1 x^2 + \lambda_2 y^2 + \lambda_3 z^2 &= 2h, \end{aligned} \tag{6.8}$$

which is a non-singular affine curve, because the differentials of the two equations are independent at any of its points (indeed, $\mathbf{F_c}$ is entirely contained in $\mathcal{U}_\mathbf{F}$). Replacing (6.8) by

$$\begin{aligned} x^2 + y^2 + z^2 &= 2k, \\ \lambda_{13} x^2 + \lambda_{23} y^2 &= 2(h - \lambda_3 k), \end{aligned} \tag{6.9}$$

we notice that the second equation in (6.9) defines a conic in $\mathbf{C}^2$, which is non-singular since $h \neq \lambda_3 k$; also the first equation in (6.8) defines $\mathbf{F_c}$ as a double cover of this conic, with 4 ramification points (there are no ramification points at infinity), yielding 1 as the genus of $\mathbf{F_c}$, by Riemann-Hurwitz.

The argument to show that the vector field is linear on the elliptic curve goes as in the first case, considered in Example 6.3. Namely, one considers the 1-form

$$\omega := \frac{\mathrm{d}x}{\lambda_{32}yz} = \frac{\mathrm{d}y}{\lambda_{13}zx} = \frac{\mathrm{d}z}{\lambda_{21}xy}$$

on $\mathbf{F_c}$, which is holomorphic because $\mathbf{F_c}$ does not intersect any of the coordinate axes. Using a local parameter it is easily verified that the 1-form is holomorphic at the 4 points at infinity. Thus, $\mathcal{V}$ is a linear vector field on the elliptic Riemann surface and the local action, corresponding to the vector field, is the action of the Riemann surface on an affine part of itself. Solving (6.8) for y and z expresses ω in terms of x only, and the integration of ω amounts to an elliptic integral.

- $\mathbf{c} = (k,h) \in l_i \setminus \{(0,0)\}$. To simplify the notation, let us assume that $i=3$. The second equation in (6.9) now reduces to

$$\lambda_{13}x^2 + \lambda_{23}y^2 = 0, \tag{6.10}$$

which describes two lines in $\mathbf{C}^2$ that pass through $(0,0)$. The fiber of $\mathbf{F}$ over (k,h) consists of two conics, intersecting in the two points that lie over $(0,0)$, i.e., in $(0,0,\sqrt{2k})$ and in $(0,0,-\sqrt{2k})$, where $\mathcal{V}$ vanishes. Each of the conics, minus these two points, is an invariant manifold $\mathbf{F}'_m$, as we show. Let $(x_0,y_0) \neq (0,0)$ and substitute this initial condition in (6.10) to find that the latter can be written as $y_0^2x^2 = x_0^2y^2$, so that $y_0x = x_0y$ (the other line does not contain the point (x_0,y_0)). Using this linear equation to eliminate y from the differential equations we arrive at the following quadratic equations on $\mathbf{C}^2$:

$$\begin{aligned} \dot{x} &= \frac{\lambda_{32}y_0}{x_0}xz = \mu_1 xz, \\ \dot{z} &= \frac{\lambda_{21}y_0}{x_0}x^2 = \mu_2 x^2. \end{aligned} \tag{6.11}$$

Notice that $\mu_1\mu_2 = \lambda_{21}\lambda_{13}$, since $(y_0/x_0)^2 = -\lambda_{13}/\lambda_{23}$. In order to solve this equation, take the derivative of the second equation,

$$\ddot{z} = 2\mu_1\mu_2x^2z = 2\mu_1 z\dot{z} = \left(\mu_1 z^2\right)^{\cdot}$$

to find that the general solution is given by

$$x(t) = \frac{4c^2\alpha(t)}{\alpha^2(t) - 4\lambda_{21}\lambda_{13}c^2}, \qquad z(t) = -\frac{c}{\mu_1}\left(\frac{\alpha^2(t) + 4\lambda_{21}\lambda_{13}c^2}{\alpha^2(t) - 4\lambda_{21}\lambda_{13}c^2}\right),$$

where $\alpha(t) := Ce^{ct}$ and where C and c are constants, which relate to the initial conditions as follows:

$$x(0) = \frac{4c^2C}{C^2 - 4\lambda_{21}\lambda_{13}c^2}, \qquad z(0) = -\frac{c}{\mu_1}\left(\frac{C^2 + 4\lambda_{21}\lambda_{13}c^2}{C^2 - 4\lambda_{21}\lambda_{13}c^2}\right).$$

Combined with $y_0x(t) = x_0y(t)$ this yields the integral curve that starts at (x_0, y_0, z_0).

In order to check how the group depends on the initial condition, eliminate y from (6.9) to find that

$$\lambda_{21}x^2(t) + \lambda_{23}z^2(t) = 2\lambda_{23}k,$$

along the solutions $(x(t), z(t))$ of (6.11). If we substitute the above general solution in it, then we find $c^2 = 2k\lambda_{13}$, so that apart from a sign, the exponential part of $\alpha(t)$ depends on the value of the constants of motion only. The sign depends on which one of the two fibers of the momentum map over $(k, \lambda_3 k)$ one picks. Thus, the algebraic group depends on the invariant manifold $\mathbf{F}'_m$ only (and not on m itself). We conclude that on $\mathbf{F}'_m$ the flow of the vector field $\mathcal{V}$ corresponds to the local action of $\mathbf{C}^*$, which again amounts to $\exp c(t_1 + t_2) = \exp ct_1 \exp ct_2$, with c constant.

- $\mathbf{c} = (k, h) = (0, 0)$. This case is a further degeneration of the previous case. Namely, both equations in (6.9) factor into two linear terms, so that the fiber consists of four lines l_i through $(0, 0, 0)$. On each of the lines l_i the vector field takes the simple form $\dot{x} = \nu_i x^2$, which is easily integrated. The fiber $\mathbf{F}_\mathbf{c}$ consist of $O := (0, 0, 0)$, where $\mathcal{V}$ vanishes, and four lines, minus O, on which $\mathcal{V}$ induces the local action of the additive group $\mathbf{C}$.

Example 6.5. We will now consider a simple but non-trivial example of a generalized a.c.i. system that is not an a.c.i. system. We take again $\mathbf{C}^3$ as phase space with coordinates (x, y, z). The (quadratic) Poisson structure $\{\cdot\,,\cdot\}$ is given by the following Poisson matrix,

$$X = \begin{pmatrix} 0 & -xy & 0 \\ xy & 0 & -yz \\ 0 & yz & 0 \end{pmatrix}.$$

We have that $K := xz$ is a Casimir and that $M_{(1)}$ is $\mathbf{C}^3$ minus the plane $y = 0$ and minus the line $x = z = 0$. As Hamiltonian we choose $H := x + y + z$, which gives the following Hamiltonian vector field $\mathcal{X}_H$:

$$\begin{aligned} \dot{x} &= -xy, \\ \dot{y} &= y(x - z), \\ \dot{z} &= yz. \end{aligned}$$

The locus where H and K are dependent is the line $x = z = 0$ so that $\mathbf{F} := (H, K)$ is integrable. Notice that the image of this line under the momentum map $\mathbf{F}$ is given by $K = 0$. To investigate the fibers of the momentum map we fix $m = (x_0, y_0, z_0) \in \mathbf{C}^3$ and we denote by h, k the values of the functions H and K at this point: $h := x_0 + y_0 + z_0$ and $k := x_0z_0$, and we define $\mathbf{c} := (h, k)$.

We need to distinguish two cases (see Figure 6.3):

(1) $k \neq 0$;
(2) $k = 0$.

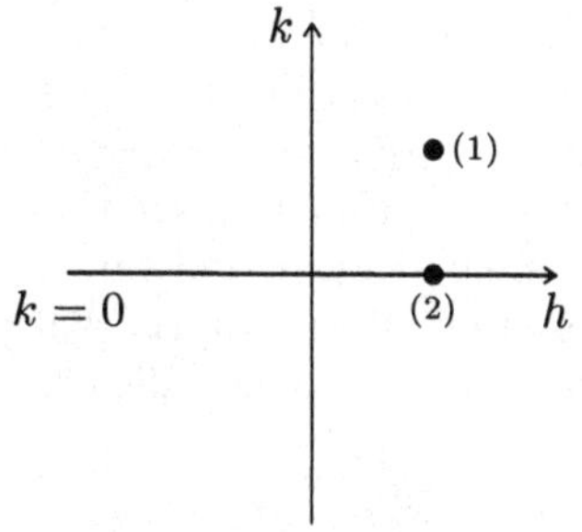

Fig. 6.3. The only special line in the HK-plane is in this case the line $K = 0$; cfr. Figure 6.1.

- $\mathbf{c} = (h, k) \in \mathbf{C}^2$, with $k \neq 0$. The fiber $\mathbf{F_c}$ is given by

$$\begin{aligned} x + y + z &= h, \\ xz &= k, \end{aligned} \tag{6.12}$$

which is a non-degenerate conic; it can be written as $x(h - x - y) = k$. Using (6.12) we rewrite the equation $\dot{x} = -xy$ as $\dot{x} = x^2 - hx + k$, or also as $\dot{u} = u^2 + C$, where $u := x - h/2$ and $C := k - h^2/4$. It follows that

$$u(t) = \sqrt{C} \tan(\sqrt{C}t + c),$$

where c is related to the initial condition by

$$x_0 = h/2 + \sqrt{C} \tan(c).$$

This gives basically the same formulas as in the last case in Example 6.3 and it follows, as in that case, that the local action is given by the action of $\mathbf{C}^*$ when $C \neq 0$ and by the action of the additive group $\mathbf{C}$ when $C = 0$. Since we have just analyzed the fibers $\mathbf{F_c}$ for generic $\mathbf{c}$ this proves that this is an example of a generalized a.c.i. system that is not an a.c.i. system. Notice that $\mathcal{X}_H$ has two zeros on every fiber, which correspond to the points in $\mathbf{F_c}$ for which $y = 0$ (these two points coincide when $C = 0$).

- $\mathbf{c} = (h, 0) \in \mathbf{C}^2$. Then $x_0 z_0 = 0$ so that $x_0 = 0$ or $z_0 = 0$. When $x_0 = z_0 = 0$ then $\mathcal{X}_H$ vanishes and the action is trivial. If, say, $z_0 = 0$ but $x_0 \neq 0$ then $\mathcal{X}_H$ reduces to (6.2), hence we get the same action (6.3) and we arrive at the same conclusion as in that case.

Notice that also in this example the vector field $\mathcal{X}_H$ induces the local action of an algebraic group in the neighborhood of *any* point in phase space.

6.2 Necessary Conditions for Algebraic Complete Integrability

In this section we give two different conditions for algebraic complete integrability. The first one, which goes back to Kowalevski, (see [102]) is based on the fact that the phase space of an a.c.i. system admits a partial compactification (at least semi-locally), to which the integrable vector fields extend to complete vector fields. The second one, which goes back to Lyapunov, (see [109]), is based on the fact that any solution to any integrable vector field of an a.c.i. system is single-valued, and similarly for any solution of the associated variational equation. Both criteria will be illustrated on several examples, allowing one to compare the specific characteristics of each one of them.

6.2.1 The Kowalevski-Painlevé Criterion

We first introduce the notion of a (convergent or formal) Laurent solution to a polynomial vector field on $\mathbf{C}^n$. We will then show that each of the integrable vector fields of an irreducible a.c.i. system on $\mathbf{C}^n$ admits one or several families of Laurent solutions (called balances), which will lead to a necessary condition for algebraic complete integrability, which we call the *Kowalevski-Painlevé Criterion*. Applications of this criterion will be given in Sections 8.1, 9.2 and 10.1.

Let us consider a polynomial vector field $\mathcal{V}$ on $\mathbf{C}^n$. In terms of linear coordinates $x_1, \dots, x_n$ it is described by a system of first order differential equations

$$\begin{aligned} \dot{x}_1 &= f_1(x_1, \dots, x_n), \\ &\vdots \\ \dot{x}_n &= f_n(x_1, \dots, x_n), \end{aligned} \tag{6.13}$$

where $f_1, \dots, f_n$ are polynomials. We often abbreviate (6.13) to

$$\dot{x} = f(x).$$

Definition 6.6. An n-tuple of Laurent series

$$x_i(t) = \frac{1}{t^{r_i}} \sum_{k=0}^{\infty} x_i^{(k)} t^k, \qquad i = 1, \dots, n, \tag{6.14}$$

with $r_i \in \mathbf{Z}$ for $i = 1, \dots, n$ is called a *formal Laurent solution* to $\dot{x} = f(x)$, or to $\mathcal{V}$, if the formal substitution of (6.14) in each of the differential equations in (6.13) yields an equality of formal series. If, in addition, there exists $\epsilon > 0$ such that $(x_1(t), \dots, x_n(t))$ converges for $0 < |t| < \epsilon$, then we call this formal

Laurent solution a *convergent Laurent solution* or simply a *Laurent solution.* Assuming in (6.14) that $x_i^{(0)} \neq 0$ we say that $x_i(t)$ has a *pole* (resp. *zero*) of order $|r_i|$ if $r_i > 0$ (resp. $r_i < 0$). We say that a (formal) Laurent solution is *strict* if $r := \max_i r_i > 0$, and we call r its *pole order.*

Every Taylor solution is of course a Laurent solution and since the vector field is holomorphic, a convergent Taylor solution exist for any initial condition. In fact, the Taylor solutions to (6.13) can collectively be written as one Taylor solution, depending on a certain number of independent parameters (in fact n); similarly, we will see that under certain conditions the Laurent solutions to (6.13) are naturally organized in families, each family depending on a certain number of independent parameters. In our case these families will always be analytic sets or algebraic sets, with the coefficients of the Laurent series being holomorphic resp. regular functions on them, so that the set of all strict Laurent solutions naturally breaks up into *irreducible* families, namely the subfamilies which correspond to the irreducible components of these analytic of algebraic sets. Each of these irreducible families of formal Laurent solutions is called a *balance.* Notice that the pole order of each balance, which is positive, depends in general on the balance.

We say that a balance depends on d parameters if the underlying family has dimension d; it is understood that e.g. in the case of an affine variety of dimension d, more than d parameters may explicitly appear in the balance, but then they are related by the polynomial equations that define the affine variety. We say that a balance is a *convergent balance* if it yields a convergent Laurent series for any particular values of the parameters, belonging to some non-empty Zariski open subset.

We will see in Section 7.1 how for an important class of differential equations the decomposition of the set of all formal Laurent solutions (of a certain type) into balances can be explicitly computed, together with as many terms in the expansion as wanted.

Definition 6.7. Suppose that $\mathcal{V}$ is a polynomial vector field on $\mathbf{C}^n$ and let $x(t)$ be a balance of $\mathcal{V}$. We say that $x(t)$ is a *principal balance* if it depends on $\dim(\mathbf{C}^n) - 1 = n - 1$ parameters. Otherwise it is called a *lower balance.*

Example 6.8. We give a simple example that shows that not every polynomial vector field on $\mathbf{C}^n$ admits a balance. Taking $n = 2$ consider

$$\begin{aligned} \dot{x} &= y, \\ \dot{y} &= y^2, \end{aligned}$$

and notice that if $(x(t), y(t))$ is a balance then $x(t)$ as well as $y(t)$ must have a pole. This implies that $y(t)$ has a simple pole, hence that $y(t)$ cannot be the derivative of a Laurent series, leading to our claim.

We will now show that an a.c.i. system always has principal balances. Intuitively is is clear why this should be the case.

Consider a generic fiber $\mathbf{F_c}$ of the momentum map of a polynomial a.c.i. system on $\mathbf{C}^n$. By assumption, $\mathbf{F_c}$ is an affine part of an Abelian variety $\mathbf{T}^r_\mathbf{c}$, so there exists a holomorphic embedding $\varphi_\mathbf{c} : \mathbf{T}^r_\mathbf{c} \to \mathbf{P}^N$, for some N, for example we may take the Kodaira map associated to a very ample line bundle on $\mathbf{T}^r_\mathbf{c}$. Let F_i be one of the components of $\mathbf{F}$ and consider the integrable vector field $\mathcal{X}_{F_i}$ which extends, by the a.c.i. assumption, to a linear vector field on $\mathbf{T}^r_\mathbf{c}$. As was shown in Section 5.3, there exists a holomorphic (quadratic) vector field $\overline{\mathcal{V}}_i$ on $\mathbf{P}^N$ which is $\varphi_\mathbf{c}$-related to $\mathcal{X}_{F_i}$, i.e., $\mathrm{d}\varphi_\mathbf{c}(\mathcal{X}_{F_i}(m)) = \overline{\mathcal{V}}_i(\varphi_\mathbf{c}(m))$ for any $m \in M$. Let $\mathcal{D}_\mathbf{c}$ be an analytic hypersurface on $\varphi_\mathbf{c}(\mathbf{T}^r_\mathbf{c})$ and let $z_1, \dots, z_N$ be local coordinates in the neighborhood of some point $p \in \mathcal{D}_\mathbf{c}$. The Taylor series $z_k \circ \varphi_\mathbf{c}(t; \mathcal{D}_\mathbf{c})$, where $k = 1, \dots, n$ will depend on $r-1$ parameters (which parameterize the generic point of $\mathcal{D}_\mathbf{c}$), yielding Laurent series $x_j(t; \mathcal{D}_\mathbf{c})$ which depend on $r-1$ parameters. If $\mathcal{D}_\mathbf{c}$ has been chosen as an irreducible component of $\mathbf{T}^r_\mathbf{c} \setminus \mathbf{F_c}$ then at least for one value of j the Laurent series $x_j(t; \mathcal{D}_\mathbf{c})$ will have a pole.

If this construction is done in a family, i.e., for $\mathbf{c}$ varying in an open subset of $\mathbf{C}^s$, then it looks reasonable to expect that the above construction can be done uniformly in $\mathbf{c}$, yielding Laurent series that depend on $s + r - 1 = n - 1$ parameters. Intuitively speaking, these $n-1$ parameters correspond to initial conditions at the divisor at infinity that is to be adjoined to $\mathbf{C}^n$ to compactify the generic fibers $\mathbf{F_c}$ at once into an Abelian variety.

We wish to make this idea precise and give a rigorous proof of the fact that irreducible a.c.i. systems admit principal balances. To do this we first prove that the phase space of an a.c.i. system admits, locally on the base space, a partial compactification, and we show that the integrable vector fields extend holomorphically to this partial compactification.

Theorem 6.9. *Let $\mathbf{F} : \mathbf{C}^n \to \mathbf{C}^s$ be a polynomial map such that for a generic $\mathbf{c} \in \mathbf{C}^s$ there exists an isomorphism $\jmath_\mathbf{c} : \mathbf{F_c} \to \mathbf{T}^r_\mathbf{c} \setminus \mathcal{D}_\mathbf{c}$, where $\mathbf{T}^r_\mathbf{c}$ is an irreducible Abelian variety, $r = n - s$, and $\mathcal{D}_\mathbf{c}$ is an analytic hypersurface of $\mathbf{T}^r_\mathbf{c}$. Suppose, in addition, that $\mathcal{V}$ is a polynomial vector field on $\mathbf{C}^n$, which has the property that its restriction to the generic fiber $\mathbf{F_c}$ is linear (i.e., it extends to a linear vector field on $\mathbf{T}^r_\mathbf{c}$). For a generic $\mathbf{c}_0 \in \mathbf{C}^s$ there exist*

(1) an open neighborhood U of $\mathbf{c}_0$ in $\mathbf{C}^s$;
(2) a non-singular algebraic variety $M(U)$;
(3) a holomorphic surjection $\pi : M(U) \to U$;
(4) an isomorphic embedding $\imath : \mathbf{F}^{-1}(U) \to M(U)$;
(5) an analytic hypersurface $\mathcal{D}$ of $M(U)$

such that

(1) the following diagram commutes,

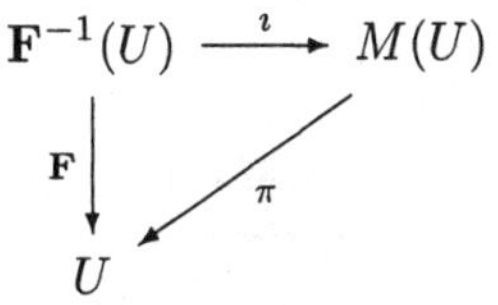

$$\begin{array}{ccc} \mathbf{F}^{-1}(U) & \xrightarrow{\imath} & M(U) \\ {\scriptstyle F}\downarrow & \swarrow {\scriptstyle \pi} & \\ U & & \end{array} \qquad (6.15)$$

(2) $\imath : \mathbf{F}^{-1}(U) \to M(U) \setminus \mathcal{D}$ is an isomorphism;

(3) for any $\mathbf{c} \in U$, the fiber $\pi^{-1}(\mathbf{c})$ is a compactification of the fiber $\mathbf{F_c}$ in the sense that there exists an isomorphism $\bar{\jmath}_\mathbf{c}$ which makes the following diagram commute,

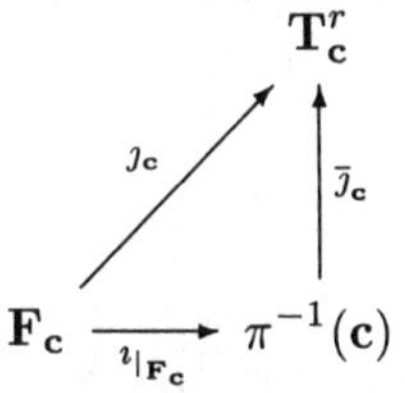

$$\begin{array}{ccc} & \mathbf{T}_\mathbf{c}^r & \\ {\scriptstyle \jmath_\mathbf{c}}\nearrow & \uparrow {\scriptstyle \bar{\jmath}_\mathbf{c}} & \\ \mathbf{F_c} & \xrightarrow{\imath_{|\mathbf{F_c}}} & \pi^{-1}(\mathbf{c}) \end{array} \qquad (6.16)$$

Moreover, $M(U)$ carries a holomorphic vector field $\overline{\mathcal{V}}$ which is $\imath$-related to the restriction of $\mathcal{V}$ to $\mathbf{F}^{-1}(U)$.

Proof. Let U be a Zariski open neighborhood of $\mathbf{c}_0$ in $\mathbf{C}^s$, such that for any $\mathbf{c} \in U$ one has that $\mathbf{F_c}$ is isomorphic to $\mathbf{T}_\mathbf{c}^r \setminus \mathcal{D}_\mathbf{c}$. In the language of Paragraph 5.2.5, we have over U a smooth family of affine parts of Abelian varieties. As we have seen in that paragraph (see Theorem 5.36), there is a partial compactification of this family over a possibly smaller Zariski open subset; in the sequel we will assume that U is small enough for the existence of a partial compactification of $\mathbf{F}$ over U. We will give in the current proof an explicit construction of this partial compactification, from which it will be clear that it has all the properties that are stated in the present theorem.

Under the isomorphism $\mathbf{F_c} \cong \mathbf{T}_\mathbf{c}^r \setminus \mathcal{D}_\mathbf{c}$ the vector field $\mathcal{V}_{|\mathbf{F_c}}$ corresponds by assumption to a linear vector field on $\mathbf{T}_\mathbf{c}^r$, whose flow will be denoted by Φ; it is obvious that this flow is complete. Consider on the affine variety $\mathbf{F}^{-1}(U) \times \mathbf{C}$ the following equivalence relation: $(m, t) \sim (m', t')$ if and only if $\mathbf{F}(m) = \mathbf{F}(m')$ and $\Phi_t(m) = \Phi_{t'}(m')$. Define $M(U)$ by

$$M(U) := \frac{\mathbf{F}^{-1}(U) \times \mathbf{C}}{\sim}$$

and define $\pi : M(U) \to U$ by $\pi(\langle m, t \rangle) := \mathbf{F}(m)$, where we denote elements of $M(U)$ by $\langle m, t \rangle$. Also, $\imath$ is defined by $\imath(m) := \langle m, 0 \rangle$. The commutativity of diagram (6.15) is obvious in view of these definitions. We have that $M(U)$ is the space of orbits of the local action of $\mathbf{C}$ on $\mathbf{F}^{-1}(U) \times \mathbf{C}$, given by

$$\begin{aligned} \Psi : V \subseteq \mathbf{C} \times (\mathbf{F}^{-1}(U) \times \mathbf{C}) &\to \mathbf{F}^{-1}(U) \times \mathbf{C} \\ (\tau, (m, t)) &\mapsto (\Phi_\tau(m), t - \tau), \end{aligned}$$

where, given any $m \in \mathbf{F}^{-1}(U)$, the open subset V can be chosen such that $(0,(m,0)) \in V$. Since $M(U)$ is the space of orbits of a local action, to show that it is a manifold we need to show that it is Hausdorff and that the local action admits everywhere local sections that are biholomorphic.

We first show that $M(U)$ is Hausdorff (in the complex topology). To do this, let $\langle m,t\rangle \neq \langle m',t'\rangle$ be two elements of $M(U)$. We need to show that there exist disjoint saturated[3] open subsets W and W' of $\mathbf{F}^{-1}(U) \times \mathbf{C}$, with $(m,t) \in W$ and $(m',t') \in W'$. If $\mathbf{F}(m) \neq \mathbf{F}(m')$ then it suffices to take $W := \mathbf{F}^{-1}(W_0) \times \mathbf{C}$ and $W' := \mathbf{F}^{-1}(W_0') \times \mathbf{C}$, where W_0 and W_0' are any two disjoint open subsets of $\mathbf{C}^s$, with $\mathbf{F}(m) \in W_0$ and $\mathbf{F}(m') \in W_0'$.

Assume therefore that $\langle m,t\rangle \neq \langle m',t'\rangle$ but $\mathbf{F}(m) = \mathbf{F}(m')$. It is easy to see that, without loss of generality, we may assume that they are of the form $\langle m,t\rangle$ and $\langle m',t\rangle$, with $m \neq m'$. Taking disjoint open neighborhoods W_0 and W_0' of (m,t) and of (m',t) in $\mathbf{F}^{-1}(U) \times \mathbf{C}$, the saturated open subsets $W := \Psi(\mathbf{C}, W_0)$ and $W' := \Psi(\mathbf{C}, W_0')$ are disjoint, hence do the job. It follows that $M(U)$ is Hausdorff.

The local action Ψ admits through any point a local section. In fact, let (m_0,t_0) be an arbitrary point of $\mathbf{F}^{-1}(U) \times \mathbf{C}$, and let V_1 and V_2 be neighborhoods of 0 in $\mathbf{C}$, resp. of (m_0,t_0) in $\mathbf{F}^{-1}(U) \times \mathbf{C}$, such that $V_1 \times V_2 \subseteq V$. Since the local action is on the second component just translation, the local action is transversal to the local hypersurface of V_2, obtained by fixing the second coordinate to t_0. This hypersurface gives a chart around m_0 by projecting on the first component, and since the local action is holomorphic (because Φ is holomorphic), the transition functions to the charts which are obtained by taking another hypersurface through (m_0,t_0), transversal to the local action, will be biholomorphic. It follows that $M(U)$ is a complex manifold.

Using either of the above local charts it is obvious that $\imath$ and π are holomorphic. It is also easy to see that $\imath$ is a biholomorphism onto its image: $\imath$ is by construction injective, and if we take any point $\langle m,0\rangle$ in the image of $\imath$ then the local hypersurface $W \times \{0\}$ containing $\langle m,0\rangle$, is biholomorphic to the neighborhood W of m. Notice that we can transport the local action of $\mathbf{C}$ on $\mathbf{F}^{-1}(U) \times \mathbf{C}$ to a global (free) action of $\mathbf{C}$ on $M(U)$.

Suppose now that $\mathbf{c} \in U$ is such that $\mathbf{T}^r_{\mathbf{c}}$ is irreducible. We show that $\pi^{-1}(\mathbf{c})$ is biholomorphic to $\mathbf{T}^r_{\mathbf{c}}$. Let us identify $\mathbf{F}_{\mathbf{c}}$ with an open subset of $\mathbf{T}^r_{\mathbf{c}}$ using the biholomorphism $\mathbf{F}_{\mathbf{c}} \cong \mathbf{T}^r_{\mathbf{c}} \setminus \mathcal{D}_{\mathbf{c}}$. The map

$$\begin{aligned} \psi_{\mathbf{c}} : \mathbf{F}_{\mathbf{c}} \times \mathbf{C} &\to \mathbf{T}^r_{\mathbf{c}} \\ (m,t) &\mapsto \Phi_t(m) \end{aligned}$$

is holomorphic and $(m,t) \sim (m',t')$ obviously implies that $\psi_{\mathbf{c}}(m,t) = \psi_{\mathbf{c}}(m',t')$. Therefore $\psi_{\mathbf{c}}$ induces a map $\tilde{\psi}_{\mathbf{c}} : \frac{\mathbf{F}_{\mathbf{c}} \times \mathbf{C}}{\sim} \to \mathbf{T}^r_{\mathbf{c}}$. Since $\mathbf{T}^r_{\mathbf{c}}$ is irreducible the (holomorphic) integral curves that start from points on $\mathcal{D}_c$ go

[3] W being saturated means that $W = \rho^{-1}(A)$ for some open subset $A \subseteq M(U)$, where $\rho : \mathbf{F}^{-1}(U) \times \mathbf{C} \to M(U)$ is the quotient map.

immediately into $\imath(\mathbf{F_c})$, i.e., for any $m \in \mathcal{D}_\mathbf{c}$ there exists $T > 0$ such that $\Phi_t(m) \in \imath(\mathbf{F_c})$ for any t such that $0 < |t| < T$. In fact, if for some m such a T would not exist then the integral curve through m would be entirely contained in $\mathcal{D}_\mathbf{c}$. The Zariski closure of this integral curve, which is a subtorus, hence an Abelian subvariety, would then also be contained in $\mathcal{D}_\mathbf{c}$. But $\mathbf{T}_\mathbf{c}^r$ is irreducible, hence it does not contain non-trivial Abelian subvarieties. Contradiction.

The fact that the integral curves that start from points on $\mathcal{D}$ go immediately into $\imath(\mathbf{F_c})$ implies that $\psi_\mathbf{c}$ is surjective. Indeed, the surjectivity of $\psi_\mathbf{c}$ is trivial for points in $\mathbf{T}_\mathbf{c}^r \setminus \mathcal{D}_\mathbf{c}$, while if $m \in \mathcal{D}$ then we can, by the above, pick t_0 such that $\Phi_{-t_0}(m) \in \mathbf{T}_\mathbf{c}^r \setminus \mathcal{D}_\mathbf{c}$, and then $\Phi_{t_0}(\Phi_{-t_0}(m)) = m$. Thus, $\psi_\mathbf{c}$ and $\tilde{\psi}_\mathbf{c}$ are surjective. Since $\tilde{\psi}_\mathbf{c}$ is injective this shows that $\tilde{\psi}_\mathbf{c}$ is a bijection. Since Φ_t is a local biholomorphism, for small $|t|$, the map $\tilde{\psi}_\mathbf{c}$ is a holomorphic bijection, i.e., a biholomorphism.

It follows from the uniqueness of the partial compactification that $M(U)$ is the partial compactification of $\mathbf{F}$ over U. In particular, $\imath$ is an open embedding and $\mathcal{D} := M(U) \setminus \imath(\mathbf{F}^{-1}(U))$ is an analytic hypersurface of $M(U)$ whose restriction to $\pi^{-1}(\mathbf{c})$ is (isomorphic to) $\mathcal{D}_\mathbf{c}$.

Finally, the easiest way to show that the vector field $\mathcal{V}$ can be transported to a vector field $\overline{\mathcal{V}}$ on $M(U)$ is by defining $\overline{\mathcal{V}}$ as the fundamental vector field that corresponds to the holomorphic action of $\mathbf{C}$ on $M(U)$, that is defined by

$$(\tau, \langle m, t\rangle) \mapsto \langle \Phi_\tau(m), t\rangle = \langle m, \tau + t\rangle.$$

It is clear that if we would have been given $r > 1$ commuting vector fields $\mathcal{V}_1, \dots, \mathcal{V}_r$ on $\mathbf{C}^n$ with the same properties as stated for $\mathcal{V}$ above, then we would have obtained r commuting vector fields $\overline{\mathcal{V}}_1, \dots, \overline{\mathcal{V}}_r$ on $M(U)$. By linearity, if they are independent in one point of $\mathbf{F_c}$ then they are independent at all points of $\mathbf{T_c}$. □

Remark 6.10. It is clear that if we would have been given $r > 1$ commuting vector fields $\mathcal{V}_1, \dots, \mathcal{V}_r$ on $\mathbf{C}^n$ with the same properties as stated for $\mathcal{V}$ above, then we would have obtained r commuting vector fields $\overline{\mathcal{V}}_1, \dots, \overline{\mathcal{V}}_r$ on $M(U)$. By linearity, if they are independent in one point of $\mathbf{F_c}$ then they are independent in all points of $\mathbf{T_c}$.

Definition 6.11. The manifold $M(U)$, constructed in the theorem, is called the *partial compactification* of $\mathbf{F}$ (over U). The analytic hypersurface $\mathcal{D}$ is called the *Painlevé building* of $\mathbf{F}$, while each of its irreducible components is called a *Painlevé wall.* For generic $\mathbf{c} \in \mathbf{C}^s$ the divisor $\mathcal{D}_\mathbf{c}$ is called the *Painlevé divisor* at $\mathbf{c}$.

The origin of the terminology "Painlevé building" and "Painlevé wall" will become more apparent when we show, in the next chapter, that in the case of weight homogeneous a.c.i. systems the Painlevé walls are affine varieties of the form $V \times \mathbf{C}^n$; a Painlevé building is then a finite collection of such walls that fit together nicely. Under the natural inclusion $\mathbf{T}_\mathbf{c}^r \hookrightarrow M(U)$ the

Painlevé divisor $\mathcal{D}_\mathbf{c}$ gets mapped to the restriction of $\mathcal{D}$ to $\pi^{-1}(\mathbf{c})$; we will usually identify $\mathcal{D}_\mathbf{c}$ with this restriction, without explicit mention. Similarly, the irreducible components of $\mathcal{D}_\mathbf{c}$ will be identified with the restriction of the irreducible components of $\mathcal{D}$ to $\pi^{-1}(\mathbf{c})$.

Remark 6.12. The conditions of the above theorem can be weakened to Abelian varieties which are not necessarily irreducible, by demanding that $\mathcal{V}$ be generically transversal to the divisor $\mathcal{D}_{\mathbf{c}_0}$, or by constructing $M(U)$ as the quotient of $\mathbf{F}^{-1}(U) \times \mathbf{C}^r$, where the action is defined by using r commuting vector fields, which are independent when restricted to $\mathbf{F}_\mathbf{c}$, for generic $\mathbf{c}$.

We can now give a first criterion for algebraic complete integrability.

Theorem 6.13 (Kowalevski-Painlevé Criterion). *Let $(\mathbf{C}^n, \{\cdot\,,\cdot\}, \mathbf{F})$ be an irreducible, polynomial a.c.i. system, where $\mathbf{F} = (F_1, \dots, F_s)$ and let $(x_1, \dots, x_n)$ be a system of linear coordinates on $\mathbf{C}^n$. Let $\mathcal{V}$ be any one of the integrable vector fields $\mathcal{X}_{F_1}, \dots, \mathcal{X}_{F_s}$. For every $1 \leqslant i \leqslant n$ such that x_i is not constant along the integral curves of $\mathcal{V}$, i.e., $\dot{x}_i := \mathcal{V}[x_i] \neq 0$, there exists a principal balance $x(t) = (x_1(t), \dots, x_n(t))$ for which $x_i(t)$ has a pole.*

Proof. Since $(\mathbf{C}^n, \{\cdot\,,\cdot\}, \mathbf{F})$ is an irreducible a.c.i. system, we may apply Theorem 6.9 on a (Zariski) open neighborhood U of a generic point $\mathbf{c}_0 \in \mathbf{C}^s$, to construct a non-singular algebraic variety $M(U)$ and a divisor $\mathcal{D}$ on it, with the properties that were stated in that theorem. The holomorphic vector field on $M(U)$ that corresponds to $\mathcal{V}$ will be denoted by $\overline{\mathcal{V}}$.

Since $M(U) \setminus \mathcal{D}$ is isomorphic to the affine variety $\mathbf{F}^{-1}(U)$ it is itself an affine variety, so that, under this isomorphism, the field of rational functions on $M(U)$ coincides with the field of rational functions of $\mathbf{F}^{-1}(U)$. As we think of $\mathbf{F}^{-1}(U)$ as being contained in $M(U)$, via the embedding $\imath$, we will use the same notations $x_1, \dots, x_n$ for the rational functions on $M(U)$ that correspond to the coordinates $x_1, \dots, x_n$ on $\mathbf{C}^n$. These functions have poles along the Painlevé walls only. In fact, each regular function on $\mathbf{C}^n$ whose restriction to a fiber $\mathbf{F}_\mathbf{c}$, with $\mathbf{c} \in U$, is non-constant, must have a pole along at least one of these irreducible components, by compactness of $\pi^{-1}(\mathbf{c})$.

Suppose now that x_i is not constant along the integral curves of $\mathcal{V}$ in $\mathbf{F}^{-1}(U)$ and let $\mathcal{D}'$ denote one of the irreducible components of $\mathcal{D}$ on which x_i experiences a pole. Pick a generic point m on $\mathcal{D}'$ and let $y_1, \dots, y_n$ be generators for the algebra of regular functions of an affine neighborhood W of m in $M(U)$. Since the algebra of rational functions of $M(U)$ and the algebra of rational functions of $\mathbf{C}^n$ coincide we can write every y_k as a rational function in $x_1, \dots, x_n$ and vice versa. The generators $y_1, \dots, y_n$ yield a system of holomorphic coordinates on W and $\overline{\mathcal{V}}$ is holomorphic on $M(U) \supseteq W$, so that $\overline{\mathcal{V}}$ is given, on W, by

$$\dot{y}_k = G_k(y_1, \dots, y_n), \qquad k = 1, \dots, n,$$

where $G_1, \dots, G_n$ are holomorphic functions on W. Since each y_k is holomorphic on W and since $\mathcal{D}$ is a divisor on $M(U)$ we can consider the Taylor series $y_k(t)$ of y_k with respect to $\overline{\mathcal{V}}$, starting at $\mathcal{D}'$,

$$y_k(t) = \sum_{l=n_k}^{\infty} \alpha_k^{(l)} t^l, \qquad \alpha_k^{(n_k)} \neq 0. \tag{6.17}$$

Since x_k is a rational function of $y_1, \dots, y_n$,

$$x_k = R_k(y_1, \dots, y_n) = \frac{p_k(y_1, \dots, y_n)}{q_k(y_1, \dots, y_n)},$$

where p_k and q_k are holomorphic functions, we find by substituting the series (6.17) in the latter

$$x_k(t) = \frac{p_k(y_1(t), \dots, y_n(t))}{q_k(y_1(t), \dots, y_n(t))} = \frac{\sum_{l \geqslant r_k} \beta_k^{(l)} t^l}{\sum_{j \geqslant s_k} \gamma_k^{(j)} t^j} = t^{r_k - s_k} \sum_{l \geqslant 0} \delta_k^{(l)} t^l, \quad \delta_k^{(0)} \neq 0, \tag{6.18}$$

yielding a Laurent solution $x_k(t)$ for $1 \leqslant k \leqslant n$. These Laurent solutions, as a whole, depend on $n-1$ free parameters. Indeed, on the one hand the Taylor expansions $y(t)$, given in (6.17) are parametrized by their initial conditions $(t = 0)$ near m, i.e., by a neighborhood of m in the $n-1$ dimensional variety $\mathcal{D}'$. On the other hand, the series $y(t)$ can be recovered from the series $x(t)$ since the functions y_k are rational functions in $x_1, \dots, x_n$. Thus, the series $x_1(t), \dots, x_n(t)$ must also depend on $n-1$ free parameters, namely the parameters that describe $\mathcal{D}'$, and hence they are principal balances.

As we will see in the next proposition the pole order of x_i along $\mathcal{D}'$ (recall that we assumed that x_i has a pole (of positive order) along $\mathcal{D}'$) equals the pole order of the Laurent series $x_i(t)$. It follows that there exists a principal balance $x(t) = (x_1(t), \dots, x_n(t))$ for which $x_i(t)$ has a pole, as claimed. □

When we want to be explicit then we will denote the principal balance (6.18) that corresponds to a Painlevé wall $\mathcal{D}'$ by $x(t;\mathcal{D}')$. More generally, for any $f \in \mathcal{F}(\mathbf{C}^n)$ we will write $f(t;\mathcal{D}')$ for its Laurent series, corresponding to $\mathcal{D}'$; the latter is computed by substituting the principal balance $x(t;\mathcal{D}')$, where $i = 1, \dots, n$, in the definition of f in terms of $x_1, \dots, x_n$.

Proposition 6.14. *Under the assumptions and notations of the previous theorem, let f be a polynomial function on $\mathbf{C}^n$. The pole order of the principal balance $f(t;\mathcal{D}')$ in t equals, for generic $\mathbf{c} \in U$, the pole order of $f_{|\mathbf{T}^r_\mathbf{c}}$ along the Painlevé divisor $\mathcal{D}'_\mathbf{c}$, where $f_{|\mathbf{T}^r_\mathbf{c}}$ is by definition $f_{|\mathbf{F}_\mathbf{c}}$, viewed as a meromorphic function on $\mathbf{T}^r_\mathbf{c}$, via $\jmath_\mathbf{c}$ (see (6.16)).*

Proof. Since $\mathcal{V}$ is transversal to $\mathcal{D}'$ at a generic point of $\mathcal{D}'$ we have that the pole that $f(t;\mathcal{D}')$ has in t equals the pole that f has along $\mathcal{D}'$, as a meromorphic function on $M(U)$. When we restrict the series $f(t;\mathcal{D}')$ to a fiber $\pi^{-1}(\mathbf{c})$, where $\mathbf{c} \in U$ is generic, then the pole order remains unchanged and the result follows. □

Notice that the corollary allows us to construct a basis of all functions with a certain pole structure along the irreducible components $\mathcal{D}'$ of $\mathcal{D}$, from knowledge of all principal balances $x(t)$. We will come back to this in Section 7.6 and in the examples.

The proof of Theorem 6.13 actually shows that there is a principal balance in the neighborhood of any point m of $\mathcal{D}'$ (any irreducible component of $\mathcal{D}$) for which $\mathcal{D}'$ is non-singular at m and $\overline{\mathcal{V}}$ is not tangent (i.e., is transversal) to $\mathcal{D}$ at m.

Notice that the points where $\mathcal{D}$ is singular or where $\overline{\mathcal{V}}$ is tangent to $\mathcal{D}$ constitute themselves a divisor in $\mathcal{D}$. What happens to the principal balances in these points?

To answer this question, let us look at the principal balances $x(t) = x(t;\mathcal{D}')$, defined in (6.18). We show that the coefficients $\delta_k^{(l)}$ are rational functions on $\mathcal{D}'$. This is so because they can be recursively defined by differentiating x_k or x_k^{-1} (depending on whether x_k is holomorphic or has a pole along $\mathcal{D}'$) sufficiently many times using the polynomial vector field $\mathcal{V}$ and restricting the result to $\mathcal{D}'$. For example, the leading coefficient of x_k is given, in case $r_k - s_k < 0$ by

$$\frac{1}{\delta_k^{(0)}} = \frac{1}{(s_k - r_k)!}\frac{\mathrm{d}^{s_k-r_k}}{\mathrm{d}t^{s_k-r_k}}_{|t=0}\frac{1}{x_k(t;\mathcal{D}')} = \frac{1}{(s_k-r_k)!}\mathcal{V}^{s_k-r_k}\left[\frac{1}{x_k}\right]_{|\mathcal{D}'},$$

and the last formula clearly shows that $\delta_k^{(0)}$ is a rational function on $\mathcal{D}'$. Similarly, the leading coefficient of x_k is given, in case $r_k - s_k \geqslant 0$ by

$$\delta_k^{(0)} = \frac{1}{(r_k - s_k)!}\frac{\mathrm{d}^{r_k-s_k}}{\mathrm{d}t^{r_k-s_k}}_{|t=0} x_k(t;\mathcal{D}') = \frac{1}{(r_k-s_k)!}\mathcal{V}^{r_k-s_k}\left[x_k\right]_{|\mathcal{D}'},$$

showing that also in this case that $\delta_k^{(0)}$ is a rational function on $\mathcal{D}'$. For the higher coefficients one proceeds in the same way, taking higher order derivatives. It follows that (6.18) is of the form

$$x_k(t;\mathcal{D}') = t^{r_k-s_k}\sum_{l\geqslant 0} R_k^{(l)} t^l,$$

where every $R_k^{(l)}$ is a rational function on $\mathcal{D}'$ for $1 \leqslant k \leqslant n$ and $l \in \mathbf{N}$.

Let us pick a generic $\mathbf{c} \in \mathbf{C}^s$ and let us consider the restriction $R_k^{(l)}{}_{|\mathcal{D}'_{\mathbf{c}}}$ of these functions to the irreducible divisor $\mathcal{D}'_{\mathbf{c}}$ of $\mathbf{T}^r_{\mathbf{c}}$. This gives us rational, hence meromorphic, functions on $\mathcal{D}'_{\mathbf{c}}$ which are, by the compactness of $\mathcal{D}'_{\mathbf{c}}$ either constant or they have a pole along some divisor $\mathcal{D}''_{\mathbf{c}}$ of $\mathcal{D}'_{\mathbf{c}}$. If, for generic $\mathbf{c}$ each element of

$$\left\{\delta_l^{(i)} \mid 1 \leqslant l \leqslant n \text{ and } i \in \mathbf{N}\right\}$$

is constant in $\mathcal{D}'_{\mathbf{c}}$ then the balance $x(t;\mathcal{D}'_{\mathbf{c}})$ cannot depend on $n-1$ free parameters, hence at least one $\delta_l^{(i)}$ must have a pole along a divisor $\mathcal{D}''_{\mathbf{c}}$ of

$\mathcal{D}'_{\mathbf{c}}$ and so the principal balance $x(t;\mathcal{D}')$ must cease to make sense along a divisor $\mathcal{D}''$ of $\mathcal{D}'$.

Let us pick an irreducible component of $\mathcal{D}''$, which for simplicity of notation we also call $\mathcal{D}''$, and let us compute $x(t;\mathcal{D}'')$ in exactly the same way as we computed $x(t;\mathcal{D}')$. Namely let us pick a genuine $m \in \mathcal{D}''$ and an appropriate chart with holomorphic coordinates $y_1, \dots, y_n$, which are of course a different set of coordinates than the one that we have used for a generic point $m \in \mathcal{D}'$.

Then once again we write the holomorphic vector field $\overline{\mathcal{V}}$ in the holomorphic coordinates $y_1, \dots, y_n$ and we compute $y(t;\mathcal{D}'')$, the Taylor series of the functions y_k with respect to $\overline{\mathcal{V}}$, starting at $\mathcal{D}''$. As before, use (6.18) to define $x(t;\mathcal{D}'')$ and conclude as before that not only the n series $y(t;\mathcal{D}'')$ but also the n series $x(t;\mathcal{D}'')$ depend on $n-2$ free parameters.

The practical moral of the story is that in order to recompute the correct $x(t;\mathcal{D}')$ series at $\mathcal{D}''$ we must use a different coordinate chart: the problem lies not in the vector field but in the coordinates that describe it.

Thus, besides principal balances, $\mathcal{V}$ must have balances which depend on $n-2$ free parameters, and by repeating the argument, balances which depend on $n-3, \dots, n-r = s$ free parameters; it stops at s because then the subvariety that corresponds to the balance is of dimension s, hence its restriction to the generic fiber $\pi^{-1}(c)$ is just a finite set of points, leading to one or several constant Laurent solutions and so the induction argument terminates at this point.

Finally we observe that we may compute the series $y(t;\mathcal{D}'')$ in two different ways. Either by grinding out the full Taylor series $y(t;\mathcal{D}')$ from the holomorphic differential equations, written in the y coordinates; or, more efficiently, using the principal balances $x(t;\mathcal{D}')$ by noting that

$$y(t;\mathcal{D}'') = y(t;\mathcal{D}')_{|\mathcal{D}''} = y(x(t;\mathcal{D}'))_{|\mathcal{D}' \to \mathcal{D}''}.$$

Namely, since $y_1 \dots, y_n$ are by assumption holomorphic coordinates near m and since $\mathcal{V}$ is holomorphic we know by Theorem 2.1 that $y(t;\mathcal{D}')$ is holomorphic in t and in $m \in \mathcal{D}'$ and so, given $m'' \in \mathcal{D}''$ we may compute $y(t;m'')$ by taking the limit of $y(t;m'_i)$, where m'_i is any sequence in $\mathcal{D}'$ that tends to m''. Indeed, by continuity of holomorphic functions we have that

$$y(t;m'') = \lim_{m'_i \to m''} y(t;m'_i) = \lim_{m'_i \to m''} y(x(t;m'_i)).$$

Since $x(t;\mathcal{D}'')$ is computed from $y(t;\mathcal{D}'')$ by (6.18) the lower balance $x(t;\mathcal{D}'')$ can completely be computed using the principal balance $x(t;\mathcal{D}')$ and holomorphic coordinates y on the neighborhood of a generic point of $\mathcal{D}''$. The same observation holds for the balances which depend on $n-3, \dots, s$ free parameters.

In summary we have shown the following theorem.

Theorem 6.15. *Suppose that* $(\mathbf{C}^n, \{\cdot,\cdot\}, \mathbf{F})$ *is an irreducible a.c.i. system, where* $\mathbf{F} = (F_1, \dots, F_s)$. *Then any of the integrable vector fields* $\mathcal{X}_{F_1}, \dots, \mathcal{X}_{F_s}$ *admits at least one balance that depends on* k *free parameters, for all* k *such that* $s \leqslant k \leqslant n-1$. *All lower balances can be computed from the principal balances.*

Although the existence of lower balances of all dimensions is also a necessary condition for algebraic complete integrability (besides the existence of principal balances), it will in most of the examples be sufficient to examine the existence of the principal balances, as stated in Theorem 6.13.

Example 6.16. Let us consider the following family of Hamiltonians

$$H = \frac{1}{2}(y_1^2 + y_2^2) + \frac{\epsilon}{3}x_1^3 + x_1 x_2^2,$$

where ϵ is an arbitrary complex parameter. The Poisson structure that we consider is the one that comes from the standard symplectic structure $\omega = \mathrm{d}x_1 \wedge \mathrm{d}y_1 + \mathrm{d}x_2 \wedge \mathrm{d}y_2$. Thus, the Hamiltonian vector field $\mathcal{X}_H$ is given by

$$\begin{aligned} \dot{x}_1 &= y_1, & \dot{y}_1 &= -\epsilon x_1^2 - x_2^2, \\ \dot{x}_2 &= y_2, & \dot{y}_2 &= -2x_1 x_2. \end{aligned} \tag{6.19}$$

We write this vector field as the following second order equation

$$\begin{aligned} \ddot{x}_1 &= -\epsilon x_1^2 - x_2^2, \\ \ddot{x}_2 &= -2x_1 x_2. \end{aligned} \tag{6.20}$$

We look for principal balances $(x_1(t), x_2(t))$ to (6.20) for which $x_1(t)$ or $x_2(t)$ have a pole. It is clear that $x_2(t)$ cannot have a pole of order larger than 2. For otherwise the second equation in (6.20) would imply that $x_1(t)$ has a double pole, which contradicts the first equation in (6.20). So the pole order of $x_2(t)$ is at most 2, and the first equation in (6.20) implies that the pole order of $x_1(t)$ is also at most 2. Hence we will look for balances to (6.20) of the form

$$x_i(t) = \frac{1}{t^2}\sum_{k \geqslant 0} x_i^{(k)} t^k, \qquad i = 1, 2. \tag{6.21}$$

Notice also that $x_1(t)$ cannot have a simple pole and that if $x_1(t)$ has no pole, then neither $x_2(t)$. So, if the above Hamiltonian H is one of the Hamiltonians of an irreducible a.c.i. system then there must exist, in view of the Kowalevski-Painlevé Criterion, a principal balance where $x_1(t)$ has a double pole and $x_2(t)$ has a double or simple pole, and there must exist a lower balance where $x_1(t)$ has a double pole (no conditions on $x_2(t)$).

In the language of Chapter 7, the vector field $\mathcal{X}_H$ is a weight homogeneous vector field with weights $\nu(x_1, x_2, y_1, y_2) = (2, 2, 3, 3)$ and the above analysis shows that all Laurent solutions to $\mathcal{X}_H$ are weight homogeneous (see Definition 7.2). The methods that we use here will be discussed in more detail in that chapter.

If we substitute (6.21) in (6.20) then we find that the leading coefficients of the series satisfy the non-linear equations

$$0 = 6x_1^{(0)} + \epsilon \left(x_1^{(0)}\right)^2 + \left(x_2^{(0)}\right)^2,$$
$$0 = x_2^{(0)}(3 + x_1^{(0)}).$$

We find three non-zero solutions, namely

$$\left(x_1^{(0)}, x_2^{(0)}\right) = \left(-\frac{6}{\epsilon}, 0\right) \tag{6.22}$$

which is valid only when $\epsilon \neq 0$, and

$$\left(x_1^{(0)}, x_2^{(0)}\right) = \left(-3, \pm 3\sqrt{2-\epsilon}\right). \tag{6.23}$$

The subsequent coefficients $x_1^{(k)}$ and $x_2^{(k)}$, with $k > 0$, must satisfy

$$\sum_{k\geqslant 0}\left((k-2)(k-3)x_1^{(k)} + \sum_{j=0}^{k}\left(\epsilon x_1^{(j)}x_1^{(k-j)} + x_2^{(j)}x_2^{(k-j)}\right)\right)t^{k-4} = 0,$$
$$\sum_{k\geqslant 0}\left((k-2)(k-3)x_2^{(k)} + 2\sum_{j=0}^{k}x_1^{(j)}x_2^{(k-j)}\right)t^{k-4} = 0,$$

Notice that $x_1^{(k)}$ and $x_2^{(k)}$ appear linearly in the coefficients of t^{k-4} and that no $x_1^{(l)}$ and $x_2^{(l)}$ with $l > k$ appear in the latter coefficients. Thus, we can solve the coefficients of t^{k-4} in these equations linearly for $x_1^{(k)}$ and $x_2^{(k)}$, and this recursively for $k = 1, 2, 3, \ldots$ In matrix form,

$$K(k)\begin{pmatrix} x_1^{(k)} \\ x_2^{(k)} \end{pmatrix} = -\sum_{j=1}^{k-1}\begin{pmatrix} \epsilon x_1^{(j)}x_1^{(k-j)} + x_2^{(j)}x_2^{(k-j)} \\ 2x_1^{(j)}x_2^{(k-j)} \end{pmatrix},$$

where

$$K(k) = \begin{pmatrix} (k-2)(k-3) + 2\epsilon x_1^{(0)} & 2x_2^{(0)} \\ 2x_2^{(0)} & (k-2)(k-3) + 2x_1^{(0)} \end{pmatrix}.$$

Notice that when k is not a root of $\det K(k)$ then the solutions $x_1^{(k)}$ and $x_2^{(k)}$ are unique, which implies that the free parameters will come in at those k where $\det K(k) = 0$, assuming that these values are non-negative integers.

Thus, for a principal (resp. lower) balance, $\det K(k)$ must have three (resp. two) non-negative integer roots. For $(x_1^{(0)}, x_2^{(0)})$ given by (6.22) we have that

$$K_1(k) = \begin{pmatrix} (k+1)(k-6) & 0 \\ 0 & (k-2)(k-3)-12/\epsilon \end{pmatrix},$$

so that

$$\det K_1(k) = (k+1)(k-6)(k^2-5k-6(2-\epsilon)/\epsilon. \tag{6.24}$$

For $(x_1^{(0)}, x_2^{(0)})$ given by (6.23) we have that

$$K_2(k) = \begin{pmatrix} (k-2)(k-3)-6\epsilon & \pm 6\sqrt{2-\epsilon} \\ \pm 6\sqrt{2-\epsilon} & (k-5)k \end{pmatrix},$$

so that

$$\det K_2(k) = (k+1)(k-6)(k^2-5k+6(2-\epsilon)). \tag{6.25}$$

A necessary condition for (6.20) to have a principal balance is that ϵ is such that $\det K_1$ or $\det K_2$ has three non-negative integer roots. Now the sum of the roots $\{\lambda_1, \lambda_2\}$ of the quadratic factors in (6.24) and in (6.25) is 5, so that these roots are either $\{0,5\}$ or $\{1,4\}$ or $\{2,3\}$. This means that $\epsilon \in \left\{1, 2, \frac{4}{3}, 6\right\}$.

Notice that if $\epsilon = 4/3$ then $\det K_1(k) = (k+1)(k-6)(k^2-5k-3)$ so that (6.22) can in this case not lead to a (lower or principal) balance. But the two solutions in (6.23) cannot lead to different balances (i.e., to a principal balance and to a lower balance), as they lead to series $(x_1(t), x_2(t))$ that only differ in the sign of $x_2(t)$. Similarly, if $\epsilon = 2$ then the solutions (6.22) and (6.23) coincide, hence lead to a single balance. Therefore, only the values $\epsilon \in \{1, 6\}$ can lead to an irreducible a.c.i. system. It is known that these two values for ϵ lead indeed to an a.c.i. system.

6.2.2 The Lyapunov Criterion

An interesting feature of a.c.i. systems is that their solutions, with any initial conditions, are single-valued. The fact that the solutions to a differential equation are not single-valued, in general, comes from the fact that analytic continuation of solutions usually depends on the (homotopy class of) the path along which the solution is analytically continued, and not just on the endpoints. Let us illustrate on a very simple example how things may go wrong.

Example 6.17. Let us consider on $\mathbf{C}^*$ the holomorphic differential equation, given by $\dot{x} = 1/(2x)$. We pick the solutions with initial condition $x = 1$ for $t = 0$. We find that these solutions satisfy $x^2(t) = t + 1$. For t in a small neighborhood of 0 (say $|t| < 1/2$) we have a holomorphic solution, namely $x(t) = \sqrt{t+1}$, where the choice of the square root is unambiguous because we have specified that $x(0) = 1$; its Taylor series is given by $x(t) = 1 + t/2 + O(t^2)$. We wish to consider the analytic continuation of this solution to $t = -2$. Therefore, consider the two half-circles, given by the following parameterizations:

$$\gamma_1 : [0,1] \to \mathbf{C} : \vartheta \mapsto t(\vartheta) = -1 + e^{\pi\sqrt{-1}\vartheta},$$
$$\gamma_2 : [0,1] \to \mathbf{C} : \vartheta \mapsto t(\vartheta) = -1 + e^{-\pi\sqrt{-1}\vartheta}.$$

Let us denote $y(\vartheta) := x(t(\vartheta))$. Along γ_1 we have that $y^2(\vartheta) = e^{\pi\sqrt{-1}\vartheta}$, so that $y(\vartheta) = e^{\pi\sqrt{-1}\vartheta/2}$, since $y(0) = 1$. Then $y(1) = \sqrt{-1}$, which means that $x(-2) = \sqrt{-1}$. However, along γ_2, $y(\vartheta) = e^{-\pi\sqrt{-1}\vartheta/2}$, so that $y(1) = -\sqrt{-1}$, which means that $x(-2) = -\sqrt{-1}$. Thus, the solution $x(t)$ is not a single-valued function of t.

It is obvious from the definition of algebraic complete integrability that the solutions are in such a case single-valued, whenever the initial conditions are taken generic, since they are meromorphic functions. We show in the following theorem, whose proof is due to Luc Haine (see [75]) that this implies that the solutions that correspond to *any* initial conditions is single-valued, and that a similar result holds for the analytic continuation of the variational equation, associated to a particular solution. This criterion was first used, without proof, by Lyapunov, who showed that the only integrable tops whose solutions have "good" analytic properties belong to the classical list, that consists of the Euler top, the Lagrange top and the Kowalevski top (see Chapter 10).

Theorem 6.18 (Lyapunov Criterion). *Let $(\mathbf{C}^n, \{\cdot\,,\cdot\}, \mathbf{F})$ be an a.c.i. system and let F be an arbitrary element of $\mathbf{F}$. All solutions to the integrable vector field $\mathcal{V} := \mathcal{X}_F$ are single-valued. Moreover, if $\gamma : [0,1] \to \mathbf{C}$ is any closed path and $x(t)$ is a solution to $\mathcal{V}$ that is holomorphic in a neighborhood of the path, then the analytic continuation along γ of the solution to the variational equations which correspond to the solution $x(t)$, is single-valued.*

Proof. Let $m_0 \in \mathbf{C}^n$ be arbitrary and consider the solution $x(t; m_0)$, defined for t sufficiently close to t_0, for which $x(t_0; m_0) = m_0$. Consider two paths $\gamma_i : [0,1] \to \mathbf{C}$ $(i = 1,2)$, with $t_0 = \gamma_1(0) = \gamma_2(0)$ and $t_1 := \gamma_1(1) = \gamma_2(1)$. Let us assume that the solution $x(t; m_0)$ can be analytically continued along both paths, and denote the analytic continuation along γ_i by $x^{(i)}(t; m_0)$. We need to show that $x^{(1)}(t_1; m_0) = x^{(2)}(t_1; m_0)$. Notice that we know that this is true if m_0 is chosen generic, because each component $x_j(t; m_0)$ of $x(t; m_0)$ is then a meromorphic function (of t), by algebraic complete integrability. Consider small tubular neighborhoods V_1 and V_2 of the two paths γ_1 and γ_2; the neighborhoods are chosen small enough so that $x^{(i)}(t; m_0)$ is holomorphic for $t \in V_i$, where $i = 1,2$. There exists a small neighborhood U of m_0 and there exist for $i = 1,2$ holomorphic functions $x^{(i)}(t; m)$ on $V_i \times U$, with $t \in V_i$

and $m \in U$ such that, for any fixed $m \in U$, $x^{(i)}(t;m)$ is a solution to $\mathcal{V}$, with initial condition $x^{(i)}(t_0;m) = m$. We have that $x^{(1)}(t_1;m) = x^{(2)}(t_1;m)$ for m in a dense subset U' of U. By continuity of $x^{(1)}$ and $x^{(2)}$ it follows that $x^{(1)}(t_1;m_0) = x^{(2)}(t_1;m_0)$.

We now turn to the variational equation, associated with a solution $x(t;m_0)$ that will be fixed throughout the rest of the proof. Define $f_i := \mathcal{V}[x_i]$, so that the differential equations for $\mathcal{V}$ are explicitly given by $\dot{x}_i = f_i(x)$. Then the *variational equation* for $\mathcal{V}$ around $x(t)$ is the linear, but non-autonomous, equation on $\mathbf{C}^n$, given by

$$\dot{\xi}_i = \sum_{j=1}^{n} \frac{\partial f_i}{\partial x_j}(x(t;m_0))\xi_j, \qquad i = 1, \dots, n. \tag{6.26}$$

They are the linearized equations of the vector field $\mathcal{V}$ in the following sense: choose any non-zero $u \in \mathbf{C}^n$ and for small $\epsilon > 0$, write the Taylor series for $x_i(t; m_0 + \epsilon u)$ and for $f_i(x(t; m_0 + \epsilon u))$,

$$x_i(t;m_0 + \epsilon u) = x_i(t;m_0) + \epsilon \sum_{j=1}^{n} \frac{\partial x_i}{\partial u_j}(t;m_0)u_j + O(\epsilon^2),$$

$$\begin{aligned} &f_i(x(t;m_0 + \epsilon u)) \\ &\quad = f_i(x(t;m_0)) + \epsilon \sum_{j,k=1}^{n} \frac{\partial f_i}{\partial x_j}(x(t;m_0))\frac{\partial x_j}{\partial u_k}(t;m_0)u_k + O(\epsilon^2). \end{aligned}$$

Substituting this in $\dot{x}_i = f_i(x)$ the term that is independent of ϵ cancels, while the linear term in ϵ gives precisely (6.26), where

$$\xi_i(t) := \sum_{j=1}^{n} \frac{\partial x_i}{\partial u_j}(t;m_0)u_j. \tag{6.27}$$

Since $x(t;m)$ is single-valued for any m in a neighborhood of m_0, the functions $\xi_i(t)$, defined by (6.27) are also single-valued, and this for any $u \in \mathbf{C}^n$. Varying u we find all solutions of the variational equation, since for fixed small t the matrix $\frac{\partial x_i}{\partial u_j}(t;m_0)$ is invertible, as $\mathcal{V}$ is holomorphic. It follows that all solutions to this variational equation are single-valued. Since the solution $x(t;m_0)$ was arbitrary, this shows that the analytic continuation of any solution to the variational equation, associated to any solution of the original equation is single-valued. □

Example 6.19. Let us treat Example 6.16 by using Theorem 6.18. Thus, we consider

$$H = \frac{1}{2}(y_1^2 + y_2^2) + \frac{\epsilon}{3}x_1^3 + x_1x_2^2$$

on $\mathbf{C}^4$, with the standard symplectic structure. In order to use Theorem 6.18 we need (a) particular solution(s) to the Hamiltonian vector field (6.19).

We start out with the simple solution

$$(x_1(t), y_1(t), x_2(t), y_2(t)) = (-6/(\epsilon t^2), 12/(\epsilon t^3), 0, 0).$$

The variational equation[4] that is associated to this solution is given by

$$\begin{aligned} \dot{\xi}_1 &= \xi_3, & \dot{\xi}_3 &= 12\xi_1/t^2, \\ \dot{\xi}_2 &= \xi_4, & \dot{\xi}_4 &= 12\xi_2/(\epsilon t^2). \end{aligned} \tag{6.28}$$

Letting $(\eta_1(t), \ldots, \eta_4(t)) = (\xi_1 t^2, \xi_2 t^2, \xi_3 t^3, \xi_4 t^3)$ we can rewrite this linear equation in the matrix form

$$\begin{pmatrix} \eta_1 \\ \eta_2 \\ \eta_3 \\ \eta_4 \end{pmatrix}^{\cdot} = \frac{1}{t} \begin{pmatrix} 2 & 0 & 1 & 0 \\ 0 & 2 & 0 & 1 \\ 12 & 0 & 3 & 0 \\ 0 & 12/\epsilon & 0 & 3 \end{pmatrix} \begin{pmatrix} \eta_1 \\ \eta_2 \\ \eta_3 \\ \eta_4 \end{pmatrix} \tag{6.29}$$

Clearly, the solutions to (6.29), and hence to (6.28) can only be single-valued when the square matrix in (6.29) is diagonalizable, with integer eigenvalues, in particular when its characteristic polynomial

$$\chi_1(k) = (k+1)(k-6)(k^2 - 5k + 6 - 12/\epsilon)$$

has only integer roots. Let us also consider the solution

$$(x_1(t), y_1(t), x_2(t), y_2(t)) = (-3t^{-2}, 6t^{-3}, 3\sqrt{2-\epsilon}\, t^{-2}, -6\sqrt{2-\epsilon}\, t^{-3}),$$

with corresponding variational equation, in the matrix form

$$\begin{pmatrix} \eta_1 \\ \eta_2 \\ \eta_3 \\ \eta_4 \end{pmatrix}^{\cdot} = \frac{1}{t} \begin{pmatrix} 2 & 0 & 1 & 0 \\ 0 & 2 & 0 & 1 \\ 6\epsilon & -6\sqrt{2-\epsilon} & 3 & 0 \\ -6\sqrt{2-\epsilon} & 6 & 0 & 3 \end{pmatrix} \begin{pmatrix} \eta_1 \\ \eta_2 \\ \eta_3 \\ \eta_4 \end{pmatrix}. \tag{6.30}$$

The solutions to (6.30) can again only be single-valued when the square matrix in (6.30) is diagonalizable, with integer eigenvalues, in particular when its characteristic polynomial

$$\chi_2(k) = (k+1)(k-6)(k^2 - 5k + 6(2-\epsilon))$$

has only integer roots[5].

---

[4] See the next chapter for more information on how to compute and integrate the variational equation, associated to a solution.

[5] Notice that while $\chi_1(k) = \det K_1(k)$ and $\chi_2(k) = \det K_2(k)$ (see (6.24) and (6.25)), as will be explained in Chapter 7, we use these polynomials in the two criteria in a very different way.

Summarizing, if we want all solutions to the variational equation associated to any solution to 6.19 to be single-valued, then ϵ must be such that the two polynomials

$$p_1(k) := k^2 - 5k + 6(1 - 2/\epsilon),$$
$$p_2(k) := k^2 - 5k + 6(2 - \epsilon),$$

have only integer roots. The discriminants of p_1 (resp. of p_2) must be non-negative, more precisely, since the two roots of p_1 (resp. of p_2) have 5 as their sum, it must be the square of an odd integer. Since $\mathrm{disc}(p_2) = 1 - 24(1-\epsilon) \in \mathbf{N}$ we have that $\epsilon \geqslant 1$. Also, $\mathrm{disc}(p_1) = 1 + 48/\epsilon \in \mathbf{N}$, so that $1 < \mathrm{disc}(p_1) \leqslant 49$, and since this discriminant must be the square of an odd integer we must have that $\mathrm{disc}(p_1) \in \{9, 25, 49\}$, which corresponds to the values $\epsilon \in \{1, 2, 6\}$. Notice that, besides the possibility of $\epsilon = 2$, these are precisely the values of ϵ that we have found by using the Kowalevski-Painlevé Criterion to this family of Hamiltonians (see Example 6.16). It is not clear how, using the Lyapunov criterion, this case can be excluded. In fact, for $\epsilon = 2$ numerical data suggest that this case is not Liouville integrable, but a rigorous proof that this case is not Liouville integrable is not known.

Remark 6.20. The Lyapunov Criterion (and its proof) is also valid for generalized a.c.i. systems, because they also have the property that the solutions that correspond to generic initial conditions are single-valued.

6.3 The Complex Liouville Theorem

In this section we will prove a complex (holomorphic) version of the Liouville Theorem (Theorem 4.28). Our assumptions (especially in Theorem 6.22) are geared towards the examples.

Theorem 6.21 (Complex Liouville Theorem I). *Suppose that M is an r-dimensional complex compact manifold and let $\mathcal{D}$ be an analytic hypersurface on M. Suppose that $M \setminus \mathcal{D}$ admits r holomorphic vector fields $\mathcal{V}_1, \dots, \mathcal{V}_r$ with the following properties.*

(1) The vector fields commute pairwise, $[\mathcal{V}_i, \mathcal{V}_j] = 0$ for $1 \leqslant i, j \leqslant r$;
(2) At every point $m \in M \setminus \mathcal{D}$ the vector fields $\mathcal{V}_1, \dots, \mathcal{V}_r$ are independent;
(3) $\mathcal{V}_1$ extends to a holomorphic vector field $\overline{\mathcal{V}}_1$ on M;
(4) The integral curves of $\overline{\mathcal{V}}_1$ that start at points $m \in \mathcal{D}$ leave $\mathcal{D}$ immediately.

Then M is a complex torus of dimension r and the vector fields $\mathcal{V}_1, \dots, \mathcal{V}_r$ extend to r commuting, everywhere independent holomorphic vector fields $\overline{\mathcal{V}}_1, \dots, \overline{\mathcal{V}}_r$ on M. If M admits, in addition, r algebraically independent meromorphic functions then M is an Abelian variety.

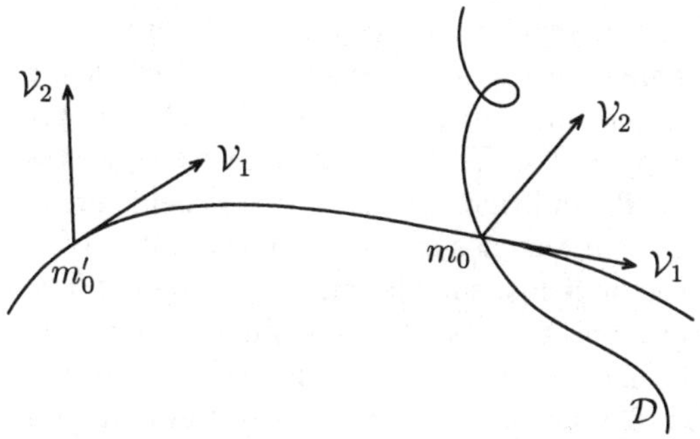

Fig. 6.4. The flow of the holomorphic vector field $\overline{\mathcal{V}}_1$ is used to extend the holomorphic vector field $\mathcal{V}_2, \ldots, \mathcal{V}_r$ in the neighborhood of any point of $\mathcal{D}$.

Proof. We first extend $\mathcal{V}_2, \ldots, \mathcal{V}_r$ to holomorphic vector fields $\overline{\mathcal{V}}_2, \ldots, \overline{\mathcal{V}}_r$ on M (see Figure 6.4). Let $m_0 \in \mathcal{D}$ and, using assumption *(4)*, let ϵ be such that if $0 < |t_1| < \epsilon$ then $\Phi^{(1)}_{t_1}(m_0) \notin \mathcal{D}$, where $\Phi^{(1)}$ denotes the flow of $\overline{\mathcal{V}}_1$. Fix t_1 such that $0 < |t_1| < \epsilon$ and let U be an open neighborhood of $m_0' := \Phi^{(1)}_{t_1}(m_0)$ in $M \setminus \mathcal{D}$ and let $V := \Phi^{(1)}_{-t_1}(U)$. We have that $\mathcal{V}_1, \ldots, \mathcal{V}_r$ are holomorphic commuting vector fields on U which are independent on U, and that $\Phi^{(1)}_{-t_1}$ is a biholomorphic map from U to V. Therefore, for $i = 1, \ldots, r$ the r vector fields $\mathcal{V}_i' := \left(\Phi^{(1)}_{-t_1}\right)_* \mathcal{V}_i$ are holomorphic commuting vector fields which are independent on V; notice that $\mathcal{V}_1' = \overline{\mathcal{V}}_1$ (on V). For any point $m \in V \setminus \mathcal{D}$ the fact that $\mathcal{V}_1$ and $\mathcal{V}_i$ commute on $V \setminus \mathcal{D}$ implies that

$$\mathcal{V}_i'(m) = \left(\Phi^{(1)}_{-t_1}\right)_* \mathcal{V}_i\left(\Phi^{(1)}_{t_1}(m)\right) = \left(\Phi^{(1)}_{-t_1}\right)_* \left(\Phi^{(1)}_{t_1}\right)_* \mathcal{V}_i(m) = \mathcal{V}_i(m)$$

so that the commuting vector fields $\mathcal{V}_i'$ extend the vector fields $\mathcal{V}_i$ holomorphically to V. By uniqueness, the vector fields $\mathcal{V}_i$ extend to commuting, everywhere independent holomorphic vector fields on all of M, which we denote by $\overline{\mathcal{V}}_i$. Since M is compact, the flow of each of the holomorphic vector fields $\overline{\mathcal{V}}_i$ on M is complete.

We can now repeat part of the proof of the Liouville theorem (Theorem 4.28) and use the completeness and commutativity of the vector fields $\overline{\mathcal{V}}_1, \ldots, \overline{\mathcal{V}}_r$ to define an action $\mathbf{C}^r \times M \to M$ by

$$((t_1, \ldots, t_r), m) \mapsto \Phi^{(1)}_{t_1} \circ \Phi^{(2)}_{t_2} \circ \cdots \circ \Phi^{(r)}_{t_r}(m),$$

thereby exhibiting M as a compact quotient of $\mathbf{C}^r$ by an additive subgroup, i.e., as a compact complex torus. If M admits n independent meromorphic functions then, by Moishezon's Theorem on Tori, (Theorem 5.13) M is an Abelian variety. $\square$

In the examples we will typically have r independent commuting vector fields on an affine variety $\mathcal{A}$ (a non-singular irreducible component of a fiber $\mathbf{F_c}$ of the momentum map), which we will embed in projective space via a regular map $\varphi : \mathcal{A} \to \mathbf{P}^N$. We will want to prove that $\mathcal{A}$ is an affine part of an Abelian variety $\mathbf{T}^r$ and that the given vector fields are linear vector fields on $\mathbf{T}^r$, by showing that the closure[6] $\overline{\varphi(\mathcal{A})}$ of the image of $\mathcal{A}$ in $\mathbf{P}^N$, together with the image vector fields, satisfies the conditions of Theorem 6.21. However, nothing guarantees that $\overline{\varphi(\mathcal{A})}$ is non-singular; moreover, it is in general difficult to check this in concrete examples. Therefore we give a version of the complex Liouville Theorem which does not assume, but instead implies that the closure is non-singular, i.e., that it is a complex manifold.

Notice that, even if $\mathcal{A}$ is indeed an affine part of an Abelian variety $\mathbf{T}^r$ there is of course no hope that φ will lead to an embedding of $\mathbf{T}^r$, unless φ is very special; we will deal with this issue in Paragraph 7.6.5.

Theorem 6.22 (Complex Liouville Theorem II). *Let $\mathcal{A} \subset \mathbf{C}^s$ be a non-singular affine variety of dimension r which supports r holomorphic vector fields $\mathcal{V}_1, \dots, \mathcal{V}_r$ and let $\varphi : \mathcal{A} \to \mathbf{C}^N \subset \mathbf{P}^N$ be a regular map; here $\mathbf{C}^N \subset \mathbf{P}^N$ is the usual inclusion of $\mathbf{C}^N$ as the complement of a hyperplane H in $\mathbf{P}^N$. We define $\Delta := \overline{\varphi(\mathcal{A})} \setminus \varphi(\mathcal{A})$ and we decompose the analytic subset Δ as $\Delta = \Delta' \cup \Delta''$, where Δ' is the union of the irreducible components of Δ of dimension $r-1$ and Δ'' is the union of the other irreducible components of Δ. Assume the following.*

(0) $\varphi : \mathcal{A} \to \mathbf{C}^N$ is an isomorphic embedding;
(1) The vector fields commute pairwise, $[\mathcal{V}_i, \mathcal{V}_j] = 0$ for $1 \leqslant i, j \leqslant r$;
(2) At every point $m \in \mathcal{A}$ the vector fields $\mathcal{V}_1, \dots, \mathcal{V}_r$ are independent;
(3) The vector field $\varphi_* \mathcal{V}_1$ extends to a vector field $\overline{\mathcal{V}}_1$ which is holomorphic on a neighborhood of Δ' in $\mathbf{P}^N$;
(4) The integral curves of $\overline{\mathcal{V}}_1$ that start at points $m \in \Delta'$ go immediately into $\varphi(\mathcal{A})$.

Then $\overline{\varphi(\mathcal{A})}$ is an Abelian variety of dimension r and $\Delta'' = \emptyset$, so that $\overline{\varphi(\mathcal{A})} = \varphi(\mathcal{A}) \cup \Delta'$. Moreover, the vector fields $\varphi_*\mathcal{V}_1, \dots, \varphi_*\mathcal{V}_r$ extend to holomorphic vector fields on $\overline{\varphi(\mathcal{A})}$.

Proof. We verify that $\overline{\varphi(\mathcal{A})}$ is a smooth manifold. Then the assumptions imply that the holomorphic vector fields $\varphi_*\mathcal{V}_1, \dots, \varphi_*\mathcal{V}_r$ satisfy the conditions of Theorem 6.21 and we may conclude that $\overline{\varphi(\mathcal{A})}$ is an Abelian variety which is equipped with r commuting holomorphic vector fields $\overline{\mathcal{V}}_1, \dots, \overline{\mathcal{V}}_r$ that extend $\varphi_*\mathcal{V}_1, \dots, \varphi_*\mathcal{V}_r$.

[6] It is a fundamental fact that we don't have to specify whether the closure is taken with respect to the Zariski topology or with respect to the usual (complex) topology, because for subsets such as $\varphi(\mathcal{A})$ they coincide ($\varphi(\mathcal{A})$ is an example of a constructible set in $\mathbf{P}^N$, and for constructible subsets of $\mathbf{P}^N$ the Zariski closure coincides with the usual closure, see [135, §I.10]).

1. Let $m_0 \in \Delta'$. We show that $\overline{\varphi(\mathcal{A})}$ is smooth at m_0. Conditions (3) and (4) imply that there exists an open neighborhood of m_0 in $\mathbf{P}^N$ on which $\overline{\mathcal{V}}_1$ is holomorphic and non-vanishing. Denoting the flow of $\overline{\mathcal{V}}_1$ by $\Phi^{(1)}$, we can choose such a neighborhood U and a (small) $t_1 \in \mathbf{C}$, such that $\Phi^{(1)}_{t_1}|_U$ is a biholomorphism onto an open neighborhood V of $\Phi^{(1)}_{t_1}(m_0)$ in $\mathbf{P}^N$, which is disjoint from Δ, and such that $\Phi^{(1)}_{t_1}(m_0) \in \varphi(\mathcal{A})$. The image of $H \cap U$ by $\Phi^{(1)}_{t_1}$ will be denoted by H'; it is an analytic hypersurface (divisor) of V. We will show that $\Phi^{(1)}_{t_1}$ restricts to a bijection between $(U \setminus H) \cap \overline{\varphi(\mathcal{A})} = U \cap \varphi(\mathcal{A})$ and $(V \setminus H') \cap \overline{\varphi(\mathcal{A})} = (V \cap \varphi(\mathcal{A})) \setminus H'$. Since $\Phi^{(1)}_{t_1}$ is a biholomorphism it restricts then also to a bijection between their closures (in U, resp. in V),

$$\Phi^{(1)}_{t_1} : \overline{U \cap \varphi(\mathcal{A})} \longleftrightarrow \overline{(V \cap \varphi(\mathcal{A})) \setminus H'}.$$

Now $\overline{(V \cap \varphi(\mathcal{A})) \setminus H'} = V \cap \varphi(\mathcal{A})$ since H' is an analytic hypersurface of V and since $\varphi(\mathcal{A})$ is closed in V (recall that $V \cap \Delta = \emptyset$). Also, $\overline{U \cap \varphi(\mathcal{A})} = U \cap \overline{\varphi(\mathcal{A})}$, since U is an open subset of $\mathbf{P}^N$. Now $V \cap \varphi(\mathcal{A})$ is a smooth submanifold of V and $\Phi^{(1)}_{t_1}$ is a biholomorphism, hence $\Phi^{(1)}_{-t_1}(V \cap \varphi(\mathcal{A})) = U \cap \overline{\varphi(\mathcal{A})}$ is a smooth submanifold of U, implying that $\overline{\varphi(\mathcal{A})}$ is smooth at m_0, as was to be shown. In order to show that $\Phi^{(1)}_{t_1}$ restricts to a bijection between $(U \setminus H) \cap \overline{\varphi(\mathcal{A})}$ and $(V \setminus H') \cap \overline{\varphi(\mathcal{A})}$ it suffices, since $\Phi^{(1)}_{t_1}$ restricts to a biholomorphism between $U \setminus H$ and $V \setminus H'$, to show that the integral curves of $\overline{\mathcal{V}}_1$ that start at points in $\varphi(\mathcal{A})$ remain in $\overline{\varphi(\mathcal{A})}$. This follows from the fact that $\overline{\mathcal{V}}_1$ is tangent to $\varphi(\mathcal{A})$ at all points of $\varphi(\mathcal{A})$: the integral curve that starts at $p \in \varphi(\mathcal{A})$ will be locally contained in $\varphi(\mathcal{A})$, hence the entire integral curve will be contained in the (Zariski) closure $\overline{\varphi(\mathcal{A})}$ of $\varphi(\mathcal{A})$.

2. We now show that $\Delta'' = \emptyset$. When $\dim \mathcal{A} = r = 1$ then Δ'' is empty for dimensional reasons, so we assume that $r \geqslant 2$. First notice that the irreducibility of $\varphi(\mathcal{A})$ implies that $\overline{\varphi(\mathcal{A})}$ is irreducible. Since φ is an isomorphic embedding we have that Δ is a hyperplane section of $\overline{\varphi(\mathcal{A})}$, namely $\Delta = \overline{\varphi(\mathcal{A})} \cap H$. Therefore, Δ is the support of an ample divisor on an irreducible projective variety of dimension at least 2, hence it is connected (see [78, Cor. III.7.9]). But in this case, there must be one of the irreducible components of Δ'' which has a non-zero intersection point m_0 with Δ'. But we know that $\overline{\varphi(\mathcal{A})}$ is smooth in a neighborhood U of $m_0 \in \Delta'$. The intersection $\Delta \cap U$ is an analytic hypersurface of the variety $\overline{\varphi(\mathcal{A})} \cap U$, and the latter is smooth at m_0. But then each of the irreducible components of $\Delta \cap U$ is of codimension 1. Contradiction.

We conclude that $\overline{\varphi(\mathcal{A})}$ is a non-singular projective manifold, and it verifies the conditions of Theorem 6.21. The conclusion follows. □

6.4 Lax Equations with a Parameter

In this section we will consider Lax equations with a parameter and we will show that they are closely related to a.c.i. systems. In particular we will prove a criterion which gives necessary and sufficient conditions for the divisor map (defined below) to transform the isospectral flow defined by the Lax equation to a linear flow on the Jacobian of the spectral curve. The proof that we give is based on arguments that first appeared in [165] and [7]. For a cohomological interpretation of this proof, see [68]. The different notions related to Lax equations with a spectral parameter and to this criterion will be illustrated, as they are developed, using one example that is at the same time very simple and very rich.

We first define the notion of a Lax equation with a parameter and we introduce the example.

Definition 6.23. Let M be a finite-dimensional affine subspace of the loop algebra $L(\mathfrak{gl}(N)) = \mathfrak{gl}(N)\left[\mathfrak{h}, \mathfrak{h}^{-1}\right]$. A *Lax equation with parameter* $\mathfrak{h}$ on M is given by a differential equation on M of the form

$$\dot{X}(\mathfrak{h}) = \left[X(\mathfrak{h}), Y(\mathfrak{h})\right], \tag{6.31}$$

where $X(\mathfrak{h})$ is a typical element of M,

$$X(\mathfrak{h}) = \sum_{i=m'}^{m} X_i \mathfrak{h}^i, \qquad X_i \in \mathfrak{gl}(N) \text{ for } m' \leqslant i \leqslant m,$$

and

$$Y(\mathfrak{h}) = \sum_{j=n'}^{n} Y_j \mathfrak{h}^j, \qquad Y_j \in \mathfrak{gl}(N) \text{ for } n' \leqslant j \leqslant n,$$

the entries of each Y_j being polynomial functions of the entries of some (or all) of the X_i.

Remark 6.24. By multiplying $X(\mathfrak{h})$ by a sufficiently high power of $\mathfrak{h}$ we may always suppose that $X(\mathfrak{h})$ is polynomial in $\mathfrak{h}$; said differently we may assume that M is a finite-dimensional subspace of $\mathfrak{gl}(N)[\mathfrak{h}]$. However, we will not use this assumption in what follows, as it is often unnatural. Notice that this property does not hold for $Y(\mathfrak{h})$. Also, often M will be contained in $\mathfrak{g}[\mathfrak{h}]$ where $\mathfrak{g}$ is a non-trivial Lie subalgebra of $\mathfrak{gl}(N)$.

Remark 6.25. The Lax equation (6.31) depends only on $Y(\mathfrak{h})$ modulo matrices that commute with $X(\mathfrak{h})$, so that we may e.g. add to $Y(\mathfrak{h})$ an arbitrary polynomial in $\mathfrak{h}$ and in $\mathfrak{h}^{-1}$, whose coefficients are powers of $X(\mathfrak{h})$.

Remark 6.26. The definition is easily adapted to Lax representations with spectral parameter (see Definition 4.42).

Example 6.27. For $d \geqslant 2$ let M_d denote the vector space of all trace-less 2×2 matrices

$$X(\mathfrak{h}) = \begin{pmatrix} v(\mathfrak{h}) & u(\mathfrak{h}) \\ w(\mathfrak{h}) & -v(\mathfrak{h}) \end{pmatrix}$$

where $u(\mathfrak{h})$, $v(\mathfrak{h})$ and $w(\mathfrak{h})$ are elements of $\mathbf{C}[\mathfrak{h}]$ of degree at most d. Letting $Y(\mathfrak{h}) := X(0)/\mathfrak{h}$ it is easy to see that

$$\dot{X}(\mathfrak{h}) = \frac{1}{\mathfrak{h}}[X(\mathfrak{h}), X(0)] \tag{6.32}$$

defines a Lax equation with parameter on M_d.

Proposition 6.28. *Given a Lax pair*

$$\dot{X}(\mathfrak{h}) = [X(\mathfrak{h}), Y(\mathfrak{h})],$$

the functions q_{kl} which are defined by the coefficients of the characteristic polynomial of $X(\mathfrak{h})$,

$$\det(\mathfrak{z}\,\mathrm{Id}_N - X(\mathfrak{h})) = \mathfrak{z}^N + \sum_{\substack{0 \leqslant l < N \\ -\ell' \leqslant k \leqslant \ell}} q_{kl}\mathfrak{h}^k \mathfrak{z}^l. \tag{6.33}$$

are constants of motion of (6.31). The plane algebraic curve, associated to each $X(\mathfrak{h})$,

$$\Gamma_X := \left\{(\mathfrak{h}, \mathfrak{z}) \in \mathbf{C} \times \mathbf{C} \mid \det(\mathfrak{z}\,\mathrm{Id}_N - X(\mathfrak{h})) = 0\right\},$$

is preserved by the flow of (6.31). Similarly, for each $X(\mathfrak{h})$ the variety of matrices $A_X \subset M$ defined by

$$A_X := \left\{X'(\mathfrak{h}) \mid X(\mathfrak{h}) \text{ and } X'(\mathfrak{h}) \text{ have the same characteristic polynomial}\right\}$$

is preserved by the flow of (6.31). For X such that Γ_X is smooth, let us denote its smooth compactification by $\overline{\Gamma_X}$ and let

$$\{p_1, \ldots, p_s\} := \overline{\Gamma_X} \setminus \Gamma_X$$

denote the points at infinity. At each of these points $\mathfrak{h}$ has a zero or a pole, i.e., (possibly after relabeling) we have that

$$\operatorname{ord}_{p_i}(\mathfrak{h}) = \begin{cases} -\mu_i & 1 \leqslant i \leqslant s' \\ \mu_i & s'+1 \leqslant i \leqslant s \end{cases} \tag{6.34}$$

where $\mu_i > 0$ for $i = 1, \ldots, s$.

Proof. The proof of the first part is formally identical to the proof in the case of Lax equations (without a parameter), given in Example 4.3. The statement about the zeros and poles of $\mathfrak{h}$ at infinity is the only remaining part. When Γ_X is smooth, it admits a unique non-singular compactification that we have denoted by $\overline{\Gamma_X}$. The parameters $\mathfrak{z}$ and $\mathfrak{h}$ in the equation $\det(\mathfrak{z}\,\mathrm{Id}_N - X(\mathfrak{h})) = 0$ for Γ_X are meromorphic functions on $\overline{\Gamma_X}$ with poles on $\{p_1, \ldots, p_s\} = \overline{\Gamma_X} \backslash \Gamma_X$ only and at each of these points at least one of $\mathfrak{z}$ and $\mathfrak{h}$ have a pole. Moreover, if $\mathfrak{z}$ has a pole at a point $p \in \overline{\Gamma_X}$ then $\mathfrak{h}$ must have a pole or a zero at that point p; indeed, using the fact that $q_{kl} = 0$ for $l \geqslant N$, we have that if $\mathfrak{h}$ would not have a zero or pole at that point, z^N would be the dominant term in (6.33) and would not be compensated by another term. This shows that if $\mathfrak{h}$ does not have a pole at some point $p_i \in \overline{\Gamma_X} \setminus \Gamma_X$ then is has a zero at that point. □

Definition 6.29. Each of the curves Γ_X (or $\overline{\Gamma_X}$) is called a *spectral curve* and each of the varieties A_X is an *isospectral variety*; when A_X is smooth, we also use *isospectral manifold.*

Notice that the spectral curve and the isospectral variety both depend on the values of the constants of motion q_{ij} only. For that reason we often write $\Gamma_{\mathbf{c}}$ and $A_{\mathbf{c}}$ instead of Γ_X and A_X, where $\mathbf{c}$ is the vector of values of q_{ij} on $X = X(\mathfrak{h})$. In the examples, discussed in this book, the spectral curve $\Gamma_{\mathbf{c}}$ is non-singular for generic values of $\mathbf{c}$.

Example 6.30. In the case of Example 6.27 we have that

$$\begin{vmatrix} \mathfrak{z} - v(\mathfrak{h}) & -u(\mathfrak{h}) \\ -w(\mathfrak{h}) & \mathfrak{z} + v(\mathfrak{h}) \end{vmatrix} = \mathfrak{z}^2 - (u(\mathfrak{h})w(\mathfrak{h}) + v^2(\mathfrak{h}))$$

so that we have $2d+1$ constants of motion, which are the coefficients of the polynomial $u(\mathfrak{h})w(\mathfrak{h}) + v^2(\mathfrak{h})$, which has degree $2d$. In fact, the leading coefficients of each of $u(\mathfrak{h})$, $v(\mathfrak{h})$ and $w(\mathfrak{h})$ are also constants of motion, since the right hand side of the Lax equation is a polynomial in $\mathfrak{h}$ of degree less than d. It is easy to see that this yields a total of $2d+3$ independent constants of motion. For any $X \in M_d$ the spectral curve Γ_X is given by

$$\mathfrak{z}^2 = u(\mathfrak{h})w(\mathfrak{h}) + v^2(\mathfrak{h}).$$

For generic X it is a non-singular affine curve of genus $g = d-1$. When $d = 2$ then it is an elliptic curve, otherwise it is hyperelliptic.

Having investigated the spectral curves and the isospectral manifolds we now turn to the eigenfunctions of $X(\mathfrak{h})$.

Proposition 6.31. *Let $X(\mathfrak{h}) \in M$ where M is a finite-dimensional affine subspace of $\mathfrak{gl}(N)[\mathfrak{h}, \mathfrak{h}^{-1}]$, and suppose that for generic $X(\mathfrak{h}) \in M$ the affine curve Γ_X is non-singular and that for generic $(\mathfrak{h}, \mathfrak{z}) \in \Gamma_X$, the eigenspace of $X(\mathfrak{h})$ with eigenvalue $\mathfrak{z}$ is one-dimensional.*

(1) *If we denote by* $\Delta_{kl}(\mathfrak{z}, X(\mathfrak{h}))$ *the cofactor of* $\mathfrak{z}\,\mathrm{Id}_N - X(\mathfrak{h})$ *corresponding to the* (k,l)*-th entry then*

$$\xi(\mathfrak{z}, X(\mathfrak{h})) := (\xi_1, \dots, \xi_N)^\top \quad \text{with } \xi_k = \frac{\Delta_{1k}(\mathfrak{z}, X(\mathfrak{h}))}{\Delta_{11}(\mathfrak{z}, X(\mathfrak{h}))}, \tag{6.35}$$

is the unique eigenvector of $X(\mathfrak{h})$ *with eigenvalue* $\mathfrak{z}$*, normalized at* $\xi_1 = 1$*. The* ξ_k*'s are rational functions in* $(\mathfrak{h}, \mathfrak{z})$ *and in the entries of the coefficients* X_i *of* $X(\mathfrak{h})$*.*

(2) *When* $X(\mathfrak{h}, t)$ *flows according to* $\dot{X}(\mathfrak{h}) = [X(\mathfrak{h}), Y(\mathfrak{h})]$*, the corresponding eigenvector* $\xi(t) := \xi(\mathfrak{z}, X(\mathfrak{h}, t))$ *satisfies the autonomous equation*

$$\dot{\xi} + Y\xi = \lambda\xi, \tag{6.36}$$

where $Y := Y(\mathfrak{h}, X(\mathfrak{h}, t))$ *and* λ *is a scalar function of* $\mathfrak{h}$*,* $\mathfrak{z}$*,* t*,*

$$\begin{aligned}\lambda := \lambda(\mathfrak{h}, \mathfrak{z}, t) = \lambda(\mathfrak{h}, \mathfrak{z}, X(\mathfrak{h}, t)) &= \Big(Y(\mathfrak{h}, X(\mathfrak{h}, t))\xi(\mathfrak{z}, X(\mathfrak{h}, t))\Big)_1 \\ &= \sum_{l=1}^N Y(\mathfrak{h}, X(\mathfrak{h}, t))_{1l} \frac{\Delta_{1l}(\mathfrak{z}, X(\mathfrak{h}, t))}{\Delta_{11}(\mathfrak{z}, X(\mathfrak{h}, t))}.\end{aligned} \tag{6.37}$$

(3) *The order of* λ*, viewed as a meromorphic function on* $\overline{\Gamma_X}$*, at the points* p_i *is given, independently of* t*, by*

$$\mathrm{ord}_{p_i}(\lambda) \geqslant \begin{cases} -n\mu_i & 1 \leqslant i \leqslant s' \\ n'\mu_i & s'+1 \leqslant i \leqslant s. \end{cases} \tag{6.38}$$

Proof. Under the assumption that for generic $(\mathfrak{h}, \mathfrak{z}) \in \Gamma_X$, the eigenspace of $X(\mathfrak{h})$ with eigenvalue $\mathfrak{z}$ is one-dimensional, we solve for the eigenfunction $\xi = \xi(\mathfrak{z}, X(\mathfrak{h}))$ by Cramer's rule, yielding (6.35). Clearly the ξ_k are rational functions in $\mathfrak{z}, \mathfrak{h}$ and thus, meromorphic on the curve $\overline{\Gamma_X}$. The meromorphic functions $\xi_k = \xi_k(\mathfrak{z}, X(\mathfrak{h}))$ evolve, as X evolves according to (6.31), but they are always meromorphic functions on the same curve. To verify the precise form of their evolution, we compute, setting $X(t) := X(\mathfrak{h}, t)$, $\xi(t) := \xi(\mathfrak{z}, X(\mathfrak{h}, t))$ and $Y(t) := Y(\mathfrak{h}, X(\mathfrak{h}, t))$, that

$$\begin{aligned} 0 &= ((X(t) - \mathfrak{z}\,\mathrm{Id}_N)\xi(t))^{\cdot} \\ &= \dot{X}(t)\xi(t) + (X(t) - \mathfrak{z}\,\mathrm{Id}_N)\dot{\xi}(t) \\ &= (X(t)Y(t) - Y(t)X(t))\xi(t) + (X(t) - \mathfrak{z}\,\mathrm{Id}_N)\dot{\xi}(t) \\ &= (X(t) - \mathfrak{z}\,\mathrm{Id}_N)(Y(t)\xi(t) + \dot{\xi}(t)), \end{aligned}$$

implying that $Y(t)\xi(t) + \dot{\xi}(t)$ is an eigenvector of $X(t)$ corresponding to the eigenvalue $\mathfrak{z}$. Since, by the assumption, $(X(t) - \mathfrak{z}\,\mathrm{Id}_N)^{-1}(0)$ is one-dimensional for generic $(\mathfrak{h}, \mathfrak{z}) \in \Gamma_X$, the vector $Y(t)\xi(t) + \dot{\xi}(t)$ must be proportional to $\xi(t)$, i.e., there exists $\lambda = \lambda(\mathfrak{h}, \mathfrak{z}, X(\mathfrak{h}, t))$ such that

$$\dot{\xi} + Y\xi = \lambda\xi. \tag{6.39}$$

To see the precise form of λ, look at the first component of the equation above; using $\xi_1 = 1$, this yields $\lambda = (\lambda\xi)_1 = (Y\xi)_1$, and, spelled out, this is (6.37). It follows that λ is a rational function in $\mathfrak{h}$ and $\mathfrak{z}$, whose coefficients are functions of the entries of $X(\mathfrak{h}, t)$.

To prove (6.38), first replace ξ by $\widehat{\xi} = \xi/\mathfrak{h}^\ell$, where ℓ is large enough so as to guarantee the holomorphy of $\widehat{\xi}$ at the points $p_i \in \overline{\Gamma_X} \setminus \Gamma_X$, where $\mathfrak{h}$ has a pole, i.e., $1 \leqslant i \leqslant s'$. Substituting $\xi = \widehat{\xi}\mathfrak{h}^\ell$ into (6.39) yields

$$\dot{\widehat{\xi}} + \sum_{i=n'}^{n} Y_i \mathfrak{h}^i \widehat{\xi} - \lambda(\mathfrak{h}, \mathfrak{z}, t)\widehat{\xi} = 0. \tag{6.40}$$

First (6.38) holds at p_i, if λ has no pole at p_i; so, let λ have a pole at p_i. Picking a component $\widehat{\xi}_k$ having no zero at p_i, or at least the one with a minimal zero of order μ at p_i, the leading contribution of (6.40) at p_i is given by

$$\mathfrak{h}^n \sum_{l=1}^{N} (Y_n)_{kl} \widehat{\xi}_l - \lambda \widehat{\xi}_k$$

and thus, the divisors at p_i read[7]

$$n \operatorname{ord}_{p_i}(\mathfrak{h}) + \nu = \operatorname{ord}_{p_i}(\lambda) + \mu, \quad \text{with } \nu \geqslant \mu,$$

implying $\operatorname{ord}_{p_i}(\lambda) \geqslant n \operatorname{ord}_{p_i}(\mathfrak{h})$. At the other points $p_i \in \{p_{s'+1}, \dots, p_s\}$, we know that $\mathfrak{h}$ vanishes, so we may consider $\widehat{\xi} = \xi\mathfrak{h}^\ell$ with ℓ sufficiently large so as to make $\xi\mathfrak{h}^\ell$ holomorphic at all those points, yielding the same equation (6.40). A similar argument leads to $\operatorname{ord}_{p_i}(\lambda) \geqslant n' \operatorname{ord}_{p_i}(\mathfrak{h})$, establishing (6.38).
□

Example 6.32. In the case of Example 6.27 we have that the generic affine curve Γ_X is non-singular and that for a generic matrix $X(\mathfrak{h})$ in M_d the eigenspace corresponding to a generic point in Γ_X is one-dimensional. ξ_2, as computed from (6.35), can be written in the following two equivalent forms:

$$\xi_2 = \frac{\mathfrak{z} - v(\mathfrak{h})}{u(\mathfrak{h})} = \frac{w(\mathfrak{h})}{\mathfrak{z} + v(\mathfrak{h})}.$$

Then (6.36) becomes

$$\begin{pmatrix} 1 \\ \frac{\mathfrak{z} - v(\mathfrak{h})}{u(\mathfrak{h})} \end{pmatrix}^{\cdot} + \frac{1}{\mathfrak{h}} \begin{pmatrix} v(0) & u(0) \\ w(0) & -v(0) \end{pmatrix} \begin{pmatrix} 1 \\ \frac{\mathfrak{z} - v(\mathfrak{h})}{u(\mathfrak{h})} \end{pmatrix} = \lambda \begin{pmatrix} 1 \\ \frac{\mathfrak{z} - v(\mathfrak{h})}{u(\mathfrak{h})} \end{pmatrix},$$

from which we find that λ is given by

$$\lambda = \frac{1}{\mathfrak{h}}(v(0) + \xi_2 u(0)) = \frac{v(0)u(\mathfrak{h}) - u(0)v(\mathfrak{h})}{\mathfrak{h}u(\mathfrak{h})} + \frac{\mathfrak{z}}{\mathfrak{h}} \frac{u(0)}{u(\mathfrak{h})}.$$

[7] If $(Y_n)_{kk} \neq 0$, we have $\mu = \nu$.

Proposition 6.33. *Let* $\dot{X}(\mathfrak{h}) = [X(\mathfrak{h}), Y(\mathfrak{h})]$ *be a Lax equation on* M*, as before. For* $t \in \mathbf{C}$*, with* $|t|$ *small, the unique integral curve* $X(\mathfrak{h}, t)$ *to*

$$\dot{X}(\mathfrak{h}) = [X(\mathfrak{h}), Y(\mathfrak{h})], \quad \text{with } X(\mathfrak{h}, 0) = X_0(\mathfrak{h}), \tag{6.41}$$

is given by

$$X(\mathfrak{h}, t) = U(\mathfrak{h}, t)X(\mathfrak{h}, 0)U(\mathfrak{h}, t)^{-1}, \tag{6.42}$$

where $U(\mathfrak{h}, t)$ *is the unique solution to*

$$\dot{U}(\mathfrak{h}, t) = -Y(\mathfrak{h}, t)U(\mathfrak{h}, t), \quad \text{with } U(\mathfrak{h}, 0) = \mathrm{Id}_N. \tag{6.43}$$

The solutions $X(\mathfrak{h}, t)$*,* $Y(\mathfrak{h}, t)$*,* $U(\mathfrak{h}, t)$ *of (6.41) and (6.43) are all holomorphic in* t *for* $|t|$ *small and in* $\mathfrak{h}$ *at all points* $(\mathfrak{h}, \mathfrak{z}) \in \Gamma_{\mathbf{c}}$*. The same holds true for the matrix* $Z(\mathfrak{h}, t)$*, defined by*

$$U(\mathfrak{h}, t) = \mathrm{Id}_N - tY(\mathfrak{h}, 0) + t^2 Z(\mathfrak{h}, t). \tag{6.44}$$

Moreover, the matrix $U(\mathfrak{h}, t)$ *acting on the normalized eigenvector* $\xi(0) := \xi(\mathfrak{z}, X(\mathfrak{h}, 0))$ *is a multiple of the normalized eigenvector* $\xi(t) := \xi(\mathfrak{z}, X(\mathfrak{h}, t))$*, namely*

$$U(\mathfrak{h}, t)\xi(0) = \left(1 - t\lambda(\mathfrak{h}, \mathfrak{z}, 0) + t^2 g(\mathfrak{h}, \mathfrak{z}, t)\right) \xi(t). \tag{6.45}$$

with

$$\lambda(\mathfrak{h}, \mathfrak{z}, 0) = \sum_{l=1}^{N} Y_{1l}(\mathfrak{h}, X(\mathfrak{h}, 0))\xi_l(\mathfrak{z}, X(\mathfrak{h}, 0))$$

$$g(\mathfrak{h}, \mathfrak{z}, t) = \sum_{l=1}^{N} Z_{1l}(\mathfrak{h}, t)\xi_l(\mathfrak{z}, X(\mathfrak{h}, 0))$$

Proof. Since equation (6.31) is a system of equations for each entry of X_i, namely

$$(\dot{X}_i)_{jk} = \text{polynomial in } (X_\ell)_{pq} \text{ for all } \ell, p, q,$$

we have that the solution of such a system (i.e., with holomorphic right hand side) is holomorphic in t for $|t|$ sufficiently small. Since the Y_i's are polynomial in the entries of the X_j's and polynomial in $\mathfrak{h}, \mathfrak{h}^{-1}$, the function $Y(\mathfrak{h}, t)$ is holomorphic in t and $\mathfrak{h}$ for $|t|$ small and $(\mathfrak{z}, \mathfrak{h}) \in \Gamma_{\mathbf{c}}$. Therefore, solving equation (6.43) also leads to a solution $U(\mathfrak{h}, t)$ holomorphic in t and $\mathfrak{h}$ for $|t|$ small and $(\mathfrak{z}, \mathfrak{h}) \in \Gamma_{\mathbf{c}}$; this yields (6.44).

We omit for the moment the dependence on $\mathfrak{h}$ from X, Y, U and ξ. Let $U(t)$ be the solution to (6.43), for $|t|$ small. Then

$$\left(U(t)^{-1}\right)^{\cdot} = -U(t)^{-1}\dot{U}(t)U(t)^{-1} = U(t)^{-1}Y(t),$$

as follows from differentiating the matrix identity $U(t)U(t)^{-1} = \mathrm{Id}_N$, and we have that

$$\begin{aligned}(U(t)^{-1}X(t)U(t))^{\cdot} &\\ &= \left(U(t)^{-1}\right)^{\cdot} X(t)U(t) + U(t)^{-1}\dot{X}(t)U(t) + U(t)^{-1}X(t)\dot{U}(t)\\ &= U(t)^{-1}Y(t)X(t)U(t) + U(t)^{-1}(X(t)Y(t) - Y(t)X(t))U(t)\\ &\qquad -U(t)^{-1}X(t)Y(t)U(t)\\ &= 0.\end{aligned}$$

So we find that

$$U(t)^{-1}X(t)U(t) = U(0)^{-1}X(0)U(0) = X(0)$$

showing (6.42). If we multiply (6.42) on the right by $U(t)\xi(0)$ then we find that

$$X(t)(U(t)\xi(0)) = U(t)X(0)\xi(0) = U(t)\mathfrak{z}\xi(0) = \mathfrak{z}(U(t)\xi(0)),$$

which means that $U(t)\xi(0)$ is an eigenvector of $X(t)$ corresponding to the eigenvalue $\mathfrak{z}$. Since it is assumed that for generic $(\mathfrak{h},\mathfrak{z}) \in \Gamma_X$, the eigenspace of $X(\mathfrak{h})$ with eigenvalue $\mathfrak{z}$ is one-dimensional, we find that

$$U(t)\xi(0) = f(t)\xi(t).$$

To compute $f(t)$, using the expression (6.44) for $U(t)$, one finds, remembering $\xi(t) := \xi(\mathfrak{z}, X(\mathfrak{h},t))$, with $\xi(t)_1 = 1$, and $Y(\mathfrak{h},t) := Y(\mathfrak{h}, X(\mathfrak{h},t))$,

$$\begin{aligned}f(t) &= (f(t)\xi(t))_1\\ &= (U(t)\xi(0))_1\\ &= \left(\left(\mathrm{Id}_N - tY(\mathfrak{h},0) + t^2 Z(\mathfrak{h},t)\right)\xi(0)\right)_1\\ &= 1 - t\left(Y(\mathfrak{h},0)\xi(0)\right)_1 + t^2\left(Z(\mathfrak{h},t)\xi(0)\right)_1\\ &= 1 - t\lambda(\mathfrak{h},\mathfrak{z},0) + t^2\sum_{l=1}^{N} Z_{1l}(\mathfrak{h},t)\xi_l(0)\end{aligned}$$

yielding (6.45), ending the proof of Proposition 6.33. □

Example 6.34. In the case of Example 6.32 we have that

$$U(\mathfrak{h},t)\xi(0) = \left(1 - t\left(\frac{v(0)u(\mathfrak{h}) - u(0)v(\mathfrak{h})}{\mathfrak{h}u(\mathfrak{h})} + \frac{\mathfrak{z}}{\mathfrak{h}}\frac{u(0)}{u(\mathfrak{h})}\right)_{|t=0} + t^2 h(t)\right)\xi(t),$$

where, for generic $\mathfrak{z}$ and $\mathfrak{h}$, the function $h(t)$ is holomorphic, for small $|t|$.

We now define the divisor map, which is for fixed values of the constants of motion $\mathbf{c}$, a map from the isospectral variety $A_{\mathbf{c}}$ to the variety of effective divisors of a certain degree on $\Gamma_{\mathbf{c}}$. For simplicity we will assume that $\mathbf{c}$ is chosen such that $\Gamma_{\mathbf{c}}$ is non-singular and such that $A_{\mathbf{c}}$ is connected. For $l = 1,\ldots,N$ and for an open subset U of $\overline{\Gamma_{\mathbf{c}}}$, denote by $(\xi_l)_U$ the divisor of zeros and poles of ξ_l, restricted to U. For a generic $X(\mathfrak{h}) \in A_{\mathbf{c}}$, with corresponding normalized eigenvector ξ, let $\mathcal{D}_X$ be the minimal effective divisor on $\Gamma_{\mathbf{c}}$ such that

$$(\xi_l)_{\Gamma_{\mathbf{c}}} \geqslant -\mathcal{D}_X, \qquad \text{for all } l = 1, \dots, N;$$

by continuity, $d := \deg(\mathcal{D}_X)$ is independent of $X = X(\mathfrak{h}) \in A_{\mathbf{c}}$ and thus, $\mathcal{D}_X$ defines an effective divisor of degree d in $\overline{\Gamma_{\mathbf{c}}}$ for any $X = X(\mathfrak{h}) \in A_{\mathbf{c}}$. The point is to study the motion of the divisor $\mathcal{D}_X$ in $\Gamma_{\mathbf{c}}$, when $X(\mathfrak{h})$ is moving in $A_{\mathbf{c}}$. Roughly speaking, $\mathcal{D}_X$ is the divisor of poles of the normalized eigenvector $\xi(\mathfrak{z}; X(\mathfrak{h}))$ on $\Gamma_{\mathbf{c}}$, not at ∞. Note for non-generic $X(\mathfrak{h})$ the divisor $\mathcal{D}_X$ may contain one or several of the points p_i at infinity.

Definition 6.35. In terms of the above notation, the *divisor map* is defined as

$$\begin{aligned} \imath_{\mathbf{c}} :\ A_{\mathbf{c}} &\to \operatorname{Div}^d(\overline{\Gamma_{\mathbf{c}}}) \\ X(\mathfrak{h}) &\mapsto \quad \mathcal{D}_X \end{aligned}$$

Example 6.36. Continuing our example, since $\xi_1 = 1$ we only need to investigate the poles of the function

$$\xi_2(\mathfrak{h}, \mathfrak{z}) = \frac{\mathfrak{z} - v(\mathfrak{h})}{u(\mathfrak{h})} = \frac{w(\mathfrak{h})}{\mathfrak{z} + v(\mathfrak{h})}.$$

If we define $x_1 \dots, x_d$ to be the d zeros of $u(\mathfrak{h})$ (with the understanding that some of the x_i may coincide, according to the multiplicity of the corresponding root of $u(\mathfrak{h})$) and we let $y_i := -v(x_i)$ then the d points $(x_1, y_1), \dots, (x_d, y_d)$ yield the minimal divisor $\mathcal{D}$ such that $(\xi_2) + \mathcal{D} \geqslant 0$ on $\Gamma_{\mathbf{c}}$.

When $X(\mathfrak{h})$ evolves according to (6.31), the image of $X(\mathfrak{h}, t)$ under $\imath_{\mathbf{c}}$ evolves on $\operatorname{Div}^d(\overline{\Gamma_{\mathbf{c}}})$ and will be denoted by $\mathcal{D}_{X(t)}$. We show in the following proposition that the transition function of the line bundle of $\mathcal{D}_{X(t)} - \mathcal{D}_{X(0)}$ is given by the function $1 - t\lambda + t^2 g(t)$, and we express this function in terms of λ, up to terms that are quadratically small.

Proposition 6.37. *Let* $\mathbf{c}$ *be such that* $\Gamma_{\mathbf{c}}$ *is non-singular and such that* $A_{\mathbf{c}}$ *is connected. Let* $X(0) = X(\mathfrak{h}, 0)$ *be a generic element in* $A_{\mathbf{c}}$ *such that* $\mathcal{D}_{X(0)} \subseteq \Gamma_{\mathbf{c}} \subset \overline{\Gamma_{\mathbf{c}}}$ *and let* (U_0, U_∞) *be an open cover of* $\overline{\Gamma_{\mathbf{c}}}$ *such that*

(1) $U_\infty \setminus U_0$ *is a neighborhood of the points* $p_1, \dots, p_s$*;*
(2) $U_0 \setminus U_\infty$ *is a neighborhood of the support of* $\mathcal{D}_{X(0)}$*.*

We have for $|t|$ *small enough and* $g(\mathfrak{h}, \mathfrak{z}, t) := \sum_{l=1}^N Z_{1l}(\mathfrak{h}, t)\xi_l(0)$*,*

$$\left(1 - t\lambda(\mathfrak{h}, \mathfrak{z}, 0) + t^2 g(\mathfrak{h}, \mathfrak{z}, t)\right)_{U_0} = \mathcal{D}_{X(t)} - \mathcal{D}_{X(0)} \tag{6.46}$$

and

$$(1 - t\lambda(\mathfrak{h}, \mathfrak{z}, 0))_{U_0} = \mathcal{D}'_{X(t)} - \mathcal{D}_{X(0)}, \tag{6.47}$$

for divisors $\mathcal{D}_{X(t)}$ and $\mathcal{D}'_{X(t)}$ in U_0, which are $O(t^2)$-close; i.e., for every holomorphic differential ω,

$$\int_{\mathcal{D}_{X(t)}}^{\mathcal{D}'_{X(t)}} \omega = O(t^2). \tag{6.48}$$

Proof. From (6.45) we have

$$U(\mathfrak{h},t)\xi(0) = \left(1 - t\lambda(\mathfrak{h},\mathfrak{z},X(\mathfrak{h},0)) + t^2 g(\mathfrak{h},\mathfrak{z},t)\right)\xi(t), \tag{6.49}$$

where $U(\mathfrak{h},t)$ is holomorphic for small $|t|$ and $(\mathfrak{h},\mathfrak{z}) \in \Gamma_{\mathbf{c}}$. Since $\mathcal{D}_{X(0)}$ is the minimal divisor of affine poles of $\xi(0)$ and since $U(\mathfrak{h},t)$ is holomorphic and near Id_N for small $|t|$ we have that $\mathcal{D}_{X(0)}$ is the minimal divisor of affine poles of $U(\mathfrak{h},t)\xi(0)$, which is the left hand side in (6.49). But $\mathcal{D}_{X(t)}$ is the minimal divisor of affine poles of $\xi(t)$. It follows that the zeros of $1 - t\lambda + t^2 g$ cancel all the poles of $\xi(t)$ that are not poles of $\mathcal{D}_{X(0)}$, leading to (6.46). Taking the difference of (6.46) and (6.47) we find

$$\begin{aligned}\mathcal{D}'_{X(t)} - \mathcal{D}_{X(t)} &= (1-t\lambda)_{U_0} - (1 - t\lambda + t^2 g(t))_{U_0} \\ &= \left(\frac{1-t\lambda(\mathfrak{h},\mathfrak{z},0)}{1-t\lambda(\mathfrak{h},\mathfrak{z},0)+t^2 g(\mathfrak{z},\mathfrak{h},t)}\right)_{U_0}.\end{aligned} \tag{6.50}$$

Consider an arbitrary holomorphic differential ω on $\overline{\Gamma_{\mathbf{c}}}$ and let $\Gamma_{\mathbf{c}}^*$ denote a canonical dissection of $\Gamma_{\mathbf{c}}$, i.e., the simply connected open subset of $\Gamma_{\mathbf{c}}$, obtained by removing a system of simple curves that generated the first homology group of $\Gamma_{\mathbf{c}}$. We may assume that these curves are disjoint from the points in $\mathcal{D}_{X(t)}$ and from the points in $\mathcal{D}'_{X(t)}$ (t is fixed). Choosing a point $p_0 \in \Gamma_{\mathbf{c}}^*$ we consider the function $\psi(p) = \int_{p_0}^{p} \omega$, which is well-defined because $\Gamma_{\mathbf{c}}^*$ is simply connected. Then, by a standard residue calculation, looking at poles and zeros of the function in bracket (6.50) in the region $U_0^* := U_0 \cap \Gamma_{\mathbf{c}}^*$, we have on the one hand,

$$\frac{1}{2\pi\sqrt{-1}} \oint_{\partial U_0^*} \psi \, \mathrm{d}\log\left(\frac{1-t\lambda(\mathfrak{h},\mathfrak{z},0)}{1-t\lambda(\mathfrak{h},\mathfrak{z},0)+t^2 g(\mathfrak{z},\mathfrak{h},t)}\right) = \int_{\mathcal{D}_{X(t)}}^{\mathcal{D}'_{X(t)}} \omega,$$

and, on the other hand, using the fact that the ratio in the log behaves like $O(t^2)$, as a function of t on ∂U_0^*, we have

$$\frac{1}{2\pi\sqrt{-1}} \oint_{\partial U_0^*} \psi \, \mathrm{d}\log\left(\frac{1-t\lambda(\mathfrak{h},\mathfrak{z},0)}{1-t\lambda(\mathfrak{h},\mathfrak{z},0)+t^2 g(\mathfrak{z},\mathfrak{h},t)}\right) = O(t^2).$$

Comparing the two formulas establishes (6.48). □

Lemma 6.38. *Consider a point $p \in \mathbf{C}$, a holomorphic differential ω in a neighborhood V of p and a holomorphic function u in $V \setminus \{p\}$ with a pole of order n at p. Consider t small enough, so that the n points $p_j(t)$, solution of*

$$u(p_j(t)) + t^{-1} = 0, \qquad j = 1, ..., n,$$

all belong to V. Then, for small t,

$$\sum_{j=1}^{n} \int_{p}^{p_j(t)} \omega = -t\,\mathrm{Res}_p(\omega u) + O(t^2).$$

Proof. Let $\omega = \mathrm{d}\psi$ with $\psi(p) = 0$. Then, for any closed path γ enclosing the zeros $p_j(t)$ of $u + t^{-1}$,

$$\begin{aligned}\frac{1}{t}\sum_{j=1}^{n} \int_{p}^{p_j(t)} \omega &= \frac{1}{t}\sum_{j=1}^{n} \psi(p_j(t)) = \frac{1}{t}\sum_{j=1}^{n} \mathrm{Res}_{p_j(t)} \frac{u'}{u+\frac{1}{t}}\psi \\ &= \sum_{j=1}^{n} \mathrm{Res}_{p_j(t)} \frac{u'}{1+tu}\psi = \frac{1}{2\pi i}\oint_{\gamma} \frac{u'}{1+tu}\psi \,\mathrm{d}z.\end{aligned}$$

From the computation above it follows that

$$\begin{aligned}\lim_{t\to 0} \frac{\mathrm{d}}{\mathrm{d}t}\sum_{j=1}^{n} \int_{p}^{p_j(t)} \omega &= \lim_{t\to 0} \frac{1}{t}\sum_{j=1}^{n} \int_{p}^{p_j(t)} \omega \\ &= \frac{1}{2\pi i}\oint_{\gamma} u'\psi \mathrm{d}z = -\frac{1}{2\pi i}\oint_{\gamma} u\omega \,\mathrm{d}z = -\,\mathrm{Res}_p(u\omega),\end{aligned}$$

ending the proof. □

Choose a divisor $\mathcal{D}_0 \in \mathrm{Div}^d(\overline{\Gamma_{\mathbf{c}}})$ and a basis $(\omega_1, \ldots, \omega_g)$ of holomorphic differentials on $\overline{\Gamma_{\mathbf{c}}}$ and let $\vec{\omega} := (\omega_1, \ldots, \omega_g)^\top$. Define the *linearizing map*

$$\begin{aligned}\jmath_{\mathbf{c}} : \mathcal{A}_{\mathbf{c}} &\to \mathrm{Jac}(\overline{\Gamma_{\mathbf{c}}}) \\ X &\mapsto \int_{\mathcal{D}_0}^{\mathcal{D}_X} \vec{\omega}.\end{aligned}$$

For example, one may choose a base point q on $\overline{\Gamma_{\mathbf{c}}}$ and take $\mathcal{D}_0 := dq$. Then the linearizing map is given by

$$\jmath_{\mathbf{c}}(X) = \sum_{i=1}^{d} \int_{q}^{q_i} \vec{\omega} \in \mathrm{Jac}(\overline{\Gamma_{\mathbf{c}}}),$$

where $\mathcal{D}_X = q_1 + \cdots + q_d$.

In what follows we will emphasize that λ only depends on t through its explicit dependence on $X(\mathfrak{h})$, as given in (6.37).

Theorem 6.39. *Along the integral curves $X(t)$ of the Lax equation $\dot{X} = [X,Y]$ the derivative of the linearizing map is given by*

$$\frac{\mathsf{d}}{\mathsf{d}t}\int_{\mathcal{D}_{X(0)}}^{\mathcal{D}_{X(t)}} \vec{\omega} = \sum_{i=1}^{s} \operatorname{Res}_{p_i} \lambda(\mathfrak{h},\mathfrak{z},t)\,\vec{\omega}. \tag{6.51}$$

Proof. For $|t|$ sufficiently small, the points $p_{ij}(t)$, which are close to p_i and are defined by

$$(\frac{1}{t} - \lambda) = \mathcal{D}'_{X(t)} - \mathcal{D}_{X(0)} + \sum_{i=1}^{s}\sum_{j=1}^{\sigma_i}(p_{ij}(t) - p_i)$$

are all contained in U_∞, by (6.47). Note the second summation is only non-zero whenever λ has a pole at p_i and that, by (6.38),

$$\sigma_i \leqslant \begin{cases} n\mu_i & 1 \leqslant i \leqslant s' \\ -n'\mu_i & s'+1 \leqslant i \leqslant s. \end{cases}$$

Let $\vec{\omega}$ be the (column) vector of holomorphic differentials $(\omega_1,\ldots,\omega_g)$ on $\overline{\Gamma_{\mathbf{c}}}$. For fixed (but small) t we have that $(\frac{1}{t} - \lambda)$ is the divisor of a meromorphic function, hence

$$\int_{\mathcal{D}_{X(0)}}^{\mathcal{D}'_{X(t)}} \vec{\omega} + \sum_{i=1}^{s}\sum_{j=1}^{\sigma_i}\int_{p_i}^{p_{ij}(t)} \vec{\omega}$$

belongs to the lattice of periods of $\overline{\Gamma_{\mathbf{c}}}$; since it vanishes for $t=0$, it does for all t, and

$$\lim_{t\to 0}\frac{\mathsf{d}}{\mathsf{d}t}\int_{\mathcal{D}_{X(0)}}^{\mathcal{D}'_{X(t)}} \vec{\omega} = -\lim_{t\to 0}\frac{\mathsf{d}}{\mathsf{d}t}\sum_{i=1}^{s}\sum_{j=1}^{\sigma_i}\int_{p_i}^{p_{ij}(t)} \vec{\omega} = \sum_{i=1}^{s}\operatorname{Res}_{p_i}\lambda\,\vec{\omega},$$

in view of Lemma 6.38. Remember that the divisor $\mathcal{D}_{X(t)}$ is the divisor of zeros of $1 - t\lambda(\mathfrak{h},\mathfrak{z},0) + t^2 g(\mathfrak{h},\mathfrak{z},t)$ in U_0 and is $O(t^2)$-close to the divisor $\mathcal{D}'_{X(t)}$, i.e., $\int_{\mathcal{D}'_{X(t)}}^{\mathcal{D}_{X(t)}} \vec{\omega} = O(t^2)$, by (6.48). Thus

$$\begin{aligned}
\left.\frac{\mathsf{d}}{\mathsf{d}t}\right|_{t=0}\int_{\mathcal{D}_{X(0)}}^{\mathcal{D}_{X(t)}} \vec{\omega} &= \left.\frac{\mathsf{d}}{\mathsf{d}t}\right|_{t=0}\int_{\mathcal{D}_{X(0)}}^{\mathcal{D}'_{X(t)}} \vec{\omega} + \left.\frac{\mathsf{d}}{\mathsf{d}t}\right|_{t=0}\int_{\mathcal{D}'_{X(t)}}^{\mathcal{D}_{X(t)}} \vec{\omega} \\
&= \left.\frac{\mathsf{d}}{\mathsf{d}t}\right|_{t=0}\int_{\mathcal{D}_{X(0)}}^{\mathcal{D}'_{X(t)}} \vec{\omega} \\
&= \sum_{i=1}^{s}\operatorname{Res}_{p_i}\lambda(\mathfrak{h},\mathfrak{z},0)\,\vec{\omega}.
\end{aligned}$$

Since the argument can be repeated at every t, the result follows, as λ depends on t only through its explicit dependence on $X(\mathfrak{h})$. □

Remark 6.40. If λ does not have poles in $\overline{\Gamma_{\mathbf{c}}}\setminus\Gamma_{\mathbf{c}}$, the right hand side of (6.51) vanishes.

Theorem 6.41 (Linearization Criterion). *The map $\jmath_{\mathbf{c}}$ linearizes the isospectral flow $\dot{X} = [X, Y]$ on $\mathcal{A}_{\mathbf{c}}$, that is to say*

$$\int_{\mathcal{D}_{X(0)}}^{\mathcal{D}_{X(t)}} \vec{\omega} = t \sum_{i=1}^{s} \operatorname{Res}_{p_i} \lambda(\mathfrak{h}, \mathfrak{z}, X(\mathfrak{h}, 0))\, \vec{\omega}, \tag{6.52}$$

if and only if there exists for each $X \in \mathcal{A}_{\mathbf{c}}$ a meromorphic function ϕ_X on $\overline{\Gamma_{\mathbf{c}}}$ with $(\phi_X)_{\overline{\Gamma_{\mathbf{c}}}} \geqslant -n \sum_{i=1}^{s'} \mu_i p_i + n' \sum_{i=s'+1}^{s} \mu_i p_i$, such that for all p_i,

$$(\textit{Laurent tail of } \frac{\mathsf{d}\lambda(\mathfrak{h}, \mathfrak{z}, X)}{\mathsf{d}t} \textit{ at } p_i) = (\textit{Laurent tail of } \phi_X \textit{ at } p_i), \tag{6.53}$$

where (remember)

$$\frac{\mathsf{d}\lambda(\mathfrak{h}, \mathfrak{z}, X)}{\mathsf{d}t} = \frac{\mathsf{d}}{\mathsf{d}t} \left(\sum_{l=1}^{N} Y_{1l}(\mathfrak{h}, \mathfrak{z}, X(\mathfrak{h}, t)) \frac{\Delta_{1l}(\mathfrak{z}, X(\mathfrak{h}, t))}{\Delta_{11}(\mathfrak{z}, X(\mathfrak{h}, t))} \right),$$

$\mathsf{d}/\mathsf{d}t$ being computed using the Lax equation $\dot{X} = [X, Y]$.

Proof. $\Longleftarrow$ Since ϕ_X is a meromorphic function with poles p_i, we have that $\phi_X\, \vec{\omega}$ is a vector of meromorphic differentials and so the sum of its residues vanishes:

$$\sum_{i=1}^{s} \operatorname{Res}_{p_i} \phi_X\, \vec{\omega} = 0.$$

Thus, using the hypothesis,

$$\begin{aligned} 0 &= \sum_{i=1}^{s} \operatorname{Res}_{p_i} \phi_X\, \vec{\omega} \\ &= \sum_{i=1}^{s} \operatorname{Res}_{p_i} \frac{\mathsf{d}\lambda(\mathfrak{h}, \mathfrak{z}, X)}{\mathsf{d}t}\, \vec{\omega} \\ &= \frac{\mathsf{d}}{\mathsf{d}t} \sum_{i=1}^{s} \operatorname{Res}_{p_i} \lambda(\mathfrak{h}, \mathfrak{z}, X)\, \vec{\omega} \\ &= \frac{\mathsf{d}^2}{\mathsf{d}t^2} \int_{\mathcal{D}_{X(0)}}^{\mathcal{D}_X} \vec{\omega}\,, \qquad \text{using Theorem 6.39.} \end{aligned}$$

It implies that $\frac{\mathsf{d}}{\mathsf{d}t} \int_{\mathcal{D}_{X(0)}}^{\mathcal{D}_{X(t)}} \vec{\omega}$ is constant along any integral curve of the differential equation $\dot{X} = [X, Y]$ and so for all t

$$\frac{\mathsf{d}}{\mathsf{d}t} \int_{\mathcal{D}_{X(0)}}^{\mathcal{D}_{X(t)}} \vec{\omega} = \sum_{i=1}^{s} \operatorname{Res}_{p_i} \lambda(\mathfrak{h}, \mathfrak{z}, X(\mathfrak{h}, t))\, \vec{\omega} = \sum_{i=1}^{s} \operatorname{Res}_{p_i} \lambda(\mathfrak{h}, \mathfrak{z}, X(\mathfrak{h}, 0))\, \vec{\omega}$$

and so

$$\int_{\mathcal{D}_{X(0)}}^{\mathcal{D}_{X(t)}} \vec{\omega} = t \sum_{i=1}^{s} \operatorname{Res}_{p_i} \lambda(\mathfrak{h}, \mathfrak{z}, X(\mathfrak{h}, 0))\, \vec{\omega} + \text{ constant},$$

with the constant equal to zero, upon evaluating both sides at $t = 0$.

$\Longrightarrow$ The relations (6.52) and (6.51) together imply that all along the trajectory

$$\text{constant } = \frac{\mathrm{d}}{\mathrm{d}t} \int_{\mathcal{D}_{X(0)}}^{\mathcal{D}_{X(t)}} \vec{\omega} = \sum_{i=1}^{s} \operatorname{Res}_{p_i} \lambda(\mathfrak{h}, \mathfrak{z}, X(\mathfrak{h}, t))\, \vec{\omega}$$

and so

$$\frac{\mathrm{d}}{\mathrm{d}t} \sum_{i=1}^{s} \operatorname{Res}_{p_i} \lambda(\mathfrak{h}, \mathfrak{z}, X(\mathfrak{h}, t))\, \vec{\omega} = \sum_{i=1}^{s} \operatorname{Res}_{p_i} \frac{\mathrm{d}\lambda(\mathfrak{h}, \mathfrak{z}, X(\mathfrak{h}, t))}{\mathrm{d}t}\, \vec{\omega} = 0.$$

Now, Lemma 5.11 guarantees the existence of a function $\phi_{X(t)}$ having the Laurent tails given by property (6.53) and holomorphic otherwise; so, by (6.38) we have the inequality $(\phi_X)_{\overline{\Gamma_{\mathbf{c}}}} \geqslant -n \sum_{i=1}^{s'} \mu_i p_i + n' \sum_{i=s'+1}^{s} \mu_i p_i$, ending the proof of Theorem 6.41. $\square$

Example 6.42. Returning to our example, we have found in Example 6.32 the following explicit expression for λ,

$$\lambda(\mathfrak{h}, \mathfrak{z}) = \frac{1}{\mathfrak{h}}(v(0) + \xi_2 u(0)) = \frac{v(0)u(\mathfrak{h}) - u(0)v(\mathfrak{h})}{\mathfrak{h}u(\mathfrak{h})} + \frac{\mathfrak{z}}{\mathfrak{h}} \frac{u(0)}{u(\mathfrak{h})}.$$

The curve $\Gamma_{\mathbf{c}}$ has the form $\mathfrak{z}^2 = c_0 \mathfrak{h}^{2d}(1 + O(\mathfrak{h}^{-1}))$, where c_0 is different from zero, for generic $\mathbf{c}$, so $\Gamma_{\mathbf{c}}$ has two points at infinity, which are given in terms of a local parameter ς by

$$\mathfrak{h} = \varsigma^{-1}, \qquad \mathfrak{z} = \pm\sqrt{c_0}\varsigma^{-d}(1 + O(\varsigma)).$$

Writing λ in terms of this local parameter yields that λ has a simple zero at $\varsigma = 0$, so it is holomorphic at infinity.

Corollary 6.43. *Suppose that $\mathfrak{h}$ has no zero at infinity and that there exists a polynomial $p(x, y, z)$ whose coefficients are arbitrary constants of the motion, and that there exists an algebraic function Ψ, whose coefficients are arbitrary constants of the motion, such that*

$$Y(\mathfrak{h}) = \Psi(p(X, \mathfrak{h}, \mathfrak{h}^{-1})) + \left(C_0 + C_1 \mathfrak{h}^{-1} + C_2 \mathfrak{h}^{-2} + \cdots\right) \tag{6.54}$$

where C_0 is a lower triangular matrix, and where the matrices $C_1, C_2, \ldots$ are arbitrary. If $\xi/\mathfrak{h}$ has no pole at the points p_i then the Linearization Criterion is satisfied by taking $\phi_X = 0$.

Proof. Since $\dot{\xi} + Y(\mathfrak{h})\xi = \lambda(\mathfrak{h},\mathfrak{z})\xi$ and since the first component of ξ is normalized to 1, λ is the first component of $Y(\mathfrak{h})\xi$. In view of (6.54), λ is the first component of $\Psi(p(A,\mathfrak{h},\mathfrak{h}^{-1}))\xi + (C_0 + \frac{C_1}{\mathfrak{h}} + \ldots)\xi$.

But

$$\Psi(p(A,\mathfrak{h},\mathfrak{h}^{-1}))\xi = \Psi(p(\mathfrak{z},\mathfrak{h},\mathfrak{h}^{-1}))\xi$$

since ξ is an eigenvector of A with eigenvalue $\mathfrak{z}$. Therefore,

$$\lambda = \Psi(p(\mathfrak{z},\mathfrak{h},\mathfrak{h}^{-1})) + (C_0)_{11} + \frac{(C_1\xi)_1}{\mathfrak{h}} + \frac{(C_2\xi)_1}{\mathfrak{h}^2} + \cdots \tag{6.55}$$

We claim that the Laurent tails of λ at each of the points p_i is independent of t. In fact, the first term in (6.55) clearly has no pole, being independent of $\mathfrak{z}$ and $\mathfrak{h}$, while the remaining terms do not have a pole at the points p_i, since we have assumed that $\xi/\mathfrak{h}$ has no pole at the points p_i. It follows that the choice $\phi_X = 0$ satisfies (6.53). □

Remark 6.44. In the examples that we will treat in this book the function Ψ is always the identity, but in some cases, such as the Neumann system, one needs a non-trivial Ψ. For a list of such examples, see [7, Theorem 4.3, page 302].

Remark 6.45. It may happen that $\jmath_{\mathbf{c}}$ not only linearizes the isospectral flows on $\mathcal{A}_{\mathbf{c}}$, but even trivializes them, i.e., the induced vector fields are zero (the corresponding flow is identity). For a non-trivial isospectral flow this cannot happen when $\jmath_{\mathbf{c}}$ is a finite map. When $\jmath_{\mathbf{c}}$ is not a finite map it suffices to check that the induced (linear) flow in one point is non-trivial to know that the linearized flow is non-trivial on the generic torus $\mathcal{A}_{\mathbf{c}}$.

Example 6.46. In the case of Example 6.42 we can also apply Corollary 6.43. Namely, if we write

$$X(\mathfrak{h}) = \sum_{i=0}^{d} X_i\mathfrak{h}^i,$$

then we see that (6.32) can also be written as

$$\dot{X}(\mathfrak{h}) = \left[X(\mathfrak{h}), \frac{X(0)}{\mathfrak{h}}\right] = \left[X(\mathfrak{h}), \frac{X(0) - X(\mathfrak{h})}{\mathfrak{h}}\right] = -\left[X(\mathfrak{h}), \sum_{i\geqslant 1} X_i\mathfrak{h}^{i-1}\right].$$

Now $\sum_{i\geqslant 1} X_i\mathfrak{h}^{i-1}$ is easily written in the form (6.54), because

$$\sum_{i\geqslant 1} X_i\mathfrak{h}^{i-1} = \frac{X(\mathfrak{h})}{\mathfrak{h}} - \frac{X(0)}{\mathfrak{h}}.$$

This gives a second proof that the example of this section satisfies the Linearization Criterion.

7 Weight Homogeneous A.c.i. Systems

In this chapter we introduce a class of a.c.i. systems for which everything can be explicitly computed. For these systems, which we will call weight homogeneous a.c.i. systems, phase space is always $\mathbf{C}^n$, and a system of linear coordinates on $\mathbf{C}^n$ can be chosen in such a way that everything (the polynomials in involution, the Poisson structure, the commuting vector fields) becomes homogeneous upon assigning weights to each of these coordinates. For these systems we will provide methods by means of which one can reveal the whole geometry of the system and prove (or disprove) algebraic complete integrability.

We show in Section 7.1 that weight homogeneous formal Laurent solutions can be effectively (algorithmically) computed: after the zeroth step (the indicial equation) each subsequent term in the Laurent solution is determined as the solution of a linear algebra problem, which we encode by introducing the Kowalevski matrix. In particular, if a free parameter enters at step k then k belongs to the spectrum of the Kowalevski matrix. We also establish a few simple, but useful, first properties of the Kowalevski matrix and of the indicial locus. We show in Section 7.2 that weight homogeneous formal Laurent solutions are always convergent. The Kowalevski matrix will be related in Section 7.4 to the grading of the algebra of constants of motion of a weight homogeneous vector field, as introduced in Section 7.3. We introduce in Section 7.5 the notion of a weight homogeneous a.c.i. system and we provide in Section 7.6 computational tools (algorithms) to compute basically everything for a weight homogeneous vector field. We finish this chapter by providing, in Section 7.7, for this class of a.c.i. systems a method which allows us to prove (or disprove) that a given Liouville integrable vector field is one of the integrable vector fields of a weight homogeneous a.c.i. system.

The implementation of the algorithm, given in Section 7.7 depends heavily on the techniques that are presented in Section 7.6. The algorithm will be illustrated several times in the examples that follow in the rest of the book. In order to make this chapter more readable we have chosen a simple but nontrivial example, which will be used to illustrate every concept and technique in this chapter, as they are introduced. This example, which we will introduce in Section 7.1, is in the literature known as the periodic 5-particle Kac-van Moerbeke lattice (see [92] and [53]).

7.1 Weight Homogeneous Vector Fields and Laurent Solutions

The integrable systems that we will deal with in the rest of this book are weight homogeneous in a sense that we will define below. First, let us recall the classical definition of a weight homogeneous polynomial. Let $\nu = (\nu_1, \dots, \nu_n)$ be a collection of positive integers without a common divisor. Such a ν is called a *weight vector*. We say that a polynomial $f \in \mathcal{F}(\mathbf{C}^n) = \mathbf{C}[x_1, \dots, x_n]$ is a *weight homogeneous polynomial* of *weight* k (with respect to ν) if

$$f(t^{\nu_1}x_1, \dots, t^{\nu_n}x_n) = t^k f(x_1, \dots, x_n)$$

for all $(x_1, \dots, x_n) \in \mathbf{C}^n$ and $t \in \mathbf{C}$. Notice that in this case $\partial f / \partial x_i$ is weight homogeneous of weight $k - \nu_i$. The weight k of f will be denoted[1] by $\varpi(f)$; moreover, when writing $\varpi(f)$ it is implicit in the notation that f is weight homogeneous. With this convention, we define for $k \in \mathbf{N}$,

$$\mathcal{F}^{(k)} := \{F \in \mathcal{F}(\mathbf{C}^n) \mid \varpi(F) = k\}.$$

Clearly, the choice of a weight vector ν induces on $\mathbf{C}[x_1, \dots, x_n]$ the structure of a graded algebra,

$$\mathbf{C}[x_1, \dots, x_n] = \bigoplus_{k=0}^{\infty} \mathcal{F}^{(k)},$$

i.e., every polynomial is in a unique way a finite sum of weight homogeneous polynomials, and $\mathcal{F}^{(k)}\mathcal{F}^{(l)} \subseteq \mathcal{F}^{(k+l)}$, for $k, l \in \mathbf{N}$. For fixed k an explicit basis for $\mathcal{F}^{(k)}$ is easily written down. The dimension of $\mathcal{F}^{(k)}$ is given by the following proposition, whose proof is an easy application of the theory of generating functions, hence is left to the reader.

Proposition 7.1. *The dimension of $\mathcal{F}^{(k)}$ has the following generating function:*

$$\prod_{i=1}^{n} \frac{1}{1 - t^{\nu_i}} = \sum_{k=0}^{\infty} \left(\dim \mathcal{F}^{(k)}\right) t^k.$$

We now define the notion of weight homogeneity for polynomial vector fields on $\mathbf{C}^n$.

Definition 7.2. A polynomial vector field on $\mathbf{C}^n$,

$$\begin{aligned} \dot{x}_1 &= f_1(x_1, \dots, x_n), \\ &\vdots \\ \dot{x}_n &= f_n(x_1, \dots, x_n), \end{aligned} \tag{7.1}$$

[1] n and ν will always be fixed, which allows us to keep the notations simple by not adding the dependence on n and ν.

is called a *weight homogeneous vector field* of weight k (with respect to ν) if each of the polynomials $f_1, \dots, f_n$ is weight homogeneous (with respect to ν) and if $\varpi(f_i) = \nu_i + k = \varpi(x_i) + k$ for $i = 1, \dots, n$. Thus, t = time has weight $-k$, so to speak. A weight homogeneous vector field of weight 1 will be simply called a *weight homogeneous vector field* (these are the most important vector fields in what follows). If (7.1) is a weight homogeneous vector field then a Laurent solution to (7.1) of the form

$$x_i(t) = \frac{1}{t^{\nu_i}} \sum_{k=0}^{\infty} x_i^{(k)} t^k, \qquad i = 1, \dots, n, \tag{7.2}$$

with $x^{(0)} \neq 0$, is called a *weight homogeneous Laurent solution.*

Remark 7.3. Weight homogeneous Laurent solutions are by definition strict Laurent solutions (meaning that $x^{(0)} \neq 0$, see Definition 6.6), so they always have a positive pole order. We will see in Section 7.2, that a weight homogeneous Laurent solution to (7.1) of the form (7.2) is automatically convergent; this is why (already) we dropped the adjective "formal".

Remark 7.4. When all weights are equal (to 1) the above terminology is simplified by replacing "weight homogeneous" by "homogeneous". Thus, our terminology is such that a vector field $\dot{x}_i = f(x)$ on $\mathbf{C}^n$ is called a homogeneous vector field if and only if each f_i is a homogeneous polynomial of degree 2.

Example 7.5. We now introduce the example that will serve as "fil rouge" throughout this chapter. We consider $\mathbf{C}^5$ with linear coordinates $x_1, \dots, x_5$ and we put $x_{i+5} = x_i$ for $i \in \mathbf{Z}$. The periodic 5-particle Kac-van Moerbeke lattice (see [92] and [53]) is given by the quadratic vector field

$$\dot{x}_i = x_i(x_{i-1} - x_{i+1}), \qquad (i = 1, \dots, 5), \tag{7.3}$$

which we will denote by $\mathcal{V}_1$, and it admits the following three independent constants of motion,

$$\begin{aligned} F_1 &= x_1 + x_2 + x_3 + x_4 + x_5, \\ F_2 &= x_1x_3 + x_2x_4 + x_3x_5 + x_4x_1 + x_5x_2, \\ F_3 &= x_1x_2x_3x_4x_5. \end{aligned} \tag{7.4}$$

If we assign to each x_i the weight 1, i.e., $\varpi(x_i) = 1$, then (7.3) becomes a weight homogeneous vector field, actually a homogeneous vector field, and the weights of the invariants (7.4) are given by $\varpi(F_1) = 1$, $\varpi(F_2) = 2$ and $\varpi(F_3) = 5$. The dimension of $\mathcal{F}^{(k)}$ can in this case be simply obtained from

$$\dim \mathcal{F}^{(k)} = \binom{5+k-1}{5-1} = \binom{4+k}{4}. \tag{7.5}$$

A vector field that commutes with (7.3) is given by

$$x_i' = x_i(x_{i+2}x_{i-1} - x_{i+1}x_{i-2}), \qquad (i = 1, \ldots, 5). \tag{7.6}$$

It is weight homogeneous of weight 2. Both vector fields and their constants of motion are invariant with respect to the order 5 automorphism σ of $\mathbf{C}^5$, which is defined by $\sigma(x_i) = x_{i+1}$. Notice that (7.3) can be written as a Lax equation (with parameter) $\dot{X}(\mathfrak{h}) = [X(\mathfrak{h}), Y(\mathfrak{h})]$, where

$$X(\mathfrak{h}) = \begin{pmatrix} 0 & x_1 & 0 & 0 & \mathfrak{h}^{-1} \\ 1 & 0 & x_2 & 0 & 0 \\ 0 & 1 & 0 & x_3 & 0 \\ 0 & 0 & 1 & 0 & x_4 \\ x_5\mathfrak{h} & 0 & 0 & 1 & 0 \end{pmatrix}, \quad Y(\mathfrak{h}) = \begin{pmatrix} 0 & 0 & x_1x_2 & 0 & 0 \\ 0 & 0 & 0 & x_2x_3 & 0 \\ 0 & 0 & 0 & 0 & x_3x_4 \\ x_4x_5\mathfrak{h} & 0 & 0 & 0 & 0 \\ 0 & x_5x_1\mathfrak{h} & 0 & 0 & 0 \end{pmatrix},$$

as is easily checked by direct computation. It is easy to fit this Lax equation in the AKS scheme, and to show, using Corollary 6.43 with $\Psi(z) = z^2$, that it satisfies the Linearization Criterion.

We show in the following proposition how, for $k \geqslant 1$, the k-th term of a weight homogeneous Laurent solution (to a weight homogeneous vector field) can be computed recursively from the previous terms $x^{(0)}, \ldots, x^{(k-1)}$. It will follow that it is sufficient to compute the first few terms of a weight homogeneous Laurent solution to be sure that these terms are indeed the first few terms of a formal Laurent solution (it will be automatically convergent, see Section 7.2). This property is precisely what makes the family of all weight homogeneous Laurent solutions computable; for Laurent solutions that are not weight homogeneous (e.g., for Laurent solutions to vector fields that are not weight homogeneous), it is neither clear with which pole order to start for each of the functions, nor how to organize the computation of the subsequent terms of the Laurent solution (see however Remark 7.9 and Paragraph 7.6.2).

Proposition 7.6. *Suppose that $\mathcal{V}$ is a weight homogeneous vector field on $\mathbf{C}^n$, given by*

$$\dot{x}_i = f_i(x_1, \ldots, x_n), \qquad (i = 1, \ldots, n),$$

and suppose that

$$x_i(t) = \frac{1}{t^{\nu_i}} \sum_{k=0}^{\infty} x_i^{(k)} t^k, \qquad (i = 1, \ldots, n) \tag{7.7}$$

is a weight homogeneous Laurent solution for this vector field. Then the leading coefficients $x_i^{(0)}$ satisfy the non-linear algebraic equations

$$\begin{aligned} \nu_1 x_1^{(0)} + f_1\left(x_1^{(0)}, \ldots, x_n^{(0)}\right) &= 0, \\ &\vdots \\ \nu_n x_n^{(0)} + f_n\left(x_1^{(0)}, \ldots, x_n^{(0)}\right) &= 0, \end{aligned} \tag{7.8}$$

while the subsequent terms $x_i^{(k)}$ *satisfy*

$$\left(k\,\mathrm{Id}_n - \mathcal{K}\left(x^{(0)}\right)\right) x^{(k)} = R^{(k)}, \tag{7.9}$$

where $x^{(k)} = \begin{pmatrix} x_1^{(k)} \\ \vdots \\ x_n^{(k)} \end{pmatrix}$, *and* $R^{(k)} = \begin{pmatrix} R_1^{(k)} \\ \vdots \\ R_n^{(k)} \end{pmatrix}$; *each* $R_i^{(k)}$ *is a polynomial, which depends on the variables* $x_1^{(l)}, \dots, x_n^{(l)}$ *with* $0 \leqslant l < k$ *only (the explicit value of* $R_i^{(k)}$ *will be given in (7.14) below). Also, the* (i,j)*-th entry of the* $(n \times n)$*-matrix* $\mathcal{K}$ *is the regular function on* $\mathbf{C}^n$, *defined by*

$$\mathcal{K}_{ij} := \frac{\partial f_i}{\partial x_j} + \nu_i \delta_{ij}, \tag{7.10}$$

where δ *is the Kronecker delta.*

Proof. Consider the weight homogeneous Laurent solution (7.7) to (7.1) and introduce for $\beta = (\beta_1, ..., \beta_n) \in \mathbf{N}^n$ the following notations $|\beta| := \sum_1^n \beta_i$ and $\beta! := \prod_{i=1}^n \beta_i!$ and

$$\frac{\partial^\beta f}{\partial x^\beta} := \frac{\partial^{\beta_1} \cdots \partial^{\beta_n} f}{\partial x_1^{\beta_1} \cdots \partial x_n^{\beta_n}}.$$

Define $U_i(t) := \sum_{k \geqslant 1} x_i^{(k)} t^k$, so that

$$x_i(t) = t^{-\nu_i}\left(x_i^{(0)} + U_i(t)\right), \tag{7.11}$$

where $1 \leqslant i \leqslant n$. Then we have on the one hand by weight homogeneity of f_i and by Taylors's Theorem that

$$\begin{aligned} t^{\nu_i+1} f_i(x(t)) &= f_i\left(x^{(0)} + U(t)\right) \qquad (7.12)\\ &= f_i\left(x^{(0)}\right) + \sum_{|\beta| \geqslant 1} \frac{1}{\beta!} \frac{\partial^\beta f_i}{\partial x^\beta}\left(x^{(0)}\right) U_1^{\beta_1}(t) \cdots U_n^{\beta_n}(t), \end{aligned}$$

while on the other hand,

$$t^{\nu_i+1} \dot{x}_i(t) = \sum_{k \geqslant 0} (k - \nu_i) x_i^{(k)} t^k, \tag{7.13}$$

as follows from (7.7). Comparing the constant terms in (7.12) and in (7.13) yields (7.8), since the constant term in the series $U_i(t)$ is zero.

In order to compare the other terms in the series (7.12) and (7.13) we expand the sum in (7.12) as a series in t, which gives for $|\beta| = 1$ the contribution

$$\sum_{j=1}^{n} \frac{\partial f_i}{\partial x_j}\left(x^{(0)}\right)\left(x_j^{(1)}t + x_j^{(2)}t^2 + \cdots\right),$$

while the terms that correspond to $|\beta| > 1$ yield $\sum_{k\geqslant 2} R^{(k)}t^k$, where

$$R_i^{(k)} := \sum_{\substack{|\beta|>1 \\ \sigma_{jl}>0 \\ \sum\sigma_{jl}=k}} \frac{1}{\beta!}\frac{\partial^\beta f_i}{\partial x^\beta}\left(x^{(0)}\right)\prod_{j=1}^{n} x_j^{(\sigma_{j1})}\cdots x_j^{(\sigma_{j\beta_j})}. \tag{7.14}$$

Noticing that R_k involves only the coefficients $x_i^{(l)}$ with $0 \leqslant l < k$, $(1 \leqslant i \leqslant n)$, we find that a comparison of the coefficients of t^k leads to n linear equations for $x_1^{(k)}, \dots, x_n^{(k)}$, to wit

$$(k - \nu_i)x_i^{(k)} = \sum_{j=1}^{n} \frac{\partial f_i}{\partial x_j}\left(x^{(0)}\right)x_j^{(k)} + R_i^{(k)}, \quad i = 1, \dots, n.$$

These equations are easily rewritten in the form (7.9), by introducing the matrix $\mathcal{K}$ (see (7.10)). □

Definition 7.7. The set of equations (7.8) is called the *indicial equation* of $\mathcal{V}$. Its solution set (which is an algebraic set in $\mathbf{C}^n$) is called the *indicial locus*, and is denoted by $\mathcal{I}$. The $n \times n$ matrix $\mathcal{K}$, defined in (7.10) is called the *Kowalevski matrix.*

Remark 7.8. $\mathcal{K}$ will only be evaluated at elements of $\mathcal{I}$; the elements of $\mathcal{K}$, restricted to $\mathcal{I}$, are regular functions on $\mathcal{I}$.

Remark 7.9. Proposition 7.6 is valid for vector fields that are almost weight homogeneous in the following sense. We assume that we are given, as before, a weight vector $\nu = (\nu_1, \dots, \nu_n)$ on $\mathbf{C}^n$ and we assume that each of the functions f_i, that appears in the differential equation $\dot{x}_i = f_i(x)$, can be written as $f_i = f_i' + f_i''$, where f_i' is weight homogeneous, of weight $\nu_i + 1$ and all terms in f_i'' have a weight that is at most ν_i. Notice that, in this case, the equations $\dot{x}_i = f_i'(x)$ define a weight homogeneous vector field on $\mathbf{C}^n$. It is easy to see that the presence of the terms f_i'' does not affect the indicial equation, and that the subsequent terms $x_i^{(k)}$ still satisfy Equation (7.9), but with a right hand side $R^{(k)}$ that is slightly different.

Example 7.10. In the case of Example (7.5) the weight homogeneous Laurent solutions of $\mathcal{V}_1$ are of the form

$$x_i(t) = \frac{1}{t}\sum_{k=0}^{\infty} x_i^{(k)} t^k, \qquad (i = 1, \dots, 5),$$

and the indicial equation is given by

$$x_i^{(0)}(1 + x_{i-1}^{(0)} - x_{i+1}^{(0)}) = 0, \qquad (i = 1, \dots, 5).$$

The following simple observations are useful for determining the indicial locus from these equations. If all $x_i^{(0)}$ are different from zero then $1+x_{i-1}^{(0)}-x_{i+1}^{(0)} = 0$ for $i = 1, \dots, 5$, which yields after summing up $5 = 0$, a contradiction. Taking into account the automorphism σ we may at first assume that $x_5^{(0)} = 0$ and that $x_1^{(0)} \neq 0$ to find one piece of the indicial locus; the entire indicial locus is then found by using the automorphism σ. The indicial equation then takes the simpler form

$$\begin{aligned} &1 - x_2^{(0)} = 0,\\ &x_2^{(0)}(1 + x_1^{(0)} - x_3^{(0)}) = 0,\\ &x_3^{(0)}(1 + x_2^{(0)} - x_4^{(0)}) = 0,\\ &x_4^{(0)}(1 + x_3^{(0)}) = 0. \end{aligned}$$

There are two solutions, depending on whether $x_4^{(0)}$ is equal to zero or not. Explicitly we find the following two points in the indicial locus, $m_4 := (-1, 1, 0, 0, 0)$ and $m_5 := (-2, 1, -1, 2, 0)$, where the labels 4 and 5 have been chosen with respect to the position of the zeros. The other eight points m_i and m_i' in the indicial locus are obtained by using the order 5 automorphism σ, for example

$$m_1 := (0, 0, -1, 1, 0) \qquad \text{and} \qquad m_1' := (0, -2, 1, -1, 2).$$

At these points the Kowalevski matrix is given by

$$\mathcal{K}(m_1) = \begin{pmatrix} 1 & 0 & 0 & 0 & 0\\ 0 & 2 & 0 & 0 & 0\\ 0 & -1 & 0 & 1 & 0\\ 0 & 0 & 1 & 0 & -1\\ 0 & 0 & 0 & 0 & 2 \end{pmatrix}, \qquad \mathcal{K}(m_1') = \begin{pmatrix} 5 & 0 & 0 & 0 & 0\\ -2 & 0 & 2 & 0 & 0\\ 0 & 1 & 0 & -1 & 0\\ 0 & 0 & -1 & 0 & 1\\ -2 & 0 & 0 & 2 & 0 \end{pmatrix}. \qquad (7.15)$$

At the other points m_i and m_i' the Kowalevski matrix is obtained by cycling the rows and the columns of the above matrices.

It is plain from the proposition that the nature of the spectrum of $\mathcal{K}$ will have pronounced implications on the algebraic complete integrability of a weight homogeneous integrable system. Indeed, if we want that a given $m \in \mathcal{I}$ leads to a principal balance (a family of Laurent solutions, depending on $n-1$ free parameters), then $\mathcal{K}(m)$ must have $n-1$ non-negative integer eigenvalues (with multiplicities; n is the dimension of phase space) to account for the $n-1$ free parameters, a strong condition! In fact, we show in the next proposition that -1 is always an eigenvalue of $\mathcal{K}$, so that, for $m \in \mathcal{I}$ as above, all other eigenvalues must be (non-negative) integers.

Proposition 7.11. *For any m which belongs to the indicial locus $\mathcal{I}$, the Kowalevski matrix $\mathcal{K}(m)$ of a weight homogeneous vector field always has -1 as an eigenvalue. The corresponding eigenspace contains $(\nu_1 m_1, \dots, \nu_n m_n)^\top$ as an eigenvector.*

Proof. The proof is similar to the proof of Euler's Formula for homogeneous polynomials. Let $\dot{x} = f(x)$ be a weight homogeneous vector field on $\mathbf{C}^n$, and let $m = (m_1, \dots, m_n) \in \mathcal{I}$ be arbitrary. We need to show that

$$(\mathcal{K}(m) + \mathrm{Id}_n) \begin{pmatrix} \nu_1 m_1 \\ \vdots \\ \nu_n m_n \end{pmatrix} = 0. \tag{7.16}$$

To show this, fix any i with $1 \leqslant i \leqslant n$ and consider

$$f_i(t^{\nu_1} x_1, \dots, t^{\nu_n} x_n) = t^{\nu_i + 1} f_i(x_1, \dots, x_n)$$

which is valid for all $t \in \mathbf{C}$ and for all $x = (x_1, \dots, x_n) \in \mathbf{C}^n$. Taking the derivative of this identity with respect to t at $t = 1$ we get

$$\sum_{j=1}^{n} \nu_j x_j \frac{\partial f_i}{\partial x_j}(x) - (\nu_i + 1) f_i(x) = 0,$$

valid for any $x \in \mathbf{C}^n$. If we substitute m for x in this equation and we use the indicial equation $f_i(m) = -\nu_i m_i$ then we get

$$\sum_{j=1}^{n} \nu_j m_j \frac{\partial f_i}{\partial x_j}(m) + (\nu_i + 1) \nu_i m_i = 0,$$

which is precisely the i-th line of (7.16). Since i was arbitrary we have shown (7.16). □

We also mention the following property of the Kowalevski matrix.

Proposition 7.12. *If $\dot{x} = f(x)$ is weight homogeneous and divergence free, the latter meaning that $\sum_{i=1}^{n} \partial f_i/\partial x_i = 0$, then the trace of its Kowalevski matrix is given by*

$$\mathrm{Trace}(\mathcal{K}(m)) = \sum_{i=1}^{n} \nu_i,$$

independently of $m \in \mathcal{I}$.

Proof. This is obvious from the definition (7.10) of $\mathcal{K}$. □

Example 7.13. In the case of Example 7.10 we easily verify from (7.15) that $\mathcal{K}(m_1)m_1^\top = -m_1^\top$ and that $\mathcal{K}(m_1')m_1'^\top = -m_1'^\top$. The vector field is divergence free, so that the trace of the matrices (7.15) is equal to 5.

Each point in the indicial locus leads to an integral curve of $\mathcal{V}$, defined for $t \neq 0$, a fact that will prove useful in the analysis of the spectrum of $\mathcal{K}$ (Section 7.4).

Proposition 7.14. *Let m be an arbitrary element of the indicial locus $\mathcal{I}$ of a weight homogeneous vector field $\dot{x} = f(x)$. Then*

$$m(t) := \left(\frac{m_1}{t^{\nu_1}}, \ldots, \frac{m_n}{t^{\nu_n}}\right)$$

is a solution to $\dot{x} = f(x)$, for $t \neq 0$.

Proof. By direct substitution, the equation at step zero is precisely the indicial equation while all equations after the indicial equations are trivially satisfied. □

For future use (see Section 7.5), we also spell out the relation between the (Zariski) tangent space at a point m in the indicial locus and the null space of $\mathcal{K}(m)$.

Proposition 7.15. *Consider the indicial locus $\mathcal{I}$ of a weight homogeneous vector field $\dot{x} = f(x)$ and let $m \in \mathcal{I}$ be arbitrary. The dimension of the Zariski tangent space $T_m\mathcal{I}$ to $\mathcal{I}$ at m equals the dimension of the null space of $\mathcal{K}(m)$.*

Proof. Let us first recall that if $M \subset \mathbf{C}^n$ is the affine variety defined by

$$G_j(x_1, \ldots, x_n) = 0, \qquad j = 1, \ldots, l,$$

where $G_1, \ldots, G_l$ are arbitrary polynomials, then the Zariski tangent space $T_m M$ to M at $m = (m_1, \ldots, m_n) \in M$ is by definition the affine subspace of $\mathbf{C}^n$ given by

$$\sum_{i=1}^{n} \frac{\partial G_j}{\partial x_i}(m)(x_i - m_i) = 0, \qquad j = 1, \ldots, l.$$

It is isomorphic to the tangent space to M at m when M is smooth at m, otherwise it is of larger dimension than the tangent space to M at smooth points of M close to m. Since $\mathcal{I}$ is given by (7.8) the Zariski tangent space $T_m\mathcal{I}$ is given by the following set of equations

$$\sum_{i=1}^{n}\left(\frac{\partial f_j}{\partial x_i}(m)+\nu_i\delta_{ij}\right)(x_i-m_i)=0, \qquad j=1,\dots,n,$$

which we can write by using the Kowalevski matrix in the following compact form: $\mathcal{K}(m)(x-m)=0$. Thus, the tangent space $T_m\mathcal{I}$ is the affine subspace of $\mathbf{C}^n$ which passes through m and whose associated vector space is the null space of $\mathcal{K}(m)$. □

Example 7.16. In the case of our example, the indicial locus consists of 10 isolated points (each taken with multiplicity 1), so that the dimension of the Zariski tangent space $T_m\mathcal{I}$ to $\mathcal{I}$ at any point $m \in \mathcal{I}$ is zero. This is consistent with the fact that $\det \mathcal{K}(m_i) = -4 \neq 0$ and $\det \mathcal{K}(m_i') = 20 \neq 0$.

In the following proposition we spell out how the family of all weight homogeneous Laurent solutions has a natural structure of an algebraic set. The family of Laurent solutions that corresponds to any of its irreducible components (which is an affine variety) is, in analogy with the terminology introduced in Paragraph 6.2.1, called a *weight homogeneous balance.*

Proposition 7.17. *Consider a weight homogeneous vector field $\mathcal{V}$ on $\mathbf{C}^n$. Suppose that for any irreducible component $\mathcal{I}'$ of the indicial locus $\mathcal{I}$ of $\mathcal{V}$ the characteristic polynomial $|k\,\mathrm{Id}_n - \mathcal{K}(m)|$ is independent of $m \in \mathcal{I}'$. Then the set of all weight homogeneous Laurent solutions to $\mathcal{V}$ is parametrized by a finite number of affine varieties (of varying dimensions) $\Gamma^{(i)}$. For any one of these affine varieties $\Gamma^{(i)}$ the coefficients that appear in the corresponding weight homogeneous balance are regular functions on $\Gamma^{(i)}$.*

Proof. Let l denote the largest integer eigenvalue of all matrices $\mathcal{K}(m)$, where m runs through $\mathcal{I}$, which is finite since $\mathcal{I}$ has only a finite number of irreducible components. Let

$$x_i(t)=\frac{1}{t^{\nu_i}}\left(x_i^{(0)}+x_i^{(1)}t+\cdots x_i^{(l)}t^l\right), \qquad i=1,\dots,n,$$

where all coefficients $x_i^{(k)}$, with $1 \leqslant i \leqslant n$ and $0 \leqslant k \leqslant l$, are undetermined. If we substitute these in the differential equations $\dot{x} = f(x)$ for $\mathcal{V}$, then we get $n(l+1)$ polynomial conditions on these $n(l+1)$ coefficients, by equating the coefficients of the first $l+1$ powers of t. This yields an algebraic subset Γ of $\mathbf{C}^{n(l+1)}$, whose points are in one-to-one correspondence with the weight homogeneous Laurent solutions to $\mathcal{V}$; in fact, since l is the largest integer eigenvalue of all matrices $\mathcal{K}(m)$, where m runs through $\mathcal{I}$, all further coefficients of the series follow uniquely from the previous ones, by Proposition 7.6.

Decomposing Γ in its irreducible components yields the affine varieties $\Gamma^{(i)}$. The first $l+1$ coefficients of each $x_i(t)$ are obviously regular functions on Γ, while for the other ones this follows from (7.9), upon using the fact that $(k\,\mathrm{Id}_n - \mathcal{K}(m))$ is invertible for $k > l$ and $m \in \mathcal{I}$. □

In order to compute the affine varieties $\Gamma^{(i)}$ in practice, it is preferable to use a different method than in the proof of the above proposition (even when one is using computer). Namely, let $\mathcal{I}'$ be an irreducible component of $\mathcal{I}$ and define, in a first step, $\Gamma := \mathcal{I}'$ (along the way, Γ will be modified). By assumption, the characteristic polynomial $\chi(k;\Gamma) := |k\,\mathrm{Id}_n - \mathcal{K}(m)|$ is independent of $m \in \Gamma$. In the following steps, as governed by (7.9) (with $k = 1, 2, \ldots$) we can uniquely solve (7.9) with $k = 1, 2, \ldots$ to find $x_i^{(1)}, x_i^{(2)}, \ldots$ as a regular function of Γ, as long as $\chi(k;\Gamma)$ is different from zero. Suppose now that k is the smallest positive integer for which $\chi(k;\Gamma) = 0$. We first enlarge Γ to $\Gamma \times \mathbf{C}^n$, where we view the unknowns $x_1^{(k)}, \ldots, x_n^{(k)}$ as coordinates on $\mathbf{C}^n$. Then the equations (7.9) define an algebraic subset of $\Gamma \times \mathbf{C}^n$, which we decompose in its irreducible components $\Gamma^{(i)}$. We pick any one of these $\Gamma^{(i)}$ and we define $\chi(k;\Gamma^{(i)}) := \chi(k;\Gamma)$; for the sake of repeating the construction, redefine Γ now by $\Gamma := \Gamma^{(i)}$ (in order to obtain all weight homogeneous balances, this and what follows will have to be done for each $\Gamma^{(i)}$ (separately)). Notice that the new Γ may have larger dimension than the old one (the dimension of Γ is the number of parameters these first terms of the balance depends on). We can now go on and solve (7.9) uniquely for $k+1, k+2, \ldots$, until we hit another zero of $\chi(k;\Gamma)$, in which case Γ will get modified again. Once we get past the largest eigenvalue, Γ will not change anymore and we are done.

Definition 7.18. The weight homogeneous balance that corresponds to the affine variety $\Gamma^{(i)}$, as given by Proposition 7.17, is denoted by $x(t;\Gamma^{(i)})$. For $m \in \Gamma^{(i)}$ the balance specializes to a Laurent series that will be denoted by $x(t;m)$. Each of the affine varieties $\Gamma^{(i)}$ that corresponds to a principal balance, i.e., depends on $n-1$ parameters, is called an *abstract Painlevé wall* of $\mathcal{V}$, while their union, which is an algebraic subset whose irreducible components are the $\Gamma^{(i)}$, is called the *abstract Painlevé building* of $\mathcal{V}$, denoted Γ.

Remark 7.19. In most of the examples the varieties $\Gamma^{(i)}$ are (some of) the original irreducible components of the indicial locus $\mathcal{I}$, multiplied by some $\mathbf{C}^N$ (where N depends on i). This is so, because in those examples the algebraic equations (7.9) can be solved polynomially for $x_1^{(k)}, \ldots x_n^{(k)}$ in terms of the previous $x_1^{(l)}, \ldots x_n^{(l)}$ (taking some of the $x_i^{(k)}$ as free parameters), because the adjoint conditions which need to be satisfied when $\chi(k;\Gamma) = 0$ turn out to be automatically satisfied, and so we merely gain free parameters at step k with no further restrictions on prior $x_i^{(l)}$, coming from the adjoint condition. See Proposition 7.22 below for a general statement.

Remark 7.20. The condition that for each irreducible component $\mathcal{I}'$ of the indicial locus the characteristic polynomial $|k\,\mathrm{Id}_n - \mathcal{K}(m)|$ is independent of $m \in \mathcal{I}'$ may seem a strong condition, but it is satisfied for the indicial locus of any a.c.i. system (see Proposition 7.32).

Example 7.21. In the case of our example, we find the following weight homogeneous balance, starting from $m_1 = (0, 0, -1, 1, 0)$, the remaining terms are uniquely determined from what is given below:

$$
\begin{aligned}
x_1(t; m_1) &= d + O(t^2),\\
x_2(t; m_1) &= bt + O(t^2),\\
x_3(t; m_1) &= -\frac{1}{t} + a - \frac{1}{3}(a^2 + 2b + c)t + O(t^2),\\
x_4(t; m_1) &= \frac{1}{t} + a + \frac{1}{3}(a^2 - b - 2c)t + O(t^2),\\
x_5(t; m_1) &= ct + O(t^2).
\end{aligned}
\tag{7.17}
$$

It is a principal balance, because there are four free parameters $a, \dots, d$. The corresponding abstract Painlevé wall is denoted by $\Gamma^{(1)}$ and is isomorphic to $\mathbf{C}^4$. Two of the free parameters, a and d, enter at step 1, while the other two, b and c, enter at step 2, which is consistent with the fact that 1 and 2 are double roots of the characteristic polynomial of $\mathcal{K}(m_1)$. Using the automorphism σ we find in total 5 abstract Painlevé walls, which constitute together the abstract Painlevé building, where the Painlevé wall that starts from m_i is denoted by $\Gamma^{(i)}$. Indeed, since the only non-negative roots of $\mathcal{K}(m_1')$ are 1, 2 and 5, with corresponding free parameters a, b and c, the weight homogeneous balance that starts at any of the m_i' is not a principal but a lower balance. We give the first few terms of the lower balance that starts at m_1', the remaining terms being uniquely determined from what is given below:

$$
\begin{aligned}
x_1(t; m_1') &= ct^4 + O(t^5),\\
x_2(t; m_1') &= -\frac{2}{t} + 2a - (a^2 + b)t + \frac{a}{5}(a^2 + 3b)t^2 + \frac{1}{10}(a^2 + b)^2 t^3 + O(t^4),\\
x_3(t; m_1') &= \frac{1}{t} + a - bt - \frac{a}{5}(a^2 + 3b)t^2 + \frac{1}{10}(a^4 + 6a^2 b + 7b^2)t^3 + O(t^4),\\
x_4(t; m_1') &= -\frac{1}{t} + a + bt - \frac{a}{5}(a^2 + 3b)t^2 - \frac{1}{10}(a^4 + 6a^2 b + 7b^2)t^3 + O(t^4),\\
x_5(t; m_1') &= \frac{2}{t} + 2a + (a^2 + b)t + \frac{a}{5}(a^2 + 3b)t^2 - \frac{1}{10}(a^2 + b)^2 t^3 + O(t^4).
\end{aligned}
$$

We have printed more terms here than for the principal balance because, when we want to compute the term in t^0 of the series, obtained by substituting the series in the constant of motion F_3, i.e., in $x_1x_2x_3x_4x_5$, then we need as many terms in the series as given. The lower balance that starts at the other points m_i' are found by using the automorphism σ.

We show in the following proposition that the affine varieties that parameterize the weight homogeneous principal balances, i.e., the abstract Painlevé walls, are of a simple, "wall-like" form. This implies, in particular, that these walls and the corresponding weight homogeneous principal balances can easily be computed.

Proposition 7.22. *Let $\mathcal{V}$ be a weight homogeneous vector field on $\mathbf{C}^n$.*

(1) For every weight homogeneous principal balance $x(t;\Gamma')$ of $\mathcal{V}$, the abstract Painlevé wall Γ' is of the form $\Gamma' = \mathcal{I}' \times \mathbf{C}^p$, where $\mathcal{I}'$ is an irreducible component of the indicial locus $\mathcal{I}$ that does not contain any singular point of $\mathcal{I}$, and where $p = n - 1 - \dim \mathcal{I}'$; accordingly, $x(t;\Gamma')$ will often be denoted by $x(t;\mathcal{I}')$.
(2) For such a $\mathcal{I}'$, the characteristic polynomial $\chi(k;m)$ is independent of $m \in \mathcal{I}'$, and the matrices $\mathcal{K}(m)$, $m \in \mathcal{I}'$ are diagonalizable.

Proof. Suppose that $x(t)$ is a weight homogeneous balance of $\mathcal{V}$ which depends on $n-1$ parameters (principal balance). Since -1 is always an eigenvalue of the Kowalevski matrix (see Proposition 7.11) and since the number of new parameters that can appear at step k are bounded by the multiplicity of k as an eigenvalue of the Kowalevski matrix, the coefficient $x^{(k)}$ of $x(t)$ must depend on a number of new parameters, which is exactly equal to the multiplicity of k as an eigenvalue of the Kowalevski matrix $\mathcal{K}(m)$, where m is any particular value of $x^{(0)}$; for otherwise we can never get at a dependence on $n-1$ parameters. Any such m is contained in the indicial locus; restricting the family, if necessary, we may assume that these values are contained in a single irreducible component $\mathcal{I}' \subseteq \mathcal{I}$. Using Proposition 7.15, we have that

$$\begin{aligned}\dim \mathcal{I}' &= \dim \text{ null space of } \mathcal{K}(m), \text{ with } m \in \mathcal{I}' \text{ generic}\\ &\leqslant \text{ mult. of } 0 \text{ in the spectrum of } \mathcal{K}(m), \text{ for } m \in \mathcal{I}' \text{ generic}\\ &\leqslant \text{ mult. of } 0 \text{ in the spectrum of } \mathcal{K}(m), \text{ for any } m \in \mathcal{I}'.\end{aligned}$$

Therefore, if a weight homogeneous principal balance contains m_0 as a value of $x^{(0)}$ then

$$\dim \mathcal{I}' = \text{ mult. of } 0 \text{ as eigenvalue of } \mathcal{K}(m_0),$$

and the number of parameters that appear at step 0 is $\dim \mathcal{I}'$, which means that the values of $x^{(0)}$ contain a neighborhood U of m_0 in $\mathcal{I}'$. We claim that every $m \in \mathcal{I}'$ appears as a value of $x^{(0)}$. In fact, by the above, the characteristic polynomial of $\mathcal{K}(m)$ is independent of m for $m \in U$, since all its roots must be integers and yet it must vary continuously with $m \in U$, hence it is independent of m for $m \in \mathcal{I}'$; in fact it is just given by $\prod_i (k-k_i)^{\rho_i}$, where ρ_i and m_i are integers. Apply now the algorithm given after the proof of Proposition 7.17, starting with $\mathcal{I}'$ and let us focus on what happens at the k-th step, where k is the first positive eigenvalue of $\mathcal{K}(m)$, $m \in \mathcal{I}'$.

We know that for generic $m \in \mathcal{I}'$ the right hand side $R^{(k)}(m)$ is contained in the image of $\mathcal{K}(m) - k\,\mathrm{Id}_n$: the adjoint condition is automatically satisfied, otherwise we will lose some of the previously gained free parameters and never end up with $n-1$ free parameters in the balance. Since $\mathcal{K}(m)$ and $R^{(k)}(m)$ depend algebraically on m it follows that the right hand side $R^{(k)}(m)$ is contained in the image of $\mathcal{K}(m) - k\,\mathrm{Id}_n$ for all $m \in \mathcal{I}'$. It means that before going to the next step, $\mathcal{I}'$ will simply be replaced by $\mathcal{I}' \times \mathbf{C}^p$, where p denotes the multiplicity of k as an eigenvalue of $\mathcal{K}(m)$, $m \in \mathcal{I}'$. Notice also that this is the situation that we described in Remark 7.19.

We now show that $\mathcal{I}'$ is non-singular. Since the multiplicity of 0 as an eigenvalue of $\mathcal{K}(m)$ is independent of $m \in \mathcal{I}'$ the dimension of the Zariski tangent space $T_m\mathcal{I}$ is also independent of $m \in \mathcal{I}'$ (again by Proposition 7.15). This means that every point m of $\mathcal{I}'$ is a non-singular point (of $\mathcal{I}$).

It is now also easy to see that $\mathcal{K}(m)$ is diagonalizable, for $m \in \mathcal{I}'$. Indeed, besides -1, which is a simple eigenvalue (with one-dimensional eigenspace), each eigenvalue admits an eigenspace whose dimension is equal to the multiplicity of the eigenvalue in the characteristic polynomial; in terms of a basis for $\mathbf{C}^n$ which is constructed from a basis of each of the eigenspaces, the matrix $\mathcal{K}(m)$ is diagonal. This diagonal matrix contains -1 and the $n-1$ other eigenvalues of $\mathcal{K}(m)$ as is entries, which are integers, independent of $m \in \mathcal{I}'$.
□

Remark 7.23. In comparing Propositions 7.17 and 7.22, notice that in Proposition 7.17 we describe *all* weight homogeneous balances, but under the assumption that for any irreducible component $\mathcal{I}'$ of the indicial locus $\mathcal{I}$ of $\mathcal{V}$ the characteristic polynomial $|k\,\mathrm{Id}_n - \mathcal{K}(m)|$ is independent of $m \in \mathcal{I}'$. In Proposition 7.22 this assumption is absent, but only the weight homogeneous principal balances are described. Moreover, in the latter proposition the description of the abstract Painlevé walls is more explicit, and is given in terms of the indicial locus.

Example 7.24. Returning to our example, we have seen that each of the Painlevé walls $\Gamma^{(1)}, \dots \Gamma^{(5)}$ is a linear space $\mathbf{C}^4$, i.e., the corresponding irreducible component $\mathcal{I}^{(i)}$ of $\mathcal{I}$ is just a point (the point m_i). The Kowalevski matrix $\mathcal{K}(m_1)$ gets diagonalized by the invertible matrix

$$\begin{pmatrix} 1 & 0 & 2 & 1 & -1 \\ 1 & 0 & 1 & 2 & 1 \\ 0 & 0 & 0 & -3 & 0 \\ 0 & 1 & 0 & 0 & 0 \\ 0 & 0 & -3 & 0 & 0 \end{pmatrix}$$

whose columns are a complete set of eigenvectors of $\mathcal{K}(m_1)$, corresponding respectively to the eigenvalues 1, 1, 2, 2 and -1.

7.2 Convergence of the Balances

The following theorem states that weight homogeneous balances are convergent (for small, non-zero $|t|$). The proof is inspired by [61].

Theorem 7.25. *Let $\mathcal{V}$ be a polynomial vector field on $\mathbf{C}^n$ which is weight homogeneous. If $x(t)$ is a weight homogeneous balance to $\mathcal{V}$, then $x(t)$ is convergent (for small, non-zero $|t|$).*

Proof. Let $x(t;\Gamma)$ be any weight homogeneous balance of $\mathcal{V}$, let $m_0 \in \Gamma$ and let Γ_0 be any compact subset of Γ that contains an open neighborhood of m_0 in Γ (for the choice of topology on Γ, see Remark 7.26 below). We prove the stronger statement that the Laurent series converges uniformly on Γ_0. To do this, define

$$N_1 := 1 + \max\left\{\left|x_i^{(k)}(m)\right| \mid m \in \Gamma_0,\ 1 \leqslant i \leqslant n,\ 0 \leqslant k \leqslant \lambda\right\}, \qquad (7.18)$$

where λ is the largest integer eigenvalue of the Kowalevski matrix $\mathcal{K}(m)$ with $m \in \Gamma_0$. Let us write the vector field $\mathcal{V}$ as before as $\dot{x}_i = f_i(x)$, where $i = 1,\dots,n$. By analyticity, there are real constants N_2 and N_3, with $N_2, N_3 > N_1$, such that for any $m \in \Gamma_0$ and for any $k \geqslant \lambda + 1$

$$\left|\frac{\partial^\beta f_i}{\partial x^\beta}(m)\right| \leqslant \beta! N_2^{|\beta|} \quad \text{and} \quad |(k \operatorname{Id}_n - \mathcal{K}(m))^{-1}| \leqslant N_3.$$

Applying these estimates to (7.9) and to (7.14) leads for $1 \leqslant i \leqslant n$ and $k \geqslant \lambda + 1$ to the recursive estimate

$$\left|x_i^{(k)}(m)\right| \leqslant N_3 \sum_{\substack{|\beta|>1 \\ \sigma_{jl}>0 \\ \sum \sigma_{jl}=k}} N_2^{|\beta|} \prod_{j=1}^{n} \left|x_j^{(\sigma_{j1})}(m)\right| \cdots \left|x_j^{(\sigma_{j\beta_j})}(m)\right|. \qquad (7.19)$$

where $m \in \Gamma_0$. Define now the series

$$V(t) := N_1 t + \sum_{k=2}^{\infty} \alpha_k t^k,$$

where the coefficients $\alpha_k \in \mathbf{R}$ are defined inductively by $\alpha_1 := N_1$ and

$$\alpha_k := N_3 \sum_{\substack{|\beta|>1 \\ \sigma_{jl}>0 \\ \sum \sigma_{jl}=k}} N_2^{|\beta|} \prod_{j=1}^{n} \alpha_{\sigma_{j1}} \dots \alpha_{\sigma_{j\beta_j}},$$

for $k \geqslant 2$.

The series $V(t)$ majorizes

$$U_i(t;m) := \sum_{k\geqslant 1} x_i^{(k)}(m)t^k$$

for all $1 \leqslant i \leqslant n$ and for any $m \in \Gamma_0$, i.e., we have that $|x_i^{(k)}(m)| \leqslant \alpha_k$ for $i = 1,\ldots,n$, for $k \in \mathbf{N}^*$ and for $m \in \Gamma_0$. Indeed, if $k \leqslant \lambda$ then (7.18) yields $|x_i^{(k)}(m)| < N_1$, and $N_1 \leqslant \alpha_k$, as is clear from the definition of α_k, N_2 and N_3. For $k \geqslant \lambda+1$ we proceed by induction: assuming $|x_i^{(j)}(m)| \leqslant \alpha_j$, for $j < k$, for $1 \leqslant i \leqslant n$, and for $m \in \Gamma_0$ we may estimate each $x_j^{(\sigma_{jl})}(m)$ in the right hand side of (7.19), which gives

$$\left|x_i^{(k)}(m)\right| \leqslant N_3 \sum_{\substack{|\beta|>1\\ \sigma_{jl}>0\\ \sum\sigma_{jl}=k}} N_2^{|\beta|} \prod_{j=1}^{n} \alpha_{\sigma_{j1}} \ldots \alpha_{\sigma_{j\beta_j}} = \alpha_k,$$

Finally from the definition of the α_k one observes that

$$\begin{aligned}
\frac{V(t) - N_1 t}{N_3} &= \sum_{k=2}^{\infty} t^k \sum_{\substack{|\beta|>1\\ \sigma_{jl}>0\\ \sum\sigma_{jl}=k}} N_2^{|\beta|} \prod_{j=1}^{n} \alpha_{\sigma_{j1}} \ldots \alpha_{\sigma_{j\beta_j}}\\
&= \sum_{i=2}^{\infty} (N_2 V(t))^i\\
&= \frac{N_2^2 V(t)^2}{1 - N_2 V(t)}
\end{aligned}$$

so that the series $V(t)$ satisfies

$$V(t) = N_1 t + N_3 N_2^2 \frac{V(t)^2}{1 - N_2 V(t)},$$

which amounts to a quadratic equation for V; solving for V yields the desired majorant for the functions U_i, which therefore converge for $|t|$ sufficiently small. Since $x_i(t;m) = t^{-\nu_i}(x^{(0)} + U_i(t;m))$, see (7.11), the Laurent series $x(t;m)$ converges for $|t|$ sufficiently small, $t \neq 0$, and $m \in \Gamma_0$. □

Remark 7.26. We have seen that if $x(t;\Gamma)$ is a principal balance then Γ is an affine variety, hence has a decent topological structure, and that usually the other balances also have this property. The compact subset Γ_0 can then be taken with respect to the complex topology. If Γ does not have a nice structure, just give it the discrete topology and take $V := m_0$; then the proof establishes simple convergence of the series, as asserted.

7.3 Weight Homogeneous Constants of Motion

Let $\mathcal{V}$ be a weight homogeneous vector field on $\mathbf{C}^n$ and write, as before, $\dot{F}$ for $\mathcal{V}[F]$, where $F \in \mathcal{F}(\mathbf{C}^n) = \mathbf{C}[x_1, \dots, x_n]$, and $\dot{x}_i = f_i(x)$, where $i = 1, \dots, n$. Define

$$\mathcal{H} := \left\{ F \in \mathcal{F}(\mathbf{C}^n) \mid \dot{F} = 0 \right\}.$$

In view of the Leibniz rule, $\mathcal{H}$ is a subalgebra of $\mathcal{F}(\mathbf{C}^n)$, which we call the *algebra of constants of motion* or the *algebra of first integrals* of $\mathcal{V}$. For $F \in \mathcal{F}(\mathbf{C}^n)$ we have that $F \in \mathcal{H}$ if and only if

$$\sum_{i=1}^{n} \frac{\partial F}{\partial x_i} f_i = 0. \tag{7.20}$$

Notice that the algebra of polynomial constants of motion of $\mathcal{V}$ is generated by weight homogeneous constants of motion. Indeed, if F is any polynomial constant of motion of $\dot{x} = f(x)$, write $F = \sum_j F_j$, where $\varpi(F_j) = j$. Then $\sum_j \dot{F}_j = \dot{F} = 0$, implies that $\dot{F}_j = 0$, because $\varpi(\dot{F}_j) = \varpi(F_j) + 1 = j + 1$. It follows that $\mathcal{H}$ is graded by weighted degree, $\mathcal{H} = \oplus_{j \in \mathbf{N}} \mathcal{H}^{(j)}$, where

$$\mathcal{H}^{(j)} := \left\{ F \in \mathcal{F}(\mathbf{C}^n) \mid \dot{F} = 0 \text{ and } \varpi(F) = j \right\}.$$

Clearly, for a given degree j a basis for $\mathcal{H}^{(j)}$ is easily calculated. In the examples we will often have weight homogeneous constants of motion $F_1, \dots, F_s$ of $\mathcal{V}$ which generate $\mathcal{H}$ (as an algebra). If they are independent (in the sense of Definition 4.11) then we can, as in Proposition 7.1, compute the dimensions $\dim \mathcal{H}^{(j)}$ as the coefficients of the following generating function:

$$\prod_{j=1}^{s} \frac{1}{1 - t^{\varpi(F_j)}} = \sum_{j=0}^{\infty} \left(\dim \mathcal{H}^{(j)} \right) t^j. \tag{7.21}$$

If $F_1, \dots, F_s$ are weight homogeneous constants of motion then we may associate, as in Section 4.1, to each $\mathbf{c} = (c_1, \dots, c_s) \in \mathbf{C}^s$ the affine variety[2]

$$\mathcal{A}_{\mathbf{c}} := \{ x \in \mathbf{C}^n \mid F_j(x) = c_j \text{ for } 1 \leqslant j \leqslant s \}.$$

By weight homogeneity, $\mathcal{A}_{\mathbf{c}}$ injects naturally in some weighted projective space $\mathbf{P}^n_\nu$, which is by definition the quotient of $\mathbf{C}^{n+1}$ by the action, given by $t \cdot (x_0, \dots, x_n) \mapsto (t^{\nu_0} x_0, \dots, t^{\nu_n} x_n)$, where $\nu_0 := 1$. When all ν_i are equal to 1 we recover the definition of the standard projective space $\mathbf{P}^n$.

[2] In the present case the functions F_j are not in involution, in fact there is no Poisson structure (yet), so there are no commuting vector fields on $\mathcal{A}_{\mathbf{c}}$ and so on. Hence the slightly different notation for these varieties than the notation $\mathbf{F}_{\mathbf{c}}$ from Section 4.1.

Thus, we may consider

$$\overline{\mathcal{A}_{\mathbf{c}}} := \left\{(z_0 : z_1 : \cdots : z_n) \in \mathbf{P}^n_\nu \mid F_j(z_1, \ldots, z_n) = c_j z_0^{\varpi(F_j)} \text{ for } 1 \leqslant j \leqslant s\right\},$$

which is well-defined by weight homogeneity. Notice that the intersection of $\overline{\mathcal{A}_{\mathbf{c}}}$ with the hyperplane at infinity (defined by $z_0 = 0$) is independent of the choice of $\mathbf{c}$. Thus, we denote

$$\mathcal{A}_\infty := \{(0 : z_1 : \cdots : z_n) \in \mathbf{P}^n_\nu \mid F_j(z_1, \ldots, z_n) = 0 \text{ for } 1 \leqslant j \leqslant s\}.$$

We show in the following proposition that the indicial locus $\mathcal{I}$ of a weight homogeneous vector field injects naturally in $\mathcal{A}_\infty$.

Proposition 7.27. *Let $\dot{x} = f(x)$ be a weight homogeneous vector field on $\mathbf{C}^n$. Let $(m_1, \ldots, m_n)$ be an element of its indicial locus $\mathcal{I}$. Then $(0 : m_1 : \cdots : m_n) \in \mathcal{A}_\infty$. The resulting map injects $\mathcal{I}$ in $\mathcal{A}_\infty \subset \mathbf{P}^n_\nu$.*

Proof. For $m \in \mathcal{I}$, consider the solution

$$m(t) := \left(\frac{m_1}{t^{\nu_1}}, \ldots, \frac{m_n}{t^{\nu_n}}\right)$$

to $\dot{x} = f(x)$, which is defined for $t \neq 0$ (See Proposition 7.14). Since each F_j is a constant of the motion, $F_j(m(t))$ is a constant (i.e., it is independent of t). Now

$$F_j(m(t)) = F_j\left(\frac{m_1}{t^{\nu_1}}, \ldots, \frac{m_n}{t^{\nu_n}}\right) = t^{-\varpi(F_j)} F_j(m_1, \ldots, m_n).$$

Since $\varpi(F_j) > 0$ and since $F_j(m(t))$ is independent of t it follows that $F_j(m_1, \ldots, m_n) = 0$, for $j = 1, \ldots, 5$, as asserted. We prove that the map

$$\begin{array}{rcl} \imath : \qquad \mathcal{I} & \to & \mathcal{A}_\infty \subset \mathbf{P}^n_\nu \\ (m_1, \ldots, m_n) & \mapsto & (0 : m_1 : \cdots : m_n) \end{array}$$

is injective; for doing this, notice that $(0 : m_1 : \cdots : m_n)$ and $(0 : t^{\nu_1} m_1 : \cdots : t^{\nu_n} m_n)$ represent the same point in $\mathbf{P}^n_\nu$, for any $t \in \mathbf{C}^*$. Suppose therefore that $(m_1, \ldots, m_n)$ and $(t^{\nu_1} m_1, \ldots, t^{\nu_n} m_n)$ both belong to the indicial locus $\mathcal{I}$, for some $t \in \mathbf{C}^*$. In view of (7.8) and since $\varpi(f_i) = \nu_i + 1$ this means that

$$\nu_i m_i + f_i(m_1, \ldots, m_n) = 0,$$
$$\nu_i m_i + t f_i(m_1, \ldots, m_n) = 0,$$

for some $t \in \mathbf{C}^*$ and for $1 \leqslant i \leqslant n$. Since at least one of the m_i is different from zero, the above equations imply $t = 1$, proving injectivity of $\imath$. □

Example 7.28. In the case of our example the constants of motion F_1, F_2, F_3 were chosen (weight) homogeneous. In this case the varieties $\mathcal{A}_{\mathbf{c}}$ embed in projective space $\mathbf{P}^5$ and $\mathcal{A}_\infty \subset \mathbf{P}^5$ is the (singular) curve, given by

$$
\begin{aligned}
0 &= x_0, \\
0 &= x_1 + x_2 + x_3 + x_4 + x_5, \\
0 &= x_1x_3 + x_2x_4 + x_3x_5 + x_4x_1 + x_5x_2, \\
0 &= x_1x_2x_3x_4x_5.
\end{aligned}
$$

We see that $\mathcal{A}_\infty$ consists of five conics, each of which is contained in one of the hyperplanes $x_i = 0$, where $i = 1, \dots, 5$ (they are all contained in the hyperplane $x_0 = 0$). We will denote the conic that is contained in $x_i = 0$ by C_i, where $i = 1, \dots, 5$. For example, the conic C_1 is given, in the plane

$$(x_0 = 0) \cap (x_1 = 0) \cap (x_2 + x_3 + x_4 + x_5 = 0)$$

by $(x_4 + x_5)^2 = x_3x_4$. Each point m_i is mapped by $\imath$ to an intersection point of three of the conics, namely the conics C_{i-1}, C_i and C_{i+1}. Moreover, the conics C_{i-1} and C_{i+1} are tangent to each other in $\imath(m_i)$. This accounts for all intersection points of the conics (see Figure 7.1). Each of the points $m_1' \dots, m_5'$ gets mapped into precisely one of the conics, namely m_i' gets mapped to the conic C_i.

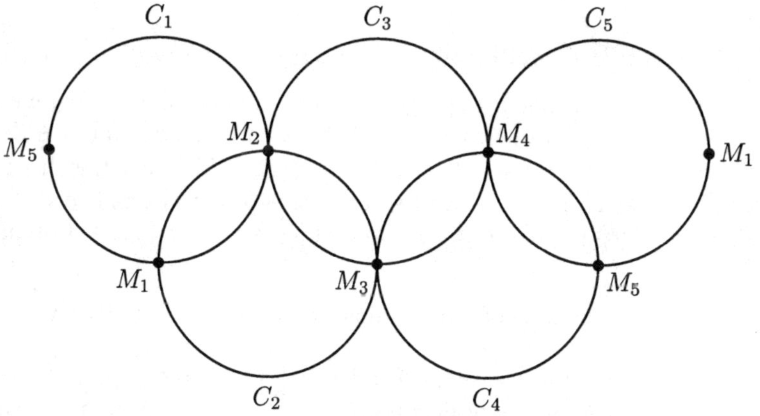

Fig. 7.1. Letting C_i denote the conic in the hyperplane $x_i = 0$ of $\mathbf{P}^5$ and $M_i := \imath(m_i)$ we have that C_i contains the points M_{i-1}, M_i and M_{i+1}, where C_i meets the other 4 curves. These conics are permuted by the projective transformation that cyclically permutes the homogeneous coordinates $x_1, \dots, x_5$ on $\mathbf{P}^5$.

7.4 The Kowalevski Matrix and its Spectrum

In this section we investigate the spectrum of the Kowalevski matrix $\mathcal{K}$, going with a weight homogeneous vector field $\mathcal{V}$. The matrix $\mathcal{K}$, which was defined in (7.10), appeared when computing, for $k > 0$, the k-th term of a weight homogeneous balance of $\mathcal{V}$. We will spell out the relation between the spectrum of $\mathcal{K}$ and the degrees of the weight homogeneous constants of motion of the vector field. Throughout the section a polynomial vector field $\mathcal{V}$ on $\mathbf{C}^n$, given by $\dot{x} = f(x)$ is fixed and we assume it to be weight homogeneous (with respect to ν). We denote, as before, by $\mathcal{H}^{(k)}$ the vector space of constants of motion of $\mathcal{V}$ that are weight homogeneous of weight k, where $k \in \mathbf{N}^*$.

We define for any $k \in \mathbf{N}$ and for any $m \in \mathcal{I}$, the following subspaces of $(\mathbf{C}^n)^*$ (the dual space of $\mathbf{C}^n$) and of $\mathbf{C}^n$:

$$\begin{aligned} \mathrm{d}\mathcal{H}_m^{(k)} &:= \left\{\mathrm{d}F(m) \mid F \in \mathcal{H}^{(k)}\right\} \subseteq T_m^*\mathbf{C}^n \cong (\mathbf{C}^n)^*, \\ E_m^{(k)} &:= \{v \in \mathbf{C}^n \mid (k\,\mathrm{Id}_n - \mathcal{K}(m))^l v = 0, \text{ for some } l \geqslant 1\} \subseteq \mathbf{C}^n, \\ \mathrm{Ann}\left(E_m^{(k)}\right) &:= \left\{\phi \in (\mathbf{C}^n)^* \mid \langle \phi, v\rangle = 0 \text{ for all } v \in E_m^{(k)}\right\}. \end{aligned}$$

By linear algebra,

$$\mathbf{C}^n = \bigoplus_{k \in \mathbf{C}} E_m^{(k)}$$

and

$$\dim E_m^{(k)} = \text{multiplicity of } k \text{ in the spectrum of } \mathcal{K}(m),$$

where $m \in \mathcal{I}$. Notice that if the algebra of constants of motion is generated by the weight homogeneous constants of motion $F_1, \dots, F_s$ and $m \in \mathcal{I}$ then $\mathrm{d}\mathcal{H}_m^{(k)} = 0$ for all k such that $k \notin \{\varpi(F_1), \dots \varpi(F_s)\}$. To prove this, it suffices to point out that if F and G are two constants of motion and $m \in \mathcal{I}$ then $\mathrm{d}(FG)(m) = F(m)\mathrm{d}G(m) + G(m)\mathrm{d}F(m) = 0$ because any constant of motion vanishes on $\mathcal{I}$.

Example 7.29. In the case of our example, we have for $m_1 \in \mathcal{I}$ that

$$\begin{aligned} \mathrm{d}F_1(m_1) &= \mathrm{d}x_1 + \mathrm{d}x_2 + \mathrm{d}x_3 + \mathrm{d}x_4 + \mathrm{d}x_5, \\ \mathrm{d}F_2(m_1) &= \mathrm{d}x_2 - \mathrm{d}x_5, \\ \mathrm{d}F_3(m_1) &= 0, \end{aligned}$$

so that $\dim \mathrm{d}\mathcal{H}_{m_1}^{(k)} = 1$ for $k \in \{1, 2\}$, and is 0 for all other values of k. The generalized eigenspaces of $\mathcal{K}(m_1)$ are trivial (see (7.15)), except for

$$E_{m_1}^{(1)} = \operatorname{span}\left\{\begin{pmatrix}0\\0\\1\\1\\0\end{pmatrix}, \begin{pmatrix}1\\0\\0\\0\\0\end{pmatrix}\right\}, \qquad E_{m_1}^{(2)} = \operatorname{span}\left\{\begin{pmatrix}0\\-3\\2\\1\\0\end{pmatrix}, \begin{pmatrix}0\\0\\1\\2\\-3\end{pmatrix}\right\}.$$

The annihilators of these subspaces are given by

$$\begin{aligned}\operatorname{Ann}\left(E_{m_1}^{(1)}\right) &= \operatorname{span}\left\{\mathrm{d}x_2,\ \mathrm{d}x_3 - \mathrm{d}x_4,\ \mathrm{d}x_5\right\},\\ \operatorname{Ann}\left(E_{m_1}^{(2)}\right) &= \operatorname{span}\left\{\mathrm{d}x_1,\ \mathrm{d}x_2 + \mathrm{d}x_3 + \mathrm{d}x_4 + \mathrm{d}x_5,\ \mathrm{d}x_2 + 2\mathrm{d}x_3 - \mathrm{d}x_4\right\}.\end{aligned}$$

At the points $m_1', \dots, m_5'$ each of the differentials is different from zero, so that $\dim \mathrm{d}\mathcal{H}_{m_i'}^{(k)} = 1$ for $k = 1, 2, 3$ and $1 \leqslant i \leqslant 5$.

We show in the following theorem that if we have p constants of motion of weight k, whose differentials are independent at $m \in \mathcal{I}$, then k is an eigenvalue of $\mathcal{K}(m)$ of multiplicity at least p. The proof is based on arguments that are due to Yoshida (see [176]).

Theorem 7.30. *Consider the weight homogeneous vector field $\dot{x} = f(x)$ on $\mathbf{C}^n$. For any $k \in \mathbf{N}$ and for any $m \in \mathcal{I}$ the following hold.*

(1) $\mathrm{d}\mathcal{H}_m^{(k)} \subseteq \bigcap_{l \neq k} \operatorname{Ann}\left(E_m^{(l)}\right)$;

(2) $\dim \mathrm{d}\mathcal{H}_m^{(k)} \leqslant \dim E_m^{(k)}$.

As a consequence, if we have p constants of motion of weight k whose differentials are independent at m, then k is an eigenvalue of $\mathcal{K}(m)$ with multiplicity at least p.

Proof. The proof of *(2)* follows easily from *(1)*, since

$$\dim E_m^{(k)} = \dim \operatorname{Ann}\left(\bigoplus_{l \neq k} E_m^{(l)}\right) - \dim\left(\bigcap_{l \neq k} \operatorname{Ann}\left(E_m^{(l)}\right)\right).$$

The proof of *(1)* will be given in a number of steps. Throughout the proof m is an arbitrary element of the indicial locus $\mathcal{I}$, and F is a weight homogeneous constant of motion of $\dot{x} = f(x)$. Recall from Proposition 7.14 that

$$m(t) := \left(\frac{m_1}{t^{\nu_1}}, \dots, \frac{m_n}{t^{\nu_n}}\right) \tag{7.22}$$

is a solution to $\dot{x} = f(x)$. We consider the variational equation

$$\dot{\xi}_i = \langle \mathrm{d}f_i(m(t)), \xi \rangle, \tag{7.23}$$

going with this solution.

1. We first show how the solutions to (7.23) are related to $\mathcal{K}(m)$, the Kowalevski matrix at m. Precisely, we claim that if, for some $\kappa \in \mathbf{C}$, the vector $\eta(t)$ satisfies for $t \in \mathbf{C}$ near 1 the linear equation

$$t\dot{\eta}(t) = (\mathcal{K}(m) - \kappa \operatorname{Id}_n)\eta(t) \tag{7.24}$$

then

$$\xi_i(t) := \eta_i(t)t^{\kappa-\nu_i}, \qquad i = 1, \dots, n, \tag{7.25}$$

is a solution to the variational equation (7.23). To see this, let us write out the latter equation explicitly in terms of coordinates as

$$\dot{\xi}_i = \sum_{j=1}^{n} \frac{\partial f_i}{\partial x_j}(m_1 t^{-\nu_1}, \dots, m_n t^{-\nu_n})\xi_j. \qquad i = 1, \dots, n. \tag{7.26}$$

Substituting (7.25) for ξ_i in (7.26) leads to

$$\begin{aligned}\dot{\eta}_i(t)t^{\kappa-\nu_i} + (\kappa - \nu_i)\eta_i(t)t^{\kappa-\nu_i-1} &= \sum_{j=1}^{n} \frac{\partial f_i}{\partial x_j}(m_1 t^{-\nu_1}, \dots, m_n t^{-\nu_n})\eta_j(t)t^{\kappa-\nu_j} \\ &= \sum_{j=1}^{n} \frac{\partial f_i}{\partial x_j}(m_1, \dots, m_n)t^{-\nu_i-1+\nu_j}\eta_j(t)t^{\kappa-\nu_j},\end{aligned}$$

where we have used that

$$\varpi\left(\frac{\partial f_i}{\partial x_j}\right) = \varpi(f_i) - \varpi(x_j) = \nu_i + 1 - \nu_j.$$

If we divide both sides by $t^{\kappa-\nu_i-1}$ then we find

$$t\dot{\eta}_i(t) = \sum_{j=1}^{n} \frac{\partial f_i}{\partial x_j}(m)\eta_j(t) + (\nu_i - \kappa)\eta_i(t),$$

and we notice that the right hand side is precisely the i-th line of (7.24), proving our claim. Notice that (7.24) can also be written as

$$\eta'(s) = (\mathcal{K}(m) - \kappa \operatorname{Id}_n)\eta(s), \tag{7.27}$$

where $s = \ln t$, and the prime denotes the derivative with respect to s. The latter equation is a homogeneous linear equation with constant coefficients, hence is easily solved. It follows that the solutions to the variational equation that is associated to a solution that comes from a point m in the indicial locus $\mathcal{I}$ can be written down explicitly in terms of the spectrum of the Kowalevski matrix $\mathcal{K}(m)$.

2. We now show if F is a constant of motion of $\dot{x} = f(x)$ then for any solution $x(t)$ of $\dot{x} = f(x)$ the linearization of F along this solution is a constant of motion of the variational equation (7.26), associated to $x(t)$. Since the linearization of F along $x(t)$ is

$$\sum_{i=1}^{n} \frac{\partial F}{\partial x_i}(x(t))\xi_i(t)$$

this means that we claim that the latter is independent of t, when F is any constant of motion of $\dot{x} = f(x)$. Indeed

$$\begin{aligned}
&\left(\sum_{i=1}^{n} \frac{\partial F}{\partial x_i}(x(t))\xi_i(t)\right)^{\cdot} \\
&= \sum_{i,j=1}^{n} \left(\frac{\partial^2 F}{\partial x_i \partial x_j}(x(t)) f_j(x(t))\xi_i(t) + \frac{\partial F}{\partial x_i}(x(t)) \frac{\partial f_i}{\partial x_j}(x(t))\xi_j(t)\right) \\
&= \sum_{i,j=1}^{n} \xi_i(t) \left(\frac{\partial^2 F}{\partial x_i \partial x_j}(x(t)) f_j(x(t)) + \frac{\partial F}{\partial x_j}(x(t)) \frac{\partial f_j}{\partial x_i}(x(t))\right) \\
&= \sum_{i=1}^{n} \xi_i(t) \frac{\partial}{\partial x_i} \left(\sum_{j=1}^{n} \frac{\partial F}{\partial x_j} f_j\right)(x(t)) \\
&= 0,
\end{aligned}$$

where the last line was obtained by using (7.20).

3. We now apply the result of step 2 to the solutions $x(t)$ and $\xi(t)$ that were considered in step 1. Namely, we take $x(t)$ as in (7.22), and we choose any η and κ that satisfy (7.24), yielding $\xi(t)$, as given by (7.25). Let F be any weight homogeneous constant of motion of $\dot{x} = f(x)$ of weight k. In view of step 2 we may conclude that

$$\sum_{i=1}^{n} \frac{\partial F}{\partial x_i}(x(t))\eta_i(t) t^{\kappa - \nu_i} \tag{7.28}$$

is independent of t. Since $\varpi(\partial F/\partial x_i) = \varpi(F) - \nu_i = k - \nu_i$ it follows that

$$\sum_{i=1}^{n} \frac{\partial F}{\partial x_i}(m)\eta_i(t) t^{\kappa - k} \tag{7.29}$$

is independent of t.

4. We specialize step 3 further by picking a particular initial condition at $t = 1$ (which in (7.27) corresponds to $s = 0$) for $\eta(t)$. Let ρ be any eigenvalue of $\mathcal{K}(m)$, different from $\varpi(F) = k$ and let η_0 be any eigenvector with eigenvalue ρ, i.e., $\eta_0 \in E_m^{(\rho)} \setminus \{0\}$.

The solution $\eta(s)$ of (7.27) for which $\eta(0) = \eta_0$ is by the spectral theorem a polynomial in s, hence the solution $\eta(t)$ of (7.24), which equals η_0 for $t = 1$, is a polynomial in $\ln t$. So, with this choice of solution, each $\eta_i(t)$ in (7.28) is a polynomial in $\ln t$ and (7.29) is independent of t. This simply means that

$$\langle \mathrm{d}F(m), \eta(t)\rangle = \sum_{i=1}^{n} \frac{\partial F}{\partial x_i}(m)\eta_i(t) = 0;$$

taking $t = 1$ we have that

$$\langle \mathrm{d}F(m), \eta_0\rangle = 0,$$

which yields $\mathrm{d}F(m) \in \mathrm{Ann}(E_m^{(\rho)})$. Since $\rho \neq \varpi(F) = k$, but is otherwise arbitrary, it follows that $\mathrm{d}F(m) \in \bigcap_{l \neq k} \mathrm{Ann}\left(E_m^{(l)}\right)$. Moreover, since $F \in \mathcal{H}^{(k)}$ is arbitrary, we have that

$$\mathrm{d}\mathcal{H}_m^{(k)} \subseteq \bigcap_{l \neq k} \mathrm{Ann}\left(E_m^{(l)}\right).$$

This finishes the proof of Theorem 7.30. □

Example 7.31. In the case of our example, the multiplicity of 1, and of 2, in the characteristic polynomial of $\mathcal{K}(m_1)$ is equal to two. This is strictly larger than the dimensions of $\mathrm{d}\mathcal{H}_{m_1}^{(1)}$ and of $\mathrm{d}\mathcal{H}_{m_1}^{(2)}$, which are both equal to one. However, the multiplicity of 1, 2 and 5 in the characteristic polynomial of $\mathcal{K}(m_1')$ is 1, which coincides with the dimensions of $\mathrm{d}\mathcal{H}_{m_1'}^{(1)}$, $\mathrm{d}\mathcal{H}_{m_1'}^{(2)}$ and $\mathrm{d}\mathcal{H}_{m_1'}^{(5)}$.

Proposition 7.32. *Let $\mathcal{V} : \dot{x} = f(x)$ be a weight homogeneous vector field and let $\mathcal{I}$ denote its indicial locus. Let $m(t)$ denote the solution to $\mathcal{V}$ that corresponds to a point $m \in \mathcal{I}$, as in Proposition 7.14. If all solutions to the variational equation corresponding to $m(t)$ are single-valued then $\mathcal{K}(m)$ is diagonalizable and all eigenvalues of $\mathcal{K}(m)$ are integers. In particular, if the weight homogeneous vector field $\mathcal{V}$ is one of the integrable vector fields of an a.c.i. system then for any $m \in \mathcal{I}$, $\mathcal{K}(m)$ is diagonalizable and all its eigenvalues are integers.*

Proof. Let us reconsider the variational equation (7.23), associated to $m(t)$. According to (7.24) and (7.25), this equation admits as solution the function $\xi(t)$, defined by $\xi_i(t) := \eta_i(t)t^{-\nu_i}$ $(i = 1, \dots, n)$, where $\eta(t)$ is any solution to

$$t\dot{\eta}(t) = \mathcal{K}(m)\eta(t). \tag{7.30}$$

To analyze the latter equation, one chooses a basis in which $\mathcal{K}(m)$ takes the Jordan form. For a block of size d, with λ on the diagonal and all ones one above the diagonal and, say, corresponding to indices $i = 1, \dots, d$ the general solution to (7.30) is given by

$$\eta_1(t) = t^\lambda P(\log(t))$$

where P is an arbitrary polynomial of degree $d-1$, and $\eta_2(t), \ldots, \eta_d(t)$ follow from $\eta_{i+1}(t) = t\dot{\eta}_i(t)$, valid for $i = 1, \ldots, d-1$. Thus, $\xi(t)$ is single-valued iff each $\eta_i(t)$ is single-valued ($i = 1, \ldots, d$) iff $d = 1$ and each λ is an integer. This proves the first part of the proposition. In the case of an a.c.i. system Theorem 6.18 implies that every solution to the variational equations are single-valued, so the first part of the proposition applies to that case. □

We finish this section with another useful theorem that allows one to express some of the free parameters of any weight homogeneous balance of $\mathcal{V}$, directly in terms of the constants of motion of $\mathcal{V}$. We first define these free parameters.

Definition 7.33. Suppose that $a_1, \ldots, a_\beta$ are free parameters that first appear in a weight homogeneous balance $x(t)$ at level k and suppose that there exist constants of motion $F_1, \ldots, F_\beta$, of order k, such that

$$F_i(x_1(t), \ldots, x_n(t)) = a_i, \qquad i = 1, \ldots, \beta.$$

Then these parameters are called *trivial parameters*, while all other free parameters are called *effective parameters*.

Proposition 7.34. *Let $\dot{x} = f(x)$ be a polynomial vector field on $\mathbf{C}^n$ which is weight homogeneous, and let $k \geqslant 1$. Let*

$$x_i(t) = \frac{1}{t^{\nu_i}}\left(x_i^{(0)} + x_i^{(1)}t + \cdots x_i^{(k-1)}t^{k-1}\right), \qquad i = 1, \ldots, n, \tag{7.31}$$

be the first k terms of a weight homogeneous balance, as constructed in Proposition 7.17. For m a generic value of $x^{(0)}$, assume that k is an eigenvalue of $\mathcal{K}(m)$, that the restriction of $k\,\mathrm{Id}_n - \mathcal{K}(m)$ to $E_m^{(k)}$ is diagonalizable and that $R^{(k)}$ is identically in the range of $k\,\mathrm{Id}_n - \mathcal{K}(m)$ (see (7.9)). Then among the $\dim E_m^{(k)}$ free parameters that enter at step k, there are precisely $\dim \mathrm{d}\mathcal{H}_m^{(k)}$ of them that can be chosen as trivial parameters.

Proof. Given (7.31), let $U(t) := \sum_{j=1}^k x^{(j)}t^j$, where $x^{(k)}$ is for the moment arbitrary, and substitute $x_i(t) = t^{-\nu_i}(m_i + U_i(t))$ into a constant of motion F for which $\varpi(F) = k$. We have by weight homogeneity of F that

$$F(x(t)) - t^{-k}F(m + U(t))$$

where the latter can be expanded by Taylor's Theorem as

$$t^{-k}\left[F(m) + \sum_{j=1}^n \frac{\partial F}{\partial x_j}(m)U_j(t) + \frac{1}{2}\sum_{j,k=1}^n \frac{\partial^2 F}{\partial x_j \partial x_k}(m)U_j(t)U_k(t) + \cdots\right].$$

Since $F(x(t))$ is independent of t, all coefficients of t^i with $i < 0$ vanish, so that

$$F(x(t)) = \sum_{i=j}^{n} \frac{\partial F}{\partial x_j}(m) x_j^{(k)} + \Delta^{(k)} + O(t),$$

where $\Delta^{(k)}$ is a polynomial which contains only terms that involve $x^{(l)}$, with $0 \leqslant l < k$. The terms in $O(t)$ will disappear when $x(t)$ is continued to a full balance, so they are irrelevant and we write

$$F(x(t)) \sim \sum_{i=1}^{n} \frac{\partial F}{\partial x_j}(m) x_j^{(k)} + \Delta^{(k)}, \tag{7.32}$$

with the understanding that $\sim$ is an equality when $x(t)$ is completed into a balance. In view of Theorem 7.30 we have that

$$\mathrm{d}\mathcal{H}_m^{(k)} \subseteq \bigcap_{l \neq k} \mathrm{Ann}\left(E_m^{(l)}\right) \cong \left(E_m^{(k)}\right)^*,$$

so we can pick $F_1 \ldots, F_\beta \in \mathcal{H}^{(k)}$, where $\beta := \dim \mathrm{d}\mathcal{H}_m^{(k)}$, as well as a basis $v^1, \ldots, v^\gamma$ of $E_m^{(k)}$, where $\gamma := \dim E_m^{(k)}$ such that

$$\sum_{l=1}^{n} \frac{\partial F_i}{\partial x_l}(m) v_l^j = \delta_{ij}, \qquad 1 \leqslant i, j \leqslant \beta. \tag{7.33}$$

Let $u^{(k)}$ be a particular solution to the k-th step equation (7.9), so that the general solution to (7.9) is given by

$$x^{(k)} = u^{(k)} + \sum_{i=1}^{\gamma} \alpha_i v^i, \tag{7.34}$$

where the α_i are arbitrary. We keep the parameters $\alpha_{\beta+1}, \ldots \alpha_\gamma$ free and we show how the other ones $(\alpha_1, \ldots, \alpha_\beta)$ can be written linearly in terms of free parameters $a_1, \ldots a_\beta$ such that $F_i(x(t)) = a_i$ for $1 \leqslant i \leqslant \beta$.

To do this, consider (7.32) for the constant of motion F_i (which has weight k), writing $\Delta_i^{(k)}$ for $\Delta^{(k)}$ and substitute (7.34) for $x^{(k)}$.

$$\begin{aligned} F_i(x(t)) &\sim \sum_{j=1}^{n} \frac{\partial F_i}{\partial x_j}(m) x_j^{(k)} + \Delta_i^{(k)} \\ &= \sum_{j=1}^{n} \frac{\partial F_i}{\partial x_j}(m) u_j^{(k)} + \sum_{j=1}^{n} \sum_{l=1}^{\gamma} \frac{\partial F_i}{\partial x_j}(m) \alpha_l v_j^l + \Delta_i^{(k)} \\ &= \sum_{j=1}^{n} \frac{\partial F_i}{\partial x_j}(m) u_j^{(k)} + \alpha_i + \sum_{j=1}^{n} \sum_{l=\beta+1}^{\gamma} \frac{\partial F_i}{\partial x_j}(m) \alpha_l v_j^l + \Delta_i^{(k)}, \end{aligned}$$

where we used (7.33) to obtain the last line. Notice that α_i appears linearly in $F_i(x(t))$ and that no other element of $\alpha_1, \dots, \alpha_\beta$ appears in it. Therefore we can make $F_i(x(t)) \sim a_i$ (which means that $F_i(x(t)) = a_i$ when the first terms for $x(t)$ are continued to a formal series) by replacing α_i by a_i plus previous data and a linear combination of $\alpha_{\beta+1}, \dots, \alpha_\gamma$ involving previous data. Finally, no other of the free parameters that show up in the k-th step can be a trivial parameter. Indeed, since for any other constant of motion F of weight k the vector $\mathrm{d}F(m)$ is a linear combination of the vectors $\mathrm{d}F_i(m)$, with $i = 1, \dots, \beta$, it suffices to see what happens to a constant of motion F of weight k for which $\mathrm{d}F(m) = 0$. In this case we have according to (7.32) that $F(x(t)) \sim \Delta^{(k)}$, so its value is independent of $x^{(k)}$ and cannot be equal to any of the free parameters that first appear at level k. □

Remark 7.35. Suppose that we have s independent constants of motion $F_1, \dots, F_s$ of weighted degrees $k_1 \leqslant k_2 \leqslant \cdots \leqslant k_s$. Suppose that $x(t; m)$ is a (lower) balance that depends on s free parameters, which enter precisely at the steps $k_1, \dots, k_s$ (including multiplicities). If for generic $\mathbf{c} = (c_1, \dots, c_s)$ the affine variety defined by $F_i(x(t; m)) = c_i$ is non-empty then it consists of a single point and all free parameters are trivial parameters. To show this, let us denote $k := k_1$ and let us assume that $k_1 = k_2 = \cdots = k_l \neq k_{l+1}$. According to (7.32)

$$\begin{pmatrix} \frac{\partial F_1}{\partial x_1}(m) & \cdots & \frac{\partial F_1}{\partial x_n}(m) \\ \vdots & & \vdots \\ \frac{\partial F_l}{\partial x_1}(m) & \cdots & \frac{\partial F_l}{\partial x_n}(m) \end{pmatrix} \begin{pmatrix} x_1^{(k)} \\ \vdots \\ x_l^{(k)} \end{pmatrix} + \begin{pmatrix} \Delta_1^{(k)} \\ \vdots \\ \Delta_l^{(k)} \end{pmatrix} = \begin{pmatrix} c_1 \\ \vdots \\ c_l \end{pmatrix}, \tag{7.35}$$

where $\Delta^{(k)}$ is constant. Since this linear system has a solution for $x^{(k)}$ when $\mathbf{c}$ is generic, the square matrix in (7.35) is invertible and the system has a (unique) solution for all $\mathbf{c}$. This allows us to express the coefficients $x_1^{(k)}, \cdots, x_l^{(k)}$ uniquely in terms of the constants $c_1, \dots, c_l$, so that each of the free parameters that enter at step k are trivial. One concludes by recursion.

Example 7.36. Still in the case of our example, the principal balance $x(t; m_1)$ depends on 4 free parameters, two of which are trivial parameters. If we substitute the balance (7.17) in $F_1 = c_1$ and in $F_2 = c_2$ then we find $a = (c_1 - d)/2$ and $c = b - c_2 - d(c_1 - d)$. If we substitute these values for a and c in (7.17) then $x(t; m_1)$ depends now on the free parameters c_1, c_2, b and d and it has the property that $F_1(x(t; m_1)) = c_1$ and $F_2(x(t; m_1)) = c_2$. In this way we have replaced c and d by the trivial parameters c_1 and c_2. The same can be done with all parameters a, b and c in the lower balances we find that the lower balances just depend on the constants of motion. Notice that this is a particular case of Remark 7.35 because the invariants have degree 1, 2 and 5 which corresponds precisely to the Kowalevski exponents of $\mathcal{K}(m_1')$.

7.5 Weight Homogeneous A.c.i. Systems

Before introducing the notion of a weight homogeneous a.c.i. system we define what it means for a Poisson structure to be weight homogeneous. Weight homogeneous Poisson structures will turn weight homogeneous Hamiltonians into weight homogeneous vector fields (of some weight).

Definition 7.37. Let $\nu = (\nu_1, \ldots, \nu_n)$ be a weight vector and let $\{\cdot\,,\cdot\}$ be a polynomial Poisson structure on $\mathbf{C}^n$. We will say that $\{\cdot\,,\cdot\}$ is a *weight homogeneous Poisson structure* of weight $\varpi(\{\cdot\,,\cdot\}) = P$ (with respect to ν) if

$$\varpi\left(\{x_i, x_j\}\right) = \nu_i + \nu_j - 2 + P.$$

for all $1 \leqslant i, j \leqslant n$.

Our definition of the weight P of a weight homogeneous Poisson structure is such that for a homogeneous Poisson structure (all ν_i equal to 1) it is equal to the (common) degree of the polynomials $\{x_i, x_j\}$, in particular it is always non-negative. Thus, if all weight are equal to 1, a (non-trivial) constant Poisson structure on $\mathbf{C}^n$ (see Example 3.3) has degree zero, a Lie-Poisson structure (see Section 3.5) has degree 1 and so on.

Remark 7.38. For general weight homogeneous Poisson structures the weight may be negative. Consider indeed on $\mathbf{C}^2$ with coordinates z_1, z_2 and with weight vector $\nu = (3, 2)$ the constant Poisson structure, given by $\{z_1, z_2\} = 1$. Its weight is -3.

Example 7.39. Consider on $\mathbf{C}^5$ the Poisson structure that is defined by

$$\{x_i, x_j\} = x_i x_j (\delta_{i,j+1} - \delta_{j,i+1}),$$

where $1 \leqslant i, j \leqslant 5$. Its Poisson matrix in terms of $x_1, \ldots, x_5$ is explicitly given by

$$\begin{pmatrix} 0 & -x_1x_2 & 0 & 0 & x_5x_1 \\ x_1x_2 & 0 & -x_2x_3 & 0 & 0 \\ 0 & x_2x_3 & 0 & -x_3x_4 & 0 \\ 0 & 0 & x_3x_4 & 0 & -x_4x_5 \\ -x_5x_1 & 0 & 0 & x_4x_5 & 0 \end{pmatrix}$$

It is obvious that $\{\cdot\,,\cdot\}$ is (weight) homogeneous of weight 2 (recall that in this example all weights are equal to 1). The rank of $\{\cdot\,,\cdot\}$ is 4. More precisely, the rank is 0 on the five 2-planes, which are defined by three non-consecutive x_i being zero (i.e., the plane $x_1 = x_2 = x_4 = 0$ and its image, under all powers of σ); the rank is 2 on the ten 3-planes which are given by $x_i = x_j = 0$, where $1 \leqslant i < j \leqslant 5$. On the remaining open subset the rank is 4.

It is easy to see that this yields a Poisson bracket with respect to which the periodic 5-particle Kac-van Moerbeke lattice is Liouville integrable. In fact we recover the vector field $\mathcal{V}_1$, given by (7.3), from $\{\cdot\,,\cdot\}$ by taking $F_1 = x_1 + x_2 + \cdots + x_5$ as Hamiltonian, while we recover the vector field (7.6), which we will denote by $\mathcal{V}_2$, by taking $F_2 = x_1x_3 + x_2x_4 + x_3x_5 + x_4x_1 + x_5x_2$ as Hamiltonian. Also, $F_3 = x_1x_2x_3x_4x_5$ is a Casimir. Since F_1, F_2 and F_3 are independent, the Kac-van Moerbeke vector field (7.3) is indeed an integrable vector field.

Notice that the image under the momentum map of the five two-planes where the rank drops to two is contained in the Zariski closed subset $c_3 = 0$. Proposition 7.56 implies that the vector fields $\mathcal{V}_1$ and $\mathcal{V}_2$ are independent at each point of the generic fiber of $\mathbf{F} = (F_1, F_2, F_3)$.

The Poisson bracket of two weight homogeneous polynomials is a weight homogeneous polynomial, when the bracket is weight homogeneous. This is shown in the following lemma.

Lemma 7.40. *Let $\{\cdot\,,\cdot\}$ be a weight homogeneous Poisson bracket of weight P on $\mathbf{C}^n$. If f and g are weight homogeneous polynomials then $\{f,g\}$ is weight homogeneous and its weight is given by*

$$\varpi\left(\{f,g\}\right) = \varpi(f) + \varpi(g) - 2 + P.$$

Proof. For any i,j such that $1 \leqslant i,j \leqslant n$ we have that

$$\varpi\left(\frac{\partial f}{\partial x_i}\frac{\partial g}{\partial x_j}\left\{x_i,x_j\right\}\right) = \varpi(f) + \varpi(g) - 2 + P.$$

Since $\{f,g\}$ is a sum of such terms the result follows. □

As a corollary, we express the condition for a weight homogeneous polynomial that its Hamiltonian vector field (with respect to a weight homogeneous Poisson structure) is weight homogeneous.

Lemma 7.41. *Let $\{\cdot\,,\cdot\}$ be a weight homogeneous Poisson structure on $\mathbf{C}^n$ and let F be a weight homogeneous polynomial. The Hamiltonian vector field $\mathcal{X}_F$ is weight homogeneous (of weight 1) if and only if*

$$\varpi(\{\cdot\,,\cdot\}) + \varpi(F) = 3. \tag{7.36}$$

Proof. For $i = 1, \ldots, n$ we have in view of the previous lemma that

$$\varpi(\dot{x}_i) = \varpi\left(\{x_i,F\}\right) = \nu_i + \varpi(F) - 2 + \varpi(\{\cdot\,,\cdot\}).$$

Thus, $\varpi(\dot{x}_i) = \nu_i + 1$ if and only if (7.36) holds. □

Thus, in order for a Hamiltonian vector field to be weight homogeneous (of weight 1), the degree of the Poisson structure (which may be negative) must be at most two; if the degree is two then the Hamiltonian must be linear, for a Lie-Poisson structure the Hamiltonian must be quadratic and so on. In view of Lemma 7.41, we define a weight homogeneous a.c.i. system as follows.

Definition 7.42. An a.c.i. system $(\mathbf{C}^n, \{\cdot\,,\cdot\}, (F_1, \dots F_s))$ is called a *weight homogeneous a.c.i. system* (with respect to ν) if $\{\cdot\,,\cdot\}$ and all the polynomials $F_1, \dots F_s$ are weight homogeneous with

(1) For at least one of the F_i, say for F_1, we have that $\varpi(\{\cdot\,,\cdot\}) + \varpi(F_1) = 3$, so that $\mathcal{X}_{F_1}$ is a weight homogeneous vector field;
(2) All principal balances of $\mathcal{X}_{F_1}$ are weight homogeneous.

Remark 7.43. It is not clear if condition (2) in Definition 7.42 follows from the other assumptions. In any case, a counterexample is unknown.

Example 7.44. We are not yet ready to show that the periodic 5-particle Kac-van Moerbeke lattice is a weight homogeneous a.c.i. system, although we have already many elements: we know already that it is Liouville integrable, we have a weight homogeneous Poisson structure $\{\cdot\,,\cdot\}$ and the vector field $\mathcal{X}_{F_1}$ is weight homogeneous. Suppose that

$$x_i(t) = \frac{1}{t^r} \sum_{j=0}^{\infty} x_i^{(j)} t^j, \qquad i = 1, \dots, 5,$$

is a Laurent solution to (7.3), with $r > 1$ and $x_i^{(0)} \neq 0$ for some i. If we denote the pole order of $x_i(t)$ by r_i (so that in particular $r = \max_i r_i$) then we have for $i = 1, \dots, 5$ and $s \in \mathbf{N}$ that

$$\operatorname{Res}_{t=0} \frac{\dot{x}_i(t)}{x_i(t)} t^s = \begin{cases} -r_i, & s = 0, \\ 0, & s > 0. \end{cases}$$

On the other hand, if we use (7.3) then we find

$$\operatorname{Res}_{t=0} \frac{\dot{x}_i(t)}{x_i(t)} t^s = \operatorname{Res}_{t=0} \left(x_{i-1}(t) - x_{i+1}(t)\right) t^s = x_{i-1}^{(r-s-1)} - x_{i+1}^{(r-s-1)}.$$

We conclude that

$$x_{i-1}^{(k)} - x_{i+1}^{(k)} = \begin{cases} -r_i, & k = r-1 \\ 0, & 0 \leq k \leq r-2. \end{cases} \tag{7.37}$$

If, say, $x_1^{(0)} \neq 0$ then (7.37), with $k = 0$, implies that $x_1^{(0)} = x_3^{(0)} = \dots$, so that all $x_i^{(0)} \neq 0$ and $r_i = r$ for $i = 1, \dots, 5$. But then (7.37), with $k = r - 1$, yields

$$x_{i-1}^{(r-1)} - x_{i+1}^{(r-1)} = -r, \qquad (i = 1, \dots, 5),$$

which is impossible, since summing these equations up yields $0 = 5r \neq 0$. This shows that all Laurent solutions to (7.3) are weight homogeneous.

7.6 Algorithms

In this section we present a few algorithms which provide for a weight homogeneous vector field on $\mathbf{C}^n$ concrete information on its algebraic complete integrability, the nature of the fibers of the momentum map, the divisor at infinity, and so on. They will be most effective in the case of two-dimensional systems (i.e., the fibers of the momentum map are complex surfaces), in which the algorithms lead to a powerful tool for proving that a given two-dimensional system is algebraic completely integrable (see Section 7.7).

All techniques given here will be illustrated on the example of the periodic 5-particle Kac-van Moerbeke lattice, that was introduced in Example 7.5, and for which a lot of data has already be computed in the previous sections. These techniques will be essential when we get to the main examples of this book.

7.6.1 The Indicial Locus $\mathcal{I}$ and the Kowalevski Matrix $\mathcal{K}$

We assume that $\mathcal{V}$ is the polynomial vector field $\dot{x} = f(x)$ on $\mathbf{C}^n$, which is weight homogeneous with respect to $\nu = (\nu_1, \dots, \nu_n)$. The indicial locus $\mathcal{I}$, which was defined in Proposition 7.6, is an algebraic subset of $\mathbf{C}^n$ which contains the leading terms of all possible weight homogeneous balances to $\mathcal{V}$. It is given by n non-linear algebraic equations

$$\nu_i x_i^{(0)} + f_i\left(x_1^{(0)}, \dots, x_n^{(0)}\right) = 0, \qquad i = 1, \dots, n,$$

where $f_1, \dots, f_n$ are the components of f. The indicial locus breaks in general up in several irreducible components of varying dimension. These irreducible components have to be determined explicitly, together with their dimension. Also, the intersection and incidence pattern of these components, as well as their singularities should be determined. From a certain point of view this step is the most difficult one; indeed no feasible algorithm is known for decomposing an algebraic set (given by explicit equations) into its irreducible components. This step should be done by hand and with a lot of care, since missing even one point in the indicial locus may mean that one is going to miss a balance, indispensable for revealing the algebraic geometry of the system and for eventually proving that it is a.c.i.

For $m \in \mathcal{I}$ the Kowalevski matrix $\mathcal{K}(m)$ is easily computed from its definition,

$$\mathcal{K}(m)_{ij} = \frac{\partial f_i}{\partial x_j}(m) + \nu_i \delta_{ij}.$$

In fact, it is natural to consider the Kowalevski matrix for a whole irreducible component $\mathcal{I}'$ of $\mathcal{I}$. Using the explicit equations for $\mathcal{I}'$, the characteristic polynomial $\chi(k; m)$ of $\mathcal{K}(m)$, with $m \in \mathcal{I}'$, follows from a direct computation. Recall that -1 is always a root of $\chi(k; m)$.

As we have seen in Theorem 7.22, $\mathcal{I}'$ can only lead to a principal balance if

(1) $\chi(k;m)$ is independent of $m \in \mathcal{I}'$;
(2) $\chi(k;m)$ has $n-1$ non-negative integer roots;
(3) $\mathcal{K}(m)$ is diagonalizable for all $m \in \mathcal{I}'$;
(4) $\mathcal{I}$ is non-singular at all points of $\mathcal{I}'$.

Thus, one should for every irreducible component $\mathcal{I}'$ of $\mathcal{I}$ check these conditions, i.e., conditions (2)–(4), since (1) is a consequence of (2) (because -1 is always an eigenvalue). Also, one should make a list of the eigenvalues (with their multiplicities) of the corresponding Kowalevski matrix. If one of the above conditions is not satisfied it does not mean that the system is not a.c.i., because this component may lead to a lower balance (see e.g. Example 7.45 below). However, if for none of the components $\mathcal{I}'$ of $\mathcal{I}$ the above four conditions are simultaneously satisfied, then $\mathcal{V}$ is not one of the integrable vector fields of a weight homogeneous a.c.i. system. Moreover, if for one of the solutions to the indicial equation the Kowalevski matrix is not diagonalizable with integer roots, then, by Proposition 7.32, $\mathcal{V}$ cannot be one of the integrable vector fields of an a.c.i. system. As we will see, these conditions are strong enough to single out the golden needles hidden in the haystack of all weight homogeneous systems of a certain type.

Example 7.45. In the case of the periodic 5-particle Kac-van Moerbeke lattice we have already determined in Example 7.10 the indicial locus and the Kowalevski matrix. We have shown that the indicial locus consists of ten points $m_1, m_1', \dots, m_5, m_5'$, where $m_1 = (-1,1,0,0,0)$ and $m_1' = (-2,1,-1,2,0)$, the other points being determined by using the automorphism σ. It is readily computed from (7.15) that $\chi(k;m_1) = (k+1)(k-1)^2(k-2)^2$ and that $\chi(k;m_1') = (k+2)(k+1)(k-1)(k-2)(k-5)$. Therefore, the points m_i satisfy the above four conditions (see Example 7.24 for condition 3), but the points m_i' don't.

7.6.2 The Principal Balances (for all Vector Fields)

Suppose that $\mathcal{I}'$ is one of the irreducible components of $\mathcal{I}$ for which the above four conditions hold. We first show how to compute the principal balance for $\mathcal{V}$ that corresponds to $\mathcal{I}'$, if it exists, and then we explain how the result is used to compute the principal balance for any other vector field on $\mathbf{C}^n$ that has "good" properties at infinity (the latter vector field needs not be weight homogeneous). Let

$$\chi(k;m) = (k+1)\prod_{i=1}^{p}(k-k_i)^{\mu_i}, \qquad m \in \mathcal{I}',$$

where for $i = 1, \dots, p$ the eigenvalue k_i is a non-negative integer and μ_i denotes its multiplicity (we assume these eigenvalues to be ordered, so that k_p is the largest eigenvalue of $\chi(k;m)$).

We explain how to compute the first few terms of the series that starts at $\mathcal{I}'$. For $l = 1, 2, \ldots$, we consider the Laurent polynomials

$$x_i(t; \mathcal{I}') = \frac{1}{t^{\nu_i}} \sum_{k=0}^{l} x_i^{(k)} t^k, \qquad i = 1, \ldots, n, \tag{7.38}$$

where $x^{(0)}$ is an arbitrary element of $\mathcal{I}'$, and where the other terms $x^{(l)}$ will be determined recursively by substituting these Laurent polynomials (up to the term in $x^{(l)}$) in the differential equations $\dot{x} = f(x)$, and equating coefficients of t in the resulting equations. Notice that since $x^{(0)} \in \mathcal{I}' \subseteq \mathcal{I}$ the coefficients of $t^{-\nu_i - 1}$ in the i-th equation cancel out, so the first non-trivial equations come from the coefficients of $t^{-\nu_i}$ in the i-th equation, where $i = 1, \ldots, n$; these equations already appear when taking $l = 1$ and they are linear in $x^{(1)}$. Their solution should depend on μ_1 free parameters if $k_1 = 1$; otherwise they are uniquely solvable for $x^{(1)}$ in terms of $x^{(0)}$. This step is called step 1. Having successfully completed step 1 we move on to step 2 where we consider now the coefficients of $t^{-\nu_i+1}$ in $\dot{x} = f(x)$, which will determine $x^{(2)}$, then we move to step 3 and so on. Notice that we could at each step also use the general formulas (7.9) and (7.14) to compute the next term in the series, but in practice it is easier to compute them by substituting the above Laurent polynomials (7.38) in $\dot{x} = f(x)$ and solving linearly for the components $x_i^{(l)}$ of $x^{(l)}$.

Remark 7.46. If we wish we can choose at the step k_i each trivial parameter equal to the value of one of the constants of motion of weight k_i, evaluated at these Laurent polynomials, as explained in Theorem 7.34. In general this does not simplify the balances, so that we advice to do it after enough terms of the series have been computed (also not all constants of motion may be known, but they will be found in a further step).

The process of computing the first terms of the balance should be at least continued until the step k_p, yielding at least the first $k_p + 1$ terms of $x(t; \mathcal{I}')$. In order to be sure that the terms that we have computed extend to a principal balance it is necessary and sufficient to check that the number of free parameters that come in at step k_i is indeed equal to μ_i. We know that if we compute further terms, then the original terms will remain unchanged and that no other free parameters will show up. In addition, we know that the coefficients of $x(t; \mathcal{I}')$ generate the algebra of regular functions on $\mathcal{I}' \times \mathbf{C}^p$, where $p = n - 1 - \dim \mathcal{I}'$. From Theorem 7.25 we know that the series $x(t; \mathcal{I}')$ is convergent for small, non-zero $|t|$.

Remark 7.47. The lower balances that are weight homogeneous are found in the same way since, by definition, they also start at points in the indicial locus. However, in some of the examples that we will treat, the lower balances will not be weight homogeneous.

For certain examples that we will consider, we will need the first few terms of the principal balance for some other vector field on $\mathbf{C}^n$. In general this vector field will not be weight homogeneous, so we cannot use the previous method. In order to justify the method that we will present, let us assume that we have a weight homogeneous a.c.i. system $(\mathbf{C}^n, \{\cdot\,,\cdot\}, \mathbf{F})$, where $\mathbf{F} = (F_1, \ldots, F_s)$ and that the vector field $\mathcal{V}_1 := \mathcal{X}_{F_1}$ is a weight homogeneous vector field. This implies that we can compute each of the principal balances $x(t; \mathcal{D}')$ algorithmically. Let $\mathcal{V}_2 := \mathcal{X}_{F_2}$ denote the vector field for which we wish to compute its principal balances $x_i(t_2; \mathcal{D}')$. According to Proposition 6.14 the pole order of $x_i(t; \mathcal{D}')$ equals the pole order of $x_i(t_2; \mathcal{D}')$, so that we know at which pole order each of the principal balances of $\mathcal{V}_2$ start. However, this is not sufficient for computing the first few terms in an efficient way, because the k-th term of the series cannot be computed from the previous terms, instead the computation of the k-th term will impose constraints on the previous terms, and this in a way that is hard to control in general. Therefore we propose a method that yields the different terms of each balance of $\mathcal{V}_2$ immediately, from the corresponding balance of $\mathcal{V}_1$.

Let us pick one of the Painlevé walls $\mathcal{D}'$ and let us first assume that $f \in \mathcal{F}(\mathbf{C}^n)$ is holomorphic in a neighborhood of a generic point of $\mathcal{D}'$, i.e., $f(x(t; \mathcal{D}'))$ is a Taylor series, where $x(t; \mathcal{D}')$ is a weight homogeneous principal balance of $\mathcal{V}_1$, that has been computed with the usual method. Since $\mathcal{V}_2$ and f are holomorphic in a neighborhood of a generic point of $\mathcal{D}'$, we have by Taylor's Theorem that

$$f(t_2; \mathcal{D}') = \sum_{j=0}^{\infty} \mathcal{V}_2^j[f]\Big|_{\mathcal{D}'} \frac{t_2^j}{j!} \tag{7.39}$$

is the Taylor series of f with respect to $\mathcal{V}_2$, starting at $\mathcal{D}'$. The computation of the coefficient of t_2^j is done as follows: first compute the repeated application of the vector field $\mathcal{V}_2$ on f. Since we know $\mathcal{V}_2 = \mathcal{X}_{F_2}$ explicitly in terms of the coordinates $x_1, \ldots, x_n$ of phase space this gives an explicit formula for $f^{(j)} := \mathcal{V}_2^j[f]$ as a rational function on phase space $\mathbf{C}^n$. Since $\mathcal{V}_2$ is holomorphic in a neighborhood of a generic point of $\mathcal{D}'$ the same is true for $f^{(j)}$. In order to compute the restriction of $f^{(j)}$ to $\mathcal{D}'$, write the first few terms of $f^{(j)}(t; \mathcal{D}')$, the Taylor series of $f^{(j)}$ with respect to $\mathcal{V}_1$, starting at $\mathcal{D}'$. This is done by substituting the first few terms of the principal balance $x(t; \mathcal{D}')$ of $\mathcal{V}_1$ in $f^{(j)}$. To restrict the series $f^{(j)}(t; \mathcal{D}')$ to $\mathcal{D}'$ it then suffices to substitute $t = 0$ in $f^{(j)}(t; \mathcal{D}')$ (the latter is holomorphic in t (for $|t|$ small) and on a neighborhood of a generic point of $\mathcal{D}'$). This yields the Taylor series $f(t_2; \mathcal{D}')$.

We have assumed that f was holomorphic in a neighborhood of a generic point of $\mathcal{D}'$; if this is not the case, but $1/f$ is holomorphic in a neighborhood of a generic point of $\mathcal{D}'$, then take $1/f$ instead of f, find the series $1/f(t_2, \mathcal{D}')$, as above, and then invert the series to find the Taylor series of f with respect to $\mathcal{V}_2$, starting at $\mathcal{D}'$. In this way, taking for f each of the functions x_i, or its inverse $1/x_i$ we find explicit formulas for the principal balance $x(t_2; \mathcal{D}')$.

In practice we will not know that we are dealing with an a.c.i. system and hence that the vector field $\mathcal{V}_2$ is holomorphic in the neighborhood of a generic point on $\mathcal{D}'$ (the Painlevé wall $\mathcal{D}'$ will not even be defined). But in this case we will actually be in the process of proving (or disproving) the algebraic integrability of the integrable system at hand. We may then perform the above computation for any of the integrable vector fields $\mathcal{X}_{F_i}$, we check if it is still a principal balance, i.e., if it formally solves the differential equations that describe the vector field, and if it depends on the right number of free parameters. If this is not the case then we know that the given integrable system is not algebraic completely integrable and we are done. In the cases that we will consider this will never happen. ; in fact we will consider the restriction of the principal balances (and of the vector fields $\mathcal{V}_i$) to the level sets $\mathbf{F_c}$ of the momentum map, and we will show that explicitly that the restriction of the vector fields $\mathcal{V}_i$ to the generic fiber of the momentum map extends to a holomorphic vector field on a neighborhood of a divisor that will play the rôle of a Painlevé divisor. The above procedure is then still justified and yields a convergent (for small $|t_2| \neq 0$) Laurent series, that is a solution to the differential equation that describes $\mathcal{V}_2$.

The reader has guessed that for computing a reasonable number (say the above $k_p + 1$) of terms of a principal balance one is almost forced to use a (powerful!) computer algebra program. In fact, even for the simplest a.c.i. systems the computation of the balances should be done with (truncated) series, because the product of two Laurent polynomials is very different (in size!) from the product of those two Laurent polynomials, viewed as truncated series. Moreover, during the computation the relations between the generators of $\mathcal{I}'$ have to be used at all times, for example to find a formula for $x^{(k)}$.

Example 7.48. For our example, the principal balances for $\mathcal{V}_1$ have already been given in Example 7.17. We give here the first few terms of $x(t_2; m_1)$, as computed from (7.17), using (7.39) with $f = x_i$ or with $f = x_i^{-1}$. For x_3, x_4 and x_5 the computation is immediate, since these do not have a pole when $x(t; m_1)$ is substituted in them; for x_1 and x_2 one first has to do the computation for the inverted series $1/x_1(t; m_1)$ and $1/x_2(t; m_1)$.

$$
\begin{aligned}
x_1(t_2; m_1) &:= d(1 - (b + c)t_2 + O(t_2^2)),\\
x_2(t_2; m_1) &:= bdt_2(1 + \frac{1}{2}(2ad - b + c)t_2 + O(t_2^2)),\\
x_3(t_2; m_1) &:= -\frac{1}{dt_2}(1 - \frac{1}{2}(2ad - b - c)t_2 + O(t_2^2)),\\
x_4(t_2; m_1) &:= \frac{1}{dt_2}(1 + \frac{1}{2}(2ad + b + c)t_2 + O(t_2^2)),\\
x_5(t_2; m_1) &:= cdt_2(1 - \frac{1}{2}(2ad - b + c)t_2 + O(t_2^2)).
\end{aligned}
$$

Remark 7.49. The above method can be used to compute the principal balances for any linear combination $\sum_{i=1}^{r}\alpha_i\mathcal{V}_i$ of the integrable vector fields $\mathcal{V}_i$. If we consider e.g. in the case of the periodic 5-particle Kac-van Moerbeke lattice, the vector field $\alpha\mathcal{V}_1+\beta\mathcal{V}_2$ then we find the following series for x_5,

$$x_2(u;m_1)=b(\alpha+\beta d)u-b((a-d)\alpha^2-(c+d^2)\alpha\beta-\frac{d}{2}(2ad-b+c)\beta^2)u^2+O(u^3).$$

Notice that, in the general case, if one substitutes $u=1$ in the principal balance that corresponds to $\sum_{i=1}^{r}\alpha_i\mathcal{V}_i$, then one gets the first few terms of the series in $\alpha_1,\ldots,\alpha_r$ of any of the phase variables; the parameters $\alpha_1,\ldots,\alpha_r$ that appear in it are dual to the vector fields $\mathcal{V}_1,\ldots,\mathcal{V}_r$, so that the series give a solution to each of these vector fields. For the above example, if the coordinates that are dual to $\mathcal{V}_1$ and $\mathcal{V}_2$ are denoted by t_1 and t_2, so that $\mathcal{V}_i=\partial/\partial t_i$, for $i=1,2$, then the above series for x_5 is, in terms of t_1 and t_2, given by

$$x_2(t_1,t_2;m_1)=b(t_1+t_2d)-\frac{b}{2}(2(a-d)t_1^2-2(c+d^2)t_1t_2-d(2ad-b+c)t_2^2)+O(t^3),$$

and similarly for $(x_1,x_3^{-1},x_4^{-1},x_5)(t_1,t_2;m_1)$, leading to a description of the meromorphic functions x_i in terms of coordinates on the universal covering space of the $\mathbf{T}_{\mathbf{c}}^r$.

7.6.3 The Constants of Motion

If a basis for the constants of motion is not known one can easily find, for every $k\geqslant 0$, the polynomial constants of motion of weight k (recall that a basis for the constants of motion can always be chosen weight homogeneous). There are in fact two ways of doing this and both have their advantages.

The first one is to write down to most general element H of $\mathcal{F}^{(k)}$, i.e., the most general weight homogeneous polynomial of weight k, and then computing

$$\dot{H}=\sum_{i=1}^{n}\frac{\partial H}{\partial x_i}f_i,$$

and expressing that the resulting polynomial (i.e., each of its coefficients) is identically zero. This gives a collection of linear equations on the $\dim\mathcal{F}^{(k)}$ unknown constants in H (a generating function for $\dim\mathcal{F}^{(k)}$ is given by Proposition 7.1). Thus, all weight homogeneous constants of motion of a given weight are easily calculated by linear algebra. If some of them are already known (for example those which are products of previously found constants of motion) they are easily excluded from the search by taking some of the $\dim\mathcal{F}^{(k)}$ coefficients in H equal to 0.

A second way of finding the constants of motion of weight k uses the (first k terms of the) balances going with all the irreducible components $\mathcal{I}^{(1)},\ldots,\mathcal{I}^{(d)}$ of $\mathcal{I}$ which lead to a principal balance (in particular which satisfy the 4 conditions above (Paragraph 7.6.1)). Take, as before, the most general weight homogeneous polynomial H of weight k. If we express that

$$H(x_1(t;\mathcal{I}^{(i)}),\ldots,x_n(t;\mathcal{I}^{(i)})) = O(t^0), \qquad i=1,\ldots,d,$$

then we get a linear system of algebraic equations in the unknown coefficients, which is easily solved. At this point, a crucial check has to be done if we find some solutions. Namely, we need to check if $\dot{H}=0$ to be sure that H is indeed a constant of motion. If so, we are fine, but if not, what does it mean? Then H would be a holomorphic function whose restriction to the fibers of the momentum map is holomorphic, is not constant, and is bounded at the divisor at infinity of these fibers ... more precisely at the components of the divisor at infinity that we have found. So, assuming that the system is a.c.i., there must be at least one principal balance that is missing, on which H has a pole. Since our algorithm gives all weight homogeneous balances it means that the system, if a.c.i., possesses principal balances which are not weight homogeneous, in particular it is not a weight homogeneous a.c.i. system. At this point no general methods are known to find such principal balances; the reader should convince himself that in the weight homogeneous case one has no idea with what order of poles to start for each function and, what is worse, all terms of the Laurent series that have been computed up to a certain step are most likely to be modified in one of the next steps (we even do not want to mention convergence). On the other hand, no weight homogeneous[3] vector field has ever been encountered with a principal balance that is not weight homogeneous (for the lower balances this however does happen). To conclude, if one is really only interested in finding the constants of motion one may just use the first (quicker, less subtle) method, but in all other cases it is better to use the second one, because one verifies at the same time if no principal balances are missing.

Example 7.50. In the case of the periodic 5-particle Kac-van Moerbeke lattice we have given in Example 7.5 three independent constants of motion and we know from Example 7.39 that there are no other, independent constants of motion. In fact, the following stronger result is true: every polynomial constant of motion must be a polynomial in F_1, F_2 and F_3. For a proposition from which this will follow easily at the end, see [169, Proposition 3.7]. As a corollary, we can use Formula 7.21 to compute the number of linearly independent constants of motion of degree $1,2,\ldots$, namely we get

$$\sum_{i=0}^{\infty}\left(\dim\mathcal{H}^{(k)}\right)t^k = \frac{1}{(1-t)(1-t^2)(1-t^5)} \tag{7.40}$$
$$= 1+t+2t^2+2t^3+3t^4+4t^5+5t^6+6t^7+7t^8+O(t^9).$$

[3] The same applies to almost weight homogeneous vector fields, see Remark 7.9.

7.6.4 The Abstract Painlevé Divisors $\Gamma_{\mathbf{c}}$

In order to motivate our definition of the abstract Painlevé divisors, which are algebraic subsets of the abstract Painlevé building, we first assume that $\mathcal{V}$ is, besides being weight homogeneous, one of the integrable vector fields of a weight homogeneous a.c.i. system $(\mathbf{C}^n, \{\cdot,\cdot\}, \mathbf{F})$, say $\mathcal{V} = \mathcal{X}_{F_1}$, where $\mathbf{F} = (F_1, \dots, F_s)$. Define

$$\lambda := \max\{\varpi(F_i) \mid i = 1, \dots, s\},$$

and suppose that the first $\lambda + 1$ terms (i.e., up to and including step λ) of a principal balance $x(t; \mathcal{D}')$ have been computed, where $\mathcal{D}'$ is a Painlevé wall of $\mathcal{V}$. By weight homogeneity,

$$\begin{aligned} F(x(t;\mathcal{D}')) &= F\left(\cdots, t^{-\nu_j}\sum_{i=0}^{\lambda} x_k^{(i)} t^i + O(t^{\lambda+1}), \cdots\right) \\ &= t^{-\lambda} F\left(\cdots, \sum_{i=0}^{\lambda} x_k^{(i)} t^i, \cdots\right) + O(t), \end{aligned}$$

so that the coefficient of t^0 can be computed by substituting only these first $\lambda + 1$ terms in F; this one coefficient yields $F(x(t; \mathcal{D}'))$ because, since F is a constant of motion, all coefficients in t of $F(x(t; \mathcal{D}'))$ vanish, except for the coefficient of t^0.

If we perform this substitution for each of $F_1, \dots, F_s$ then we find s algebraic relations $F_i(x(t; \mathcal{D}')) = c_i$ between the $n - 1$ free parameters and the s values c_i of the constants of motion F_i. Their common solution gives (if $\mathbf{c} = (c_1, \dots, c_r)$ is chosen generically) an affine variety of possible initial conditions for a solution of $\mathcal{V}$ that lives for non-zero small t in the fiber $\mathbf{F}_{\mathbf{c}}$ of the momentum map. Therefore, this affine variety is a Zariski open subset of one of the irreducible components of the Painlevé divisor $\mathcal{D}_{\mathbf{c}}$. The explicit equations of the irreducible components of $\mathcal{D}_{\mathbf{c}}$ are very useful: for example, if the fibers $\mathbf{F}_{\mathbf{c}}$ of the momentum map are two-dimensional, then this divisor will be an algebraic curve and, as we have seen in Chapter 5, much of the geometry of an Abelian variety is encoded in the hyperplane sections that it contains. If we repeat this for each Painlevé wall then we will find all irreducible components of $\mathcal{D}_{\mathbf{c}}$. We will see in the next paragraph how to determine explicitly how the different irreducible components of the Painlevé divisors $\mathcal{D}_{\mathbf{c}}$ intersect (for generic $\mathbf{c}$) and how they live in the Abelian variety; but by now we are only able to compute explicit equations for the irreducible components of the Painlevé divisors.

Suppose now that we do not know whether or not we are dealing with an a.c.i. system. So all that we have at our disposal is a weight homogeneous vector field on $\mathbf{C}^n$, for which we have determined all weight homogeneous principal balances.

Let $x(t;\Gamma')$ be such a balance and suppose that the first $\lambda+1$ terms of it have been computed, as in the case of an a.c.i. system. If we substitute these series in the equations $F_i = c_i$, where the constants c_i are fixed, but generic, then we find again s algebraic relations between the $n-1$ free parameters and the s values c_i of the constants of motion F_i. This gives us for generic $\mathbf{c} = (c_1, \dots, c_s)$ an algebraic set in the abstract Painlevé building, which is similar to the Painlevé divisor at $\mathbf{c}$, so we call it the *abstract Painlevé divisor* at $\mathbf{c}$; in the case of an a.c.i. system each irreducible component of the abstract Painlevé divisor at $\mathbf{c}$ is isomorphic to a Zariski open subset of an irreducible component of the Painlevé divisor at $\mathbf{c}$. We denote the abstract Painlevé divisor at $\mathbf{c}$ by $\Gamma_{\mathbf{c}}$ and the irreducible component of $\Gamma_{\mathbf{c}}$ that corresponds to a Painlevé wall Γ' is denoted by $\Gamma'_{\mathbf{c}}$. The restriction of the principal balance $x(t;\Gamma')$ to $\Gamma_{\mathbf{c}}$, which is obtained by imposing the algebraic relations on the free parameters, that define $\Gamma_{\mathbf{c}}$, is denoted by $x(t;\Gamma'_{\mathbf{c}})$.

Example 7.51. In the case of the periodic Kac-van Moerbeke system we find an equation for the abstract Painlevé divisor at $\mathbf{c}$ by substituting (7.17) in the equations $F_i = c_i$, $i = 1, \dots, 3$, to wit

$$\begin{aligned} 2a + d &= c_1, \\ b - c + 2ad &= c_2, \\ bcd &= -c_3. \end{aligned}$$

This yields an affine curve $\Gamma_{\mathbf{c}}^{(1)}$, which is contained in the abstract Painlevé building $\mathbf{C}^4$. By linearly eliminating a and c from these equations we find that $\Gamma_{\mathbf{c}}^{(1)}$ is isomorphic to the plane curve

$$bd(b - c_2 + d(c_1 - d)) + c_3 = 0,$$

which is non-singular for generic values of $c_1, \dots, c_3$. Viewing the completion of this curve as a double cover of $\mathbf{P}^1$, via the map $(b,d) \mapsto d$, we find by Riemann-Hurwitz that the genus of $\Gamma_{\mathbf{c}}^{(1)}$ is two. $\Gamma_{\mathbf{c}}^{(1)}$ has three points at infinity, corresponding to $d = 0$ or $d = \infty$, denoted by ∞, ∞' and ∞''. In terms of a local parameter ς a neighborhood of these points is given by

$$\begin{aligned} \infty: &\quad d = \varsigma^{-1}, \quad b = \varsigma^{-2} - c_1\varsigma^{-1} + c_2 - c_3\varsigma^3 + O(\varsigma^4), \\ \infty': &\quad b = \varsigma^{-1}, \quad d = -c_3\varsigma^2 - c_2c_3\varsigma^3 - c_2^2c_3\varsigma^4 + O(\varsigma^5), \\ \infty'': &\quad d = \varsigma^{-1}, \quad b = c_3\varsigma^3 + c_1c_3\varsigma^4 + O(\varsigma^5). \end{aligned} \tag{7.41}$$

It is clear that, if we use any other principal balance $x(t;m_i)$ then we get an affine curve $\Gamma_{\mathbf{c}}^{(i)}$, which is isomorphic to $\Gamma_{\mathbf{c}}^{(1)}$. Thus, the abstract Painlevé divisor $\Gamma_{\mathbf{c}}$ consists in this case of five isomorphic curves of genus two; these non-singular curves are affine curves and each of them is compactified into a Riemann surface by adding three points.

7.6.5 Embedding the Tori $\mathbf{T}^r_{\mathbf{c}}$

We now show how the balances are used to embed the tori $\mathbf{T}^r_{\mathbf{c}}$ explicitly into projective space. As in the previous paragraph, we first assume that we are dealing with a weight homogeneous a.c.i. system $(\mathbf{C}^n, \{\cdot\,,\cdot\}, \mathbf{F})$, where $\mathbf{F} = (F_1, \ldots, F_s)$, and that $\mathcal{V}$ is the weight homogeneous vector field $\mathcal{X}_{F_1}$. Let $x(t; \mathcal{D}^{(1)}), \ldots, x(t; \mathcal{D}^{(d)})$ be the principal balances, where $\mathcal{D}^{(1)}, \ldots, \mathcal{D}^{(d)}$ are the Painlevé walls. We recall from Proposition 6.14 that for any polynomial $H \in \mathbf{C}[x_1, \ldots, x_n]$ the pole (in t) of the series $H(x(t; \mathcal{D}^{(i)}))$ for generic $\mathbf{c} \in \mathbf{C}^s$ is precisely the pole that $H_{|\mathbf{F_c}}$ has, viewed as a meromorphic function on the completion of $\mathbf{F_c}$ (which is an Abelian variety). Thus, for any polynomial of weight (at most) k we can find out exactly what is the pole structure of its restriction to the generic fiber of the momentum map, by using the first $k+1$ terms in the balances $x(t; \mathcal{D}^{(1)}), \ldots, x(t; \mathcal{D}^{(d)})$. We will specify the (maximal) pole structure by a vector $\rho = (\rho_1, \ldots, \rho_d)$, where each $\rho_i \in \mathbf{N}$; we call such a vector a *pole vector*. The idea is then to search, for $k = 0, 1, 2, \ldots$, all weight homogeneous polynomials $z_0, \ldots, z_N$ of weight k such that $z_j(x(t; \mathcal{D}^{(i)}))$ has a pole of order at most ρ_i, for $i = 1, \ldots, d$ and $j = 0, \ldots, N$. For a given weight k this is a finite process, as the vector space of all polynomials of a given weight k is finite-dimensional; we will see later for which k to stop.

Let us consider, for generic $\mathbf{c}$, the completion $\mathbf{T}^r_{\mathbf{c}}$ of the fiber $\mathbf{F_c}$, where we denote the irreducible components of the Painlevé divisor $\mathcal{D}_{\mathbf{c}}$ by $\mathcal{D}^{(i)}$, for $i = 1, \ldots, d$. According to Proposition 6.14 the found polynomials, restricted to $\mathbf{F_c}$, correspond to sections of the line bundle

$$\left[\rho_1 \mathcal{D}^{(1)}_{\mathbf{c}}\right] \otimes \cdots \otimes \left[\rho_d \mathcal{D}^{(d)}_{\mathbf{c}}\right], \tag{7.42}$$

on the completion of $\mathbf{F_c}$ (which is an Abelian variety).

Now $\mathcal{D}^{(1)}_{\mathbf{c}} \cup \cdots \cup \mathcal{D}^{(d)}_{\mathbf{c}}$ is the divisor to be adjoined to $\mathbf{F_c}$ to complete it into an Abelian variety, so that the line bundle will be very ample as soon as it is the cube of the line bundle of an effective divisor (Theorem 5.17). Since for irreducible Abelian varieties any effective divisor is ample, it suffices to take ρ such that $\sum_i \rho_i \geqslant 3$ to be sure that (7.42) is very ample. Moreover, the number of independent polynomials (independent when restricted to the completion of $\mathbf{F_c}$) can be computed from (5.15) when the polarization type that each of the $\mathcal{D}^{(i)}_{\mathbf{c}}$ induces on the tori is known.

If we have found a collection of independent functions with at most a simple pole along $\mathcal{D}_{\mathbf{c}}$ then we can easily construct from them several functions with at most a double pole along $\mathcal{D}_{\mathbf{c}}$. One obvious way of getting such functions is just by taking the product of two functions (not necessarily different) with a simple pole along $\mathcal{D}_{\mathbf{c}}$. Another way is by taking Wronskians: as we explained in Section 5.3 the Wronskian of two functions with a simple pole at worst along $\mathcal{D}_{\mathbf{c}}$ is a function with a double pole at worst along $\mathcal{D}_{\mathbf{c}}$.

Recall in this respect Piovan's Theorem (Theorem 5.37) which says that, in the case of a symmetric divisor on a Jacobi surface, which induces twice the principal polarization, these two constructions lead to a basis of $L(2\mathcal{D}_{\mathbf{c}})$, when the vector field $\mathcal{V}$ is generic. Notice that the fact that this is only true for a generic vector field is not a restriction in practice, because in the examples one always has explicit formulas for all (two in the case of Jacobi surfaces) vector fields.

Things are quite different if we do not know whether or not we are dealing with an a.c.i. system (which will usually be the case). We will discuss this in detail in Section 7.7, in particular in Paragraph 7.7.1. Therefore, in the next example, we will just give a list of functions that embeds the tori $\mathbf{T}^r_{\mathbf{c}}$, and we explain later how this list was found and why it provides an embedding.

Example 7.52. In the case of the periodic 5-particle Kac-van Moerbeke lattice we have that the five principal balances $x_i(t;\Gamma^{(i)})$ $(i=1,\dots,5)$ have at worst a simple pole in t. One obvious choice of ρ would be $\rho=(1,1,1,1,1)$ but it leads to an embedding of the tori in a very large projective space, namely in $\mathbf{P}^{24}$. Therefore, we take a smaller (but less symmetric) pole vector, namely $\rho=(3,0,0,0,0)$. It leads to the following independent functions z_i, which we give together with the first few terms of their series $z_i(t;m_1)$.

$$
\begin{aligned}
z_0 &:= 1,\\
z_1 &:= x_3x_4 = -\frac{1}{t^2}+\frac{1}{3}(a^2-b+c)+O(t),\\
z_2 &:= x_1x_3x_4 = -\frac{d}{t^2}+\frac{d}{6}(2a^2+b-c)+O(t),\\
z_3 &:= x_3x_4(x_2+x_3) = \frac{1}{t^3}-\frac{a}{t^2}+O(1),\\
z_4 &:= x_1x_3^2x_4 = \frac{d}{t^3}-\frac{ad}{t^2}+\frac{d}{2t}(b+c)+O(1),\\
z_5 &:= x_1x_3x_4^2 = -\frac{d}{t^3}-\frac{ad}{t^2}+\frac{d}{2t}(b+c)+O(1),\\
z_6 &:= x_1x_3x_4((x_1+x_5)x_3-(x_1+x_2)x_4) = \frac{2d^2}{t^3}+\frac{d}{t^2}(b+c)+O(1),\\
z_7 &:= x_1x_2x_3^2x_4^2 = \frac{bd}{t^3}+\frac{bd}{t^2}(d-a)+O(t^{-1}),\\
z_8 &:= x_1x_3x_4^2((x_1+x_2)^2+x_1x_5) = -\frac{d^3}{t^3}-\frac{d^2}{t^2}(ad+2b+c)+O(t^{-1}).
\end{aligned}
\tag{7.43}
$$

As we will see later, for generic $\mathbf{c}$ the map defined by

$$(x_1,\dots,x_5)\in\mathbf{F}_{\mathbf{c}}\mapsto(1:z_1:\dots:z_8)$$

extends to a holomorphic embedding of $\mathbf{T}^2_{\mathbf{c}}$ into $\mathbf{P}^8$.

7.6.6 The Quadratic Differential Equations

We know from Section 5.2.2 that if $z_0, \ldots, z_N$ are meromorphic functions that correspond to a basis of sections of a projectively normal line bundle on an Abelian variety $\mathbf{T}^r$, then any holomorphic vector field on $\mathbf{T}^r$ extends to a vector field on the embedding space $\mathbf{P}^N$, and that moreover, this vector field is quadratic (in any of the standard affine charts $z_i \neq 0$). Recall also from that section that for any effective divisor $\mathcal{D}$ on an Abelian variety one has that $[3\mathcal{D}]$ is always projectively normal, though $[\mathcal{D}]$ or $[2\mathcal{D}]$ may already be projectively normal. When analyzing a weight homogeneous vector field, hoping to prove that it is a.c.i. one is therefore tempted to try to construct these quadratic differential equations. Indeed, if one shows that they do exist one has actually shown that the vector field extends to a holomorphic vector field on $\mathbf{P}^N$, in particular to a neighborhood of the divisor at infinity, one of the essential conditions in the complex Liouville Theorem (Theorem 6.22).

Let us consider as above a weight homogeneous vector field $\mathcal{V}$ on $\mathbf{C}^n$ which is described by $\dot{x} = f(x)$. Suppose that the Wronskian $W(z_i, z_j) = \dot{z}_i z_j - z_i \dot{z}_j$, with z_i, z_j polynomials in the x_k is expressible as a quadratic polynomial,

$$W(z_i, z_j) = \sum_{k,l=0}^{N} \alpha_{ij}^{kl} z_k z_l \tag{7.44}$$

where each α_{ij}^{kl} depends on the values $\mathbf{c} = (c_1, \ldots, c_s)$ of the constants of motion $F_1, \ldots, F_s$ (only); upon restriction to the generic fiber $\mathbf{F_c}$ of the momentum map the a_{ij}^{kl} are just constants and the above quadratic polynomials have constant coefficients. Each α_{ij}^{kl} is weight homogeneous of weight $\nu_i + \nu_j - \nu_k - \nu_l + 1$, since

$$\varpi(W(z_i, z_j)) = \varpi(\dot{z}_i z_j - z_i \dot{z}_j) = \nu_i + \nu_j + 1,$$

while $\varpi(\alpha_{ij}^{kl} z_k z_l) = \varpi(\alpha_{ij}^{kl}) + \nu_k + \nu_l$. On the coordinate chart $(z_j \neq 0) \subset \mathbf{P}^N$ the functions $z_0/z_j, \ldots, z_N/z_j$ are a system of coordinates, and the quadratic differential equations are given in terms of these coordinates by

$$\left(\frac{z_i}{z_j}\right)^{\cdot} = \frac{W(z_i, z_j)}{z_j^2} = \sum_{k,l=0}^{N} \alpha_{ij}^{kl} \frac{z_k}{z_j} \frac{z_l}{z_j}.$$

If the coefficients α_{ij}^{kl} are polynomials (in $F_1, \ldots, F_s$) then they can easily be found from (7.44) by substituting for each α_{ij}^{kl} the most general weight homogeneous polynomial of degree $\nu_i + \nu_j - \nu_k - \nu_l + 1$ in $F_1, \ldots, F_s$, i.e.,

$$\alpha_{ij}^{kl} = \sum \beta_{i,j;i_1 \ldots i_s}^{kl} F_1^{i_1} \ldots F_s^{i_s},$$

where the sum runs over all s-tuples of non-negative integers $(i_1, \ldots, i_s)$ for which

$$\sum_{p=1}^{s} i_p \varpi(F_p) = \nu_i + \nu_j - \nu_k - \nu_l + 1,$$

and where each $\beta^{kl}_{i,j;i_1\ldots i_s} \in \mathbf{C}$ is to be determined. This is done by writing both sides of (7.44) in terms of the original variables x_i which yields, by equating the coefficients of the x_i in both sides, a (big) system of linear equations, which is easily solved. If the first few terms of one of the principal balances $x(t;\mathcal{I}')$ has a rather simple form, then it may be simpler in the last step to replace the x_i by that principal balance and the constants c_i in terms of the free parameters and to equate powers in t in the result (each coefficient in t will be a polynomial which depends on a number of free parameters, hence they have to be equated as polynomials in the free parameters). Alternatively, we can also work directly with the z_i and deduce the coefficients α^{kl}_{ij} in (7.44) by substituting the principal balances for the z_i, following the above procedure for the x_i.

For proving algebraic complete integrability it is, as we will see, enough to prove the existence of the quadratic differential equations, without actually computing them. We will give in Paragraph 7.7.2 a practical way to check their existence.

Example 7.53. We write the quadratic differential equations in two different charts in $\mathbf{P}^8$. We start with a first chart, which is the one in which we can take the original embedding variables $z_1, \ldots, z_8$ as affine coordinates (recall that $z_0 = 1$). Then $\mathcal{V}_1$ can be written as the following quadratic[4] differential equations.

$$\begin{aligned}
\dot z_1 &= z_2 + 2z_3 - c_1 z_1,\\
\dot z_2 &= z_4 - z_5,\\
\dot z_3 &= z_4 + 2z_5 - 3z_1^2 + c_1 z_3 - 2c_2 z_1,\\
\dot z_4 &= -3z_1 z_2 + c_1 z_4 + \frac{1}{2}(c_3 - c_2 z_2 - z_6),\\
\dot z_5 &= 3z_1 z_2 - c_1 z_5 - \frac{1}{2}(c_3 - c_2 z_2 + z_6),\\
\dot z_6 &= (2z_1 + c_2)(z_4 + z_5) - 4z_2^2 - 2c_1 z_1 z_2,\\
\dot z_7 &= 3z_2 z_5 + c_1 z_7 + \frac{1}{2}(3z_1 z_6 - 3c_2 z_1 z_2 - c_3 z_1),\\
\dot z_8 &= -3z_2(z_4 + z_5 - c_1 z_2) + c_3 z_1 + \frac{1}{2}c_2(z_6 - c_2 z_2 + c_3).
\end{aligned}$$

For the second chart, which corresponds to $z_1 \neq 0$, we define $y_i := z_i/z_1$, for $i = 0, \ldots, 8$, and we obtain the following quadratic differential equations.

[4] In order to make these equations purely quadratic, use z_0 (which equals 1); the first equation e.g. should be written as $\dot z_1 = z_0(z_2 + 2z_3 - c_1 z_1)$. In the other chart, use y_1 (which also equals 1).

$$\begin{aligned}
\dot{y}_0 &= -y_2 - 2y_3 + c_1,\\
\dot{y}_1 &= 0,\\
\dot{y}_2 &= -(y_4 + y_5) - 2y_0y_7 - y_2(y_2 - c_1),\\
\dot{y}_3 &= y_3^2 + 2y_0y_7 - y_5 - c_1y_3 + c_2,\\
\dot{y}_4 &= y_2(c_2 - y_4 - y_5) - y_3y_5 - c_3y_0,\\
\dot{y}_5 &= y_6 + y_5(2y_2 + y_3) - c_2y_2 + c_3y_0,\\
\dot{y}_6 &= 2(y_4^2 + y_4y_5 + y_5^2) - (y_4 + y_5)(2c_1y_2 + c_2) + (c_2y_2 - c_3y_0)(c_1 + y_2),\\
\dot{y}_7 &= y_7(y_3 + 2y_2 - c_1y_0) + c_3,\\
\dot{y}_8 &= y_2y_7 + y_5y_6 - y_8(y_2 + y_3) - c_1y_5(y_4 + y_5) + c_1^2y_2y_5 - c_3y_0y_5.
\end{aligned}$$

7.6.7 The Holomorphic Differentials on $\mathcal{D}_c$

In this section we show, following an idea by Luc Haine (see [74]), how the holomorphic one-forms $\mathrm{d}t_{1|\mathcal{D}}, \dots, \mathrm{d}t_{r|\mathcal{D}}$ on an irreducible analytic hypersurface $\mathcal{D}$ (which is singular or not) in an Abelian variety $\mathbf{T}^r$ can be computed; the t_i in this expression are any system of linear coordinates on $\mathbf{T}^r$. We will mainly be interested in the $(r-1)$-forms

$$\omega_i := \mathrm{d}t_1 \wedge \ldots \wedge \widehat{\mathrm{d}t_i} \wedge \ldots \wedge \mathrm{d}t_r \Big|_{\mathcal{D}}. \tag{7.45}$$

These have several applications: the locus of tangency of $\partial/\partial t_i$ on $\mathcal{D}$ is the locus where $\omega_i = 0$. Indeed, $\partial/\partial t_i$ is tangent to $\mathcal{D}$ at a smooth point m precisely if $\frac{\partial f}{\partial t_i}(m) = 0$, where $f(t_1, \dots, t_r) = 0$ locally defines $\mathcal{D}$, but then

$$0 = \mathrm{d}f_{|\mathcal{D}}(m) = \sum_{j=1}^{r} \frac{\partial f}{\partial t_j}(m)\, \mathrm{d}t_{j|\mathcal{D}}(m) = \sum_{j\neq i} \frac{\partial f}{\partial t_j}(m)\, \mathrm{d}t_{j|\mathcal{D}}(m),$$

so that $\omega_i(m) = 0$. Also, we have seen in Section 5.3 that in order to compute a basis of the holomorphic differentials on a divisor $\mathcal{D}$ on an Abelian variety $\mathbf{T}^r$, we need to compute precisely the holomorphic differentials ω_i.

Our computation, which is done in the context of integrable systems, depends in an essential way on the fact that we have at our disposal explicit equations for the vector fields. We assume that $\mathcal{D}$ is a very ample divisor, with each component having multiplicity 1 (i.e., $\mathcal{D}$ is an analytic hypersurface).

Let $(y_0, \dots, y_N)$ be a basis of $L(\mathcal{D})$, which is chosen such that y_0 is one of the functions that has the largest pole along $\mathcal{D}$, among all functions of this basis (such a choice of basis exists, even if $\mathcal{D}$ is not irreducible; in fact a generic basis of $L(\mathcal{D})$ has this property). We write the Laurent solutions of the functions y_i with respect to all of the irreducible components $\mathcal{D}'$ of $\mathcal{D}$ collectively as one Laurent series, i.e., we write $y_i(t;\mathcal{D}) = \frac{1}{t}\sum_{i\geqslant 0} y_i^{(j)} t^j$, where each of the $y_j^{(j)}$ is a regular function on $\mathcal{D}$. Because of the above assumption on y_0 we have that the restriction of $y_0^{(0)}$ to any of the irreducible components of $\mathcal{D}$ is non-zero, but this needs not be true for $y_i^{(0)}$ when $i > 0$; the fact that the series start at t^{-1} at worst comes from the fact that each component of $\mathcal{D}$ has multiplicity 1. We will denote the r independent integrable vector fields by $\mathcal{V}_1, \dots, \mathcal{V}_r$. The functions $y_1/y_0, \dots, y_N/y_0$ are holomorphic, so that the functions $y_1/y_0, \dots, y_r/y_0$ can be picked to define a holomorphic chart for generic $m \in \mathcal{D}$, since $L(\mathcal{D})$ embeds $\mathbf{T}^r$, and we have

$$\begin{pmatrix} \mathsf{d}\left(\frac{y_1}{y_0}\right) \\ \vdots \\ \mathsf{d}\left(\frac{y_r}{y_0}\right) \end{pmatrix} = \begin{pmatrix} \mathcal{V}_1\left[\frac{y_1}{y_0}\right] & \cdots & \mathcal{V}_r\left[\frac{y_1}{y_0}\right] \\ \vdots & & \vdots \\ \mathcal{V}_1\left[\frac{y_r}{y_0}\right] & \cdots & \mathcal{V}_r\left[\frac{y_r}{y_0}\right] \end{pmatrix} \begin{pmatrix} \mathsf{d}t_1 \\ \vdots \\ \mathsf{d}t_r \end{pmatrix}. \tag{7.46}$$

We restrict (7.46) to $\mathcal{D}$. For the left hand side we have,

$$\mathsf{d}\left(\frac{y_i}{y_0}\right)\Big|_{\mathcal{D}} = \mathsf{d}\left(\frac{y_i^{(0)}}{y_0^{(0)}}\right),$$

as follows by putting $t = 0$ in the Laurent series $\frac{y_i}{y_0}(t;\mathcal{D})$ of the function y_i/y_0, which is holomorphic in a neighborhood of a generic point of $\mathcal{D}$, using $\mathsf{d}t_{|\mathcal{D}} = 0$. For the entries of the square matrix in (7.46) we first compute the functions $\mathcal{V}_j[y_i/y_0]$ in terms of the original phase variables $x_1 \dots, x_n$, which can easily be done because we know explicit formulas for the vector fields $\mathcal{V}_j$ and for the embedding functions y_i in terms of the phase variables. Since these vector fields and these functions are holomorphic in a neighborhood of a generic point of $\mathcal{D}$ we can then compute the restriction of $\mathcal{V}_j[y_i/y_0]$ by first substituting the Laurent series $x_1(t;\mathcal{D}), \dots, x_n(t;\mathcal{D})$ for $x_1, \dots, x_n$ in $\mathcal{V}_j[y_i/y_0]$ and then putting $t = 0$. For $\mathcal{V}_1$ this can be done in a slightly simpler way, namely we can use the fact that $\mathcal{V}_1[f] = \frac{\partial f}{\partial t}(t)$, giving

$$\mathcal{V}_1\left[\frac{y_i}{y_0}\right]\Big|_{\mathcal{D}} = \frac{\mathsf{d}}{\mathsf{d}t}_{|t=0}\frac{y_i}{y_0}(t) = \frac{y_0^{(0)}y_i^{(1)} - y_i^{(0)}y_0^{(1)}}{(y_0^{(0)})^2}.$$

This leads to explicit formulas for the one-forms $\mathsf{d}t_{1|\mathcal{D}}, \dots \mathsf{d}t_{r|\mathcal{D}}$ on $\mathcal{D}$, hence also for their wedges, such as the top-forms $\omega_1, \dots, \omega_r$. Of course the fact that we picked the $y_1/y_0, \dots, y_r/y_0$ correctly is equivalent to the determinant of the matrix in (7.46) being different from zero for generic $m \in \mathcal{D}$.

Example 7.54. Suppose that $r = 2$, with $\mathcal{D}$ a curve in $\mathbf{T}^2$, where all components have multiplicity 1 when $\mathcal{D}$ is viewed as a divisor. We use the above method to compute $\omega_1 = \mathsf{d}t_2\Big|_{\mathcal{D}}$ and $\omega_2 = \mathsf{d}t_1\Big|_{\mathcal{D}}$. Let us suppose that y_0, $y_1 = 1$ and $y_2 = y$ have been chosen as above with y_0 having a simple pole on each of the irreducible components of $\mathcal{D}$ and such that $1/y_0$ and y/y_0 define a holomorphic chart around a generic point of $\mathcal{D}$. We denote the residues of y and of y_0 as series in t by $y^{(0)}$ resp. $y_0^{(0)}$, as before. Applying the methods explained above, we consider

$$\begin{pmatrix} \mathrm{d}\left(\frac{1}{y_0}\right) \\ \mathrm{d}\left(\frac{y}{y_0}\right) \end{pmatrix} = \begin{pmatrix} \mathcal{V}_1\left[\frac{1}{y_0}\right] & \mathcal{V}_2\left[\frac{1}{y_0}\right] \\ \mathcal{V}_1\left[\frac{y}{y_0}\right] & \mathcal{V}_2\left[\frac{y}{y_0}\right] \end{pmatrix} \begin{pmatrix} \mathrm{d}t_1 \\ \mathrm{d}t_2 \end{pmatrix}, \tag{7.47}$$

which we solve for $\mathrm{d}t_1$ and $\mathrm{d}t_2$, and which we restrict to $\mathcal{D}$, to find

$$\begin{pmatrix} \omega_2 \\ \omega_1 \end{pmatrix} = \frac{1}{\delta} \begin{pmatrix} \mathcal{V}_2[y/y_0]\big|_{\mathcal{D}} & -\mathcal{V}_2[1/y_0]\big|_{\mathcal{D}} \\ -\mathcal{V}_1[y/y_0]\big|_{\mathcal{D}} & 1/y_0^{(0)} \end{pmatrix} \begin{pmatrix} 0 \\ \mathrm{d}\left(y^{(0)}/y_0^{(0)}\right) \end{pmatrix},$$

where δ is the determinant of the square matrix in (7.47), restricted to $\mathcal{D}$,

$$\delta = \frac{1}{(y_0^{(0)})^2} \begin{vmatrix} y_0^{(0)} & \mathcal{V}_2[1/y_0]\big|_{\mathcal{D}} \\ y_0^{(0)}y^{(1)} - y^{(0)}y_0^{(1)} & \mathcal{V}_2[y/y_0]\big|_{\mathcal{D}} \end{vmatrix},$$

which is non-zero by the above assumptions on $y^{(0)}$ and y. It follows that

$$\begin{aligned} \omega_1 &= \frac{1}{\delta y_0^{(0)}} \mathrm{d}\left(\frac{y^{(0)}}{y_0^{(0)}}\right) \\ \omega_2 &= -\frac{1}{\delta}\mathcal{V}_2[1/y_0]\big|_{\mathcal{D}} \mathrm{d}\left(\frac{y^{(0)}}{y_0^{(0)}}\right). \end{aligned} \tag{7.48}$$

The zeros of ω_1 resp. ω_2 provide the points of tangency of the vector fields $\mathcal{V}_1$ and $\mathcal{V}_2$ respectively. Since the degree of the canonical bundle on $\mathcal{D}$ is $2g-2$, the tangency locus of $\mathcal{V}_1$ consist of $2g-2$ points, including multiplicities, and similarly for the tangency locus of $\mathcal{V}_2$.

Example 7.55. In the case of the periodic 5-particle Kac-van Moerbeke lattice we have chosen $(3, 0, 0, 0, 0)$ as pole vector, so in this case the divisor is not an analytic hypersurface, but it is three times an analytic hypersurface and the above method does not work. The holomorphic differentials on the curve can in this case be constructed at from the equation of the curve because the curve is hyperelliptic of genus two.

7.7 Proving Algebraic Complete Integrability

We will now present an effective method for proving the algebraic complete integrability of a weight homogeneous a.c.i. system, by using the principal balances to (one of) the weight homogeneous vector field(s). The algorithms that we will give are very effective in the case of two dimensions, but they may be less so in the case of higher dimensions because the dimension of the smallest projective space in which a higher dimensional Abelian variety can be embedded is very large ($3^g - 1$ for a g-dimensional Jacobian).

We suppose that we are given on $\mathbf{C}^n$ a weight homogeneous Poisson structure $\{\cdot\,,\cdot\}$ of rank $2r$ and $s := n - r$ independent weight homogeneous polynomials $\mathbf{F} = (F_1, \dots, F_s)$, which are in involution with respect to $\{\cdot\,,\cdot\}$. Then $(\mathbf{C}^n, \{\cdot\,,\cdot\}, \mathbf{F})$ is integrable in the sense of Liouville. The independent integrable vector fields are denoted by $\mathcal{V}_1, \dots, \mathcal{V}_r$, where we assume that $\mathcal{V}_1$ is weight homogeneous (i.e., it has weight 1). As usual we will denote the weight vector by $\nu = (\nu_1, \dots, \nu_n)$. For generic $\mathbf{c} \in \mathbf{C}^s$ we wish to check if the fiber $\mathbf{F_c}$ of the momentum map is (up to isomorphism) an affine part of an Abelian variety $\mathbf{T}^r_{\mathbf{c}}$ and if the Hamiltonian vector fields $\mathcal{V}_1, \dots, \mathcal{V}_r$ are linear on $\mathbf{T}^r_{\mathbf{c}}$.

The method that we give consists in using the weight homogeneous principal balances to $\mathcal{V}_1$ to show that the conditions of the complex Liouville Theorem (Theorem 6.22) are satisfied. Let us first see which of these conditions are automatically satisfied; then we will discuss in detail the other conditions and we will point out what exactly has to be checked in each of the examples. Since $\mathbf{c}$ is generic we have by Sard's Theorem that $\mathbf{F_c}$ is non-singular. Assume now that for generic $\mathbf{c}$ the fiber $\mathbf{F_c}$ is, in addition, irreducible, so that it is a non-singular affine variety; if $\mathbf{F_c}$ is not irreducible then what we explain next has to be done for each of its irreducible components separately. On $\mathbf{F_c}$ we have by Liouville integrability r commuting vector fields $\mathcal{V}_1, \dots, \mathcal{V}_r$.

If the Poisson structure is symplectic, so $n = 2r$, then these vector fields are independent at every point of the generic fiber of the momentum map since a symplectic structure yields an isomorphism between tangent vectors and covectors in each point of phase space, and since the differentials $\mathrm{d}F_1, \dots, \mathrm{d}F_r$ are independent at all points of a generic fiber of the momentum map.

If the Poisson structure is not symplectic (which may happen even if $n = 2r$) then things are very different. In fact, the integrable vector fields may become dependent at one or several points of *every* fiber of the momentum map. For the example of the periodic Kac-van Moerbeke system we have shown in Example 7.39 that the two vector fields are independent at every point of the generic fiber of the momentum map. However, many systems that are not a.c.i. are different in this respect; for example, each of the integrable vector fields of the non-periodic Toda lattice and of the non-periodic Kac-van Moerbeke system (see Example 6.5) have a non-empty zero locus, when restricted to any of the fibers of the momentum map.

In the following proposition we give a sufficient condition for the independence of the integrable vector fields at every point of the generic fiber of the momentum map. This condition will be satisfied in all the examples that we will study. In the case of a regular Poisson manifold (e.g. a symplectic manifold) one has that $M_{(r)} = M$ so that the given condition is automatically satisfied.

Proposition 7.56. *Let* $(\mathbf{C}^n, \{\cdot,\cdot\}, \mathbf{F})$ *be a Liouville integrable system of rank* $2r$*, where each* F_i *in* $\mathbf{F} = (F_1, \dots, F_r)$ *is assumed to be a polynomial. If the image of* $\mathbf{F}(M \setminus M_{(r)})$ *is contained in a Zariski closed subset of* $\mathbf{C}^s$*, different from* $\mathbf{C}^s$*, then for generic* $\mathbf{c} \in \mathbf{C}^s$ *the vector fields* $\mathcal{X}_{F_1}, \dots, \mathcal{X}_{F_s}$ *are independent at each point of the fiber* $\mathbf{F_c}$*.*

Proof. Let C denote a Zariski closed subset of $\mathbf{C}^s$ such that $\mathbf{F}(M \setminus M_{(r)}) \subset C$, with $C \neq \mathbf{C}^s$. Then $\mathbf{F}^{-1}(\mathbf{C}^s \setminus C) \subset M_{(r)}$, so that for generic $\mathbf{c} \in \mathbf{C}^s$ one has $\mathbf{F_c} \subset M_{(r)}$. Since for generic $\mathbf{c} \in \mathbf{C}^s$ one also has that $\mathbf{F_c} \subset \mathcal{U}_\mathbf{F}$ it follows that for generic $\mathbf{c} \in \mathbf{C}^s, \mathbf{F_c} \subset \mathcal{U}_\mathbf{F} \cap M_{(r)}$. Proposition 4.12 implies that the vector fields $\mathcal{X}_{F_1}, \dots, \mathcal{X}_{F_s}$ are independent on the fibers of $\mathbf{F}$ over such $\mathbf{c}$. □

Thus, we suppose that conditions *(1)* and *(2)* in Theorem 6.22 are verified. The first thing to be done is then to construct, for generic $\mathbf{c}$, an embedding $\varphi_\mathbf{c} : \mathbf{F_c} \to \mathbf{P}^N$, which satisfies the adjunction formula, thereby making its extendibility to an embedding of $\mathbf{T}^r_\mathbf{c}$ plausible (as long as the adjunction formula is not satisfied the embedding surely does not extend to an embedding of $\mathbf{T}^r_\mathbf{c}$ in $\mathbf{P}^N$). This will be explained in Paragraph 7.7.1.

Given the "to be" embedding, we proceed to write down the quadratic equations for one of the vector fields, say for $\overline{\mathcal{V}} := (\varphi_\mathbf{c})_* \mathcal{V}$. To do this it may be necessary to enlarge the constructed embedding, taking functions that have higher poles at infinity; see for example the Kowalevski top (Section 10.3). Having found the quadratic differential equations we are done with condition *(3)* because these vector fields are holomorphic on all of $\mathbf{P}^N$. As we will prove in Paragraph 7.7.2, it is enough to construct the quadratic differential equations in two different charts $z_i \neq 0$ of $\mathbf{P}^N$.

The last thing to be checked is that the flow of $\overline{\mathcal{V}}$ which starts at any point of $\Delta'_\mathbf{c}$ (in the notation of Theorem 6.22) goes into $\varphi_\mathbf{c}(\mathbf{F_c})$ immediately. For this, see Paragraph 7.7.3. Then we can conclude that the projective closure $\overline{\varphi_\mathbf{c}(\mathbf{F_c})}$ of $\varphi_\mathbf{c}(\mathbf{F_c})$ is an Abelian variety and that the commuting vector fields $(\varphi_\mathbf{c})_* \mathcal{V}_1, \dots, (\varphi_\mathbf{c})_* \mathcal{V}_r$ extend to holomorphic vector fields $\overline{\mathcal{V}}_1, \dots, \overline{\mathcal{V}}_r$ on $\overline{\varphi_\mathbf{c}(\mathbf{F_c})}$. The polarization type that is induced by the divisor at infinity can in certain cases be read off from the divisor itself (for example, if the divisor is a non-singular curve of genus 2 then it induces a principal polarization on the Abelian surface in which it lives, its Jacobi surface), in other cases we use an extra argument, based on Proposition 5.20. This gives a complete description of the algebraic integrability of the integrable system. Notice that, at this point, everything that we have constructed formally (the balances, Painlevé architecture, embedding, ...) can be identified with the corresponding object that can be constructed for the a.c.i. system.

7.7.1 Embedding the Tori $\mathbf{T}_{\mathbf{c}}^r$ and Adjunction

Given an integrable system $(\mathbf{C}^n, \{\cdot\,,\cdot\}, \mathbf{F})$, where $\mathbf{F} = (F_1, \dots, F_s)$ we wish to construct, for generic $\mathbf{c}$, a map $\varphi_{\mathbf{c}}$ from $\mathbf{F_c}$ into some projective space $\mathbf{P}^N$ which, in case the integrable systems turns out to be a.c.i., embeds the Abelian variety of which $\mathbf{F_c}$ is an affine part; the latter means in practice that the projective closure $\overline{\varphi_{\mathbf{c}}(\mathbf{F_c})}$ of $\varphi_{\mathbf{c}}(\mathbf{F_c})$ is an Abelian variety (embedded in projective space). In order to construct such a map $\varphi_{\mathbf{c}}$ (if it exists) we construct a polynomial map $\mathbf{C}^n \to \mathbf{C}^N$ whose restriction to a generic $\mathbf{F_c}$ has the main properties of the Kodaira map associated to a very ample line bundle on an Abelian variety.

Let $x(t;\Gamma^{(1)}), \dots, x(t;\Gamma^{(d)})$ denote the weight homogeneous principal balances. We know from Proposition 7.22 that each of the Painlevé walls $\Gamma^{(i)}$ is of the form $\mathcal{I}^{(i)} \times \mathbf{C}^{p_i}$, where $\mathcal{I}^{(i)}$ is an irreducible component of the indicial locus $\mathcal{I}$, and where $p_i = n-1-\dim \mathcal{I}^{(i)}$. By the same proposition, each of the coefficients of the principal balance $x(t;\Gamma^{(i)})$ is a regular function on $\Gamma^{(i)}$. We choose a pole vector $(\rho_1, \dots, \rho_d)$, with $\rho_i \geqslant 0$ for $i = 1, \dots, d$. We introduce the vector space of polynomials in $x_1, \dots, x_n$ with no harder pole than $t^{-\rho_i}$ when evaluated on $x(t;\Gamma^{(i)})$, where $1 \leqslant i \leqslant d$,

$$\mathcal{Z}_\rho := \left\{ F \in \mathbf{C}[x_1, \dots, x_n] \mid \operatorname{ord}_{t=0} F(x(t;\Gamma^{(i)})) \geqslant -\rho_i \text{ for } 1 \leqslant i \leqslant d \right\}.$$

$\mathcal{Z}_\rho$ is an $\mathcal{H}$-module, i.e., if $z \in \mathcal{Z}_\rho$ and $H \in \mathcal{H}$ ($\mathcal{H}$ is the algebra generated by $F_1, \dots, F_s$) then $zH \in \mathcal{Z}_\rho$. Moreover, if $S \subseteq \mathcal{Z}_\rho$ and if $\mathcal{H}(S)$ denotes the $\mathcal{H}$-module generated by S, then

$$\operatorname{span}\left\{ z_{|\mathbf{F_c}} \mid z \in S \right\} = \operatorname{span}\left\{ z_{|\mathbf{F_c}} \mid z \in \mathcal{H}(S) \right\}.$$

Since we are looking for polynomials $z \in \mathcal{Z}_\rho$ whose restrictions to $\mathbf{F_c}$ are independent, we will content ourselves in finding elements of $\mathcal{Z}_\rho$ that are not merely independent over $\mathbf{C}$, but that are independent over $\mathcal{H}$. In order to organize the computation — which tends to be enormous, even for simple systems — one naturally searches such elements by weight. To do this we use the grading

$$\mathbf{C}[x_1, \dots, x_n] = \bigoplus_{k=0}^{\infty} \mathcal{F}^{(k)}.$$

Namely, denote $\mathcal{Z}_\rho^{(k)} := \mathcal{Z}_\rho \cap \mathcal{F}^{(k)}$ and suppose that we have constructed a maximal set of elements $z_0, \dots, z_j$ of $\oplus_{k=0}^{l-1} \mathcal{Z}_\rho^{(k)}$ that are independent (over $\mathcal{H}$). By multiplying these elements with a basis of $\oplus_{k=0}^{l} \mathcal{H}^{(k)}$, in all possible ways that yield a product of weight l, we construct a maximal set of linearly independent (over $\mathbf{C}$) elements in $\mathcal{Z}_\rho^{(l)}$, where each of them is dependent over $\mathcal{H}$ on the elements of $\mathcal{Z}_\rho$ that we have already obtained. These elements can, by direct computation, easily be extended to a basis (over $\mathbf{C}$) of $\mathcal{Z}_\rho^{(l)}$, by adding elements $z_{j+1}, \dots, z_{j'}$. Then $z_0, \dots, z_{j'}$ forms a maximal set of elements of $\oplus_{k=0}^{l}\mathcal{Z}_\rho^{(k)}$, which are independent over $\mathcal{H}$. The number $j'-j$ of independent elements that are added at level l is given by:

$$\zeta_l := \dim\left(\mathcal{Z}_\rho^{(l)} / \oplus_{k=0}^{l-1} \mathcal{H}^{(l-k)} \mathcal{Z}_\rho^{(k)}\right) = \dim \mathcal{Z}_\rho^{(l)} - \sum_{k=0}^{l-1} \zeta_k \dim \mathcal{H}^{(l-k)}. \quad (7.49)$$

This formula allows one to compute recursively the ζ_l, because the dimension of the space $\mathcal{H}^{(l)}$ follows from (7.21) and the dimension of $\mathcal{Z}_\rho^{(l)}$ is found by taking the most general general polynomial of weight l and expressing that it belongs to $\mathcal{Z}_\rho$.

The main problem is then: for which l do we stop? We answer this in a negative way, indicating when we should certainly *not* stop; see also Paragraph 7.7.2 for an alternative indication of when to stop. Let us suppose that we have constructed $N+1$ elements $z_0=1, z_1, \dots, z_N$ of $\mathcal{Z}_\rho$ which are independent over $\mathcal{H}$ and that the $\mathcal{H}$-module spanned by them contains all elements in $\mathcal{Z}_\rho$ of weight at most l. We consider for generic $\mathbf{c}$ the regular map

$$\begin{aligned} \varphi_{\mathbf{c}} : \mathbf{F}_{\mathbf{c}} &\to \mathbf{P}^N \\ m &\mapsto (1 : z_1(m) : \dots : z_N(m)). \end{aligned}$$

To begin with, we will of course increase l until $\varphi_{\mathbf{c}}$ is an isomorphic embedding of $\mathbf{F}_{\mathbf{c}}$. Once this has been achieved we substitute each of the Laurent series $x(t;\Gamma^{(1)}), \dots, x(t;\Gamma^{(d)})$, restricted to $\mathbf{F}_{\mathbf{c}}$ in this embedding and we let $t\to 0$. This yields, in the case of $\Gamma^{(i)}$, an injective map

$$\varphi_{\mathbf{c}}^{(i)} : \Gamma_{\mathbf{c}}^{(i)} \longrightarrow \mathbf{P}^N.$$

The closure of the image of this map, $\overline{\varphi_{\mathbf{c}}^{(i)}(\Gamma^{(i)})}$ will be denoted by $\mathcal{D}_{\mathbf{c}}^{(i)}$. Notice that the extension of $\varphi_{\mathbf{c}}^{(i)}$ to the smooth compactification of $\Gamma_{\mathbf{c}}^{(i)}$ is not necessarily an embedding, it does not even need to be injective (we will see examples of this). As we have seen in Section 5.3, if $\varphi_{\mathbf{c}}$ extends to an embedding of $\mathbf{T}_{\mathbf{c}}^r$ then $\mathcal{D}^{(i)}$ is one of the irreducible components of the Painlevé divisor on $\mathbf{T}_{\mathbf{c}}^r$, which corresponds to the linear vector field $(\varphi_{\mathbf{c}})_*\mathcal{V}$ on $\mathbf{T}_{\mathbf{c}}^r$. Thus, we must first increase l until the dimension of the image of each of these divisors is $r-1$. Next we must make sure that the total divisor $\sum_{i=1}^d \rho_i \mathcal{D}_{\mathbf{c}}^{(i)}$ satisfies the adjunction formula, i.e., we must have that

$$g_a\left(\sum_{i=1}^d \rho_i \mathcal{D}_{\mathbf{c}}^{(i)}\right) = N + r, \quad (7.50)$$

Indeed, according to Theorem 5.38, if the $N+1$ functions that we have constructed provide the Kodaira map that corresponds to $\sum_{i=1}^d \mathcal{D}_{\mathbf{c}}^{(i)}$, so that $\dim L\left(\sum_{i=1}^d \rho_i \mathcal{D}_{\mathbf{c}}^{(i)}\right) = N+1$, then

$$g_a\left(\sum_{i=1}^{d}\rho_i\mathcal{D}_{\mathbf{c}}^{(i)}\right)-r+1=\dim L\left(\sum_{i=1}^{d}\rho_i\mathcal{D}_{\mathbf{c}}^{(i)}\right).$$

Therefore we must increase l until (7.50) is satisfied. When this occurs we stop, because experience indicates that increasing l will not lead to new functions (over $\mathcal{H}$).

Example 7.57. In the case of the periodic 5-particle Kac-van Moerbeke lattice, we have chosen $\rho=(3,0,0,0,0)$ in order to keep the size of N down. Let us first construct the functions that were given in (7.43). Thus, we look for a basis (over $\mathcal{H}$) of the polynomials which have a triple pole at most when $x(t;m_1)$ is substituted in them, and no pole at all when $x(t;m_i)$ for $2\leqslant i\leqslant 5$ is substituted in them. The result is given in the Table 7.1, where we compute for small k the following data, corresponding to the different columns (in that order).

1. $\dim\mathcal{F}^{(k)}$, the number of independent polynomials of weight k, which is computed from (7.5);
2. $\dim\mathcal{H}^{(k)}$, the number of independent constants of motion of weight k, which are the coefficients of the series (7.40);
3. $\dim\mathcal{Z}_\rho^{(k)}$, the number of linearly independent polynomials which have a triple pole at most when $x(t;m_1)$ is substituted in them, and no pole at all when $x(t;m_i)$ for $2\leqslant i\leqslant 5$ is substituted in them (i.e., $\rho=(3,0,0,0,0)$). This is done by explicitly substituting the series $(x;m_i)$ in the most general polynomial of weight k; for doing this a computer program is very useful;
4. The number of elements in $\mathcal{Z}_\rho^{(k)}$ that are dependent of the previous ones over $\mathcal{H}$. This is computed from the previous data by the formula $\sum_{j=0}^{i-1}\zeta_j\dim\mathcal{H}^{(i-j)}$;
5. ζ_k is the number of new elements at level k, and is computed as the difference of the two previous columns, see (7.49);
6. The last column gives a choice of these new functions; their explicit expressions were given in Example 7.52.

We fix a generic $\mathbf{c}=(c_1,c_2,c_3)\in\mathbf{C}^3$ and we consider the map $\varphi_{\mathbf{c}}:\mathbf{F}_{\mathbf{c}}\to\mathbf{P}^8$ given by $(x_1,\dots,x_5)\mapsto(z_0:\dots:z_8)$. Notice that since $\mathbf{c}$ is generic we may assume that $c_3\neq 0$. Then $\mathbf{F}_{\mathbf{c}}$ does not intersect any of the hyperplanes $x_i=0$. Then the map $\varphi_{\mathbf{c}}$ is birational (not biregular!) on its image, since we have

$$x_1=z_4/z_2,\quad x_2=z_5/z_2,\quad x_4=z_2/z_1,\quad x_5=z_7/(z_1z_2),\tag{7.51}$$

while x_3 is recovered from $F_1=c_1$. The leading terms of the series (7.43) of the functions z_i lead to an embedding of the abstract Painlevé divisor $\Gamma_{\mathbf{c}}^{(1)}$ (see Example 7.51), given by

$$\varphi_{\mathbf{c}}^{(1)}:(b,d)\mapsto(0:0:0:1:d:-d:2d^2:bd:-d^3),\tag{7.52}$$

Table 7.1. The polynomials of weight at most 7 which have a triple pole at most when $x(t;m_1)$ is substituted in them, and which have no pole when any of $x(t;m_2),\dots,x(t;m_5)$ are substituted in them.

k	$\dim\mathcal{F}^{(k)}$	$\dim\mathcal{H}^{(k)}$	$\dim\mathcal{Z}_\rho^{(k)}$	# dep	ζ_k	indep. functions
0	1	1	1	0	1	z_0
1	5	1	1	1	0	—
2	15	2	3	2	1	z_1
3	35	2	5	3	2	z_2, z_3
4	70	3	9	7	2	z_4, z_5
5	126	4	13	12	1	z_6
6	210	5	19	17	2	z_7, z_8
7	330	6	24	24	0	—

which is obviously also an embedding of the *affine* curve $\Gamma_{\mathbf{c}}^{(1)}$. Similarly, the Taylor (!) series $z(t;m_2),\dots,z(t;m_5)$ lead to four different embeddings of the (isomorphic) affine curves $\Gamma_{\mathbf{c}}^{(2)},\dots,\Gamma_{\mathbf{c}}^{(5)}$, to wit

$$\begin{aligned}
\varphi_{\mathbf{c}}^{(2)} &: (b,d)\mapsto (1:-b:0:-bd:0:bc:c_3:0:bc(c+d^2)),\\
\varphi_{\mathbf{c}}^{(3)} &: (b,d)\mapsto (1:0:bd:0:bd^2:0:bd(2ad-b):0:2ab^2d),\\
\varphi_{\mathbf{c}}^{(4)} &: (b,d)\mapsto (1:0:-cd:cd:0:-cd^2:cd(c+2ad):\\
&\qquad\qquad\qquad\qquad -(cd)^2:-cd^2(4a^2-b)),\\
\varphi_{\mathbf{c}}^{(5)} &: (b,d)\mapsto (1:c:0:2ac:bc:0:-c_3:-bc^2:bc^2),
\end{aligned}\tag{7.53}$$

where we recall that $a=(c_1-d)/2$ and $c=b-c_2-d(c_1-d)$. Since $bd\neq 0$ (in fact, $bcd=-c_3$) we see at once that the five image curves are disjoint. However, these images are not complete, being just the embedding of the affine curves $\Gamma_{\mathbf{c}}^{(i)}$, so we check if maybe their closures (in projective space) intersect. In order to do this, we compute the image of ∞,∞' and ∞'' under each of the above five embeddings. This is done by substituting the local parameterizations (7.41) in the embeddings (7.52) and (7.53). The result is given in Table 7.2.

In this table the points P_i are the following points in $\mathbf{P}^8$ (some of them depend on $\mathbf{c}$, but we do not add this dependence in the notation).

$$\begin{aligned}
P_1 &= (0:0:0:1:0:0:0:0:0),\\
P_2 &= (0:0:0:0:0:0:0:0:1),\\
P_3 &= (1:0:0:0:0:0:c_3:0:-c_1c_3),\\
P_4 &= (1:0:0:0:0:0:-c_3:0:0),\\
P_5 &= (0:0:0:0:0:0:0:1:-1).
\end{aligned}$$

Table 7.2. The images of the points at infinity on the curves $\Gamma_{\mathbf{c}}^{(i)}$ under the five embeddings of these curves, given by (7.52) and (7.53).

	$\varphi_{\mathbf{c}}^{(1)}$	$\varphi_{\mathbf{c}}^{(2)}$	$\varphi_{\mathbf{c}}^{(3)}$	$\varphi_{\mathbf{c}}^{(4)}$	$\varphi_{\mathbf{c}}^{(5)}$
∞	P_5	P_1	P_2	P_3	P_4
∞'	P_1	P_2	P_3	P_4	P_5
∞''	P_2	P_3	P_4	P_5	P_1

Denoting $\overline{\varphi_{\mathbf{c}}^{(i)}(\Gamma_{\mathbf{c}}^{(i)})}$ by $\mathcal{D}_{\mathbf{c}}^{(i)}$, we see that $\mathcal{D}_{\mathbf{c}}^{(i)}$ contains the points P_{i-1}, P_i and P_{i+1} and that each $\mathcal{D}_{\mathbf{c}}^{(i)}$ intersects its neighbor $\mathcal{D}_{\mathbf{c}}^{(i+1)}$ in two different points P_i and P_{i+1}, while being simply tangent to the divisors $\mathcal{D}_{\mathbf{c}}^{(i-2)}$ and $\mathcal{D}_{\mathbf{c}}^{(i+2)}$ at P_{i-1} and P_{i+1} respectively. The resulting Painlevé divisor is represented in Figure 7.2. Notice that the geometric divisor of the five genus 2 curves is very similar to the one in Figure 7.1, where the curves are conics in $\mathbf{P}^5$. In fact, the Abelian surfaces $\mathbf{T}_{\mathbf{c}}^2$ that we have constructed are the singular surfaces $\overline{\mathcal{A}_{\mathbf{c}}} \subset \mathbf{P}^5$, blown up along $\mathcal{A}_\infty$ (see Section 7.3).

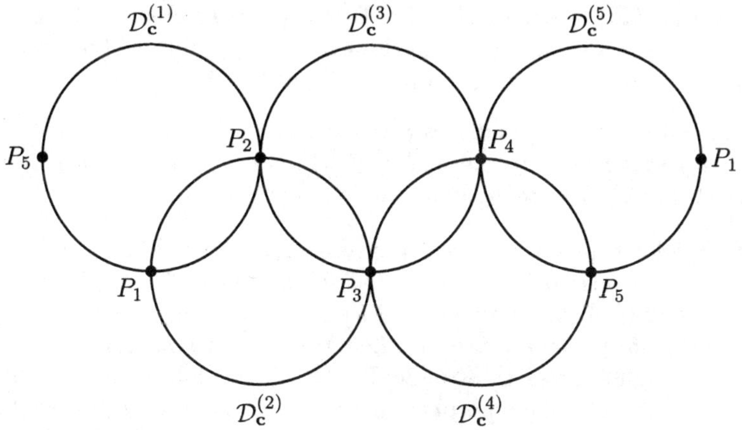

Fig. 7.2. The Painlevé divisor $\mathcal{D}_{\mathbf{c}}$ of the Kac-van Moerbeke system consists of five genus two curves that intersect in five points, each of which is a triple point of $\mathcal{D}_{\mathbf{c}}$, with two of the curves being simply tangent and the third transversal to the other two. To make the picture exact the two points labeled P_1 need to be identified, as well as the two points labeled P_5, and this should be done in such a way that the curves $\mathcal{D}_{\mathbf{c}}^{(5)}$ and $\mathcal{D}_{\mathbf{c}}^{(2)}$ are tangent, as well as the curves $\mathcal{D}_{\mathbf{c}}^{(4)}$ and $\mathcal{D}_{\mathbf{c}}^{(1)}$ at these points. It follows from Proposition 5.42 that $g_a(\mathcal{D}_{\mathbf{c}}) = 26$.

Let us now verify the adjunction formula for the present embedding, given by the functions $z_0, \dots, z_8$. The genus of $\mathcal{D}_{\mathbf{c}}^{(1)}$ is two, because it is the image of a genus two curve under an embedding. Since the pole vector that we have chosen is $(3,0,0,0,0)$ we need to compute the arithmetic genus of $3\mathcal{D}_{\mathbf{c}}^{(1)}$, which is, according to Example 5.41, given by $g_a(3\mathcal{D}_{\mathbf{c}}^{(1)}) = 3^2(2-1)+1 = 10$. This is indeed $N + r = 8 + 2$, as demanded by the adjunction formula. Therefore, this is a good place to stop (in fact the last line of Table 7.1 indicates that no new functions are found at weight 7).

7.7.2 Extending One of the Vector Fields $\mathcal{X}_F$

We suppose now that an embedding of $\mathbf{F_c}$ in $\mathbf{P}^N$ has been constructed, as in the previous paragraph. In order to apply the Liouville Theorem (Theorem 6.22), we wish to extend one of the vector fields, say $(\varphi_{\mathbf{c}})_*\mathcal{V}$, to a neighborhood of $\mathcal{D}'_{\mathbf{c}}$, which is defined, as in Theorem 6.22, as the union of all $(r-1)$-dimensional components of $\mathcal{D}_{\mathbf{c}}$, where the latter is in our case given by $\mathcal{D}_{\mathbf{c}} := \overline{\varphi_{\mathbf{c}}(\mathbf{F_c})} \setminus \varphi_{\mathbf{c}}(\mathbf{F_c})$. If the pole vector is chosen large enough and $\mathbf{F_c}$ is indeed an affine part of an irreducible Abelian variety then we know from Section 5.3 that any linear vector field on $\mathbf{F_c}$ extends to a quadratic (hence holomorphic) vector field on all of $\mathbf{P}^N$; to be precise, it extends to a vector field which is quadratic in each of the standard charts $z_i \neq 0$ of $\mathbf{P}^N$. Since we are trying to prove algebraic complete integrability we will therefore try to prove that $(\varphi_{\mathbf{c}})_*\mathcal{V}$ extends to all of $\mathbf{P}^N$. A priori holomorphicity has to be shown in each of the charts $z_i \neq 0$. However, in view of the following lemma, it is sufficient to do this for two different charts $z_i \neq 0$, $z_j \neq 0$.

Lemma 7.58. *Suppose that $\overline{\mathcal{V}}$ is a vector field on $\mathbf{P}^N$ which is holomorphic in two different charts $z_i \neq 0$ and $z_j \neq 0$. Then $\overline{\mathcal{V}}$ is a holomorphic vector field on $\mathbf{P}^N$, i.e., it is holomorphic in any chart $z_k \neq 0$.*

Proof. A holomorphic vector field $\overline{\mathcal{V}}$ on a smooth manifold M is a derivation which is characterized by the fact that for any $m \in M$ and for any open neighborhood U of m, the function $\overline{\mathcal{V}}[F]$ is a holomorphic function on U, for any holomorphic function F on U. Let $M = \mathbf{P}^N$, let $m \in \mathbf{P}^N$ and choose any open neighborhood U of m in $\mathbf{P}^N$. If $\overline{\mathcal{V}}$ is holomorphic on the open subset $z_i \neq 0$ and is holomorphic on the open subset $z_j \neq 0$ then $\overline{\mathcal{V}}[F]$ is holomorphic, away from the codimension two subset $z_i = z_j = 0$.

But recall that Hartog's Theorem says that if $V \subseteq U$ are open subsets such that the codimension of $U \setminus V$ in U is at least two, and G is a holomorphic function on V, then G extends (uniquely) to a holomorphic function on U. Applied to our case, with $V = U \setminus \{z_i = z_j = 0\}$ we find that $\overline{\mathcal{V}}[F]$ extends to a holomorphic function on U, so that, by the above characterization, $\overline{\mathcal{V}}$ is a holomorphic vector field on U. Since U is an arbitrary open subset of $\mathbf{P}^N$ the vector field $\overline{\mathcal{V}}$ is holomorphic on $\mathbf{P}^N$. □

Example 7.59. In the case of the periodic 5-particle Kac-van Moerbeke lattice, we have already written the vector field $\overline{\mathcal{V}}$ as a quadratic differential equation in two different charts in Example 7.53. It follows from Lemma 7.58 that $\overline{\mathcal{V}}$ is a holomorphic vector field on all of $\mathbf{P}^8$.

In order to apply the complex Liouville Theorem it is, strictly speaking, not necessary to construct the quadratic vector fields explicitly in two such charts; it is sufficient to prove their existence, which is in certain cases easier. The following proposition shows how this can be done in certain cases.

Proposition 7.60. *Let $\dot{x} = f(x)$ be a weight homogeneous vector field on $\mathbf{C}^n$ and suppose that each of its d principal balances is weight homogeneous. We suppose that $z_0 = 1, z_1, \dots, z_N \in \mathcal{Z}_\rho$, where $\rho = (1, 1, \dots, 1)$ and we write the d principal balances collectively as one family of Laurent solutions*

$$z_i(t) = \frac{1}{t}\left(z_i^{(0)} + z_i^{(1)}t + O(t^2)\right).$$

For $0 \leqslant i, j \leqslant N$ the Wronskian $W(z_i, z_j)$ is a quadratic polynomial in $z_0, \dots, z_N$ with coefficients in the algebra $\mathcal{H}$ of weighted homogeneous polynomial constants of motion if the following two conditions are satisfied.

(1) The $\mathcal{H}$-module generated by $(z_0, \dots, z_N)$ contains all elements of $\mathcal{Z}_\rho$ of weight at most $2S + 1$, where S is the maximum of the weights of $z_1, \dots, z_N$;
(2) The expression $z_i^{(0)} z_j^{(1)} - z_j^{(0)} z_i^{(1)}$ can be written as a quadratic polynomial in $z_0^{(0)}, \dots z_N^{(0)}$, with coefficients in $\mathcal{H}$.

Proof. With i, j as above we have that

$$\dot{z}_i(t) z_j(t) - z_i(t)\dot{z}_j(t) = \left(z_j^{(0)} z_i^{(1)} - z_i^{(0)} z_j^{(1)}\right) t^{-2} + O\left(t^{-1}\right).$$

In view of *(2)* there exist elements $a_{ij}^{kl} \in \mathcal{H}$, where $0 \leqslant k, l \leqslant N$ such that

$$z_j^{(0)} z_i^{(1)} - z_i^{(0)} z_j^{(1)} = \sum_{k,l=0}^{N} a_{ij}^{kl} z_k^{(0)} z_l^{(0)}.$$

It follows that the Laurent series

$$\dot{z}_i(t) z_j(t) - z_i(t)\dot{z}_j(t) - \sum_{k,l=0}^{N} a_{ij}^{kl} z_k(t) z_l(t),$$

which is the Laurent series of $\dot{z}_i z_j - z_i \dot{z}_j - \sum_{k,l=0}^{N} a_{ij}^{kl} z_k z_l$, has at most a simple pole in t. Since S is, by definition, the maximum of the weights of $z_1, \dots, z_N$, we have that

$$\dot{z}_i z_j - z_i \dot{z}_j - \sum_{k,l=0}^{N} a_{ij}^{kl} z_k z_l \in \bigoplus_{s=0}^{2S+1} \mathcal{Z}_\rho^{(s)}.$$

Using *(1)*, the latter is a linear combination of the z_i, with coefficients in $\mathcal{H}$. Using $z_0 = 1$ to rewrite this as a quadratic polynomial in the z_i the result follows. □

Remark 7.61. By Lemma (7.60), in order to explain that $\overline{\mathcal{V}}$ extends to a holomorphic vector field on all of $\mathbf{P}^N$ it suffices to check that $W(z_i, z_j)$ is a quadratic polynomial in terms of the z_k and this for $0 \leqslant i \leqslant N$ and for two different values of j and thus, it suffices to check this condition for $2N-1$ values of (i,j).

7.7.3 Going into the Affine

We will now show how to check condition *(4)* in Theorem 6.22. We suppose as before that we have constructed a map

$$\begin{aligned} \varphi_{\mathbf{c}} : \mathbf{C}^n &\to \mathbf{P}^N \\ m &\mapsto (1 : z_1(m) : \cdots : z_N(m)), \end{aligned}$$

which restricts for generic $\mathbf{c} \in \mathbf{C}^s$ to an isomorphic embedding of the fiber $\mathbf{F}_{\mathbf{c}}$, which we assume for simplicity to be connected (otherwise the procedure below has to be repeated for every irreducible component of $\mathbf{F}_{\mathbf{c}}$). Remember that $\varphi_{\mathbf{c}}$ gave rise, for each of the d weight homogeneous principal balances $x(t;\Gamma^{(i)})$, to a holomorphic map $\varphi_{\mathbf{c}}^{(i)} : \Gamma_{\mathbf{c}}^{(i)} \to \mathbf{P}^N$, whose image is $(r-1)$-dimensional. We denote, as before, by $\mathcal{D}_{\mathbf{c}}^{(i)}$ the analytic hypersurface $\overline{\varphi_{\mathbf{c}}^{(i)}(\Gamma_{\mathbf{c}}^{(i)})}$ of $\overline{\varphi_{\mathbf{c}}(\mathbf{F}_{\mathbf{c}})}$, where $i = 1, \ldots, d$. As we already pointed out these hypersurfaces have the following properties. First, since $\varphi_{\mathbf{c}}$ is a regular map, the above closure can be taken with respect to the complex topology or with respect to the Zariski topology, because both yield the same result (see [135, §I.10]). Second, when $r \geqslant 2$ then $\mathcal{D}_{\mathbf{c}}$, which was defined as $\mathcal{D}_{\mathbf{c}} := \overline{\varphi_{\mathbf{c}}(\mathbf{F}_{\mathbf{c}})} \setminus \varphi_{\mathbf{c}}(\mathbf{F}_{\mathbf{c}})$, is connected (see [78, Cor. III.7.9]); the case $r = 1$, in which $\mathcal{D}_{\mathbf{c}}$ simply consists of points, will be excluded in the present discussion, see Remark 7.65 below. Since $\varphi_{\mathbf{c}}$ is an isomorphic embedding of $\mathbf{F}_{\mathbf{c}}$, we have, in addition, that

$$\mathcal{D}_{\mathbf{c}} = \overline{\varphi_{\mathbf{c}}(\mathbf{F}_{\mathbf{c}})} \setminus \varphi_{\mathbf{c}}(\mathbf{F}_{\mathbf{c}}) = \overline{\varphi_{\mathbf{c}}(\mathbf{F}_{\mathbf{c}})} \cap (z_0 = 0),$$

in particular $\mathcal{D}_{\mathbf{c}}$ is contained in the subspace $z_0 = 0$. Thus, if $z_j(m)$, $1 \leqslant j \leqslant N$ are finite, we must be in the affine $\varphi_{\mathbf{c}}(\mathbf{F}_{\mathbf{c}})$.

Let us denote the distinct irreducible components of $\mathcal{D}_{\mathbf{c}}$ of maximal dimension, that are different from the d components $\mathcal{D}_{\mathbf{c}}^{(1)},\dots,\mathcal{D}_{\mathbf{c}}^{(d)}$, if any, by $\mathcal{D}_{\mathbf{c}}^{(d+1)},\dots,\mathcal{D}_{\mathbf{c}}^{(\tilde{d})}$. Such components may indeed exist, i.e., it may happen that $d<\tilde{d}$: nothing guarantees that each divisor $\mathcal{D}_{\mathbf{c}}^{(i)}$ is found from a weight homogeneous balance.

We first show that we flow immediately into the affine, $\varphi_{\mathbf{c}}(\mathbf{F}_{\mathbf{c}})$, when we start from points in $\cup_{i=1}^{d}\varphi_{\mathbf{c}}^{(i)}(\Gamma_{\mathbf{c}}^{(i)})$. Let us consider a principal balance $x(t;\Gamma^{(i)})$, let P be any point in $\varphi_{\mathbf{c}}^{(i)}(\Gamma_{\mathbf{c}}^{(i)})$ and let $p\in\Gamma_{\mathbf{c}}^{(i)}$ be such that $\varphi_{\mathbf{c}}^{(i)}(p)=P$. We need to show that $\Phi_t(P)\in\varphi_{\mathbf{c}}(\mathbf{F}_{\mathbf{c}})$ for $|t|$ small and non-zero, where Φ denotes, as before, the flow of the holomorphic vector field $\overline{\mathcal{V}}$ on $\mathbf{P}^N$. To prove this, recall on the one hand from Section 7.2 that the weight homogeneous Laurent series $x_j(t;p)$ are convergent for small and non-zero $|t|$, and on the other hand that the embedding functions z_j are polynomials in the phase variables $x_1,\dots,x_n$. Thus, we may pick $\epsilon>0$ such that $z_j(t;p)$ is finite for $j=1,\dots,N$ and for t such that $0<|t|<\epsilon$. For such a t we therefore have that $\Phi_t(P)$ does not belong to the hyperplane $z_0=0$, hence it gets mapped into $\varphi_{\mathbf{c}}(\mathbf{F}_{\mathbf{c}})$, as was claimed.

Remark 7.62. In the special case where at least one of the ρ_i is zero, as in our Kac-van Moerbeke example, where $\rho=(3,0,0,0,0)$, $\varphi_{\mathbf{c}}$ is an embedding but not an isomorphic embedding. By construction, for $2\leqslant i\leqslant 5$ all the functions z_j are finite on $\overline{\varphi_{\mathbf{c}}^{(i)}(\Gamma_{\mathbf{c}}^{(i)})}$, since $\rho_i=0$, and so $\mathcal{D}_{\mathbf{c}}\not\subseteq\{z_0=0\}$. Thus, to check in such a case that the flow starting at $\mathcal{D}_{\mathbf{c}}$ immediately goes into the affine $\varphi_{\mathbf{c}}(\mathbf{F}_{\mathbf{c}})$, we must actually check using the series $z(t;\Gamma_{\mathbf{c}}^{(i)})$ that the $x_j(t;\Gamma_{\mathbf{c}}^{(i)})$ are finite, for $|t|\neq 0$ small, which is about the same effort as above.

The varieties $\Gamma_{\mathbf{c}}^{(i)}$ are not compact, so $K^{(i)}:=\mathcal{D}_{\mathbf{c}}^{(i)}\setminus\varphi_{\mathbf{c}}^{(i)}(\Gamma_{\mathbf{c}}^{(i)})$ is not empty, where $i=1,\dots,d$. For points in $K^{(i)}$ no general argument guarantees that the flow goes into the affine immediately, hence this has to be checked in every example. We first explain how to do this in the case of two-dimensional integrable systems ($r=2$), where clearly $K:=\cup_{i=1}^{d}K^{(i)}$ consists of a finite number of points. Let P_0 be one of the points of K, say $P_0\in K^{(i)}$, and let us consider the Laurent series $z_j(t;\Gamma_{\mathbf{c}}^{(i)})$, which is obtained by substituting the principal balance $x(t;\Gamma^{(i)})$, restricted to $\mathbf{F}_{\mathbf{c}}$, in z_j. We denote by p_0 one of the points of the smooth compactification $\overline{\Gamma_{\mathbf{c}}^{(i)}}$ of $\Gamma_{\mathbf{c}}^{(i)}$ for which

$$\lim_{p\to p_0}\varphi_{\mathbf{c}}^{(i)}(p)=P_0;$$

notice that p_0 needs not be unique, since $\varphi_{\mathbf{c}}^{(i)}$ needs not extend to an embedding of its smooth[5] compactification.

[5] Recall from Theorem 7.22 that all Painlevé walls $\Gamma_{\mathbf{c}}^{(i)}$ are non-singular, hence they admit a smooth compactification. However, the divisors $\mathcal{D}_{\mathbf{c}}^{(i)}=\overline{\varphi_{\mathbf{c}}^{(i)}(\Gamma_{\mathbf{c}}^{(i)})}$ may be singular, as we will see in the examples.

For $j = 1, \dots, N$, write

$$z_j(t; \Gamma_{\mathbf{c}}^{(i)}) = \frac{1}{t^l} \sum_{k=0}^{\infty} z_j^{(k)} t^k,$$

where l is chosen minimal, i.e., $z_j^{(0)} \neq 0$ for at least one j. Among the j for which $z_j^{(0)} \neq 0$, choose the index α for which $z_\alpha^{(0)}(\Gamma_{\mathbf{c}}^{(i)})$ has the largest pole as $p \to p_0$ in $\overline{\Gamma_{\mathbf{c}}^{(i)}}$. Then, for any j the limit

$$\lim_{p \to p_0} \lim_{t \to 0} \frac{z_j(t; \Gamma_{\mathbf{c}}^{(i)})}{z_\alpha(t; \Gamma_{\mathbf{c}}^{(i)})}$$

is finite, and thus, $z_\alpha \neq 0$ defines a chart about P_0. Since $\overline{\mathcal{V}}$ is holomorphic on $\mathbf{P}^N$ it follows that $\overline{\mathcal{V}}$ is holomorphic in the coordinates $y_j = z_j/z_\alpha$, $0 \leqslant j \leqslant N$. Now the series $y_j(t; \Gamma_{\mathbf{c}}^{(i)})$ solve, for $|t|$ small and non-zero, the differential equations that describe $\mathcal{V}$ in the chart $z_0 \neq 0$, hence we may identify these series with the (Taylor series) solutions to $\overline{\mathcal{V}}$ in the chart $z_\alpha \neq 0$.

Explicitly, we have for $p \in \Gamma_{\mathbf{c}}^{(i)}$ close to p_0 that the series $y_j(t;p)$ equals the series $y_j(t; \varphi_{\mathbf{c}}(p))$, where the latter is the Taylor series of y_j with respect to $\overline{\mathcal{V}}$, starting at $\varphi_{\mathbf{c}}(p)$. Since the latter series admits a limit, which is nothing but $y_j(t; P_0)$ the series $y_j(t;p)$ admits a limit as $p \to p_0$; we will denote it by $y_j(t;p_0)$. Suppose now that we can show that $y_0(t;p_0) \neq 0$ (as a series in t) then it follows that all $z_j(t;p_0)$ are finite (for $|t|$ small and non-zero), because $z_j = y_j/y_0$ and we are done. Hence, we only need to check in the two-dimensional examples that for any principal balance $x(t; \Gamma^{(i)})$ the Taylor series

$$y_0(t; \Gamma_{\mathbf{c}}^{(i)}) := \frac{1}{z_\alpha(t; \Gamma_{\mathbf{c}}^{(i)})} = \sum_{k \geqslant 0} a_k(p) t^k, \tag{7.54}$$

has a non-zero limit when $p \in \Gamma_{\mathbf{c}}^{(i)}$ tends to p_0, and this for enough points p_0 so as to cover all points in K. Notice that, as a bonus, the above computation yields the lower balance that goes with $P_0 \in K$.

When $r > 2$ one takes instead of m an irreducible component of K and one performs the same check, where p tends now to a generic point of the boundary of $\varphi_{\mathbf{c}}(\Gamma_{\mathbf{c}}^{(i)})$. This will guarantee that the flow which starts from any point on $\mathcal{D}_{\mathbf{c}}^{(i)}$ goes into the affine immediately, except for a subvariety of codimension at least two (i.e., if the test is successful). Since the zero locus of a holomorphic function is of codimension one this property extends, by Hartog's Theorem, over that subset and the flow of $\overline{\mathcal{V}}$, starting from any point on $\mathcal{D}_{\mathbf{c}}^{(i)}$ goes into the affine immediately. It suffices therefore, when $r > 2$ to do the check for a generic boundary point of $\varphi_{\mathbf{c}}(\Gamma_{\mathbf{c}}^{(i)})$, which makes the test also in the higher dimensional case feasible.

Example 7.63. As we pointed out in Example 7.57, we need in the case of the periodic 5-particle Kac-van Moerbeke lattice do a little bit more work, because the map $\varphi_{\mathbf{c}}$ is not an isomorphic embedding. We will do the check for one of the points, say for $P_1 = (0:0:0:1:0:0:0:0:0)$, see Table 7.2. Since the only non-zero entry corresponds to z_3 we must substitute the parameterization (7.41) of a neighborhood of ∞' in $1/z_3(t;m_1)$ and we must show that its limit, for $\varsigma \to 0$, is non-zero. Since $z_3(t;m_1) = \frac{1}{t^3} - \frac{a}{t^2} + O(1)$ we have that

$$z_3^{-1}(t;m_1) = t^3(1+at+O(t^2)) = t^3\left(1+\frac{1}{2}(c_1+c_3\varsigma^2+O(\varsigma^3))t+O(t^2)\right), \tag{7.55}$$

which yields, in the limit $\varsigma \to 0$, the series $t^3(1+c_1t/2+O(t^2))$, which is different from zero. This means that, starting from P_1 we do not flow into the divisor $\mathcal{D}_{\mathbf{c}}^{(1)}$. In order to show that we flow into the affine $\varphi_{\mathbf{c}}(\mathbf{F}_{\mathbf{c}})$ we show that $z_1/z_3(t;m_1)$ and $z_2/z_3(t;m_1)$ have also a non-zero limit, as $\varsigma \to 0$, and so by (7.51) all the $x_i(t;m_1)$ are finite for $|t| \neq 0$ small, showing explicitly that we flow into the finite from P_1. For the first one this is trivial since

$$\frac{z_1}{z_3}(t;m_1) = \left(-\frac{1}{t^2}+O(1)\right)(t^3+O(t^4)) = -t+O(t^2),$$

as follows from (7.43). For the second one the computation is longer, since the first terms of the series $z_2/z_3(t;m_1)$ vanish when taking the limit, in fact three more terms are needed in (7.43) in order to find a non-zero term; besides this fact, which only makes the computation longer, the calculation is trivial and we only state the final formula,

$$\lim_{\varsigma\to 0}\frac{z_2}{z_3}(t;m_1) = -\frac{c_3}{4}t^5+O(t^6).$$

Repeating the argument for the other points P_i proves that we flow into the affine, starting from any of the points in $\mathcal{D}_{\mathbf{c}}^{(1)} \cup \cdots \cup \mathcal{D}_{\mathbf{c}}^{(5)}$.

Let us assume now that we have successfully completed the above check, so that we know that the flow of $\overline{\mathcal{V}}$ goes into the affine immediately, when starting from any of the points of $\mathcal{D}_{\mathbf{c}}^{(i)}$, where $i = 1, \ldots, d$. If we can show that there are no other points in $\mathcal{D}_{\mathbf{c}}$, i.e., that $d = \tilde{d}$ then we are done. Since $\mathcal{D}_{\mathbf{c}}$ is connected (being the hyperplane section $\overline{\varphi_{\mathbf{c}}(\mathbf{F}_{\mathbf{c}})} \cap \{z_0 = 0\}$ of the irreducible variety $\overline{\varphi_{\mathbf{c}}(\mathbf{F}_{\mathbf{c}})}$), there must be at least one point P_0 in $\cup_{i=1}^{d}\mathcal{D}_{\mathbf{c}}^{(i)}$ which belongs to $\cup_{i=d+1}^{\tilde{d}}\mathcal{D}_{\mathbf{c}}^{(i)}$, if the latter is non-empty; therefore $P_0 \in \mathcal{D}_{\mathbf{c}}^{(i)}$ for some $i = 1, \ldots, d$. We first show that there exists a neighborhood U of P_0 in $\overline{\varphi_{\mathbf{c}}(\mathbf{F}_{\mathbf{c}})}$ and an $\epsilon > 0$ such that

$$U \subseteq \left\{\Phi_t(\mathcal{D}_{\mathbf{c}}^{(i)}) \mid 0 \leqslant |t| < \epsilon\right\}, \tag{7.56}$$

where $\varPhi$ denotes the flow of $\overline{\mathcal{V}}$ on $\mathbf{P}^N$. Since P_0 belongs to $\mathcal{D}_{\mathbf{c}}^{(i)}$ it flows into the affine immediately. By precisely the same argument as in the first part of the proof of Theorem 6.22 we have that $\overline{\varphi_{\mathbf{c}}(\mathbf{F_c})}$ is smooth at P_0, and hence in a neighborhood V of P_0 in $\overline{\varphi_{\mathbf{c}}(\mathbf{F_c})}$. We consider the restriction of $\overline{\mathcal{V}}$ to V (recall that $\overline{\mathcal{V}}$ is tangent to $\varphi_{\mathbf{c}}(\mathbf{F_c})$, hence it is tangent to its closure $\overline{\varphi_{\mathbf{c}}(\mathbf{F_c})}$ at smooth points) and we consider local coordinates $(t, s_2, \ldots, s_r) = (t, s)$ in a neighborhood $V' \subseteq V$ of P_0 such that $\overline{\mathcal{V}} = \partial/\partial t$ in this neighborhood, which can be done since $\overline{\mathcal{V}}$ does not vanish at P_0. Let $V'' \subseteq V'$ be an open neighborhood of P_0 in $\overline{\varphi_{\mathbf{c}}(\mathbf{F_c})}$ and let $\epsilon > 0$, both chosen small enough to guarantee that $\varPhi_t(V'') \subseteq V'$ for any t with $|t| < \epsilon$. Notice that the flow $\varPhi$ takes on V'' the simple form $\varPhi_t(t_0, s) = (t + t_0, s)$. The subset $\mathcal{D}_{\mathbf{c}}^{(i)}$, which is a divisor of $\overline{\varphi_{\mathbf{c}}(\mathbf{F_c})}$, is locally — say on V'' — given as the zero locus of a holomorphic function, which is a monic polynomial in t, i.e., $\mathcal{D}_{\mathbf{c}}^{(i)}$ is given by $Q_k = 0$ where

$$Q_k(t, s) := t^k + a_{k-1}(s)t^{k-1} + \ldots + a_0(s),$$

the coefficients a_l being holomorphic functions in $s = (s_2, \ldots, s_r)$ (only), that vanish at $s = 0$. Indeed, this is an immediate application of the Weierstrass Preparation Theorem (Theorem 5.7): we only need to check that the t-axis (given by $s = 0$) is not contained (locally) in $\mathcal{D}_{\mathbf{c}}^{(i)}$, but this is so since P_0 flows into the affine immediately. Let $\delta > 0$ be chosen such that for any s in the polydisk $B_\delta := \{s \mid |s_l| < \delta \text{ for } 2 \leqslant l \leqslant r\}$ the k solutions t to $Q_k(t, s) = 0$ all satisfy $|t| < \epsilon/2$, which can be done since the coefficients of Q_k are continuous functions of s. Consider

$$U := \{(t, s) \in V'' \mid |t| < \epsilon/2,\ s \in B_\delta\}.$$

If $(t, s) \in U$ then, by definition, there exists $t_0 \in \mathbf{C}$ with $|t_0| < \epsilon/2$ such that $Q_k(t_0, s) = 0$, i.e., such that $(t_0, s) \in \mathcal{D}_{\mathbf{c}}^{(i)}$, and $|t| < \epsilon/2$. Then $(t, s) \in \varPhi_{t-t_0}(t_0, s)$, where $|t - t_0| < \epsilon$ and $(t_0, s) \in \mathcal{D}_{\mathbf{c}}^{(i)}$. Thus, if $(t, s) \in U$ then

$$(t, s) \in \left\{\varPhi_t(\mathcal{D}_{\mathbf{c}}^{(i)}) \mid 0 \leqslant |t| < \epsilon)\right\},$$

showing (7.56).

Let U be a neighborhood of P_0 in $\overline{\varphi_{\mathbf{c}}(\mathbf{F_c})}$, as in (7.56), and let us suppose that $\overline{\mathcal{V}}$ is transversal to one of the branches of $\mathcal{D}_{\mathbf{c}}^{(i)}$ at P_0, where $1 \leqslant i \leqslant d$. If this is not the case one looks for a linear combination of the vector fields $\overline{\mathcal{V}}_1, \ldots, \overline{\mathcal{V}}_r$, with this property; in this case, i.e., when we need to replace $\overline{\mathcal{V}}$ by this linear combination we need to check for the latter that it is holomorphic (by exhibiting the quadratic differential equations), as we did for $\mathcal{V}$. In fact, such a linear combination does not need to exist, but it does exist in all the examples that we will treat; for the case in which such a $\mathcal{V}$ does not exist, see Remark 7.66 below.

Defining α as before and letting $y_j := z_j/z_\alpha$ for $j = 0, \ldots, N$ the above transversality condition at P_0 can be checked by the following rank condition.

$$\mathrm{Rk}\begin{pmatrix} \frac{\partial y_0}{\partial t}(t;\Gamma^{(i)}) & \frac{\partial y_0}{\partial s_2}(t;\Gamma^{(i)}) & \cdots & \frac{\partial y_0}{\partial s_r}(t;\Gamma^{(i)}) \\ \vdots & \vdots & & \vdots \\ \frac{\partial y_N}{\partial t}(t;\Gamma^{(i)}) & \frac{\partial y_N}{\partial s_2}(t;\Gamma^{(i)}) & \cdots & \frac{\partial y_N}{\partial s_r}(t;\Gamma^{(i)}) \end{pmatrix}_{|t=s=0} = r.$$

For example, if $r = 2$ then it is enough to show that there exist $j \neq k$ such that

$$\begin{vmatrix} \frac{\partial y_j}{\partial t}(t;\Gamma^{(i)}) & \frac{\partial y_k}{\partial s}(t;\Gamma^{(i)}) \\ \frac{\partial y_j}{\partial t}(t;\Gamma^{(i)}) & \frac{\partial y_k}{\partial s}(t;\Gamma^{(i)}) \end{vmatrix}_{|t=s=0} \neq 0,$$

where we have written s_2 simply as s.

By the implicit function theorem, the fact that $\overline{\mathcal{V}}$ is transversal to one of the branches of $\mathcal{D}_{\mathbf{c}}^{(i)}$ at P_0 implies that $(t, s_2, \ldots, s_r)$ is a system of holomorphic coordinates on a neighborhood of P_0 in $\varphi_{\mathbf{c}}(\mathbf{F}_{\mathbf{c}})$, where the coordinates $s_2, \ldots, s_r$ come from coordinates on the chosen branch of $\mathcal{D}_{\mathbf{c}}^{(i)}$ at P_0. As before, the solution to $\overline{\mathcal{V}}$ starting on the branch of $\mathcal{D}_{\mathbf{c}}^{(i)}$ will consist of holomorphic functions $y_j(t, s_2, \ldots, s_r)$ of their arguments. Note that since $z_j = y_j/y_0$, $0 \leqslant j \leqslant N$, we have that z_j is finite whenever y_0 is finite and then we are in the affine (as usual in the case when $\varphi_{\mathbf{c}}$ is not an isomorphic embedding we must work slightly harder to check we are in the affine). If we write

$$y_0(t, s_2, \ldots, s_r) = a_\mu(t, s) + a_{\mu+1}(t, s) + \cdots,$$

where a_l is the sum of all monomials of degree l in the series, and where $a_\mu(t, s)$ is not identically zero, then $\mu = \mu(y_0; P_0)$ is the multiplicity of $\mathcal{D}_{\mathbf{c}}$ at P_0, since y_0 is the locally defining function of $\mathcal{D}_{\mathbf{c}}$. Since $\overline{\mathcal{V}}$ is transversal to $\mathcal{D}_{\mathbf{c}}$ at a generic point of each irreducible component of $\mathcal{D}_{\mathbf{c}}$ that meets at P_0 (since they all flow into the affine $\varphi_{\mathbf{c}}(\mathbf{F}_{\mathbf{c}})$), we can compute the multiplicity of y_0 restricted to these components, which is just the generic pole of z_α along each of these components (the ones that we have found from the principal balances, $\mathcal{D}^{(1)}, \ldots, \mathcal{D}^{(d)}$, as well as the other ones, $\mathcal{D}^{(d+1)}, \ldots, \mathcal{D}^{(\tilde{d})}$). Thus, if we only consider the first group and we check that

$$\mu(y_0; P_0) = \sum_{1 \leqslant i \leqslant d} \mu(\mathcal{D}_{\mathbf{c}}^{(i)}, P_0)\left(\text{pole order of } z_\alpha(t; \Gamma_{\mathbf{c}}^{(i)})\right), \tag{7.57}$$

where $\mu(\mathcal{D}_{\mathbf{c}}^{(i)}, P_0)$ is the multiplicity of P_0 on $\mathcal{D}_{\mathbf{c}}^{(i)}$, then we have shown that there cannot be any more components of $\mathcal{D}_{\mathbf{c}}$ of codimension 1 which contain P_0 other than those corresponding to the d principal balances that were found (otherwise such a component would raise the multiplicity above $\mu(y_0, P_0)$).

For a point $P \in \varphi_{\mathbf{c}}^{(i)}(\Gamma_{\mathbf{c}}^{(i)})$ this condition holds automatically, since the defining z_α one picks at such a point will look like

$$z_\alpha(t;\Gamma_{\mathbf{c}}^{(i)}) = \frac{c}{t^l} + \cdots, \qquad y_0(t;\Gamma_{\mathbf{c}}^{(i)}) = \frac{1}{z_\alpha} = \frac{t^l}{c} + \cdots,$$

where $c \neq 0$, so that $\mu = l$ and our condition is satisfied. It follows that we only need to check the multiplicity condition (7.57) for the infamous set of points

$$K := \bigcup_{i=1}^{d} K^{(i)} = \bigcup_{i=1}^{d} \mathcal{D}_{\mathbf{c}}^{(i)} \setminus \varphi_{\mathbf{c}}^{(i)}(\Gamma_{\mathbf{c}}^{(i)}),$$

to conclude that $\mathcal{D}_{\mathbf{c}}^{(i)}$ with $d+1 \leqslant i \leqslant d'$ does not exist and so $\mathcal{D}_{\mathbf{c}} = \sum_{i=1}^{d} \mathcal{D}_{\mathbf{c}}^{(i)}$.

Remark 7.64. In the case when we have an embedding $\varphi_{\mathbf{c}} : \mathbf{F}_{\mathbf{c}} \to \mathbf{P}^N$ which is not an isomorphic embedding, the latter arguments need to be trivially modified. For example in the Kac-van Moerbeke lattice we have that $\mathcal{D}_{\mathbf{c}} \subseteq \{z_0 = 0\} \cap \overline{\varphi_{\mathbf{c}}(\mathbf{F}_{\mathbf{c}})}$ is replaced by

$$\mathcal{D}_{\mathbf{c}} \subseteq (\{z_0 = 0\} \cup \{z_1/z_3 = 0\} \cup \{z_2/z_3\} = 0) \cap \overline{\varphi_{\mathbf{c}}(\mathbf{F}_{\mathbf{c}})},$$

and thus, if we are in a chart U defined e.g. by z_3, so that $y_j = z_j/z_3$, $0 \leqslant j \leqslant 8$, then

$$U \cap \mathcal{D}_{\mathbf{c}} \subseteq \{y_0 = 0\} \cup \{y_1 = 0\} \cup y_2 = 0,$$

and $y_0 y_1 y_2 = 0$ is a local defining function for the divisor at P_0. The multiplicity condition (7.57) would now become

$$\mu(y_0; P_1) + \mu(y_1; P_1) + \mu(y_2, P_1) = \sum_{j=0}^{2} \sum_{\substack{1 \leqslant i \leqslant 5 \\ \mathcal{D}_{\mathbf{c}}^{(i)} \ni P_1}} \left(\text{zero order of } y_j(t;\Gamma_{\mathbf{c}}^{(i)})\right), \tag{7.58}$$

i.e., the multiplicities of the y_0, y_1, y_2, due to their generic polar behavior on the Painlevé divisors containing P_1 must agree with the multiplicity coming from $y_j(t, s_2)$, $0 \leqslant j \leqslant 2$. In order to do this check, first compute the left hand side in (7.58) by writing for $j = 0, 1, 2$ the series $y_j(t;\Gamma_{\mathbf{c}}^{(1)})$ in terms of the local parameter ς, given in (7.41), that parameterizes a neighborhood of ∞'. This gives

$$\begin{aligned} y_0(t;\Gamma_{\mathbf{c}}^{(1)}) &= -t^3 + a_4(t,\varsigma) + \cdots \\ y_1(t;\Gamma_{\mathbf{c}}^{(1)}) &= -t + a_2(t,\varsigma) + \cdots \\ y_2(t;\Gamma_{\mathbf{c}}^{(1)}) &= -c_3\varsigma^2 t + a_4'(t,\varsigma) + \cdots, \end{aligned}$$

so that the left hand side in (7.58) evaluates to $3+1+3=7$. On the other hand, the functions y_0, y_1 and y_2 have a respective zero of order 3, 1 and 1 along $\mathcal{D}_{\mathbf{c}}^{(1)}$, as is read off at once from the principal balances $z_j(t;\Gamma_{\mathbf{c}}^{(1)})$, giving a contribution 5 to the right hand side in (7.58).

Similarly one reads off from the principal balances $z_j(t; \Gamma_{\mathbf{c}}^{(2)})$ and $z_j(t; \Gamma_{\mathbf{c}}^{(5)})$ that neither y_0 nor y_1 have a zero along $\mathcal{D}_{\mathbf{c}}^{(2)}$ and $\mathcal{D}_{\mathbf{c}}^{(5)}$, while y_2 has a simple zero along $\mathcal{D}_{\mathbf{c}}^{(2)}$ as well as along $\mathcal{D}_{\mathbf{c}}^{(5)}$, so that the right hand side in (7.58) adds up to 7, as required.

Remark 7.65. We have explained how to check that we go into the affine when $r \neq 1$ (r is the dimension of the fibers of the momentum map). In the one-dimensional case this is however simpler. Suppose that we have an equation for an affine curve and an embedding for this curve into projective space. In this case we can determine all points that will form the divisor $\mathcal{D}_{\mathbf{c}}$ by computing the image of the points at infinity, precisely as we did in the general case. Then we must check that the number of (principal) balances equals the number of points in $\mathcal{D}_{\mathbf{c}}$.

Remark 7.66. We have assumed that there exists a linear combination $\overline{\mathcal{V}}$ of the vector fields $\overline{\mathcal{V}}_1, \ldots, \overline{\mathcal{V}}_r$, which is transversal to one of the branches of $\mathcal{D}_{\mathbf{c}}^{(j)}$ at P_0. It is however possible that such a $\mathcal{V}$ does not exist. Consider for example the case in which $\mathcal{D}_{\mathbf{c}}^{(j)}$ is a curve which is singular at P_0, but with only one branch (a cusp $y^2 = x^3$ is the easiest example) and that this is the only irreducible component of $\mathcal{D}_{\mathbf{c}}$ through P_0 that we have found using the principal balances. In this case we cannot use the parameters on the divisor to build a system of holomorphic coordinates on a neighborhood of P_0. In this case, take the parameters $t_1, \ldots, t_r$ that are dual to the vector fields $\overline{\mathcal{V}}_1, \ldots, \overline{\mathcal{V}}_r$, which are certainly holomorphic coordinates in a neighborhood of P_0, if the system is a.c.i., and write y_0 as a series in these parameters (we explained how to do this in Remark 7.49; it should in this case also be checked in our usual way that any linear combination of the vector fields $\overline{\mathcal{V}}_1, \ldots, \overline{\mathcal{V}}_r$ is holomorphic). Then the multiplicity of $\mathcal{D}_{\mathbf{c}}$ at P_0 can again be read off as the degree of the first non-zero term in the series $y_0(t_1, \ldots, t_r)$.

Part III

Examples

8 Integrable Geodesic Flow on SO(4)

8.1 Geodesic Flow on SO(4)

8.1.1 From Geodesic Flow on G to a Hamiltonian Flow on $\mathfrak{g}$

Let $\mathbf{G}$ be a (real) Lie group. A left invariant metric on $\mathbf{G}$ is completely specified by an inner product $(\cdot\,,\cdot)$ on its Lie algebra $\mathfrak{g}$. Namely, if we denote by L_g left translation in $\mathbf{G}$ over $g \in \mathbf{G}$ and if we denote the inner product on $T_g\mathbf{G}$ by $(\cdot\,,\cdot)_g$, then left invariance means that

$$(X_g, Y_g)_g = \left(\mathrm{d}L_{g^{-1}}(X_g), \mathrm{d}L_{g^{-1}}(Y_g)\right)_e, \tag{8.1}$$

for any X_g, $Y_g \in T_g\mathbf{G}$, where $e \in \mathbf{G}$ is the identity element in $\mathbf{G}$. Identifying $T_e\mathbf{G}$ with $\mathfrak{g}$ we will write in the sequel $(\cdot\,,\cdot)$ for $(\cdot\,,\cdot)_e$. We consider geodesic flow on $\mathbf{G}$ with respect to $(\cdot\,,\cdot)_g$ and we use the calculus of variations to rewrite the variational equations of geodesic flow as a Hamiltonian vector field on $\mathfrak{g}$. We assume for the computation that follows that the reader is familiar with the basics of the calculus of variations (basically the definition of the variational derivative and the Euler-Lagrange equations); see [1] for a quick introduction. In order to make the computation more readable we will assume in the sequel that $\mathbf{G}$ is a semi-simple Lie group, embedded in $\mathbf{GL}(\mathfrak{g})$ by the adjoint representation Ad. Then (8.1) can be written in the simpler form

$$(X_g, Y_g)_g = \left(g^{-1}X_g, g^{-1}Y_g\right).$$

and the Killing form $\langle\cdot\,|\,\cdot\rangle$ on $\mathfrak{g} \subseteq \mathfrak{gl}(\mathfrak{g})$ is given by

$$\langle X\,|\,Y\rangle = \mathrm{Trace}(XY).$$

Let us consider an arbitrary curve $g : [a,b] \to \mathbf{G} : t \mapsto g(t)$ in $\mathbf{G}$. Then g is a geodesic if and only if it minimizes (locally) the length functional, given by

$$l(g) = \int_a^b \sqrt{(\dot{g}(t), \dot{g}(t))_{g(t)}}\,\mathrm{d}t,$$

which is the same as minimizing the energy functional

$$\tilde{l}(g) = \frac{1}{2}\int_a^b (\dot{g}(t), \dot{g}(t))_{g(t)}\,\mathrm{d}t = \frac{1}{2}\int_a^b \left(g^{-1}(t)\dot{g}(t), g^{-1}(t)\dot{g}(t)\right)\mathrm{d}t,$$

(the 1/2 is irrelevant but comes in handy later). Let I denote the symmetric endomorphism of $\mathfrak{g}$, which is defined by

$$\langle I(X) \mid Y \rangle = (X, Y), \qquad X, Y \in \mathfrak{g}.$$

Then we have to minimize

$$\tilde{l}(g) = \frac{1}{2} \int_a^b \langle I(g^{-1}(t)\dot{g}(t)) \mid g^{-1}(t)\dot{g}(t) \rangle \mathrm{d}t.$$

The Euler-Lagrange equations are given by

$$\frac{\delta F}{\delta g} = \frac{\mathrm{d}}{\mathrm{d}t}\frac{\delta F}{\delta \dot{g}}, \qquad F(g, \dot{g}) = \frac{1}{2} \langle I(g^{-1}\dot{g}) \mid g^{-1}\dot{g} \rangle, \tag{8.2}$$

where $\frac{\delta F}{\delta g}$ and $\frac{\delta F}{\delta \dot{g}}$ are the variational derivatives in the direction of g, resp. $\dot{g}$. Hence we compute, using the fact that I is symmetric, and using the properties of Trace,

$$\begin{aligned} \delta_g F &= \frac{1}{2}\delta_g \langle g^{-1}\dot{g} \mid I(g^{-1}\dot{g}) \rangle \\ &= -\langle g^{-1}\,\delta g\, g^{-1}\dot{g} \mid I(g^{-1}\dot{g}) \rangle \\ &= -\langle \delta g \mid g^{-1}\dot{g} I(g^{-1}\dot{g}) g^{-1} \rangle, \end{aligned}$$

so that

$$\frac{\delta F}{\delta g} = -g^{-1}\dot{g}\, I(g^{-1}\dot{g})\, g^{-1} = -\omega X g^{-1}, \tag{8.3}$$

where we have defined

$$\omega := g^{-1}\dot{g}, \quad \text{and} \quad X := I(\omega).$$

Similarly,

$$\delta_{\dot{g}} F = \frac{1}{2}\delta_{\dot{g}} \langle \dot{g} \mid I(g^{-1}\dot{g})g^{-1} \rangle = \langle \delta\dot{g} \mid I(g^{-1}\dot{g})g^{-1} \rangle,$$

which gives in terms of the above notation

$$\frac{\delta F}{\delta \dot{g}} = I(g^{-1}\dot{g})g^{-1} = Xg^{-1}. \tag{8.4}$$

If we substitute (8.3) and (8.4) in the Euler-Lagrange equations (8.2) then we find that

$$\begin{aligned} 0 &= \left(-\frac{\delta F}{\delta g} + \frac{\mathrm{d}}{\mathrm{d}t}\frac{\delta F}{\delta \dot{g}}\right) g, \\ &= \left(\omega X g^{-1} + \frac{\mathrm{d}}{\mathrm{d}t}(Xg^{-1})\right) g \\ &= \omega X + \dot{X} - X\omega. \end{aligned}$$

We conclude that geodesic flow on **G** is governed by the following two equations:

$$\dot{X} = [X, \omega],$$
$$\dot{g} = g\omega,$$

where $X = I(\omega)$. The first one of these equations is non-linear, but autonomous and should be solved first for X, which also gives ω; the second equation becomes then a linear, non-autonomous equation, which can be solved for g, giving the geodesic. Thus, the first equation, which defines a vector field $\mathcal{V}$ on $\mathfrak{g}$, is the more fundamental one and is the one that will be studied here in a few cases. In fact, in the previous discussion, $\mathfrak{g}$ was a real Lie algebra, but we will study the complexification[1] of $\mathcal{V}$ (on the complexification of $\mathfrak{g}$). Somewhat imprecise, we will refer to the vector field $\mathcal{V}$ (or its flow) on $\mathfrak{g}$ as *geodesic flow* on $\mathfrak{g}$ (whether $\mathcal{V}$ and $\mathfrak{g}$ are real or complex).

Suppose now that H is an arbitrary quadratic form on $\mathfrak{g}$. Recall from Section 3.5 that its gradient at X (with respect to $\langle\cdot\,|\,\cdot\rangle$) is the element $\nabla H(X) \in \mathfrak{g}$, defined by

$$\langle \nabla H(X) \,|\, Y \rangle = \langle \mathrm{d}H(X), Y \rangle .$$

By non-degeneracy of $\langle\cdot\,|\,\cdot\rangle$ we can find, for any given ω (which is a linear function of X, namely $\omega = I^{-1}(X)$), a quadratic form H such that $\nabla H(X) = \omega = I^{-1}(X)$. Therefore, we may restate the problem of studying geodesic flow on $\mathfrak{g}$ by the problem of studying the vector field given by the Lax equation

$$\dot{X} = [X, \nabla H(X)], \tag{8.5}$$

where H is a quadratic form on $\mathfrak{g}$. As we pointed out in Section 3.5 such a vector field is Hamiltonian with respect to the Lie-Poisson structure on $\mathfrak{g}$.

8.1.2 Half-diagonal Metrics on $\mathfrak{so}(4)$

We now turn to the case of $\mathfrak{so}(4)$. We represent an element X of $\mathfrak{so}(4)$ by a skew-symmetric matrix, $X = (X_{ij})_{1\leqslant i,j\leqslant 4}$, $X_{ij} = -X_{ji}$, and we identify $\mathfrak{so}(4)$ with its dual by using $\langle X \,|\, Y\rangle = \mathrm{Trace}(XY)$. If we denote by E_{ij} the matrix which has a 1 at position (i,j) and zeros elsewhere, then (3.33) yields

$$\nabla X_{ij}(X) = \frac{1}{2}(E_{ji} - E_{ij}).$$

Notice that the gradient is independent of X, because the function X_{ij} is a linear function.

[1] Since we are dealing with a quadratic vector field on a real vector space, there is a natural way to complexify the vector space and the vector field.

It follows that the Lie-Poisson bracket on $\mathfrak{so}(4)$ is given by

$$\begin{aligned}\{X_{ij}, X_{kl}\}(X) &= \frac{1}{4}\operatorname{Trace}(X[E_{ji} - E_{ij}, E_{lk} - E_{kl}])\\ &= \frac{1}{2}\operatorname{Trace}(X[E_{ji}, E_{lk}] - X[E_{ij}, E_{lk}]),\end{aligned}$$

where $1 \leqslant i, j, k, l \leqslant 4$. It follows that the Poisson matrix of the Lie-Poisson structure on $\mathfrak{so}(4)$ admits the following Poisson matrix,

$$\frac{1}{2}\begin{pmatrix} 0 & X_{23} & X_{31} & X_{24} & X_{41} & 0 \\ X_{32} & 0 & X_{12} & X_{34} & 0 & X_{41} \\ X_{13} & X_{21} & 0 & 0 & X_{34} & X_{42} \\ X_{42} & X_{43} & 0 & 0 & X_{12} & X_{13} \\ X_{14} & 0 & X_{43} & X_{21} & 0 & X_{23} \\ 0 & X_{14} & X_{24} & X_{31} & X_{32} & 0 \end{pmatrix}, \tag{8.6}$$

where we have ordered the coordinates X_{ij} as follows: X_{12}, X_{13}, X_{23}, X_{14}, X_{24}, X_{34}. A basis for the Casimirs for the Lie-Poisson structure on $\mathfrak{so}(4)$ is given by

$$\begin{aligned}Q_0 &:= \sqrt{\det X} = X_{12}X_{34} + X_{23}X_{14} + X_{31}X_{24},\\ Q_1 &:= -\tfrac{1}{2}\operatorname{Trace}(X^2) = \textstyle\sum_{1\leqslant i<j\leqslant 4} X_{ij}^2.\end{aligned} \tag{8.7}$$

Now $\mathfrak{so}(4) \cong \mathfrak{so}(3) \oplus \mathfrak{so}(3)$, leading to another system of natural coordinates on $\mathfrak{so}(4)$, namely define $x = (x_1, \dots, x_6)$ by

$$X_{ij} = -\frac{1}{2}(x_k + x_{k+3}),\ X_{k4} = -\frac{1}{2}(x_k - x_{k+3}),\ 1 \leqslant i, j, k \leqslant 3, \tag{8.8}$$

where (i, j, k) denotes any cyclic permutations of $(1, 2, 3)$. In these new coordinates, the Poisson bracket takes the following simple form,

$$\begin{aligned}\{x_i, x_j\} &= -\epsilon_{ijk}x_k,\\ \{x_{i+3}, x_{j+3}\} &= -\epsilon_{ijk}x_{k+3},\\ \{x_i, x_{j+3}\} &= \{x_{i+3}, x_j\} = 0,\end{aligned} \tag{8.9}$$

where $1 \leqslant i, j, k \leqslant 3$, and where ϵ_{ijk} is the skew-symmetric tensor for which $\epsilon_{123} = 1$. In these coordinates $x_1, \dots, x_6$ the Poisson matrix is given by

$$\begin{pmatrix} 0 & -x_3 & x_2 \\ x_3 & 0 & -x_1 \\ -x_2 & x_1 & 0 \end{pmatrix} \oplus \begin{pmatrix} 0 & -x_6 & x_5 \\ x_6 & 0 & -x_4 \\ -x_5 & x_4 & 0 \end{pmatrix}. \tag{8.10}$$

It follows from (8.9) that the rank of $\{\cdot\,,\cdot\}$ is 4, except on the two three-planes $x_1 = x_2 = x_3 = 0$ and $x_4 = x_5 = x_6 = 0$, where the rank drops to 2 or to 0 (at the origin).

In the x_i coordinates, a basis for the the Casimirs is given by

$$Q_0' := x_1^2 + x_2^2 + x_3^2,$$
$$Q_1' := x_4^2 + x_5^2 + x_6^2.$$

Since Q_0' resp. Q_1' evaluate to zero on the above three-planes, Proposition 7.56 implies that for any integrable system that we consider on $(\mathfrak{so}(4), \{\cdot\,,\cdot\})$ the two commuting vector fields will be independent on the generic fiber $\mathbf{F_c}$ of the momentum map.

Let us consider the most general quadratic form H on $\mathfrak{so}(4)$, in terms of these coordinates. By means of the adjoint action (which respects the direct sum decomposition of $\mathfrak{so}(4)$), we may diagonalize each $\mathfrak{so}(3)$ piece separately and so we may assume that H has the form

$$H = \frac{1}{2}\sum_{i=1}^{6} \lambda_i x_i^2 + \sum_{1\leqslant i,j\leqslant 3} \mu_{ij} x_i x_{j+3} \tag{8.11}$$

so that H depends on 15 parameters. We will in the sequel only consider a special class, that we introduce now.

Definition 8.1. A metric on $\mathfrak{so}(4)$ is called a *half-diagonal metric* if it is given by a quadratic form

$$H = \frac{1}{2}\sum_{1}^{6} \lambda_i x_i^2 + \sum_{j=1}^{3} \mu_j x_j x_{j+3}, \tag{8.12}$$

where $\lambda_1, \dots, \lambda_6, \mu_1, \dots, \mu_3 \in \mathbf{C}$. Setting $\lambda_{ij} := \lambda_i - \lambda_j$ the above half-diagonal metric is called *non-degenerate* when

$$\lambda_{12}\lambda_{23}\lambda_{31}\lambda_{45}\lambda_{56}\lambda_{64}\mu_1\mu_2\mu_3 \neq 0.$$

Remark 8.2. This condition excludes the product of two Euler tops, because this corresponds to $\mu_1 = \mu_2 = \mu_3 = 0$ (see Example 6.4).

For the half-diagonal metric, given by the quadratic form (8.12), the Hamiltonian vector field $\mathcal{X}_H := \{\cdot\,, X\}$ takes the following form, as computed from (8.9).

$$\begin{aligned}
\dot{x}_1 &= x_2x_3\lambda_{32} + x_2x_6\mu_3 - x_3x_5\mu_2,\\
\dot{x}_2 &= x_3x_1\lambda_{13} + x_3x_4\mu_1 - x_1x_6\mu_3,\\
\dot{x}_3 &= x_1x_2\lambda_{21} + x_1x_5\mu_2 - x_2x_4\mu_1,\\
\dot{x}_4 &= x_5x_6\lambda_{65} + x_3x_5\mu_3 - x_2x_6\mu_2,\\
\dot{x}_5 &= x_6x_4\lambda_{46} + x_1x_6\mu_1 - x_3x_4\mu_3,\\
\dot{x}_6 &= x_4x_5\lambda_{54} + x_2x_4\mu_2 - x_1x_5\mu_1.
\end{aligned}$$

Since the symplectic leaves are 4-dimensional, for the system to be Liouville integrable it must have a constant of motion, that is independent of the Hamiltonian H and the Casimirs Q_0 and Q_1 (equivalently, the Casimirs Q_0' and Q_1').

8.1.3 The Kowalevski-Painlevé Criterion

The purpose of this section is to use the Kowalevski-Painlevé Criterion (Theorem 6.13) to determine which non-degenerate half-diagonal metrics (see Definition 8.1) on $\mathfrak{so}(4)$ lead to an irreducible weight homogeneous a.c.i. system. This criterion selects three families of metrics, and each of them will be shown in one of the three subsequent sections to be algebraic completely integrable.

Theorem 8.3. *Let H be a quadratic form on $\mathfrak{so}(4)$ which defines a non-degenerate half-diagonal metric on $\mathfrak{so}(4)$, i.e., H is of the form*

$$H = \frac{1}{2}\sum_{1}^{6} \lambda_i x_i^2 + \sum_{j=1}^{3} \mu_j x_j x_{j+3},$$

with

$$\lambda_{12}\lambda_{23}\lambda_{31}\lambda_{45}\lambda_{56}\lambda_{64}\mu_1\mu_2\mu_3 \neq 0, \tag{8.13}$$

where $\lambda_{ij} := \lambda_i - \lambda_j$. If $\mathcal{X}_H$ is one of the integrable vector fields of an irreducible weight homogeneous a.c.i. system on $(\mathfrak{so}(4), \{\cdot\,,\cdot\})$ then the metric satisfies one of the following conditions:

(1) $\lambda_{14} = \lambda_{25} = \lambda_{36}$ *and* $\mu_1^2\lambda_{23} + \mu_2^2\lambda_{31} + \mu_3^2\lambda_{12} + \lambda_{12}\lambda_{23}\lambda_{31} = 0$;

(2)

$$(\mu_1^2, \mu_2^2, \mu_3^2) = \frac{E\bar{E}}{F^2}\left(\frac{(\lambda_{23}-\lambda_{56})^2}{\lambda_{23}\lambda_{56}}, \frac{(\lambda_{31}-\lambda_{64})^2}{\lambda_{31}\lambda_{64}}, \frac{(\lambda_{12}-\lambda_{45})^2}{\lambda_{12}\lambda_{45}}\right)$$

with the following sign specification

$$\mu_1\mu_2\mu_3 = \frac{E\bar{E}}{F^3}(\lambda_{12}-\lambda_{45})(\lambda_{23}-\lambda_{56})(\lambda_{31}-\lambda_{64}),$$

where $E := \lambda_{12}\lambda_{23}\lambda_{31}$ *and* $\bar{E} := \lambda_{45}\lambda_{56}\lambda_{64}$ *and* $F := \lambda_{46}\lambda_{32} - \lambda_{65}\lambda_{13} \neq 0$.

(3) There exist constants ρ, q and square roots ϵ_1 and ϵ_2 of 1, with $\rho \in \mathbf{C}^*$ and $q \in \mathbf{C} \setminus \{0, \pm 1, \pm 1/3\}$, such that, up to a possible transposition $(\lambda_{12}, \lambda_{23}, \lambda_{31}) \leftrightarrow (\lambda_{45}, \lambda_{56}, \lambda_{64})$,

$$\begin{aligned} \lambda_{12} &= \rho(q-1)^3(3q+1), & \lambda_{45} &= \rho(q-1)(3q+1)^3,\\ \lambda_{23} &= 16\rho q^3, & \lambda_{56} &= 16\rho q,\\ \lambda_{31} &= \rho(1-3q)(q+1)^3, & \lambda_{64} &= \rho(1-3q)^3(q+1), \end{aligned}$$

$$\begin{aligned} \mu_1 &= \epsilon_1\rho(q^2-1)(9q^2-1),\\ \mu_2 &= 4\epsilon_2\rho q(q-1)(3q+1),\\ \mu_3 &= 4\epsilon_1\epsilon_2\rho q(q+1)(3q-1). \end{aligned}$$

The constant ρ is a common factor, which can be put equal to 1 by rescaling the metric.

Our proof is based on the following proposition, which is a consequence of the Kowalevski-Painlevé Criterion (Theorem 6.13).

Proposition 8.4. *Let H be as in Theorem 8.3, namely H is a quadratic form on $\mathfrak{so}(4)$ which defines a non-degenerate half-diagonal metric on $\mathfrak{so}(4)$. If $\mathcal{X}_H$ is one of the integrable vector fields of an irreducible weight homogeneous a.c.i. system on $(\mathfrak{so}(4),\{\cdot\,,\cdot\})$ then the indicial locus of $\mathcal{X}_H$ contains at least one curve. Moreover, a curve that is contained in the indicial locus cannot be contained in one of the coordinate hyperplanes $x_i^{(0)}=0$.*

Proof. If $\mathcal{X}_H$ is to be one of the integrable vector fields of an irreducible weight homogeneous a.c.i. system then, according to the Kowalevski-Painlevé Criterion, $\mathcal{X}_H$ must have a weight homogeneous principal balance. $\mathcal{X}_H$ is a homogeneous vector field (i.e., the weight of each of the variables x_i as well as of $\mathcal{X}_H$ is 1), so that the indicial locus $\mathcal{I}\subset\mathbf{C}^6$ is given by the equations

$$\begin{aligned}
-x_1^{(0)} &= x_2^{(0)}x_3^{(0)}\lambda_{32}+x_2^{(0)}x_6^{(0)}\mu_3-x_3^{(0)}x_5^{(0)}\mu_2,\\
-x_2^{(0)} &= x_3^{(0)}x_1^{(0)}\lambda_{13}+x_3^{(0)}x_4^{(0)}\mu_1-x_1^{(0)}x_6^{(0)}\mu_3,\\
-x_3^{(0)} &= x_1^{(0)}x_2^{(0)}\lambda_{21}+x_1^{(0)}x_5^{(0)}\mu_2-x_2^{(0)}x_4^{(0)}\mu_1,\\
-x_4^{(0)} &= x_5^{(0)}x_6^{(0)}\lambda_{65}+x_3^{(0)}x_5^{(0)}\mu_3-x_2^{(0)}x_6^{(0)}\mu_2,\\
-x_5^{(0)} &= x_6^{(0)}x_4^{(0)}\lambda_{46}+x_1^{(0)}x_6^{(0)}\mu_1-x_3^{(0)}x_4^{(0)}\mu_3,\\
-x_6^{(0)} &= x_4^{(0)}x_5^{(0)}\lambda_{54}+x_2^{(0)}x_4^{(0)}\mu_2-x_1^{(0)}x_5^{(0)}\mu_1.
\end{aligned}\tag{8.14}$$

Proposition 7.11 implies that for each point $x^{(0)}$ in the indicial locus $\mathcal{I}$ the Kowalevski matrix $\mathcal{K}(x^{(0)})$ has -1 as an eigenvalue. Thus, if we want a weight homogeneous principal balance, then for some $x^{(0)}\in\mathcal{I}$, all other eigenvalues must be non-negative integers. Since $\mathcal{X}_H$ is divergence free, Proposition 7.12 implies that the sum of the eigenvalues of $\mathcal{K}(x^{(0)})$ must be 6 (the sum of the weights of all the variables).

We claim that $\mathcal{K}(x^{(0)})$ has 2 as a triple eigenvalue (at least), for any $x^{(0)}\in\mathcal{I}$ for which $\left(x_1^{(0)},x_2^{(0)},x_3^{(0)}\right)\neq(0,0,0)$ and $\left(x_4^{(0)},x_5^{(0)},x_6^{(0)}\right)\neq(0,0,0)$. In view of Theorem 7.30 this can be shown by proving that the three quadratic polynomials Q_0', Q_1' and H have independent differentials at such a $x^{(0)}$. Indeed, let $x^{(0)}\in\mathcal{I}\backslash\{0\}$, and rewrite (8.14) as $-x^{(0)}=J(x^{(0)})[\mathrm{d}H(x^{(0)})]$, where J denotes the Poisson matrix of $(\mathfrak{so}(4),\{\cdot\,,\cdot\})$ in the coordinates $x_1,\dots,x_6$, and $[\mathrm{d}H(x^{(0)})]$ denotes the column vector which represents $\mathrm{d}H(x^{(0)})$ in terms of these coordinates. Suppose that a_1, a_2 and a_3 are constants such that $a_1\mathrm{d}Q_0'(x^{(0)})+a_2\mathrm{d}Q_1'(x^{(0)})+a_3\mathrm{d}H(x^{(0)})=0$. Since Q_0' and Q_1' are Casimirs we have that $J(x^{(0)})[\mathrm{d}Q_0'(x^{(0)})]=J(x^{(0)})[\mathrm{d}Q_1'(x^{(0)})]=0$, giving

$$0=a_3J(x^{(0)})[\mathrm{d}H(x^{(0)})]=-a_3x^{(0)},$$

which implies that $a_3=0$ since $x^{(0)}\neq 0$. Moreover, if $x^{(0)}\in\mathcal{I}$ is such that $\left(x_1^{(0)},x_2^{(0)},x_3^{(0)}\right)\neq(0,0,0)$ and $\left(x_4^{(0)},x_5^{(0)},x_6^{(0)}\right)\neq(0,0,0)$ then $\mathrm{d}Q_0'(x^{(0)})$ and $\mathrm{d}Q_1'(x^{(0)})$ are obviously independent, showing our claim.

Moreover, if at least for one of the principal balances the leading terms satisfy $\left(x_1^{(0)}, x_2^{(0)}, x_3^{(0)}\right) = (0,0,0)$ or $\left(x_4^{(0)}, x_5^{(0)}, x_6^{(0)}\right) = (0,0,0)$ then $\mu_1 = \mu_2 = \mu_3 = 0$, which is excluded by non-degeneracy. Let us show this in case $\left(x_4^{(0)}, x_5^{(0)}, x_6^{(0)}\right) = (0,0,0)$. The indicial equation can then be solved to yield 4 points, solutions to

$$(x_1^{(0)})^2 = \frac{1}{\lambda_{31}\lambda_{12}}, \qquad (x_2^{(0)})^2 = \frac{1}{\lambda_{12}\lambda_{23}}, \qquad (x_3^{(0)})^2 = \frac{1}{\lambda_{23}\lambda_{31}}, \tag{8.15}$$

with

$$x_1^{(0)} x_2^{(0)} x_3^{(0)} = \frac{1}{\lambda_{12}\lambda_{23}\lambda_{31}}. \tag{8.16}$$

At these 4 points the Kowalevski matrix is given by

$$\begin{pmatrix} 1 & x_3^{(0)}\lambda_{32} & x_2^{(0)}\lambda_{32} & 0 & -x_3^{(0)}\mu_2 & x_2^{(0)}\mu_3 \\ x_3^{(0)}\lambda_{13} & 1 & x_1^{(0)}\lambda_{13} & x_3^{(0)}\mu_1 & 0 & -x_1^{(0)}\mu_3 \\ x_2^{(0)}\lambda_{21} & x_1^{(0)}\lambda_{21} & 1 & -x_2^{(0)}\mu_1 & x_1^{(0)}\mu_2 & 0 \\ 0 & 0 & 0 & 1 & x_3^{(0)}\mu_3 & -x_2^{(0)}\mu_2 \\ 0 & 0 & 0 & -x_3^{(0)}\mu_3 & 1 & x_1^{(0)}\mu_1 \\ 0 & 0 & 0 & x_2^{(0)}\mu_2 & -x_1^{(0)}\mu_1 & 1 \end{pmatrix}. \tag{8.17}$$

Using (8.15) and (8.16) it is easily computed that the north-west block of this matrix has characteristic polynomial $(k-2)^2(k+1)$, while the south-east block has characteristic polynomial

$$(k-1)\left((k-1)^2 + \frac{\mu_1^2}{\lambda_{13}\lambda_{21}} + \frac{\mu_2^2}{\lambda_{21}\lambda_{32}} + \frac{\mu_3^2}{\lambda_{32}\lambda_{13}}\right). \tag{8.18}$$

It follows that $|\mathcal{K}(x^{(0)}) - k\,\mathrm{Id}_6\,|$ has -1, 1, 2, 2 as roots and two extra roots, which are the roots of the quadratic factor in (8.18). But these eigenvalues must be non-negative integers and their sum must be $6-4=2$, leaving $1,1$ and $0,2$ as the only possibilities. If 0 is an eigenvalue then a free parameter must occur in the solution to the indicial equation, but we just saw that the indicial locus consists in this case of 4 points only. Hence we must have that 1 is a triple eigenvalue of the Kowalevski matrix, which entails that

$$\mu_1^2\lambda_{23} + \mu_2^2\lambda_{31} + \mu_3^2\lambda_{12} = 0.$$

If we want three free parameters to appear at step 1 then the south-east block of (8.17) must be the identity matrix (since the north-west block does not have 1 as an eigenvalue), by Proposition 7.22. But this is impossible by (8.15) and by non-degeneracy.

Summarizing, if the metric is non-degenerate then $\mathcal{X}_H$ can only have principal balances if for some $x^{(0)} \in \mathcal{I}$ the spectrum of $\mathcal{K}(x^{(0)})$ has -1 as an eigenvalue and 2 as a triple eigenvalue. Since the sum of the other eigenvalues of $\mathcal{K}(x^{(0)})$, which are non-negative integers, is $6-(-1)-3.2=1$, this means that 0 and 1 are also eigenvalues, and that the indicial locus contains a curve (corresponding to the free parameter that enters at step 0, i.e., the indicial equation).

Finally, let us show that the hyperplane section $\left\{x_1^{(0)}=0\right\} \cap \mathcal{I}$ cannot contain a curve if the metric is non-degenerate. If we substitute $x_1^{(0)}=0$ in (8.14) then the second and third equations yield $x_3^{(0)}\left(1+(\mu_1 x_4^{(0)})^2\right)=0$ and $x_3^{(0)}=\mu_1 x_2^{(0)} x_4^{(0)}$. It is easy to see that $x_3^{(0)}=0$ implies either $x_2^{(0)}=0$ or $x_4^{(0)}=x_5^{(0)}=x_6^{(0)}=0$, which only leads to points. Therefore, let us assume that $x_3^{(0)} \neq 0$ so that $x_3^{(0)}=\pm\sqrt{-1}x_2^{(0)}$ and $x_4^{(0)}=\pm\sqrt{-1}/\mu_1$. Upon substituting in the first and in the two last equations of (8.14) we get the following linear system in $x_2^{(0)}$, $x_5^{(0)}$ and $x_6^{(0)}$,

$$\begin{pmatrix} \lambda_{32} & -\mu_2 & \mp\sqrt{-1}\mu_3 \\ \mu_3 & \mu_1 & \pm\sqrt{-1}\lambda_{46} \\ \mu_2 & \lambda_{54} & \mp\sqrt{-1}\mu_1 \end{pmatrix} \begin{pmatrix} x_2^{(0)} \\ x_5^{(0)} \\ x_6^{(0)} \end{pmatrix} = \begin{pmatrix} 0 \\ 0 \\ 0 \end{pmatrix}. \tag{8.19}$$

Since at least one of the determinants $\begin{vmatrix} -\mu_2 & \mp\sqrt{-1}\mu_3 \\ \lambda_{54} & \mp\sqrt{-1}\mu_1 \end{vmatrix}$ and $\begin{vmatrix} \mu_3 & \mu_1 \\ \mu_2 & \lambda_{54} \end{vmatrix}$ is different from zero (by the non-degeneracy condition (8.13)) the rank of (8.19) is at least two and $x_5^{(0)}$ and $x_6^{(0)}$ are proportional to $x_2^{(0)}$ (the factors of proportionality are rational functions of the λ_{ij} and of the μ_i). It follows that the fourth equation in (8.14) is of the form $\alpha\left(x_2^{(0)}\right)^2=x_4^{(0)}=\pm\sqrt{-1}/\mu_1$, which either has two solutions (if $\alpha \neq 0$) or no solutions at all (if $\alpha=0$). In any case, this shows that a solution to (8.14) with $x_1^{(0)}=0$ cannot depend on a parameter. This proves that the hyperplane section $\left\{x_1^{(0)}=0\right\} \cap \mathcal{I}$ cannot contain a curve. By symmetry, the result follows for all hyperplane sections $\left\{x_i^{(0)}=0\right\} \cap \mathcal{I}$, where $i=1,\dots,6$. □

Let us assume that the parameters λ_{ij} and μ_i that define the non-degenerate metric are fixed, and that these parameters are such that the indicial locus contains an algebraic curve, which is not contained in any of the hyperplanes $x_i^{(0)}=0$.

Then the algebraic set, defined by

$$
\begin{aligned}
-x_1^{(0)} &= x_2^{(0)}x_3^{(0)}(\lambda_{32} + \mu_3 w - \mu_2 v),\\
-x_2^{(0)} &= x_3^{(0)}x_1^{(0)}(\lambda_{13} + \mu_1 u - \mu_3 w),\\
-x_3^{(0)} &= x_1^{(0)}x_2^{(0)}(\lambda_{21} + \mu_2 v - \mu_1 u),\\
-x_1^{(0)} &= x_2^{(0)}x_3^{(0)}u^{-1}(\lambda_{65} vw + \mu_3 v - \mu_2 w),\\
-x_2^{(0)} &= x_3^{(0)}x_1^{(0)}v^{-1}(\lambda_{46} wu + \mu_1 w - \mu_3 u),\\
-x_3^{(0)} &= x_1^{(0)}x_2^{(0)}w^{-1}(\lambda_{54} uv + \mu_2 u - \mu_1 v),
\end{aligned}
\tag{8.20}
$$

contains a curve, which we denote by Σ; (8.20) has been obtained from (8.14) by putting $u := x_4^{(0)}/x_1^{(0)}$, $v := x_5^{(0)}/x_2^{(0)}$ and $w := x_6^{(0)}/x_3^{(0)}$. In the sequel, we view u, v and w as rational functions on Σ. We wish to express the fact that Σ is a curve in terms of u, v and w only. To do this, we compare the first and fourth equations in (8.20) to find a formula for u as a rational function of v and w; we also find such a formula for u by comparing the second and fifth equations in (8.20), and another one by comparing the third and sixth equations in (8.20). Altogether this gives the following three expressions for u in terms of v and w (as rational functions on Σ).

$$
u = \frac{\mu_3 v - \mu_2 w + \lambda_{65} vw}{\mu_3 w - \mu_2 v + \lambda_{32}} = \frac{\lambda_{13} v - \mu_1 w - \mu_3 vw}{\lambda_{46} w - \mu_1 v - \mu_3} = \frac{\mu_1 v + \lambda_{21} w + \mu_2 vw}{\mu_1 w + \lambda_{54} v + \mu_2}.
\tag{8.21}
$$

Note that, in (8.21), both the numerator and the denominator of the first equation cannot be identically zero on Σ, since we have shown that the curve cannot be contained in any of the hyperplanes $x_i = 0$. However, that could happen, say in the second equality. But then v and w would be determined and the first equation would yield u; from the first three equations in (8.20) we would only get a finite number of possibilities for $x_1^{(0)}, \dots, x_3^{(0)}$, so we would not have a curve. Thus, as functions on Σ, all equations in (8.21) are meaningful as written.

Comparing in (8.21) the third fraction with the first and the second yields a pair of equations in v and w,

$$
v^2(A_i w + B_i) + v(C_i w^2 + D_i w - \bar{C}_i) - w(\bar{B}_i w + \bar{A}_i) = 0, \quad i = 1, 2, \tag{8.22}
$$

where

$$
\begin{aligned}
A_1 &:= \mu_2^2 + \lambda_{54}\lambda_{65}, & A_2 &:= -\mu_1\mu_2 + \lambda_{54}\mu_3,\\
B_1 &:= \mu_1\mu_2 + \lambda_{54}\mu_3, & B_2 &:= -\mu_1^2 - \lambda_{13}\lambda_{54},\\
C_1 &:= -\mu_2\mu_3 + \lambda_{65}\mu_1, & C_2 &:= \mu_1\mu_3 + \lambda_{46}\mu_2,\\
D_1 &:= \mu_2(2\lambda_{25} - \lambda_{14} - \lambda_{36}), & D_2 &:= \mu_1(\lambda_{36} - \lambda_{25}),
\end{aligned}
\tag{8.23}
$$

and where $\bar{\ }$ denotes the involution that permutes λ_1 with λ_4, and λ_2 with λ_5 and λ_3 with λ_6 (recall that $\lambda_{ij} = \lambda_i - \lambda_j$).

For future use we also define

$$P := \mu_1^2\lambda_{32} + \mu_2^2\lambda_{13} + \mu_3^2\lambda_{21} + \lambda_{32}\lambda_{13}\lambda_{21}, \tag{8.24}$$

which is invariant under the cyclic permutation $1 \Rightarrow 2 \Rightarrow 3 \Rightarrow 1$. Its barred analogue is given by

$$\bar{P} := \mu_1^2\lambda_{65} + \mu_2^2\lambda_{46} + \mu_3^2\lambda_{54} + \lambda_{65}\lambda_{46}\lambda_{54}. \tag{8.25}$$

Lemma 8.5. *The indicial locus $\mathcal{I}$ is, away from the hyperplanes $x_i^{(0)} = 0$, a finite (ramified) cover of the plane algebraic subset, defined by the two equations (8.22). Therefore, $\mathcal{I}$ can only contain a curve if (8.22) contains a curve.*

Proof. The value of u in terms of v and w can be computed from one of the equations in (8.21). If we multiply the first three equations in (8.20) then we see that $x_1^{(0)}x_2^{(0)}x_3^{(0)}$ can be expressed rationally in terms of u, v and w. Given this fact it is also seen that $(x_1^{(0)})^2$ can be expressed rationally in terms of u, v and w, simply by multiplying the first equation in (8.21) by $x_1^{(0)}$. It follows in the same way from the second and third equations that $(x_2^{(0)})^2$ and $(x_3^{(0)})^2$ can be expressed rationally in terms of u, v and w. It follows that, at least away from the hyperplanes $x_i^{(0)} = 0$, the indicial locus is a finite cover of the locus defined by (8.22). Since by Proposition 8.4 none of these hyperplanes can contain a curve, the claim follows. □

It follows that we need to analyze under which conditions on the coefficients $\lambda_1, \dots, \lambda_6, \mu_1, \dots, \mu_3$ the algebraic subset, defined by the two equations in (8.22), contains a curve. Since each of the equations in (8.22) describes an algebraic curve (unless all of its coefficients are zero) the two equations can only define a curve in the following two cases.

(1) The two equations in (8.22) are proportional;
(2) The two equations in (8.22) have a common factor without being proportional.

We will first analyze the first case.

Proposition 8.6. *The following three conditions are equivalent.*

(i) The two equations in (8.22) are proportional;
(ii) $\lambda_{14} = \lambda_{25} = \lambda_{36}$ and $P = 0$;
(iii) $\lambda_{14} = \lambda_{25} = \lambda_{36}$ and $\bar{P} = 0$.

Proof. The equivalence of *(ii)* and *(iii)* is an immediate consequence of the fact that $\lambda_{14} = \lambda_{25} = \lambda_{36}$ is equivalent to the three equalities $\lambda_{12} = \lambda_{45}$ and $\lambda_{23} = \lambda_{56}$ and $\lambda_{31} = \lambda_{64}$ (P and $\bar{P}$ were defined in (8.24) and in (8.25)).

Suppose now that the two equations in (8.22) are proportional. Then the rank of

$$\begin{pmatrix} A_1 & B_1 & C_1 & D_1 & \bar{A}_1 & \bar{B}_1 & \bar{C}_1 \\ A_2 & B_2 & C_2 & D_2 & \bar{A}_2 & \bar{B}_2 & \bar{C}_2 \end{pmatrix} \tag{8.26}$$

is less than two. If we define P by (8.24) then we have the following identities, as follows at once from (8.23):

$$\begin{aligned} \mu_2^{-1}\begin{vmatrix} \bar{A}_1 & \bar{C}_1 \\ \bar{A}_2 & \bar{C}_2 \end{vmatrix} &= P, \\ \lambda_{21}^{-1}\begin{vmatrix} \bar{B}_1 & \bar{A}_1 \\ \bar{B}_2 & \bar{A}_2 \end{vmatrix} &= P + \bar{A}_1(\lambda_{36} - \lambda_{14}), \\ \mu_1^{-1}\begin{vmatrix} \bar{B}_1 & \bar{C}_1 \\ \bar{B}_2 & \bar{C}_2 \end{vmatrix} &= P + \bar{A}_1'(\lambda_{14} - \lambda_{25}), \end{aligned} \tag{8.27}$$

where $\bar{A}_1' := \mu_3^2 + \lambda_{32}\lambda_{13}$. Thus, we have that $P = 0$ and that $\bar{A}_1(\lambda_{36} - \lambda_{14}) = 0 = \bar{A}_1'(\lambda_{14} - \lambda_{25})$. Similarly, considering the barred analogues of these equations we have that $\bar{P} = 0$ and that $A_1(\lambda_{36} - \lambda_{14}) = 0 = A_1'(\lambda_{14} - \lambda_{25})$. Thus, $P = 0 = \bar{P}$ and we only need to show that $\lambda_{14} = \lambda_{25} = \lambda_{36}$. This is obvious from the previous equations, except if $A_1 = 0 = \bar{A}_1$, or if $A_1' = 0 = \bar{A}_1'$. If $A_1' = 0 = \bar{A}_1'$, but $A_1 \neq 0$ or $\bar{A}_1 \neq 0$ then we still have that $\lambda_{13} = \lambda_{46}$, while $A_1' = 0 = \bar{A}_1'$ yields that $\mu_3^2 + \lambda_{32}\lambda_{13} = 0 = \mu_3^2 + \lambda_{65}\lambda_{46}$. Thus, $\lambda_{32} = \lambda_{65}$ and we are done. Similarly, if $A_1 = 0 = \bar{A}_1$, but $\bar{A}_1' \neq 0$ or $A_1' \neq 0$ we are done. So we only need to analyze the case

$$A_1 = \bar{A}_1 = \bar{A}_1' = A_1' = 0.$$

Then $\mu_2^2 = \lambda_{54}\lambda_{56} = \lambda_{21}\lambda_{23}$ and $\mu_3^2 = \lambda_{56}\lambda_{46} = \lambda_{23}\lambda_{13}$, which upon substitution in $P = \bar{P} = 0$ yields $\mu_1^2 = \lambda_{64}\lambda_{54} = \lambda_{31}\lambda_{21}$. Substituting $\lambda_{46} = \lambda_{45} + \lambda_{56}$ and $\lambda_{13} = \lambda_{12} + \lambda_{23}$ in these equations we find that $(\lambda_{12}, \lambda_{23}, \lambda_{31}) = \pm(\lambda_{45}, \lambda_{56}, \lambda_{64})$. We show that the minus sign is impossible. Suppose that $(\lambda_{12}, \lambda_{23}, \lambda_{31}) = -(\lambda_{45}, \lambda_{56}, \lambda_{64})$. Then $\bar{B}_2 = 2\lambda_{13}\lambda_{21} = B_2$ ($\neq 0$) so that, by the rank condition, $B_1 = \bar{B}_1$. But this is impossible by non-degeneracy.

In order to show that *(ii)* implies *(i)*, notice that if $\lambda_{14} = \lambda_{25} = \lambda_{36}$ then $D_1 = D_2 = 0$ and that all 2×2 determinants of (8.26) are zero or divisible by P. □

Having analyzed the case in which the two equations (8.22) are proportional, we now get to the case in which they are not proportional, but have a common factor.

Proposition 8.7. *Suppose that the two equations in (8.22) have a common factor of the form $v - \delta w$, without being proportional, and with $\delta \neq 0$, then*

$$(\mu_1^2, \mu_2^2, \mu_3^2) = \frac{E\bar{E}}{F^2}\left(\frac{(\lambda_{23}-\lambda_{56})^2}{\lambda_{23}\lambda_{56}}, \frac{(\lambda_{31}-\lambda_{64})^2}{\lambda_{31}\lambda_{64}}, \frac{(\lambda_{12}-\lambda_{45})^2}{\lambda_{12}\lambda_{45}}\right) \tag{8.28}$$

with the following sign specification

$$\mu_1\mu_2\mu_3 = \frac{E\bar{E}}{F^3}(\lambda_{12}-\lambda_{45})(\lambda_{23}-\lambda_{56})(\lambda_{31}-\lambda_{64}), \tag{8.29}$$

where $E := \lambda_{12}\lambda_{23}\lambda_{31}$ and $F := \lambda_{46}\lambda_{32} - \lambda_{65}\lambda_{13}$.

Proof. Substitute δw for v in (8.22) and express that the resulting polynomial in w is zero, which expresses that the equations in (8.22) have a common factor $v - \delta w$. The resulting equations on the coefficients are given by

$$\begin{aligned} &\delta A_i + C_i = 0, \\ &\delta^2 B_i + \delta D_i - \bar{B}_i = 0, \qquad i = 1, 2. \\ &\delta \bar{C}_i + \bar{A}_i = 0. \end{aligned} \tag{8.30}$$

If we substitute (8.23) into three of these equations, namely in $\delta A_i + C_i = 0$ for $i = 1, 2$ and in $\delta\bar{C}_1 + \bar{A}_1 = 0$ then we find three equations which are linear in μ_1, λ_{13} and λ_{46}, when λ_{54} is written as $\lambda_{56} - \lambda_{46}$. We write them as the following linear system:

$$\begin{pmatrix} \mu_3 - \delta\mu_2 & 0 & \mu_2 - \delta\mu_3 \\ -\lambda_{56} & 0 & \delta\lambda_{56} \\ \delta\lambda_{23} & -\lambda_{23} & 0 \end{pmatrix}\begin{pmatrix} \mu_1 \\ \lambda_{13} \\ \lambda_{46} \end{pmatrix} = \begin{pmatrix} -\delta\mu_3\lambda_{56} \\ \mu_2\mu_3 + \delta(\lambda_{56}^2 - \mu_2^2) \\ \mu_2^2 - \lambda_{23}^2 - \delta\mu_2\mu_3 \end{pmatrix}. \tag{8.31}$$

The determinant of the above matrix is $\lambda_{23}\lambda_{56}\mu_2(1-\delta^2)$. Let us assume for now that $\delta \neq \pm 1$ so that, by non-degeneracy, this determinant is non-zero and we can solve (8.31) linearly for μ_1, λ_{13} and λ_{46}. If we substitute the result in two other equations of (8.30), namely in $\delta\bar{C}_2 + \bar{A}_2 = 0$ and in $\delta^2 B_1 + \delta D_1 - \bar{B}_1 = 0$, then these equations reduce to

$$(\lambda_{56} \pm \lambda_{23})(\delta\lambda_{23}\lambda_{56} + (\mu_3\delta - \mu_2)(\mu_2\delta - \mu_3)) = 0,$$

so that, again by non-degeneracy, we may conclude that

$$\lambda_{23} = -\frac{(\mu_3\delta - \mu_2)(\mu_2\delta - \mu_3)}{\delta\lambda_{56}}. \tag{8.32}$$

Substituting this value in (8.31) we can express μ_1, λ_{13} and λ_{46} in the following compact form

$$\begin{aligned}\mu_1 &= \frac{\delta\lambda_{56}^2 + (\delta\mu_2 - \mu_3)(\delta\mu_3 - \mu_2)}{\lambda_{56}(\delta^2 - 1)},\\ \lambda_{46} &= -\frac{(\delta\mu_2 - \mu_3)^2 - \delta^2\lambda_{56}^2}{\lambda_{56}(\delta^2 - 1)},\\ \lambda_{13} &= \lambda_{46}\frac{\mu_2 - \delta\mu_3}{\delta(\delta\mu_2 - \mu_3)}.\end{aligned} \tag{8.33}$$

If we substitute (8.32) and (8.33) in the remaining equation of (8.30), namely in the equation $\delta^2 B_2 + \delta D_2 - \bar{B}_2 = 0$, then we find that that equation is automatically satisfied, so that we have the complete solution to (8.30). Again by direct substitution of (8.32) and (8.33), the following identities are seen to be satisfied:

$$\begin{aligned}\mu_1^2(\lambda_{46}\lambda_{32} - \lambda_{65}\lambda_{13})^2 &= \lambda_{12}\lambda_{31}\lambda_{45}\lambda_{64}(\lambda_{23} - \lambda_{56})^2,\\ \mu_2^2(\lambda_{54}\lambda_{13} - \lambda_{46}\lambda_{21})^2 &= \lambda_{23}\lambda_{12}\lambda_{56}\lambda_{45}(\lambda_{31} - \lambda_{64})^2,\\ \mu_3^2(\lambda_{65}\lambda_{21} - \lambda_{54}\lambda_{32})^2 &= \lambda_{31}\lambda_{23}\lambda_{64}\lambda_{56}(\lambda_{12} - \lambda_{45})^2,\end{aligned}$$

which proves (8.28). Finally, substitute in the relation $\begin{vmatrix} A_2 & C_2 \\ \bar{C}_2 & \bar{A}_2 \end{vmatrix} = 0$, which is a consequence of (8.30), the found values of $\mu_1^2, \ldots, \mu_3^2$ to obtain the relation (8.29).

We have assumed that $\delta \neq \pm 1$. But when $\delta = \pm 1$ then $\lambda_{14} = \lambda_{25} = \lambda_{36}$ and $P = 0$, as follows by direct computation from (8.30). By Proposition 8.6 the two equations (8.22) are proportional, contrary to our assumption. □

Remark 8.8. As in the prior case, $P = \bar{P} = 0$, as follows from $\bar{A}_1\bar{C}_2 - \bar{A}_2\bar{C}_1 = 0$ and $A_1C_2 - A_2C_1 = 0$, themselves a consequence of (8.30).

Proposition 8.9. *Suppose that two equations in (8.22) have a common factor of the form $v - (\alpha w + \beta)$ without being proportional, and with $\alpha\beta \neq 0$. There exist constants ρ, q and square roots ϵ_1 and ϵ_2 of 1, with $\rho \in \mathbf{C}^*$ and $q \in \mathbf{C} \setminus \{0, \pm 1, \pm 1/3\}$, such that*

$$\begin{aligned}\lambda_{12} &= \rho(q-1)^3(3q+1) & \lambda_{45} &= \rho(q-1)(3q+1)^3\\ \lambda_{23} &= 16\rho q^3 & \lambda_{56} &= 16\rho q\\ \lambda_{31} &= \rho(1-3q)(q+1)^3 & \lambda_{64} &= \rho(1-3q)^3(q+1)\end{aligned}$$

$$\begin{aligned}\mu_1 &= \epsilon_1\rho(q^2-1)(9q^2-1)\\ \mu_2 &= 4\epsilon_2\rho q(q-1)(3q+1)\\ \mu_3 &= 4\epsilon_1\epsilon_2\rho q(q+1)(3q-1).\end{aligned}$$

Proof. Let us suppose that both equations in (8.22) are divisible by the same factor $v-(\alpha w+\beta)$, where α and β are independent of $i=1,2$, and $\alpha\beta\neq 0$. By looking at the top degrees we conclude that this means that these two equations can be written as

$$(v-(\alpha w+\beta))(v(A_iw+B_i)+\delta_iw),$$

where $\delta_i=\bar{B}_i/\alpha=\bar{A}_i/\beta$, leading to the following equations

$$\begin{aligned} \bar{A}_i &= \beta\delta_i, & C_i &= -\alpha A_i, & & \\ \bar{B}_i &= \alpha\delta_i, & D_i &= \delta_i-\beta A_i-\alpha B_i, & i&=1,2. \\ \bar{C}_i &= \beta B_i, & & & & \end{aligned} \tag{8.34}$$

We deduce from them the following relations which are the ones that will be used in what follows:

$$\begin{vmatrix} \bar{A}_1 & \bar{A}_2 \\ \bar{B}_1 & \bar{B}_2 \end{vmatrix}=0, \qquad \begin{vmatrix} A_1 & A_2 \\ C_1 & C_2 \end{vmatrix}=0, \qquad \begin{vmatrix} B_1 & \bar{C}_2 \\ B_2 & \bar{C}_2 \end{vmatrix}=0, \tag{8.35}$$

$$\bar{A}_iB_iC_i=-A_i\bar{B}_i\bar{C}_i, \tag{8.36}$$

$$\bar{C}_i^2(A_i\bar{A}_i+B_i\bar{B}_i)-\bar{A}_i^2B_i^2=-\bar{A}_iB_i\bar{C}_iD_i. \tag{8.37}$$

To obtain the last equation, deduce from (8.34) that $\alpha\bar{A}_iB_i=\bar{B}_i\bar{C}_i$ and that $\beta B_i=\bar{C}_i$. Substituting this into

$$D_i=\frac{\bar{B}_i}{\alpha}-\beta A_i-\alpha B_i,$$

first multiplied by $-\bar{A}_iB_i\bar{C}_i$ yields the desired expression.

We will now investigate the equations in uw-space of the curve that is common to the two curves defined by (8.22). Since this common curve is given in vw-space by $v=\alpha w+\beta$, we get its equation in uw-space by substituting $\alpha w+\beta$ for v in the first equation of (8.21). This gives

$$u=\frac{\mu_3(\alpha w+\beta)-\mu_2w+\lambda_{65}(\alpha w+\beta)w}{\mu_3w-\mu_2(\alpha w+\beta)+\lambda_{32}},$$

which we write as

$$ul_1(w)-(\lambda_{65}\alpha w^2+\kappa w+\mu_3\beta)=0, \tag{8.38}$$

where it is important that the linear term $l_1(w)$ is not identically zero and that the coefficients $\lambda_{65}\alpha$ and $\mu_3\beta$ are different from zero (by non-degeneracy and since $\alpha\beta\neq 0$). We claim that (8.38) must be divisible by a linear function $\xi w+\eta$, with $\xi\neq 0$.

To show this, let us first point out that if any relation holds between u, v, w, the cycled relation holds between u, v, w by cycling $u \Rightarrow v \Rightarrow w \Rightarrow u, 1 \Rightarrow 2 \Rightarrow 3 \Rightarrow 1$ and $4 \Rightarrow 5 \Rightarrow 6 \Rightarrow 4$, yielding similar, but equivalent equations, for v in terms of u and w, to wit

$$w^2(A_i'u + B_i') + w(C_i'u^2 + D_i'u - \bar{C}_i') - u(\bar{B}_i'u + \bar{A}_1') = 0, \tag{8.39}$$

where $i = 1, 2$. The primed expressions $A_i', B_i', \ldots$ refer to the expressions but cycled once, i.e., $1 \Rightarrow 2 \Rightarrow 3 \Rightarrow 1, 4 \Rightarrow 5 \Rightarrow 6 \Rightarrow 4$.

We have now three different equations (8.38) and (8.39) (with $i = 1, 2$) for the curve in uw space. If (8.38) is irreducible then (8.39) must for $i = 1, 2$ be divisible by (8.38), since the latter is linear in u. But notice first that (8.39) does not have a constant term (a term that is independent of u and w), while (8.38) has such a term (which is different from zero). And notice also that the quotient must be independent of w because of the non-zero coefficient of w^2 in (8.38). Therefore, the quotient must be a multiple of u, which means that the two equations in (8.39) are proportional. But then we are in the case of Proposition 8.6, so that the two equations in (8.22) are themselves proportional, contrary to our assumption. It follows that (8.38) cannot be irreducible. Since it is linear in u, with a non-zero coefficient which is independent of u, (8.38) must be divisible by a linear function $\xi w + \eta$, with $\xi \neq 0$.

Upon dividing (8.38) by the factor $\xi w + \eta$ we find that the common factor of the two equations (8.39) is of the form $u - (\alpha' w + \beta')$, with $\alpha'\beta' \neq 0$. But this means that we are in the same situation as in the beginning of this proof, but with respect to the shifted equation (8.39). This implies that the relations (8.35), (8.36) and (8.37), imply the same relations but shifted. Applying this to

$$0 = \begin{vmatrix} \bar{A}_1 & \bar{A}_2 \\ \bar{B}_1 & \bar{B}_2 \end{vmatrix} = \lambda_{21}(\lambda_{23}\mu_1^2 + \lambda_{64}\mu_2^2 + \lambda_{12}\mu_3^2 + \lambda_{23}\lambda_{64}\lambda_{12})$$

we get the following three equations

$$\begin{aligned} 0 &= \lambda_{23}\mu_1^2 + \lambda_{64}\mu_2^2 + \lambda_{12}\mu_3^2 + \lambda_{12}\lambda_{23}\lambda_{64}, \\ 0 &= \lambda_{23}\mu_1^2 + \lambda_{31}\mu_2^2 + \lambda_{45}\mu_3^2 + \lambda_{23}\lambda_{31}\lambda_{45}, \\ 0 &= \lambda_{56}\mu_1^2 + \lambda_{31}\mu_2^2 + \lambda_{12}\mu_3^2 + \lambda_{31}\lambda_{12}\lambda_{56}. \end{aligned} \tag{8.40}$$

A fourth equation, which is shift-invariant, is found from $\begin{vmatrix} A_1 & C_1 \\ A_2 & C_2 \end{vmatrix} = 0$, and is given by

$$-\bar{P} = \lambda_{56}\mu_1^2 + \lambda_{64}\mu_2^2 + \lambda_{45}\mu_3^2 + \lambda_{56}\lambda_{64}\lambda_{45} = 0. \tag{8.41}$$

The consistency of the four equations (8.40) and (8.41) implies that $\Delta = 0$, with

$$\Delta := \begin{vmatrix} \lambda_{23} & \lambda_{64} & \lambda_{12} & \lambda_{12}\lambda_{23}\lambda_{64} \\ \lambda_{23} & \lambda_{31} & \lambda_{45} & \lambda_{23}\lambda_{31}\lambda_{45} \\ \lambda_{56} & \lambda_{31} & \lambda_{12} & \lambda_{31}\lambda_{12}\lambda_{56} \\ \lambda_{56} & \lambda_{64} & \lambda_{45} & \lambda_{56}\lambda_{64}\lambda_{45} \end{vmatrix} \lambda_{45}^{-2}\lambda_{56}^{-2}\lambda_{64}^{-2}$$

$$= (1 - xy - yz - zx)^2 - 4xyz(x + y + z - 2),$$

where

$$x := \frac{\lambda_{12}}{\lambda_{45}}, \qquad y := \frac{\lambda_{23}}{\lambda_{56}}, \qquad z := \frac{\lambda_{31}}{\lambda_{64}}.$$

Moreover, the consistency implies that three of the four equations (8.40) and (8.41) can be solved linearly for $(\mu_1^2, \mu_2^2, \mu_3^2)$. If, say, the three equations (8.40) can be solved linearly for $(\mu_1^2, \mu_2^2, \mu_3^2)$, i.e.,

$$\Delta_0 := \begin{vmatrix} \lambda_{23} & \lambda_{64} & \lambda_{12} \\ \lambda_{23} & \lambda_{31} & \lambda_{45} \\ \lambda_{56} & \lambda_{31} & \lambda_{12}. \end{vmatrix} \neq 0,$$

then (8.36), for $i = 1$, can be written after (partly) substituting the found values for $(\mu_1^2, \mu_2^2, \mu_3^2)$, in the following simple form:

$$\mu_1\mu_3(\mu_2^4 - \lambda_{12}\lambda_{23}\lambda_{45}\lambda_{56}) + \mu_2 E\bar{E}^2 F\Delta\Delta_0^{-2} = 0,$$

where E and F were defined in Proposition 8.7. Since $\Delta = 0$ this gives us an expression for μ_2^4, which, together with its cycled forms, is given by

$$\mu_1^4 = \lambda_{12}\lambda_{31}\lambda_{45}\lambda_{64}, \qquad \mu_2^4 = \lambda_{23}\lambda_{12}\lambda_{56}\lambda_{45}, \qquad \mu_3^4 = \lambda_{31}\lambda_{23}\lambda_{64}\lambda_{56}. \quad (8.42)$$

For any other three of the four equations (8.40) and (8.41) one finds the same result. The fact that the discriminant of Δ in x is given by

$$\mathrm{disc}(\Delta, x) = 16yz(z-1)^2(y-1)^2$$

suggests to choose p, q, r such that $p^2 = x$, $q^2 = y$ and $r^2 = z$. This allows us to replace in all formulas the parameters λ_{12}, λ_{23} and λ_{31} by the parameters p, q and r:

$$\lambda_{12} = p^2\lambda_{45}, \qquad \lambda_{23} = q^2\lambda_{56}, \qquad \lambda_{31} = r^2\lambda_{64}. \quad (8.43)$$

The signs of p, q and r are chosen as follows. If we substitute (8.43) in (8.42) then we find that

$$\mu_1^4 = (pr\lambda_{45}\lambda_{64})^2, \qquad \mu_2^4 = (qp\lambda_{56}\lambda_{45})^2, \qquad \mu_3^4 = (rq\lambda_{64}\lambda_{56})^2,$$

and the signs of p, q and r can be chosen such that

$$\mu_1^2 = pr\lambda_{45}\lambda_{64}, \qquad \mu_2^2 = qp\lambda_{56}\lambda_{45}, \qquad \mu_3^2 = rq\lambda_{64}\lambda_{56}. \quad (8.44)$$

A priori there is also the possibility of a common minus sign for each of the three formulas in (8.44), but this possibility is excluded by substituting (8.43) and (8.44) (or its alternative with a common minus sign) into the two equations in (8.36): only when substituting (8.44) are those equations satisfied. This fixes the signs of p, q and r up to a common $\pm$ sign. To fix the latter, multiply the three formulas in (8.44) to obtain that

$$(\mu_1\mu_2\mu_3)^2 = (pqr\lambda_{45}\lambda_{56}\lambda_{64})^2.$$

We now fix the common $\pm$ sign for p, q and r by demanding that

$$\mu_1\mu_2\mu_3 = pqr\lambda_{45}\lambda_{56}\lambda_{64}. \tag{8.45}$$

As we already pointed out, if we substitute (8.43) and (8.44) in (8.36) then we find that these two equations are automatically satisfied. On the other hand, if we substitute them in any of the three equations in (8.35) then we find that each of them is (by non-degeneracy) equivalent to

$$pq + qr + rp + 1 = 0. \tag{8.46}$$

If we substitute (8.43) and (8.44) in the two equations in (8.37) we find formulas that depend on $\mu_1\mu_2\mu_3$ only, hence we can eliminate the dependence on the μ_i completely by using (8.45). The resulting equations are given by

$$(p+q)(pq+1)((q+1)^2\lambda_{56} + (p-1)^2\lambda_{45}) - r(p-1)^2(q+1)^2\lambda_{64} = 0, \tag{8.47}$$

and

$$\begin{aligned}(p+r)^2((r-1)^2\lambda_{64} - (1+p)^2\lambda_{45}) +& \\ (r-1)(1+p)(q(p+1)(r-1) + (q^2-1)(p+r))\lambda_{56} &= 0.\end{aligned} \tag{8.48}$$

If we add to these two linear equations in λ_{45}, λ_{56} and λ_{64} the following obvious equation

$$\lambda_{45} + \lambda_{56} + \lambda_{64} = 0 \tag{8.49}$$

then the determinant of the resulting linear system admits the following compact representation

$$2\frac{r(r+1)^2(r-1)^3(q-1)^2(q+1)^4(3qr+q-r+1)}{(q+r)^5},$$

where we have used (8.46) to eliminate p (linearly). By non-degeneracy the linear system admits a non-trivial solution for λ_{45}, λ_{56} and $\lambda_{64} = 0$, hence the determinant vanishes. By non-degeneracy and since we assumed that the two equations in (8.22) are not proportional, we conclude that $3qr+q-r+1=0$. Indeed, if $r=0$ then $\lambda_{31}=0$, which contradicts non-degeneracy; if $r^2=1$ then (8.46) implies that $p^2=1$ or that $q^2=1$, which in either case implies, by (8.43), that $\lambda_{14}=\lambda_{25}=\lambda_{36}$, so that by Proposition 8.6 the two equations in (8.22) are proportional (recall from (8.41) that $\bar{P}=0$); if $q^2=1$ we arrive at the same conclusion.

Thus, $3qr + q - r + 1 = 0$ which gives, together with (8.46) the following expressions for p and r in terms of q,

$$p = \frac{q-1}{3q+1}, \qquad r = -\frac{q+1}{3q-1}.$$

It follows that we can express all the parameters of the metric in terms of q. To do this, first solve the homogeneous linear system (8.47), (8.48), (8.49) for λ_{45}, λ_{56} and λ_{64}. Up to a non-zero constant factor (which we called ρ in the statement of the proposition) the solution is given by

$$\begin{aligned}
\lambda_{45} &= (q-1)(3q+1)^3, \\
\lambda_{56} &= 16q, \\
\lambda_{64} &= (1-3q)^3(1+q).
\end{aligned}$$

The values of λ_{12}, λ_{23} and λ_{31} then follow at once from it and from (8.43). Also, we get from (8.44) that

$$\begin{aligned}
\mu_1^2 &= (q^2-1)^2(9q^2-1)^2, \\
\mu_2^2 &= 16q^2(q-1)^2(3q+1)^2, \\
\mu_3^2 &= 16q^2(q+1)^2(3q-1)^2.
\end{aligned}$$

Up to signs this lead to the values of the μ_i by taking square roots. A constraint on the signs is found from (8.45), which, again by direct substitution, yields

$$\mu_1\mu_2\mu_3 = 16q^2(9q^2-1)^2(q^2-1)^2.$$

This leads to the formulas that were stated in the proposition. Conversely, it follows from direct substitution that the given formulas satisfy (8.35), (8.36) and (8.37) for any value of q (for $q \in \{0, \pm 1, \pm 1/3\}$ one of the parameters λ_{ij} or μ_i vanishes, which is forbidden by non-degeneracy). □

Proposition 8.10. *Suppose that two equations in (8.22) are not proportional, but have a common factor of the form $w - \alpha$, or of the form $v - \alpha$, where α is constant. Then the indicial locus $\mathcal{I}$ cannot contain a curve.*

Proof. Let us assume that that we have a common factor of the form $w - \alpha$. Notice that $\alpha \neq 0$ in view of Proposition 8.4. If we substitute α for w in (8.22) then we find the following relations on the coefficients.

$$\begin{aligned}
&A_i\alpha + B_i = 0, \\
&C_i\alpha^2 + D_i\alpha - \bar{C}_i = 0, \qquad (i = 1, 2) \\
&\bar{B}_i\alpha + \bar{A}_i = 0.
\end{aligned} \tag{8.50}$$

This leads to

$$
\begin{aligned}
0 &= \begin{vmatrix} A_1 & B_1 \\ A_2 & B_2 \end{vmatrix} = \lambda_{65}\mu_1^2 + \lambda_{13}\mu_2^2 + \lambda_{54}\mu_3^2 + \lambda_{13}\lambda_{65}\lambda_{54}, \\
0 &= \begin{vmatrix} \bar{A}_1 & \bar{B}_1 \\ \bar{A}_2 & \bar{B}_2 \end{vmatrix} = \lambda_{32}\mu_1^2 + \lambda_{46}\mu_2^2 + \lambda_{21}\mu_3^2 + \lambda_{46}\lambda_{32}\lambda_{21}.
\end{aligned} \tag{8.51}
$$

Let us first assume that $\lambda_{65}\lambda_{46} - \lambda_{32}\lambda_{13} \neq 0$. Then we may solve (8.51) for μ_1^2 and μ_2^2. When we substitute the result in

$$
\begin{aligned}
0 &= \begin{vmatrix} A_1 & B_1 \\ \bar{B}_1 & \bar{A}_1 \end{vmatrix} - \begin{vmatrix} A_2 & B_2 \\ \bar{B}_2 & \bar{A}_2 \end{vmatrix} \\
&= (\mu_1^2 - \mu_2^2)^2 - (\lambda_{13}\lambda_{45} + \lambda_{12}\lambda_{46})\mu_1^2 + (\lambda_{12}\lambda_{23} + \lambda_{45}\lambda_{56})\mu_2^2 \\
&\qquad + \lambda_{12}\lambda_{45}(\lambda_{23}\lambda_{56} + \lambda_{13}\lambda_{46} - 2\mu_3^2),
\end{aligned} \tag{8.52}
$$

then we find a quadratic polynomial in μ_3^2, whose solutions are given by

$$
\mu_3^2 = \lambda_{23}\lambda_{56}, \quad \text{and} \quad \mu_3^2 = \lambda_{13}\lambda_{46}\left(\frac{\lambda_{23} - \lambda_{56}}{\lambda_{13} - \lambda_{46}}\right)^2. \tag{8.53}
$$

If we substitute the first possibility, $\mu_3^2 = \lambda_{23}\lambda_{56}$ back into (8.51), then it simplifies to

$$
\begin{aligned}
0 &= \lambda_{65}\mu_1^2 + \lambda_{13}\mu_2^2 + \lambda_{12}\lambda_{65}\lambda_{54}, \\
0 &= \lambda_{32}\mu_1^2 + \lambda_{46}\mu_2^2 + \lambda_{45}\lambda_{32}\lambda_{21},
\end{aligned}
$$

which is easily seen to have $\mu_1^2 = \lambda_{54}\lambda_{21}$ and $\mu_2 = 0$ as its only solutions. But this possibility is excluded by non-degeneracy of the metric. Consider therefore the second possibility in (8.53), and substitute it into (8.51). This gives

$$
\mu_1^2 = \left(\frac{\lambda_{45} - \lambda_{12}}{\lambda_{13} - \lambda_{46}}\right)^2 \lambda_{13}\lambda_{46}, \qquad \mu_2 = \pm\frac{\lambda_{23}\lambda_{46} - \lambda_{13}\lambda_{56}}{\lambda_{13} - \lambda_{46}}.
$$

There is one restriction on the signs of the μ_i, which follows from substituting these solutions in

$$
0 = \begin{vmatrix} A_1 & B_1 \\ \bar{B}_1 & \bar{A}_1 \end{vmatrix} + \begin{vmatrix} A_2 & B_2 \\ \bar{B}_2 & \bar{A}_2 \end{vmatrix},
$$

namely

$$
\mu_1\mu_2\mu_3 = \lambda_{13}\lambda_{46}\frac{(\lambda_{12} - \lambda_{45})(\lambda_{23} - \lambda_{56})(\lambda_{23}\lambda_{46} - \lambda_{13}\lambda_{56})}{(\lambda_{13} - \lambda_{46})^3} \tag{8.54}
$$

so that

$$
\mu_1\mu_3 = \pm\lambda_{13}\lambda_{46}\frac{(\lambda_{12} - \lambda_{45})(\lambda_{23} - \lambda_{56})}{(\lambda_{13} - \lambda_{46})^2}. \tag{8.55}
$$

It then follows, by using $A_1\alpha+B_1=0$ and (8.53) and (8.54) to solve for $\alpha\mu_3$, and then using (8.53) again to solve for α^2 we find

$$\alpha\mu_3=-\frac{\mu_1\mu_2\mu_3+\lambda_{54}\mu_3^2}{\lambda_{54}\lambda_{65}+\mu_2^2}=\lambda_{13}\frac{\lambda_{23}-\lambda_{56}}{\lambda_{13}-\lambda_{46}},\qquad\text{and}\qquad\alpha^2=\frac{\lambda_{13}}{\lambda_{46}}.\tag{8.56}$$

If we substitute these formulas and (8.55) in the identity $\alpha\mu_1\alpha\mu_3=\alpha^2\mu_1\mu_3$ then we find

$$\alpha\mu_1=\mp\lambda_{13}\frac{\lambda_{45}-\lambda_{12}}{\lambda_{13}-\lambda_{46}}.$$

Consider now u, as given in the first equality in (8.21) (recall that we are looking for a curve in uvw-space), and notice that $\mu_1 u$ is expressible in terms of $\alpha\mu_1$, $\alpha\mu_3$, $\mu_1\mu_3$, μ_2 and v only, for each of which we have found an explicit expression in terms of the λ_{ij} (recall that v is free and that $w=\alpha$). Therefore we can compute $\mu_1 u$ by a direct substitution in (8.21), and we find that the result is independent of v, namely

$$\mu_1 u=-\lambda_{13}\frac{\lambda_{12}-\lambda_{45}}{\lambda_{13}-\lambda_{46}}.$$

Finally we plug this and (8.56) into the right hand side of the second equation in (8.20) to find

$$\lambda_{13}+\mu_1u-\mu_3\alpha=\lambda_{13}\left(1-\frac{\lambda_{12}-\lambda_{45}}{\lambda_{13}-\lambda_{46}}-\frac{\lambda_{23}-\lambda_{56}}{\lambda_{13}-\lambda_{46}}\right)=0,$$

and so $x_2^{(0)}=0$. In view of Proposition 8.4 this means that if $w-\alpha$ is a common factor of the two polynomials in (8.22) and $\lambda_{65}\lambda_{46}-\lambda_{32}\lambda_{13}\neq 0$ then the indicial locus does not contain a curve.

Let us now assume that $\lambda_{65}\lambda_{46}-\lambda_{32}\lambda_{13}=0$, but that $\lambda_{65}\lambda_{21}-\lambda_{32}\lambda_{54}\neq 0$. Then we proceed as in the previous case, except that we solve (8.51) for μ_1^2 and μ_3^2 and we substitute the result in (8.52). This gives

$$\mu_1^2=\lambda_{13}\frac{(\lambda_{23}-\lambda_{64})^2}{\lambda_{46}},\mu_2=\pm\lambda_{23}\frac{\lambda_{13}-\lambda_{64}}{\lambda_{46}},\mu_3^2=\frac{\lambda_{13}\lambda_{23}^2}{\lambda_{46}}.$$

A sign specification is again found from (8.54), namely

$$\mu_1\mu_3=\pm\lambda_{13}\lambda_{23}\frac{\lambda_{23}-\lambda_{64}}{\lambda_{46}}.$$

We find $\mu_1\alpha$ and $\mu_3\alpha$ from $A_1\alpha+B_1=0$, as above, yielding also $\mu_1 u$, to wit

$$\mu_1\alpha=\mp\lambda_{13}\frac{\lambda_{23}-\lambda_{64}}{\lambda_{46}}=\pm\mu_1u,\quad \mu_3\alpha=\frac{\lambda_{13}\lambda_{23}}{\lambda_{64}},$$

where $\alpha^2=\lambda_{13}/\lambda_{46}$, as in (8.56). Then again $x_2^{(0)}=0$, since

$$\lambda_{13}+\mu_1u-\mu_3\alpha=\lambda_{13}\left(1+\frac{\lambda_{23}-\lambda_{64}}{\lambda_{64}}-\frac{\lambda_{23}}{\lambda_{64}}\right)=0,$$

excluding also this case.

We are left with the possibility

$$\lambda_{65}\lambda_{46} - \lambda_{32}\lambda_{13} = 0 = \lambda_{65}\lambda_{21} - \lambda_{32}\lambda_{54}. \tag{8.57}$$

If we write $\lambda_{54} = \lambda_{64} - \lambda_{65}$ then (8.57) implies, upon eliminating λ_{65}, that $\lambda_{13} = \lambda_{46}\epsilon$ and that $\lambda_{65} = \lambda_{32}\epsilon$, where $\epsilon^2 = 1$. If $\epsilon = 1$ then it follows that $\lambda_{14} = \lambda_{25} = \lambda_{36}$, and both equations in (8.51) reduce to $P = \bar{P} = 0$. In view of Proposition (8.6) the two equations in (8.22) are proportional, which is contrary to our assumption. Therefore, $\epsilon = -1$, which means that $\lambda_{13} = -\lambda_{46}$ and $\lambda_{65} = -\lambda_{32}$. In this case, first solve for any of the (proportional) equations in (8.51) for μ_3^2 and notice that, upon substituting the found value in (8.52), the latter becomes a complete square,

$$(\mu_1^2 - \mu_2^2 + \lambda_{12}^2)^2 = 0. \tag{8.58}$$

This gives us a formula for μ_2^2. Substitute now $\alpha = -B_1/A_1$, as follows from (8.50) in each of the two equations on the second line of (8.50) to find that $\mu_1^2 + \lambda_{12}^2 = 0$, or that $\mu_3 = \mu_1\mu_2/(\lambda_2 - \lambda_1)$. If we substitute the second possibility in $A_i\alpha + B_i$, for $i = 1, 2$, we find $\alpha\mu_1\mu_2 = 0$, which is impossible by non-degeneracy of the metric. Comparing the first possibility $\mu_1^2 + \lambda_{12}^2 = 0$ to (8.58) implies that $\mu_2 = 0$, which is also excluded by non-degeneracy.

This shows that we cannot have a common factor of the form $w - \alpha$, except possibly when we are in the case of Proposition 8.6. In other words, if the parameters that define the non-degenerate metric do not satisfy both $\lambda_{14} = \lambda_{25} = \lambda_{36}$ and $P = 0$, then a curve which is contained in the indicial locus cannot be contained in a hyperplane of the form $w = \alpha$. By symmetry, it can then also not be contained in a hyperplane of the form $v = \alpha$ or of the form $u = \alpha$. In particular, if the two equations in (8.22) are not proportional then they also cannot have a common factor of the form $v - \alpha$. □

We now assemble the previous propositions to prove the main theorem.

Proof of Theorem 8.3. As we pointed out the two equations in (8.22) must define a curve, which means that they are proportional, or have a common factor (without being proportional). We dealt with the case of proportional equations in Proposition 8.6, which lead to the first case in Theorem 8.3. Suppose now that the two equations in (8.22) have a common factor $\phi(v, w)$, without being proportional. Since both equations are quadratic in v the factor is of degree 0, 1 or 2 in v. The case in which the degree in v is zero, i.e., the factor is of the form $w - \alpha$, was treated in Proposition 8.10; we have seen that in this case the indicial locus cannot contain a curve. The next possibility is that $\phi(v, w)$ is linear and monic in v. Looking at the equations (8.22) one sees that there are only the following possibilities:

(1) $\phi(v, w) = v - \alpha$, with $\alpha \neq 0$;
(2) $\phi(v, w) = v - \alpha w$, with $\alpha \neq 0$;
(3) $\phi(v, w) = v - (\alpha w + \beta)$, with $\alpha\beta \neq 0$;
(4) $\phi(v, w) = v - w(\alpha w + \beta)$, with $\alpha \neq 0$,

where α and β are constants.

We dealt with case (1)–(3) in Propositions 8.7, 8.9 and 8.10. Let us analyze case (4), i.e., we suppose that the two equations in (8.22) have a common factor of the form $v - w(\alpha w + \beta)$, with $\alpha \neq 0$, without being proportional. If we substitute $v = w(\alpha w + \beta)$ in these equations and we express that the resulting polynomial in w is zero, and so its coefficients are identically zero, then we find the following equations.

$$
\begin{aligned}
&A_i = 0,\\
&\alpha B_i + C_i = 0,\\
&2\alpha\beta B_i + \beta C_i + \alpha D_i = 0, \qquad\qquad (i = 1, 2)\\
&\beta^2 B_i + \beta D_i - \alpha\bar{C}_i - \bar{B}_i = 0,\\
&\beta\bar{C}_i + \bar{A}_i = 0.
\end{aligned}
\tag{8.59}
$$

Notice that the last equation implies that $P = 0$ (see (8.27)). Since $A_1 = A_2 = 0$ we have (see (8.23)) that $\mu_2^2 = \lambda_{54}\lambda_{56}$ and $\mu_1 = \lambda_{54}\mu_3/\mu_2$. Then the second equation in (8.59) implies that

$$
0 = \begin{vmatrix} B_1 & C_1 \\ B_2 & C_2 \end{vmatrix} = 2\lambda_{45}(\lambda_{13} - \lambda_{46})\mu_2\mu_3,
$$

so that $\lambda_{14} = \lambda_{36}$, by non-degeneracy. If we substitute this and $\alpha = -C_1/B_1$, $\beta = -\bar{A}_1/\bar{C}_1$ in $\alpha D_1 - \beta C_1 = 0$, gotten by subtracting 2β times the second equation from the third in (8.59), then we get

$$
\lambda_{45}^2\lambda_{56}\lambda_{13}(\lambda_{23} - \lambda_{56})\mu_2\mu_3^2 = 0,
$$

so that $\lambda_{25} = \lambda_{36}$, again by non-degeneracy. Thus, we have that $P = 0$ and that $\lambda_{14} = \lambda_{25} = \lambda_{36}$, which implies, according to Proposition 8.6, that the two equations in (8.22) are proportional. Which is contrary to our assumptions, showing that the two equations in (8.22) cannot have a common factor of the form $v - w(\alpha w + \beta)$, with $\alpha \neq 0$ (except when they are proportional).

In principle we need to consider now the cases in which $\phi(v, w)$ is linear in v but not monic, i.e., its leading term is linear in w, leading to four cases, as above. We claim that each of these four cases corresponds to one of the cases (1) – (4). Let us show this for

$$
\phi(v, w) = (\beta' w + \alpha')v - w.
$$

If the latter is a common factor of the two equations in (8.22) then they can be written as

$$
((\beta' w + \alpha')v - w)(v\delta_i + \bar{B}_i w + \bar{A}_i).
$$

This leads to the following equations, coming from the coefficients of $v^i w^j$:

$$\begin{aligned} A_i &= \beta'\delta_i', & \bar{C}_i &= -\alpha'\bar{A}_i, & \\ B_i &= \alpha'\delta_i', & -D_i &= \delta_i' - \beta'\bar{A}_i - \alpha'\bar{B}_i, & i = 1, 2. \\ C_i &= \beta'\bar{B}_i. & & \end{aligned}$$

A simple comparison of these equations with (8.34) shows that these equations are the barred duals of each other, hence leading to the same solutions, but with $\lambda_{12} \leftrightarrow \lambda_{45}$ and $\lambda_{23} \leftrightarrow \lambda_{56}$ and $\lambda_{31} \leftrightarrow \lambda_{64}$. Notice that this leads to nothing new in Propositions 8.6 and 8.7 because the conditions that are given there on the metric are invariant under this transposition. By the same means the case in which $\phi(v, w)$ is a quadratic polynomial in v is shown to be dual to the case in which $\phi(v, w)$ is independent of v. Therefore we have analyzed all possible cases and Theorem 8.3 is proven.

Remark 8.11. When searching for half-diagonal non-degenerate metrics on $\mathfrak{so}(4)$ which lead to an irreducible a.c.i. system we have only singled out those irreducible systems which admit a weight homogeneous principal balance. A priori there might also exist such metrics which admit only principal balances that are not (weight) homogeneous, but no method is at present known to single out these cases. Interestingly enough, it can however be shown by direct computation that the mere existence of balances (not necessarily principal balances) already singles out a particular class of metrics. This is a consequence of the fact that if

$$x(t) = \frac{1}{t^r} \sum_{k \in \mathbf{N}} x^{(k)} t^k$$

is a balance to (8.13), with $x^{(0)} \neq 0$ and $r \geqslant 2$, then $x^{(0)}$ must satisfy $f_i(x_1^{(0)}, \dots, x_6^{(0)}) = 0$, for $i = 1, \dots, 6$, where we have written the i-th equation in (8.13) as $\dot{x}_i = f_i(x_1, \dots, x_6)$. The existence of a non-zero solution to the system of algebraic equation $f_i(x_1^{(0)}, \dots, x_6^{(0)}) = 0$, where $i = 1, \dots, 6$, implies, by direct computation, that the metric necessarily satisfies one of the following seven conditions.

(1) $\lambda_{23}\lambda_{56} = (\mu_2 \pm \mu_3)^2$,
(2) $\lambda_{31}\lambda_{64} = (\mu_3 \pm \mu_1)^2$,
(3) $\lambda_{12}\lambda_{45} = (\mu_1 \pm \mu_2)^2$,
(4) $\sum_{i=1}^3 a_i\mu_i^2(\mu_i^2 + b_i) + b_i\mu_{i+1}^2\mu_{i-1}^2 = 0$,

where $a_1 = \lambda_{65}\lambda_{32}$ and $b_1 = \lambda_{64}\lambda_{12} - \lambda_{45}\lambda_{31}$, and a_2, b_2, a_3, b_3 are obtained from a_1 and b_1 by the cyclic permutation

$$(\lambda_1, \dots, \lambda_6, \mu_1, \dots, \mu_3) \longrightarrow (\lambda_2, \lambda_3, \lambda_1, \lambda_5, \lambda_6, \lambda_4, \mu_2, \mu_3, \mu_1).$$

8.2 Geodesic Flow on SO(4) for the Manakov Metric

8.2.1 From Metric I to the Manakov Metric

In this paragraph we show that, except for some limiting cases, the case *(1)* of Theorem 8.3 corresponds to the Manakov metric. Recall that in case *(1)*, H is given by

$$H = \frac{1}{2}\sum_{i=1}^{6} \lambda_i x_i^2 + \sum_{l=1}^{3} \mu_l x_l x_{l+3},$$

where the parameters defining the metric satisfy

$$\begin{gathered} \lambda_{14} = \lambda_{25} = \lambda_{36}, \\ \mu_1^2\lambda_{23} + \mu_2^2\lambda_{31} + \mu_3^2\lambda_{12} + \lambda_{12}\lambda_{23}\lambda_{31} = 0, \end{gathered}$$

with $\lambda_{ij} := \lambda_i - \lambda_j$. We have that, up to a Casimir, the metric is diagonal in the natural $\mathfrak{so}(4)$ coordinates X_{ij}. Indeed, solving (8.8) for $x_1, \dots, x_6$ and substituting the result in H yields[2]

$$\begin{aligned} H &= \frac{1}{2}\sum_{i=1}^{3} \left(\lambda_i (X_{jk} + X_{i4})^2 + \lambda_{i+3}(X_{jk} - X_{i4})^2 + 2\mu_i (X_{jk}^2 - X_{i4}^2)\right) \\ &= \frac{1}{2}\sum_{i=1}^{3}((\lambda_i + \lambda_{i+3} + 2\mu_i)X_{jk}^2 + (\lambda_i + \lambda_{i+3} - 2\mu_i)X_{i4}^2) \\ &\quad + \sum_{i=1}^{3} \lambda_{i,i+3} X_{jk} X_{i4}, \end{aligned}$$

and we see that the last term is a multiple of the Casimir Q_0 (see (8.7)), since $\lambda_{i,i+3}$ is independent of i (see (8.60)). In the sequel we drop this Casimir from H and we write

$$H = \frac{1}{2}\sum_{1 \leqslant i < j \leqslant 4} \Lambda_{ij} X_{ij}^2,$$

which means that the Λ_{ij} are defined by

$$\begin{aligned} \Lambda_{12} &= 2(\lambda_3 + \mu_3) - c, & \Lambda_{14} &= 2(\lambda_1 - \mu_1) - c, \\ \Lambda_{23} &= 2(\lambda_1 + \mu_1) - c, & \Lambda_{24} &= 2(\lambda_2 - \mu_2) - c, \\ \Lambda_{13} &= 2(\lambda_2 + \mu_2) - c, & \Lambda_{34} &= 2(\lambda_3 - \mu_3) - c, \end{aligned}$$

where $c = \lambda_{14} = \lambda_{25} = \lambda_{36}$. If we solve these equations linearly for λ_i, μ_i, $i = 1, \dots, 3$ in terms of the Λ_{ij}, while treating c as a given constant, and we substitute the solution in condition (8.60), then the constant c disappears

[2] We use here again the convention that (i, j, k) is any cyclic permutation of $(1, 2, 3)$.

and we find that the constants Λ_{ij} satisfy $K(\Lambda) = 0$, where $K(\Lambda)$ is the cubic polynomial, defined by

$$\begin{aligned} K(\Lambda) := &\Lambda_{12}\Lambda_{34}(\Lambda_{23} + \Lambda_{14} - \Lambda_{13} - \Lambda_{24}) \\ &+\Lambda_{23}\Lambda_{14}(\Lambda_{13} + \Lambda_{24} - \Lambda_{12} - \Lambda_{34}) \\ &+\Lambda_{13}\Lambda_{24}(\Lambda_{12} + \Lambda_{34} - \Lambda_{23} - \Lambda_{14}). \end{aligned} \tag{8.60}$$

This defines a cubic hypersurface $\mathcal{C}$ in the six-dimensional vector space of all diagonal metrics. We claim that $\mathcal{C}$ contains a Zariski open subset $\mathcal{U}$ such that if $\Lambda \in \mathcal{U}$ then there exist constants A_i, B_i, where $i = 1, \dots, 4$, such that

$$\Lambda_{ij} = \frac{B_i - B_j}{A_i - A_j}. \tag{8.61}$$

To see this, notice first that if such constants exist, then there also exist such constants with $A_4 = B_4 = 0$ and $A_3 = 1$. It is then easy to solve (8.61) with $1 \leqslant i < j \leqslant 4$, but $(i, j) \neq (1, 2)$, for the A_i and B_i in terms of the Λ_{ij}, giving

$$A_1 = \frac{\Lambda_{34} - \Lambda_{13}}{\Lambda_{14} - \Lambda_{13}}, \qquad A_2 = \frac{\Lambda_{34} - \Lambda_{23}}{\Lambda_{24} - \Lambda_{23}}, \qquad A_3 = 1, \qquad A_4 = 0,$$

$$B_1 = \frac{\Lambda_{34} - \Lambda_{13}}{\Lambda_{14} - \Lambda_{13}}\Lambda_{14}, \quad B_2 = \frac{\Lambda_{34} - \Lambda_{23}}{\Lambda_{24} - \Lambda_{23}}\Lambda_{24}, \quad B_3 = \Lambda_{34}, \quad B_4 = 0.$$

The remaining equation, (8.61) with $i = 1$ and $j = 2$, is then equivalent to (8.60), so we have parametrized a Zariski open subset $\mathcal{U}$ of $\mathcal{C}$, showing our claim. Since a metric on $\mathfrak{so}(4)$ which is of the form

$$H = \frac{1}{2} \sum_{1 \leqslant i < j \leqslant 4} \frac{B_i - B_j}{A_i - A_j} X_{ij}^2,$$

is called a *Manakov metric*, it shows that, except for some limiting cases, the case *(1)* of Theorem 8.3 corresponds to the Manakov metric.

The Liouville integrability of the Manakov metric follows from the following observations, due to Manakov (see [110]; his statement is the n-dimensional generalization of the statements that we give here). First, let $A_1 \dots, A_4$ be pairwise different, but otherwise arbitrary, and similarly for $B_1, \dots, B_4$. If we define the symmetric matrix Λ by

$$\Lambda_{ij} = \frac{B_i - B_j}{A_i - A_j},$$

then the following equality holds, for any 4×4 matrix X:

$$[X, B] + [A, \Lambda \cdot X] = 0, \tag{8.62}$$

where $A := \text{diag}(A_1, \dots, A_4)$ and $B := \text{diag}(B_1, \dots, B_4)$ and where $\Lambda \cdot X$ denotes the *Kronecker product*: $(\Lambda \cdot X)_{ij} = \Lambda_{ij} X_{ij}$. The proof of this statement is immediate.

His second observation is that, as a consequence of (8.62), the Lax equation $\dot{X} = [X, \Lambda \cdot X]$, which is the geodesic flow equation (8.5) with H the Manakov metric, is equivalent to the Lax equation with parameter

$$(A\mathfrak{h} + X)^{\cdot} = [A\mathfrak{h} + X, B\mathfrak{h} + \Lambda \cdot X]. \tag{8.63}$$

Again the proof is immediate, but notice that the combination of his two observations yields, in view of Proposition 6.28, the integrability of the geodesic flow on $\mathfrak{so}(4)$, defined by the Manakov metric. Indeed, the characteristic polynomial of $X + A\mathfrak{h}$ is given by

$$|A\mathfrak{h} + X - \mathfrak{z}\mathrm{Id}_4| = \prod_{i=1}^{4}(A_i\mathfrak{h} - \mathfrak{z}) + Q_1\mathfrak{z}^2 + Q_2\mathfrak{z}\mathfrak{h} + Q_3\mathfrak{h}^2 + Q_0^2, \tag{8.64}$$

where the two Casimirs Q_0 and Q_1 are given by (8.7), and Q_2 and Q_3 are given by

$$\begin{aligned} Q_2 &:= -\textstyle\sum_{1\leqslant i<j\leqslant 4}(A_k + A_l)X_{ij}^2, \\ Q_3 &:= \textstyle\sum_{1\leqslant i<j\leqslant 4} A_k A_l X_{ij}^2, \end{aligned} \qquad \text{where} \qquad \{i,j,k,l\} = \{1,2,3,4\}.$$

The four quadratic polynomials $Q_0, \dots, Q_3$ provide four independent constants of motion for $\mathcal{V}_1$, which is sufficient for its integrability. Indeed, the phase space is $\mathbf{C}^6$, and since we have two independent Casimirs Q_0 and Q_1 only, the rank of the Poisson structure being 4 at a generic point of $\mathbf{C}^6$, there can be at most 4 independent constants of motion and this number is the right number for Liouville integrability, see Definition 4.13. This means that the Hamiltonian H of $\mathcal{V}_1$ is necessarily dependent on these quadratic polynomials; in fact one easily verifies that

$$H = \frac{1}{2}\sum_{i=1}^{3} Q_i \sum_{j=1}^{4} A_j^{3-i} B_j \prod_{k\neq j} \frac{1}{A_j - A_k}. \tag{8.65}$$

We now show that the Lax equation (8.63) satisfies the Linearization Criterion. We do this by verifying the conditions in Corollary 6.43. Let $f(x) = \sum_{i=0}^{3} c_i x^i$ be the cubic polynomial, for which $f(A_k) = B_k$, where $k = 1, \dots, 4$. We claim that

$$B\mathfrak{h} + \Lambda \cdot X = \left[\mathfrak{h} f\left(\frac{A\mathfrak{h} + X}{\mathfrak{h}}\right)\right]_+, \tag{8.66}$$

which verifies condition (6.54). In fact,

$$\begin{aligned} \left[\mathfrak{h} f\left(\frac{A\mathfrak{h} + X}{\mathfrak{h}}\right)\right]_+ &= \left[\sum_{i=0}^{3} c_i \mathfrak{h}\left(\frac{A\mathfrak{h} + X}{\mathfrak{h}}\right)^i\right]_+ \\ &= \sum_{i=0}^{3} c_i \left(A^i\mathfrak{h} + \sum_{j=0}^{i-1} A^j X A^{i-j-1}\right) \end{aligned}$$

$$= B\mathfrak{h} + \sum_{i=0}^{3} c_i \sum_{j=0}^{i-1} A^j X A^{i-j-1},$$

since $f(A) = \sum_{i=0}^{3} c_i A^i = B$, and for $1 \leqslant k, l \leqslant 4$ we have

$$\begin{aligned}
\sum_{i=0}^{3} c_i \sum_{j=0}^{i-1} \left(A^j X A^{i-j-1}\right)_{kl} &= \sum_{i=0}^{3} c_i \sum_{j=0}^{i-1} A_k^j X_{kl} A_l^{i-j-1} \\
&= X_{kl} \sum_{i=0}^{3} c_i \sum_{j=0}^{i-1} A_k^j A_l^{i-j-1} \\
&= X_{kl} \sum_{i=0}^{3} c_i \frac{A_k^i - A_l^i}{A_k - A_l} \\
&= X_{kl} \frac{f(A_k) - f(A_l)}{A_k - A_l} \\
&= \Lambda_{kl} X_{kl},
\end{aligned}$$

since $f(A_k) = B_k$ and $\Lambda_{kl} = (B_k - B_l)/(A_k - A_l)$. This proves (8.66).

In order to verify the remaining conditions in Corollary 6.43, let us fix generic constants $\bar{c}_0, \dots, \bar{c}_3$ and let us consider the smooth plane algebraic curve, defined by

$$\prod_{i=1}^{4} (A_i \mathfrak{h} - \mathfrak{z}) + \bar{c}_1 \mathfrak{z}^2 + \bar{c}_2 \mathfrak{z}\mathfrak{h} + \bar{c}_3 \mathfrak{h}^2 + \bar{c}_0^2 = 0, \tag{8.67}$$

and let $X \in \mathfrak{so}(4)$ be such that the characteristic polynomial of $A\mathfrak{h} + X$ is given by the left hand side of (8.67), see (8.64). Four points $p_1, \dots, p_4$ need to be added to this curve in order to complete it into a compact Riemann surface. Explicitly, these points are given, in terms of a local parameter ς, by

$$p_i: \quad \mathfrak{h} = \frac{1}{\varsigma}, \qquad \mathfrak{z} = \frac{A_i}{\varsigma} + u_i \varsigma + O(\varsigma^2), \tag{8.68}$$

where each u_i depends on the values $\bar{c}_1, \dots, \bar{c}_3$ of the constants of motion $Q_1, \dots, Q_3$ only, namely

$$u_i = -\frac{\bar{c}_1 A_i^2 + \bar{c}_2 A_i + \bar{c}_3}{\prod_{j \neq i} (A_i - A_j)},$$

in particular $u_i \neq 0$, since X (and hence $\bar{c}_0, \dots, \bar{c}_3$) is assumed generic. It follows that that $\mathfrak{h}$ does not vanish at any of the points at infinity p_i, which is one of the conditions that needed to be checked.

The final condition to be checked is that $\xi_i/\mathfrak{h}$ is finite at each of these points. Denoting, as before, the (i,j)-th cofactor of $A\mathfrak{h}+X-\mathfrak{z}\,\mathrm{Id}_4$ by Δ_{ij}, we have that $\xi_i = \Delta_{1i}/\Delta_{11}$. For ξ_2 at direct substitution of (8.68) in Δ_{12}/Δ_{11} yields the following leading behavior at the points p_i.

$$\xi_2 = \begin{cases} -\dfrac{X_{12}}{A_1-A_2}\varsigma + O(\varsigma^2) & \text{at } p_1, \\[2mm] -\dfrac{1}{\varsigma}\dfrac{X_{12}}{u_2 + X_{23}^2/(A_2-A_3) + X_{24}^2/(A_2-A_4)} + O(\varsigma^0) & \text{at } p_2, \\[2mm] \dfrac{X_{13}X_{23}/(A_2-A_3)}{u_3 - X_{23}^2/(A_2-A_3) + X_{34}^2/(A_3-A_4)} + O(\varsigma) & \text{at } p_3, \\[2mm] \dfrac{X_{14}X_{24}/(A_2-A_4)}{u_4 - X_{24}^2/(A_2-A_4) - X_{34}^2/(A_3-A_4)} + O(\varsigma) & \text{at } p_4. \end{cases}$$

Thus, ξ_2 has a simple pole at worst at each of the point p_i and $\xi_2/\mathfrak{h}$ has no pole at all at these points. By symmetry the same holds true for $\xi_3/\mathfrak{h}$ and for $\xi_4/\mathfrak{h}$ so that, by Corollary 6.43, we may conclude that (8.63) satisfies the conditions of Theorem 6.41.

8.2.2 A Curve of Rank Three Quadrics

In this section we want to exhibit a remarkable property of the linear span of the quadratic constants of motion $Q_0,\dots,Q_3$ of the Manakov metric. Recall that these polynomials, which appear as coefficients in the characteristic polynomial of the matrix $X + A\mathfrak{h}$, are given by

$$\begin{aligned} Q_0 &= X_{12}X_{34} + X_{23}X_{14} + X_{31}X_{24}, \\ Q_1 &= \textstyle\sum_{1\leqslant i<j\leqslant 4} X_{ij}^2, \\ Q_2 &= -\textstyle\sum_{1\leqslant i<j\leqslant 4}(A_k + A_l)X_{ij}^2, \\ Q_3 &= \textstyle\sum_{1\leqslant i<j\leqslant 4} A_kA_lX_{ij}^2. \end{aligned} \tag{8.69}$$

Consider the quadratic polynomial $Q_\kappa := \sum_{i=0}^{3}\kappa_iQ_i$, which is a linear combination of the Q_i, where $\kappa = (\kappa_0 : \kappa_1 : \kappa_2 : \kappa_3)$ is arbitrary[3]. For generic κ we have that Q_κ defines a non-degenerate quadratic form, which will be identified with Q_κ. The matrix of Q_κ with respect to the coordinates $x_1,\dots,x_6$ (see (8.8)) is a direct sum of three blocks, corresponding to the coordinate pairs x_i, x_{i+3} for $i = 1,2,3$. Namely, the block K_i that corresponds to x_i, x_{i+3} is given by

$$K_i = \frac{1}{4}\begin{pmatrix} \kappa_0 + 2\kappa_1 - \kappa_2\sum_l A_l + \kappa_3A_i^+ & \kappa_2A_i' - \kappa_3A_i^- \\ \kappa_2A_i' - \kappa_3A_i^- & -\kappa_0 + 2\kappa_1 - \kappa_2\sum_l A_l + \kappa_3A_i^+ \end{pmatrix},$$

[3] It is most natural to view κ as an element of $\mathbf{P}^3$, which we will do in the sequel.

where $A_i^{\pm} = A_j A_k \pm A_i A_4$ and $A_i' = A_j + A_k - A_i - A_4$ (recall that $\{i,j,k\} = \{1,2,3\}$). It follows that $\det(Q_\kappa) = \prod_{i=1}^3 \det(K_i) = 0$ defines a surface of degree 6 in $\mathbf{P}^3$, which consists of three quadratic cones $C_1 \cup C_2 \cup C_3$ in $\mathbf{P}^3$, where C_i is given, for $i = 1,2,3$, by $\det(K_i) = 0$, i.e.,

$$C_i : \kappa_0^2 = 4(\kappa_1 - (A_i + A_4)\kappa_2 + A_i A_4 \kappa_3)(\kappa_1 - (A_j + A_k)\kappa_2 + A_j A_k \kappa_3), \quad (8.70)$$

where $\{i,j,k\} = \{1,2,3\}$. The remarkable property, alluded to above, is that the intersection of these three quadratic cones is a curve, more precisely an elliptic curve, and that for each point of this curve the corresponding quadric has rank three (at most). To see this, it suffices to take the difference of any two of the equations (8.70) for the cones C_1, C_2, C_3, which always yields, up to a non-zero constant, $\kappa_1\kappa_3 - \kappa_2^2 = 0$. Thus, the intersection $C_1 \cap C_2 \cap C_3$ consists of the projective curve $C_1 \cap C_0$, where C_0 is the smooth quadric given by $\kappa_1\kappa_3 - \kappa_2^2 = 0$. It is easily verified that the resulting projective curve $\overline{\mathcal{E}}$ is non-singular. In order to determine its genus, substitute the equation for C_0 in the one for C_1 to find that $\overline{\mathcal{E}}$ is birational to the projective curve

$$\kappa_0^2\kappa_3^2 = 4\prod_{i=1}^4(\kappa_2 - A_i\kappa_3), \qquad \kappa_1\kappa_3 = \kappa_2^2. \quad (8.71)$$

Notice that this curve is singular at $(1:0:0:0)$; in fact this curve has two irreducible components, one of which is $\overline{\mathcal{E}}$, and the other one is the line $\kappa_2 = \kappa_3 = 0$. We read off from (8.71) that $\overline{\mathcal{E}}$ is a double cover of $\mathbf{P}^1$, ramified in four points (the points $(\kappa_2 : \kappa_3) = (A_i : 1)$), hence $\overline{\mathcal{E}}$ is an elliptic curve. For any point $\kappa \in \mathcal{E}$, with $\kappa_3 = 1$, i.e., $\kappa = (\kappa_0 : \kappa_2^2 : \kappa_2 : 1)$ with $\kappa_0^2 = 4\prod_{i=1}^4(\kappa_2 - A_i)$ we have that

$$\begin{aligned} Q_\kappa &= \kappa_2^2 Q_1 + \kappa_2 Q_2 + Q_3 \pm 2\textstyle\prod_{i=1}^4 \sqrt{\kappa_2 - A_i} Q_0 \\ &= \textstyle\sum_{i=1}^3 \left(\sqrt{(\kappa_2 - A_i)(\kappa_2 - A_4)}X_{jk} \pm \sqrt{(\kappa_2 - A_j)(\kappa_2 - A_k)}X_{i4}\right)^2, \end{aligned} \quad (8.72)$$

so that each of these quadrics Q_κ has rank three. Similarly, the two points of $\overline{\mathcal{E}}$ for which $\kappa_3 = 0$, namely $\kappa = (\pm 2 : 1 : 0 : 0)$ lead to two rank three quadrics, to wit

$$Q_\kappa = (X_{12} \pm X_{34})^2 + (X_{23} \pm X_{14})^2 + (X_{31} \pm X_{24})^2.$$

It is clear that it is very exceptional that a three-dimensional family of (generically) rank six quadrics contains a whole curve of rank three quadrics. Still, we will observe such a phenomenon for each of the three integrable geodesic flows that we will discuss.

Four particularly simple quadrics are obtained by substituting $\kappa_2 = A_i$ in (8.72), where $i = 1, \ldots, 4$, and dividing by $\prod_{k \neq j}(A_j - A_k)$.

Namely we obtain the following four quadrics:

$$\begin{aligned}
R_1 &:= \frac{X_{12}^2}{A_1 - A_2} + \frac{X_{13}^2}{A_1 - A_3} + \frac{X_{14}^2}{A_1 - A_4}, \\
R_2 &:= \frac{X_{23}^2}{A_2 - A_3} + \frac{X_{24}^2}{A_2 - A_4} + \frac{X_{12}^2}{A_2 - A_1}, \\
R_3 &:= \frac{X_{34}^2}{A_3 - A_4} + \frac{X_{13}^2}{A_3 - A_1} + \frac{X_{23}^2}{A_3 - A_2}, \\
R_4 &:= \frac{X_{14}^2}{A_4 - A_1} + \frac{X_{24}^2}{A_4 - A_2} + \frac{X_{34}^2}{A_4 - A_3}.
\end{aligned} \tag{8.73}$$

Their sum is zero, but any three of them are independent, having the same span as the span of Q_1, Q_2 and Q_3. Notice that, in terms of these polynomials, Formula (8.65) can also be written in the following form:

$$H = \frac{1}{2}\sum_{j=1}^{4} B_j \sum_{i=1}^{3} \frac{A_j^{3-i} Q_i}{\prod_{k\neq j}(A_j - A_k)} = \frac{1}{2}\sum_{j=1}^{4} B_j R_j.$$

Since the sum of the R_i is zero, H also be simply written in terms of the first three of them as

$$H = \frac{1}{2}\sum_{j=1}^{3}(B_j - B_4)R_j.$$

8.2.3 A Normal Form for the Manakov Metric

It is easy to get rid of the constants in the polynomials R_j by rescaling the variables: for $i = 1, \ldots, 3$, let

$$x_i := \frac{X_{jk}}{\sqrt{A_k - A_j}}, \qquad x_{i+3} := \frac{X_{i4}}{\sqrt{A_i - A_4}},$$

in which formulas we use again the convention that (i, j, k) is any cyclic permutation of $(1, 2, 3)$, and the choice of sign of each of the square roots is irrelevant. In the new coordinates, the quadratic polynomials R_1, R_2 and R_3, together with Q_0, take the following form.

$$\begin{aligned}
F_1 &= x_2^2 - x_3^2 + x_4^2, \\
F_2 &= x_3^2 - x_1^2 + x_5^2, \\
F_3 &= x_1^2 - x_2^2 + x_6^2, \\
F_4 &= \alpha x_1 x_4 + \beta x_2 x_5 + \gamma x_3 x_6,
\end{aligned} \tag{8.74}$$

where $\alpha^2 + \beta^2 + \gamma^2 = 0$.

The constants α, β, γ are given by

$$\begin{aligned} \alpha &= \lambda_0 \sqrt{A_2 - A_1}\sqrt{A_3 - A_4}, \\ \beta &= \lambda_0 \sqrt{A_1 - A_3}\sqrt{A_2 - A_4}, \\ \gamma &= \lambda_0 \sqrt{A_3 - A_2}\sqrt{A_1 - A_4}, \end{aligned} \tag{8.75}$$

where λ_0 is a constant that can be scaled in such a way that $\alpha\beta\gamma = 1$ (at the cost of a simple rescaling of F_4); in the sequel we will suppose that this scaling has been done.

The vector field $\mathcal{X}_H$ and its commuting vector field are, in terms of the coordinates $x_1, \ldots, x_6$, a linear combination of the following two commuting vector fields:

$$\begin{aligned} \dot{x}_1 &= \alpha x_5 x_6, & x'_1 &= \alpha x_2 x_3, \\ \dot{x}_2 &= \beta x_6 x_4, & x'_2 &= \gamma x_4 x_6 + \alpha x_1 x_3, \\ \dot{x}_3 &= \gamma x_4 x_5, & x'_3 &= -\beta x_4 x_5 + \alpha x_1 x_2, \\ \dot{x}_4 &= -\beta x_2 x_6 + \gamma x_3 x_5, & x'_4 &= -\gamma x_2 x_6 - \beta x_3 x_5, \\ \dot{x}_5 &= -\gamma x_3 x_4 + \alpha x_1 x_6, & x'_5 &= \beta x_3 x_4, \\ \dot{x}_6 &= -\alpha x_1 x_5 + \beta x_2 x_4, & x'_6 &= \gamma x_2 x_4. \end{aligned} \tag{8.76}$$

In the sequel we will call these $\mathcal{V}_1$ and $\mathcal{V}_2$ (in that order). They are Hamiltonian with respect to the Poisson structure $\{\cdot\,,\cdot\}$, defined by the following matrix:

$$\begin{pmatrix} 0 & \frac{1}{\alpha}(\beta-\gamma)\gamma x_3 & \frac{1}{\alpha}(\beta+\gamma)\beta x_2 & 0 & -\frac{1}{\alpha}(\beta+\gamma)\gamma x_6 & \frac{1}{\alpha}(\gamma-\beta)\beta x_5 \\ -\frac{1}{\alpha}(\beta-\gamma)\gamma x_3 & 0 & -\alpha x_1 & \gamma x_6 & 0 & (\beta-\gamma)x_4 \\ -\frac{1}{\alpha}(\beta+\gamma)\beta x_2 & \alpha x_1 & 0 & -\beta x_5 & (\beta+\gamma)x_4 & 0 \\ 0 & -\gamma x_6 & \beta x_5 & 0 & \gamma x_3 & -\beta x_2 \\ \frac{1}{\alpha}(\beta+\gamma)\gamma x_6 & 0 & -(\gamma+\beta)x_4 & -\gamma x_3 & 0 & \alpha x_1 \\ \frac{1}{\alpha}(\beta-\gamma)\beta x_5 & (\gamma-\beta)x_4 & 0 & \beta x_2 & -\alpha x_1 & 0 \end{pmatrix}. \tag{8.77}$$

With respect to this Poisson structure, $2\mathcal{V}_1$ corresponds to the Hamiltonian $F_1 + F_2 + F_3$, while $2\mathcal{V}_2$ corresponds to the Hamiltonian F_1; two independent Casimirs are given by F_4 and $(\gamma^2 - \beta^2)F_1 + \beta(\gamma - \beta)F_2 + \gamma(\gamma + \beta)F_3$. As we pointed out in Paragraph 8.1.2 these two commuting vector fields are independent on the generic fiber of the momentum map.

The integrable system $(\mathbf{C}^6, \{\cdot\,,\cdot\}, \mathbf{F})$, where $F := (F_1, F_2, F_3, F_4)$ provides a normal form for the integrable system defined by all Manakov metrics on $\mathfrak{so}(4)$. Since the parameters (α, β, γ) that define the system satisfy $\alpha^2 + \beta^2 + \gamma^2 = 0$ and $\alpha\beta\gamma = 1$ this family of integrable systems depends only on one parameter. Since (α, β, γ) will be fixed in the sequel we do not indicate the dependence of $\{\cdot\,,\cdot\}$ and $\mathbf{F}$ on (α, β, γ) in the notation.

To finish this paragraph we make a few obvious but useful observations about this integrable system. The group of involutions on $\mathbf{C}^6$, generated by

$$\begin{aligned}\sigma_1(x_1,\ldots,x_6) &= (x_1,-x_2,-x_3,x_4,-x_5,-x_6),\\ \sigma_2(x_1,\ldots,x_6) &= (-x_1,x_2,-x_3,-x_4,x_5,-x_6),\end{aligned} \tag{8.78}$$

leaves the vector fields $\mathcal{V}_1$ and $\mathcal{V}_2$ invariant, as well as all fibers $\mathbf{F_c}$ of the momentum map. The involution τ on $\mathbf{C}^6$, which is defined by

$$\tau(x_1,\ldots,x_6) = (-x_1,-x_2,-x_3,-x_4,-x_5,-x_6),$$

has the property of reversing the sign of both vector fields, still leaving all fibers $\mathbf{F_c}$ of the momentum map invariant. Of course, the composition of τ with any of σ_1 or σ_2 (or both) has the same properties.

There is also another involution η that we will use. It is defined by

$$\eta(x_1,\ldots,x_6) = (x_1,x_5,-\sqrt{-1}x_6,x_4,x_2,\sqrt{-1}x_3), \tag{8.79}$$

and its main property is that it permutes the two vector fields $\mathcal{V}_1$ and $\mathcal{V}_2$ (up to a constant). It has the following effect on the constants of motion:

$$\eta(F_1,F_2,F_3,F_4) = (F_1+F_2+F_3,-F_3,-F_2,F_4), \tag{8.80}$$

so that η does not leave the fibers of the momentum map invariant. There is also another automorphism, which has order three, which is not an automorphism of the integrable system itself (with parameters α,β,γ), but with acts on the whole family of integrable systems, i.e., it permutes the parameters α,β,γ. This automorphism, which will be denoted by π, is defined by

$$\begin{aligned}\pi(x_1,\ldots,x_6) &= (x_2,x_3,x_1,x_5,x_6,x_4),\\ \pi(\alpha,\beta,\gamma) &= (\beta,\gamma,\alpha).\end{aligned} \tag{8.81}$$

π acts on the constants of motion in the following way:

$$\pi(F_1,F_2,F_3,F_4) = (F_2,F_3,F_1,F_4). \tag{8.82}$$

π does not change the vector field $\mathcal{V}_1$, but it changes the vector field $\mathcal{V}_2$ to $(\beta\mathcal{V}_2-\gamma\mathcal{V}_1)/\alpha$. As we shall see the involutions represent additional algebraic structure which can be used to reduce the work when carrying out the algorithms of Section 7.6, a point to keep in mind in general.

8.2.4 Algebraic Complete Integrability of the Manakov Metric

In this paragraph we will show that if the parameters α,β and γ satisfy $\alpha^2+\beta^2+\gamma^2=0$ and $\alpha\beta\gamma=1$, but are otherwise arbitrary, then the integrable system $(\mathbf{C}^6,\{\cdot,\cdot\},\mathbf{F})$, defined in the previous paragraph, is algebraic completely integrable. We do this by using the methods that were developed in Chapter 7; see especially Sections 7.6 and 7.7.

The vector field $\mathcal{V}_1$ is homogeneous, so that the indicial equation is given by

$$\begin{aligned}
0 &= x_1^{(0)} + \alpha x_5^{(0)} x_6^{(0)},\\
0 &= x_2^{(0)} + \beta x_6^{(0)} x_4^{(0)},\\
0 &= x_3^{(0)} + \gamma x_4^{(0)} x_5^{(0)},\\
0 &= x_4^{(0)} - \beta x_2^{(0)} x_6^{(0)} + \gamma x_3^{(0)} x_5^{(0)},\\
0 &= x_5^{(0)} - \gamma x_3^{(0)} x_4^{(0)} + \alpha x_1^{(0)} x_6^{(0)},\\
0 &= x_6^{(0)} - \alpha x_1^{(0)} x_5^{(0)} + \beta x_2^{(0)} x_4^{(0)}.
\end{aligned}$$

If we put $x_4^{(0)} = \alpha a$ and $x_5^{(0)} = \beta b$ and $x_6^{(0)} = \gamma c$ then we find that the indicial solution is given by

$$x^{(0)} = (-bc, -ca, -ab, \alpha a, \beta b, \gamma c),$$

where a, b and c satisfy

$$a(b^2 - c^2 - \alpha^2) = b(c^2 - a^2 - \beta^2) = c(a^2 - b^2 - \gamma^2) = 0.$$

If $a = b = 0$ then $c = 0$, since $\gamma \neq 0$, which leads to $x^{(0)} = 0$. By symmetry, it follows that for a strict Laurent solution at most one of a, b, c is equal to zero, so that, in any case,

$$a^2 = c^2 - \beta^2, \qquad b^2 = c^2 + \alpha^2. \tag{8.83}$$

The equations (8.83) define a non-singular elliptic curve $\Gamma^{(0)}$ in $\mathbf{C}^3$. Indeed, $\Gamma^{(0)}$ is a 4 : 1 cover of $\mathbf{C}$ (with coordinate c), which is ramified at the eight points covering $c = \pm\beta$ and $c = \pm\sqrt{-1}\alpha$, and which is unramified at infinity, hence it has genus 1 by Riemann-Hurwitz. Notice that the conditions on (α, β, γ) imply that these 8 points are all different. The involutions σ_1 and σ_2 induce two involutions on $\Gamma^{(0)}$, which are given by the 4 : 1 cover

$$\begin{aligned}
\sigma_1(a, b, c) &= (a, -b, -c),\\
\sigma_2(a, b, c) &= (-a, b, -c),
\end{aligned} \tag{8.84}$$

and the quotient of $\Gamma^{(0)}$ by the group, generated by these two involutions is another elliptic curve $\mathcal{E}$, given by

$$w^2 = v(v - \beta^2)(v + \alpha^2), \tag{8.85}$$

where the quotient map, which is unramified, even at infinity, is given by

$$\begin{aligned}
\rho : \ \Gamma^{(0)} &\to \mathcal{E}\\
(a, b, c) &\mapsto (v, w) = (c^2, abc).
\end{aligned}$$

Notice that we used the same letter $\mathcal{E}$ for the elliptic curve (8.85) as for the curve $\overline{\mathcal{E}}$ of rank three quadrics (8.71). This is on purpose: $\overline{\mathcal{E}}$ is a smooth compactification of $\mathcal{E}$, as is shown in the following lemma.

Lemma 8.12. *The Riemann surface that compactifies $\mathcal{E}$, defined by (8.85), is isomorphic to the curve $\overline{\mathcal{E}}$ of rank three quadrics (8.71).*

Proof. Both Riemann surfaces are double covers of $\mathbf{P}^1$, ramified over four points, namely $0, \beta^2, -\alpha^2$ and ∞ for the Riemann surfaces that compactifies $\mathcal{E}$ and $A_1, \dots, A_4$ for $\overline{\mathcal{E}}$. These elliptic Riemann surfaces will be isomorphic if and only if these four-tuples have the same cross-ratio, when taken in some order, because four-tuples on $\mathbf{P}^1$ with the same cross-ratio correspond under a homography (projective transformation). Now the cross-ratio of $A_1, \dots, A_4$ is given by

$$(A_1, A_2, A_3, A_4) = \frac{A_1 - A_3}{A_1 - A_4} : \frac{A_2 - A_3}{A_2 - A_4},$$

while

$$(0, \alpha^2, -\beta^2, \infty) = \frac{\begin{vmatrix} 1 & 0 \\ 1 & -\beta^2 \end{vmatrix}}{\begin{vmatrix} 1 & 0 \\ 0 & 1 \end{vmatrix}} : \frac{\begin{vmatrix} 1 & \alpha^2 \\ 1 & -\beta^2 \end{vmatrix}}{\begin{vmatrix} 1 & \alpha^2 \\ 0 & 1 \end{vmatrix}} = -\frac{\beta^2}{-\alpha^2 - \beta^2} = -\frac{\beta^2}{\gamma^2}$$

Using (8.75) we see that

$$-\frac{\beta^2}{\gamma^2} = \frac{(A_1 - A_3)(A_2 - A_4)}{(A_1 - A_4)(A_2 - A_3)}$$

so that indeed

$$(A_1, A_2, A_3, A_4) = (0, \alpha^2, -\beta^2, \infty),$$

which needed to be shown. □

The covers that we have obtained can be summarized in the following commuting diagram.

$$\begin{array}{ccc} \Gamma^{(0)} & \xrightarrow[\text{unram}]{4:1} & \mathcal{E} \\ {\scriptstyle 4:1}\Big\downarrow{\scriptstyle \text{ram}} & & {\scriptstyle 2:1}\Big\downarrow{\scriptstyle \text{ram}} \\ \mathbf{P}^1 & \xrightarrow[\text{ram}]{2:1} & \mathbf{P}^1 \end{array}$$

The $4:1$ cover $\rho : \Gamma^{(0)} \to \mathcal{E}$ maps the eight ramification points of $c : \Gamma^{(0)} \to \mathbf{P}^1$ to two of the ramification points of $v : \mathcal{E} \to \mathbf{P}^1$; the two other ramification points of the latter cover correspond to $v = 0, \infty$.

If we denote the Kowalevski matrix that corresponds to the point with parameters (a,b,c) by $\mathcal{K}(a,b,c)$, then we have that

$$\mathcal{K}(a,b,c)=\begin{pmatrix} 1 & 0 & 0 & 0 & \alpha\gamma c & \alpha\beta b\\ 0 & 1 & 0 & \beta\gamma c & 0 & \alpha\beta a\\ 0 & 0 & 1 & \beta\gamma b & \alpha\gamma a & 0\\ 0 & -\beta\gamma c & \beta\gamma b & 1 & -\gamma ab & \beta ac\\ \alpha\gamma c & 0 & -\alpha\gamma a & \gamma ab & 1 & -\alpha bc\\ -\alpha\beta b & \alpha\beta a & 0 & -\beta ac & abc & 1 \end{pmatrix}$$

Its characteristic polynomial is given by

$$|k\operatorname{Id}_6-\mathcal{K}(a,b,c)|=k(k^2-1)(k-2)^3,$$

independently of a, b and c. The first three terms of the principal balance are given by

$$\begin{aligned}
x_1(t)&=-\frac{bc}{t}-da+(bg/\gamma+cf/\beta)t+O(t^2),\\
x_2(t)&=-\frac{ca}{t}-db+(ce/\alpha+ag/\gamma)t+O(t^2),\\
x_3(t)&=-\frac{ab}{t}-dc+(af/\beta+be/\alpha)t+O(t^2),\\
x_4(t)&=\frac{\alpha a}{t}+et+O(t^2),\\
x_5(t)&=\frac{\beta b}{t}+ft+O(t^2),\\
x_6(t)&=\frac{\gamma c}{t}+gt+O(t^2),
\end{aligned}\tag{8.86}$$

where $d,\dots,g$ are the new free parameters that come in at steps 1 and 2. Since 2 is the largest eigenvalue of the Kowalevski matrix these first few terms extend to a weight homogeneous principal balance. As we pointed out in Section 8.1, the three free parameters that appear in step 2 are trivial parameters, so they can be expressed linearly in terms of the values of three of the constants of motion, which is most easily done by substituting the series $x(t)$ in $F_i(x(t))=c_i$, for $i=1,\dots,3$. Solving the resulting equations linearly for e, f and g we get

$$\begin{aligned}
e&:=\frac{\alpha a}{6}(2c_1(c^2-\gamma^2)+c_2(2c^2-\gamma^2)+c_3(2b^2+\beta^2))+\frac{V\alpha bc}{4},\\
f&:=\pi(e),\qquad g:=\pi^2(e),
\end{aligned}$$

where

$$V:=-\frac{1}{abc}(d^2+c_1b^2c^2+c_2c^2a^2+c_3a^2b^2).$$

In this formula, the action of π on the parameters $a, \dots, d$ is defined by

$$\pi(a,b,c,d) := (b,c,a,d),$$

so that in particular $\pi(V) = V$ (see (8.82) for the action of π on the constants of motion). By direct substitution of the balance in F_4 we see that $F_4(x(t)) = V/2$, so that the equation of the abstract Painlevé divisor of the fiber $\mathbf{F_c}$, where $\mathbf{c} = (c_1, \dots, c_4)$ is given by the algebraic curve $\Gamma_{\mathbf{c}}$ in $\mathbf{C}^4$, defined by

$$\begin{aligned} a^2 &= c^2 - \beta^2, \\ b^2 &= c^2 + \alpha^2, \\ d^2 &+ c_1 b^2 c^2 + c_2 c^2 a^2 + c_3 a^2 b^2 + 2abcc_4 = 0, \end{aligned} \tag{8.87}$$

where the first two equations are the ones that define the elliptic curve $\Gamma^{(0)}$ in $\mathbf{C}^3$, which appeared as the indicial locus in step 0, and the third equation is $V/2 = c_4$. Notice that $\Gamma^{(0)}$ is independent of the values of the constants of motion, unlike $\Gamma_{\mathbf{c}}$, which explains our notation.

Lemma 8.13. *For generic* $\mathbf{c}$ *the affine curve* $\Gamma_{\mathbf{c}}$ *can be completed into a compact Riemann surface* $\overline{\Gamma_{\mathbf{c}}}$ *of genus* 9 *by adding* 8 *points* $\infty^{\epsilon_1\epsilon_2\epsilon_3}$ *at infinity, where* $\epsilon_1^2 = \epsilon_2^2 = \epsilon_3^2 = 1$. *The map* $\overline{\Gamma_{\mathbf{c}}} \to \overline{\Gamma^{(0)}}$, *which is defined for* $(a,b,c,d) \in \Gamma_{\mathbf{c}}$ *by* $(a,b,c,d) \mapsto (a,b,c)$, *is a double cover, whose* 16 *ramification points belong to* $\Gamma_{\mathbf{c}}$ *(they are different from the points* $\infty^{\epsilon_1\epsilon_2\epsilon_3}$*).*

Proof. The affine ramification points are found by substituting $d = 0$ in (8.87). By using the other two equations, the third equation in (8.87) can then be written in the form $P_2(c^2) = abc$, where P_2 is a quadratic polynomial. Squaring the latter equation we get $P_2^2(c^2) = (c^2 - \beta^2)(c^2 + \alpha^2)c^2$, which has 8 different solutions c, yielding 32 points (a,b,c) on $\Gamma^{(0)}$, but only for half of them will the sign of abc match the sign of $P_2(c^2)$. Thus, we have 16 affine ramification points. The map is unramified at infinity. Indeed, in terms of a local parameter ς the 4 points at infinity of $\Gamma^{(0)}$ are given by

$$a = \frac{\epsilon_1}{\varsigma} + O(\varsigma), \qquad b = \frac{\epsilon_2}{\varsigma} + O(\varsigma), \qquad c = \frac{1}{\varsigma}, \tag{8.88}$$

where $\epsilon_1^2 = \epsilon_2^2 = 1$, yielding that d is given in terms of a local parameter by

$$d = \epsilon_3 \frac{\sqrt{-c_1 - c_2 - c_3}}{\varsigma^2} \left(1 + \frac{\epsilon_1 \epsilon_2 c_4}{c_1 + c_2 + c_3} \varsigma + O(\varsigma^2)\right), \tag{8.89}$$

where $\epsilon_3^2 = 1$, which shows that $\Gamma_{\mathbf{c}}$ has eight points at infinity and that the projection map $\overline{\Gamma_{\mathbf{c}}} \to \overline{\Gamma^{(0)}}$ is unramified at infinity. In conclusion, $\overline{\Gamma_{\mathbf{c}}}$ is a double cover of an elliptic curve, which is ramified at 16 points, so Formula (5.10) implies that the genus of $\overline{\Gamma_{\mathbf{c}}}$ is given by

$$g(\overline{\Gamma_{\mathbf{c}}}) = 2g(\overline{\Gamma_0}) - 1 + 16/2 = 9,$$

as was asserted. □

The eight points in $\overline{\Gamma_{\mathbf{c}}}\backslash\Gamma_{\mathbf{c}}$ will show up again later; we will denote them by $\infty^{\epsilon_1\epsilon_2\epsilon_3}$, where the latter corresponds by definition to the parameterizations, given in (8.88) and (8.89).

We now look for (homogeneous) polynomials which have a simple pole in t when the principal balance is substituted in them. In the notation of Paragraph 7.7.1 this means that we wish to construct a basis of $\mathcal{Z}_\rho$ as a $\mathcal{H}$-module, where the pole vector ρ is chosen as $\rho := (1)$. If we look for polynomials of weight 1 first then we find nothing but the (span of the) functions $x_1, \ldots, x_6$; we define $z_0 := 1$ and $z_i := x_i$ for $i = 1, \ldots, 6$. Clearly, the adjunction formula will not be satisfied if we only consider the functions $z_0, \ldots, z_6$, because their residues are independent of d. Therefore we look for quadratic functions and we find by simple inspection that

$$\begin{aligned} z_7 &:= \alpha z_2 z_5 - \beta z_1 z_4, \\ z_7' &:= \beta z_3 z_6 - \gamma z_2 z_5, \\ z_7'' &:= \gamma z_1 z_4 - \alpha z_3 z_6, \end{aligned}$$

are three functions that have the required property. Notice that z_7' and z_7'' can be expressed in terms of z_7 by

$$\gamma z_7' = \beta F_4 + \alpha z_7, \qquad \gamma z_7'' = \beta z_7 - \alpha F_4,$$

so that, as far as the embedding is concerned, we may restrict ourselves to z_7. It is clear that, for generic $\mathbf{c}$, the map

$$\begin{aligned} \varphi_{\mathbf{c}} : \quad & \mathbf{F_c} \to \mathbf{P}^7 \\ & (x_1, \ldots, x_6) \mapsto (1 : z_1 : \cdots : z_7) \end{aligned}$$

is an isomorphic embedding of the smooth surface $\mathbf{F_c}$ into $\mathbf{P}^7$. Since the residue of the Laurent series of z_7 is given by γd, we now consider the following map:

$$\begin{aligned} \varphi_{\mathbf{c}}' : \quad & \Gamma_{\mathbf{c}} \to \mathbf{P}^7 \\ & (a, b, c, d) \mapsto (0 : -bc : -ca : -ab : \alpha a : \beta b : \gamma c : \gamma d) \end{aligned} \tag{8.90}$$

It is obviously an embedding of the non-singular abstract Painlevé divisor $\Gamma_{\mathbf{c}}$ into $\mathbf{P}^7$; to see what happens at the eight points $\infty^{\epsilon_1\epsilon_2\epsilon_3}$, substitute (8.88) and (8.89) into (8.90), for small ς to find that a small neighborhood of the point $\infty^{\epsilon_1\epsilon_2\epsilon_3}$ is mapped to

$$(0 : \epsilon_2 : \epsilon_1 : \epsilon_1\epsilon_2 : 0 : 0 : 0 : -\gamma\epsilon_3\sqrt{-c_1 - c_2 - c_3}(1 + \star\varsigma) \in \mathbf{P}^7,$$

where $\star = \epsilon_1\epsilon_2 c_4/(c_1 + c_2 + c_3) \neq 0$.

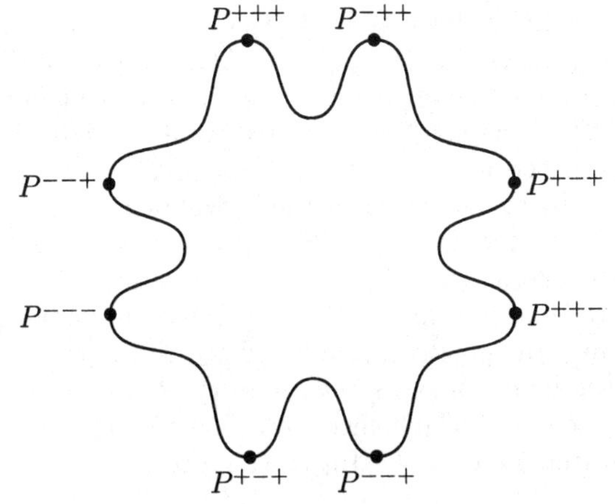

Fig. 8.1. The abstract Painlevé divisor Γ_c embeds in $\mathbf{P}^7$ and the closure of its image is a smooth curve $\mathcal{D}_c$ of genus 9, which is a double cover of the elliptic curve $\Gamma^{(0)}$ that appears as the indicial locus of the vector field $\mathcal{V}_1$. The points of tangency of $\overline{\mathcal{V}}_1$ are the eight points $P^{\epsilon_1\epsilon_2\epsilon_3}$ at infinity.

It follows that the images of the points $\infty^{\epsilon_1\epsilon_2\epsilon_3}$ are all different, and that they are non-singular points of the curve $\mathcal{D}_\mathbf{c} := \overline{\varphi'_\mathbf{c}(\Gamma_\mathbf{c})}$ in $\mathbf{P}^7$. We will denote these points by $P^{\epsilon_1\epsilon_2\epsilon_3}$. It follows that the non-singular curve $\mathcal{D}_\mathbf{c}$ has genus 9, just like $\overline{\Gamma_\mathbf{c}}$; since $9 = 7+2$ the adjunction formula (7.50) is satisfied. See Figure 8.1. This suggests that this is a good place to stop, and indeed we will see later that the functions $z_0, \dots, z_7$ lead to an embedding of the compactified generic fiber $\mathbf{F_c}$, which will turn out to be an Abelian surface. But in the next step, when we try to write down the quadratic differential equations we get stuck: already $\dot{z}_7$ cannot be written as a quadratic polynomial in $z_0, \dots, z_7$, with coefficients in $\mathcal{H}$, as is easily checked. The reason is that the Painlevé divisor $\mathcal{D}_\mathbf{c}$ is very ample but is not normally generated (i.e., its line bundle $[\mathcal{D}_\mathbf{c}]$ is not normally generated). Therefore we look at a basis (over $\mathcal{H}$) for the space of polynomials which have a double pole at most when the principal balance is substituted in them. These are all obtained by considering the 36 products $z_{ij} := z_i z_j$, where $0 \leqslant i \leqslant j \leqslant 7$ and the following two natural Wronskians

$$z_8 := W(z_1, z_4) = \dot{z}_1 z_4 - z_1 \dot{z}_4 = \alpha x_4 x_5 x_6 + \beta x_1 x_2 x_6 - \gamma x_1 x_3 x_5,$$
$$z_9 := W(z_2, z_5) = \dot{z}_2 z_5 - z_2 \dot{z}_5 = \beta x_4 x_5 x_6 + \gamma x_2 x_3 x_4 - \alpha x_1 x_2 x_6.$$

Their Laurent series have the following leading terms:

$$z_8(t) = -\alpha \frac{a^2 d}{t^2} + O(t^{-1}), \qquad z_9(t) = -\beta \frac{b^2 d}{t^2} + O(t^{-1}). \tag{8.91}$$

Notice that the similar Wronskian $W(z_3, z_6)$ leads to nothing new because $\alpha W(z_1, z_4) + \beta W(z_2, z_5) + \gamma W(z_3, z_6) = 0$. Also, only 30 of the products z_{ij} are independent over $\mathcal{H}$, giving a total of 32 independent functions, but for the quadratic differential equations it is irrelevant which of the products are dependent on the other ones. Notice that if $\dim L(\mathcal{D}_{\mathbf{c}}) = 8$ then $\dim L(\mathcal{D}_{\mathbf{c}}) = 4 \dim L(\mathcal{D}_{\mathbf{c}}) = 32$, by (5.38), so this is the logical place to stop. For generic $\mathbf{c}$ we will denote the embedding $\mathbf{F}_{\mathbf{c}} \to \mathbf{P}^{31}$ by $\psi_{\mathbf{c}}$. It is checked as above that $\psi_{\mathbf{c}}$ leads to an embedding $\psi'_{\mathbf{c}}$ of $\Gamma_{\mathbf{c}}$, and that the closure of $\psi'_{\mathbf{c}}(\Gamma_{\mathbf{c}})$ is smooth. Since this closure is isomorphic to $\mathcal{D}_{\mathbf{c}}$ we will use the same notation for it, $\mathcal{D}_{\mathbf{c}} := \overline{\psi'_{\mathbf{c}}(\Gamma_{\mathbf{c}})}$. Also, the points at infinity of $\mathcal{D}_{\mathbf{c}}$, which are the images of the points $\infty^{\epsilon_1\epsilon_2\epsilon_3}$ will be denoted, as before, by $P^{\epsilon_1\epsilon_2\epsilon_3}$.

According to Lemma 7.58 it suffices now to verify that the following Wronskians are quadratic elements of $\mathcal{H}[z_{00}, z_{01}, \dots, z_{77}, z_8, z_9]$:

$$W(z_k, z_{ij}) \qquad W(z_k, z_8) \qquad W(z_k, z_9),$$

where $0 \leqslant i \leqslant j \leqslant 7$ and $k = 0, 1$. But notice that for the Wronskians in the first line this is obvious, because besides the Wronskian $W(z_0, z_{77})$ they are all of degree 4 at most in $x_1, \dots, x_6$, hence they are all quadratic in $z_{00}, \dots, z_{77}$; also $W(z_0, z_{77}) = -2z_7\dot{z}_7$ and $\dot{z}_7$ is cubic in $x_1, \dots, x_6$, so that the latter Wronskian is also quadratic in $z_{00}, \dots, z_{77}$. By the same argument the simplest ones of the $W(z_1, z_{ij})$ are of degree at most 4 in $x_1, \dots, x_6$, hence they are quadratic in $z_{00}, \dots, z_{77}$, and we only have to verify

$$W(z_1, z_{i7}) \qquad W(z_1, z_8) \qquad W(z_1, z_9),$$

where $i = 1, \dots, 7$. After some work these Wronskians can be written in the following form, as is easily verified,

$$\begin{aligned}
W(z_1, z_1z_7) &= z_1^2(\alpha z_9 - \beta z_8),\\
W(z_1, z_2z_7) &= z_1z_2(\alpha z_9 - \beta z_8) + z_6z_7^2,\\
W(z_1, z_3z_7) &= z_1z_3(\alpha z_9 - \beta z_8) - z_5z_7z_7'',\\
W(z_1, z_4z_7) &= z_1z_4(\alpha z_9 - \beta z_8) + z_7z_8,\\
W(z_1, z_5z_7) &= z_1z_5(\alpha z_9 - \beta z_8) + z_7(\alpha c_2z_6 + z_3z_7''),\\
W(z_1, z_6z_7) &= z_1z_6(\alpha z_9 - \beta z_8) + z_7(\alpha c_3z_5 + z_2z_7),\\
W(z_1, z_7^2) &= 2z_1z_7(\alpha z_9 - \beta z_8) + \alpha z_5z_6z_7^2,\\
W(z_1, z_8) &= \alpha z_5z_6z_8 + z_1((\alpha^2c_2 - \beta^2c_1)z_1z_4 + \alpha\gamma(c_1 + c_2 + c_3)z_3z_6\\
&\qquad + \alpha c_4(z_1^2 - z_5^2) - \beta z_7(z_4^2 - z_5^2) + \gamma z_6^2z_7''),\\
W(z_1, z_9) &= z_8(\beta z_5z_6 + \gamma z_2z_3) + z_1(\alpha^2(c_1 + c_2)z_2z_5 - \alpha(z_2^2 + z_5^2)z_7).
\end{aligned}$$

The above Wronskians are written in such a way that it is clear that each of them is a quadratic polynomial in $z_{00}, \dots, z_{77}, z_8, z_9$. This shows that $\mathcal{V}_1$ extends to a holomorphic vector field $\overline{\mathcal{V}}_1$ on $\mathbf{P}^{31}$. Since $\mathcal{V}_2$ is obtained from $\mathcal{V}_1$ by using the involution ρ, the same is true for $\mathcal{V}_2$, and hence any linear combination of $\mathcal{V}_1$ and $\mathcal{V}_2$ extends to a holomorphic vector field on $\mathbf{P}^{31}$.

Let us now check that the flow of $\mathcal{V}_1$, starting from the eight points $P^{\epsilon_1\epsilon_2\epsilon_3}$, goes immediately into the affine. Recall from Paragraph 7.7.3 that in order to check this we must check for each of these points that the Taylor series $1/z_i(t)$ is non-zero at this point, where i is chosen as follows. First, the pole order of z_i along $\mathcal{D}_{\mathbf{c}}$ should be maximal among the pole orders, along $\mathcal{D}_{\mathbf{c}}$, of all functions z_i, so that $i \neq 0, \dots, 7$ (this pole order is then 2). Second, the leading coefficient $z_i^{(0)}$, which is a rational function on $\mathcal{D}_{\mathbf{c}}$, should have a pole order at $P^{\epsilon_1\epsilon_2\epsilon_3}$, which is maximal among all the pole orders of the leading coefficients $z_j^{(0)}$, with $j \neq 0, \dots, 7$. It follows from (8.88) that the residue of z_i has for $i = 1, 2, 3, 7$ a double pole at the points $P^{\epsilon_1\epsilon_2\epsilon_3}$, and has a simple pole at these points for $i = 4, 5, 6$; also the pole order of the leading terms of z_8 and z_9 at these points is 4, as is read off from (8.91). Therefore, the function $1/z_1^2$ satisfies the above conditions, and it suffices to check that the Taylor series $1/z_1^2(t)$ has a non-zero limit at the points $P^{\epsilon_1\epsilon_2\epsilon_3}$. In fact, it suffices to show that the Taylor series $1/z_1(t)$ has a non-zero limit at these points: indeed, the latter is holomorphic in a neighborhood of these points since $1/z_1(t) = z_1(t)/z_1^2(t)$, which allows us to compute the limiting series of $1/z_1(t)$; then the limiting series of $1/z_1^2(t)$ is just its square. It follows from (8.86) that

$$\frac{1}{z_1(t)} = -\frac{t}{bc}\left(1 - \frac{ad}{bc}t + \left(\frac{a^2d^2}{b^2c^2} + \frac{f}{b\beta} + \frac{g}{c\gamma}\right)t^2\right) + O(t^4).$$

Using (8.88) and (8.89) to express each coefficient in terms of the local parameter ς and letting $\varsigma \to 0$ we find

$$\lim_{\varsigma\to 0}\frac{1}{bc} = \lim_{\varsigma\to 0}\epsilon_2\varsigma^2 = 0,$$

$$\lim_{\varsigma\to 0}\frac{ad}{b^2c^2} = \lim_{\varsigma\to 0}\frac{\epsilon_1\epsilon_3\sqrt{-c_1-c_2-c_3}\varsigma^4}{\varsigma^3} = 0,$$

$$\lim_{\varsigma\to 0}\left(\frac{a^2d^2}{b^3c^3} + \frac{f}{b^2c\beta} + \frac{g}{bc^2\gamma}\right) = -\epsilon_2(c_1+c_2+c_3) \neq 0.$$

It follows that the series of $1/z_1$ has a non-zero limit at the points $P^{\epsilon_1\epsilon_2\epsilon_3}$. Moreover, inverting the series we find that the series for $x_1 = z_1$ that starts at these points is given by

$$x_1(t; P^{\epsilon_1\epsilon_2\epsilon_3}) = \frac{1}{t^3}\left(\frac{1}{\epsilon_2(c_1+c_2+c_3)} + O(t)\right).$$

Computing in a similar way the limits of $x_i(t)/x_1(t)$ one finds the lower balances. Notice that these are not weight homogeneous.

In order to complete the proof of the algebraic complete integrability of the Manakov metrics on $\mathfrak{so}(4)$ we need to show that $\mathcal{D}_{\mathbf{c}}$ is the only divisor in $\overline{\psi_{\mathbf{c}}(\mathbf{F}_{\mathbf{c}})} \setminus \psi_{\mathbf{c}}(\mathbf{F}_{\mathbf{c}})$.

As we explained in Paragraph 7.7.3 we do this in general by a degree count. In this case there is a simpler way, namely we can use the involution η, which interchanges the two vector fields $\overline{\mathcal{V}}_1$ and $\overline{\mathcal{V}}_2$, up to a factor (see (8.79) for the definition of η). To do this, we first extend the involution η to the free parameters that appear in the principal balance. Notice that this does not lead to an involution on $\overline{\psi_{\mathbf{c}}(\mathbf{F_c})} \setminus \psi_{\mathbf{c}}(\mathbf{F_c})$, because η itself did not preserve the constants of motion, hence does not preserve the fibers of the momentum map (see (8.80)). Instead, it will lead to a map from $\mathcal{D}_{\mathbf{c}}$ to $\mathcal{D}_{\eta(c)}$. In order to find this map, compare the embedding of $\Gamma_{\mathbf{c}}$ in $\mathbf{P}^7$, given by the residues of the functions $z_0 = 1, z_1, \dots, z_7$ with the embedding of $\Gamma_{\eta(c)}$, given by the residues of the functions $\eta(z_0) = 1, \eta(z_1), \dots, \eta(z_7)$. We get, consistent with the definition of the action of η on the z_i,

$$\eta(a) = -\frac{\sqrt{-1}}{\alpha a}, \quad \eta(b) = \frac{\sqrt{-1}\gamma c}{a}, \quad \eta(c) = -\frac{\beta b}{a}, \quad \eta(d) = -\frac{\sqrt{-1}d}{\alpha a^2}.$$

The main thing to be observed from these formulas is that all points that are at infinity on $\mathcal{D}_{\mathbf{c}}$ are mapped to points that are in the finite part of $\mathcal{D}_{\eta(c)}$, i.e., to points which are reached by the principal balance to $\mathcal{V}_1$. But we know that such points always satisfy the multiplicity condition (7.7.3), hence no other components of $\overline{\psi_{\mathbf{c}}(\mathbf{F_c})} \setminus \psi_{\mathbf{c}}(\mathbf{F_c})$ pass through these points.

This leads to the following theorem, that states the algebraic complete integrability of the Manakov metric on $\mathfrak{so}(4)$. We also give a first description of the tori $\mathbf{T}^2_{\mathbf{c}}$ that complete the generic fibers $\mathbf{F_c}$ of its momentum map.

Theorem 8.14. *For any fixed $(\alpha, \beta, \gamma) \in \mathbf{C}^3$ satisfying $\alpha\beta\gamma = 1$ as well as $\alpha^2 + \beta^2 + \gamma^2 = 0$, consider the integrable system $(\mathbf{C}^6, \{\cdot\,,\cdot\}, \mathbf{F})$, which provides a normal form for the Manakov geodesic flow on $\mathfrak{so}(4)$, where $F = (F_1, F_2, F_3, F_4)$ is given by (8.74) and the matrix of $\{\cdot\,,\cdot\}$ is given by (8.77); a basis of its integrable vector fields is given by (8.76).*

(1) $(\mathbf{C}^6, \{\cdot\,,\cdot\}, \mathbf{F})$ is a weight homogeneous a.c.i. system;
(2) For generic $\mathbf{c}$ the fiber $\mathbf{F_c}$ of its momentum map completes into an Abelian surface $\mathbf{T}^2_{\mathbf{c}}$ by adding a smooth genus 9 curve $\mathcal{D}_{\mathbf{c}}$;
(3) $\mathcal{D}_{\mathbf{c}}$ is a ramified double cover of an elliptic curve $\overline{\Gamma^{(0)}}$, which is itself a fourfold unramified cover of $\overline{\mathcal{E}}$, the curve of rank three quadrics;
(4) The line bundle $[\mathcal{D}_{\mathbf{c}}]$ defines a polarization of type $(2,4)$ on $\mathbf{T}^2_{\mathbf{c}}$ and leads to an embedding of $\mathbf{T}^2_{\mathbf{c}}$ in $\mathbf{P}^7$;
(5) The involutions σ_1 and σ_2 define on $\mathbf{T}^2_{\mathbf{c}}$ two translations, over a half period, that leave $\mathcal{D}_{\mathbf{c}}$ invariant.

Proof. We have verified that the conditions of the Complex Liouville Theorem (Theorem 6.22) are satisfied. It follows that, for generic $\mathbf{c}$ the projective variety

$$\overline{\psi_{\mathbf{c}}(\mathbf{F_c})} = \psi_{\mathbf{c}}(\mathbf{F_c}) \cup \mathcal{D}_{\mathbf{c}}$$

is an Abelian variety and that the restriction of the vector fields $\overline{\mathcal{V}}_1$ and $\overline{\mathcal{V}}_2$ to these Abelian varieties is linear.

The smooth genus 9 curve $\mathcal{D}_{\mathbf{c}}$ is isomorphic to $\overline{\Gamma_{\mathbf{c}}}$, which was shown in Lemma 8.13 to be a double ramified cover of $\Gamma^{(0)}$. The latter is, according to Lemma 8.12 an unramified fourfold cover of the curve of rank three quadrics. Since $\mathcal{D}_{\mathbf{c}}$ has genus 9 we have in view of (5.25) that the elementary divisors δ_1, δ_2 of the polarization that $\mathcal{D}_{\mathbf{c}}$ induces on $\mathbf{T}^2_{\mathbf{c}}$ satisfy $\delta_1\delta_2 = 8$. In view of Ramanan's Theorem (Theorem 5.18), $\mathcal{D}_{\mathbf{c}}$ is very ample; notice however that, as we have seen, $[\mathcal{D}_{\mathbf{c}}]$ is not normally generated. The embedding that we have constructed is precisely the Kodaire map, associated to $\mathcal{D}_{\mathbf{c}}$. Indeed, the functions $z_0, \dots, z_7$ provide independent sections of $[\mathcal{D}_{\mathbf{c}}]$ and they form a basis of $L(\mathcal{D}_{\mathbf{c}})$ because $\dim L(\mathcal{D}_{\mathbf{c}}) = 8$, again by (5.25). Among these 8 functions, 6 are odd with respect to the time involution τ (namely $z_1, \dots, z_6$) and 2 are even (z_0 and z_7), so that $(\delta_1, \delta_2) = (1, 8)$ is impossible, by Proposition 5.20. It follows that $(\delta_1, \delta_2) = (2, 4)$. Finally, since σ_1 and σ_2 leave the vector fields invariant, their restriction to the tori $\mathbf{T}^2_{\mathbf{c}}$ are translations; being involutions they are translations over half periods. Since they leave the affine part $\mathbf{F}_{\mathbf{c}}$ of $\mathbf{T}^2_{\mathbf{c}}$ invariant they leave also its complement $\mathcal{D}_{\mathbf{c}}$ invariant. □

A different description of the Abelian surfaces $\mathbf{T}^2_{\mathbf{c}}$ will be given in the next paragraph.

For future use, we compute the holomorphic differentials on $\mathcal{D}_{\mathbf{c}}$ that come from the natural differentials $\mathrm{d}t_1$ and $\mathrm{d}t_2$ on $\mathbf{T}^2_{\mathbf{c}}$ and we show that the points $P^{\epsilon_1\epsilon_2\epsilon_3}$ are precisely the points on $\mathcal{D}_{\mathbf{c}}$ where the vector field $(\varphi_{\mathbf{c}})_*\mathcal{V}_1$ is tangent to $\mathcal{D}_{\mathbf{c}}$. See Paragraph 7.6.7 for an explanation[4] and for an algorithm that allows us to compute this restriction. In the notations of Paragraph 7.6.7, Example 7.54, we choose $y_0 := z_1$ and $y := z_2$, so that

$$\begin{aligned} y_0^{(0)} &= -bc, & y_0^{(1)} &= -ad,\\ y^{(0)} &= -ac, & y^{(1)} &= -bd. \end{aligned}$$

Also, (see (8.76)),

$$\begin{aligned} \mathcal{V}_2[1/y_0] &= -z_1'/z_1^2 = -\alpha x_2 x_3/x_1^2,\\ \mathcal{V}_2[y/y_0] &= (z_2' z_1 - z_2 z_1')/z_1^2 = (\gamma x_1 x_4 x_6 + \alpha x_3(x_1^2 - x_2^2))/x_1^2, \end{aligned}$$

so that $\mathcal{V}_2[1/y_0]_{|\mathcal{D}_{\mathbf{c}}} = -\alpha a^2/(bc)$ and $\mathcal{V}_2[y/y_0]_{|\mathcal{D}_{\mathbf{c}}} = d/(\alpha b^2 c)$. It follows that

$$\delta = \frac{1}{b^2c^2}\begin{vmatrix} -bc & -\alpha a^2/(bc)\\ (b^2-a^2)cd & d/(\alpha b^2 c)\end{vmatrix} = -\frac{\gamma d}{\beta b^3}.$$

[4] The vector fields $(\varphi_{\mathbf{c}})_*\mathcal{V}_1$ and $(\varphi_{\mathbf{c}})_*\mathcal{V}_2$ do not extend to all of $\mathbf{P}^7$, but to a neighborhood of a generic point, which suffices for computing the holomorphic differentials.

Substituting this in (7.48) and using $a\,\mathrm{d}a = b\,\mathrm{d}b$, a consequence of $a^2 - b^2 = \gamma^2$, we find

$$\omega_1 = \frac{\beta b^3}{\gamma bcd}\mathrm{d}\left(\frac{a}{b}\right) = -\frac{\mathrm{d}a}{\alpha bcd}, \tag{8.92}$$

$$\omega_2 = -\frac{\alpha\beta a^2 b^2}{\gamma cd}\mathrm{d}\left(\frac{a}{b}\right) = \frac{a^2\mathrm{d}a}{bcd}. \tag{8.93}$$

It follows then, by a direct substitution of (8.88) and (8.89) in (8.92) that ω_1 has a double zero at the eight points $P^{\epsilon_1\epsilon_2\epsilon_3}$. So the vector field $\overline{\mathcal{V}}_1$ is doubly tangent to $\mathcal{D}_{\mathbf{c}}$ at these points and is nowhere else tangent to $\mathcal{D}_{\mathbf{c}}$ since $2(g(\mathcal{D}_{\mathbf{c}}) - 1) = 16$. Similarly, ω_2 is non-zero at these points, so that $\overline{\mathcal{V}}_2$ is transversal to $\mathcal{D}_{\mathbf{c}}$ at these points, as follows also from the independence of these vector fields.

8.2.5 The Invariant Manifolds as Prym Varieties

The purpose of this paragraph is to show that, for generic $\mathbf{c}$, the tori $\mathbf{T}^2_{\mathbf{c}}$ which compactify the smooth fibers $\mathbf{F}_{\mathbf{c}}$ of the momentum map $\mathbf{F}$ of the Manakov geodesic flow are Prym varieties and to relate them to the Abelian surfaces, themselves also Prym varieties, on which the Lax equation (8.63) linearizes. The main results that are presented here are due to Luc Haine [74]. See also [26].

We first introduce the notation that we will use; most of it has already been introduced earlier. We fix a generic $\mathbf{c} = (c_1, \dots, c_4)$ and we denote by $\mathbf{F}_{\mathbf{c}}$ the fiber of the momentum map $\mathbf{F} = (F_1, \dots, F_4) : \mathbf{C}^6 \to \mathbf{C}^4$, where $F_1, \dots, F_4$ are given by (8.74). We have seen that $\mathbf{F}_{\mathbf{c}}$ compactifies into an Abelian surface $\mathbf{T}^2_{\mathbf{c}}$ by adding a smooth irreducible curve $\mathcal{D}_{\mathbf{c}}$ of genus 9. This curve appeared in Paragraph 8.2.4 as a ramified double cover of the elliptic curve $\overline{\Gamma^{(0)}}$, where the latter is itself a fourfold unramified cover of $\overline{\mathcal{E}}$, the curve of rank three quadrics (see Lemma 8.12). The restriction of σ_1 and σ_2 to $\mathbf{F}_{\mathbf{c}}$ extends to two translations on $\mathbf{T}^2_{\mathbf{c}}$ over a half period; we will denote these two translations by the same letters σ_1 and σ_2. Since these translations leave $\mathbf{F}_{\mathbf{c}}$ invariant they leave also $\mathcal{D}_{\mathbf{c}}$ invariant and we may consider $\mathcal{C}_{\mathbf{c}} := \mathcal{D}_{\mathbf{c}} / \langle \sigma_1, \sigma_2 \rangle$. The group of translations generated by σ_1 and σ_2 has order 4 and has no fixed points, hence the cover $\mathcal{D}_{\mathbf{c}} \to \mathcal{C}_{\mathbf{c}}$ is unramified and the genus of $\mathcal{C}_{\mathbf{c}}$ is $(9-1)/4 + 1 = 3$, by Riemann-Hurwitz. If we define $\widetilde{\mathbf{T}}^2_{\mathbf{c}} := \mathbf{T}^2_{\mathbf{c}} / \langle \sigma_1, \sigma_2 \rangle$ then we get the following commutative diagram.

$$\begin{array}{ccccc}
\mathbf{F}_{\mathbf{c}} & \xrightarrow{\subset} & \mathbf{T}^2_{\mathbf{c}} & \xleftarrow{\supset} & \mathcal{D}_{\mathbf{c}} \\
 & & {\scriptstyle 4:1}\big\downarrow{\scriptstyle \text{unram}} & & {\scriptstyle 4:1}\big\downarrow{\scriptstyle \text{unram}} \\
 & & \widetilde{\mathbf{T}}^2_{\mathbf{c}} & \xleftarrow{\supset} & \mathcal{C}_{\mathbf{c}}
\end{array}$$

Since $\mathcal{C}_{\mathbf{c}}$ is a smooth curve of genus 3 it induces in view of (5.25) on the Abelian surface $\widetilde{\mathbf{T}}_{\mathbf{c}}^2$ a polarization of type $(1,2)$. Moreover, in view of Example 5.32, $\widetilde{\mathbf{T}}_{\mathbf{c}}^2$ is isomorphic to the Prym variety $\operatorname{Prym}(\mathcal{C}_{\mathbf{c}}/\widetilde{\mathcal{E}}_{\mathbf{c}})$, where $\widetilde{\mathcal{E}}_{\mathbf{c}}$ is an elliptic curve, being a quotient of $\mathcal{C}_{\mathbf{c}}$ by a double cover $\mathcal{C}_{\mathbf{c}} \to \widetilde{\mathcal{E}}_{\mathbf{c}}$ which is ramified in 4 points. We will see later that $\widetilde{\mathcal{E}}_{\mathbf{c}}$ is isomorphic to $\overline{\mathcal{E}}$, the curve of rank three quadrics (which is independent of $\mathbf{c}$). We first show that $\mathcal{C}_{\mathbf{c}}$ is indeed a cover of $\overline{\mathcal{E}}$, making the following diagram commutative.

$$\begin{array}{ccc}
\mathcal{D}_{\mathbf{c}} & \xrightarrow[\text{ram}]{2:1} & \Gamma^{(0)} \\
{\scriptstyle 4:1}\downarrow{\scriptstyle \text{unram}} & & {\scriptstyle 4:1}\downarrow{\scriptstyle \text{unram}} \\
\mathcal{C}_{\mathbf{c}} & \xrightarrow[\text{ram}]{2:1} & \overline{\mathcal{E}}
\end{array} \tag{8.94}$$

We do this by writing down explicit equations for the maps. Namely,

$$\begin{array}{ccccc}
\mathcal{D}_{\mathbf{c}} & \xrightarrow{2:1} & \overline{\Gamma^{(0)}} & \xrightarrow{4:1} & \overline{\mathcal{E}} \\
(a,b,c,d) & \mapsto & (a,b,c) & \mapsto & (c^2, abc)
\end{array}$$

where we recall that affine equations of $\mathcal{D}_{\mathbf{c}}$, $\overline{\Gamma^{(0)}}$ and $\overline{\mathcal{E}}$ are given by (8.87), (8.83) and (8.85). The action of σ_1 and σ_2 on $\mathcal{D}_{\mathbf{c}}$ is given by

$$\begin{aligned}
\sigma_1(a,b,c,d) &= (a,-b,-c,d), \\
\sigma_2(a,b,c,d) &= (-a,b,-c,d),
\end{aligned}$$

as follows from (8.78) and (8.90) (the action on a,b,c was already determined in (8.84)). It follows that σ_1 and σ_2 leave the fibers of $\mathcal{D}_{\mathbf{c}} \to \mathcal{E}$ invariant, so that there is an induced morphism $\mathcal{C}_{\mathbf{c}} = \mathcal{D}_{\mathbf{c}}/\langle\sigma_1,\sigma_2\rangle \to \mathcal{E}$. Explicitly, an affine equation for $\mathcal{C}_{\mathbf{c}}$ is given by

$$\begin{aligned}
&w^2 = v(v-\beta^2)(v+\alpha^2), \\
&d^2 + c_1(v+\alpha^2)v + c_2(v-\beta^2)v + c_3(v+\alpha^2)(v-\beta^2) + 2wc_4 = 0,
\end{aligned}$$

where the second equation exhibits $\mathcal{C}_{\mathbf{c}}$ as a double cover of $\mathcal{E}$ (which is given by the first equation), namely the maps are given as follows.

$$\begin{array}{ccccc}
\mathcal{D}_{\mathbf{c}} & \xrightarrow{4:1} & \mathcal{C}_{\mathbf{c}} & \xrightarrow{2:1} & \overline{\mathcal{E}} \\
(a,b,c,d) & \mapsto & (v,w,d) = (c^2, abc, d) & \mapsto & (v,w) = (c^2, abc)
\end{array} \tag{8.95}$$

We will also need the spectral curve $|A\mathfrak{h} + X - \mathfrak{z}\operatorname{Id}_4| = 0$ (see (8.64)), which we rewrite, using $(v,u) := (1/\mathfrak{h}, \mathfrak{z}/\mathfrak{h})$ as

$$\mathcal{K}_{\mathbf{c}} : \prod_{i=1}^{4}(A_i - u) + (\bar{c}_1 u^2 + \bar{c}_2 u + \bar{c}_3)v^2 + \bar{c}_0^2 v^4 = 0, \tag{8.96}$$

where $(\bar{c}_0, \ldots, \bar{c}_3)$ are the values of the constants of motion (8.69); they depend on the constants $\mathbf{c} = (c_1, \ldots, c_4)$ only, hence the notation $\mathcal{K}_{\mathbf{c}}$ rather than $\mathcal{K}_{\bar{\mathbf{c}}}$ for the curve. $\mathcal{K}_{\mathbf{c}}$ admits the involution defined by $\imath(u,v) := (u,-v)$, leading to the quotient curve

$$\mathcal{K}_{\mathbf{c}}^{(0)} : \prod_{i=1}^{4}(A_i - u) + (\bar{c}_1 u^2 + \bar{c}_2 u + \bar{c}_3)t + \bar{c}_0^2 t^2 = 0. \tag{8.97}$$

which is an elliptic curve, since it is a double cover of $\mathbf{P}^1$, ramified at the four points (u_i, t_i) on $\mathcal{K}_{\mathbf{c}}^{(0)}$ that correspond to the 4 zeros u_i of $(\bar{c}_1 u^2 + \bar{c}_2 u + \bar{c}_3)^2 - 4\bar{c}_0^2 \prod_{i=1}^{4}(A_i - u)$. The double cover $\mathcal{K}_{\mathbf{c}} \to \mathcal{K}_{\mathbf{c}}^{(0)}$ has four ramification points, which are the points $p_i := (A_i, 0)$. It is unramified at infinity: $\mathcal{K}_{\mathbf{c}}$ has four points at infinity, which we will denote by $q_1, q_2, q_3 = \imath(q_1)$ and $q_4 = \imath(q_2)$. Taking a local parameter at these points one verifies that the divisor of zeros and poles of u and v is given by

$$(u) = 4 \text{ zeros } - \sum_{i=1}^{4} q_i, \qquad (v) = \sum_{i=1}^{4} p_i - \sum_{i=1}^{4} q_i.$$

Summarizing, the covers that are related to the spectral curve $\mathcal{K}_{\mathbf{c}}$ are given by

$$\begin{array}{ccccc} \mathcal{K}_{\mathbf{c}} & \xrightarrow{2:1} & \mathcal{K}_{\mathbf{c}}^{(0)} & \xrightarrow{2:1} & \mathbf{P}^1 \\ (u,v) & \mapsto & (u,v^2) & \mapsto & u \end{array}$$

Observe that images of the branch points p_i of $\mathcal{K}_{\mathbf{c}} \to \mathcal{K}_{\mathbf{c}}^{(0)}$ in $\mathbf{P}^1$ are precisely the branch points A_i of the cover $\overline{\mathcal{E}} \to \mathbf{P}^1$.

We are now ready to give a precise description of the generic fibers $\mathbf{F}_{\mathbf{c}}$, the Abelian surfaces $\mathbf{T}_{\mathbf{c}}^2$ into which they compactify, and of the Abelian surfaces on which the Lax equation (8.63) linearizes.

Proposition 8.15 (Haine). *Let* $\mathbf{c} = (c_1, \ldots, c_4)$ *be generic and let* $\jmath_{\mathbf{c}} : \mathbf{F}_{\mathbf{c}} \to \mathrm{Jac}(\mathcal{K}_{\mathbf{c}})$ *be the linearizing map as defined in Section 6.4.*

(1) $\jmath_{\mathbf{c}}$ *is onto* $\mathrm{Prym}(\mathcal{K}_{\mathbf{c}}/\mathcal{K}_{\mathbf{c}}^{(0)}) \setminus \Theta_{\mathbf{c}}$*, where* $\Theta_{\mathbf{c}}$ *is (a translate of) the theta divisor of* $\mathrm{Prym}(\mathcal{K}_{\mathbf{c}}/\mathcal{K}_{\mathbf{c}}^{(0)})$*;*

(2) $\jmath_{\mathbf{c}}$ *extends to an isogeny* $\overline{\jmath_{\mathbf{c}}} : \mathbf{T}_{\mathbf{c}}^2 \to \mathrm{Prym}(\mathcal{K}_{\mathbf{c}}/\mathcal{K}_{\mathbf{c}}^{(0)})$ *of degree 4, which induces an isomorphism* $\widetilde{\mathbf{T}}_{\mathbf{c}}^2 \to \mathrm{Prym}(\mathcal{K}_{\mathbf{c}}/\mathcal{K}_{\mathbf{c}}^{(0)})$*;*

(3) $\mathbf{T}_{\mathbf{c}}^2$ *is isomorphic to the Prym surface* $\mathrm{Prym}(\mathcal{C}_{\mathbf{c}}/\mathcal{E})$*;*

(4) $\mathbf{T}_{\mathbf{c}}^2 \cong \mathrm{Prym}(\mathcal{C}_{\mathbf{c}}/\mathcal{E})$ *and* $\widetilde{\mathbf{T}}_{\mathbf{c}}^2 \cong \mathrm{Prym}(\mathcal{K}_{\mathbf{c}}/\mathcal{K}_{\mathbf{c}}^{(0)})$ *are dual Abelian surfaces.*

Proof. *(1)* Let $\mathbf{c}$ be generic, but fixed and let $X \in \mathbf{F}_{\mathbf{c}}$ be any element. We consider the divisor map, as in Section 6.4. Namely, as the Lax equation (8.63) defines an isospectral flow we consider, as in Section 6.4 the eigenvector $\xi = (\xi_1, \ldots, \xi_4)^\top$, normalized at $\xi_1 = 1$, to the spectral problem

$$(A\mathfrak{h} + X)\xi = \mathfrak{z}\xi,$$

which, under the above change of variables $(v, u) := (1/\mathfrak{h}, \mathfrak{z}/\mathfrak{h})$ on the spectral curve is equivalent to

$$(A + Xv)\xi = u\xi.$$

In terms of the cofactors Δ_{ij} of the matrix $A + Xv$, the components of $\xi = \xi(u, v)$ are given by $\xi_k = \Delta_{1k}/\Delta_{11}$. Taking a local parameter at the points p_i and at the points q_i we find that Δ_{1k} has the following divisor of zeros and poles,

$$\Delta_{1k} = 2\sum_{i=1}^{4} p_i - 3\sum_{i=1}^{4} q_i - p_1 - p_k + E_k,$$

where E_k is a positive divisor of degree 6. Thus,

$$(\xi_k) = p_1 - p_k + E_k - E_1. \tag{8.98}$$

We define, as in Section 6.4[5] $\phi_{\mathbf{c}}(X)$ to be the minimal positive divisor such that $(\xi_k) + \phi_{\mathbf{c}}(X) \geqslant 0$. Notice that, in view of (8.98), $\phi_{\mathbf{c}}(X) \in \text{Div}^6(\mathcal{K}_{\mathbf{c}})$ is forced to contain the points p_2, p_3 and p_4, that do not move with X (when X moves according to (8.63)). Notice also that $\phi_{\mathbf{c}}(X)$ does not contain the point p_1, since Δ_{11} does not vanish at p_1. By the antisymmetry of X, the zeros of the principal minors of X come in pairs, in particular $E_1 = \phi_{\mathbf{c}}(X) + \imath(\phi_{\mathbf{c}}(X)) - 2(p_2 + p_3 + p_4)$ and

$$\begin{aligned}\left(\frac{\Delta_{11}}{v^2}\right) &= \phi_{\mathbf{c}}(X) + \imath(\phi_{\mathbf{c}}(X)) - \sum_{i=1}^{4} q_i - 2\sum_{i=1}^{4} p_i \\ &= \left(\phi_{\mathbf{c}}(X) - q_1 - q_2 - \sum_{i=1}^{4} p_i\right) + \imath\left(\phi_{\mathbf{c}}(X) - q_1 - q_2 - \sum_{i=1}^{4} p_i\right).\end{aligned}$$

This shows that $\phi_{\mathbf{c}}(X) - q_1 - q_2 - \sum_{i=1}^{4} p_i$ is an odd divisor, so that its equivalence class belongs to $\text{Prym}(\mathcal{K}_{\mathbf{c}}/\mathcal{K}_{\mathbf{c}}^{(0)})$. Explicitly, if we let $\vec{\omega} := (\eta, \chi_1, \chi_2)^\top$ be a basis of holomorphic differentials on $\mathcal{K}_{\mathbf{c}}$ with $\imath^*\eta = \eta$ and $\imath^*\chi_i = -\chi_i$, for $i = 1, 2$ (see Paragraph 5.2.4) then the linearizing map, with base divisor $\mathbf{0} = q_1 + q_2 + \sum_{i=1}^{4} p_i$ is given by

$$\begin{aligned}\jmath_{\mathbf{c}} : \mathbf{F}_{\mathbf{c}} &\to \text{Prym}(\mathcal{K}_{\mathbf{c}}/\mathcal{K}_{\mathbf{c}}^{(0)}) \subset \text{Jac}(\mathcal{K}_{\mathbf{c}}) \\ X &\mapsto \int_{\mathbf{0}}^{\phi_{\mathbf{c}}(X)} \vec{\omega}.\end{aligned} \tag{8.99}$$

[5] In that section we denoted this divisor by $\mathcal{D}_X$, omitting the reference to $\mathbf{c}$.

We now show that the image of $\phi_{\mathbf{c}}$ consists precisely of the set of divisors of degree 6, containing p_2, p_3 and p_4, but not containing p_1. The image of $\jmath_{\mathbf{c}}$ is then precisely those points of $\text{Prym}(\mathcal{K}_{\mathbf{c}}/\mathcal{K}_{\mathbf{c}}^{(0)})$ that do not lie on (a certain translate of) the theta divisor of $\text{Jac}(\mathcal{K}_{\mathbf{c}})$, in view of Riemann's Theorem (Theorem 5.24).

Let $D - q_1 - q_2 - \sum_{i=1}^{4} p_i$ be an odd divisor of degree 0, where D is effective (of degree 6), and contains p_2, p_3 and p_4, but not p_1. It is assumed that $D' := D - p_2 - p_3 - p_4$ is a general divisor on $\mathcal{K}_{\mathbf{c}}$, so that D' is not linearly equivalent to a divisor of this type, but containing also p_1. Then it follows that

$$1 \leqslant \dim L(D' - p_1 + p_k) \leqslant L(D' - p_1) + 1 = 1,$$

where the first inequality is a consequence of the Riemann-Roch Theorem, the second one a consequence of the fact that adding a pole increases the dimension at most by one and the final equality follows because D' is a general divisor that does not contain p_1. Setting $\xi_1 := 1$ we can now reconstruct the functions ξ_i by demanding that ξ_i be the unique function, up to a constant, in $L(D' - p_1 + p_k)$. Comparing the pole orders of the functions ξ_j at the points p_i it follows that there exist X_{ij}, such that

$$\mathfrak{z}\xi_k = A_k \mathfrak{h} \xi_k + \sum_{i \neq k} X_{ki} \xi_i,$$

where the freedom of picking the constants in the ξ_i (for $i = 2, 3, 4$) allows us to pick them such that $X_{1i} + X_{i1} = 0$ for $i = 2, 3, 4$. Moreover, since $D - q_1 - q_2 - \sum_{i=1}^{4} p_i$ is odd, the matrix X must be skew-symmetric. Notice that the above equation does not determine X completely since we can still multiply the each ξ_i (except for $\xi_1 = 1$) by ± 1. However, among the 8 remaining possibilities, only 4 of them are compatible with $Q_0 = c_0$. Notice that these four possibilities correspond exactly under the group of translations generated by σ_1 and σ_2, as expected. This shows that, away from the (above translate of the) theta divisor, $\phi_{\mathbf{c}}$ is surjective, as was to be shown.

(2) We know that both $\mathbf{T}_{\mathbf{c}}^2$ and $\text{Prym}(\mathcal{K}_{\mathbf{c}}/\mathcal{K}_{\mathbf{c}}^{(0)})$ are Abelian surfaces and, from the preceding paragraph, that on the affine part $\mathbf{F}_{\mathbf{c}}$ the morphism $\jmath_{\mathbf{c}}$ corresponds to taking the quotient with respect to the group of translations, generated by σ_1 and σ_2. Thus, $\jmath_{\mathbf{c}}$ extends to

$$\overline{\jmath_{\mathbf{c}}} : \mathbf{T}_{\mathbf{c}}^2 \to \text{Prym}(\mathcal{K}_{\mathbf{c}}/\mathcal{K}_{\mathbf{c}}^{(0)}),$$

which is the isogeny of degree 4, that corresponds to taking the quotient of $\mathbf{T}_{\mathbf{c}}^2$ with respect to the group of translations, generated by σ_1 and σ_2. The conclusion follows, as $\widetilde{\mathbf{T}}_{\mathbf{c}}^2$ is the precisely the quotient of $\mathbf{T}_{\mathbf{c}}^2$ by this group of translations.

(3) We now consider the holomorphic differentials on the curves $\mathcal{D}_{\mathbf{c}}$, $\mathcal{C}_{\mathbf{c}}$ and $\overline{\mathcal{E}}$ and on the Abelian surfaces that contain these curves. Clearly, a basis for the holomorphic differentials on $\mathbf{T}^2_{\mathbf{c}}$ is given by $\mathsf{d}t_1|_{\mathbf{T}^2_{\mathbf{c}}}$ and $\mathsf{d}t_2|_{\mathbf{T}^2_{\mathbf{c}}}$. Since $\widetilde{\mathbf{T}}^2_{\mathbf{c}}$ is obtained from $\mathbf{T}^2_{\mathbf{c}}$ by dividing by two translations (over half periods) these differentials descend to a basis for the holomorphic differentials on $\widetilde{\mathbf{T}}^2_{\mathbf{c}}$, that we denote by $\mathsf{d}t_1|_{\widetilde{\mathbf{T}}^2_{\mathbf{c}}}$ and $\mathsf{d}t_2|_{\widetilde{\mathbf{T}}^2_{\mathbf{c}}}$.

We have already computed the restrictions ω_2 and ω_1 of the former to $\mathcal{D}_{\mathbf{c}}$: recall from (8.92) that

$$\omega_1 = -\frac{\mathsf{d}a}{abcd}, \qquad \omega_2 = \frac{a^2\mathsf{d}a}{bcd}.$$

Pushing these differentials down to $\mathcal{C}_{\mathbf{c}}$ we find two independent holomorphic differentials on $\mathcal{C}_{\mathbf{c}}$, to wit

$$\psi_1 := \frac{\mathsf{d}v}{wd}, \quad \psi_2 := \frac{v\mathsf{d}v}{wd}. \tag{8.100}$$

In fact, if we denote the covering map $\mathcal{D}_{\mathbf{c}} \to \mathcal{C}_{\mathbf{c}}$ by π then

$$\begin{aligned}
\pi^*\psi_1 &= \frac{2c\mathsf{d}c}{abcd} = \frac{2\mathsf{d}a}{bcd} = -2\alpha\omega_1,\\
\pi^*\psi_2 &= \frac{2c^2\mathsf{d}a}{bcd} = \frac{2(a^2+\beta^2)\mathsf{d}a}{bcd} = 2\omega_2 - 2\alpha\beta^2\omega_1.
\end{aligned}$$

These differentials ψ_1 and ψ_2 are odd with respect to the involution on $\mathcal{C}_{\mathbf{c}}$, defined by $\imath : (v, w, d) \mapsto (v, w, -d)$. If we add the pull-back θ of a holomorphic differential on $\overline{\mathcal{E}}$, then (ψ_1, ψ_2, θ) will be a basis for the holomorphic differentials on $\mathcal{C}_{\mathbf{c}}$, with θ even. We will use these differentials to compute the period matrix of $\mathbf{T}^2_{\mathbf{c}}$ and of $\widetilde{\mathbf{T}}^2_{\mathbf{c}}$. The computation is based on the following three ideas:

1. As we have seen when discussing the Albanese variety (see Paragraph 5.2.3), we can compute the period matrix of a complex torus by integrating a basis of the holomorphic differentials on it over a basis for the first homology of the torus. In the present case this will lead to a lattice in $\mathbf{C}^2$, spanned by 4 vector
2. By the Lefschetz hyperplane section theorem (see [69, pp. 156–159]) the first homology group of an algebraic variety is generated by the first homology of any ample divisor on it. In the present case this means that $H_1(\mathbf{T}^2_{\mathbf{c}}, \mathbf{Z})$ is generated by 4 elements of $H_1(\mathcal{D}_{\mathbf{c}}, \mathbf{Z})$ and that $H_1(\widetilde{\mathbf{T}}^2_{\mathbf{c}}, \mathbf{Z})$ is generated by 4 elements of $H_1(\mathcal{C}_{\mathbf{c}}, \mathbf{Z})$;
3. Integrating these differentials over cycles on these loops amounts to integrating the restrictions of these differentials to the embedded curves, over these same loops. This is obvious, but it is crucial, since it replaces integration over loops in the Abelian variety to integration over loops on a Riemann surface.

Let us first do this for $\widetilde{\mathbf{T}}^2_{\mathbf{c}}$. We integrate the basis $\mathrm{d}t_{1|_{\widetilde{\mathbf{T}}^2_{\mathbf{c}}}}$ and $\mathrm{d}t_{2|_{\widetilde{\mathbf{T}}^2_{\mathbf{c}}}}$ over a set of generators for $H_1(\mathcal{C}_{\mathbf{c}}, \mathbf{Z})$. As we just pointed out this amounts to integrating ψ_1 and ψ_2 over these generators. Now $H_1(\mathcal{C}_{\mathbf{c}}, \mathbf{Z})$ is generated by two even generators A^+ and B^+ that come from generators A and B on the elliptic curve $\mathcal{E}$ and four generators A^-, C^-, B^-, D^- that are odd with respect to $\imath$ (see Paragraph 5.2.4, in particular Figure 5.1). Since the differentials ψ_1 and ψ_2 are odd their integrals over the even generators A^+ and B^+ vanish and hence the period matrix of $\widetilde{\mathbf{T}}^2_{\mathbf{c}}$ is given by

$$\begin{pmatrix} \oint_{A^-} \psi_1 & \oint_{C^-} \psi_1 & \oint_{B^-} \psi_1 & \oint_{D^-} \psi_1 \\ \oint_{A^-} \psi_2 & \oint_{C^-} \psi_2 & \oint_{B^-} \psi_2 & \oint_{D^-} \psi_2 \end{pmatrix}. \tag{8.101}$$

Let us compute now the period matrix for $\mathbf{T}^2_{\mathbf{c}}$. To do this we need to relate carefully the homology of $\mathcal{D}_{\mathbf{c}} \subset \mathbf{T}^2_{\mathbf{c}}$ and of $\mathcal{C}_{\mathbf{c}} \subset \widetilde{\mathbf{T}}^2_{\mathbf{c}}$. The commutative diagram (8.94) induces the following commutative diagram in homology:

$$\begin{array}{ccc} H_1(\mathcal{D}_{\mathbf{c}}, \mathbf{Z}) & \xrightarrow{surj} & H_1(\Gamma^{(0)}, \mathbf{Z}) \\ \downarrow & & \downarrow \\ H_1(\mathcal{C}_{\mathbf{c}}, \mathbf{Z}) & \xrightarrow{surj} & H_1(\overline{\mathcal{E}}, \mathbf{Z}) \end{array} \tag{8.102}$$

In this diagram each of the horizontal arrows is induced by a ramified double covering map. It is an elementary fact from topology that such a homomorphism is surjective, which means that every loop is homologous to a loop that can be lifted; we simply say that the loop can be lifted. The vertical arrows are induced by unramified covering maps of degree 4, corresponding to two involutions, hence both homomorphisms will have a cokernel that is isomorphic to $\mathbf{Z}/2\mathbf{Z} \oplus \mathbf{Z}/2\mathbf{Z}$. The cycles A and B that generate $H_1(\overline{\mathcal{E}}, \mathbf{Z})$ are generators for the cokernel of $H_1(\Gamma^{(0)}, \mathbf{Z}) \to H_1(\overline{\mathcal{E}}, \mathbf{Z})$, hence they cannot be lifted to $\Gamma^{(0)}$, but $2A$ and $2B$ can. By commutativity of (8.102) the liftings $\widetilde{A}$ and $\widetilde{B}$ of A and B to $\mathcal{C}_{\mathbf{c}}$ cannot be lifted to $\mathcal{D}_{\mathbf{c}}$, being generators of the cokernel of $H_1(\mathcal{D}_{\mathbf{c}}, \mathbf{Z}) \to H_1(\mathcal{C}_{\mathbf{c}}, \mathbf{Z})$, but $2\widetilde{A}$ and $2\widetilde{B}$ and any other generator of $H_1(\mathcal{C}_{\mathbf{c}}, \mathbf{Z})$ can. Thus, we can lift $2A^-$, C^-, $2B^-$ and D^- to cycles on $\mathcal{D}_{\mathbf{c}}$, which gives, by integration of ω_1 and ω_2 the following matrix of periods,

$$\begin{pmatrix} 2\oint_{A^-} \psi_1 & \oint_{C^-} \psi_1 & 2\oint_{B^-} \psi_1 & \oint_{D^-} \psi_1 \\ 2\oint_{A^-} \psi_2 & \oint_{C^-} \psi_2 & 2\oint_{B^-} \psi_2 & \oint_{D^-} \psi_2 \end{pmatrix}. \tag{8.103}$$

We show that this is the period matrix of $\mathbf{T}^2_{\mathbf{c}}$. Notice first that this matrix is indeed the matrix of periods of an Abelian surface, as it is the period matrix of the Prym surface $\mathrm{Prym}(\mathcal{C}_{\mathbf{c}}/\mathcal{E})$ (see (5.24)). This does not prove yet that it is the period matrix of $\mathbf{T}^2_{\mathbf{c}}$, as the chosen cycles may not generate $H_1(\mathbf{T}^2_{\mathbf{c}}, \mathbf{Z})$. However, we know that the degree of the isogeny $\mathbf{T}^2_{\mathbf{c}} \to \widetilde{\mathbf{T}}^2_{\mathbf{c}}$ is four, as we devided by a group of four translations to obtain $\widetilde{\mathbf{T}}^2_{\mathbf{c}}$ from $\mathbf{T}^2_{\mathbf{c}}$. This corresponds precisely to the index of the period lattice defined by (8.103) as a subgroup of the the one defined by (8.101). Thus, the chosen loops generate $H_1(\mathbf{T}^2_{\mathbf{c}}, \mathbf{Z})$ and (8.103) is the period matrix of $\mathbf{T}^2_{\mathbf{c}}$.

(4) Let us write the period matrix (8.101) of $\widetilde{\mathbf{T}}^2_{\mathbf{c}}$ as $\begin{pmatrix}\tilde{E} & \tilde{F}\end{pmatrix}$ and the period matrix (8.103) of $\mathbf{T}^2_{\mathbf{c}}$ as $(E\ F)$. Denoting $\Delta := \begin{pmatrix} 2 & 0 \\ 0 & 1 \end{pmatrix}$ we have on the one hand that

$$E = \tilde{E}\Delta \qquad \text{and} \qquad F = \tilde{F}\Delta,$$

and on the other hand that, upon normalizing ψ_1 and ψ_2, an alternative period matrix for $\mathbf{T}^2_{\mathbf{c}}$ is obtained, namely $(\Delta\ Z)$, where Z is given by $Z = \Delta E^{-1}F$. According to (5.16), the matrix for the dual of $\mathbf{T}^2_{\mathbf{c}}$ is given by

$$\left(2\Delta^{-1},\quad 2\Delta^{-1}\Delta E^{-1}F\Delta^{-1}\right) = \left(2\Delta^{-1},\quad 2\Delta^{-1}\tilde{E}^{-1}\tilde{F}\right)$$

which amounts to $\begin{pmatrix}\tilde{E} & \tilde{F}\end{pmatrix}$, as was to be shown. □

8.2.6 A.c.i. Diagonal Metrics on $\mathfrak{so}(4)$

In this paragraph we show that the only diagonal metrics on $\mathfrak{so}(4)$ that define an a.c.i. system correspond either to metric I (roughly speaking the Manakov metric, see Paragraph 8.2.1) or to a product of Euler tops. This result was first proven by Adler and van Moerbeke (see [9]) by using the Kowalevski-Painlevé Criterion; their proof is (essentially) contained in the proof of Theorem 8.3. The proof that we give here is based on a alternative proof, due to Luc Haine (see [75]), which uses the fact that all solutions to each of the variational equations of an a.c.i. system must be single-valued (Theorem 6.18). This proof has the advantage of admitting a generalization to $\mathfrak{so}(n)$ and of not assuming

(1) any non-degeneracy of the metric;
(2) irreducibility of the a.c.i. system;
(3) weight homogeneity of the principal balances.

In the proof that we give here the crucial step in the argument is clarified by the use of the Kowalevski matrix.

Theorem 8.16. *Let H be a quadratic form on $\mathfrak{so}(4)$ which defines a diagonal metric on $\mathfrak{so}(4)$,*

$$H = \frac{1}{2} \sum_{1 \leqslant i < j \leqslant 4} \Lambda_{ij} X_{ij}^2.$$

If $\mathcal{X}_H$ is one of the integrable vector fields of an a.c.i. system then the constants Λ_{ij} satisfy $K(\Lambda) = 0$ ($K(\Lambda)$ was defined in (8.60)), or

$$\Lambda_{12} - \Lambda_{34} = \Lambda_{23} - \Lambda_{14} = \Lambda_{13} - \Lambda_{24} = 0. \tag{8.104}$$

Proof. The Hamiltonian vector field $\mathcal{X}_H$ is given by

$$\dot{X} = [X, \Lambda \cdot X],$$

where Λ is the symmetric matrix with entries Λ_{ij} and $\Lambda \cdot X$ denotes, as before, the Kronecker product of Λ and X, i.e., $(\Lambda \cdot X)_{ij} = \Lambda_{ij} X_{ij}$. The indicial locus $\mathcal{I}$ is given by the $X^{(0)} \in \mathfrak{so}(4)$ for which

$$X^{(0)} + [X^{(0)}, \Lambda \cdot X^{(0)}].$$

Let us assume that

$$\Lambda_1 := (\Lambda_{42} - \Lambda_{23})(\Lambda_{23} - \Lambda_{34})(\Lambda_{34} - \Lambda_{42}) \neq 0. \tag{8.105}$$

Then $\mathcal{I}$ contains the 4 points $X^{(0)}$ for which

$$\begin{aligned} X_{12}^{(0)} = X_{13}^{(0)} = X_{14}^{(0)} &= 0, \\ \left(X_{ij}^{(0)}\right)^2 &= \frac{\Lambda_{jk} - \Lambda_{ik}}{\Lambda_1}, \\ X_{23}^{(0)} X_{34}^{(0)} X_{42}^{(0)} &= -\Lambda_1^{-1}. \end{aligned}$$

where (i, j, k) is any cyclic permutation of $(2, 3, 4)$. For any of these points $X^{(0)}$ the Kowalevski matrix is given by $\mathcal{K}(X^{(0)}) = \begin{pmatrix} \mathcal{K}_1 & 0 \\ 0 & \mathcal{K}_2 \end{pmatrix}$, where

$$\mathcal{K}_1 = \begin{pmatrix} 1 & X_{23}(\Lambda_{13} - \Lambda_{23}) & X_{24}(\Lambda_{14} - \Lambda_{24}) \\ X_{23}(\Lambda_{23} - \Lambda_{12}) & 1 & X_{34}(\Lambda_{14} - \Lambda_{34}) \\ X_{24}(\Lambda_{24} - \Lambda_{12}) & X_{34}(\Lambda_{34} - \Lambda_{13}) & 1 \end{pmatrix}$$

and

$$\mathcal{K}_2 = \begin{pmatrix} 1 & X_{34}(\Lambda_{24} - \Lambda_{34}) & X_{24}(\Lambda_{24} - \Lambda_{34}) \\ X_{34}(\Lambda_{34} - \Lambda_{23}) & 1 & X_{23}(\Lambda_{34} - \Lambda_{23}) \\ X_{24}(\Lambda_{23} - \Lambda_{24}) & X_{23}(\Lambda_{23} - \Lambda_{24}) & 1 \end{pmatrix}.$$

It follows that the characteristic polynomial of $\mathcal{K}(X^{(0)})$ is given by

$$\chi(\mathcal{K}(X^{(0)}), \mu) = (\mu - 2)^3 (\mu + 1)(\mu^2 - \mu - K(\Lambda)\Lambda_1^{-1}).$$

Proposition 7.32 implies that, if $\mathcal{X}_H$ defines an a.c.i. system, then all roots of this polynomial are integers. Thus,

$$\Lambda_1 = 0 \text{ or } \exists r_1 \in \mathbf{Z} \text{ such that } K(\Lambda) = r_1(r_1 - 1)\Lambda_1.$$

By performing all possible cyclic permutations of $(1, 2, 3, 4)$ in the definition (8.105) of Λ_1, yielding[6] $\Lambda_2 = (-1)(\Lambda_{13} - \Lambda_{34})(\Lambda_{34} - \Lambda_{41})(\Lambda_{41} - \Lambda_{13})$, and so on, we find that if $\mathcal{X}_H$ defines an a.c.i. system, then for any $i \in \{1, \ldots, 4\}$,

$$\Lambda_i = 0 \text{ or } \exists r_i \in \mathbf{Z} \text{ such that } K(\Lambda) = r_i(r_i - 1)\Lambda_i. \tag{8.106}$$

Similarly we consider the 4 points in $\mathcal{I}$ for which

$$\begin{aligned} X_{1i}^{(0)} &= X_{jk}^{(0)}, \\ \left(X_{1i}^{(0)}\right)^2 &= \frac{\Lambda_{1k} - \Lambda_{ki} + \Lambda_{ij} - \Lambda_{j1}}{\Lambda_0}, \\ X_{12}^{(0)} X_{13}^{(0)} X_{14}^{(0)} &= \Lambda_0^{-1}, \end{aligned} \tag{8.107}$$

where (i, j, k) is any cyclic permutation of $(2, 3, 4)$ and where

$$\Lambda_0 := (\Lambda_{23} + \Lambda_{14} - \Lambda_{13} - \Lambda_{24})(\Lambda_{13} + \Lambda_{24} - \Lambda_{12} - \Lambda_{34})(\Lambda_{12} + \Lambda_{34} - \Lambda_{23} - \Lambda_{14}).$$

The characteristic polynomial of the Kowalevski matrix that corresponds to these points is given by

$$\chi(\mathcal{K}(X^{(0)}), \mu) = (\mu - 2)^2(\mu - 1)(\mu + 1)(\mu^2 - 2\mu - 4K(\Lambda)\Lambda_0^{-1}). \tag{8.108}$$

It follows that

$$\Lambda_0 = 0 \text{ or } \exists r_0 \in \mathbf{Z} \text{ such that } 4K(\Lambda) = r_0(r_0 - 2)\Lambda_0. \tag{8.109}$$

The theorem follows by cleverly combining (8.106) and (8.109). Namely, assume first that $\Lambda_0, \Lambda_1, \ldots, \Lambda_4$ are all different from zero. Then either $K(\Lambda) = 0$, as was to be shown, or we can solve each Λ_i in terms of $K(\Lambda)$ by using (8.106) and (8.109); substituted in the identity

$$2K(\Lambda) + \sum_{i=0}^{4} \Lambda_i = 0$$

this yields

$$2 \mid \frac{4}{r_0(r_0 - 2)} + \sum_{i=1}^{4} \frac{1}{r_i(r_i - 1)} = 0. \tag{8.110}$$

Since all r_i are integers all terms in this sum are strictly positive, except when $r_0 = 1$; for future use, notice that this also implies that all $r_i(r_i - 1)$ are equal to 2.

[6] Since $K(\Lambda) \mapsto -K(\Lambda)$ when doing such a permutation we add a minus sign in the definition of Λ_2, hence also in the definition of Λ_4.

Now $r_0 = 1$ implies that $4K(\Lambda) = -\Lambda_0$, so that (8.108) has 1 as a triple root. After a simple conjugation the Kowalevski matrix of case (8.107) takes the block form $\mathcal{K}(X^{(0)}) = \begin{pmatrix} \mathcal{K}_1 & \mathcal{K}_2 \\ 0 & \mathcal{K}_0 \end{pmatrix}$, for some matrices $\mathcal{K}_0$, $\mathcal{K}_1$ and $\mathcal{K}_2$. The matrix $\mathcal{K}_0$ has 1 as a triple eigenvalue and is given by

$$\mathcal{K}_0 = \begin{pmatrix} 1 & X_{12}^{(0)}(\Lambda_{34} - \Lambda_{12}) & X_{23}^{(0)}(\Lambda_{14} - \Lambda_{23}) \\ X_{12}^{(0)}(\Lambda_{12} - \Lambda_{34}) & 1 & X_{13}^{(0)}(\Lambda_{13} - \Lambda_{24}) \\ X_{23}^{(0)}(\Lambda_{23} - \Lambda_{14}) & X_{13}^{(0)}(\Lambda_{24} - \Lambda_{13}) & 1 \end{pmatrix},$$

where $X_{12}^{(0)}$, $X_{23}^{(0)}$ and $X_{13}^{(0)}$ are non-zero, being given by (8.107). Proposition 7.32 implies that this matrix must be diagonalizable, i.e., that it is the identity matrix, which is so if and only if

$$\Lambda_{12} = \Lambda_{34} \text{ and } \Lambda_{23} = \Lambda_{14} \text{ and } \Lambda_{34} = \Lambda_{12},$$

which is precisely (8.104). For the case in which one of several of the Λ_i are zero one uses the equation $\Lambda_i = 0$, rather than the equation that expresses $K(\Lambda)$ in terms of Λ_i. If $K(\Lambda) \neq 0$ then we will still have (8.110), but with one or several of the terms containing the r_i missing. But such an equation does not admit any integer solutions. Therefore we conclude that, in this case, $K(\Lambda) = 0$. □

8.2.7 From the Manakov Flow to the Clebsch Flow

In this paragraph we show how the Clebsch Hamiltonian can be obtained from the Manakov Hamiltonian by taking a proper limit. The *Clebsch* Hamiltonian (see [47]) is given (on $\mathbf{C}^6$, with coordinates $l_1, \dots, l_3, p_1, \dots, p_3$) by

$$H := \frac{1}{2}\left(\kappa_1 l_1^2 + \kappa_2 l_2^2 + \kappa_3 l_3^2 + \lambda_1 p_1^2 + \lambda_2 p_2^2 + \lambda_3 p_3^2\right), \tag{8.111}$$

where the constants $\kappa_1, \dots, \lambda_3$ satisfy

$$\frac{\lambda_2 - \lambda_3}{\kappa_1} + \frac{\lambda_3 - \lambda_1}{\kappa_2} + \frac{\lambda_1 - \lambda_2}{\kappa_3} = 0. \tag{8.112}$$

It is one of the known integrable cases of geodesic flow on the Lie algebra $\mathfrak{e}(3) = \mathfrak{so}(3) \times \mathbf{R}^3$ (see Paragraph 8.1.1). The Lie-Poisson structure on $\mathfrak{e}(3)$ is completely specified by the following brackets,

$$\{l_i, l_j\} = \epsilon_{ijk} l_k, \qquad \{l_i, p_j\} = \epsilon_{ijk} p_k, \qquad \{p_i, p_j\} = 0, \tag{8.113}$$

where ϵ_{ijk} is an skew-symmetric tensor, with $\epsilon_{123} = 1$.

Therefore, the geodesic flow that corresponds to a given quadratic form H on $\mathfrak{e}(3) = \mathfrak{so}(3) \times \mathbf{C}^3 \cong \mathbf{C}^6$ is given by

$$\begin{aligned} \dot{l} &= p \wedge \frac{\partial H}{\partial p} - l \wedge \frac{\partial H}{\partial l}, \\ \dot{p} &= p \wedge \frac{\partial H}{\partial l}, \end{aligned}$$

where, $p = (p_1, p_2, p_3)^\top$ and $l = (l_1, l_2, l_3)^\top$. Also, $\wedge$ denotes the cross product on $\mathbf{R}^3$ and $\frac{\partial H}{\partial p} = \left(\frac{\partial H}{\partial p_1}, \frac{\partial H}{\partial p_2}, \frac{\partial H}{\partial p_3}\right)^\top$, and similarly for $\frac{\partial H}{\partial l}$. The physical importance of these equations stems from the fact that they describe the motion of a solid in a perfect fluid (see [104, pp. 117–124]). It was found by Clebsch that K, defined by

$$K := c(l_1^2 + l_2^2 + l_3^2) + \kappa_1 p_1^2 + \kappa_2 p_2^2 + \kappa_3 p_3^2$$

where

$$c := \frac{\kappa_1(\kappa_2 - \kappa_3)}{\lambda_2 - \lambda_3} = \frac{\kappa_2(\kappa_3 - \kappa_1)}{\lambda_3 - \lambda_1} = \frac{\kappa_3(\kappa_1 - \kappa_2)}{\lambda_1 - \lambda_2},$$

is an extra constant of motion, making the Clebsch Hamiltonian integrable; this is easily verified by direct computation, in particular it is easily verified that (8.112) implies that the three given formulas for c are equivalent.

In order to connect geodesic flow on $\mathfrak{so}(4)$ with geodesic flow on $\mathfrak{e}(3)$ we contract the Lie-Poisson structure of $\mathfrak{so}(4)$ to the Lie-Poisson structure of $\mathfrak{e}(3)$. We do this as follows. For fixed non-zero ϵ, define the following linear change of variables:

$$l_k := X_{ij}, \qquad p_i := \epsilon X_{i4}, \tag{8.114}$$

where (i, j, k) denotes any cyclic permutation of $(1, 2, 3)$. In terms of these new variables, the Poisson matrix (8.6) takes the following form.

$$\frac{1}{2}\begin{pmatrix} 0 & l_3 & -l_2 & 0 & p_3 & -p_2 \\ l_3 & 0 & l_1 & -p_3 & 0 & p_1 \\ l_2 & -l_1 & 0 & p_2 & -p_1 & 0 \\ 0 & p_3 & -p_2 & 0 & \epsilon^2 l_3 & -\epsilon^2 l_2 \\ -p_3 & 0 & p_1 & -\epsilon^2 l_3 & 0 & \epsilon^2 l_1 \\ p_2 & -p_1 & 0 & \epsilon^2 l_2 & -\epsilon^2 l_1 & 0 \end{pmatrix}, \tag{8.115}$$

where we have ordered the coordinates as follows: $l_1, \ldots, l_3, p_1, \ldots, p_3$. If we let $\epsilon \to 0$ then the resulting Poisson structure still makes sense, and it is precisely (i.e., up to a factor 2) the Lie-Poisson structure of $\mathfrak{e}(3)$, as given by (8.113). It follows easily from (8.113) or from (8.115) that a basis for the Casimirs of $\mathfrak{e}(3)$ is given by $p_1^2 + p_2^2 + p_3^2$ and $p_1 l_1 + p_2 l_2 + p_3 l_3$.

In fact, these Casimirs can be obtained from the Casimirs Q_0 and Q_1 of $\mathfrak{so}(4)$ (see (8.7)) as follows. Let Q_0^ϵ denote Q_0, expressed in the coordinates $l_1, \ldots, p_3$ by using (8.114), and define Q_1^ϵ similarly. Then

$$Q_0^\epsilon = \frac{1}{\epsilon}(l_1p_1 + l_2p_2 + l_3p_3),$$

$$Q_1^\epsilon = l_1^2 + l_2^2 + l_3^2 + \frac{1}{\epsilon^2}(p_1^2 + p_2^2 + p_3^2),$$

and we see that ϵQ_0^ϵ and $\epsilon^2 Q_1^\epsilon$ admit a limit as $\epsilon \to 0$. These limits are precisely the two Casimirs that were given above.

Let us apply the same procedure to the Manakov Hamiltonian

$$H = \frac{1}{2} \sum_{1 \leqslant i < j \leqslant 4} \frac{B_i - B_j}{A_i - A_j} X_{ij}^2.$$

The parameters that appear in this Hamiltonian are written as follows in terms of ϵ in order to get the desired limit:

$$A_i = 1 + \epsilon^2 a_i, \quad B_i = b_i, \quad A_4 = B_4 = 0.$$

We substitute this in H, which we write in terms of the coordinates $l_1, \ldots, p_3$ by using (8.114), and we denote the result by H^ϵ. Then $\lim_{\epsilon \to 0} \epsilon^2 H^\epsilon$ exists and is given by

$$\lim_{\epsilon \to 0} \epsilon^2 H^\epsilon = \frac{1}{2}\left(\frac{b_2 - b_3}{a_2 - a_3} l_1^2 + \frac{b_3 - b_1}{a_3 - a_1} l_2^2 + \frac{b_1 - b_2}{a_1 - a_2} l_3^2 + b_1p_1^2 + b_2p_2^2 + b_3p_3^2\right)$$

Obviously, this quadratic form is of the form (8.111), and it is easy to see that the coefficients satisfy (8.112). In fact, for generic values of $\kappa_1, \ldots, \lambda_3$ the Clebsch Hamiltonian (8.111) is obtained in this way.

The similar limits of the constants of motion $R_1, \ldots, R_4$ of the Manakov flow, given in (8.73), give the following constants of motion of the Clebsch flow.

$$\hat{R}_1 := \lim_{\epsilon \to 0} \epsilon^2 R_1^\epsilon = p_1^2 + \frac{l_2^2}{a_1 - a_3} + \frac{l_3^2}{a_1 - a_2},$$

$$\hat{R}_2 := \lim_{\epsilon \to 0} \epsilon^2 R_2^\epsilon = p_2^2 + \frac{l_1^2}{a_2 - a_3} + \frac{l_3^2}{a_2 - a_1},$$

$$\hat{R}_3 := \lim_{\epsilon \to 0} \epsilon^2 R_3^\epsilon = p_3^2 + \frac{l_1^2}{a_3 - a_2} + \frac{l_2^2}{a_3 - a_1},$$

$$\hat{R}_4 := \lim_{\epsilon \to 0} \epsilon^2 R_4^\epsilon = -p_1^2 - p_2^2 - p_3^2.$$

Notice that Clebsch's constant of motion K can be written in terms of the polynomials $\hat{R}_i$ as

$$K = \kappa_1 \hat{R}_1 + \kappa_2 \hat{R}_2 + \kappa_3 \hat{R}_3.$$

Another integrable flow on $\mathfrak{e}(3)$ will be given in Paragraph 8.3.4.

8.3 Geodesic Flow on SO(4) for Metric II and Hyperelliptic Jacobians

8.3.1 A Normal Form for Metric II

In this section we look at case *(2)* of Theorem 8.3 and we show that it leads to a single integrable system on $\mathbf{C}^6$ (i.e., without parameters). Let us recall that in this case the Hamiltonian is given by

$$H = \frac{1}{2}\sum_1^6 \lambda_i x_i^2 + \sum_{j=1}^3 \mu_j x_j x_{j+3},$$

it is non-degenerate in the sense that

$$\lambda_{12}\lambda_{23}\lambda_{31}\lambda_{45}\lambda_{56}\lambda_{64}\mu_1\mu_2\mu_3 \neq 0,$$

where $\lambda_{ij} := \lambda_i - \lambda_j$, and the parameters satisfy the following algebraic relations

$$(\mu_1^2, \mu_2^2, \mu_3^2) = \frac{E\bar{E}}{F^2}\left(\frac{(\lambda_{23}-\lambda_{56})^2}{\lambda_{23}\lambda_{56}}, \frac{(\lambda_{31}-\lambda_{64})^2}{\lambda_{31}\lambda_{64}}, \frac{(\lambda_{12}-\lambda_{45})^2}{\lambda_{12}\lambda_{45}}\right) \tag{8.116}$$

with the following sign specification

$$\mu_1\mu_2\mu_3 = \frac{E\bar{E}}{F^3}(\lambda_{12}-\lambda_{45})(\lambda_{23}-\lambda_{56})(\lambda_{31}-\lambda_{64}),$$

where $E := \lambda_{12}\lambda_{23}\lambda_{31}$ and $\bar{E} := \lambda_{45}\lambda_{56}\lambda_{64}$ and $F := \lambda_{46}\lambda_{32} - \lambda_{65}\lambda_{13}$. Roughly speaking, the λ_i may be fixed arbitrarily, and then there are four possibilities for (μ_1, μ_2, μ_3). In a first step, we consider the linear change of variables

$$\begin{aligned} y_i &= e_i(e_i x_i + e_{i+3}x_{i+3}), \\ y_{i+3} &= e_{i+3}(e_{i+3}x_i + e_i x_{i+3}), \end{aligned} \tag{8.117}$$

where

$$\begin{aligned} (e_1^2, e_2^2, e_3^2) &= (\lambda_{12}\lambda_{46}, \lambda_{23}\lambda_{54}, \lambda_{31}\lambda_{65}), \\ (e_4^2, e_5^2, e_6^2) &= (\lambda_{45}\lambda_{13}, \lambda_{56}\lambda_{21}, \lambda_{64}\lambda_{32}), \end{aligned} \tag{8.118}$$

and with the following sign specification

$$\begin{aligned} e_1e_4 &= \mu_1\frac{\lambda_{46}\lambda_{23} - \lambda_{56}\lambda_{13}}{\lambda_{23}-\lambda_{56}}, \\ e_2e_5 &= \mu_2\frac{\lambda_{54}\lambda_{31} - \lambda_{64}\lambda_{21}}{\lambda_{31}-\lambda_{64}}, \\ e_3e_6 &= \mu_3\frac{\lambda_{65}\lambda_{12} - \lambda_{45}\lambda_{32}}{\lambda_{12}-\lambda_{45}}. \end{aligned} \tag{8.119}$$

Indeed, it follows by direct computation from (8.116) that (8.118) and (8.119) are consistent, leaving three signs unspecified, but the choice of these signs is irrelevant for the change of variables (8.117). Note that $e_i^2 - e_{i+3}^2 = F \neq 0$, for $i = 1,2,3$ and that $\prod_{i=1}^3 e_i^2 = \prod_{i=1}^3 e_{i+3}^2 = E\bar{E} \neq 0$, so that the coordinate transformation is invertible. Consider now the four (linearly independent) quadratic polynomials

$$\begin{aligned} P_1 &:= \lambda_{12} y_3^2 - \lambda_{31} y_5^2, \\ P_2 &:= \lambda_{23} y_1^2 - \lambda_{12} y_6^2, \\ P_3 &:= \lambda_{31} y_2^2 - \lambda_{23} y_4^2, \\ P_4 &:= (y_1 - y_4)^2 + (y_2 - y_5)^2 + (y_3 - y_6)^2. \end{aligned} \tag{8.120}$$

A direct computation shows that each of them is a constant of motion of the vector field $\mathcal{X}_H$, which in the x_i coordinates is given by (8.13). In fact, the span of these quadrics contains the two quadrics $\sum_{i=1}^3 x_i^2$ and $\sum_{i=1}^3 x_{i+3}^2$, the Hamiltonian H and an extra, independent, Hamiltonian, thereby proving that we have a Liouville integrable system. The parameters in $P_1, \dots, P_3$ are easily scaled away by taking

$$\begin{aligned} z_1 &= f_1^{-1}\sqrt{\frac{\lambda_{31}}{\lambda_{12}}}\, y_1, & z_2 &= f_2^{-1}\sqrt{\frac{\lambda_{12}}{\lambda_{23}}}\, y_2, & z_3 &= f_3^{-1}\sqrt{\frac{\lambda_{23}}{\lambda_{31}}}\, y_3, \\ z_4 &= f_4^{-1}\sqrt{\frac{\lambda_{12}}{\lambda_{31}}}\, y_4, & z_5 &= f_5^{-1}\sqrt{\frac{\lambda_{23}}{\lambda_{12}}}\, y_5, & z_6 &= f_6^{-1}\sqrt{\frac{\lambda_{31}}{\lambda_{23}}}\, y_6, \end{aligned}$$

where the f_i are arbitrary constants, satisfying

$$f_1^2 = f_6^2, \qquad f_2^2 = f_4^2, \qquad f_3^2 = f_5^2, \tag{8.121}$$

and where the signs of the square roots are irrelevant, but with the understanding that $f_i f_{i+3} z_i z_{i+3} = y_i y_{i+3}$ (the latter fact does not play a rôle for scaling away the parameters in $P_1, \dots, P_3$, but it will be relevant when we deal with P_4). Now plug these into $P_4 + \kappa_1 P_1 + \kappa_2 P_2 + \kappa_3 P_3$, where the κ_i are chosen such that the resulting quadratic polynomial contains (among others) the following three terms: $a_1 z_1^2$, $a_2 z_2^2$ and $a_3 z_3^2$, namely

$$\kappa_i = \frac{a_k \lambda_{jk} + f_k^2 \lambda_{ik}}{f_k^2 \lambda_{ij} \lambda_{ki}}, \tag{8.122}$$

where $i = (1,2,3)$ and (i,j,k) is any cyclic permutation of $(1,2,3)$. Then the full polynomial is given by

$$\begin{aligned} & a_1 z_1^2 - (a_2 + f_2^2) z_4^2 - 2 f_1 f_4 z_1 z_4 \\ + {} & a_2 z_2^2 - (a_3 + f_3^2) z_5^2 - 2 f_2 f_5 z_2 z_5 \\ + {} & a_3 z_3^2 - (a_1 + f_1^2) z_6^2 - 2 f_3 f_6 z_3 z_6. \end{aligned}$$

We see by using (8.121) that this polynomial is of the form $a_1(z_1 - z_4)^2 + a_2(z_2 - z_5)^2 + a_3(z_3 - z_6)^2$ if and only if a_1, a_2, a_3 satisfy

$$\begin{aligned} a_1^2 &= (a_1 + a_2)(a_3 + a_1), \\ a_2^2 &= (a_2 + a_3)(a_1 + a_2), \\ a_3^2 &= (a_3 + a_1)(a_2 + a_3). \end{aligned}$$

Each of these three equations can be written as $\frac{1}{a_1} + \frac{1}{a_2} + \frac{1}{a_3} = 0$, so that its general solution, with $a_1 a_2 a_3 \neq 0$, is given by $a_1 = -\frac{a_2 a_3}{a_2 + a_3}$, with a_2 and a_3 arbitrary, but such that $a_2 a_3 \neq 0$ and $a_2 + a_3 \neq 0$. By choosing κ_2, κ_3 and κ_1 as given by (8.122) we can respectively choose a_1, a_2 and a_3 as we wish. Taking $a_2 = a_3 = -2$ we get $a_1 = 1$, and the constants f_i can be chosen as

$$f_1 = f_2 = f_4 = f_6 = 1, \qquad f_3 = f_5 = -2.$$

The fourth invariant takes then the simple form $(z_1 - z_4)^2 - 2(z_2 - z_5)^2 - 2(z_3 - z_6)^2$, independently of the parameters of the metric! Summarizing, after a linear change of variables[7] four independent quadratic invariants of the geodesic flow are given by

$$\begin{aligned} F_1 &:= x_3^2 - x_5^2, \\ F_2 &:= x_1^2 - x_6^2, \\ F_3 &:= x_2^2 - x_4^2, \\ F_4 &:= (x_1 - x_4)^2 - 2(x_2 - x_5)^2 - 2(x_3 - x_6)^2. \end{aligned} \tag{8.123}$$

As in the case of the Manakov metric the span $\mathbf{F}_\kappa := \sum_{i=1}^4 \kappa_i F_i$ of the four quadrics $F_1, \dots, F_4$ contains a curve of rank three quadrics. The matrix of F_κ in the coordinates $x_1, \dots, x_6$ is given by

$$\begin{pmatrix} \kappa_4 + \kappa_2 & 0 & 0 & -\kappa_4 & 0 & 0 \\ 0 & \kappa_3 - 2\kappa_4 & 0 & 0 & 2\kappa_4 & 0 \\ 0 & 0 & \kappa_1 - 2\kappa_4 & 0 & 0 & 2\kappa_4 \\ -\kappa_4 & 0 & 0 & \kappa_4 - \kappa_3 & 0 & 0 \\ 0 & 2\kappa_4 & 0 & 0 & -2\kappa_4 - \kappa_1 & 0 \\ 0 & 0 & 2\kappa_4 & 0 & 0 & -2\kappa_4 - \kappa_2 \end{pmatrix}.$$

It is clear that this matrix decomposes into three blocks, and by using elementary row and column matrices one sees easily that the determinant of this matrix is given by

$$(\kappa_4(\kappa_3 - \kappa_2) + \kappa_2\kappa_3)(2\kappa_4(\kappa_1 - \kappa_3) - \kappa_1\kappa_3)(2\kappa_4(\kappa_1 - \kappa_2) + \kappa_1\kappa_2).$$

[7] We do a final relabeling $z_i \to x_i$, because our favorite phase variables are called x_i.

The rank of $\mathbf{F}_\kappa$ is smaller than 6 on the union of three quadratic surfaces $C_1 \cup C_2 \cup C_3$. Eliminating κ_4 from their equations we find that they all pass through the rational curve C, given by

$$(\kappa_1 : \ldots : \kappa_4) = (2uv(u-v) : u(u^2-v^2) : v(u^2-v^2) : uv(u+v)), \quad (8.124)$$

where $(u:v) \in \mathbf{P}^1$ is arbitrary. For such an $(u:v)$ the quadric $\mathbf{F}_\kappa$, with κ defined by (8.124) can be written as

$$\mathbf{F}_\kappa = (u+v)(ux_1 - vx_4)^2 - v((u+v)x_2 - 2ux_5)^2 - u(2vx_3 - (u+v)x_6)^2,$$

showing that each point κ on the rational curve C leads to a quadric $\mathbf{F}_\kappa$ of rank three, and thus, F_1, F_2 and F_3 correspond to the distinguished points on C where the rank drops to 2.

We now come to the commuting vector fields. One of them can be computed from the original Hamiltonian, by going through the above linear changes of variables, which is a rather tedious procedure and it yields only one of the vector fields. However, we know that both vector fields must be quadratic (and homogeneous), since the above invariants are quadratic and since the Poisson structure is linear (in the original variables, hence in the new variables). Notice moreover that it follows from $\dot{F}_1 = 0$ (where the dot refers to any vector field on $\mathbf{C}^6$ for which $F_1, \ldots, F_4$ are constant of motion) that $\dot{x}_5$ must be divisible by x_3 and that $\dot{x}_3$ must be divisible by x_5; moreover, the remaining factor must be the same. The same reasoning applies to F_2 and to F_3. This leads to a simple formula for the vector space of all quadratic vector fields on $\mathbf{C}^6$ for which $F_1, \ldots, F_3$ are constants of motion; expressing that F_4 is also a constant of motion leads, by elementary linear algebra, to the two-dimensional space, spanned by the following two vector fields.

$$\begin{aligned}
\dot{x}_1 &= 2x_5x_6, & x_1' &= x_2x_6,\\
\dot{x}_2 &= 2x_3x_4, & x_2' &= x_4(2x_3 - x_6),\\
\dot{x}_3 &= x_5(x_1 + x_4), & x_3' &= x_4x_5,\\
\dot{x}_4 &= 2x_2x_3, & x_4' &= x_2(2x_3 - x_6),\\
\dot{x}_5 &= x_3(x_1 + x_4), & x_5' &= x_3x_4,\\
\dot{x}_6 &= 2x_1x_5, & x_6' &= x_1x_2.
\end{aligned} \quad (8.125)$$

We denote these two vector fields, in that order, by $\mathcal{V}_1$ and $\mathcal{V}_2$. As we pointed out in Paragraph 8.1.2, $\mathcal{V}_1$ and $\mathcal{V}_2$ are independent on the generic fiber of the momentum map.

The Poisson structure can be found by going through the above linear changes of coordinates, but it is much simpler to look for all possible linear Poisson structures on $\mathbf{C}^6$ with respect to which $\mathcal{V}_1$ and $\mathcal{V}_2$ are Hamiltonian. The advantage of this procedure is that, as it turns out, we find many such Poisson structures, which are all compatible, yielding a multi-Hamiltonian structure for this system.

Consider, for $(\alpha,\beta,\gamma)\in\mathbf{C}^3$, the matrix

$$\begin{pmatrix} 0 & \alpha x_6 & -\beta x_5 & 0 & -\beta x_3-2\gamma x_6 & \beta(x_2-2x_5) \\ -\alpha x_6 & 0 & 2\gamma x_4 & \alpha(x_6-2x_3) & 0 & -\alpha x_1-\beta x_4 \\ \beta x_5 & -2\gamma x_4 & 0 & -\alpha x_5-2\gamma x_2 & -\gamma(x_1+x_4) & 0 \\ 0 & \alpha(2x_3-x_6) & \alpha x_5+2\gamma x_2 & 0 & \alpha x_3 & -\beta x_2 \\ \beta x_3+2\gamma x_6 & 0 & \gamma(x_1+x_4) & -\alpha x_3 & 0 & 2\gamma x_1 \\ \beta(2x_5-x_2) & \alpha x_1+\beta x_4 & 0 & \beta x_2 & -2\gamma x_1 & 0 \end{pmatrix}. \tag{8.126}$$

For any $(\alpha,\beta,\gamma)\in\mathbf{C}^3$ it is the Poisson matrix of a Poisson structure $P_{\alpha\beta\gamma}$ on $\mathbf{C}^6$. If $(\alpha,\beta,\gamma)\neq(0,0,0)$ then $P_{\alpha\beta\gamma}$ generates the Hamiltonian vector fields $\mathcal{V}_1$ and $\mathcal{V}_2$ as described in Table 8.1; generators for the algebra of Casimirs of these structures $P_{\alpha\beta\gamma}$ also follow from the table.

Table 8.1. The tri-Hamiltonian structure of the integrable system that corresponds to geodesic flow on $\mathbf{SO}(4)$ for metric II.

	F_1	F_2	F_3	F_4
P_{100}	0	0	$2\mathcal{V}_2$	$2(\mathcal{V}_1-2\mathcal{V}_2)$
P_{010}	0	$2(\mathcal{V}_1-\mathcal{V}_2)$	0	$2(\mathcal{V}_1-2\mathcal{V}_2)$
P_{001}	$2\mathcal{V}_1$	0	0	$4(\mathcal{V}_1-2\mathcal{V}_2)$

We will refer to the integrable system $(\mathbf{C}^6,\{\cdot\,,\cdot\}_{\alpha\beta\gamma},\mathbf{F})$, where $\mathbf{F}:=(F_1,F_2,F_3,F_4)$, as the integrable system that corresponds to geodesic flow on $\mathbf{SO}(4)$ for *metric II.*

8.3.2 Algebraic Complete Integrability

Let us first point out a few features of this integrable system. The group of involutions on $\mathbf{C}^6$, generated by

$$\begin{aligned} \sigma_1(x_1,\dots,x_6)&=(x_1,-x_2,-x_3,x_4,-x_5,-x_6),\\ \sigma_2(x_1,\dots,x_6)&=(-x_1,x_2,-x_3,-x_4,x_5,-x_6), \end{aligned} \tag{8.127}$$

leaves the vector fields $\mathcal{V}_1$ and $\mathcal{V}_2$ invariant, as well as all fibers $\mathbf{F_c}$ of the momentum map. We also consider the involution τ on $\mathbf{C}^6$, which is defined by

$$\tau(x_1,\dots,x_6)=(-x_1,-x_2,-x_3,-x_4,-x_5,-x_6). \tag{8.128}$$

It has the property of reversing the sign of both vector fields, still leaving all fibers $\mathbf{F_c}$ of the momentum map invariant. Of course, the composition of τ with any of σ_1 or σ_2 (or both) has the same properties. Another type of involution is given by

$$\pi(x_1, \ldots, x_6) = (x_4, x_6, x_5, x_1, x_3, x_2).$$

It does not leave the generic fiber $\mathbf{F_c}$ of the momentum map invariant, because it acts on the constants of motion in the following way:

$$\pi(F_1, F_2, F_3, F_4) = (-F_1, -F_3, -F_2, F_4).$$

It leaves only the vector field $\mathcal{V}_1$ invariant, because

$$\pi_*\mathcal{V}_1 = \mathcal{V}_1, \qquad \pi_*\mathcal{V}_2 = \mathcal{V}_1 - \mathcal{V}_2.$$

Let us now look for the weight homogeneous Laurent solutions of the vector field $\mathcal{V}_1$. Since the weights of all variables are equal to 1, the indicial equation is given by

$$\begin{aligned}
0 &= x_1^{(0)} + 2x_5^{(0)}x_6^{(0)},\\
0 &= x_2^{(0)} + 2x_3^{(0)}x_4^{(0)},\\
0 &= x_3^{(0)} + x_5^{(0)}(x_1^{(0)} + x_4^{(0)}),\\
0 &= x_4^{(0)} + 2x_2^{(0)}x_3^{(0)},\\
0 &= x_5^{(0)} + x_3^{(0)}(x_1^{(0)} + x_4^{(0)}),\\
0 &= x_6^{(0)} + 2x_1^{(0)}x_5^{(0)}.
\end{aligned}$$

It follows that the indicial locus is given by four lines (rational curves) $\mathcal{I}^{\epsilon_1\epsilon_2}$, where $\epsilon_1^2 = \epsilon_2^2 = 1$, which are given in parametric form by

$$(x_1^{(0)}, \ldots, x_6^{(0)}) = \left((a-1)\epsilon_2,\, a\epsilon_1,\, \frac{\epsilon_1\epsilon_2}{2},\, -a\epsilon_2,\, \frac{\epsilon_1}{2},\, (1-a)\epsilon_1\epsilon_2\right),$$

where $a \in \mathbf{C}$ is the parameter. If we denote the Kowalevski matrix that corresponds to the point with parameter a on the line $\mathcal{I}^{\epsilon_1\epsilon_2}$ by $\mathcal{K}_{\epsilon_1\epsilon_2}(a)$, then we have that

$$\mathcal{K}_{\epsilon_1\epsilon_2}(a) = \begin{pmatrix}
1 & 0 & 0 & 0 & 2(1-a)\epsilon_1\epsilon_2 & \epsilon_1\\
0 & 1 & -2a\epsilon_2 & \epsilon_1\epsilon_2 & 0 & 0\\
\frac{1}{2}\epsilon_1 & 0 & 1 & \frac{1}{2}\epsilon_1 & -\epsilon_2 & 0\\
0 & \epsilon_1\epsilon_2 & 2a\epsilon_1 & 1 & 0 & 0\\
\frac{1}{2}\epsilon_1\epsilon_2 & 0 & -\epsilon_2 & \frac{1}{2}\epsilon_1\epsilon_2 & 1 & 0\\
\epsilon_1 & 0 & 0 & 0 & 2(a-1)\epsilon_2 & 1
\end{pmatrix}$$

Its characteristic polynomial is given by

$$|k\,\mathrm{Id}_6 - \mathcal{K}_{\epsilon_1\epsilon_2}(a)| = k(k^2-1)(k-2)^3,$$

independently of ϵ_1, ϵ_2 and a.

We now establish the existence of principal balances. Indeed, as in the case of the Manakov metric, we need to show that these balances do exist. Since $\mathcal{K}_{\epsilon_1\epsilon_2}(a)$ has 5 non-negative integer eigenvalues, the largest of which is equal to 2, it suffices to compute the first three terms of the series and to verify that it depends indeed on 5 free parameters, to know that they are the first terms of a principal balance. It is easily checked that the following Laurent polynomials

$$\begin{aligned}
x_1(t) &= \frac{(a-1)\epsilon_2}{t}\left(1 - bt + (b^2-d-e)t^2 + O(t^3)\right),\\
x_2(t) &= \frac{\epsilon_1}{t}\left(a - abt + ((a-1)(ae-c-ab^2)+a^2d)t^2 + O(t^3)\right),\\
x_3(t) &= \frac{\epsilon_1\epsilon_2}{2t}\left(1 + bt - ((a-1)e + ad - c - ab^2)t^2 + O(t^3)\right),\\
x_4(t) &= \frac{\epsilon_2}{t}\left(-a + abt + ct^2 + O(t^3)\right),\\
x_5(t) &= \frac{\epsilon_1}{2t}\left(1 + bt + dt^2 + O(t^3)\right),\\
x_6(t) &= \frac{(a-1)\epsilon_1\epsilon_2}{t}\left(-1 + bt - et^2 + O(t^3)\right),
\end{aligned} \tag{8.129}$$

which depend on the 5 free parameters $a, \dots, e$, satisfy the differential equations which describe $\mathcal{V}_1$, up to higher order terms, hence lead to four principal balances (indexed by (ϵ_1, ϵ_2)), where $\epsilon_1^2 = \epsilon_2^2 = 1$; when we need to be more precise about the balance we will write it as $x(t; \mathcal{I}^{\epsilon_1\epsilon_2})$. The next terms are uniquely (linearly) determined by the given ones, so they can easily be computed, when needed (in essentially all computations that follow the above terms suffice). Notice that the Painlevé building consists in this case of 4 Painlevé walls, each of which isomorphic to $\mathbf{C}^5$.

We now look for independent homogeneous polynomials which have a simple pole at most when any of the principal balances is substituted in them. In the notation of Paragraph 7.7.1, we look for a basis of $\mathcal{Z}_\rho$ as an $\mathcal{H}$-module, where the pole vector ρ is chosen as $\rho = (1,1,1,1)$. The result is summarized in Table 8.2, which suggests that a basis of this space of polynomials is given by the last column in the table. These polynomials are defined by $z_i := x_i$, for $i = 1, \dots, 6$ and for $i = 7, \dots, 15$ they are given, together with their polar part as follows.

$$\begin{aligned}
z_7 &= x_5(2x_3 - x_6) - x_2x_3 \sim \frac{\epsilon_2 b}{t},\\
z_8 &= x_1(2x_3 - x_6) - x_4x_6 \sim \frac{2\epsilon_1 b(a-1)}{t},
\end{aligned}$$

Table 8.2. The polynomials of degree at most 4 which have a simple pole at most when any of the principal balances is substituted in them. In order to verify that $\dim \mathcal{Z}_\rho^4 = 30$, as is asserted in the last line, one extra term in (8.129) needs to be computed.

k	$\dim \mathcal{F}^k$	$\dim \mathcal{H}^k$	$\dim \mathcal{Z}_\rho^k$	# dep	ζ_k	indep. functions
0	1	1	1	0	1	$z_0 = 1$
1	6	0	6	0	6	$z_1, \dots, z_6$
2	21	4	9	4	5	$z_7, \dots, z_{11}$
3	56	0	28	24	4	$z_{12}, \dots, z_{15}$
4	126	10	30	30	0	—

$$
\begin{aligned}
z_9 &= x_4(2x_5 - x_2) - x_1x_2 \sim -\frac{2\epsilon_1\epsilon_2 ab}{t},\\
z_{10} &= (2x_5 - x_2)^2 - x_6^2 \sim \frac{4b(1-a)}{t},\\
z_{11} &= (2x_3 - x_6)^2 - x_2^2 \sim \frac{4ab}{t},\\
z_{12} &= x_1x_2x_3 - x_4x_5x_6 \sim \frac{(a-1)(a(b^2-d-e)+c)}{t},\\
z_{13} &= x_2x_3x_6 - x_1x_4x_5 \sim \frac{\epsilon_1(1-a)(ae+c)}{t},\\
z_{14} &= x_2x_5x_6 - x_1x_3x_4 \sim \frac{\epsilon_1\epsilon_2(a-1)(a^2(b^2-d-e)+a(c-d)-c)}{t},\\
z_{15} &= x_1x_2x_5 - x_3x_4x_6 \sim \frac{\epsilon_2(a-1)((1-a)(a(b^2-d-e)+c)+a(d-e))}{t}.
\end{aligned}
\tag{8.130}
$$

The action of the involutions σ_1 and σ_2 on these functions is given by

$$
\begin{aligned}
\sigma_1(z_7, \dots, z_{15}) &= (z_7, -z_8, , -z_9, z_{10}, z_{11}, z_{12}, -z_{13}, -z_{14}, z_{15}),\\
\sigma_2(z_7, \dots, z_{15}) &= (-z_7, z_8, , -z_9, z_{10}, z_{11}, z_{12}, z_{13}, -z_{14}, -z_{15}),
\end{aligned}
\tag{8.131}
$$

as follows from (8.127). For generic $\mathbf{c} = (c_1, \dots, c_4)$ we consider the following map,

$$
\begin{aligned}
\varphi_{\mathbf{c}} : \quad & \mathbf{F_c} \to \mathbf{P}^7\\
& (x_1, \dots, x_6) \mapsto (1 : z_1 : \cdots : z_{15}),
\end{aligned}
$$

which is an isomorphic embedding of $\mathbf{F_c}$ into projective space. We use it to compute an equation for each of the irreducible components of the abstract Painlevé divisor, which is done by substituting (8.129) in the equations $F_1 = c_1, \dots, F_4 = c_4$, where the F_i are the constants of motion, defined in (8.123) and $(c_1, \dots, c_4) \in \mathbf{C}^4$ is generic.

The first three equations yield

$$\begin{aligned}
&a(b^2-d-e)+c+e-d=2c_1,\\
&2(a-1)^2(b^2-d-2e)=c_2,\\
&-2a(a^2(b^2-d-e)+a(c+e-b^2)-2c)=c_3.
\end{aligned}$$

These equations are linear in the parameters c, d and e, and we can solve linearly for these in terms of the values of the constants of motion, giving

$$\begin{aligned}
c&=-\frac{ab^2}{3}+\frac{a(a+1)}{3}c_1+\frac{a}{12(a-1)}c_2+\frac{a+3}{12a}c_3,\\
d&=\frac{b^2}{3}+\frac{2(a-2)}{3}c_1+\frac{c_2}{6(a-1)}+\frac{c_3}{6a},\\
e&=\frac{b^2}{3}-\frac{a-2}{3}c_1-\frac{a+2}{12(a-1)^2}c_2-\frac{c_3}{12a}.
\end{aligned}$$

This is not surprising: they appear at step 2 and we already pointed out that we have three quadratic constants of motion, which are independent when evaluated at an arbitrary point of the indicial locus; Proposition 7.34 therefore guarantees that we can express these three parameters linearly in terms of the constants of motion. The fourth equation then reduces, for any $(c_1,\dots,c_4)\in\mathbf{C}^4$, to the following equation of an affine curve in $\mathbf{C}^2$,

$$4b^2=\frac{c_3}{a}+\frac{c_2}{a-1}-4ac_1+2c_1+2c_2-2c_3-c_4, \tag{8.132}$$

which is a hyperelliptic curve of genus 2. Notice that the curve is independent of (ϵ_1,ϵ_2); therefore we simply denote it by $\Gamma_{\mathbf{c}}$. In order to compactify $\Gamma_{\mathbf{c}}$, for generic c, into a compact Riemann surface, denoted $\overline{\Gamma_{\mathbf{c}}}$, we need to add three points to it, corresponding to $a=0$, $a=1$ and $a=\infty$; we denote these points of $\overline{\Gamma_{\mathbf{c}}}$ by $\mathbf{0}$, $\mathbf{1}$ and ∞. Notice that each of these three points is a Weierstrass point of $\overline{\Gamma_{\mathbf{c}}}$. A local parameter ς at each of these point is given as follows:

$$\begin{aligned}
&\mathbf{0}: && a=\varsigma^2, && b=\frac{\sqrt{c_3}}{2\varsigma}\left(1+\frac{2c_1+c_2-2c_3-c_4}{2c_3}\varsigma^2+O(\varsigma^3)\right),\\
&\mathbf{1}: && a=1+\varsigma^2, && b=\frac{\sqrt{c_2}}{2\varsigma}\left(1-\frac{2c_1-2c_2+c_3+c_4}{2c_2}\varsigma^2+O(\varsigma^3)\right),\\
&\infty: && a=\varsigma^{-2}, && b=\frac{\sqrt{-c_1}}{\varsigma}\left(1-\frac{2c_1+2c_2-2c_3-c_4}{8c_1}\varsigma^2+O(\varsigma^3)\right).
\end{aligned} \tag{8.133}$$

For fixed generic $\mathbf{c}$ the series $z_i(t;\mathcal{I}^{\epsilon_1\epsilon_2})$ restrict to series that we denote by $z_i(t;\Gamma_{\mathbf{c}}^{\epsilon_1\epsilon_2})$. The residues of these series lead, for fixed $\epsilon_1=\pm1$ and $\epsilon_2=\pm1$, to a map, which we will denote by $\varphi_{\mathbf{c}}^{\epsilon_1\epsilon_2}$; the projective closure of the image of this map will be denoted by $\mathcal{D}_{\mathbf{c}}^{\epsilon_1\epsilon_2}$, where we write $\mathcal{D}_{\mathbf{c}}^{+-}$ for $\mathcal{D}_{\mathbf{c}}^{1,-1}$ and so on.

This gives us four divisors on $\overline{\varphi_{\mathbf{c}}(\mathbf{F_c})}$, and verifying the adjunction formula (7.50) means that we must check that the curve, formed by these divisors, is a (singular) curve of genus $15 + 2 = 17$. A direct substitution of these parameterizations[8] in the residues, yields, up to terms of order ς^2, and up to a rescaling of ς, the following:

$$
\begin{aligned}
\mathbf{0} &: (0 : \cdots : 0 : \varsigma : -2\epsilon_1\epsilon_2\varsigma : 0 : 4\epsilon_2\varsigma : 0 : 1 : -\epsilon_1 : -\epsilon_1\epsilon_2 : \epsilon_2),\\
\mathbf{1} &: (0 : \cdots : 0 : \epsilon_2\varsigma : 0 : -2\epsilon_1\epsilon_2\varsigma : 0 : 4\varsigma : 1 : \epsilon_1 : \epsilon_1\epsilon_2 : \epsilon_2),\\
\infty &: (0 : \cdots : 0 : \epsilon_2\varsigma : -\varsigma : -2\epsilon_1\epsilon_2\varsigma : 2\epsilon_1\epsilon_2\varsigma : 1 : -\epsilon_1 : \epsilon_1\epsilon_2 : -\epsilon_2).
\end{aligned}
\tag{8.134}
$$

It follows that these three points are mapped to the following four points in $\mathbf{P}^{15}$:

$$
\begin{aligned}
P^{+-} &:= (0 : \cdots : 0 : 1 : 1 : 1 : 1),\\
P^{-+} &:= (0 : \cdots : 0 : 1 : -1 : 1 : -1),\\
P^{--} &:= (0 : \cdots : 0 : 1 : -1 : -1 : 1),\\
P^{++} &:= (0 : \cdots : 0 : 1 : 1 : -1 : -1).
\end{aligned}
\tag{8.135}
$$

Notice that these points are independent of the constants of motion $\mathbf{c}$. The involutions σ_1 and σ_2 permute the points $P^{\epsilon_1\epsilon_2}$ and the divisors $\mathcal{D}_{\mathbf{c}}^{\epsilon_1\epsilon_2}$ in the following way, as follows from (8.131).

$$
\begin{array}{ccc}
P^{++} & \overset{\sigma_1}{\longleftrightarrow} & P^{-+}\\
\sigma_2 \updownarrow & & \updownarrow \sigma_2\\
P^{+-} & \underset{\sigma_1}{\longleftrightarrow} & P^{--}
\end{array}
\qquad\qquad
\begin{array}{ccc}
\mathcal{D}_{\mathbf{c}}^{++} & \overset{\sigma_1}{\longleftrightarrow} & \mathcal{D}_{\mathbf{c}}^{-+}\\
\sigma_2 \updownarrow & & \updownarrow \sigma_2\\
\mathcal{D}_{\mathbf{c}}^{+-} & \underset{\sigma_1}{\longleftrightarrow} & \mathcal{D}_{\mathbf{c}}^{--}
\end{array}
$$

The precise correspondence between the points $\mathbf{0}$, $\mathbf{1}$ and ∞ on $\overline{\Gamma_{\mathbf{c}}}$ and the four points $P^{\epsilon_1\epsilon_2}$ under the different embeddings is given in Table 8.3.

Looking at the terms in ς in (8.134) we see that all tangents at the intersection points are different, so that the Painlevé divisor has four ordinary triple points, the points $P^{\epsilon_1\epsilon_2}$, and has no other singularities. Thus, we find the intersection pattern, given in Figure 8.2. Since the Euler characteristic of an ordinary triple point is 3 and since the genus of each of the four curves equals 2, Proposition 5.42 gives $g = 2 \times 4 + 1 - 4 + 12 = 17$, as required by the adjunction formula.

Since the adjunction formula is satisfied when considering the functions $z_0, \ldots, z_{15}$ we now verify that $(\varphi_{\mathbf{c}})_* \mathcal{V}_1$ extends to a holomorphic vector field $\overline{\mathcal{V}_1}$ in $\mathbf{P}^{15}$. It is easy to very that the functions z_i themselves satisfy the following quadratic equations.

[8] The given terms are sufficient to do the computation.

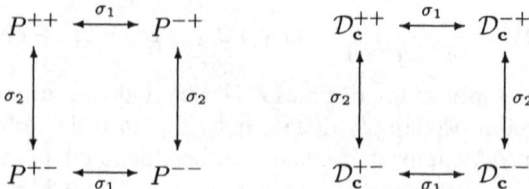

Table 8.3. By the four embeddings $\varphi_{\mathbf{c}}^{\epsilon_1\epsilon_2}$ of $\Gamma_{\mathbf{c}}$ into $\mathbf{P}^{15}$ the three points at infinity of $\Gamma_{\mathbf{c}}$ are mapped to four different points, making each of the image points an intersection point of three copies of $\overline{\Gamma_{\mathbf{c}}}$. Our labeling of these points $P^{\epsilon_1\epsilon_2}$ and these image curves $\mathcal{D}_{\mathbf{c}}^{\epsilon_1\epsilon_2} := \overline{\varphi_{\mathbf{c}}^{\epsilon_1\epsilon_2}(\Gamma_{\mathbf{c}})}$ is such that $\mathcal{D}_{\mathbf{c}}^{\epsilon_1\epsilon_2}$ does not contain $P^{\epsilon_1\epsilon_2}$, but contains all other triple points of $\mathcal{D}_{\mathbf{c}}$. See also Figure 8.2.

	$\mathcal{D}_{\mathbf{c}}^{++}$	$\mathcal{D}_{\mathbf{c}}^{--}$	$\mathcal{D}_{\mathbf{c}}^{-+}$	$\mathcal{D}_{\mathbf{c}}^{+-}$
0	P^{--}	P^{++}	P^{+-}	P^{-+}
1	P^{+-}	P^{-+}	P^{--}	P^{++}
∞	P^{-+}	P^{+-}	P^{++}	P^{--}

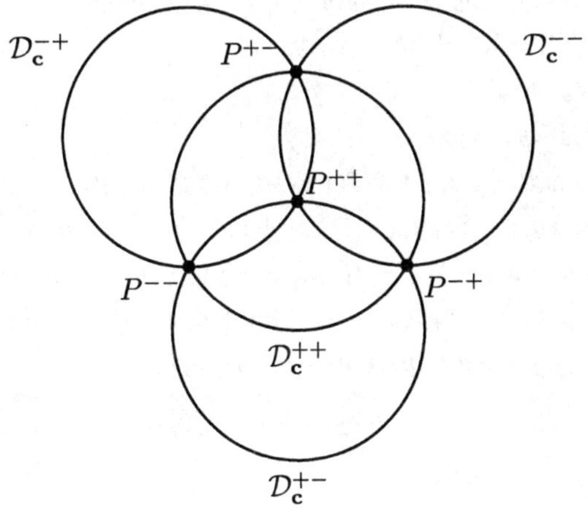

Fig. 8.2. The Painlevé divisor $\mathcal{D}_{\mathbf{c}}$ consists of four genus two curves $\mathcal{D}_{\mathbf{c}}^{\epsilon_1\epsilon_2}$ that intersect in four points, each of which is an ordinary triple point of $\mathcal{D}_{\mathbf{c}}$, the three branches coming from three different curves. The divisor contains all 16 half periods of $\mathrm{Jac}(\overline{\Gamma_{\mathbf{c}}})$: four of them are the singular points $P^{\epsilon_1\epsilon_2}$, while each curve contains three other half periods, that are its other Weierstrass points. It follows from Proposition 5.42 that $g_a(\mathcal{D}_{\mathbf{c}}) = 17$.

$$
\begin{aligned}
\dot z_1 &= 2z_5z_6, & \dot z_9 &= 2z_2z_7,\\
\dot z_2 &= 2z_3z_4, & \dot z_{10} &= 4z_1z_7,\\
\dot z_3 &= z_5(z_1+z_4), & \dot z_{11} &= 4z_4z_7,\\
\dot z_4 &= 2z_2z_3, & \dot z_{12} &= (z_1+z_4)z_{15} + 2c_1z_1z_4,\\
\dot z_5 &= z_3(z_1+z_4), & \dot z_{13} &= (z_1+z_4)z_{14} + 2c_1z_4z_6,\\
\dot z_6 &= 2z_1z_5, & \dot z_{14} &= (z_1+z_4)z_{13} - 2c_1z_1z_2,\\
\dot z_7 &= z_5z_9 + z_3z_8, & \dot z_{15} &= (z_1+z_4)z_{12} - 2c_1z_2z_6,\\
\dot z_8 &= 2z_6z_7,
\end{aligned}
$$

This establishes the fact that $(\varphi_{\mathbf{c}})_* \mathcal{V}_1$ extends to a quadratic vector field in the chart $Z_0 \neq 0$ in $\mathbf{P}^{15}$. In order to show that $(\varphi_{\mathbf{c}})_* \mathcal{V}_1$ also extends to a quadratic vector field in the chart $Z_1 \neq 0$ in $\mathbf{P}^{15}$ we define $y_i := z_i/z_1$, for $i = 0, \ldots, 15$. In terms of the y_i it is easy to verify the following equations.

$$
\begin{aligned}
\dot{y}_0 &= -2y_5y_6,\\
\dot{y}_1 &= 0,\\
\dot{y}_2 &= -2y_{14}y_0,\\
\dot{y}_3 &= y_0(c_2y_5 - y_{13}) - y_6y_7,\\
\dot{y}_4 &= 2y_0y_{12},\\
\dot{y}_5 &= y_7(2y_5 - y_2) - y_3y_{10} - y_0y_{14} + c_2y_0y_3,\\
\dot{y}_6 &= 2c_2y_0y_5,\\
\dot{y}_7 &= 2c_2(2y_3^2 - y_5^2) - 2y_5y_{13} - 2c_1 - (1 + y_4)y_{15}\\
&\quad +(2c_1 - 2c_2 + 2c_3 + c_4)y_3y_6,\\
\dot{y}_8 &= -2y_6y_{12},\\
\dot{y}_9 &= 2y_{12}(2y_5 - y_2),\\
\dot{y}_{10} &= 8c_2y_3y_5 - 4y_{12} + 2(2c_1 - 3c_2 + c_3 + c_4)y_5y_6,\\
\dot{y}_{11} &= -4y_4y_{12} - 8y_5y_{14} - 2(2c_1 + c_2 - 3c_3 - c_4)y_5y_6,\\
\dot{y}_{12} &= (2c_2 - 6c_1 - 2c_3 - c_4)y_0y_{15} - y_9y_{13} + (4c_1 + c_3)y_3y_8\\
&\quad -2y_7y_{12} - 2c_1y_6y_8 - 2c_1(2c_1 + c_2 - c_4)y_0 + 6c_1c_2y_0y_4,\\
\dot{y}_{13} &= c_2(y_0y_{14} + c_3y_0y_3 + y_2y_7) - y_8y_{15},\\
\dot{y}_{14} &= \frac{1}{2}c_2y_5(y_{11} - (2c_1 + c_2 - 3c_3 - c_4)y_0) - y_8y_{12},\\
\dot{y}_{15} &= -y_8y_{13} - c_2y_3y_9.
\end{aligned}
$$

In view of Lemma 7.58 this proves that $(\varphi_{\mathbf{c}})_* \mathcal{V}_1$ extends to a holomorphic vector field $\overline{\mathcal{V}}_1$ on $\mathbf{P}^{15}$.

Remark 8.17. Though the 16 functions $z_0, \ldots, z_{15}$ are simple, being very symmetric, it is sometimes more convenient to consider the subspace of their span, which consists of the functions which do not have a pole when one of the principal balances is substituted in them. The idea is that, if this integrable system is indeed an a.c.i. system (as we will see) then the embedding which is given by the above 16 functions with a simple pole along the 4 Painlevé divisors, can be replaced by the lower-dimensional embedding which is given by a basis for the functions with a simple pole along 3 of the Painlevé divisors. Indeed, according to Lefschetz's Theorem, the third power of any ample line bundel on an Abelian variety is always very ample (see Theorem 5.17). One such a basis is easily read off from (8.129) and (8.130), by looking for 9 independent polynomials that have no residue for $\epsilon_1 = -\epsilon_2 = 1$.

One possible basis is given by the following.

$$
\begin{aligned}
z'_0 &= 1,\\
z'_1 &= z_2 - z_4 = x_2 - x_4,\\
z'_2 &= z_3 + z_5 = x_3 + x_5,\\
z'_3 &= z_1 + z_6 = x_1 + x_6,\\
z'_4 &= z_1 + z_2 + 2z_3 = x_1 + x_2 + 2x_3,\\
z'_5 &= 2z_8 + z_{10} = (2x_5 - x_2)^2 - 2x_6(x_1 + x_4) + 4x_1x_3 - x_6^2,\\
z'_6 &= 2z_9 - z_{11} = -(2x_3 - x_6)^2 - 2x_2(x_1 + x_4) + 4x_4x_5 + x_2^2,\\
z'_7 &= 2z_7 - z_8 + z_9\\
&= x_1x_6 - (2x_3 + x_2)x_1 - (2x_5 - x_4)x_6 + (2x_3 + x_4)(2x_5 - x_2),\\
z'_8 &= z_{12} + z_{13} + z_{14} + z_{15} = (x_2 - x_4)(x_3 + x_5)(x_1 + x_6).
\end{aligned}
$$

It can be verified as above that, for generic constant of the motion, the residues of these functions map the three curves onto three curves in $\mathbf{P}^8$, which all meet in an ordinary triple point and which pairwise intersect in another point, which is an ordinary double point (the picture is Figure 8.2 with the divisor $\mathcal{D}_{\mathbf{c}}^{+-}$ removed). Since each double point has Euler characteristic 2, while the triple point has Euler characteristic 3, Proposition 5.42 gives $g = 3 \times 2 + 1 - 3 + 3 + 3 = 10$, which is again in agreement with the adjunction formula. It can be verified that in these variables the equations of $\mathcal{V}_1$ are also quadratic, but they are more complicated (and less symmetric) than the above ones for the functions z_i, which explains why we have given preference to prove algebraic integrability by using the larger set of variables.

We now check, following Paragraph 7.7.3, that we flow into the affine immediately when starting from any point in $\overline{\varphi_{\mathbf{c}}(\mathbf{F_c})} \setminus \varphi_{\mathbf{c}}(\mathbf{F_c})$. Notice that the four embeddings $\varphi_{\mathbf{c}}^{\epsilon_1\epsilon_2}$ map the point $\mathbf{0} \in \overline{\Gamma}_{\mathbf{c}}$ to the four points $P^{\epsilon_1\epsilon_2}$. Therefore, we first check that, in a neighborhood of $\mathbf{0}$, the series $1/z_{15}(t; \Gamma_{\mathbf{c}}^{\epsilon_1\epsilon_2})$ admits, for generic $\mathbf{c}$, a non-zero limit as $\varsigma \to 0$, where the parameterization of a neighborhood of $\mathbf{0}$ is given by (8.133); the choice of z_{15} is based on the fact that $z_{15} \neq 0$ defines a chart in a neighborhood of $\mathbf{0}$, as can be read off from (8.135). For this, a few more terms in the principal balance need to be computed (one needs the terms up to t^5). The result is that

$$
\lim_{\varsigma \to 0} \frac{1}{z_{15}}(t; \Gamma_{\mathbf{c}}^{\epsilon_1\epsilon_2}) = \frac{\epsilon_2 c_1}{3} t^5 + O(t^6),
$$

and so the series is not identically zero and we go into the affine immediately, when starting from any of the points $P^{\epsilon_1\epsilon_2}$.

Finally we need to check that there are no other divisors in $\overline{\varphi_{\mathbf{c}}(\mathbf{F_c})} \setminus \varphi_{\mathbf{c}}(\mathbf{F_c})$. By substituting (8.133), we have that

$$
\frac{1}{z_{15}}(t; \Gamma_{\mathbf{c}}^{\epsilon_1\epsilon_2}) = -\frac{2\epsilon_2}{c_3}(2\varsigma - \sqrt{c_3}t)\varsigma t + \cdots,
$$

where the dots stand for terms of total degree at least 4 in ς and t.

It follows that the multiplicity of $1/z_{15}$ at each of the points $P^{\epsilon_1\epsilon_2}$ equals three, which coincides with the sum of the order of zero of $1/z_{15}$ on each of the three intersecting branches, giving $1+1+1=3$. Notice that the flow is transversal at any $P^{\epsilon_1\epsilon_2}$ to one of the divisors, so the function $1/z_{15}(t,s)$ is expanded in holomorphic coordinates and hence its multiplicity is meaningful. Thus, we have verified the conditions of the Complex Liouville Theorem (Theorem 6.22) and we may conclude that $\overline{\varphi_{\mathbf{c}}(\mathbf{F_c})}$ is an Abelian surface, for generic $\mathbf{c}$, and that the vector fields $\overline{\mathcal{V}}_1$ and $\overline{\mathcal{V}}_2$ restrict to linear vector fields on these tori. Our conclusions are summarized in the following theorem.

Theorem 8.18. *Let $(\mathbf{C}^6, \{\cdot\,,\cdot\}, \mathbf{F})$ be the integrable system which provides a normal form for the geodesic flow on $\mathfrak{so}(4)$ with respect to metric II, where $F = (F_1, F_2, F_3, F_4)$ is given by (8.123) and $\{\cdot\,,\cdot\}$ is any of the Poisson structures (8.126), with multi-Hamiltonian vector fields (8.125).*

(1) $(\mathbf{C}^6, \{\cdot\,,\cdot\}, \mathbf{F})$ is a weight homogeneous a.c.i. system;
(2) For generic $\mathbf{c}$ the fiber $\mathbf{F_c}$ of its momentum map completes into a Jacobi surface $\mathrm{Jac}(\overline{\Gamma_{\mathbf{c}}})$ by adding a singular divisor $\mathcal{D}_{\mathbf{c}}$;
(3) $\mathcal{D}_{\mathbf{c}}$ consists of 4 copies of the genus two curve $\overline{\Gamma_{\mathbf{c}}}$ that intersect according to the pattern, indicated in Figure 8.2;
(4) The line bundle $[\mathcal{D}_{\mathbf{c}}]$ defines a polarization of type $(4,4)$ on $\mathbf{T}^2_{\mathbf{c}}$ and leads to an embedding of $\mathrm{Jac}(\overline{\Gamma_{\mathbf{c}}})$ in $\mathbf{P}^{15}$;
(5) The involutions σ_1 and σ_2 define on $\mathrm{Jac}(\overline{\Gamma_{\mathbf{c}}})$ two translations, over a half period, that permute the components of $\mathcal{D}_{\mathbf{c}}$.

The fact that the Abelian surface $\overline{\varphi_{\mathbf{c}}(\mathbf{F_c})}$ is a Jacobi surface follows simply from the fact that this surface contains a smooth genus two curve, hence it is its Jacobian. Notice also that the divisor $\mathcal{D}_{\mathbf{c}}$ contains all sixteen fixed points of the time-involution on $\mathrm{Jac}(\overline{\Gamma_{\mathbf{c}}})$ that is induced by τ (see (8.128)), and that these points are precisely the 16 Weierstrass points of the four curves that make up $\mathcal{D}_{\mathbf{c}}$. Indeed, τ induces on the parameters (a,b) that appear in the principal balances the action $(a,b) \mapsto (a,-b)$, as is read off from (8.129), which amounts to the hyperelliptic involution on each of the curves.

8.3.3 A Lax Equation for Metric II

In this section we give a Lax equation for the metric II. To do this, we first construct, following [38], an explicit map from the generic fiber $\mathbf{F_c}$ of the momentum map to the Jacobian of the Riemann surface $\overline{\Gamma_{\mathbf{c}}}$.

Let $\mathbf{c}$ be generic, such that the affine curve $\Gamma_{\mathbf{c}}$, given by (8.132) is a non-singular curve of genus 2; in particular we assume that $c_1c_2c_3 \neq 0$. We consider the functions which have at worst a double pole along one of the components of the divisor $\mathcal{D}_{\mathbf{c}}$ on $\mathrm{Jac}(\overline{\Gamma_{\mathbf{c}}})$, and no other poles; say that the component that we pick is $\mathcal{D}_{\mathbf{c}}^{-+}$. These functions are obtained by constructing those polynomials on $\mathbf{C}^6$ which have at worst a double pole in t when $x(t;\Gamma_{\mathbf{c}}^{1,-1})$ is substituted into them, and no poles when the other principal balances are substituted.

From (8.129) we easily find the following basis,

$$\begin{aligned} \theta_0 &:= 1, \\ \theta_1 &:= (x_2 + x_4)(x_3 + x_5), \\ \theta_2 &:= (x_3 + x_5)(x_1 + x_6), \\ \theta_3 &:= (x_1 + x_6)(x_2 + x_4), \end{aligned} \tag{8.136}$$

where we think of these polynomials as being restricted to $\mathbf{F_c}$. Consider now the Kodaira map, corresponding to these functions,

$$\begin{aligned} \varphi_{\mathbf{c}} : \ \operatorname{Jac}(\overline{\Gamma_{\mathbf{c}}}) \setminus \mathcal{D}_{\mathbf{c}}^{-+} &\to \mathbf{P}^3 \\ P = (x_1, \dots, x_6) &\mapsto (\theta_0(P) : \theta_1(P) : \theta_2(P) : \theta_3(P)). \end{aligned}$$

Since the functions θ_i correspond to the sections of $[2\mathcal{D}_{\mathbf{c}}^{-+}]$, which defines twice the principal polarization on $\operatorname{Jac}(\overline{\Gamma_{\mathbf{c}}})$, the map $\varphi_{\mathbf{c}}$ maps the surface $\operatorname{Jac}(\overline{\Gamma_{\mathbf{c}}})$ to its Kummer surface, which is a singular quartic in $\mathbf{P}^3$ (see Paragraph 5.2.3). An equation for this quartic surface can be computed by eliminating the variables $x_1, \dots, x_6$ from the equations (8.123) and (8.136). For doing this it is useful to first rewrite these equations using the six coordinates $x_2 \pm x_4$, $x_3 \pm x_5$ and $x_1 \pm x_6$: solving the equations (8.136) and the first three equations in (8.123) for these variables and substituting these values in the remaining equation, the equation for the Kummer surface of $\operatorname{Jac}(\overline{\Gamma_{\mathbf{c}}})$ can be written in the form

$$\theta_3^2((\theta_1 + \theta_2 - 2c_1)^2 + 8c_1\theta_1) + 2f_3(\theta_1, \theta_2)\theta_3 + f_4(\theta_1, \theta_2) = 0, \tag{8.137}$$

where

$$\begin{aligned} f_3 &= 2(\theta_1 c_2 + \theta_2 c_3)c_1 + (\theta_1 + \theta_2)(\theta_1 c_2 - \theta_2 c_3 - 2\theta_1\theta_2) + 2\theta_1\theta_2(2c_1 - c_4), \\ f_4 &= (\theta_1 c_2 - \theta_2 c_3 + 2\theta_1\theta_2)^2. \end{aligned}$$

Looking at the leading term in (8.137) (see [169, Section V.4]) a system of linearizing variables (ξ_1, ξ_2) is given by the equations

$$-2c_1(\xi_1 + \xi_2) = \theta_1 + \theta_2 - 2c_1, \quad -2c_1\xi_1\xi_2 = \theta_1. \tag{8.138}$$

The fact that this provides linearizing variables is checked in the present case as follows. First make use of (8.136), to rewrite the equations (8.138) as

$$\begin{aligned} (x_3 + x_5)(x_2 + x_4) &= -2c_1\xi_1\xi_2, \\ (x_3 + x_5)(x_1 + x_6) &= 2c_1(\xi_1 - 1)(\xi_2 - 1). \end{aligned} \tag{8.139}$$

Since $\mathbf{c}$ is generic the variables ξ_1 and ξ_2 are both different from 1 and from 0; in fact, the two left hand sides of (8.139) are different from zero if $c_1c_2c_3 \neq 0$, as follows from (8.123). Therefore, we can divide in the computations that follow by ξ_i and by $\xi_i - 1$, when necessary.

Differentiating the equations (8.139) with respect to the vector field $\mathcal{V}_1$ given by (8.125), we find that

$$\begin{aligned} \xi_1^{-1}\dot{\xi}_1 + \xi_2^{-1}\dot{\xi}_2 &= x_1 + x_4 + 2x_3, \\ (\xi_1 - 1)^{-1}\dot{\xi}_1 + (\xi_2 - 1)^{-1}\dot{\xi}_2 &= x_1 + x_4 + 2x_5. \end{aligned} \tag{8.140}$$

Then we can solve the first three equations of (8.123), together with (8.139) and the difference of the two equations in (8.140) for $x_1, \ldots, x_6$, using the above suggested change of coordinates. Substituting these values in the second equation of (8.140) we find that

$$\left(\frac{\dot{\xi}_1}{\xi_1(\xi_1-1)}\right)^2 - \left(\frac{\dot{\xi}_2}{\xi_2(\xi_2-1)}\right)^2 = \frac{1}{\xi_1 - \xi_2}\left[4c_1 + \frac{c_2}{(\xi_1-1)(\xi_2-1)} + \frac{c_3}{\xi_1\xi_2}\right]. \tag{8.141}$$

Notice that this equation is linear in $\dot{\xi}_1^2$ and $\dot{\xi}_2^2$. Finally we substitute the above solved values for $x_1, \ldots, x_6$ in the fourth equation of (8.123) to find another equation in $\dot{\xi}_1$ and $\dot{\xi}_2$ which, together with (8.141) leads to

$$\dot{\xi}_i^2 = \frac{f(\xi_i)}{(\xi_1 - \xi_2)^2}, \qquad i = 1, 2,$$

where

$$f(x) = x(1-x)[4c_1x^3 - (6c_1 + 2c_2 - 2c_3 - c_4)x^2 + (2c_1 + c_2 - 3c_3 - c_4)x + c_3].$$

Notice that the affine curve $y^2 = f(x)$ is the affine curve $\Gamma_{\mathbf{c}}$, with the Weierstrass points, lying over $x = 0$ and $x = 1$ added (see (8.132)). It follows that, in terms of the coordinates ξ_1, ξ_2 given by (8.138), the differential equations (8.125) lead to

$$\frac{\dot{\xi}_1}{\sqrt{f(\xi_1)}} + \frac{\dot{\xi}_2}{\sqrt{f(\xi_2)}} = 0, \qquad \frac{\xi_1\dot{\xi}_1}{\sqrt{f(\xi_1)}} + \frac{\xi_2\dot{\xi}_2}{\sqrt{f(\xi_2)}} = 1. \tag{8.142}$$

This means that when we view $P_1 + P_2 := \left(\xi_1, \sqrt{f(\xi_1)}\right) + \left(\xi_2, \sqrt{f(\xi_2)}\right)$ as a divisor on the hyperelliptic curve $y^2 = f(x)$, then the $\mathcal{V}_1$-flow of its linear equivalence class is linear on the Jacobian of this curve, as (8.142) is equivalent to

$$\frac{\mathsf{d}}{\mathsf{d}t}\left(\sum_{i=1}^{2}\int_{0_i}^{P_i}\vec{\omega}\right) = \begin{pmatrix}0\\1\end{pmatrix},$$

with $\vec{\omega} = (\omega_1, \omega_2)^\top = \left(\frac{dx}{\sqrt{f(x)}}, \frac{x\,dx}{\sqrt{f(x)}}\right)^\top$, a basis for the holomorphic differentials on $\overline{\Gamma}_{\mathbf{c}}$. According to Mumford's description of hyperelliptic Jacobians (see [133, Section 3.1]), if Γ is a hyperelliptic curve of genus two then $\overline{\Gamma}$ can be embedded in its Jacobian in such a way that $\mathrm{Jac}(\overline{\Gamma}) \setminus \Gamma$ is isomorphic to the space of pairs of polynomials $(u(\mathfrak{h}), v(\mathfrak{h}))$ such that $u(\mathfrak{h})$ is monic of degree two, $v(\mathfrak{h})$ is of degree less than two and $f(\mathfrak{h}) - v^2(\mathfrak{h})$ is divisible by $u(\mathfrak{h})$. Let us describe the map from $\mathbf{F_c}$ into $\mathrm{Jac}(\overline{\Gamma_{\mathbf{c}}})$ in terms of these polynomials. We define the polynomial $u(\mathfrak{h})$ by demanding that its roots are ξ_1 and ξ_2, i.e., from (8.140) and (8.123) conclude

$$u(\mathfrak{h}) = \mathfrak{h}^2 + \left(\frac{x_1 + x_2 + x_4 + x_6}{2(x_3 - x_5)} - 1\right)\mathfrak{h} - \frac{x_2 + x_4}{2(x_3 - x_5)}.$$

The polynomial $v(\mathfrak{h})$ is defined as the derivative of $u(\mathfrak{h})$ in the direction of $\mathcal{V}_1$ and can be most easily described by the following formulas, gotten from (8.140):

$$v(0) = u(0)(x_1 + x_4 + 2x_3), \quad v(1) = u(1)(x_1 + x_4 + 2x_5).$$
$$v(\mathfrak{h}) = (v(1) - v(0))\mathfrak{h} + v(0).$$

It is easy to check that $f(\mathfrak{h}) - v^2(\mathfrak{h})$ is divisible by $u(\mathfrak{h})$ so that the above formulas indeed define a point of $\mathrm{Jac}(\overline{\Gamma_{\mathbf{c}}}) \setminus \mathcal{D}_{\mathbf{c}}^{-+}$. This leads to a Lax equation $\dot{X}(\mathfrak{h}) = [X(\mathfrak{h}), Y(\mathfrak{h})]$ for the metric II by taking

$$X(\mathfrak{h}) = \begin{pmatrix} v(\mathfrak{h}) & u(\mathfrak{h}) \\ w(\mathfrak{h}) & -v(\mathfrak{h}) \end{pmatrix}, \qquad Y(\mathfrak{h}) = \begin{pmatrix} 0 & 1 \\ w_3\mathfrak{h} + w_2 - u_1 w_3 & 0 \end{pmatrix},$$

where $w(\mathfrak{h})$ is the cubic polynomial $(f(\mathfrak{h}) - v^2(\mathfrak{h}))/u(\mathfrak{h})$, and where we have written $u(\mathfrak{h}) = \mathfrak{h}^2 + u_1\mathfrak{h} + u_0$ and $v(\mathfrak{h}) = v_1\mathfrak{h} + v_0$ and $w(\mathfrak{h}) = w_3\mathfrak{h}^3 + \cdots + w_0$. The verification is done by direct substitution; notice that the characteristic polynomial of $X(\mathfrak{h})$ is precisely the polynomial which defines the curve $\Gamma_{\mathbf{c}}$. For more information on this Lax equation and its relation to the Mumford system, see [169, Chapter VIII].

8.3.4 From Metric II to the Lyapunov-Steklov Flow

In Paragraph 8.2.7 we have shown that the limit of the Manakov Hamiltonian yields the Clebsch Hamiltonian, which yields one of the integrable cases of geodesic flow on $\mathfrak{e}(3)$. We will now show that metric II leads to another integrable case, which is known as the Lyapunov-Steklov case.

Recall that in the case of metric II, the Hamiltonian is given by

$$H = \frac{1}{2}\sum_{1}^{6} \lambda_i x_i^2 + \sum_{j=1}^{3} \mu_j x_j x_{j+3},$$

where the coefficients satisfy the following relations ($\lambda_{ij} := \lambda_i - \lambda_j$).

$$(\mu_1^2, \mu_2^2, \mu_3^2) = \frac{E\bar{E}}{F^2}\left(\frac{(\lambda_{23} - \lambda_{56})^2}{\lambda_{23}\lambda_{56}}, \frac{(\lambda_{31} - \lambda_{64})^2}{\lambda_{31}\lambda_{64}}, \frac{(\lambda_{12} - \lambda_{45})^2}{\lambda_{12}\lambda_{45}}\right) \qquad (8.143)$$

$$\mu_1\mu_2\mu_3 = \frac{E\bar{E}}{F^3}(\lambda_{12}-\lambda_{45})(\lambda_{23}-\lambda_{56})(\lambda_{31}-\lambda_{64}),$$

where $E := \lambda_{12}\lambda_{23}\lambda_{31}$ and $\bar{E} := \lambda_{45}\lambda_{56}\lambda_{64}$ and $F := \lambda_{46}\lambda_{32} - \lambda_{65}\lambda_{13}$. For $i = 1, \dots, 3$, define

$$\begin{aligned}
\lambda_i &= \frac{a_i}{4\epsilon^2} + \frac{a_1a_2a_3}{2\epsilon a_i}, \qquad \lambda_{i+3} = \frac{a_i}{4\epsilon^2} - \frac{a_1a_2a_3}{2\epsilon a_i}, \\
\mu_i &= \frac{a_i}{4\epsilon^2}\prod_{j\neq i}\sqrt{1-4\epsilon^2a_j^2} \\
&= \frac{a_i}{4\epsilon^2} - \frac{a_i}{2}(a_1^2+a_2^2+a_3^2-a_i^2) + 0(\epsilon^2).
\end{aligned} \tag{8.144}$$

Then it is easy to verify that these values for λ_i and μ_i satisfy (8.143) (for the precise definition of the μ_i in (8.144), choose the sign of the square roots for μ_1 and μ_2 arbitrary and choose the sign of the square root for μ_3 such that the third equation in (8.143) is satisfied). As in the Clebsch case, we substitute the values (8.144) in H, besides substituting $X_{ij} = l_k$ and $X_{i4} = p_i/\epsilon$, where (i,j,k) is any cyclic permutation of $(1,2,3)$ and where the coordinates x_i are expressed in terms of the coordinates X_{ij} by (8.8). Denoting the result by H^ϵ it is easy to see that ϵ^2H^ϵ has a non-zero limit as $\epsilon \to 0$, namely

$$\lim_{\epsilon\to 0}\epsilon^2H^\epsilon = \frac{1}{2}\sum_{i=1}^{3}a_i(l_i-(a_1+a_2+a_3-a_i)p_i)^2 + C,$$

where C is the following Casimir of $\mathfrak{e}(3)$,

$$C := (a_1a_2+a_2a_3+a_3a_1)(l_1p_1+l_2p_2+l_3p_3) - a_1a_2a_3(p_1^2+p_2^2+p_3^2).$$

The constants of motion of the Lyapunov-Steklov Hamiltonian are most easily found by taking the limit of the constants of motion (8.120). If we compute y_i^ϵ in the same way as we computed H^ϵ above, where the y_i are defined in terms of the x_i in (8.117), then we find that $\epsilon^4y_i^\epsilon$ has, for $i = 1, \dots, 6$ a non-zero limit as $\epsilon \to 0$, namely for $i = 1, \dots, 3$ we find that

$$\begin{aligned}
\lim_{\epsilon\to 0}\epsilon^4y_i^\epsilon &= \frac{1}{8}(a_1-a_2)(a_2-a_3)(a_3-a_1)\left(\frac{l_i}{a_j-a_k}+p_i\right), \\
\lim_{\epsilon\to 0}\epsilon^4y_{i+3}^\epsilon &= \frac{1}{8}(a_1-a_2)(a_2-a_3)(a_3-a_1)\left(\frac{l_i}{a_j-a_k}-p_i\right).
\end{aligned}$$

It follows that the three first constants of motion in (8.120) get transformed (up to a constant) into

$$(a_i-a_j)\left(\frac{l_k}{a_i-a_j}+p_k\right)^2 + (a_i-a_k)\left(\frac{l_j}{a_i-a_k}+p_j\right)^2,$$

where (i,j,k) is any cyclic permutation of $(1,2,3)$. Similarly, the last constant of motion in (8.120) gets transformed into the Casimir $p_1^2+p_2^2+p_3^2$.

8.4 Geodesic Flow on SO(4) for Metric III and Abelian Surfaces of Type (1, 6)

8.4.1 A Normal Form for Metric III

We now turn to case *(3)* in Theorem 8.3. By a trivial rescaling of the coordinates, so as to get rid of the square roots ϵ_1 and ϵ_2 of 1, and by a transposition of the coordinates

$$y_1 \leftrightarrow y_4, \qquad y_2 \leftrightarrow y_5, \qquad y_3 \leftrightarrow y_6,$$

we may assume that in this case the Hamiltonian H is of the form

$$H = \frac{1}{2}\sum_{i=1}^{6} \lambda_i y_i^2 + \sum_{j=1}^{3} \mu_j y_j y_{j+3},$$

where the constants λ_i and μ_i can be expressed in terms of constants ρ, q with $\rho \in \mathbf{C}^*$ and $q \in \mathbf{C} \setminus \{0, \pm 1, \pm 1/3\}$

$$\begin{aligned} \lambda_{12} &= \rho(q-1)^3(3q+1), & \lambda_{45} &= \rho(q-1)(3q+1)^3, \\ \lambda_{23} &= 16\rho q^3, & \lambda_{56} &= 16\rho q, \\ \lambda_{31} &= \rho(1-3q)(q+1)^3, & \lambda_{64} &= \rho(1-3q)^3(q+1), \end{aligned}$$

$$\begin{aligned} \mu_1 &= \rho(1-q^2)(1-9q^2), \\ \mu_2 &= 4\rho q(1-q)(3q+1), \\ \mu_3 &= 4\rho q(q+1)(1-3q). \end{aligned}$$

The constant ρ is a common factor, which can be given any non-zero value; we take

$$\rho := \frac{1}{(q-1)(3q+1)^3}.$$

In order to simplify the formulas that follow, we define

$$\alpha := \frac{q-1}{3q+1}, \qquad \text{so that} \qquad q = \frac{\alpha+1}{1-3\alpha},$$

and notice that $\alpha \in \mathbf{C} \setminus \{0, \pm 1, \pm 1/3\}$, just like q. If we remove from H a multiple of the Casimirs $y_1^2 + y_2^2 + y_3^2$ and $y_4^2 + y_5^2 + y_6^2$, so as to kill the coefficients of y_2^2 and y_5^2 then H takes, with the above choices of ρ and α the following form.

$$\begin{aligned} H = \alpha^2 y_1^2 + y_4^2 &+ \frac{(1+\alpha)^3(3\alpha-1)}{16\alpha} y_3^2 + \frac{(1+\alpha)(3\alpha-1)^3}{16\alpha} y_6^2 \\ &- \frac{(\alpha-1)(3\alpha+1)}{2} y_1 y_4 + \frac{(\alpha+1)(3\alpha-1)}{2} y_2 y_5 - \frac{(9\alpha^2-1)(\alpha^2-1)}{8\alpha} y_3 y_6. \end{aligned}$$

Finally, we do one more (linear) change of variables, namely let $x_1, \dots, x_6$ be new coordinates[9] on $\mathbf{C}^6$, which are defined by

$$\begin{pmatrix} y_1 \\ y_4 \end{pmatrix} = \sqrt{-1}\begin{pmatrix} 3\alpha+1 & 1 \\ \alpha-1 & -1 \end{pmatrix}\begin{pmatrix} (\alpha-1)x_1 \\ (3\alpha-1)(\alpha+1)x_4 \end{pmatrix},$$
$$\begin{pmatrix} y_2 \\ y_5 \end{pmatrix} = \sqrt{-1}\begin{pmatrix} 3\alpha-1 & 1 \\ \alpha+1 & -1 \end{pmatrix}\begin{pmatrix} (\alpha+1)x_2 \\ (3\alpha+1)(\alpha-1)x_5 \end{pmatrix},$$
$$\begin{pmatrix} y_3 \\ y_6 \end{pmatrix} = -\begin{pmatrix} 3\alpha+1 & 3\alpha-1 \\ \alpha-1 & \alpha+1 \end{pmatrix}\begin{pmatrix} (\alpha-1)x_3 \\ (\alpha+1)x_6 \end{pmatrix}.$$

Notice that each of these matrices is invertible, since $\alpha \in \mathbf{C} \setminus \{0, \pm 1, \pm 1/3\}$. In terms of the coordinates x_i the Hamiltonian vector field $\mathcal{X}_H$ is (up to the constant factor $(9\alpha^2-1)(\alpha^2-1)$) given by

$$\begin{aligned}
\dot{x}_1 &= x_3 x_5 & \dot{x}_4 &= \frac{2\alpha x_5 x_6 + (\alpha-1)x_2 x_3}{3\alpha-1} \\
\dot{x}_2 &= x_4 x_6 & \dot{x}_5 &= \frac{2\alpha x_3 x_4 + (\alpha+1)x_1 x_6}{3\alpha+1} \\
\dot{x}_3 &= \frac{1-\alpha}{2}x_4 x_5 + x_1 x_5 + \frac{1+\alpha}{2}x_1 x_2 & \dot{x}_6 &= \frac{1+\alpha}{2}x_4 x_5 + x_2 x_4 + \frac{1-\alpha}{2}x_1 x_2
\end{aligned}$$

This (homogeneous) vector field will in the sequel be denoted by $\mathcal{V}_1$. Four independent constants of motion of $\mathcal{V}_1$ are given by

$$\begin{aligned}
F_1 &:= \alpha G_2 + \frac{1-\alpha}{1+3\alpha}G_7, \\
F_2 &:= -\alpha G_1 + \frac{1+\alpha}{1-3\alpha}G_8, \\
F_3 &:= \frac{2G_6}{9\alpha^2-1} - \frac{G_1}{1+3\alpha} - \frac{G_2}{1-3\alpha}, \\
F_4 &:= \frac{\alpha-1}{3\alpha+1}(G_1^2+G_4^2) + \frac{\alpha+1}{3\alpha-1}(G_2^2+G_5^2) - 3\frac{\alpha^2-1}{9\alpha^2-1}(2G_1G_2 - G_3^2) \\
&\quad + \frac{4}{9\alpha^2-1}\left[(\alpha+1)G_2(G_6+G_8) - (\alpha-1)G_1(G_6+G_7)\right],
\end{aligned} \tag{8.145}$$

where

$$\begin{aligned}
G_1 &:= x_4^2 - x_2 x_5, & G_5 &:= -\frac{2}{1+3\alpha}(x_1 x_6 - x_3 x_4), \\
G_2 &:= x_5^2 - x_1 x_4, & G_6 &:= x_1 x_4 + x_2 x_5 - x_3 x_6, \\
G_3 &:= x_1 x_2 - x_4 x_5, & G_7 &:= x_1^2 - x_3^2 + x_1 x_4, \\
G_4 &:= -\frac{2}{1-3\alpha}(x_2 x_3 - x_5 x_6), & G_8 &:= x_2^2 - x_6^2 + x_2 x_5.
\end{aligned}$$

[9] These new coordinates are motivated by the equations of a curve of rank three quartics, as explained in [12, Chapter 8].

The fact that $F_1, \ldots, F_3$ are constants of motion of $\mathcal{V}_1$ is checked at once. In order to check this for F_4 it is useful to first point out that the derivatives of the quadratic polynomials $G_1, \ldots, G_5$, in the direction of $\mathcal{V}_1$, are linear functions of $G_1, \ldots, G_5$, with linear coefficients in the x_i, namely

$$\begin{aligned}
\dot{G}_1 &= \frac{1+\alpha}{2}(x_2G_5 - x_4G_4),\\
\dot{G}_2 &= \frac{1-\alpha}{2}(x_1G_4 - x_5G_5),\\
\dot{G}_3 &= \alpha(x_5G_4 - x_4G_5),\\
\dot{G}_4 &= -\frac{1+\alpha}{1-3\alpha}(x_6G_5 + (x_2+x_5)G_3),\\
\dot{G}_5 &= -\frac{1-\alpha}{1+3\alpha}(x_3G_4 + (x_1+x_4)G_3).
\end{aligned} \tag{8.146}$$

The derivatives of $G_6, \ldots, G_8$ are expressed in terms of $\dot{G}_1$ and $\dot{G}_2$ by using $\dot{F}_1 = \dot{F}_2 = \dot{F}_3 = 0$. From these formulas for the $\dot{G}_i$ one concludes easily that $\dot{F}_4 = 0$, while the quadratic Hamiltonian H is easily written as a linear combination of F_1, F_2 and F_3. In terms of these three constants of motion, two generators for the algebra of Casimirs are given by

$$\begin{aligned}
F_1' &:= F_1(1-\alpha)(1+3\alpha)^3 + F_2(1+\alpha)(1-3\alpha)^3 - F_3(1-\alpha^2)(1-9\alpha^2)^2,\\
F_2' &:= F_1(1-\alpha)^3(1+3\alpha) + F_2(1+\alpha)^3(1-3\alpha) - F_3(1-\alpha^2)^2(1-9\alpha^2).
\end{aligned} \tag{8.147}$$

Since the functions $F_1, \ldots, F_4$ are clearly independent, this shows that the geodesic flow on $\mathfrak{so}(4)$ given by metric III is Liouville integrable. In the sequel $\mathbf{F} := (F_1, F_2, F_3, F_4)$ denotes its momentum map. The vector field $\mathcal{X}_{F_4}$ that commutes with $\mathcal{V}_1$ would be a natural choice for $\mathcal{V}_2$, but we will prefer to use $\mathcal{V}_2 := \mathcal{X}_{F_4'}$, where F_4' is defined by

$$F_4' := (3\alpha^2+1)^2F_4 - 32\alpha^2(3\alpha^2-1)F_3^2 - (F_1' + 3F_2')F_3. \tag{8.148}$$

$\mathcal{V}_2$ is a weight homogeneous vector field of weight 3; we will give an expression of it later, when we need it. As we pointed out in Paragraph 8.1.2 the two commuting vector fields $\mathcal{V}_1$ and $\mathcal{V}_2$ are independent on the generic fiber of the momentum map.

We finish this paragraph by listing a few involutions that are relevant for this integrable system. The group of involutions on $\mathbf{C}^6$, generated by

$$\begin{aligned}
\sigma_1(x_1, \ldots, x_6) &= (x_1, -x_2, -x_3, x_4, -x_5, -x_6),\\
\sigma_2(x_1, \ldots, x_6) &= (-x_1, x_2, -x_3, -x_4, x_5, -x_6),
\end{aligned} \tag{8.149}$$

leaves the vector fields $\mathcal{V}_1$ and $\mathcal{V}_2$ invariant, as well as all fibers $\mathbf{F_c}$ ($\mathbf{c} \in \mathbf{C}^4$) of the momentum map.

This is easily seen by using the fact that the action on the functions G_i is as follows,

$$\sigma_1(G_1,\ldots,G_8) = (G_1,G_2,-G_3,G_4,-G_5,G_6,G_7,G_8),$$
$$\sigma_2(G_1,\ldots,G_8) = (G_1,G_2,-G_3,-G_4,G_5,G_6,G_7,G_8),$$

The involution τ on $\mathbf{C}^6$, which is defined by

$$\tau(x_1,\ldots,x_6) = (-x_1,-x_2,-x_3,-x_4,-x_5,-x_6), \tag{8.150}$$

has the property of reversing the sign of both vector fields, still leaving all fibers $\mathbf{F_c}$ of the momentum map invariant. Of course, the composition of τ with any of σ_1 or σ_2 (or both) has the same properties. An involution on the whole family of integrable systems is defined by

$$\pi(x_1,\ldots,x_6,\alpha) := (x_2,x_1,x_6,x_5,x_4,x_3,-\alpha). \tag{8.151}$$

It leaves the two vector fields invariant, but it permutes the fibers of the momentum map, since

$$\pi(F_1,F_2,F_3,F_4) = (F_2,F_1,F_3,F_4)$$

as follows from

$$\pi(G_1,\ldots,G_8) = (G_2,G_1,G_3,G_5,G_4,G_6,G_8,G_7).$$

In order to compactify the notation, we will often replace π by an overline bar. In this notation, $\bar{\alpha} = -\alpha$ and $\bar{G}_1 = G_2$ and so on.

8.4.2 A Lax Equation for Metric III

In this paragraph we give a Lax equation for metric III, which was discovered by Reiman and Semenov-Tian-Shanskii (see [148]). Consider the following quadratic form on $\mathfrak{so}(4)$,

$$H := \sum_{i=1}^{3}\left((\frac{c_i}{3}+d_i)x_i^2 + (3c_i+d_i)x_{i+3}^2 + 2(d_i-c_i)x_ix_{i+3}\right) \tag{8.152}$$

with

$$c_i := \frac{b_i}{a_i}, \qquad \text{and} \qquad d_i := \frac{b_j-b_k}{a_j-a_k},$$

where (i,j,k) is any cyclic permutation of $(1,2,3)$, and where the parameters a_i and b_i satisfy

$$a_1+a_2+a_3=0, \qquad \text{and} \qquad b_1+b_2+b_3=0,$$

but are otherwise arbitrary. It is easy to see that each of the non-degenerate metrics that is defined by H is, for any value of the parameters, a special case of metric III.

To see this, it suffices to make in *(3)* of Theorem 8.3 the following choices for q, ρ, ϵ_1 and ϵ_2:

$$q := \frac{a_2 - a_3}{3a_1}, \qquad \text{and} \qquad \rho := \frac{9}{8}\frac{(b_2a_1 - b_1a_2)a_1^3}{a_2a_3(a_1 - a_2)(a_2 - a_3)(a_3 - a_1)}.$$

and $\epsilon_1 := -\epsilon_2 := -1$. For example

$$\frac{c_1 - c_2}{3} + d_1 - d_2 = \lambda_{12} = \rho(q-1)^3(3q+1),$$

and so on. Consider the following parameterization of the (15-dimensional) Lie algebra $\mathfrak{g}_2$

$$M(u, v, a, y, z) :=$$

$$\begin{pmatrix} 0 & -\frac{u_3+v_3}{2} & \frac{u_2+v_2}{2} & \frac{v_1-u_1}{2} & -y_2 & y_3 & a_1 \\ \frac{u_3+v_3}{2} & 0 & -\frac{u_1+v_1}{2} & \frac{v_2-u_2}{2} & y_1 & a_2 & z_3 \\ -\frac{u_2+v_2}{2} & \frac{u_1+v_1}{2} & 0 & \frac{v_3-u_3}{2} & a_3 & z_1 & -z_2 \\ \frac{u_1-v_1}{2} & \frac{u_2-v_2}{2} & \frac{u_3-v_3}{2} & 0 & y_3-z_3 & y_2-z_2 & y_1-z_1 \\ -y_2 & y_1 & a_3 & y_3-z_3 & 0 & v_1 & -v_2 \\ y_3 & a_2 & z_1 & y_2-z_2 & -v_1 & 0 & v_3 \\ a_1 & z_3 & -z_2 & y_1-z_1 & v_2 & -v_3 & 0 \end{pmatrix}$$

where $u = (u_1, u_2, u_3)$ and so on. Define, in terms of this parameterization,

$$\begin{aligned} X &:= M((x_4, x_5, x_6), (x_1/3, x_2/3, x_3/3), 0, 0, 0), \\ Y &:= M((y_4, y_5, y_6), (y_1/3, y_2/3, y_3/3), 0, 0, 0), \\ A &:= M(0, 0, (a_1, a_2, a_3), 0, 0), \\ B &:= M(0, 0, (b_1, b_2, b_3), 0, 0), \end{aligned}$$

where the y_i are linear combinations of the x_i, which are defined by:

$$\begin{aligned} y_i + y_{i+3} &= d_i(x_i + x_{i+3}), \\ y_i - 3y_{i+3} &= c_i(x_i - 3x_{i+3}), \end{aligned}$$

where $i = 1, \ldots, 3$. Then it is easy to check, using (8.10) that the Hamiltonian vector field $\{\cdot\,, H\}$, with H defined in (8.152), is given, up to a constant, by the following Lax equation with parameter,

$$(X + A\mathfrak{h})^{\cdot} = [X + A\mathfrak{h}, Y + B\mathfrak{h}].$$

Note that, using $[A,Y]=[B,X]$, we see as before (in (8.66)) that

$$B\mathfrak{h}+Y=\left[\mathfrak{h}f\left(\frac{A\mathfrak{h}+X}{\mathfrak{h}}\right)\right]_+,$$

where $f(x)=x(e_1x^2+e_2x^4+e_3x^6)$ is such that $f(a_i)=b_i$ for $1\leqslant i\leqslant 3$ and from that and on analysis of the spectral curve $\mathcal{C}:\det(X+A\mathfrak{h}-\xi\,\mathrm{Id}_7)=0$, ($\xi/\mathfrak{h}$ is finite at infinity) we see that the Linearization Criterion (Corollary 6.43) is satisfied and so the Lax equation induces linear flow on $\mathrm{Jac}(\mathcal{C})$.

8.4.3 Algebraic Complete Integrability

We now study the geometry of the integrable systems which is defined by metric III and we establish its algebraic complete integrability. The latter is done, as in the case of metrics I and II, by going through the algorithm, given in Section 7.7. In the case of metric III some of the formulas get really huge, for example the formulas for the quadratic vector fields (which, for that reason, we will only give in one of the charts). However, the geometry is in this case even richer than in the previous cases. For these two reasons we will rather concentrate on the geometry, and we will not always give display all the terms of the series that were used in the computations.

We first compute the indicial locus of $\mathcal{V}_1$, which is the algebraic subset $\mathcal{I}$ of $\mathbf{C}^6$, defined by the following equations.

$$\begin{aligned}
&x_1^{(0)}+x_3^{(0)}x_5^{(0)}=0,\\
&x_2^{(0)}+x_4^{(0)}x_6^{(0)}=0,\\
&x_3^{(0)}+\frac{1-\alpha}{2}x_4^{(0)}x_5^{(0)}+x_1^{(0)}x_5^{(0)}+\frac{1+\alpha}{2}x_1^{(0)}x_2^{(0)}=0,\\
&x_4^{(0)}+\frac{2\alpha x_5^{(0)}x_6^{(0)}+(\alpha-1)x_2^{(0)}x_3^{(0)}}{3\alpha-1}=0,\\
&x_5^{(0)}+\frac{2\alpha x_3^{(0)}x_4^{(0)}+(\alpha+1)x_1^{(0)}x_6^{(0)}}{3\alpha+1}=0,\\
&x_6^{(0)}+\frac{1+\alpha}{2}x_4^{(0)}x_5^{(0)}+x_2^{(0)}x_4^{(0)}+\frac{1-\alpha}{2}x_1^{(0)}x_2^{(0)}=0.
\end{aligned}$$

Solving the first two equations,

$$x_1^{(0)}=-x_3^{(0)}x_5^{(0)},\qquad\text{and}\qquad x_2^{(0)}=-x_4^{(0)}x_6^{(0)},$$

and substituting the result in the remaining equations, we get four linear homogeneous equations in $(x_3^{(0)},x_6^{(0)},x_3^{(0)}x_6^{(0)},1)$ with coefficients in $x_4^{(0)}$ and $x_5^{(0)}$, which we call a and b respectively.

They can be written as

$$\begin{pmatrix} 2(1-b^2) & 0 & (1+\alpha)ab & (1-\alpha)ab \\ 0 & 2\alpha b & (1-\alpha)a & (3\alpha-1)a \\ 2\alpha a & 0 & -(1+\alpha)b & (1+3\alpha)b \\ 0 & 2(1-a^2) & (1-\alpha)ab & (1+\alpha)ab \end{pmatrix} \begin{pmatrix} x_3^{(0)} \\ x_6^{(0)} \\ x_3^{(0)}x_6^{(0)} \\ 1 \end{pmatrix} = 0. \tag{8.153}$$

The determinant of this matrix must vanish, because otherwise (8.153) would have no solution. Thus

$$8\alpha ab(1-a^2-b^2)\left((\alpha^2+2\alpha-1)a^2+(\alpha^2-2\alpha-1)b^2+2\right)=0. \tag{8.154}$$

If $a=0$ or $b=0$ then $x^{(0)}=0$, hence this does not lead to a balance. Suppose now that $a^2+b^2=1$. Then the unique solution to (8.153) is given by $x_3^{(0)}=-b/a=1/x_6^{(0)}$, leading to the following rational curve $\Gamma^{(0)}$ in $\mathbf{C}^6$,

$$x^{(0)}=\left(\frac{b^2}{a},\frac{a^2}{b},-\frac{b}{a},a,b,-\frac{a}{b}\right), \qquad a^2+b^2=1. \tag{8.155}$$

From (8.154), there is one other alternative, namely $(\alpha^2+2\alpha-1)a^2+(\alpha^2-2\alpha-1)b^2+2=0$, but it is easy to see that this does not lead to a curve in the indicial locus, but to a finite number of points. As we have seen in the proof of Proposition 8.4, in the case of geodesic flow on $\mathfrak{so}(4)$ (with respect to a non-degenerate half-diagonal metric) the principal balances start out from 1-dimensional irreducible components of the indicial locus. Therefore, we have only one weight homogeneous principal balance, and its leading term $x^{(0)}$ is given by (8.155). The Kowalevski matrix, evaluated at $x^{(0)}$, reads

$$\mathcal{K}(x^{(0)})=\begin{pmatrix} 1 & 0 & b & 0 & -\frac{b}{a} & 0 \\ 0 & 1 & 0 & -\frac{a}{b} & 0 & a \\ \frac{2-a^2(1-\alpha)}{2b} & \frac{(1+\alpha)b^2}{2a} & 1 & \frac{(1-\alpha)b}{2} & \frac{2-a^2(1+\alpha)}{2a} & 0 \\ 0 & \frac{(1-\alpha)b}{(3\alpha-1)a} & \frac{(\alpha-1)a^2}{(3\alpha-1)b} & 1 & \frac{2\alpha a}{(1-3\alpha)b} & \frac{2\alpha b}{3\alpha-1} \\ -\frac{(1+\alpha)a}{(3\alpha+1)b} & 0 & \frac{2\alpha a}{3\alpha+1} & -\frac{2\alpha b}{(3\alpha+1)a} & 1 & \frac{(1+\alpha)b^2}{(3\alpha+1)a} \\ \frac{(1-\alpha)a^2}{2b} & \frac{1+a^2-\alpha b^2}{2a} & 0 & \frac{1+a^2+\alpha b^2}{2b} & \frac{(1+\alpha)a}{2} & 1 \end{pmatrix},$$

and its characteristic polynomial is given by

$$|k\,\mathrm{Id}_6-\mathcal{K}(x^{(0)})|=(k+1)k(k-1)(k-2)^3.$$

Thus, we are expected to find, besides the free parameter that enters in the indicial locus, a free parameter, c, at step 1 and 3 free parameters d, e and f, at step 2.

As we have seen in Paragraph 8.1.3, the three parameters that enter at step 2 are trivial parameters; since there is no linear constant of motion, the parameters that enter at levels 0 and 1 are effective parameters. We now give the first three terms of the principal balance. The subsequent terms (one more term is needed for some of the computations that we will do) follow uniquely from the given ones, since the largest eigenvalue of $\mathcal{K}(x^{(0)})$ is 2.

$$
\begin{aligned}
x_1(t) &= \frac{b^2}{at}\left(1-\frac{c}{2}\left(4\alpha^2a^2+(1-\alpha)^2\right)t+((1+\alpha)e+f)t^2+O(t^3)\right),\\
x_2(t) &= \frac{a^2}{bt}\left(1+\frac{c}{2}\left(4\alpha^2b^2+(1+\alpha)^2\right)t+x_2^{(2)}t^2+O(t^3)\right),\\
x_3(t) &= -\frac{b}{at}\left(1+\frac{c}{2}\left(4\alpha^2a^2-(1-\alpha)^2\right)t-2e\alpha t^2+O(t^3)\right),\\
x_4(t) &= -\frac{a}{t}\left(1+\frac{c}{2}\left(4\alpha^2b^2-(1+\alpha)^2\right)t+x_2^{(4)}t^2+O(t^3)\right),\\
x_5(t) &= -\frac{b}{t}\left(1-\frac{c}{2}\left(4\alpha^2a^2-(1-\alpha)^2\right)t++x_2^{(5)}t^2+O(t^3)\right),\\
x_6(t) &= --\frac{a}{bt}\left(1-\frac{c}{2}\left(4\alpha^2b^2-(1+\alpha)^2\right)t-2d\alpha t^2+O(t^3)\right),
\end{aligned}
\tag{8.156}
$$

where

$$
\begin{aligned}
x_2^{(2)} &= 4a^4c^2\alpha^4+c^2(\alpha-1)^2((3\alpha+1)^2-8(2\alpha+1)\alpha a^2)/4+d(\alpha-1)-f,\\
x_4^{(2)} &= 2a^2c^2\alpha(1-\alpha)(1+\alpha)^2+d(\alpha+1)+f,\\
x_5^{(2)} &= c^2\left(4a^2\alpha^2-(\alpha-1)^2\right)^2/4+e(\alpha-1)-f.
\end{aligned}
$$

Since the free parameters d, e, f are trivial parameters they can be expressed linearly in terms of the values c_1, c_2, c_3 of the constants of motion in a such a way that $F_i(x(t)) = c_i$ for $i = 1, 2, 3$. Notice that when the above series are substituted in the functions $G_1, \ldots, G_8$ then the resulting series have a simple pole. In fact, the leading term of these is given as follows.

$$
\begin{aligned}
(G_1,\ldots,G_8)(t) = -\frac{2}{t}(&a^2c(\alpha+1), b^2c(\alpha-1), -2\alpha abc, ac(1+\alpha),\\
&bc(\alpha-1), c((3\alpha^2-1)(a^2-b^2)+2\alpha)/2,\\
&ab^2c(3\alpha+1), \alpha(1-3\alpha)a^2c)+O(1).
\end{aligned}
\tag{8.157}
$$

Since F_4 is a quadratic function in $G_1, \ldots, G_8$, we only need the terms up to order t in the latter to compute the constant term in F_4, which has the useful consequence that it suffices to compute the terms up to order t^2 in $x_1(t), \ldots, x_6(t)$ (since F_4 is of the fourth degree, one needs a priori the terms up to order t^3 to do this).

For generic[10] $\mathbf{c} = (c_1, \dots, c_4)$ the abstract Painlevé divisor $\Gamma_\mathbf{c}$, as computed from $F_4(x(t)) = c_4$, is given by the following curve in $\mathbf{C}^3$:

$$P_\mathbf{c}(a^2, b^2, c^2) = 0, \qquad a^2 + b^2 = 1,$$

where

$$\begin{aligned} &P_\mathbf{c}(A, B, C) \\ &\quad = (ABC)^2(\alpha_1 A + \alpha_2)^2 - 2ABC\left((\alpha_1 A + \alpha_2)P_\mathbf{c}^-(A, B) + 2ABc_2'\right) \\ &\qquad - AB\left((\alpha_3 A + \alpha_4)c_4 + 4c_1c_2 - \alpha_4\bar{\alpha}_4 c_3^2\right) + (P_\mathbf{c}^+(A, B))^2, \end{aligned}$$

with

$$P_\mathbf{c}^\pm(A, B) := c_1 A + c_2 B \pm (c_3(3\alpha^2 + 1) - c_1 - c_2)AB,$$

and where the α_i are given by

$$\alpha_1 := 16\alpha^3, \quad \alpha_2 := (\alpha - 1)^3(3\alpha + 1), \quad \alpha_3 := 4\alpha, \quad \alpha_4 := (\alpha - 1)(3\alpha + 1),$$

and where $\bar{\alpha}_i$ is obtained from α_i by changing the sign in α, for example $\bar{\alpha}_2 = (\alpha + 1)^3(3\alpha - 1)$. In order to compute the genus of the abstract Painlevé divisor $\Gamma_\mathbf{c}$ we use a few related curves, that are obtained by considering the quotient of $\Gamma_\mathbf{c}$ by certain symmetries. In fact, changing the sign of any of the parameters a, b or c leads to an involution on $\Gamma_\mathbf{c}$, and hence to a quotient curve.

Table 8.4. The abstract Painlevé divisor $\Gamma_\mathbf{c}$ admits three obvious involutions. We denote them by σ_1, σ_2 and τ because they correspond to the involutions that were defined in (8.149) and (8.150), as follows by comparing the residues of the functions $x_i(t)$ and $G_j(t)$ (see (8.156) and (8.157)).

	σ_1	σ_2	τ
a	a	$-a$	a
b	$-b$	b	b
c	c	c	$-c$

Proposition 8.19. *For generic* $\mathbf{c}$ *the abstract Painlevé divisor* $\Gamma_\mathbf{c}$ *is non-singular and its smooth compactification* $\overline{\Gamma_\mathbf{c}}$ *has genus* 17. *The quotient of* $\overline{\Gamma_\mathbf{c}}$ *by the group generated by* σ_1 *and* σ_2 *is a smooth curve* $\overline{\Delta_\mathbf{c}}$ *of genus* 5 *and the quotient map* $\overline{\Gamma_\mathbf{c}} \to \overline{\Delta_\mathbf{c}}$ *is an unramified* 4 : 1 *map. The quotient of* $\overline{\Delta_\mathbf{c}}$ *by* τ *is a hyperelliptic curve* $\overline{\mathcal{H}_\mathbf{c}}$ *of genus* 2 *and the quotient map* $\overline{\Delta_\mathbf{c}} \to \overline{\mathcal{H}_\mathbf{c}}$ *is ramified at* 4 *points.*

[10] Fixing the values c_i of the constants of motion F_i also fixes the values of the constants of motion F_i' (see (8.147) and (8.148)); the value of F_i' is denoted by c_i'.

Proof. It is easy to give explicit equations for the two quotient curves $\Delta_{\mathbf{c}}$ and $\mathcal{H}_{\mathbf{c}}$, and to describe the two quotient maps. The result is summarized in the following diagram.

$$\begin{array}{ccc}
\Gamma_{\mathbf{c}} : P_{\mathbf{c}}(a^2, b^2, c^2) = 0,\ a^2 + b^2 = 1 & (a, b, c) & \text{g=17} \\
\downarrow & \downarrow & \\
\Delta_{\mathbf{c}} : P_{\mathbf{c}}(A, 1 - A, c^2) = 0 & (A, c) = (a^2, c) & \text{g=5} \\
\downarrow & \downarrow & \\
\mathcal{H}_{\mathbf{c}} : P_{\mathbf{c}}(A, 1 - A, W/(A(1 - A))) = 0 & (A, W) = (A, c^2 A(1 - A)) & \text{g=2} \\
\downarrow & \downarrow & \\
\mathcal{C} : \mathbf{P}^1 & A & \text{g=0}
\end{array}$$

The genera of these curves were computed as follows. Let p_3 and p_4 be the polynomials, defined by

$$P_{\mathbf{c}}(A, 1 - A, C) = (A(1 - A)C)^2(\alpha_1 A + \alpha_2)^2 - 2A(1 - A)Cp_3(A) + p_4(A), \tag{8.158}$$

so that $\Delta_{\mathbf{c}}$ can be written as

$$(A(1 - A)c^2)^2(\alpha_1 A + \alpha_2)^2 - 2A(1 - A)c^2 p_3(A) + p_4(A) = 0, \tag{8.159}$$

and $\mathcal{H}_{\mathbf{c}}$ as

$$W^2(\alpha_1 A + \alpha_2)^2 - 2Wp_3(A) + p_4(A) = 0. \tag{8.160}$$

Solving (8.160) for W we find

$$W = \frac{p_3(A) \pm \sqrt{p_3^2(A) - p_4(A)(\alpha_1 A + \alpha_2)^2}}{(\alpha_1 A + \alpha_2)^2}, \tag{8.161}$$

where it is important to notice that the discriminant $p_3^2(A) - p_4(A)(\alpha_1 A + \alpha_2)^2$ is of degree 5 only (the leading term cancels out), which implies that the cover $\mathcal{H}_{\mathbf{c}} \to \mathbf{P}^1$ is ramified at infinity. In fact, writing A in terms of a local parameter ς as $A = 1/\varsigma^2$ we find from (8.161) that

$$W = -\varsigma^{-2}\left(\frac{c_1 + c_2 - c_3(3\alpha^2 + 1)}{\alpha_1} + O(\varsigma)\right),$$

where $O(\varsigma)$ is a holomorphic series in ς, without constant term, so that the cover is indeed ramified at infinity. Thus, the double cover $\overline{\mathcal{H}_{\mathbf{c}}} \to \mathbf{P}^1$ which is defined by $(A, W) \mapsto A$ has six ramification points, so that the genus of $\mathcal{H}_{\mathbf{c}}$ is 2, by Riemann-Hurwitz.

The ramification points of the double cover $\Delta_{\mathbf{c}} \to \mathcal{H}_{\mathbf{c}}$ are found by taking $c = 0$, which yields the zeros of $p_4(A)$, i.e., 4 points. This cover is unramified at infinity.

To show this, solve $W = c^2A(1-A)$ for c, which gives, in view of (8.161),

$$c = \pm\frac{1}{\alpha_1 A + \alpha_2}\sqrt{\frac{p_3(A) \pm \sqrt{p_3^2(A) - p_4(A)(\alpha_1 A + \alpha_2)^2}}{A(1-A)}}.$$

Thus, writing $A = 1/\varsigma^2$ we find that $c = \pm c_0\varsigma + O(\varsigma^2)$, where c_0 is a non-zero constant, so we find two points and the map is unramified at these points. Similarly, it is checked by writing $A = -\alpha_2/\alpha_1 + \varsigma$ that the map is unramified over the points for which $A = -\alpha_2/\alpha_1$. We conclude by Riemann-Hurwitz that the genus of $\overline{\Delta_{\mathbf{c}}}$ is 5.

Finally, the genus of $\Gamma_{\mathbf{c}}$ is 17 because the 4 : 1 cover $\overline{\Gamma_{\mathbf{c}}} \to \overline{\Delta_{\mathbf{c}}}$ is unramified. In fact, the only affine points where ramification can occur are those for which $a = 0$ or $b = 0$. But $a = 0$ implies $b^2 = 1$, while

$$P_{\mathbf{c}}(0,1,c^2) = (P_{\mathbf{c}}^+(0,1))^2 = c_2^2,$$

which is different from zero because $\mathbf{c}$ is generic, leading to no ramification points on $\Gamma_{\mathbf{c}}$. Similarly $b = 0$ leads to no points. Thus, there are no affine ramification points. Now we saw earlier in this proof that, near infinity, $A = 1/\varsigma^2$, in terms of a local parameter ς. It follows that $a = \pm 1/\varsigma$, leading to two points, and similarly for b. Thus, the 4 : 1 cover $\overline{\Gamma_{\mathbf{c}}} \to \overline{\Delta_{\mathbf{c}}}$ is unramified, as asserted. Once we know that $\mathbf{F}_{\mathbf{c}}$ is an affine part of an Abelian surface the fact that this cover is unramified becomes geometrically clear: since the two involutions σ_1 and σ_2 preserve the two vector fields they are translations (over half periods), so they cannot have fixed points. □

The curves that appear in Proposition 8.19 are also naturally related to a certain curve of rank four quartics, as we explain now. Picking generic values of $\mathbf{c} = (c_1, c_2, c_3, c_4)$ we define $\bar{F}_i := F_i - c_i x_0^2$, where $i = 1, \ldots, 3$ and $\bar{F}_4 := F_4 - c_4 x_0^4$. Here, x_0 is a new variable that we use to make the polynomials $F_i - c_i$ homogeneous. We will say that a quartic polynomial in the variables $x_0, x_1, \ldots, x_6$ has rank four if it has, possibly after a linear change of variables in $x_1, \ldots, x_6$, the form (up to a constant factor)

$$\sum_{i=1}^{3} x_i^2 Q_i(x) + x_0^4,$$

where each of the $Q_i(x)$ is a quadratic polynomial in $x_0, \ldots, x_6$. In the proposition that follows we will consider the projective surface $\overline{I_{\mathbf{c}}} \subset \mathbf{C}^6$ which is defined by

$$\overline{I_{\mathbf{c}}} = \bigcap_{i=1}^{4} \left\{ x = (x_0 : x_1 : \cdots : x_6) \mid \bar{F}_i(x) = 0 \right\},$$

which is the dumb (singular) compactification of the level set $\mathbf{F}_{\mathbf{c}}$ of the momentum map. For a proof of this proposition we refer to [12, Chapter 8].

Proposition 8.20. *The rational curve C of Proposition 8.19 can be realized in the linear space of quartics*

$$\sum_{1\leqslant i\leqslant j\leqslant 4} \kappa_{ij}(F_iF_j - c_ic_jx_0^4) + \kappa\bar{F}_4$$

as a curve of rank four quartics

$$\sum_{i=1}^{3}(\mu_i(A)x_i + \mu_{i+3}(A)x_{i+3})^2G_i(x) - \mu(A)x_0^4,$$

where each $G_i(x)$ is a quadratic polynomial, and where A is the parameter on the curve; precisely this expression is obtained by substituting F_i for c_i in $p_4(A)$ (see (8.160)). Conversely, there is a $4:1$ map from this curve of rank four quadrics to one of the components C' of the locus[11] $\overline{I}\cap(x_0=0)$, given by (8.155), along which the surface $\overline{I}\subset\mathbf{P}^6$ has a four-fold normal crossing with a number of pinch points. Blowing up $\overline{I}$ along C' turns C' into the curve $\overline{\Gamma_{\mathbf{c}}}$.

We proceed to construct an embedding of the manifolds $\mathbf{F_c}$ in projective space. To do this we look for the polynomials which have a simple pole at most when the principal balance is substituted in them. The result is given in Table 8.5; it suggests that the 24 found functions provide a basis for $\mathcal{Z}_\rho$ (here, $\rho=(1)$). The 24 functions z_i that appear in the table are defined as

Table 8.5. The polynomials of degree at most 5 which have a simple pole at most when the principal balance is substituted in them.

k	$\dim\mathcal{F}^k$	$\dim\mathcal{H}^k$	$\dim\mathcal{Z}_\rho^k$	# dep	ζ_k	indep. functions
0	1	1	1	0	1	$z_0=1$
1	6	0	6	0	6	$z_1,\dots,z_6$
2	21	3	8	3	5	$z_7,\dots,z_{11}$
3	56	0	26	18	8	$z_{12},\dots,z_{19}$
4	126	7	26	22	4	$z_{20},\dots,z_{23}$
5	252	0	66	66	0	—

follows. For $i=1,\dots,6$ we take $z_i:=x_i$. Then come the functions G_i; we have seen that the $G_1(t),\dots,G_8(t)$ have a simple pole; since the functions $G_6,\dots,G_8$ depend over $\mathcal{H}$ on the other ones, as is clear from (8.145), we define $z_{i+6}:=G_i$, for $i=1,\dots,5$.

[11] This locus can also be described as the projectivization of $\mathbf{F_c}$, where $\mathbf{c}=(0,0,0,0)$.

The cubic polynomials that appear in the table are defined by

$$
\begin{aligned}
z_{12} &:= 2\alpha x_2 G_2 - (1-\alpha)x_4 G_3, & z_{16} &:= (1-\alpha)x_4 G_4 + (1+\alpha)x_2 G_5,\\
z_{13} &:= -2\alpha x_1 G_1 - (1+\alpha)x_5 G_3, & z_{17} &:= (\alpha-1)x_3 G_1 - (1+\alpha)x_6 G_2,\\
z_{14} &:= (1-\alpha)x_5 G_4 + (1+\alpha)x_4 G_5, & z_{18} &:= 2\alpha x_5 G_2 - (1-\alpha)x_1 G_3,\\
z_{15} &:= (1+\alpha)x_5 G_5 + (1-\alpha)x_1 G_4, & z_{19} &:= -2\alpha x_4 G_1 - (1+\alpha)x_2 G_3.
\end{aligned}
$$

They are permuted by the involution π in the following way,

$$
\pi(z_{12},\ldots,z_{19}) = (z_{13}, z_{12}, z_{14}, z_{16}, z_{15}, z_{17}, z_{19}, z_{18}). \tag{8.162}
$$

Finally, the four polynomials of degree 4 are defined in terms of the polynomials G_i, to wit,

$$
\begin{aligned}
z_{20} &:= 4\alpha^2 G_1 G_2 + (1-\alpha^2)G_3^2\\
z_{21} &:= 2\alpha G_1 G_5 - (1-\alpha)G_3 G_4,\\
z_{22} &:= -2\alpha G_2 G_4 - (1+\alpha)G_3 G_5,\\
z_{23} &:= 2\alpha G_4 G_5 + G_3((1+\alpha)G_2 - (1-\alpha)G_1).
\end{aligned}
$$

They are also permuted by π, namely

$$
\pi(z_{20},\ldots,z_{23}) = (z_{20}, z_{22}, z_{21}, -z_{23}). \tag{8.163}
$$

It is clear that these functions provide an isomorphic embedding of $\mathbf{F_c}$ into $\mathbf{P}^{23}$, because the functions $z_1,\ldots,z_6$ are precisely the original phase variables. Similarly, we see from (8.156) and from (8.157) that the parameters a, b and c appear separately in the residues of the functions $z_1(t),\ldots,z_{11}(t)$, so that the map $\varphi'_{\mathbf{c}} : \Gamma_{\mathbf{c}} \to \mathbf{P}^{23}$, defined by the residues of the series $z_i(t)$ provides an embedding of the Painlevé divisor $\Gamma_{\mathbf{c}}$. Let us denote the closure of $\varphi'_{\mathbf{c}}(\Gamma_{\mathbf{c}})$ by $\mathcal{D}_{\mathbf{c}}$. We use a local parameter around the points at infinity of $\Gamma_{\mathbf{c}}$ to determine the singularities of $\mathcal{D}_{\mathbf{c}}$ and to determine its genus. We first start with the eight points $\mathbf{0}^{\epsilon_1\epsilon_2\epsilon_3}$ for which $a=0$, where $\epsilon_1^2=\epsilon_2^2=\epsilon_3^2=1$. The point $\mathbf{0}^{\epsilon_1\epsilon_2\epsilon_3}$ is given, in terms of a local parameter ς, by

$$
a = \varsigma, \qquad b = \epsilon_3\left(1 - \frac{\varsigma^2}{2} + O(\varsigma^4)\right), \qquad c = \frac{\epsilon_1}{\varsigma}\sqrt{\frac{c_2}{\alpha_2}}(1 + \star\epsilon_2\varsigma + O(\varsigma^2)), \tag{8.164}
$$

where it is important that $\star$ is non-zero and independent of the ϵ_i, but its precise value will be irrelevant. If we substitute these local parameterizations in the map $\varphi'_{\mathbf{c}} : \Gamma_{\mathbf{c}} \to \mathbf{P}^{23}$, defined by the residues of the series $z_i(t)$ then we find, after letting $\varsigma \to 0$, the following 8 points in $\mathbf{P}^{23}$,

$$
\begin{aligned}
&\Big(\, 0:1:0:-\epsilon_3:0:0:0:0:2\epsilon_1(\alpha-1)\sqrt{\tfrac{c_2}{\alpha_2}}:0:0:2\epsilon_1\epsilon_3(\alpha-1)\sqrt{\tfrac{c_2}{\alpha_2}}:\\
&\quad 0:\tfrac{8c_2\alpha}{(1-\alpha)(3\alpha+1)}:\tfrac{4\epsilon_3c_2(\alpha+1)}{(1-\alpha)(3\alpha+1)}:-\tfrac{4\epsilon_2c_2\star}{3\alpha+1}:0:\tfrac{2\epsilon_3c_2(3\alpha^2+1)}{(\alpha-1)(3\alpha+1)}:\\
&\quad -\tfrac{2\epsilon_2\epsilon_3c_2(3\alpha-1)}{3\alpha+1}:0:-32\tfrac{c_2\epsilon_1\alpha^2}{3\alpha+1}\sqrt{\tfrac{c_2}{\alpha_2}}:-16\tfrac{\epsilon_1\epsilon_3c_2\alpha}{3\alpha+1}\sqrt{\tfrac{c_2}{\alpha_2}}:\\
&\quad 4\epsilon_1\epsilon_2(1-\alpha)c_2\sqrt{\tfrac{c_2}{\alpha_2}}\star:-4\epsilon_1\epsilon_2\epsilon_3(1-\alpha)c_2\sqrt{\tfrac{c_2}{\alpha_2}}\star\,\Big).
\end{aligned}
$$

It is clear that this leads to 8 different points in $\mathbf{P}^{23}$, as ϵ_1, ϵ_2 and ϵ_3 take, independently, the values ± 1. For the 8 points at infinity that correspond to $b = 0$, which we denote by $\mathbf{1}^{\epsilon_1\epsilon_2\epsilon_3}$, we arrive at the same conclusion by using the involution π (see (8.151)): they map to 8 distinct points in $\mathbf{P}^{23}$. They are moreover distinct from the image points of the points $\mathbf{0}^{\epsilon_1\epsilon_2\epsilon_3}$ because π permutes x_1 and x_2.

We now investigate the points where a (and hence b) has a pole, to wit the points $\infty^{\epsilon_1\epsilon_2\epsilon_3}$ that are given in terms of a local parameter ς by

$$
\begin{aligned}
a &= \frac{1}{\varsigma}, \qquad b = \frac{\epsilon_3\sqrt{-1}}{\varsigma}\left(1 + O(\varsigma^2)\right), \\
c &= \frac{\epsilon_1}{4}\sqrt{\frac{c_1 + c_2 - c_3(3\alpha^2+1)}{\alpha^3}}\,\varsigma(1 + \star\epsilon_2\varsigma + O(\varsigma^2)) \\
&= \frac{\epsilon_1\beta_0}{4}\varsigma(1 + \star\epsilon_2\varsigma + O(\varsigma^2)),
\end{aligned}
\tag{8.165}
$$

where the last equality is a definition (of β_0), and where, as before, the precise value of $\star$, which is independent of the e_i and different from zero, is irrelevant. If we substitute these local parameterizations as before in the map $\varphi'_{\mathbf{c}} : \Gamma_{\mathbf{c}} \to \mathbf{P}^{23}$, and we let $\varsigma \to 0$ then we find the following 8 distinct points in $\mathbf{P}^{23}$,

$$
\begin{aligned}
(\, &0 : 1 : \epsilon_3\sqrt{-1} : 0 : -1 : -\epsilon_3\sqrt{-1} : 0 : \beta_0\epsilon_1\tfrac{1+\alpha}{2} : \beta_0\epsilon_1\tfrac{1-\alpha}{2} : -\beta_0\epsilon_1\epsilon_3\alpha : \\
&0 : 0 : -\tfrac{\sqrt{-1}\epsilon_3\beta_1}{4\alpha^2} : \tfrac{\beta_1}{4\alpha^2} : -2\epsilon_2\epsilon_3\sqrt{-1}\star\alpha^3\beta_0^2 : 2\epsilon_2\star\alpha^3\beta_0^2 : -2\epsilon_2\star\alpha^3\beta_0^2 : \\
&0 : \tfrac{\epsilon_3\sqrt{-1}\beta_2}{4\alpha^2} : -\tfrac{\beta_2}{4\alpha^2} : -\epsilon_1\beta_0\beta_3 : -2\sqrt{-1}\epsilon_1\epsilon_3\beta_0^3\alpha : 2\epsilon_1\beta_0^3\alpha : \tfrac{\epsilon_1\epsilon_3\sqrt{-1}\beta_0}{2}\beta_4\,),
\end{aligned}
$$

where

$$
\begin{aligned}
\tfrac{\beta_1}{\alpha^2-1} &= \tfrac{3\alpha^2+1}{1-\alpha}c_1 + \tfrac{(1-\alpha)(3\alpha+1)}{1+\alpha}c_2 + (3\alpha^2(3\alpha-1) - (\alpha+1))c_3, \\
\tfrac{\beta_2}{(\alpha-1)(3\alpha-1)} &= \tfrac{\alpha^2(3\alpha+1)+5\alpha-1}{(1-\alpha)(3\alpha-1)}c_1 + (1-\alpha)c_2 + (\alpha^2(3\alpha+1) + \alpha - 1)c_3, \\
\tfrac{\beta_3}{\alpha^2-1} &= \tfrac{1+\alpha}{1-\alpha}c_1 + \tfrac{1-\alpha}{1+\alpha}c_2 + (3\alpha^2 - 1)c_3, \\
\tfrac{\beta_4}{\alpha^2-1} &= -\tfrac{1-3\alpha}{1-\alpha}c_1 - \tfrac{1+3\alpha}{1+\alpha}c_2 + (9\alpha^2 + 1)c_3.
\end{aligned}
$$

It is clear that this leads to 8 different points in $\mathbf{P}^{23}$, which are different from the 16 points at infinity that we have found so far.

Next, we move to the points at infinity satisfying $\alpha_1 a^2 + \alpha_2 = 0$. A local parameterization of a neighborhood of these points is given by

$$
a = a_0 + \varsigma, \qquad b = b_0 + O(\varsigma), \qquad c = \frac{c_0}{\alpha_1 a_0 \varsigma} + O(\varsigma^0),
$$

where

$$\begin{aligned}
a_0^2 &= -\frac{(\alpha-1)^3(3\alpha+1)}{16\alpha^3} = -\frac{\alpha_2}{\alpha_1}, \\
b_0^2 &= \frac{(\alpha+1)^3(3\alpha-1)}{16\alpha^3} = \frac{\bar{\alpha}_2}{\alpha_1}, \\
c_0^2 &= -\alpha_2 c_1 - \bar{\alpha}_2 c_2 + \alpha_4\bar{\alpha}_4(\alpha^2-1)c_3 = c_2',
\end{aligned}$$

accounting for eight points on $\overline{\Gamma_{\mathbf{c}}}$, that we denote by $s_{\epsilon_1\epsilon_2\epsilon_3}$, where $(\epsilon_1, \epsilon_2, \epsilon_3)$ satisfies $\epsilon_1^2 = \epsilon_2^2 = \epsilon_3^2 = 1$ and refers to the signs of a_0, b_0 and c_0. Again, the given terms will suffice to do the computation. We substitute these local parameterizations as before in the map $\varphi'_{\mathbf{c}} : \Gamma_{\mathbf{c}} \to \mathbf{P}^{23}$, and we let $\varsigma \to 0$. This leads to the following points in $\mathbf{P}^{23}$.

$$\begin{aligned}
&\left(0:\cdots:0:1:-\frac{\alpha_1 a_0 b_0}{\alpha_4(\alpha^2-1)}:-\frac{8\alpha^2 a_0}{\alpha_4(\alpha-1)}:-\frac{8\alpha^2 b_0}{\alpha_4(\alpha+1)}:-\frac{8\alpha^2 b_0}{\bar{\alpha}_4(\alpha+1)}:\right.\\
&\quad\frac{8\alpha^2 a_0}{\alpha_4(\alpha-1)}:-\frac{\bar{\alpha}_4}{\alpha_4}:\frac{\alpha_1 a_0 b_0}{\bar{\alpha}_4(\alpha^2-1)}:\frac{16\alpha^2 b_0 c_0}{\alpha_4\bar{\alpha}_4(\alpha+1)}:\frac{2c_0}{\alpha_4(\alpha+1)}: \qquad (8.166)\\
&\quad\left.\frac{2b_0 c_0}{\bar{\alpha}_4(\alpha+1)a_0}:-\frac{c_0}{2(\alpha+1)\alpha a_0}\right).
\end{aligned}$$

It is easy to see that this leads again to eight distinct points, that are different from the 24 points that we have obtained so far.

Finally, we move to the points satisfying $\alpha_3 a^2 + \alpha_4 = 0$. A local parameterization of a neighborhood of these points is given by

$$a = a_1 + \varsigma, \qquad b = b_1 + O(\varsigma), \qquad c = \frac{c_0}{\alpha_4\bar{\alpha}_4} + O(\varsigma), \tag{8.167}$$

where

$$\begin{aligned}
a_1^2 &= -\frac{(\alpha-1)(3\alpha+1)}{4\alpha} = -\frac{\alpha_4}{\alpha_3}, \\
b_1^2 &= \frac{(\alpha+1)(3\alpha-1)}{4\alpha} = \frac{\bar{\alpha}_4}{\alpha_3},
\end{aligned}$$

and c_0^2 is as before, accouting for an another set of eight points $t_{\epsilon_1\epsilon_2\epsilon_3}$ on $\overline{\Gamma_{\mathbf{c}}}$. Once more the given terms will suffice to do the computation. We substitute these local parameterizations in the map $\varphi'_{\mathbf{c}} : \Gamma_{\mathbf{c}} \to \mathbf{P}^{23}$, and we let $\varsigma \to 0$. This leads to the following points in $\mathbf{P}^{23}$.

$$\begin{aligned}
&\left(0:\cdots:0:1:-\frac{\alpha_3 a_1 b_1}{\alpha_4}:\frac{\alpha_3 a_1}{\alpha_4}:\frac{\alpha_3 b_1}{\alpha_4}:-\frac{\alpha_3 b_1}{\bar{\alpha}_4}:-\frac{\alpha_3 a_1}{\alpha_4}:\right.\\
&\quad\left.-\frac{\bar{\alpha}_4}{\alpha_4}:\frac{\alpha_3 a_1 b_1}{\bar{\alpha}_4}:-\frac{2\alpha_3 b_1 c_0}{\alpha_4\bar{\alpha}_4}:\frac{2c_0}{\alpha_4(\alpha+1)}:\frac{2\alpha_3 a_1 b_1 c_0}{\alpha_4\bar{\alpha}_4(1-\alpha)}:\frac{\alpha_3 a_1 c_0}{\alpha_4(1-\alpha^2)}\right).
\end{aligned}$$

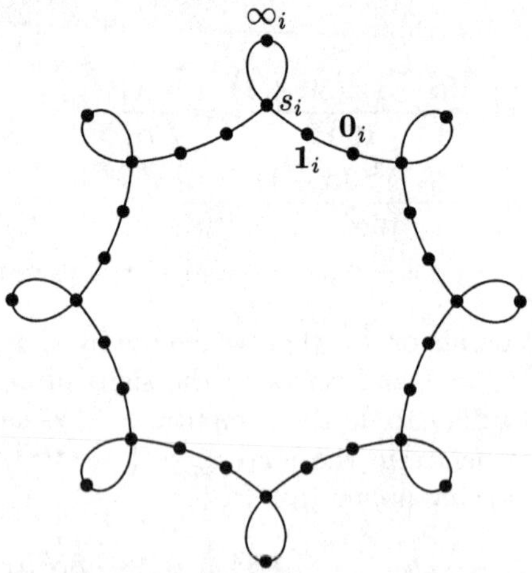

Fig. 8.3. The image $\mathcal{D}_{\mathbf{c}} := \overline{\varphi'_{\mathbf{c}}(\Gamma_{\mathbf{c}})}$ of the abstract Painlevé divisor of metric III is a curve of genus 25 with 8 ordinary double points (nodes) and no other singularities. The vector field $\overline{\mathcal{V}}_1$ is tangent to one of the branches of each of the 8 nodes of $\mathcal{D}_{\mathbf{c}}$ (these nodes are the images of the points $s_{\epsilon_1\epsilon_2\epsilon_3}$, which are also the images of the points $t_{\epsilon_1\epsilon_2\epsilon_3}$) and at 24 of the smooth points of $\mathcal{D}_{\mathbf{c}}$ (the image points of the points $\mathbf{0}^{\epsilon_1\epsilon_2\epsilon_3}$, $\mathbf{1}^{\epsilon_1\epsilon_2\epsilon_3}$ and $\infty^{\epsilon_1\epsilon_2\epsilon_3}$). We mark a typical representative of these points simply by s_i, $\mathbf{0}_i$, $\mathbf{1}_i$ and ∞_i.

These points coincide with the eight points (8.166). To see this, use the following sign conventions:

$$2\alpha a_0 = (1-\alpha)a_1, \qquad 2\alpha b_0 = -(1+\alpha)b_1.$$

Summarizing, we have shown that $\mathcal{D}_{\mathbf{c}}$ is singular, having eight double points; they come from 8 points on $\Gamma_{\mathbf{c}}$ for which $a^2 = -\alpha_2/\alpha_1$, which get identified in pairs, by $\varphi'_{\mathbf{c}}$, with 8 points on $\Gamma_{\mathbf{c}}$ for which $a^2 = -\alpha_4/\alpha_3$. Computing an extra term in the series it follows easily that these points are actually ordinary double points, i.e., the two branches of $\mathcal{D}_{\mathbf{c}}$ meet transversally at these points, as depicted in Figure (8.3). Since the genus of $\overline{\Gamma_{\mathbf{c}}}$ is 17, and since the Euler characteristic of an ordinary double point is 1, we find that the genus of $\mathcal{D}_{\mathbf{c}}$ is $17+8=25$, in particular the embedding that we constructed satisfies the adjunction formula (7.50).

We now exhibit the quadratic differential equations for $\mathcal{V}_1$ in the chart $Z_0 \neq 0$. For the chart $Z_1 \neq 0$, the algorithm that we gave in Paragraph 7.6.6 also yields that $\mathcal{V}_1$ is given by quadratic equations, but the formulas get rather long, so they will not be given here. Since the original equations for $\mathcal{V}_1$ are quadratic, $\dot{z}_1, \dots, \dot{z}_6$ is given by quadratic equations. For the functions $z_7 = G_1, \dots, z_{11} = G_5$ this follows from (8.146), which asserts that $\dot{G}_i$ is, for $i = 1, \dots, 5$ linear in $x_1 \dots, x_6$ on the one hand, and linear in $G_1, \dots, G_5$ on the other hand.

Since the involution π preserves the vector field $\mathcal{V}_1$, it is, in view of (8.162) sufficient to give the formulas for the following 5 functions, to show that $z_{12}, \ldots, z_{19}$ also satisfy quadratic differential equations.

$$\begin{aligned}
\dot{z}_{12} &= \frac{\alpha^2-1}{2}(z_7 z_{11} - z_9 z_{10}) - z_4 z_{17},\\
\dot{z}_{14} &= z_3 z_{16} + (1-\alpha^2) c_3 z_9 - (1+\alpha) z_{10} z_{11},\\
\dot{z}_{15} &= \frac{1+\alpha}{3\alpha-1}\left((1-\alpha)(z_8^2+z_9^2) - 2c_1 z_7 - 2\alpha z_{11}^2\right) + (1-\alpha^2)c_3(z_1 z_4 - z_5^2)\\
&\quad +(1+\alpha) z_7 z_8 + z_3 z_{14},\\
\dot{z}_{17} &= \frac{1}{2}\left((1+\alpha)^2 z_8 z_9 + (1-\alpha^2) z_{10} z_{11} + (1-\alpha)^2 z_7 z_9\right) - z_1 z_{12} - z_2 z_{13},\\
\dot{z}_{19} &= \frac{3\alpha^2+1}{2} z_8 z_{11} - z_1 z_{17}.
\end{aligned}$$

Finally, for the remaining functions $z_{20}, \ldots, z_{23}$ we need to do, in view of (8.163), only the following three functions.

$$\begin{aligned}
\dot{z}_{20} &= \frac{2\alpha}{1+\alpha}\left(2\alpha z_8 z_{16} - 2 z_{10} z_{13} + (\alpha-1) z_9 z_{14}\right),\\
\dot{z}_{21} &= \frac{1-\alpha}{9\alpha^2-1}\left(4\alpha c_1 z_6 z_{10} - (1-\alpha) z_3 z_{22}\right) + \frac{(\alpha-1)(9\alpha^2-4\alpha-1)}{4(3\alpha-1)} z_{15} z_{11}\\
&\quad +\frac{1}{2}(\alpha^2-1)(3\alpha+1)c_3(z_7 z_5 + z_2 z_8) + \frac{1}{4}(1+\alpha)(3\alpha-1) z_{16} z_{11}\\
&\quad -(\alpha-1)c_2 z_1 z_9 + \frac{(\alpha-1)(3\alpha-1)}{2(3\alpha+1)}\left((\alpha+1)(3\alpha-1)c_3 z_9 - z_{23}\right) z_1\\
&\quad -\frac{3\alpha^2+1}{2(3\alpha+1)} z_4 z_{23} - \frac{\alpha}{3\alpha+1}\left((\alpha+1) z_{14} + 2(\alpha-1)^2 c_3 z_3\right) z_{10}\\
&\quad -\frac{(1+\alpha)(3\alpha-1)}{3\alpha+1} c_1 (z_5 z_7 + z_2 z_8),\\
\dot{z}_{23} &= \frac{1}{2(3\alpha-1)}\left((3\alpha+1)(\alpha-1) z_9 z_{15} + 8\alpha c_1 z_2 z_{10} - (3\alpha^2+1) z_5 z_{22}\right)\\
&\quad +\frac{1}{2(3\alpha+1)}\left((3\alpha-1)(\alpha+1) z_9 z_{16} - 8\alpha c_2 z_1 z_{11} - (3\alpha^2+1) z_4 z_{21}\right)\\
&\quad -\frac{1+\alpha}{2}(z_{10} z_{12} + 4\alpha c_3 z_4 z_{11}) + \frac{1-\alpha}{2}(z_{11} z_{13} - 4\alpha c_3 z_5 z_{10}).
\end{aligned}$$

Let us check that the flow that starts from the points $\varphi'_{\mathbf{c}}(\mathbf{0}^{\epsilon_1\epsilon_2\epsilon_3})$ goes into the affine immediately. Since z_1 defines a chart around any of these points it suffices to show that $1/z_1(t,\varsigma)$ has a non-zero limit when $\varsigma \to 0$. Substituting (8.164) in the first terms of the series for z_1 we find that for ς small

$$\frac{1}{z_1}(t,\varsigma) = \frac{\varsigma}{1-\varsigma^2} t + \frac{\epsilon_1}{2} \frac{(4\alpha^2\varsigma^2 + (1-\alpha)^2)(1+\star\epsilon_2\varsigma)}{1-\varsigma^2}\sqrt{\frac{c_2}{\alpha_2}} t^2 + O(t^3),$$

so that

$$\lim_{\varsigma\to 0}\frac{1}{z_1}(t,\varsigma)=\frac{\epsilon_1}{2}(1-\alpha)^2\sqrt{\frac{c_2}{\alpha_2}}t^2+O(t^3),$$

which shows that the flow that starts from the points $\varphi_{\mathbf{c}}'(\mathbf{0}^{\epsilon_1\epsilon_2\epsilon_3})$ goes indeed into the affine immediately. Since the involution π permutes the points $\varphi_{\mathbf{c}}'(\mathbf{0}^{\epsilon_1\epsilon_2\epsilon_3})$ and $\varphi_{\mathbf{c}}'(\mathbf{1}^{\epsilon_1\epsilon_2\epsilon_3})$ this property follows also for the latter points. The function z_1 also defines a chart around the points $\infty^{\epsilon_1\epsilon_2\epsilon_3}$ and we find by substituting (8.165) in the first terms of the series for z_1 that

$$\frac{1}{z_1}(t,\varsigma)=\frac{\varsigma}{1-\varsigma^2}t-\frac{\epsilon_1}{8}\frac{(4\alpha^2+(1-\alpha)^2\varsigma^2)(1+\star\epsilon_2\varsigma)}{1-\varsigma^2}\beta_0t^2+O(t^3),$$

(for the definition of β_0, see (8.165)) so that

$$\lim_{\varsigma\to 0}\frac{1}{z_1}(t,\varsigma)=\frac{\epsilon_1\alpha^2}{2}\beta_0t^2+O(t^3),$$

showing that the flow that starts from the points $\varphi_{\mathbf{c}}'(\infty^{\epsilon_1\epsilon_2\epsilon_3})$ goes into the affine immediately. For the remaining eight points, which are the double points on $\mathcal{D}_{\mathbf{c}}$ that were obtained from the points $s_{\epsilon_1\epsilon_2\epsilon_3}$, let us merely[12] check that the degree of $\mathcal{D}_{\mathbf{c}}$ at these special points, which is two, agrees with the degree of $\overline{\varphi_{\mathbf{c}}(\mathbf{F}_{\mathbf{c}})}\setminus\varphi_{\mathbf{c}}(\mathbf{F}_{\mathbf{c}})$. Since z_{12} defines a chart at these points it suffices to substitute (8.167) in the series for $1/z_{12}$, which yields,

$$\frac{1}{z_{12}}(t,\varsigma)=\frac{16a_1b_1\alpha^2\varsigma t}{(\alpha^2-1)^2((3\alpha+1)c_1+(3\alpha-1)c_2-\alpha(9\alpha^2-1)c_3)}+O(t^2,t\varsigma^2),$$

showing that the degree is indeed two. Notice that this confirms also our earlier claim that the double points are ordinary double points of $\mathcal{D}_{\mathbf{c}}$. Note that the degree calculation is meaningful, as the vector field $\mathcal{V}_1$ is only tangent to one of the branches of $\mathcal{D}_{\mathbf{c}}$ which cross at these points and we have taken the development along the non-tangent branch. Similarly, the other special points that come from the points $\mathbf{0}^{\epsilon_1\epsilon_2\epsilon_3}$, $\mathbf{1}^{\epsilon_1\epsilon_2\epsilon_3}$ and $\infty^{\epsilon_1\epsilon_2\epsilon_3}$ are simple (i.e., smooth points) of $\overline{\varphi_{\mathbf{c}}(\mathbf{F}_{\mathbf{c}})}\setminus\varphi_{\mathbf{c}}(\mathbf{F}_{\mathbf{c}})$. This leads to the following theorem, stating the algebraic complete integrability of metric III, and describing some of its algebraic geometric features.

Theorem 8.21. *For any fixed* $\alpha\in\mathbf{C}\setminus\{0,\pm1,\pm1/3\}$ *consider the integrable system* $(\mathbf{C}^6,\{\cdot\,,\cdot\},\mathbf{F})$, *which provides a normal form for the geodesic flow on* $\mathfrak{so}(4)$ *with respect to metric III, where* $F=(F_1,F_2,F_3,F_4)$ *is given by (8.145) and the matrix of* $\{\cdot\,,\cdot\}$ *is given by (8.9).*

[12] To check that the flow of $\mathcal{V}_1$, starting from these points, goes into the affine immediately, two more terms in the series (8.156) need to be computed.

(1) $(\mathbf{C}^6, \{\cdot,\cdot\}, \mathbf{F})$ *is a weight homogeneous a.c.i. system;*
(2) For generic $\mathbf{c}$ *the fiber* $\mathbf{F_c}$ *of its momentum map completes into an Abelian surface* $\mathbf{T}^2_\mathbf{c}$ *by adding a curve* $\mathcal{D}_\mathbf{c}$ *of genus* 25 *that has* 8 *nodes and is smooth elsewhere;*
(3) The line bundle $[\mathcal{D}_\mathbf{c}]$ *defines a polarization of type* $(2,12)$ *on* $\mathbf{T}^2_\mathbf{c}$ *and it leads of an embedding of the latter in* $\mathbf{P}^{23}$*;*
(4) The involutions σ_1 *and* σ_2 *define on* $\mathbf{T}^2_\mathbf{c}$ *two translations, over a half period, that leave* $\mathcal{D}_\mathbf{c}$ *invariant;*
(5) The quotient curve $\widetilde{\mathcal{D}}_\mathbf{c} := \mathcal{D}_\mathbf{c}/\langle\sigma_1,\sigma_2\rangle$ *is of genus* 7 *and has two nodes. It defines on the quotient Abelian surface* $\widetilde{\mathbf{T}}^2_\mathbf{c} := \mathbf{T}^2_\mathbf{c}/\langle\sigma_1,\sigma_2\rangle$ *a polarization of type* $(1,6)$*, leading to an embedding of* $\widetilde{\mathbf{T}}^2_\mathbf{c}$ *into* $\mathbf{P}^5$.

Proof. We have verified all the conditions that appear in Theorem 6.22, so that we can conclude that, for generic $\mathbf{c}$, the fiber $\mathbf{F_c}$ of the momentum map completes into an Abelian surface $\mathbf{T}^2_\mathbf{c}$ by adjoining the singular irrecucible divisor $\mathcal{D}_\mathbf{c}$ to it; also, the latter theorem implies that the commuting vector fields $\mathcal{V}_1$ and $\mathcal{V}_2$ are linear vector fields on these tori. This proves that we have indeed an a.c.i. system.

Since the involutions σ_1 and σ_2 preserve these vector fields we can now identify their restriction to these tori as translations over half periods. Similarly, we can now identify the restriction of τ to $\mathbf{T}^2_\mathbf{c}$ as the (-1)-involution on $\mathbf{T}^2_\mathbf{c}$. As we have embedded the tori $\mathbf{T}^2_\mathbf{c}$ in $\mathbf{P}^{23}$ the polarization type that is induced by $\mathcal{D}_\mathbf{c}$ on $\mathbf{T}^2_\mathbf{c}$ must be $(1,24)$ or $(2,12)$. Since τ changes the sign of all the phase variables $z_1,\dots,z_6$, which have degree (weight) one, the embedding variables of even weight are fixed for τ, while the ones of odd weight change signs when acted upon by τ. It follows from Table 8.5 that there are 10 of the z_i that are even for τ and 14 that are odd for τ.

In view of Proposition 5.20 the number of even/odd sections of $\mathcal{D}_\mathbf{c}$ must be $12+\epsilon$ when the polarization type is $(1,24)$, where $\epsilon\in\{0,-1,1\}$, which does not agree with our count. However, the same proposition predicts that the number of even/odd sections of $\mathcal{D}_\mathbf{c}$ must be $12+2\epsilon$ when the polarization type is $(2,12)$, where $\epsilon\in\{0,-1,1\}$, which is precisely what we found (with $\epsilon=\pm1$).

Finally, the quotient of $\mathbf{T}^2_\mathbf{c}$ by the translations σ_1 and σ_2 yields a new Abelian surface $\widetilde{\mathbf{T}}^2_\mathbf{c}$. Since $\mathbf{F_c}$ is invariant for these involutions the same is true for the divisor at infinity $\mathcal{D}_\mathbf{c}$. Thus, $\widetilde{\mathbf{T}}^2_\mathbf{c}$ contains the curve $\mathcal{D}_\mathbf{c}/\langle\sigma_1,\sigma_2\rangle$, which has two nodes, as the eight nodes form two orbits under the group generated by the two translations. The Kodaira map that corresponds to this quotient curve is found by selecting in $\varphi_\mathbf{c}$ those elements that are σ_1 and σ_2-invariant, since the action of each of these translations is diagonal in the basis $z_0,\dots,z_{23}$. We find the following functions: z_0, z_7, z_8, z_{15}, z_{16} and z_{20}. Consider now the smooth curve $\Delta_\mathbf{c}$, that appeared as an unramified quotient of the abstract Painlevé divisor $\overline{\Gamma_\mathbf{c}}$.

In view of Table 8.4 we have a commutative diagram

$$
\begin{array}{ccccc}
\overline{\Gamma_{\mathbf{c}}} & \xrightarrow{\text{birat}} & \mathcal{D}_{\mathbf{c}} & \longleftarrow & 8 \text{ nodes} \\
\downarrow {\scriptstyle 4:1 \text{ unram}} & & \downarrow {\scriptstyle 4:1 \text{ unram}} & & \downarrow {\scriptstyle 4:1 \text{ unram}} \\
\Delta_{\mathbf{c}} & \xrightarrow[\text{birat}]{} & \widetilde{\mathcal{D}}_{\mathbf{c}} & \longleftarrow & 2 \text{ nodes}
\end{array}
$$

As we know that $\Delta_{\mathbf{c}}$ has genus 5 we may conclude that $\widetilde{\mathcal{D}}_{\mathbf{c}}$ has genus 7, since it has two nodes and no other singularities. Thus, $\widetilde{\mathcal{D}}_{\mathbf{c}}$ induces on $\mathbf{T}^2_{\mathbf{c}}$ a polarization of type $(1,6)$ and we conclude from Ramanan's Theorem (Theorem 5.18) that $\widetilde{\mathcal{D}}_{\mathbf{c}}$ is very ample, and hence that the above Kodaira map of $\tilde{\mathbf{T}}^2_{\mathbf{c}}$ into $\mathbf{P}^5$ is an embedding, as expected. □

We can now use the methods that are explained in Paragraph 7.6.7 to compute the holomorphic differentials ω_1 and ω_2 on $\mathcal{D}_{\mathbf{c}}$ that come from the differentials $\mathrm{d}t_1$ and $\mathrm{d}t_2$ on $\mathbf{T}^2_{\mathbf{c}}$. In the notation of that paragraph we choose $y_0 := x_1$ and $y := x_2$. We need to compute explicitly $\mathcal{V}_2[x_1]$ and $\mathcal{V}_2[x_2]$; notice that these can be expressed as quadratic polynomials in terms of the embedding variables x_i because, restricted to these tori, x_1 and x_2 have a simple pole along $\mathcal{D}_{\mathbf{c}}$ and because $[\mathcal{D}_{\mathbf{c}}]$ is normally generated. For x_1 the result is (up to a constant) as follows.

$$
\begin{aligned}
x_1' = 2(\alpha-1)z_1z_{15} - 3(\alpha^2-1)z_3(z_{12}+z_{18}) - 2(\alpha+1)z_5z_{14} + 8\alpha z_{10}F_1 + {} \\
(\alpha+1)(3\alpha-1)\left(2(3\alpha^2-3\alpha-2)z_8z_{10} + (\alpha+1)(3\alpha-1)z_9z_{11}\right)/2 - {} \\
2z_3z_5\left((\alpha-1)(3\alpha+1)F_1 + (\alpha+1)(3\alpha-1)F_2 - (\alpha^2-1)(9\alpha^2+1)F_3\right).
\end{aligned}
$$

For x_2 the result is found immediately from it by using the involution π (which conserves both vector fields). By substituting the principal balances (8.156) in $\mathcal{V}_1[1/x_1] = -x_3x_5/x_1^2$, in $\mathcal{V}_1[x_2/x_1] = (x_1x_4x_6 - x_2x_3x_5)/x_1^2$, in $\mathcal{V}_2[1/x_1] = -x_1'/x_1^2$ and in $\mathcal{V}_2[x_2/x_1] = (x_2'x_1 - x_1'x_2)/x_1^2$, where x_i' is given above, one finds

$$
\begin{aligned}
\mathcal{V}_1\left[\frac{1}{x_1}\right]_{|\mathcal{D}_{\mathbf{c}}} &= \frac{a}{b^2}, \\
\mathcal{V}_1\left[\frac{x_2}{x_1}\right]_{|\mathcal{D}_{\mathbf{c}}} &= (3\alpha^2+1)\frac{a^3c}{b^3}, \\
\mathcal{V}_2\left[\frac{1}{x_1}\right]_{|\mathcal{D}_{\mathbf{c}}} &= -\frac{\alpha_3a^2+\alpha_4}{ab^4}\left((\alpha_1a^2+\alpha_2)a^2b^2c^2 - P_{\mathbf{c}}^-(a^2,b^2)\right), \\
\mathcal{V}_2\left[\frac{x_2}{x_1}\right]_{|\mathcal{D}_{\mathbf{c}}} &= -(3\alpha^2+1)\frac{ac}{b^5}\Big(2a^2b^2c_2' + {} \\
&\qquad \left((\alpha_3-\alpha_1)a^2+\alpha_4-\alpha_2\right)\left((\alpha_1a^2+\alpha_2)a^2b^2c^2 - P_{\mathbf{c}}^-(a^2,b^2)\right)\Big),
\end{aligned}
$$

where we recall that c_2' is the value of the constant of motion F_2'. In the notation of Paragraph 7.6.7 this means that, up to multiplicative constants,

$$\begin{aligned}\delta &= \frac{a^2c}{b^7}\left((\alpha_1a^2+\alpha_2)^2a^2b^2c^2-(\alpha_1a^2+\alpha_2)P_{\mathbf{c}}^-(a^2,b^2)-2a^2b^2c_2'\right)\\ &= \frac{a^2c}{b^7}\left((\alpha_1a^2+\alpha_2)^2a^2b^2c^2-p_3(a^2)\right),\end{aligned}$$

upon using the definition (8.158) of p_3. Noticing that the latter expression between bracket is the square root of the discriminant of $P_{\mathbf{c}}(a^2,b^2,c^2)$, as a polynomial in $a^2b^2c^2$ (with coefficients in $\mathbf{C}[a,b]$), we get

$$\delta = \frac{a^2c}{b^7}\sqrt{p_3^2(a^2)-p_4(a^2)(\alpha_1a^2+\alpha_2)^2}.$$

We compute now ω_1 and ω_2 from (7.48) using $\mathsf{d}(a/b) = \mathsf{d}a/b^3$, as follows from $a^2+b^2=1$, and the residues $y^{(0)} = a^2/b$ and $y_0^{(0)} = b^2/a$. We get

$$\begin{aligned}\omega_1 &= \frac{b^5\mathsf{d}(a^3/b^3)}{ac\sqrt{p_3^2(a^2)-p_4(a^2)(\alpha_1a^2+\alpha_2)^2}}\\ &= \frac{3a\mathsf{d}a}{c\sqrt{p_3^2(a^2)-p_4(a^2)(\alpha_1a^2+\alpha_2)^2}},\\ \omega_2 &= -\frac{1}{y_0^{(0)}}\mathcal{V}_2\left[\frac{1}{y_0}\right]\omega_1\\ &= \frac{3a(\alpha_3a^2+\alpha_4)\left((\alpha_1a^2+\alpha_2)a^2b^2c^2-P_{\mathbf{c}}^-(a^2,b^2)\right)\mathsf{d}a}{b^6c\sqrt{p_3^2(a^2)-p_4(a^2)(\alpha_1a^2+\alpha_2)^2}}.\end{aligned}$$

Notice that these differentials descend to the curve $\Delta_{\mathbf{c}}$, namely ω_1 descends to the differential form ψ_1, given by

$$\begin{aligned}\psi_1 &= \frac{3\,\mathsf{d}A}{c\sqrt{p_3^2(A)-p_4(A)(\alpha_1A+\alpha_2)^2}}\\ &= \frac{3\,\mathsf{d}A}{c((\alpha_1A+\alpha_2)A(1-A)c^2-p_3(A))},\end{aligned}$$

and similarly for ω_2. Notice that ψ_1 and ψ_2 do not descend to $\mathcal{H}_{\mathbf{c}}$, in fact both of them are odd with respect to the covering involution on $\Delta_{\mathbf{c}}$, which is given by $(A,c)\mapsto(A,-c)$.

It is now easy to verify the earlier announced tangency locus of the vector field $\mathcal{V}_1$ along $\mathcal{D}_{\mathbf{c}}$. Since this locus is given by the zeros of ω_1 it is already clear that the only possible candidates are the points at infinity, i.e., the images of the points $\mathbf{0}^{\epsilon_1\epsilon_2\epsilon_3}$, $\mathbf{1}^{\epsilon_1\epsilon_2\epsilon_3}$ and $\infty^{\epsilon_1\epsilon_2\epsilon_3}$ as well as the nodes on $\mathcal{D}_{\mathbf{c}}^{\epsilon_1\epsilon_2\epsilon_3}$.

This is done by using the parameterizations around these points, given above. For example, for the points $\infty^{\epsilon_1\epsilon_2\epsilon_3}$ the parameterization is given by (8.165); we have that $p_3^2(A) - p_4(A)(\alpha_1 A + \alpha_2)^2 \sim A^5 = a^{10}$ since, as we noticed before, the degree of $p_3^2(A) - p_4(A)(\alpha_1 A + \alpha_2)^2$ is 5, so that, around $\infty^{\epsilon_1\epsilon_2\epsilon_3}$,

$$\omega_1 \sim \frac{\varsigma^{-1}\mathrm{d}(\varsigma^{-1})}{\varsigma\sqrt{\varsigma^{-10}}} + \cdots \sim \varsigma \mathrm{d}\varsigma + \cdots,$$

showing that at the 8 points on $\mathcal{D}_{\mathbf{c}}$ that correspond to the points $\infty^{\epsilon_1\epsilon_2\epsilon_3}$ the vector field $\mathcal{V}_1$ is tangent to $\mathcal{D}_{\mathbf{c}}$ and that this tangency is simple. For the other points at infinity one arrives at precisely the same conclusion (for the nodes the tangency is, of course, only along one of the branches). See Figure 8.3.

9 Periodic Toda Lattices Associated to Cartan Matrices

9.1 Different Forms of the Periodic Toda Lattice

In its original form, the n-particle periodic Toda lattice is given by the Hamiltonian on $\mathbf{R}^{2n}$

$$H = \frac{1}{2}\sum_{i=1}^{n} p_i^2 + \sum_{i=1}^{n} e^{q_i - q_{i+1}},$$

where $q_{n+1} = q_1$; the symplectic structure is the canonical one, $\{q_i, q_j\} = \{p_i, p_j\} = 0$ and $\{q_i, p_j\} = \delta_{ij}$, where $1 \leqslant i, j \leqslant n$. For a mechanical interpretation, consider n unit mass particles on a circle that are connected by exponential springs. In [33], Bogoyavlensky proposed a Lie algebraic generalization, where the original Toda lattice corresponds to the root system $\mathfrak{a}_{n-1}$. Denoting by l the rank of the root system, the general form of the Hamiltonian is

$$H = \frac{1}{2}\sum_{i=1}^{n} p_i^2 + V_\bullet,$$

where $n = l + 1$ for the root systems $\mathfrak{a}_l, \mathfrak{e}_6, \mathfrak{e}_7, \mathfrak{g}_2$ and $n = l$ for the other root systems. Denoting

$$V_k := \sum_{i=1}^{k} e^{q_i - q_{i+1}}$$

the potential $V_\bullet$ is given for the root systems that correspond to the classical Lie algebras by the following expressions:

$$\begin{aligned}
V_{\mathfrak{a}_l} &= V_l + \exp(q_{l+1} - q_1), \qquad l \geqslant 2,\\
V_{\mathfrak{b}_l} &= V_{l-1} + \exp(q_l) + \exp(-q_1 - q_2), \qquad l \geqslant 2,\\
V_{\mathfrak{c}_l} &= V_{l-1} + \exp(2q_l) + \exp(-2q_l), \qquad l \geqslant 3,\\
V_{\mathfrak{d}_l} &= V_{l-1} + \exp(q_{l-1} + q_l) + \exp(-q_1 - q_2), \qquad n \geqslant 4.
\end{aligned}$$

For the exceptional Lie algebras they are given by

$$\begin{aligned}
V_{\mathfrak{e}_6} &= V_5 + \exp((-q_1 - q_2 - q_3 + q_4 + q_5 + q_6)/2 + q_7/\sqrt{2}) + \exp(-\sqrt{2}q_7,)\\
V_{\mathfrak{e}_7} &= V_5 + \exp((q_2 + \cdots + q_7 - q_8 - q_1)/2) + \exp(-q_1 - q_2) + \exp(q_8 - q_7),\\
V_{\mathfrak{e}_8} &= V_6 + \exp((-q_1 + q_2 + \cdots + q_7 - q_8)/2) + \exp(-q_1 - q_2)\\
&\quad + \exp(q_7 + q_8),
\end{aligned}$$

$$V_{\mathfrak{f}_4} = V_2 + \exp(q_3) + \exp((q_4 - q_1 - q_2 - q_3)/2) + \exp(-q_1 - q_4),$$
$$V_{\mathfrak{g}_2} = \exp(q_1 - q_2) + \exp(-2q_1 + q_2 + q_3) + \exp(q_1 + q_2 - 2q_3).$$

We will investigate the algebraic integrability of a class of polynomial Hamiltonians, that generalize these systems, up to a change of variables. In order to describe these systems we first generalize the above Hamiltonians, considering on $\mathbf{R}^{2l}$ Hamiltonians of the form

$$H = \frac{1}{2}\sum_{i=1}^{l} p_i^2 + \sum_{i=0}^{l} \exp\left(\sum_{j=1}^{l} N_{ij} q_j\right),$$

where N is a matrix of size $(l+1, l)$. The coordinates q_i, p_i are assumed to be canonical coordinates, as above, and we denote the standard inner product on $\mathbf{R}^l$ (or $\mathbf{R}^{l+1}$) by $\langle \cdot \,|\, \cdot \rangle$. For now our only assumption on the matrix N is that $N^\top$ has a unique normalized null vector $\xi = (\xi_0, \xi_1, \dots, \xi_l)^\top = \begin{pmatrix} 1 \\ \bar{\xi} \end{pmatrix}$. Notice that this condition implies that the lower $l \times l$ block of N, which we will denote by $\bar{N}$, is invertible. Thus, $\bar{N} = (N_{ij})_{1 \leqslant i,j \leqslant l}$ and the first line of N is given by $-\bar{\xi}^\top \bar{N}$. We also define the symmetric $l \times l$ matrix $\bar{R} := \bar{N}\bar{N}^\top$. Consider the linear change of variables

$$\bar{q} = \bar{N} q, \qquad \bar{p} = \left(\bar{N}^\top\right)^{-1} p$$

and notice that in terms of the new variables $\bar{q}$ and $\bar{p}$ the Poisson structure is still the canonical one, $\{\bar{q}_i, \bar{q}_j\} = \{\bar{p}_i, \bar{p}_j\} = 0$ and $\{\bar{q}_i, \bar{p}_j\} = \delta_{ij}$, where $1 \leqslant i, j \leqslant n$. In these new coordinates the Hamiltonian $H = H(\bar{q}, \bar{p})$ is given by

$$H = \frac{1}{2}\langle \bar{R}\bar{p} \,|\, \bar{p} \rangle + \sum_{i=1}^{l} \exp(\bar{q}_i) + \exp(-\langle \bar{\xi} \,|\, \bar{q} \rangle).$$

We make everything polynomial by putting

$$a_0 = \exp(-\langle \bar{\xi} \,|\, \bar{q} \rangle), \qquad a_i = \exp(\bar{q}_i), \qquad b_i = \bar{p}_i,$$

where $i = 1 \dots, l$. Then the Hamiltonian $H = H(a, b)$ is given by

$$H = \frac{1}{2}\langle \bar{R} b \,|\, b \rangle + \sum_{i=0}^{l} a_i \tag{9.1}$$

while the Poisson structure has become a linear Poisson structure, to wit

$$\begin{aligned} \{a_i, a_k\} &= 0, \qquad & \{a_i, b_j\} &= a_i \delta_{ij}, \\ \{b_i, b_j\} &= 0, \qquad & \{a_0, b_j\} &= -\xi_j a_0, \end{aligned} \tag{9.2}$$

where $i, j = 1, \dots, l$ and $k = 0, \dots, l$.

Notice that the definition of the a_i implies that $\prod_{i=0}^{l} a_i^{\xi_i} = 1$, but the above Hamiltonian H and the Poisson structure $\{\cdot\,,\cdot\}$ make perfectly sense on all of $\mathbf{R}^{2l+1}$, with $\prod_{i=0}^{l} a_i^{\xi_i}$ being a Casimir. In the sequel it will be irrelevant if we fix this Casimir to the value 1 or if we work on the bigger affine space, which is simpler, hence we will work on the latter. From (9.1) and (9.2) one easily computes the following equations for the vector field $\mathcal{X}_H$:

$$\begin{aligned} \dot{a}_i &= a_i \textstyle\sum_{j=1}^{l} \bar{R}_{ij} b_j, \\ \dot{a}_0 &= -a_0 \textstyle\sum_{i,j=1}^{l} \xi_i \bar{R}_{ij} b_j, \end{aligned} \qquad \dot{b}_i = -a_i + a_0 \xi_i, \tag{9.3}$$

where $1 \leqslant i, j \leqslant l$.

We wish to write these equations in a simpler form. To this, consider the symmetric matrix $R := NN^\top$ which is of rank l since ξ is a null vector of R and since

$$R = \begin{pmatrix} -\xi^\top \bar{N} \\ \bar{N} \end{pmatrix} \begin{pmatrix} -\bar{N}^\top \xi \ , \ \bar{N}^\top \end{pmatrix} = \begin{pmatrix} \langle \bar{R}\xi \,|\, \xi \rangle & -(\bar{R}\xi)^\top \\ -\bar{R}\xi & \bar{R} \end{pmatrix}.$$

R can be diagonalized by an orthogonal matrix so we may factorize R as $R = EE^\top$, where E is a square matrix of size $l+1$ and of rank l. The $l+1$ columns of E are dependent of rank l, hence ξ is (up to a multiple) the unique null vector of $E^\top$. Let α denote a non-zero null vector of E. We define another set of coordinates by letting $u := E^\top \begin{pmatrix} 0 \\ b \end{pmatrix}$ and $v := -a$, where $b := (b_1, \dots, b_l)^\top$ and $a := (a_0, \dots, a_l)^\top$. Notice that the u coordinates are dependent since $\alpha^\top u = 0$. The Hamiltonian takes the simple form

$$H = \frac{1}{2} \langle u \,|\, u \rangle - \sum_{i=0}^{l} v_i, \tag{9.4}$$

while the vector field $\mathcal{X}_H$ takes the symmetric form

$$\dot{v}_i = v_i \sum_{j=0}^{l} e_{ij} u_j, \qquad \dot{u}_i = \sum_{j=0}^{l} e_{ji} v_j.$$

This can be shown directly from (9.3), by performing the change of variables, or by computing $\mathcal{X}_H$ from (9.4), by first checking that the Poisson structure is given, on the hyperplane $\alpha^\top u = 0$, by $\{u_i, u_j\} = \{v_i, v_j\} = 0$ and $\{v_i, u_j\} = e_{ij} v_i$, where $0 \leqslant i, j \leqslant l$. For the last change of variables, let

$$x_i := \frac{1}{2} \langle e_i \,|\, e_i \rangle \, v_i, \qquad y_i := \sum_{j=0}^{l} e_{ij} u_j,$$

for $i = 0, \dots, l$, where e_i is the i-th row of E. Notice that, again, the y-coordinates are dependent, as E does not have maximal rank.

Precisely, since ξ generates the null space of E, we have that $\sum_{j=0}^{l}\xi_j y_j = 0$. If we denote for $x, y \in \mathbf{R}^{l+1}$ by $x \cdot y$ the vector in $\mathbf{R}^{l+1}$, defined by $(x \cdot y)_i := x_i y_i$, where $i = 0, \dots, l$, then in terms of the variables x_i, y_i the vector field $\mathcal{X}_H$ takes on the hyperplane $\sum_{j=0}^{l}\xi_j y_j = 0$ its simplest form

$$\dot{x} = x \cdot y, \qquad \dot{y} = Ax, \tag{9.5}$$

where A is the square matrix $A = (a_{ij})_{0 \leqslant i,j \leqslant l}$, whose entries are computed from the rows of the matrix E by

$$a_{ij} := 2\frac{\langle e_i \,|\, e_j\rangle}{\langle e_j \,|\, e_j\rangle}, \qquad 0 \leqslant i, j \leqslant l.$$

The Poisson bracket is, on this hyperplane, given by $\{x_i, x_j\} = \{y_i, y_j\} = 0$ and

$$\{x_i, y_j\} = \langle e_i \,|\, e_j\rangle\, x_i = \frac{1}{2}\langle e_j \,|\, e_j\rangle\, a_{ij} x_i. \tag{9.6}$$

It has the following Casimir,

$$C_1 := \prod_{i=0}^{l} x_i^{\xi_i}. \tag{9.7}$$

Indeed $\{y_i, C_1\} = C_1 \left\langle e_i \,|\, \sum_{j=0}^{l}\xi_j e_j\right\rangle = 0$ and obviously $\{x_i, C_1\} = 0$, for any $i \in \{0, \dots, l\}$, showing that C_1 is a Casimir. Notice that the Poisson structure is indeed well-defined on the hyperplane $\sum_{j=0}^{l}\xi_j y_j = 0$, since for $i \in \{0, \dots, l\}$ we have that $\{x_i, C_0\} = x_i \left\langle e_i \,|\, \sum_{j=0}^{l}\xi_j e_j\right\rangle = 0$, while $\{y_i, C_0\}$ is obviously zero.

Equation (9.5) is the form of the generalized Toda lattice with which we shall work. As before we will consider these equations on a complex, rather than a real space, namely we consider them on the complex hyperplane M of $\mathbf{C}^{2(l+1)}$, defined by

$$M := \left\{ (x, y) \in \mathbf{C}^{2(l+1)} \mid \sum_{j=0}^{l}\xi_j y_j = 0 \right\}. \tag{9.8}$$

The matrices E and A, however, will always be taken real.

9.2 The Kowalevski-Painlevé Criterion

In this section we select from the Hamiltonian vector fields (9.5) those that satisfy the Kowalevski-Painlevé Criterion. Recall from Section 6.2.1 that this criterion gives necessary conditions for algebraic complete integrability (with the generic fiber of the moment map not containing an elliptic curve). The proof of this fact was first[1] given in [10]; the proof that we give here follows the notations and setup of [15].

Motivated by the previous section we consider a generic system of linearly dependent vectors in a real Euclidean space $(\mathbf{R}^{l+1}, \langle \cdot \,|\, \cdot \rangle)$, defined as follows.

Definition 9.1. For $l \geqslant 1$, let $e_0, \dots, e_l$ be vectors in $\mathbf{R}^{l+1}$. Suppose that

1. The $l+1$ vectors $e_0, \dots, e_l$ are linearly dependent;
2. Any l of the vectors $e_0, \dots, e_l$ are linearly independent;
3. The real numbers $\xi_0, \xi_1, \dots, \xi_l \in \mathbf{R}\setminus\{0\}$ which satisfy $\sum_{i=0}^{l} \xi_i e_i = 0$ have a non-zero sum, i.e., $\sum_{i=0}^{l} \xi_i \neq 0$.

Then we call $(e_0, \dots, e_l)$ a *generic* $(l+1)$*-tuple of rank* l in $\mathbf{R}^{l+1}$.

Given vectors $e_0, \dots, e_l$ in $\mathbf{R}^{l+1}$, let E denote the matrix whose i-th row consists of the coordinates of e_i with respect to some fixed orthonormal basis $\mathcal{B}$. Then the above three conditions for the rows of E to be a generic $(l+1)$-tuple of rank l in $\mathbf{R}^{l+1}$ are expressed in terms of E as follows.

1. $\det(E) = 0$;
2. $E^\top$ has a unique null vector of the form $\xi = (1, \xi_1, \dots, \xi_l)^\top$; all entries ξ_i of ξ are different from zero;
3. $\sum_{i=0}^{l} \xi_i \neq 0$.

Suppose now that E satisfies those conditions. Since each of the vectors e_i is different from 0, we may define a $(l+1, l+1)$ matrix A by

$$a_{ij} := \frac{2\,\langle e_i \,|\, e_j \rangle}{\langle e_j \,|\, e_j \rangle}, \qquad (0 \leqslant i, j \leqslant l).$$

We will call this matrix the *Cartan matrix* of E. Since $\mathcal{B}$ is an orthonormal basis, the relation between A and E is given by

$$A = EE^\top D, \qquad D \text{ diagonal, invertible.}$$

We list a few useful consequences for A. Since every row of E is a linear combination of the other rows of E, which are independent, every principal minor of A is different from zero. In particular, A has rank l and the null space of $A^\top$ is spanned by ξ, where ξ is the above null vector of $E^\top$. Also, the null space of A is spanned by $D^{-1}\xi$.

[1] A similar result was proven independently by Yoshida [175], but under stronger hypothesis.

Notice that all entries of $D^{-1}\xi$ are different from zero, because all entries of ξ are different from zero; we normalize $D^{-1}\xi$ so as to obtain that its first element is 1 and we call the resulting vector $\hat{\xi} = (1, \hat{\xi}_1, \dots, \hat{\xi}_l)^\top$. The fact that $\xi^\top E = 0$ also implies that $\xi^\top u = 0$, for any u in the image of A; in particular, condition 3 above implies that $\delta := (1, \dots, 1)^\top$ does not belong to the image of A. Finally, all diagonal entries of A are equal to 2 and $a_{ij} = 0$ if and only if $a_{ji} = 0$, for $0 \leqslant i, j \leqslant l$.

Example 9.2. Suppose that $\mathfrak{g}$ is a simple Lie algebra of rank l and that ν is an outer automorphism of $\mathfrak{g}$. The $l+1$ simple roots

$$\bar{\Pi} = \{\bar{\alpha}_0 = (\alpha_0, 1), \bar{\alpha}_1 = (\alpha_1, 0), \dots, \bar{\alpha}_l = (\alpha_l, 0)\}$$

of the twisted affine Lie algebra $L(\mathfrak{g}, \nu)$ define a generic system of dependent vectors in $\mathfrak{h}^*_{\mathbf{R}} \times \mathbf{R}$, as follows easily from the properties stated in Section 2.4. The Cartan matrix of the matrix E formed by these roots is, of course, the Cartan matrix of $L(\mathfrak{g}, \nu)$ with respect to $\bar{\Pi}$, as defined in Section 2.4. See Tables 2.4 and 2.5 for the list of all Cartan matrices that correspond to a twisted affine Lie algebra.

Theorem 9.3. *For $l > 1$, let the $(l+1) \times (l+1)$ matrix E define a generic system of dependent vectors in $\mathbf{R}^{l+1}$ and let A be its Cartan matrix. Consider the vector field $\mathcal{V}$ on $M \subset \mathbf{C}^{2(l+1)}$ which is given by*

$$\begin{aligned} \dot{x} &= x \cdot y, \\ \dot{y} &= Ax, \end{aligned} \tag{9.9}$$

where $x, y \in \mathbf{C}^{l+1}$ and where $x \cdot y$ is a shorthand for the element of $\mathbf{C}^{l+1}$ for which $(x \cdot y)_i = x_i y_i$, $i = 0, \dots, l$. If the vector field $\mathcal{V}$ is one of the integrable vector fields of an irreducible a.c.i. system then A is the Cartan matrix of a twisted affine Lie algebra.

We first prove two lemmas that will be used in the proof of the theorem. The first one describes the leading behavior of all possible Laurent solutions to (9.9), while the second one gives a more detailed description of the first few terms of the principal balances.

Lemma 9.4. *Let A be the Cartan matrix of an $(l+1) \times (l+1)$ matrix E, which defines a generic system of dependent vectors in $\mathbf{R}^{l+1}$, where $l > 1$. All formal Laurent solutions to (9.9) are weight homogeneous, where all variables x_i have weight 2 and all variables y_i have weight 1: these Laurent solutions have at most a double pole for all variables x_i and at most a simple pole for all variables y_i. Each of these formal Laurent solutions is convergent (for small values of $|t|$).*

Proof. The equations (9.9) for the vector field $\mathcal{V}$ imply that if one of the x_i variables has a pole then also one of the y variables and vice-versa. Thus, we look for Laurent solutions to (9.9) of the form

$$\begin{aligned} x(t) &= \frac{1}{t^m}\left(x^{(m)} + O(t)\right), \\ y(t) &= \frac{1}{t^k}\left(y^{(k)} + O(t)\right), \end{aligned} \tag{9.10}$$

with $m, k > 0$ and $x^{(m)} \neq 0$ and $y^{(k)} \neq 0$.

By a direct substitution of (9.10) and (9.10) in (9.9) we find

$$-m\frac{x^{(m)}}{t^{m+1}} + \cdots = \frac{x^{(m)} \cdot y^{(k)}}{t^{m+k}} + \cdots, \tag{9.11}$$

$$-k\frac{y^{(k)}}{t^{k+1}} + \cdots = \frac{Ax^{(m)}}{t^m} + \cdots \tag{9.12}$$

We show that $m = 2$ and $k = 1$. To do this, assume first that $m > k + 1$. Then (9.12) implies that $x^{(m)}$ is a non-zero multiple of $\hat{\xi}$, so that all entries of $x^{(m)}$ are different from zero. Since $y^{(k)} \neq 0$ this implies that $k = 1$ and that $y^{(k)} = -m\delta$. But (9.12) says that $y^{(k)}$ must be in the image of A, which is impossible since we have seen that δ is not in the image of A. Thus, $m > k + 1$ is impossible. But $m < k + 1$ is also impossible, as follows from (9.12). Therefore, we conclude that $m = k + 1$.

We proceed to show that $k = 1$, so that $m = 2$. To do this, assume that $k > 1$ and let $m = k + 1$. Then (9.11) and (9.12) imply that

$$x^{(k+1)} \cdot y^{(k)} = 0, \tag{9.13}$$

$$-ky^{(k)} = Ax^{(k+1)}. \tag{9.14}$$

Let us denote, just for the sake of the present computation, the complex conjugate of a vector $v \in \mathbf{C}^{l+1}$ by $\bar{v}$. The matrices A, E and D are real but the vectors $x^{(k+1)}$ and $y^{(k)}$ need not be real. However, notice that $x^{(k+1)} \cdot y^{(k)} = 0$ implies that for any $i = 0, \ldots, l$ one has $x_i^{(k+1)} = 0$ or $y_i^{(k)} = 0$. Therefore, using (9.14), we have that

$$(Dx^{(k+1)})^\top \overline{Ax^{(k+1)}} = -k(Dx^{(k+1)})^\top \overline{y^{(k)}} = 0.$$

Since $A = EE^\top D$ this implies that

$$0 = (Dx^{(k+1)})^\top EE^\top \overline{Dx^{(k+1)}} = (E^\top Dx^{(k+1)})^\top \overline{E^\top Dx^{(k+1)}}.$$

Thus, $E^\top Dx^{(k+1)} = 0$ and, by (9.14) we conclude that

$$y^{(k)} = -\frac{1}{k}Ax^{(k+1)} = -\frac{1}{k}EE^\top Dx^{(k+1)} = 0,$$

a contradiction. This shows that $k > 1$ is impossible, so that $k = 1$ and $m = 2$.

Since the vector field (9.9) is weight homogeneous, with weight 2 for all x_i and weight 1 for all y_i, this means that all Laurent solutions to (9.9) are weight homogeneous. By Theorem 7.25 all formal solutions to (9.9) are convergent (for small values of $|t|$). □

Lemma 9.5. *Under the hypothesis of Theorem 9.3, suppose that*

$$x(t) = \frac{1}{t^2}\sum_{k\geqslant 0} x^{(k)}t^k, \qquad y(t) = \frac{1}{t}\sum_{k\geqslant 0} y^{(k)}t^k, \tag{9.15}$$

is a balance for (9.9). Let $S \subset \{0,1,\dots,l\}$ denote the set of indices i for which $x_i^{(0)} \neq 0$ and denote $s := \#S$. Then $0 < s \leqslant l$ and $y_i^{(0)} = -2$ whenever $i \in S$. Moreover, if (9.15) is a principal balance then $s = 1$ and $x^{(1)} = 0$.

Proof. By direct substitution of (9.15) in (9.9) one finds that the indicial equation yields

$$\begin{aligned} -2x^{(0)} &= x^{(0)} \cdot y^{(0)}, \\ -y^{(0)} &= Ax^{(0)}. \end{aligned} \tag{9.16}$$

Notice that if all $x_i^{(0)}$ are different from zero then $y^{(0)} = -2\delta$, which is impossible since, as we have seen, δ is not in the image of A. So at least one of the $x_i^{(0)}$ is equal to zero, $s \leqslant l$. Also, $x^{(0)} = 0$ leads to $y^{(0)} = 0$, which does not correspond to a balance, so that $s > 0$.

Let us denote by S the subset of $\{0,1,\dots,l\}$ that contains those i for which $x_i^{(0)} \neq 0$. This set S will be supposed fixed throughout the proof and we denote $s := \#S$. Notice that the first equation in (9.16) implies that $y_i^{(0)} = -2$ if $i \in S$.

For a vector $v = (v_0, v_1, \dots, v_l)^\top \in \mathbf{C}^{l+1}$, let us denote[2] by $\overline{v}$ (resp. by $\underline{v}$) the vector of $\mathbf{C}^s$ (resp. of $\mathbf{C}^{l+1-s}$) whose components are the components v_i of v for which $i \in S$ (resp. $i \notin S$). Also, $\overline{A}$ denotes the $s \times s$ matrix that contains the entries a_{ij} with $i, j \in S$.

With these notations the (partially solved) indicial equation can be rewritten in the following (redundant) form.

$$\begin{aligned} \underline{x^{(0)}} &= 0, \\ \underline{y^{(0)}} &= -\underline{Ax^{(0)}}, \\ \overline{y^{(0)}} &= -2\overline{\delta}, \\ \overline{A}\,\overline{x^{(0)}} &= 2\overline{\delta}. \end{aligned} \tag{9.17}$$

[2] After permuting the variables x_i (and the y_i at the same time) one may assume that $x_0^{(0)}, \dots, x_{s-1}^{(0)}$ are all different from zero while $x_s^{(0)} = \cdots = x_l^{(0)} = 0$, but we will not do this, because that is not what one wants to do in practice when computing all balances.

We see from these equations that the solutions to the indicial equation do not contain any free parameters: since any principal minor of A is different from zero, $\bar{A}$ is invertible and $\overline{x^{(0)}}$ is determined from these equations, hence they have a unique solution for $x^{(0)}$. Then $y^{(0)}$ follows from $y^{(0)} = -Ax^{(0)}$. We now move to the equations that come after the indicial equation. First, using (9.17), the equations at step 1 are given by

$$\begin{aligned} Ax^{(1)} &= 0, \\ \underline{x^{(1)}} \cdot \left(\underline{y^{(0)}} + \underline{\delta}\right) &= 0, \\ \overline{x^{(0)}} \cdot \overline{y^{(1)}} &= \overline{x^{(1)}}. \end{aligned} \tag{9.18}$$

For $k \geqslant 2$ we find the following equations.

$$\begin{aligned} (k-2)x^{(k)} &= x^{(k)} \cdot y^{(0)} + x^{(0)} \cdot y^{(k)} + R^{(k-1)} \\ (k-1)y^{(k)} &= Ax^{(k)}, \end{aligned}$$

where $R^{(k-1)}$ involves only the variables $x_i^{(l)}$ and $y_i^{(l)}$ with $l < k$. Splitting these vector equations up in the S part and the remaining part, using (9.17) and substituting the second into the first we can rewrite these equations as follows.

$$\begin{aligned} &(k-2)\underline{x^{(k)}} + \underline{x^{(k)}} \cdot \underline{Ax^{(0)}} = \underline{R^{(k-1)}}, \\ &k(k-1)\overline{x^{(k)}} - \overline{x^{(0)}} \cdot \overline{Ax^{(k)}} = (k-1)\overline{R^{(k-1)}}, \\ &(k-1)y^{(k)} = Ax^{(k)}. \end{aligned} \tag{9.19}$$

We show that if the Laurent solution (9.15) contains $2l = \dim M - 1$ free parameters then $x^{(1)} = 0$. Indeed, if $x^{(1)} \neq 0$ then (9.18) implies that $x^{(1)} = cq$, where c is possibly a free parameter and q spans the null space of A, as before. In addition, (9.18) does not impose any constraints on $\underline{y^{(1)}}$, leading to $l - s$ free parameters, since $\sum_{j=0}^{l} \xi_j y_j = 0$ (see (9.8)). In total we have, after the first step, at most $l+1-s$ free parameters. Since all $x_i^{(1)}$ are different from zero, (9.18) implies that $\underline{Ax^{(0)}} = -\underline{y^{(0)}} - \underline{\delta}$. Using this, the first equation in (9.19) becomes

$$(k-1)\underline{x^{(k)}} = \underline{R^{(k-1)}},$$

which can be solved uniquely for $\underline{x^{(k)}}$, since $k \geqslant 2$. The resulting value of $\underline{x^{(k)}}$ depends on the values of $x^{(l)}$ and $y^{(l)}$ with $0 \leqslant l < k$ only. Since for $k \geqslant 2$, $y^{(k)}$ follows uniquely from $x^{(k)}$, the only free parameters that can appear after the first step must come from the equations that determine $\overline{x^{(k)}}$. But since $\underline{x^{(k)}}$ depends on the values of $x^{(l)}$ and $y^{(l)}$ with $0 \leqslant l < k$ only, we can rewrite the second equation in (9.19) as

$$k(k-1)\overline{x^{(k)}} - \overline{x^{(0)}} \cdot \left(\overline{A}\,\overline{x^{(k)}}\right) = \bar{R}^{(k-1)} \tag{9.20}$$

where $\bar{R}^{(k-1)}$ depends on the values of $x^{(l)}$ and $y^{(l)}$ with $0 \leqslant l < k$ only.

Since the spectrum of the $s \times s$ matrix, whose (i,j)-th entry (with $0 \leqslant i,j < s$) is given by $x_i^{(0)} a_{ij}$, has at most s elements, in particular at most s elements of the form $k(k-1)$, the values of all x_i^k with $k \geqslant 2$ are uniquely determined, except for at most s of them, which yield free parameters for $k \geqslant 2$. Combined with the $l+1-s$ parameters that we found (at most) for $k < 2$ we get at most $l+1$ free parameters, which is smaller that $2l$ since $l > 1$. Our claim, that if the Laurent solution contains $2l$ free parameters then $x^{(1)} = 0$, follows.

By a similar counting argument we find that if the Laurent solution contains $2l$ free parameters then $s = 1$. Indeed, we find as before no free parameters in the indicial equation, $l - s$ free parameters in the first step and at most $l - s$ in the determination of $x^{(k)}$ in the following steps; but the determination of $x^{(k)}$ in the steps that follow the first step yields also at most s free parameters. In total we get at most $2l + 1 - s$ free parameters, which is less that $2l$ unless $s = 1$. □

We are now ready to prove Theorem 9.3.

Proof. (of Theorem 9.3) We apply the Kowalevski-Painlevé Criterion (see Theorem 6.13) to the weight homogeneous vector field $\mathcal{V}$. According to that criterion there must exist for $0 \leqslant i \leqslant l$ a principal balance for which $x_i(t)$ has a pole. According to Lemma 9.5 $x_i(t)$ must then have a double pole, zero residue, and all other $x_j(t)$ (with $j \neq i$) have no pole. We will show that this implies that a_{ji} is a non-positive integer for all $j \neq i$ (recall that $a_{ii} = 2$). Repeating the argument for every $0 \leqslant i \leqslant l$ yields that

$$a_{ij} = \frac{2\langle e_i \,|\, e_j\rangle}{\langle e_j \,|\, e_j\rangle} \in \mathbf{Z}_- = \{0, -1, -2, \dots\} \quad \text{for} \quad 0 \leqslant i < j \leqslant l.$$

In view of Proposition 2.19 A is the Cartan matrix of a twisted affine Lie algebra.

By relabeling, it suffices to show the above implication for $i = 0$. Let

$$x(t) = \frac{1}{t^2}\left(x^{(0)} + x^{(1)}t + \sum_{k \geqslant 2} x^{(k)} t^k\right)$$

$$y(t) = \frac{1}{t}\left(y^{(0)} + y^{(1)}t + \sum_{k \geqslant 2} y^{(k)} t^k\right)$$

be a principal balance for (9.9), with $x^{(1)} = 0$ and $x^{(0)} = x_0^{(0)}(1, 0, \dots, 0)^\top$. Then $y^{(0)} = -Ax^{(0)} = x_0^{(0)}(-a_{00}, \dots, -a_{l0})^\top$, which implies that $x_0^{(0)} = 1$, since $y_0^{(0)} = -2 = -a_{00}$. The equations (9.18) which correspond to step 1 simply yield that $y_0^{(1)} = 0$ and that all other $y_i^{(1)}$ are free parameters, $l - 1$ in total (since $\sum_{j=0}^{l} \xi_j y_j = 0$).

For the other steps ($k \geqslant 2$) we investigate (9.19). The last equation says that all $y^{(k)}$ are determined by the $x^{(k)}$, hence all free parameters must come from the $x^{(k)}$. Since $a_{00} = 2$ Equation (9.20) implies

$$k(k-1)x_0^{(k)} - 2x_0^{(k)} = (k-1)\bar{R}^{(k-1)}$$

which yields that $x_0^{(2)}$ is a free parameter (it is easy to check that $\bar{R}^{(1)} = 0$). The first equation in (9.19) leads to l equations

$$(k-2+a_{i0})x_i^{(k)} = (R^{(k-1)})_i$$

which provides l free parameters (which is precisely the number of free parameters needed to arrive at the right number $2l$) only if for any $i = 1, \dots, l$ the equation $k - 2 + a_{i0} = 0$ admits an integral solution for k with $k \geqslant 2$. In other words, all entries a_{i0} with $1 \leqslant i \leqslant l$ must be non-positive and integral. □

Remark 9.6. The positive (but singular) symmetric matrix $EE^\top$ will have the same property as A that all its off-diagonal entries are negative. According to the Frobenius-Perron Theorem (see [40, p. 162]) all entries of the null vector $\xi = (1, \xi_1, \dots, \xi_l)^\top$ of $A^\top$ are positive. For each of the twisted affine Lie algebras the corresponding normalized vector ξ is given in Tables 2.4 and 2.5.

9.3 A Lax Equation for the Periodic Toda Lattice

We will now give a Lax equation for the Toda lattice associated to any of the twisted affine Lie algebras $L(\mathfrak{g}, \nu)$. Recall from Section 2.4 that $L(\mathfrak{g}, \nu)$ admits a system of simple roots $\bar{\Pi} = \{\bar{\alpha}_0 = (\alpha_0, 1), \bar{\alpha}_1 = (\alpha_1, 0), \dots, \bar{\alpha}_l = (\alpha_l, 0)\}$, where $\{\alpha_1, \dots, \alpha_l\}$ is a system of simple roots for $\mathfrak{g}_0$ (the subalgebra of $\mathfrak{g}$ that is fixed under ν), and where α_0 is, in the untwisted case, minus the highest long root of $\mathfrak{g} = \mathfrak{g}_0$, while α_0 is given by (2.32) in the twisted case. Starting from root vectors and coroots of $\mathfrak{g}_0$ we can choose root vectors $E_{\bar{\alpha}} \in L(\mathfrak{g}, \nu)$ and coroots H_α in $\mathfrak{h}_0 = \mathfrak{h} \cap \mathfrak{g}_0$ such that

$$\begin{aligned} [H, E_{\bar{\alpha}}] &= \langle \alpha, H \rangle E_{\bar{\alpha}}, \qquad H \in \mathfrak{h}_0, \\ [E_{\bar{\alpha}}, E_{-\bar{\alpha}}] &= H_\alpha. \end{aligned} \tag{9.21}$$

We also introduce, for $i = 1, \dots, l$ the *coweights* $\hat{\lambda}_1, \dots, \hat{\lambda}_l \in \mathfrak{h}_0$ by

$$\left\langle \alpha_j, \hat{\lambda}_i \right\rangle = \delta_{ij}, \qquad 1 \leqslant i, j \leqslant l. \tag{9.22}$$

Consider now the elements of $L(\mathfrak{g},\nu)$, defined by

$$X := \sum_{i=1}^{l} y_i\hat{\lambda}_i - \sum_{i=0}^{l} x_i E_{\alpha_i} + \sum_{i=0}^{l} E_{-\alpha_i}, \qquad Y := -\sum_{j=0}^{l} x_j E_{\alpha_j}, \tag{9.23}$$

where Y is naturally interpreted as the projection of X on the subspace generated by the positive root vectors. We claim that the equation

$$\dot{X} = [X, Y]$$

is a Lax equation (with parameter, because of $E_{\pm\alpha_0}$) for the Toda vector field

$$\begin{aligned} \dot{x} &= x \cdot y, \\ \dot{y} &= Ax, \end{aligned} \tag{9.24}$$

on the hyperplane M of $\mathbf{C}^{2(l+1)}$, defined by (9.8). To show this, we compute the three contributions to $[X, Y]$, and we compare them to

$$\begin{aligned} \dot{X} &= \sum_{i=1}^{l} \dot{y}_i\hat{\lambda}_i - \sum_{i=0}^{l} \dot{x}_i E_{\alpha_i} \\ &= \sum_{i=1}^{l}\sum_{j=0}^{l} a_{ij}x_j\hat{\lambda}_i - \sum_{i=0}^{l} x_i y_i E_{\alpha_i}. \end{aligned}$$

The first term in $[X, Y]$ is given by

$$-\sum_{i=1}^{l}\sum_{j=0}^{l} x_j y_i \left[\hat{\lambda}_i, E_{\alpha_j}\right] = -\sum_{i=1}^{l}\sum_{j=0}^{l} x_j y_i \left\langle \alpha_j, \hat{\lambda}_i \right\rangle E_{\alpha_j},$$

where we used (9.21) for $\hat{\lambda}_i \in \mathfrak{h}_0$. Now

$$\left\langle \alpha_j, \hat{\lambda}_i \right\rangle = \begin{cases} \delta_{ij} & \text{if} \quad j \neq 0; \\ -\xi_i & \text{if} \quad j = 0, \end{cases}$$

where the last equality is obtained from $\alpha_0 = -\sum_{j=1}^{l} \xi_i\alpha_i$. Summing up and using $\sum_{j=0}^{l} \xi_i y_i = 0$ we conclude that the first term in $[X, Y]$ matches exactly the second term in $\dot{X}$. The second term in $[X, Y]$ is given by

$$-\sum_{i,j=0}^{l} x_i x_j \left[E_{\alpha_i}, E_{\alpha_j}\right] = 0,$$

by skew-symmetry.

For the last term in $[X, Y]$ we find

$$-\sum_{i,j=0}^{l} x_j \left[E_{-\alpha_i}, E_{\alpha_j}\right] = \sum_{j=0}^{l} x_j H_{\alpha_j},$$

but (9.22) and (2.23) yield

$$H_{\alpha_j} = \sum_{i=1}^{l} a_{ij} \hat{\lambda}_i, \tag{9.25}$$

and so if we substitute this into the previous formula then we find precisely the first term in $\dot{X}$. This show that $\dot{X} = [X, Y]$ is a Lax equation for the Toda vector field (9.24).

In an arbitrary faithful representation ρ of $\mathfrak{g}$ into $\mathfrak{gl}(N)$ we get, in view of the AKS Theorem, commuting Hamiltonians, given by the coefficients of the equation

$$\det(\mathfrak{z}\,\mathrm{Id}_N - \rho(L)) = \sum_{i,j} \mathfrak{z}^i \mathfrak{h}^j H_{ij} = 0,$$

where the dependence of $\rho(L)$ on $\mathfrak{h}$, polynomial in $\mathfrak{h}$ and $\mathfrak{h}^{-1}$ comes from the root vectors E_{α_0} and $E_{-\alpha_0}$. Setting $x_0 = 0$ the polynomials H_{i0} are invariant under the coadjoint action of $\mathfrak{g}_0$. By Chevalley's Theorem they are extensions of the Weyl polynomials of $\mathfrak{g}_0$, in particular they are independent, yielding l independent polynomials in involution on M, in addition to the Casimir C_1. Thus, the Toda lattice associated to any of the affine Lie algebras is Liouville integrable.

In the examples that follow, we give the explicit Hamiltonians for the periodic Toda lattices that involve precisely three (connected) particles. There are precisely six cases of them, going with the extended root systems $\mathfrak{a}_2^{(1)}$, $\mathfrak{c}_2^{(1)}$, $\mathfrak{g}_2^{(1)}$, $\mathfrak{a}_4^{(2)}$, $\mathfrak{d}_3^{(2)}$ and $\mathfrak{d}_4^{(3)}$. We give in each case the Cartan matrix A, the Casimir $C_0 = \sum_{i=0}^{2} \xi_i y_i$ that defines the subspace, $C_0 = 0$, the Casimir $C_1 = \prod_{i=0}^{2} x_i^{\xi_i}$, the energy H_1 and the extra constant of motion H_2.

Example 9.7. For the classical periodic 3-body Toda lattice the Cartan matrix is given by

$$A = \begin{pmatrix} 2 & -1 & -1 \\ -1 & 2 & -1 \\ -1 & -1 & 2 \end{pmatrix}.$$

The Casimir and commuting Hamiltonians are given by

$$\begin{aligned} C_0 &= y_0 + y_1 + y_2 = 0, \\ C_1 &= x_0 x_1 x_2, \\ H_1 &= (y_0 - y_1)^2 + (y_1 - y_2)^2 + (y_2 - y_0)^2 - 18(x_0 + x_1 + x_2), \\ H_2 &= \prod_{i<j} (y_i - y_j) - 9(x_0(y_1 - y_2) + x_1(y_2 - y_0) + x_2(y_0 - y_1)). \end{aligned}$$

Example 9.8. For the $\mathfrak{c}_2^{(1)}$ Toda lattice the Cartan matrix is given by

$$A = \begin{pmatrix} 2 & -2 & 0 \\ -1 & 2 & -1 \\ 0 & -2 & 2 \end{pmatrix}.$$

The Casimir and commuting Hamiltonians are given by

$$\begin{aligned} C_0 &= y_0 + 2y_1 + y_2 = 0, \\ C_1 &= x_0 x_1^2 x_2, \\ H_1 &= y_0^2 - 4x_0 + y_2^2 - 4x_2 - 8x_1, \\ H_2 &= (y_0^2 - 4x_0)(y_2^2 - 4x_2) - 8x_1(y_0 y_2 - 2x_1). \end{aligned}$$

Example 9.9. For the third and last untwisted case, the $\mathfrak{g}_2^{(1)}$ Toda lattice, the Cartan matrix is given by

$$A = \begin{pmatrix} 2 & 0 & -1 \\ 0 & 2 & -1 \\ -1 & -3 & 2 \end{pmatrix}.$$

Denoting $\phi_i := y_i^2 - 4x_i$, the Casimir and commuting Hamiltonians are in this case given by

$$\begin{aligned} C_0 &= y_0 + 3y_1 + 2y_2 = 0, \\ C_1 &= x_0 x_1^3 x_2^2, \\ H_1 &= y_0^2 - 4x_0 + 3(y_1^2 - 4x_1) - 4x_2, \\ H_2 &= 27(\phi_0 - \phi_1)^2 \phi_1 + 16x_2[18x_0 y_1(y_2 + 2y_0) + 3x_2(y_2 + 2y_0)^2 - \\ &\qquad 108x_1(x_0 + x_1) + 6x_1(y_2 + 2y_0)(4y_2 + 5y_0) + 2y_1(y_2 - y_0)(y_2 + 2y_0)^2]. \end{aligned}$$

Example 9.10. There are two twisted cases that go with an involution (automorphism of order two). The first one is the $\mathfrak{a}_4^{(2)}$ Toda lattice, whose Cartan matrix is given by

$$A = \begin{pmatrix} 2 & -2 & 0 \\ -1 & 2 & -2 \\ 0 & -1 & 2 \end{pmatrix}.$$

The Casimir and commuting Hamiltonians are explicitly given by

$$\begin{aligned} C_0 &= y_0 + 2y_1 + 2y_2 = 0, \\ C_1 &= x_0 x_1^2 x_2^2, \\ H_1 &= y_0^2 - 4x_0 + 4(y_2^2 - 4x_2) - 8x_1, \\ H_2 &= (y_0^2 - 4x_0)(y_2^2 - 4x_2) - 4x_1(y_0 y_2 - x_1 - 4x_2). \end{aligned}$$

Example 9.11. The second one is the $\mathfrak{d}_3^{(2)}$ Toda lattice, whose Cartan matrix is given by

$$A = \begin{pmatrix} 2 & -1 & 0 \\ -2 & 2 & -2 \\ 0 & -1 & 2 \end{pmatrix}.$$

The Casimir and commuting Hamiltonians are explicitly given by

$$\begin{aligned} C_0 &= y_0 + y_1 + y_2 = 0, \\ C_1 &= x_0 x_1 x_2, \\ H_1 &= y_0^2 - 4x_0 + y_2^2 - 4x_2 - 2x_1, \\ H_2 &= (y_0^2 - 4x_0)(y_2^2 - 4x_2) - x_1(2y_0y_2 - 4x_0 - x_1 - 4x_2). \end{aligned}$$

Example 9.12. There is one twisted case that goes with an automorphism of order three. It is the $\mathfrak{d}_4^{(3)}$ Toda lattice, whose Cartan matrix is given by

$$A = \begin{pmatrix} 2 & -1 & 0 \\ -1 & 2 & -1 \\ 0 & -3 & 2 \end{pmatrix}.$$

For this final case the Casimir and commuting Hamiltonians take the following form.

$$\begin{aligned} C_0 &= y_0 + 2y_1 + y_2 = 0, \\ C_1 &= x_0 x_1^2 x_2, \\ H_1 &= y_0^2 - 4x_0 + \frac{1}{3}(y_2^2 - 4x_2) - 4x_1, \\ H_2 &= ((y_0(y_1 + y_2) - x_1)y_1 + x_2y_0)^2 - 4x_1((y_0(y_1 + y_2) - x_1)^2 - x_2(y_0^2 - x_1)) \\ &\quad -4x_0\left((x_2 - x_0)^2 + 3x_1(x_0 + x_1) + 5x_1x_2 - \psi/2\right), \end{aligned}$$

where

$$\begin{aligned} \psi = {} & x_0(2y_0^2 - 8y_2^2) + 2x_1(12y_1^2 + 21y_1y_2 + 10y_2^2) \\ & -4x_2(y_0^2 - 2y_2^2) - y_1(y_1 + y_2)(3y_0^2 - 7y_2^2). \end{aligned}$$

Remark 9.13. Taking $x_0 = 0$ and computing y_0 from $C_0 = 0$ yields a *non-periodic Toda lattice.* The equations of motion can be written down directly by taking the Cartan matrix A of a (semi-)simple Lie algebra, giving

$$\dot{x} = x \cdot y, \qquad \dot{y} = Ax.$$

We will need the commuting Hamiltonians in the case of two particles. They are easily found from the periodic 3-particle Toda lattice, and are given in Table 9.1.

Table 9.1. For each of the four non-periodic 2-particle Toda lattices we display the commuting Hamiltonians. They are obtained, up to a constant, from the periodic 3-particle Toda lattice (see Examples 9.7 – 9.12) by setting $x_0 = 0$ and by substituting y_0 from $C_0 = y_0 + \xi_1 y_1 + \xi_2 y_2 = 0$.

	Non-periodic commuting Hamiltonians
$\mathfrak{a}_1 \oplus \mathfrak{a}_1$	$H_1 = y_\alpha^2 + y_\beta^2 - 4(x_\alpha + x_\beta)$
α β	$H_2 = y_\alpha^2 - 4x_\alpha$
$\mathfrak{a}_2$	$H_1 = y_\alpha^2 + y_\alpha y_\beta + y_\beta^2 - 3(x_\alpha + x_\beta)$
α β	$H_2 = (2y_\alpha + y_\beta)(y_\beta - y_\alpha)(y_\alpha + 2y_\beta) + 9x_\alpha(y_\alpha + 2y_\beta) - 9x_\beta(y_\alpha + y_\beta)$
$\mathfrak{c}_2$	$H_1 = (y_\alpha + 2y_\beta)^2 + y_\alpha - 4x_\alpha - 8x_\beta$
α β	$H_2 = (y_\alpha^2 - 4x_\alpha)(y_\alpha + 2y_\beta)^2 + 8x_\beta(y_\alpha(y_\alpha + 2y_\beta) + 2x_\beta)$
$\mathfrak{g}_2$	$H_1 = (2y_\alpha + 3y_\beta)^2 + 3y_\beta^2 - 4(x_\alpha + 3x_\beta)$
α β	$H_2 = [((y_\alpha + 2y_\beta)(y_\alpha + y_\beta) + x_\beta)y_\beta + x_\alpha(y_\alpha + 2y_\beta)]^2$ $-4x_\beta[(y_\alpha + 2y_\beta)(y_\alpha + y_\beta) + x_\beta]^2 + 4x_\alpha x_\beta[(y_\alpha + 2y_\beta)^2 - x_\beta]$

9.4 Algebraic Integrability of the $\mathfrak{a}_2^{(1)}$ Toda Lattice

We will now show that the usual periodic 3-body Toda lattice is a.c.i. We will do this by using the methods of Chapter 7. For an alternative proof, see [169, Ch. VII.7].

The equations of motion are given by

$$\begin{aligned} \dot{x} &= x \cdot y, \\ \dot{y} &= Ax, \end{aligned} \tag{9.26}$$

where $x = (x_0, x_1, x_2)^\top$ and $y = (y_0, y_1, y_2)^\top$, with $y_0 + y_1 + y_2 = 0$. The Cartan matrix A is in this case given by

$$A = \begin{pmatrix} 2 & -1 & -1 \\ -1 & 2 & -1 \\ -1 & -1 & 2 \end{pmatrix},$$

and $\xi = (1,1,1)^\top$ is the normalized null vector of $A^\top$. In order to simplify the formulas and computations that follow we perform a linear change of coordinates on $y_0, \dots, y_2$, namely we define

$$x_3 := \frac{y_1 - y_2}{3}, \qquad x_4 := \frac{y_2 - y_0}{3}, \qquad x_5 := \frac{y_0 - y_1}{3}. \tag{9.27}$$

Then the vector field (9.26) and a vector field that commutes with it are given by

$$
\begin{aligned}
\dot{x}_0 &= x_0(x_5 - x_4), & x_0' &= x_0x_3(x_4 - x_5) + x_0(x_2 - x_1),\\
\dot{x}_1 &= x_1(x_3 - x_5), & x_1' &= x_1x_4(x_5 - x_3) + x_1(x_0 - x_2),\\
\dot{x}_2 &= x_2(x_4 - x_3), & x_2' &= x_2x_5(x_3 - x_4) + x_2(x_1 - x_0),\\
\dot{x}_3 &= x_1 - x_2, & x_3' &= x_2x_5 - x_1x_4,\\
\dot{x}_4 &= x_2 - x_0, & x_4' &= x_0x_3 - x_2x_5,\\
\dot{x}_5 &= x_0 - x_1, & x_5' &= x_1x_4 - x_0x_3.
\end{aligned}
\tag{9.28}
$$

We will denote these vector fields by $\mathcal{V}_1$ and $\mathcal{V}_2$ respectively. Notice that (9.27) implies that $x_3 + x_4 + x_5 = 0$, while the vector fields (9.28) make perfect sense (and commute) on the whole of $\mathbf{C}^6$. We will therefore drop in the sequel the condition that $x_3 + x_4 + x_5 = 0$; this is even very natural from the Hamiltonian point of view, as the linear function $x_3 + x_4 + x_5$ is a Casimir for a natural Lie-Poisson structure on $\mathbf{C}^6$ with respect to which both $\mathcal{V}_1$ and $\mathcal{V}_2$ are Hamiltonian. Namely, consider the Poisson structure $\{\cdot\,,\cdot\}$ on $\mathbf{C}^6$, defined by the skew-symmetric matrix

$$
\begin{pmatrix} 0 & -X \\ X^\top & 0 \end{pmatrix}, \quad \text{where} \quad X := \begin{pmatrix} 0 & x_0 & -x_0 \\ -x_1 & 0 & x_1 \\ x_2 & -x_2 & 0 \end{pmatrix}.
$$

and consider the following integrals of the vector fields $\mathcal{V}_1$ and $\mathcal{V}_2$:

$$
\begin{aligned}
F_1 &:= x_0x_1x_2,\\
F_2 &:= x_3 + x_4 + x_5,\\
F_3 &:= \frac{1}{2}(x_3^2 + x_4^2 + x_5^2) - x_0 - x_1 - x_2,\\
F_4 &:= x_3x_4x_5 + x_0x_3 + x_1x_4 + x_2x_5.
\end{aligned}
\tag{9.29}
$$

Then it is easy to check that F_1 and F_2 are Casimirs of $\{\cdot\,,\cdot\}$, and that $\mathcal{X}_{F_{i+2}} = \mathcal{V}_i$ for $i = 1, 2$. Denoting $\mathbf{F} := (F_1, \dots, F_4)$ we have that $\mathbf{F}$ is involutive and independent, hence $(\mathbf{C}^6, \{\cdot\,,\cdot\}, \mathbf{F})$ is Liouville integrable. The rank of $\{\cdot\,,\cdot\}$ is 4 at all points of $\mathbf{C}^6$, except at the three four-planes $x_0 = x_1 = 0$ and $x_1 = x_2 = 0$ and $x_2 = x_0 = 0$. Since F_1 is constant on these four-planes, Proposition 7.56 implies that the commuting vector fields $\mathcal{V}_1$ and $\mathcal{V}_2$ are independent on the generic fiber $\mathbf{F}_\mathbf{c}$ of the momentum map.

We assign the weight 2 to x_0, x_1 and x_2 and the weight 1 to x_3, x_4 and x_5. Then $\mathcal{V}_1$ is a weight homogeneous vector field (of weight 1), while $\mathcal{V}_2$ is weight homogeneous of weight 2. Also, $\{\cdot\,,\cdot\}$ has weight 1 and the weights of the constants of motion F_i are given by

$$
\varpi(F_1, F_2, F_3, F_4) = (6, 1, 2, 3).
$$

For future use, we also introduce the order three automorphism of $\mathbf{C}^6$, which is given by

$$\pi(x_0,\ldots,x_5) := (x_1,x_2,x_0,x_4,x_5,x_3).$$

π is Poisson automorphism which preserves the functions F_i, hence it also preserves the vector fields $\mathcal{V}_1$ and $\mathcal{V}_2$.

We now turn to the principal balances of $\mathcal{V}_1$. The indicial locus $\mathcal{I}$ is the subset of $\mathbf{C}^6$, given by

$$\begin{aligned}
0 &= x_0^{(0)}(2+x_5^{(0)}-x_4^{(0)}),\\
0 &= x_1^{(0)}(2+x_3^{(0)}-x_5^{(0)}),\\
0 &= x_2^{(0)}(2+x_4^{(0)}-x_3^{(0)}),\\
0 &= x_3^{(0)}+x_1^{(0)}-x_2^{(0)},\\
0 &= x_4^{(0)}+x_2^{(0)}-x_0^{(0)},\\
0 &= x_5^{(0)}+x_0^{(0)}-x_1^{(0)}.
\end{aligned}$$

It consists of six points, two of which are given by

$$(x_0^{(0)},\ldots,x_5^{(0)}) = \begin{cases}(1,0,0,0,1,-1),\\ (2,2,0,-2,2,0),\end{cases}$$

and four other points, obtained by applying π and π^2 to these points. For the point $w_0 := (1,0,0,0,1,-1)$ the Kowalevski matrix is given by

$$\mathcal{K}(w_0) = \begin{pmatrix}
0 & 0 & 0 & 0 & -1 & 1\\
0 & 3 & 0 & 0 & 0 & 0\\
0 & 0 & 3 & 0 & 0 & 0\\
0 & 1 & -1 & 1 & 0 & 0\\
-1 & 0 & 1 & 0 & 1 & 0\\
1 & -1 & 0 & 0 & 0 & 1
\end{pmatrix}.$$

Its characteristic polynomial is given by

$$|\mu\,\mathrm{Id}_6 - \mathcal{K}(w_0)| = (\mu+1)(\mu-2)(\mu-1)^2(\mu-3)^2,$$

so it has 5 non-negative integer eigenvalues, as required for a principal balance. The fact that 1, 2 and 3 are roots of the characteristic polynomial follows from Theorem 7.30, namely the differentials $\mathrm{d}F_1(w_0),\ldots,\mathrm{d}F_4(w_0)$ are given by

$$\begin{aligned}
\mathrm{d}F_1(w_0) &:= 0,\\
\mathrm{d}F_2(w_0) &:= \mathrm{d}x_3+\mathrm{d}x_4+\mathrm{d}x_5,\\
\mathrm{d}F_3(w_0) &:= -\mathrm{d}x_0-\mathrm{d}x_1-\mathrm{d}x_2+\mathrm{d}x_4-\mathrm{d}x_5,\\
\mathrm{d}F_4(w_0) &:= \mathrm{d}x_1-\mathrm{d}x_2,
\end{aligned}$$

yielding a non-zero $\mathrm{d}\mathcal{H}^{(k)}$ for $k=1$, 2 and 3.

Also, the fact that -1 is a root follows from Proposition 7.11 and, obviously, the sum of all roots is $9 = \operatorname{Trace}(\mathcal{K}(w_0))$. In view of Proposition 7.34 we will get a trivial parameter entering at levels 1, 2 and 3, if w_0 leads indeed to a principal balance.

For the point $w_0' := (2,2,0,-2,2,0)$ the Kowalevski matrix is given by

$$\mathcal{K}(w_0') = \begin{pmatrix} 0 & 0 & 0 & 0 & -2 & 2 \\ 0 & 0 & 0 & 2 & 0 & -2 \\ 0 & 0 & 6 & 0 & 0 & 0 \\ 0 & 1 & -1 & 1 & 0 & 0 \\ -1 & 0 & 1 & 0 & 1 & 0 \\ 1 & -1 & 0 & 0 & 0 & 1 \end{pmatrix}.$$

and the differentials $\mathrm{d}H_i(w_0')$ take the form

$$\begin{aligned}
\mathrm{d}F_1(w_0') &:= 4\mathrm{d}x_2, \\
\mathrm{d}F_2(w_0') &:= \mathrm{d}x_3 + \mathrm{d}x_4 + \mathrm{d}x_5, \\
\mathrm{d}F_3(w_0') &:= -\mathrm{d}x_0 - \mathrm{d}x_1 - \mathrm{d}x_2 - 2\mathrm{d}x_3 + 2\mathrm{d}x_4, \\
\mathrm{d}F_4(w_0') &:= 2(-\mathrm{d}x_0 + \mathrm{d}x_1 + \mathrm{d}x_3 + \mathrm{d}x_4 - 2\mathrm{d}x_5),
\end{aligned}$$

so that 1, 2, 3 and 6 are eigenvalues of $\mathcal{K}(w_0')$, besides -1 (Proposition 7.11). Since the sum of all eigenvalues of $\mathcal{K}(w_0')$ is 9 it follows that

$$|\operatorname{Id}_6 - \mathcal{K}(w_0')| = (\mu+2)(\mu+1)(\mu-1)(\mu-2)(\mu-3)(\mu-6),$$

so this matrix has only 4 non-negative integer eigenvalues, and w_0' cannot lead to a principal balance. We show that w_0 leads indeed to a principal balance by exhibiting the first terms (the five free parameters, going with steps 1, 1, 2, 3, 3 respectively, are denoted by $a, \dots, e$, all subsequent terms are uniquely determined by the given ones).

$$\begin{aligned}
x_0(t;w_0) &= \frac{1}{t^2}\left(1 + ct^2 + dt^3 + O(t^4)\right), \\
x_1(t;w_0) &= -(5d+2e)t + O(t^2), \\
x_2(t;w_0) &= (d+2e)t + O(t^2), \\
x_3(t;w_0) &= a - (3d+2e)t^2 + O(t^3), \\
x_4(t;w_0) &= \frac{1}{t} + b - ct + et^2 + O(t^3), \\
x_5(t;w_0) &= -\frac{1}{t} + b + ct + (3d+e)t^2 + O(t^3).
\end{aligned} \tag{9.30}$$

Letting $w_1 := \pi(w_0) = (0,0,1,1,-1,0)$ and $w_2 := \pi(w_1) = (0,1,0,-1,0,1)$ the principal balances $x(t;w_1)$ and $x(t;w_2)$ are obtained from the above formulas for $x(t;w_0)$ by applying the automorphism π.

As we pointed out, we can pick a trivial parameter at positions 1, 2 and 3: in order to have that $F_i(x(t;w_0)) = c_i$ for $i = 2,3,4$ the parameters a, c and e should be chosen as follows:

$$\begin{aligned} a &= c_2 - 2b, \\ c &= b^2 - \frac{2}{3}c_2 b + \frac{c_2^2 - 2c_3}{6}, \\ e &= -\frac{3}{2}d - 2b^3 + 2c_2b^2 + \frac{1}{4}(2c_3 - 3c_2^2)b + \frac{1}{8}(c_2^3 - 2c_2c_3 - 2c_4). \end{aligned} \tag{9.31}$$

An equation for one of the irreducible components $\Gamma_{\mathbf{c}}^{(0)}$ of the abstract Painlevé divisor is then given by $F_1(x(t;w_0)) = c_1$, which yields an algebraic relation between b and d, namely

$$64d^2 = \left(16b^3 - 16c_2b^2 + (6c_2^2 - 4c_3)b + 2c_2c_3 + 2c_4 - c_2^3\right)^2 + 16c_1.$$

The affine curve $\Gamma_{\mathbf{c}}^{(0)}$ is, for generic $c_1, \dots, c_4$, a smooth hyperelliptic curve of genus 2. Upon computing the abstract Painlevé divisors $\Gamma_{\mathbf{c}}^{(1)}$ and $\Gamma_{\mathbf{c}}^{(2)}$ which correspond to the other principal balances $x(t;w_1)$ and $x(t;w_2)$ we arrive at the same equation since π preserves the constants of motion; when it is irrelevant which one of the isomorphic $\Gamma_{\mathbf{c}}^{(i)}$ we are talking about the curve will just be denoted by $\Gamma_{\mathbf{c}}$. In order to complete the non-singular curve $\Gamma_{\mathbf{c}}$ into a compact Riemann surface we need to adjoin two points ∞^+ and ∞^-, which are given in terms of a local parameter ς by

$$b = \varsigma^{-1}, \qquad d = \pm\frac{2}{\varsigma^3}\left(1 - c_2\varsigma + \frac{1}{8}\left(3c_2^2 - 2c_3\right)\varsigma^2 + O(\varsigma^3)\right). \tag{9.32}$$

We now look for polynomials $P(x)$ which have the property that each of the series $P(x(t;w_j))$ has at most a simple pole, where $j = 0,1,2$. The fact that $\Gamma_{\mathbf{c}}$ has genus 2 suggests that the invariant manifolds $\mathbf{F}_{\mathbf{c}}$ are Jacobians of genus 2 curves, and these can be embedded in $\mathbf{P}^8$ by using the sections of the third power of the line bundle that corresponds to the theta divisor (which is isomorphic to the Riemann surfaces that underlies the Jacobian). Indeed, if we denote by Θ the theta divisor of a Jacobi surface, then by (5.25),

$$\dim L(3\Theta) = \frac{(3\Theta)\cdot(3\Theta)}{2} = 9\frac{\Theta\cdot\Theta}{2} = 9\dim L(\Theta) = 9(g(\Theta) - 1) = 9,$$

and probably

$$\mathcal{D}_{\mathbf{c}}^{(0)} + \mathcal{D}_{\mathbf{c}}^{(1)} + \mathcal{D}_{\mathbf{c}}^{(2)} \sim 3\Theta,$$

with $\Gamma_{\mathbf{c}}^{(i)} \hookrightarrow \mathcal{D}_{\mathbf{c}}^{(i)}$. This suggests that there should only be 9 independent polynomials with the above property, and that they should suffice for proving algebraic integrability (recall that the third power of any ample line bundle on an Abelian variety is very ample and normally generated).

Table 9.2. A list of 9 independent polynomials which have a simple pole in t at worst, when any of the 3 principal balances $x(t; w_j)$ is substituted in them. For each of these polynomials z_i we give its weight and the residues of $z_i(x(t; w_j))$ for $j = 0, \ldots, 2$.

z_i	$\varpi(z_i)$	$\text{Res}_t\, z_i(x(t;w_0))$	$\text{Res}_t\, z_i(x(t;w_1))$	$\text{Res}_t\, z_i(x(t;w_2))$
$z_0 := 1$	0	0	0	0
$z_1 := x_3$	1	0	1	-1
$z_2 := x_5$	1	-1	0	1
$z_3 := x_0 + x_4x_5$	2	0	$-a$	a
$z_4 := x_1 + x_3x_5$	2	$-a$	a	0
$z_5 := x_1x_4 - x_5z_3$	3	$b^2 + 3c$	a^2	$-2ab$
$z_6 := x_0x_1$	4	$-5d - 2e$	0	$d + 2e$
$z_7 := x_1x_2$	4	0	$d + 2e$	$-5d - 2e$
$z_8 := x_0x_2$	4	$d + 2e$	$-5d - 2e$	0

We give a list of 9 such functions in Table 9.2, together with the leading behavior when any of the principal balances $x(t; w_j)$ is substituted in them. These functions define a map $\varphi_{\mathbf{c}} : \mathbf{F}_{\mathbf{c}} \to \mathbf{P}^8$ which is, for generic $\mathbf{c}$, an embedding. We consider the three maps $\varphi_{\mathbf{c}}^{(j)} : \Gamma_{\mathbf{c}} \to \mathbf{P}^8$, given by the residues $\text{Res}_t\, z_i(x(t; w_j))$, where $j = 0, \ldots, 2$. Explicitly, $\varphi_{\mathbf{c}}^{(0)}$ is for $(b, d) \in \Gamma_{\mathbf{c}}$ given by

$$\varphi_{\mathbf{c}}^{(0)}(b,d) = (0 : 0 : -1 : 0 : -a : b^2 + 3c : -5d - 2e : 0 : d + 2e), \qquad (9.33)$$

where the values of a, c and e in terms of b and d are given by (9.31). Similarly the other two maps $\varphi_{\mathbf{c}}^{(1)}$ and $\varphi_{\mathbf{c}}^{(2)}$ are written down from Table 9.2. We see at once that each of them yields an isomorphic embedding of the curve, and that all image curves are disjoint. In order to compute the image of the compact Riemann surface $\overline{\Gamma_{\mathbf{c}}}$ that corresponds to $\Gamma_{\mathbf{c}}$, write (9.33) and the other two embeddings in terms of a local parameter ς around each of the two points ∞^+ et ∞^- in $\overline{\Gamma_{\mathbf{c}}} \setminus \Gamma_{\mathbf{c}}$, by using (9.32) and let $\varsigma \to 0$. This results in the following three image points

$$P^0 := (0:0:0:0:0:0:0:1:0),$$
$$P^1 := (0:0:0:0:0:0:1:0:0),$$
$$P^2 := (0:0:0:0:0:0:0:0:1),$$

namely the precise correspondence is given as follows:

$$\varphi_{\mathbf{c}}^{(0)}(\infty^+) = \varphi_{\mathbf{c}}^{(1)}(\infty_{\mathbf{c}}^-) = P^2,$$
$$\varphi_{\mathbf{c}}^{(1)}(\infty^+) = \varphi_{\mathbf{c}}^{(2)}(\infty_{\mathbf{c}}^-) = P^0,$$
$$\varphi_{\mathbf{c}}^{(2)}(\infty^+) = \varphi_{\mathbf{c}}^{(0)}(\infty_{\mathbf{c}}^-) = P^1.$$

This means that each pair of image curves has an intersection point; it is clear that the three intersection points P_0, P_1 and P_3 are all different.

Notice that these points are independent of $\mathbf{c}$ and that they are cyclically permuted by the automorphism π. With some more effort one finds, up to order 2, the following description of the embedded curves along this tangency point:

$$
\begin{aligned}
\varphi_{\mathbf{c}}^{(0)}(\infty^+) &\sim (0:0:0:0:2\varsigma^2:2\varsigma(2-c_2\varsigma):0:0:\star),\\
\varphi_{\mathbf{c}}^{(1)}(\infty^+) &\sim (0:0:0:2\varsigma^2:-2\varsigma^2:2\varsigma(2-2c_2\varsigma):0:\star:0),\\
\varphi_{\mathbf{c}}^{(2)}(\infty^+) &\sim (0:0:0:-2\varsigma^2:0:2\varsigma(2-c_2\varsigma):\star:0:0),\\
\varphi_{\mathbf{c}}^{(0)}(\infty^-) &\sim (0:0:0:0:2\varsigma^2:2\varsigma(2-c_2\varsigma):-\star:0:0),\\
\varphi_{\mathbf{c}}^{(1)}(\infty^-) &\sim (0:0:0:2\varsigma^2:-2\varsigma^2:2\varsigma(2-2c_2\varsigma):0:0:-\star),\\
\varphi_{\mathbf{c}}^{(2)}(\infty^-) &\sim (0:0:0:-2\varsigma^2:0:2\varsigma(2-c_2\varsigma):0:-\star:0),
\end{aligned}
$$

where $\star = -8 + 8c_2\varsigma + \varsigma^2(2c_3 - 3c_2^2)$. It follows that the curves are tangent at their common points, and that the tangency is simple. The configuration made up by the three image curves is represented in Figure 9.1. With this

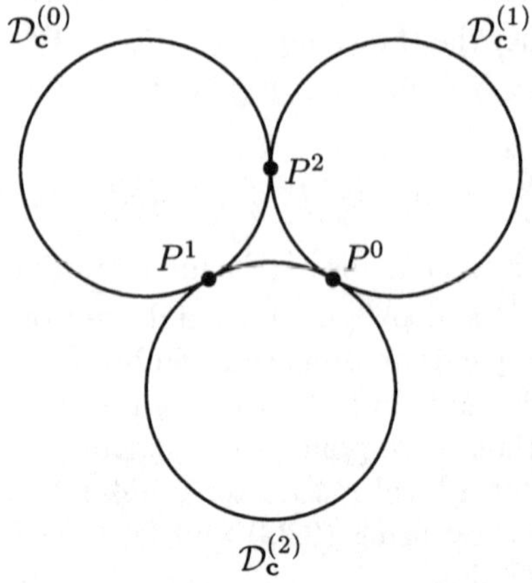

Fig. 9.1. For generic $\mathbf{c}$ the Toda divisor $\mathcal{D}_{\mathbf{c}}$ consists of 3 isomorphic non-singular hyperelliptic curves of genus 2. These curves are translates of each other by one third of a period. They intersect pairwise in a single point, which is a point of simple tangency (tacnode); moreover, the tangency locus of the vector field $\overline{\mathcal{V}}_1$, constructed below, to $\mathcal{D}_{\mathbf{c}}$, consists of the three tacnodes of $\mathcal{D}_{\mathbf{c}}$.

information it is easy to verify the adjunction formula: each of the three singular points is a tacnode, hence has Euler characteristic 2, so that the arithmetic genus of the image, as computed from Proposition (5.42), is $3 \times 2+1-3+3\times 2 = 10$. Since the embedding space is $\mathbf{P}^8$ the adjunction formula (7.50) is satisfied.

We now give the quadratic differential equations in two charts $Z_0 \neq 0$ and $Z_1 \neq 0$. In order to simplify the formulas we will put $c_2 = 0$, where we recall that c_2 is the value of one of the Casimirs, and that it is equal to zero in the original Toda lattice (see the beginning of this section). In the chart $Z_0 \neq 0$ we just use the z_i as coordinates. It is easily checked that $\mathcal{V}_1$ can be written in this chart the following form.

$$\begin{aligned}
\dot z_1 &= z_3 + 2z_4 - z_1(z_1 + 2z_2) + c_3, \\
\dot z_2 &= z_3 - z_4 + z_2(2z_1 + z_2), & \dot z_6 &= z_6(2z_1 + z_2), \\
\dot z_3 &= -z_3(z_1 + 2z_2) - 2z_5 + c_4, & \dot z_7 &= -z_7(z_1 + 2z_2), \\
\dot z_4 &= z_1 z_3 + z_2(z_3 + z_4 + c_3), & \dot z_8 &= -z_8(z_1 - z_2). \\
\dot z_5 &= z_5(z_1 + z_2) + z_3(z_4 - z_3) - z_6 + z_7 - c_4 z_2,
\end{aligned}$$

For the second chart, $Z_1 \neq 0$, define $y_i := z_i/z_1$. Then the quadratic differential equations take the following form.

$$\begin{aligned}
\dot y_0 &= 2y_2 + 1 - y_0(2y_4 + y_3 + c_3 y_0), \\
\dot y_1 &= 0, \\
\dot y_2 &= y_2(y_3 + y_4) + y_0 y_5 + y_4 + c_4 y_0^2, \\
\dot y_3 &= y_0(y_8 - y_6) - y_5 - c_4 y_0 y_2, \\
\dot y_4 &= y_0(y_6 + 2y_7) - y_2 y_5 + c_3(y_2^2 + y_0 y_4) + c_4 y_0(y_2 + 1), \\
\dot y_5 &= y_8(y_2 - 1) - y_6(y_2 + 2) - c_3 y_3 - c_4(y_2^2 + y_0(y_4 + 2y_3) - 1), \\
\dot y_6 &= y_3 y_6 + y_4 y_8 + \star, \\
\dot y_7 &= -y_7(y_3 + 2y_4 + c_3 y_0), \\
\dot y_8 &= y_6(y_3 + y_4) - y_3 y_8 + c_3 y_0(2y_6 - y_8) - \star,
\end{aligned}$$

where

$$\star := y_3 y_7 + y_5^2 + c_3(y_3 y_4 - y_2 y_5) - c_4(y_2 y_4 - y_3) - c_3 c_4 y_0 y_2 - c_4^2 y_0^2.$$

This shows that $(\varphi_{\mathbf{c}})_*(\mathcal{V}_1)$ extends to a holomorphic vector field $\overline{\mathcal{V}}_1$ on $\mathbf{P}^8$.

We now show that the integral curves of $\overline{\mathcal{V}}_1$ that start at the three singular points go into the affine immediately. In view of the automorphism π it suffices to do this for one of these points. We do this for $P^2 = \lim_{P \to \infty^+} \varphi_{\mathbf{c}}^{(0)}(P)$. Among the residues of the Laurent series $z_0(t; \Gamma_{\mathbf{c}}^{(0)}), \dots, z_8(t; \Gamma_{\mathbf{c}}^{(0)})$ the one which has the highest pole at this point is the residue of $z_8(t; \Gamma_{\mathbf{c}}^{(0)})$ (see Table 9.2). We need to show that (see (7.54))

$$\lim_{P \to \infty^+} \frac{1}{z_8(t; \Gamma_{\mathbf{c}}^{(0)})} \neq 0,$$

which is computed by writing the series $z_8(t; \Gamma_{\mathbf{c}}^{(0)})$ in terms of a local parameter ς around ∞^+ by using (9.32), and letting $\varsigma \to 0$.

To do this, we need the four leading terms of z_8:

$$z_8(t;w_0)=\frac{d+2e}{t}+(b-a)(d+2e)+((a-b)^2+c)(d+2e)t$$
$$+\frac{1}{6}(d+2e)((b-a)^3+3c(b-a)+12d+6e)t^2+O(t^3).$$

The first few terms of the inverse of this series are given by

$$\frac{1}{z_8(t;w_0)}=\frac{t}{d+2e}-\frac{b-a}{d+2e}t^2+\frac{1}{2}\frac{(a-b)^2-c}{d+2e}t^3$$
$$-\frac{1}{6}\frac{(b-a)^3+3c(a-b)+12d+6e}{d+2e}t^4+O(t^5).$$

It suffices now to compute to substitute (9.31) and (9.32) in each of these terms: letting $\varsigma\to 0$ one easily finds

$$\lim_{\varsigma\to 0}\frac{1}{z_8(t;\varsigma)}=\frac{1}{4}t^4+O(t^5)\neq 0,$$

which shows that the integral curves of $\overline{\mathcal{V}}_1$ that start at $\varphi_{\mathbf{c}}^{(0)}(\infty^+)$ go into the affine immediately. Notice that the fact that the leading term $4/t^4$ of $z_8=x_0x_2$ is in agreement with the fact that there is a lower balance for which $x_0(t)=2/t^2+O(t^{-1})$ and $x_2(t)=2/t^2+O(t^{-1})$.

The last thing that we need to check is that there are no other divisors in $\overline{\varphi_{\mathbf{c}}(\mathbf{F}_{\mathbf{c}})}\setminus\varphi_{\mathbf{c}}(\mathbf{F}_{\mathbf{c}})$ that pass through the three singular points. Again it is sufficient to do this for one of the three points; we do it for the same point as above. Since the vector field $\overline{\mathcal{V}}_1$ is tangent to both $\overline{\varphi_{\mathbf{c}}^{(0)}(\Gamma_{\mathbf{c}}^{(0)})}$ and $\overline{\varphi_{\mathbf{c}}^{(1)}(\Gamma_{\mathbf{c}}^{(1)})}$ at this point, we need to consider the vector field $\mathcal{V}_2$. We show that $(\varphi_c)_*\mathcal{V}_2$ exists and extends to a holomorphic vector field on $\mathbf{P}^8$. This is done by exhibiting the quadratic differential equations in two of the charts. Below we just give them for the chart $Z_0\neq 0$; as before we put $c_2=0$ in order to simplify the formulas (the prime denotes the time derivative with respect to $\mathcal{V}_2$; the corresponding time variable will be denoted by t_2).

$$\begin{aligned}
z_1'&=z_1z_4-z_2(z_3+c_3),\\
z_2'&=-(z_1+z_2)z_4-z_1z_3,\\
z_3'&=z_8-z_6-z_1z_5-c_4z_2,\\
z_4'&=z_5(z_1+z_2)+z_6-z_7-c_3z_2(z_1+z_2)-c_4z_1,\\
z_5'&=z_2(z_6-z_8)+z_1(z_7-z_8)-c_3z_1z_3+c_4(z_1^2+z_1z_2+z_2^2),\\
z_6'&=z_6(z_3-z_4),\\
z_7'&=z_7(z_3+2z_4+c_3),\\
z_8'&=-z_8(2z_3+z_4+c_3).
\end{aligned}$$

The reader is welcomed to compute the vector field in the chart $Z_1 \neq 0$ as an exercice, which amounts to computing the coefficients α_{i1}^{jk} in

$$W(z_i, z_1) = z_i' z_1 - z_i z_1' = \sum_{j,k=0}^{8} \alpha_{i1}^{jk} z_j z_k, \qquad 0 \leqslant i \leqslant 8.$$

We need to compute the first few terms of the Taylor series of $1/z_8(t_2; \Gamma_{\mathbf{c}}^{(0)})$, and express the free parameters in terms of a local parameter ς in a neighborhood of ∞^+. The resulting series in t_2 and in ς should start at degree 2, by (7.57) since we have found two divisors that pass through $\lim_{P\to\infty^+} \varphi_{\mathbf{c}}^{(0)}(P)$, on which $1/z_8$ has a simple zero. As explained in Paragraph 7.6.2 we compute these first few terms as follows:

$$\frac{1}{z_8(t_2; \Gamma_{\mathbf{c}}^{(0)})} = \frac{1}{z_8(t)}\bigg|_{t=0} + \left(\frac{1}{z_8(t)}\right)'\bigg|_{t=0} t_2 + \left(\frac{1}{z_8(t)}\right)''\bigg|_{t=0} \frac{t_2^2}{2} + \cdots,$$

where $z_8(t) := z_8(t; \Gamma_{\mathbf{c}}^{(0)})$. The first term yields no contribution. For the second term we have that

$$\left(\frac{1}{z_8}\right)' = -\frac{x_2 - x_0 + (x_3 - x_5)x_4}{x_0 x_2},$$

and we find by substituting the Laurent series $x(t; \Gamma_{\mathbf{c}}^{(0)})$ that in a neighborhood of ∞^+,

$$\left(\frac{1}{z_8}\right)'(t; \Gamma_{\mathbf{c}}^{(0)})\bigg|_{t=0} = -\frac{\varsigma^2}{4} + O(\varsigma^3),$$

which yields only terms of order 3 and higher in ς, t_2. The next term however yields a term of order 2 since

$$\left(\frac{1}{z_8}\right)''(t; \Gamma_{\mathbf{c}}^{(0)})\bigg|_{t=0} = -2 + O(\varsigma).$$

It follows that there are indeed no other divisors in $\overline{\varphi_{\mathbf{c}}(\mathbf{F_c})} \setminus \varphi_{\mathbf{c}}(\mathbf{F_c})$ than the three divisors that we have found. In conclusion, we have shown that the $\mathfrak{a}_2^{(1)}$ Toda lattice is a.c.i. Since the torus that compactifies the generic $\mathbf{F_c}$ contains a non-singular curve of genus 2 it is equal to the Jacobian of this curve; moreover, $\mathbf{F_c}$ is an affine part of this Jacobian, obtained by removing three copies of this curve that are translates of each other by a third of a period.

9.5 The Geometry of the Periodic Toda Lattices

In the previous section we have shown that the $\mathfrak{a}_2^{(1)}$ Toda lattice is an irreducible a.c.i. system, and we could have similarly shown this for the other surface cases ($l = 2$). We will derive in the present section a complete description of the geometry of any periodic Toda lattice, assuming that is an irreducible a.c.i. system. This is a reasonable assumption, since it was shown in [8] that the $\mathfrak{a}_l^{(1)}, \mathfrak{b}_l^{(1)}, \mathfrak{c}_l^{(1)}, \mathfrak{d}_l^{(1)}$ Toda lattices all satisfy the Linearization Criterion (Theorem 6.41). Moreover, a proof of the algebraic integrability of all $\mathfrak{a}_l^{(1)}$ Toda lattices was given in [165]. The point being that, while it is never trivial to show that a system is a.c.i., for higher dimensional integrable systems it is a natural *starting point* from which to derive the specific algebraic geometry of the system.

9.5.1 Notation

We first introduce the Lie algebra notation that will be used throughout this section; see Section 2.4 for some background on the concepts that will be used. We consider the integrable vector field $\mathcal{V}$ on $\mathbf{C}^{2l+2}$, given by

$$\dot{x} = x \cdot y, \qquad \dot{y} = Ax,$$

where A is the Cartan matrix of a twisted affine Lie algebra $L(\mathfrak{g}, \nu)$, and $x, y \in \mathbf{C}^{l+1}$, $x = (x_0, x_1, \dots, x_l)^\top$, $y = (y_0, y_1, \dots, y_l)$ and $x \cdot y = (x_0y_0, x_1y_1, \dots, x_ly_l)^\top$. The integer l is the rank of the simple Lie algebra $\mathfrak{g}_0$, which is the fixed point set of the automorphism ν; recall that ν may have order 1, 2 or 3 (see Tables 2.3, 2.4 and 2.5). We choose a Cartan subalgebra $\mathfrak{h}$ of $\mathfrak{g}$ and we define $\mathfrak{h}_0 := \mathfrak{g}_0 \cap \mathfrak{h}$, which is a Cartan subalgebra of $\mathfrak{g}_0$. From a chosen basis $(\alpha_1, \dots, \alpha_l)$ of simple roots for $\mathfrak{g}_0$ (with respect to $\mathfrak{h}_0$) we construct a basis of simple roots[3] $\Pi = (\alpha_0, \alpha_1, \dots, \alpha_l)$ for $L(\mathfrak{g}, \nu)$. We recall that

$$\alpha_0 = \begin{cases} -\text{ (highest long root of } \mathfrak{g}) & (\nu = \mathrm{Id}) \\ -2 \text{ (highest short root of } \mathfrak{b}_l) & (\text{case } \alpha_{2l}^{(2)}) \\ -\text{ (highest short root of } \mathfrak{g}_0) & (\text{otherwise}) \end{cases}$$

We will often deal with subsets $\Pi' \subset \Pi$. Such a subset, always assumed non-empty, corresponds to a semi-simple Lie algebra, which we denote by $\mathfrak{g}'$. Since Π' determines in a canonical way a Dynkin subdiagram of the Dynkin diagram of $L(\mathfrak{g}, \nu)$ we often refer to Π' as being a Dynkin subdiagram of Π; notice that such subdiagrams are not necessarily connected, in fact they are connected precisely when $\mathfrak{g}'$ is simple (rather than just semi-simple).

[3] In the notation of Section 2.4, $\Pi = (\bar{\alpha}_0, \bar{\alpha}_1, \dots, \bar{\alpha}_l)$; for readability of the formulas we will, in this section, omit the bars, as it will not lead to any ambiguity.

The *boundary* $\partial\Pi'$ of Π' is defined by

$$\partial\Pi' := \{\beta \in \Pi \setminus \Pi' \mid \exists \alpha \in \Pi' \text{ such that } a_{\alpha\beta} \neq 0\},$$

and the *exterior* as

$$\Pi'' := \Pi \setminus (\Pi' \cup \partial\Pi'),$$

yielding a partition of Π as $\Pi = \Pi' \cup \partial\Pi' \cup \Pi''$. In the Dynkin diagram each element of $\partial\Pi'$ is connected to at least one element of Π', but there is no edge between elements of Π' and elements of Π''. Notice that $\partial\Pi'' \subseteq \partial\Pi'$.

The set of positive roots of $\mathfrak{g}'$ is denoted by Φ'_+, and the corresponding dual Weyl element (see (2.26)) is denoted by h', with dual Weyl integers n'_i,

$$h' = \sum_{\alpha\in\Phi'_+} H_\alpha = \sum_{\alpha_i\in\Pi'} n'_i H_i. \tag{9.34}$$

We will often deal with subsets $\Pi' \subset \Pi$ that have l or 2 elements.

(1) Subsets $\Pi' \subset \Pi$ which are obtained by removing one root are often denoted by $\Pi^{(\alpha)}$ when $\Pi' = \Pi \setminus \{\alpha\}$ and we often abbreviate $\Pi^{(\alpha_i)}$ to $\Pi^{(i)}$, where $i = 0, \dots, l$. The Weyl group of these root systems is denoted by $\mathbf{W}'$, resp. $\mathbf{W}^{(\alpha)}$ and $\mathbf{W}^{(i)}$. The same applies to the Cartan matrix A, where $A^{(i)}$ means that the i-th row and column[4] of A have been removed; for example $A^{(0)}$ is the Cartan matrix of $\mathfrak{g}_0$.
(2) Subsets $\Pi' \subset \Pi$ that consist of precisely two elements, $\Pi' = \{\alpha, \beta\}$. The corresponding Weyl group is denoted by $\mathbf{W}^{(\alpha\beta)}$ and the corresponding Cartan matrix by $A^{(\alpha\beta)}$.

Notice that, when $l = 2$ and $\Pi = \{\alpha, \beta, \gamma\}$, then these notations are such that $\mathbf{W}^{(\alpha\beta)} = \mathbf{W}^{(\gamma)}$ and $A^{(\alpha\beta)} = A^{(\gamma)}$.

Also, we will sometimes be tempted to view the indices i that appear in x_i and y_i as coming from the roots $\alpha_i \in \Pi$, so that we will sometimes write x_α and y_α for $\alpha \in \Pi$, meaning x_i and y_i when $\alpha = \alpha_i$. These notations are not just of a practical nature, in fact they highlight the Lie algebraic nature of the different entities that constitute the periodic Toda lattices.

For the Lie algebra $\mathfrak{g}_0$, with Cartan subalgebra $\mathfrak{h}_0$, the following notation will also be used. Recall that for any root α the corresponding coroot H_α is defined by

$$H_\alpha = 2\frac{h_\alpha}{\langle h_\alpha \mid h_\alpha\rangle}.$$

where $\langle \alpha, \cdot\rangle = \langle h_\alpha \mid \cdot\rangle$ as elements of $\mathfrak{h}_0^*$ and where $\langle \cdot \mid \cdot \rangle$ denotes the Killing form of $\mathfrak{g}_0$; as before, we write H_i for H_{α_i}. The coefficients of the highest long root of $\mathfrak{g}_0$, with respect to $\Pi^{(0)}$, will be denoted by $\eta_1, \dots, \eta_l$.

[4] Recall that the labeling of the entries of the Cartan matrix A start at 0.

Therefore, by definition,

$$(\text{highest long root of } \mathfrak{g}_0) = \sum_{i=1}^{l} \eta_i \alpha_i. \tag{9.35}$$

The fundamental weights λ_i (belonging to $\mathfrak{h}_0^*$) and coweights $\hat{\lambda}_j$ (belonging to $\mathfrak{h}_0$) of $\mathfrak{g}_0$ are defined by the dual statements ($a_{ij} = \langle \alpha_i, H_j \rangle$)

$$\langle \lambda_i, H_j \rangle = \delta_{ij}, \qquad \left\langle \alpha_i, \hat{\lambda}_j \right\rangle = \delta_{ij}, \qquad 1 \leqslant i, j \leqslant l.$$

We have, according to (2.23), that

$$(\alpha_1, \ldots, \alpha_l)^\top = A^{(0)} (\lambda_1, \ldots, \lambda_l)^\top, \tag{9.36}$$

and also the dual statement (see (9.25))

$$(H_1, \ldots, H_l) = (\hat{\lambda}_1, \ldots, \hat{\lambda}_l) A^{(0)}.$$

Associated to the Cartan matrix A there are a few vectors of integers (generally speaking) that will show up several times. The normalized null vector of $A^\top$ is denoted by $\xi = (\xi_0 = 1, \xi_1, \ldots, \xi_l)^\top$ while the normalized null vector of A is denoted by $\hat{\xi} = (\hat{\xi}_0 = 1, \hat{\xi}_1, \ldots, \hat{\xi}_l)^\top$. Then we have the dual relations

$$\sum_{i=0}^{l} \xi_i \alpha_i = 0, \qquad \text{and} \qquad \sum_{i=0}^{l} \hat{\xi}_i H_i = 0.$$

If we denote the (i,j)-th cofactor of A by Δ_{ij} then

$$\xi_i = \frac{\Delta_{ij}}{\Delta_{0j}}, \qquad \text{and} \qquad \hat{\xi}_i = \frac{\Delta_{ji}}{\Delta_{j0}},$$

independently of j, and $\det A^{(i)} = \Delta_{ii}$. It follows that

$$\xi_i \hat{\xi}_i = \frac{\Delta_{ii}}{\Delta_{0i}} \frac{\Delta_{0i}}{\Delta_{00}} = \frac{\det A^{(i)}}{\det A^{(0)}}, \tag{9.37}$$

for $i = 0, \ldots, l$. A final set of integers is obtained from A as follows. Since $A^{(0)}$ is an invertible matrix, with entries in $\mathbf{Z}$, there exists a matrix R, with entries in $\mathbf{Z}$, such that $RA^{(0)} = \operatorname{diag}(\rho_1, \ldots, \rho_l)$, where each ρ_i is a positive integer; we will assume that R has been chosen such that each of the positive integers ρ_i is as small as possible. Notice that, multiplying (9.36) by R, this amounts to saying that R is chosen such that each ρ_i is the smallest value for which $\rho_i \lambda_i$ belongs to the root lattice of $\mathfrak{g}_0$, where λ_i denotes the i-th fundamental weight of $\mathfrak{g}_0$. We also put $\rho_0 := 1$.

For any of the vectors $\xi, \hat{\xi}, n, \eta$ and ρ it will sometimes be convenient to label their components by elements of Π, writing e.g. ξ_α for ξ_i when $\alpha = \alpha_i$.

9.5.2 The Balances of the Periodic Toda Lattice

We will first discuss the balances (principal and lower) of the periodic Toda lattice, associated to a twisted affine Lie algebra $L(\mathfrak{g},\nu)$. We will do this by refining the proof of Lemma 9.5, the extra information that we have now being the fact that the matrix A, that determines the Toda vector field, is the Cartan matrix of a twisted affine Lie algebra. Part of our final conclusion will depend on our fundamental assumption that the periodic Toda lattice under consideration leads to an a.c.i. system.

Let us recall that we have seen that the Toda vector field

$$\dot{x} = x \cdot y, \qquad \dot{y} = Ax, \qquad x, y \in \mathbf{C}^{l+1} \tag{9.38}$$

is a weight homogeneous vector field on $M \subset \mathbf{C}^{2(l+1)}$, where M is the hyperplane that is defined by (9.8), where the weights are given by $\varpi(x_i) = 2$ and $\varpi(y_i) = 1$ for $i = 0, \dots, l$, and that all the balances of the vector field are weight homogeneous, i.e., they are of the form

$$x(t) = \frac{1}{t^2} \sum_{k \geqslant 0} x^{(k)} t^k, \qquad y(t) = \frac{1}{t} \sum_{k \geqslant 0} y^{(k)} t^k,$$

with $x^{(0)} \neq 0$ and $y^{(0)} \neq 0$. We fix a subset $S \subset \{0, 1, \dots, l\}$, with $s := \#S$ satisfying $1 \leqslant s \leqslant l$ and we denote, as in the proof of Lemma 9.5, by $\overline{v}$ (resp. by $\underline{v}$) the vector of $\mathbf{C}^s$ (resp. of $\mathbf{C}^{l+1-s}$) whose components are the components v_i of v for which $i \in S$ (resp. $i \notin S$). As we have seen in the proof of the latter lemma, there is a unique solution $(x^{(0)}, y^{(0)})$ to the indicial equation (9.16), satisfying $\underline{x^{(0)}} = 0$ and $x_i^{(0)} \neq 0$ for all $i \in S$. Namely, we have that $y^{(0)} = -Ax^{(0)}$ while such an $x^{(0)}$ is entirely determined by $\underline{x^{(0)}} = 0$ and $\overline{A}\, \overline{x^{(0)}} = 2\overline{\delta}$, where $\overline{A}$ is the non-singular $(s \times s)$-matrix that contains the entries a_{ij} with $i, j \in S$. Recall also that for $i \in S$ we have that $y_i^{(0)} = -2$. Using the fact that A is the Cartan matrix of a twisted affine Lie algebra we can explicitly solve $\overline{A}\, \overline{x^{(0)}} = 2\overline{\delta}$ for $x^{(0)}$. To do this, let Π' consist of the simple roots α_i for which $i \in S$; using the conventions of Paragraph 9.5.1 the corresponding Lie algebra is denoted by $\mathfrak{g}'$, the dual Weyl element by h', with dual Weyl integers n_i'. For $i \in S$, let $x_i^{(0)} := n_i'$. Then $\overline{A}\, \overline{x^{(0)}} = 2\overline{\delta}$ follows from the definition (2.22) of the Cartan integers, the definition (9.34) of the Weyl element and (2.29): for any $i \in S$,

$$\sum_{j \in S} a_{ij} n_j' = \sum_{j \in S} \langle \alpha_i, H_j \rangle\, n_j' = \langle \alpha_i, h' \rangle = 2.$$

The explicit value for $y_i^{(0)}$, with $i \notin S$, can then be written as follows

$$y_i^{(0)} := -\sum_{j \in S} a_{ij} n_j' = -\sum_{j \in S} \langle \alpha_i, H_j \rangle\, n_j' = -\langle \alpha_i, h' \rangle\,.$$

Let us denote by $\mathcal{K}(x^{(0)}, y^{(0)})$ the Kowalevski matrix that corresponds to $(x^{(0)}, y^{(0)})$.

It follows easily from the definition (7.10) of the Kowalevski matrix, applied to the vector field (9.38), that $\mathcal{K}(x^{(0)}, y^{(0)})$ is given by the $(2l+2) \times (2l+2)$ matrix

$$\mathcal{K}(x^{(0)}, y^{(0)}) = \begin{pmatrix} 2+y_0^{(0)} & & 0 & x_0^{(0)} & & 0 \\ & \ddots & & & \ddots & \\ 0 & & 2+y_l^{(0)} & 0 & & x_l^{(0)} \\ & & & 1 & & 0 \\ & A & & & \ddots & \\ & & & 0 & & 1 \end{pmatrix}$$

We claim that 1 is an eigenvalue of multiplicity $l+1-s$ of $\mathcal{K}(x^{(0)}, y^{(0)})$, when $(x^{(0)}, y^{(0)})$ is the above solution to the indicial equation. To see this, write

$$\mathcal{K}(x^{(0)}, y^{(0)}) - \mathrm{Id}_{2l+2} = \begin{pmatrix} B & C \\ A & 0 \end{pmatrix}$$

and notice that the diagonal matrix C has rank s, since $x_i^{(0)} \neq 0 \leftrightarrow i \in S$. We may conclude that $\mathcal{K}(x^{(0)}, y^{(0)}) - \mathrm{Id}_{2l+2}$ has rank $l+1+s$ if it has no null vector of the form $(v, w)^\top$, with $v \neq 0$. In fact, if $(v, w)^\top$ is such a null vector then v is a non-trivial null vector of A, hence $v = \hat{\xi}$, up to a non-zero constant. But then $y_i^{(0)} = -1$ for any $i \notin S$ since all $\hat{\xi}_i \neq 0$. But this is absurd since, if $i \notin S$, then $y_i^{(0)} = -\sum_{j \in S} a_{ij} n'_j \geqslant 0$ since $a_{ij} \geqslant 0$ when $i \notin S$ and $j \in S$, and all n'_j are positive integers. It follows that 1 is an eigenvalue of multiplicity $l+1-s$ of $\mathcal{K}(x^{(0)}, y^{(0)})$; taking into account that $\sum_{j=0}^{l} \xi_j y_j = 0$, at most $l-s$ free parameters enter at step 1. For future reference, note also that $y_i^{(1)} = 0$ for all $i \in S$ and $x_i^{(1)} = 0$ for $i = 0, 1, \ldots, l$, while the free parameters come from the $y_i^{(1)}$, with $i \notin S$.

In order to analyze what happens at step $k \geqslant 2$ we determine the rank of $k\,\mathrm{Id}_{2l+2} - \mathcal{K}(x^{(0)}, y^{(0)})$ for $k \geqslant 2$. After a number of row and column transformations we can rewrite this matrix in the block form

$$\begin{pmatrix} -k\,\mathrm{Id}_s & 0 & \overline{x^{(0)}} & 0 \\ 0 & (2-k)\,\mathrm{Id}_{l+1-s} + \underline{y^{(0)}} & 0 & 0 \\ \overline{A} & \star_1 & (1-k)\,\mathrm{Id}_s & 0 \\ \star_2 & \star_3 & 0 & (1-k)\,\mathrm{Id}_{l+1-s} \end{pmatrix}$$

where the $\star$'s are blocks, taken from A, whose precise value is irrelevant, and the matrices such as $\overline{x^{(0)}}$ are the above defined vectors, but written as diagonal matrices.

As a consequence, $k \geqslant 2$ will be a Kowalevski exponent in the following two cases: (1) when $k = 2 + y_i^{(0)}$ for some $i \notin S$ or (2) when k is a zero of

$$\begin{vmatrix} -k\,\mathrm{Id}_s & \overline{x^{(0)}} \\ \overline{A} & (1-k)\,\mathrm{Id}_s \end{vmatrix} = \det(k(k-1)\,\mathrm{Id}_s - N'\overline{A}),$$

where we have denoted $N' := \overline{x^{(0)}}$, the diagonal matrix that contains the integers n'_i with $i \in S$. For case (1), since $y_i^{(0)}$ is a non-negative integer whenever $i \notin S$, as we remarked above, we have $l+1-s$ Kowalevski exponents, given by

$$\left\{2 + y_i^{(0)} \mid i \notin S\right\} = \{2 - \langle \alpha_i, h' \rangle \mid i \notin S\}.$$

For (2), recall from Proposition 2.16 that

$$\mathrm{Spec}(N'\overline{A}) = \{m'_1(m'_1+1), \dots, m'_s(m'_s+1)\},$$

where $m'_1, \dots, m'_s$ are the exponents of $\mathfrak{g}'$. It follows that the integers $m'_i + 1$ are roots of $\det(k(k-1)\,\mathrm{Id}_s - N'\overline{A})$; but notice that when k is a root of the latter, then also $1-k$. This leads to the following complete description of the spectrum of the Kowalevski matrix $\mathcal{K}(x^{(0)}, y^{(0)})$.

Proposition 9.14. *The indicial locus $\mathcal{I}$ of the Toda vector field (9.38) is parametrized by the subsets $\Pi' \subset \Pi$, where $0 < s = \#\Pi' \leqslant l$: for such a set Π' there is a unique $(x^{(0)}, y^{(0)}) \in \mathcal{I}$ such that $x_i^{(0)} \neq 0 \leftrightarrow \alpha_i \in \Pi'$. Let $m'_1, \dots, m'_s$ denote the exponents, h' the dual Weyl element and $n'_1, \dots, n'_s$ the dual Weyl integers of the semi-simple Lie algebra $\mathfrak{g}'$ with simple roots Π'. Then*

$$\begin{aligned}\mathrm{Spec}(\mathcal{K}(x^{(0)}, y^{(0)})) = \{&-m'_s, \dots, -m'_1, 1, \dots, 1, 1+m'_1, \dots, 1+m'_s\} \\ &\bigcup \{2 - \langle \alpha_i, h' \rangle \mid i \notin S\},\end{aligned}$$

where the eigenvalue 1 has multiplicity $l+1-s$. In particular, if $\Pi = \{\alpha\}$ then

$$\mathrm{Spec}(\mathcal{K}(x^{(0)}, y^{(0)})) = \{-1, 1, 1, \dots, 1, 2\} \bigcup \{2 - a_{\gamma\alpha} \mid \gamma \neq \alpha\}.$$

This leads to the first few terms of all the balances to the Toda lattice.

Lemma 9.15. *Under the a.c.i. assumption, for any Π' such that $\emptyset \neq \Pi' \subset \Pi$ the first few terms of the corresponding balance are given by*

$$\begin{aligned}
x_i(t; \Pi') &= \frac{n'_i}{t^2}(1 + O(t^2)) \text{ for } \alpha_i \in \Pi', \\
x_i(t; \Pi') &= b_i t^{-\langle \alpha_i, h' \rangle}(1 + c_i t + O(t^2)) \text{ for } \alpha_i \notin \Pi', \\
y_i(t; \Pi') &= -\frac{2}{t}(1 + O(t^2)) \text{ for } \alpha_i \in \Pi', \\
y_i(t; \Pi') &= \frac{1}{t}(-\langle \alpha_i, h' \rangle + c_i t + O(t^2)) \text{ for } \alpha_i \notin \Pi',
\end{aligned} \tag{9.39}$$

where the b_i and c_i are free parameters, with $\sum_{\alpha_i \notin \Pi'} \xi_i c_i = 0$.

In particular, taking $\Pi' = \{\alpha_i\}$ the corresponding principal balance takes the following form,

$$\begin{aligned}
x_i(t;\alpha_i) &= \frac{1}{t^2}(1 + b_i t^2 + O(t^3)),\\
x_j(t;\alpha_i) &= \frac{b_j}{t^{a_{ji}}}(1 + c_j t + O(t^2)) \text{ for } j \neq i,\\
y_i(t;\alpha_i) &= -\frac{2}{t}(1 - b_i t^2 + O(t^2)),\\
y_j(t;\alpha_i) &= -\frac{a_{ji}}{t} + c_j + O(t) \text{ for } j \neq i,
\end{aligned} \tag{9.40}$$

with $\sum_{j\neq i} \xi_j c_j = 0$.

Proof. We first show that every balance is necessarily of the above form (9.39). In fact, we have already obtained complete information about the first two terms of all balances (steps 0 and 1), so it suffices to prove that the series for $x_i(t;\Pi')$, with $\alpha_i \notin \Pi'$ is the one that is given above. To do this, consider for $\alpha_i \notin \Pi'$ the differential equation $\dot{x}_i = x_i y_i$ for x_i, given y_i as above, and substitute in it

$$\begin{aligned}
x_i(t;\Pi') &= \frac{1}{t^2}\left(x_i^{(k)} t^k + x_i^{(k+1)} t^{k+1} + O(t^{k+2})\right)\\
y_i(t;\Pi') &= \frac{1}{t}(-\langle \alpha_i, h'\rangle + c_i t + O(t^2)),
\end{aligned}$$

where k is chosen such that $x^{(k)} \neq 0$. We get

$$\begin{aligned}
t^{k-3}\left((k-2)x_i^{(k)} + (k-1)x_i^{(k+1)} t + O(t^2)\right) &=\\
t^{k-3}\left(x_i^{(k)} + x_i^{(k+1)} t + O(t^2)\right)\left(-\langle \alpha_i, h'\rangle + c_i t + O(t^2)\right)&,
\end{aligned}$$

from which $k = 2 - \langle \alpha_i, h'\rangle$ and $x_i^{(k+1)} = x_i^{(k)} c_i$. This yield the proposed first few terms for $x_i(t;\Pi')$, with $\alpha_i \notin \Pi'$. Specializing (9.39) to $\Pi' = \{\alpha_i\}$ we get (9.40), except that we still have prove that $x_i^{(2)}$ (or $y_i^{(2)}$) is a free parameter (which we called b_i). To do this, first notice that $\dot{x}_k = x_k y_k$ only involves $x_i^{(2)}$ and/or $y_i^{(2)}$ when $k = i$, in which case we simply get $y_i^{(2)} = 2x_i^{(2)}$, as the equation at step 2. Next, notice that since $\dot{y} = Ax$ is linear in x and y together, and since y starts with a pole of order 1, the resulting equations at step 2 are simply given by $y_j^{(2)} = \sum a_{jk} x_k^{(2)}$. This equation involves $x_i^{(2)}$ and/or $y_i^{(2)}$ only when $j = i$, since $a_{ji} \neq 0$ for $j \neq i$ implies that $x_j^{(2)} = 0$, as follows from the second line in (9.40). For the same reason, the equation $y_i^{(2)} = \sum a_{ik} x_k^{(2)}$ reduces to the equation $y_i^{(2)} = 2x_i^{(2)}$ that we have before. Thus, $x_i^{(2)}$ is a free parameter and $y_i^{(2)} = 2x_i^{(2)}$, as we needed to prove.

From the above analysis we find that the balance $(x(t;\Pi'), y(t;\Pi'))$ depends on $2l+1-\#\Pi'$ free parameters *at most*, with $x_i(t;\Pi')$ having a pole precisely for those i for which $\alpha_i \in \Pi'$. It follows that we can only get a principal balance when $\#\Pi' = 1$. Using the algebraic integrability assumption the Kowalevski-Painlevé Criterion implies that for each $i \in \{0, 1, \ldots, l\}$ we *must* have a principal balance $(x(t;\Pi'), y(t;\Pi'))$ with the property that $x_i(t;\Pi')$ has a pole. This means for any $\alpha \in \Pi$ the series $(x(t;\alpha), y(t;\alpha))$ is a principal balance. Notice that $x_j(t;\alpha_i)$ has a pole when $j = i$, it has a zero when $a_{ji} < 0$ (α_j and α_i are joined in the Dynkin diagram) and it is finite, non-zero otherwise.

For generic $\mathbf{c} \in \mathbf{C}^{l+1}$ let $\mathbf{T}^l_{\mathbf{c}}$ denote the Abelian variety that compactifies the generic level manifold $\mathbf{F_c}$ of the momentum map and let $\mathcal{D}_{\mathbf{c}}$ denote the analytic hypersurface (divisor) of $\mathbf{T}^l_{\mathbf{c}}$ that was added to do the compactification. Since every principal balances corresponds to precisely one irreducible component of $\mathcal{D}_{\mathbf{c}}$ it is natural to label the irreducible component of $\mathcal{D}_{\mathbf{c}}$ that corresponds to $\{\alpha\} \subset \Pi$ by $\mathcal{D}^{(\alpha)}_{\mathbf{c}}$; again, we usually write $\mathcal{D}^{(i)}_{\mathbf{c}}$ for $\mathcal{D}^{(\alpha_i)}_{\mathbf{c}}$. Thus,

$$\mathcal{D}_{\mathbf{c}} = \sum_{i=0}^{l} \mathcal{D}^{(i)}_{\mathbf{c}} = \sum_{\alpha \in \Pi} \mathcal{D}^{(\alpha)}_{\mathbf{c}}.$$

Also, for generic $\mathbf{c} \in \mathbf{C}^{l+1}$ the restriction of the series $x(t;\alpha)$ and $y(t;\alpha)$ to $\mathbf{T}^l_{\mathbf{c}}$ is denoted by $x(t;\mathcal{D}^{(\alpha)}_{\mathbf{c}})$ and $y(t;\mathcal{D}^{(\alpha)}_{\mathbf{c}})$ (or $x(t;\mathcal{D}^{(i)}_{\mathbf{c}})$ and $y(t;\mathcal{D}^{(i)}_{\mathbf{c}})$ when $\alpha = \alpha_i$). It follows from Proposition 6.14 that, as a meromorphic function on $\mathbf{T}^l_{\mathbf{c}}$, x_i has a double pole along $\mathcal{D}^{(i)}_{\mathbf{c}}$, and it has no other poles. For $\{\alpha_i, \alpha_j\} \subset \Pi$, with $i \neq j$, it follows that $\frac{1}{x_i} = 0$ on $\mathcal{D}^{(i)}_{\mathbf{c}} \cap \mathcal{D}^{(j)}_{\mathbf{c}}$, and also that $\frac{1}{x_j} = 0$ on $\mathcal{D}^{(i)}_{\mathbf{c}} \cap \mathcal{D}^{(j)}_{\mathbf{c}}$. Since two divisors in $\mathbf{T}^l_{\mathbf{c}}$, such as $\mathcal{D}^{(i)}_{\mathbf{c}}$ and $\mathcal{D}^{(j)}_{\mathbf{c}}$, always intersect (in a subvariety of codimension 2) there must however be balances, depending on $2l-1$ free parameters for which both x_i and x_j, simultaneously have a pole. Since for no one of the principal balances, x_i and x_j have simultaneously a pole there is precisely one candidate, namely the balance $(x(t;\{\alpha_i,\alpha_j\}), y(t;\{\alpha_i,\alpha_j\}))$ is the unique balance which depends on at most $2l$ free parameters and which is such that $1/x_i(t;\{\alpha_i,\alpha_j\})$ and $1/x_j(t;\{\alpha_i,\alpha_j\})$ both vanish along $\mathcal{D}^{(i)}_{\mathbf{c}} \cap \mathcal{D}^{(j)}_{\mathbf{c}}$. We conclude that the series $(x(t;\{\alpha_i,\alpha_j\}), y(t;\{\alpha_i,\alpha_j\}))$ depends indeed on $2l-1$ free parameters, i.e., every Kowalevski exponent (including multiplicities) leads to a free parameter. By repeating the same argument for larger subsets Π' of Π it follows that for any subset $\Pi' \subset \Pi$, with $\Pi' \neq \emptyset$ and $\Pi' \neq \Pi$ there is a unique balance to (9.38) which depends on $2l+1-\#\Pi'$ free parameters and whose first terms are given by (9.39). $\square$

In particular we have shown the following theorem about the balances of the Toda lattice.

Theorem 9.16. *Consider the Toda lattice, associated to the affine Lie algebra $L(\mathfrak{g},\nu)$. Assuming it is a.c.i., there is a one-to-one correspondence between proper, non-empty subdiagrams Π' of the Dynkin diagram Π of $L(\mathfrak{g},\nu)$ and balances of the Toda vector field*

$$\dot{x} = x \cdot y, \qquad \dot{y} = Ax, \qquad (x,y) \in M \subset \mathbf{C}^{2(l+1)},$$

where A is the Cartan matrix of $L(\mathfrak{g},\nu)$, corresponding to the simple roots $\alpha_0,\dots,\alpha_l$ of $L(\mathfrak{g},\nu)$. The balance that corresponds to Π' has the form (9.39), where h' denotes the Weyl element of Π' and the n_i' are the corresponding Weyl integers, and it depends on $2l+1-\#\Pi'$ free parameters. In particular, there is a natural one-to-one correspondence between the $l+1$ irreducible components of the Painlevé divisor $\mathcal{D}_{\mathbf{c}}$ and the dots in the Dynkin diagram of $L(\mathfrak{g},\nu)$.

For future reference we rewrite the balances (9.39) in a slightly different form, where we use root labels $\{\alpha_0,\dots,\alpha_l\}$ as indices, instead of using the integers $\{0,\dots,l\}$ (see Paragraph 9.5.1). As before, Π' denotes any non-empty proper subset of Π.

$$\begin{aligned}
x_\alpha(t;\Pi') &= \frac{n'_\alpha}{t^2}(1+O(t^2)) \text{ for } \alpha \in \Pi',\\
x_\alpha(t;\Pi') &= b_\alpha t^{-\langle \alpha,h'\rangle}(1+c_\alpha t+O(t^2)) \text{ for } \alpha \notin \Pi',\\
y_\alpha(t;\Pi') &= -\frac{2}{t}(1+O(t^2)) \text{ for } \alpha \in \Pi',\\
y_\alpha(t;\Pi') &= \frac{1}{t}(-\langle \alpha,h'\rangle + c_\alpha t+O(t^2)) \text{ for } \alpha \notin \Pi'.
\end{aligned} \tag{9.41}$$

The variables b_α and c_α are free parameters and h' is the dual Weyl element, corresponding to Π', with dual Weyl integers n'_α ($\alpha \in \Pi'$).

9.5.3 Equivalence of Painlevé Divisors

In the previous paragraph we have seen that for generic $\mathbf{c}$ the Painlevé divisor $\mathcal{D}_{\mathbf{c}}$ of the a.c.i. Toda lattice, associated to the affine Lie algebra $L(\mathfrak{g},\nu)$, consists of $l+1$ irreducible components $\mathcal{D}_{\mathbf{c}}^{(0)},\dots,\mathcal{D}_{\mathbf{c}}^{(l)}$, which are in a natural one-to-one correspondence with the dots in the Dynkin diagram of $L(\mathfrak{g},\nu)$. We show in the following theorem that these irreducible components are, up to a rational multiple, linearly equivalent. It follows that they define, up to a multiple, the same polarization on $\mathbf{T}_{\mathbf{c}}^l$. We fix a generic $\mathbf{c} \in \mathbf{C}^{l+1}$ and we denote, for $i \in \{0,1,\dots,l\}$ by (x_i) the divisor of zeros and poles of x_i, viewed as a meromorphic function on $\mathbf{T}_{\mathbf{c}}^l$.

Theorem 9.17. *Under the a.c.i. assumption the divisor of zeros and poles of the meromorphic functions* $x_0,\ldots,x_l$*, restricted to* $\mathbf{T}^l_{\mathbf{c}}$*, is given by*

$$\begin{pmatrix}(x_0)\\(x_1)\\ \vdots \\ (x_l)\end{pmatrix} = -A\begin{pmatrix}\mathcal{D}^{(0)}_{\mathbf{c}}\\ \mathcal{D}^{(1)}_{\mathbf{c}}\\ \vdots \\ \mathcal{D}^{(l)}_{\mathbf{c}}\end{pmatrix}, \tag{9.42}$$

leading to the following linear equivalence

$$\rho_i\mathcal{D}^{(i)}_{\mathbf{c}} \sim \rho_i\hat{\xi}_i\mathcal{D}^{(0)}_{\mathbf{c}}, \tag{9.43}$$

where we recall from Paragraph 9.5.1 that ρ_i *is the smallest positive integer such that* $\rho_i\lambda_i$ *belongs to the root lattice of* $\mathfrak{g}_0$ *and that* $\lambda_i \in (\mathfrak{h}_0)^*$ *is the* i*-th fundamental weight of* $\mathfrak{g}_0$*.*

Proof. For generic $\mathbf{c}$, each of the polynomials $x_0,\ldots,x_l$, restricted to $\mathbf{F}_{\mathbf{c}}$, is everywhere different from zero since $\prod_{i=0}^l x_i^{\xi_i}$ is a Casimir (see (9.7)), where all ξ_i are positive integers. Thus, all zeros and poles of these functions can be read off from the principal balances to the Toda vector field, using Proposition 6.14. From (9.39) we get that if $0 \leqslant i \neq j \leqslant l$ then x_i has a double pole along $\mathcal{D}^{(i)}_{\mathbf{c}}$ and has, for $j \neq i$, along $\mathcal{D}^{(j)}_{\mathbf{c}}$ a zero of order

$$-\left\langle \alpha_i, h'\right\rangle = -\left\langle \alpha_i, h_{\alpha_j}\right\rangle = -a_{ij},$$

where we have used that $h' = h_{\alpha_j}$ if $\Pi' = \{\alpha_j\}$. Formula (9.42) follows.

The fact that A has corank 1 and that all principal minors have maximal rank suggests that one may express each of the divisors $\mathcal{D}^{(i)}_{\mathbf{c}}$ as a multiple of one of them, say of $\mathcal{D}^{(0)}_{\mathbf{c}}$. We do this by using the integer matrix R, introduced in Paragraph 9.5.1, with entries r_{ij}, $1 \leqslant i,j \leqslant l$. We let $\Gamma := \operatorname{diag}(1,\rho_1,\ldots,\rho_l)$, where $RA^{(0)} = \operatorname{diag}(\rho_1,\ldots,\rho_l)$. Since the divisor of zeros and poles of any meromorphic function is linearly equivalent to zero, $(x_i) \sim 0$, we deduce from (9.42) that

$$A\begin{pmatrix}\mathcal{D}^{(0)}_{\mathbf{c}}\\ \vdots \\ \mathcal{D}^{(l)}_{\mathbf{c}}\end{pmatrix} \sim \begin{pmatrix}0\\ \vdots\\ 0\end{pmatrix} \qquad \text{so that} \qquad A^{(0)}\begin{pmatrix}\mathcal{D}^{(1)}_{\mathbf{c}}\\ \vdots \\ \mathcal{D}^{(l)}_{\mathbf{c}}\end{pmatrix} \sim \begin{pmatrix}0\\ \vdots\\ -a_{k0}\mathcal{D}^{(0)}_{\mathbf{c}}\\ \vdots\\ 0\end{pmatrix}$$

where $k \neq 0$ is such that $a_{k0} \neq 0$. There is precisely one such a k in $\{1,\ldots,l\}$, except for $\mathfrak{a}_l^{(1)}$, a case that we exclude at first.

If we multiply the latter equation on the left by R, the i-th line gives a linear equivalence formula for each of the $\mathcal{D}_{\mathbf{c}}^{(i)}$ in terms of $\mathcal{D}_{\mathbf{c}}^{(0)}$, to wit

$$\rho_i \mathcal{D}_{\mathbf{c}}^{(i)} \sim -r_{ik} a_{k0} \mathcal{D}_{\mathbf{c}}^{(0)}. \tag{9.44}$$

It follows that

$$\begin{pmatrix} 0 \\ 0 \\ \vdots \\ 0 \end{pmatrix} \sim A \begin{pmatrix} \mathcal{D}_{\mathbf{c}}^{(0)} \\ \mathcal{D}_{\mathbf{c}}^{(1)} \\ \vdots \\ \mathcal{D}_{\mathbf{c}}^{(l)} \end{pmatrix} = A\Gamma^{-1} \begin{pmatrix} \mathcal{D}_{\mathbf{c}}^{(0)} \\ \rho_1 \mathcal{D}_{\mathbf{c}}^{(1)} \\ \vdots \\ \rho_l \mathcal{D}_{\mathbf{c}}^{(l)} \end{pmatrix} \sim -A\Gamma^{-1} \begin{pmatrix} \mathcal{D}_{\mathbf{c}}^{(0)} \\ r_{1k} a_{k0} \mathcal{D}_{\mathbf{c}}^{(0)} \\ \vdots \\ r_{lk} a_{k0} \mathcal{D}_{\mathbf{c}}^{(0)} \end{pmatrix}.$$

Since $\mathcal{D}_{\mathbf{c}}^{(0)}$ is an effective divisor on $\mathbf{T}_{\mathbf{c}}^l$ the latter linear equivalence implies that $\Gamma^{-1} \begin{pmatrix} 1 \\ r_{1k} a_{k0} \\ \vdots \\ r_{lk} a_{k0} \end{pmatrix}$ is a (normalized) null-vector of A, hence equals $\hat{\xi}$. We conclude that

$$\begin{pmatrix} 1 \\ r_{1k} a_{k0} \\ \vdots \\ r_{lk} a_{k0} \end{pmatrix} = \Gamma \hat{\xi} = \begin{pmatrix} 1 \\ \rho_1 \hat{\xi}_1 \\ \vdots \\ \rho_l \hat{\xi}_l \end{pmatrix},$$

which, substituted in (9.44), leads at once to (9.43). The argument works for $\mathfrak{a}_l^{(1)}$ with $r_{ik} a_{k0}$ replaced by $r_{ik} a_{k0} + r_{ik'} a_{k'0}$. □

It follows that we can express the intersection of any of the divisors $\mathcal{D}_{\mathbf{c}}^{(i)}$ in terms of the self-intersection of one of them, say of $\mathcal{D}_{\mathbf{c}}^{(0)}$. By the general theory of Abelian varieties the arithmetic genus of any one of them, say $g_a(\mathcal{D}_{\mathbf{c}}^{(0)})$, and the dimension of $L(\mathcal{D}_{\mathbf{c}}^{(0)})$ can be expressed as the intersection of any l of the $l+1$ irreducible components of $\mathcal{D}_{\mathbf{c}}$. This is the content of the following proposition.

Proposition 9.18. *Under the a.c.i. assumption,*

$$\mathcal{D}_{\mathbf{c}}^{(s_1)} \cdot \mathcal{D}_{\mathbf{c}}^{(s_2)} \cdots \mathcal{D}_{\mathbf{c}}^{(s_k)} = \prod_{i=1}^{k} \hat{\xi}_{s_i} \left(\mathcal{D}_{\mathbf{c}}^{(0)}\right)^k, \tag{9.45}$$

where $\{s_1, \dots, s_k\} \subset \{0, 1, \dots, l\}$ *is arbitrary.*

As a consequence,

$$g_a(\mathcal{D}_{\mathbf{c}}^{(0)}) - l + 1 = \frac{\hat{\xi}_i}{l! \prod_{j=0}^{l} \hat{\xi}_j} \mathcal{D}_{\mathbf{c}}^{(0)} \cdots \widehat{\mathcal{D}_{\mathbf{c}}^{(i)}} \cdots \mathcal{D}_{\mathbf{c}}^{(l)}, \tag{9.46}$$

where $i \in \{0, 1, \ldots, l\}$ is arbitrary. The latter formula yields the dimension of $L(\mathcal{D}_{\mathbf{c}}^{(0)})$, in view of

$$\dim L(\mathcal{D}_{\mathbf{c}}^{(0)}) = \prod_{i=1}^{l} \delta_i = g_a(\mathcal{D}_{\mathbf{c}}^{(0)}) - l + 1, \tag{9.47}$$

where $(\delta_1, \ldots, \delta_l)$ is the type of the polarization that $\mathcal{D}_{\mathbf{c}}^{(0)}$ induces on $\mathbf{T}_{\mathbf{c}}^l$. More generally, for any non-negative integers $(r_0, r_1, \ldots, r_l)$,

$$\begin{aligned} g_a\left(\sum_{i=0}^{l} r_i \mathcal{D}_{\mathbf{c}}^{(i)}\right) - l + 1 &= \dim L\left(\sum_{i=0}^{l} r_i \mathcal{D}_{\mathbf{c}}^{(i)}\right) \\ &= \left(\sum_{i=0}^{l} r_i \hat{\xi}_i\right)^l \dim L(\mathcal{D}_{\mathbf{c}}^{(0)}) \\ &= \left(\sum_{i=0}^{l} r_i \hat{\xi}_i\right)^l \left(g_a\left(\mathcal{D}_{\mathbf{c}}^{(0)}\right) - l + 1\right). \end{aligned}$$

Proof. Since the homology class of the intersection of divisors depends on the linear equivalence class of the divisors only, (9.43) implies that

$$\begin{aligned} \left(\prod_{i=1}^{k} \rho_{s_i}\right) \mathcal{D}_{\mathbf{c}}^{(s_1)} \cdots \mathcal{D}_{\mathbf{c}}^{(s_k)} &= \rho_{s_1} \mathcal{D}_{\mathbf{c}}^{(s_1)} \cdot \rho_{s_2} \mathcal{D}_{\mathbf{c}}^{(s_2)} \cdots \rho_{s_k} \mathcal{D}_{\mathbf{c}}^{(s_k)} \\ &= \rho_{s_1} \hat{\xi}_{s_1} \mathcal{D}_{\mathbf{c}}^{(0)} \cdot \rho_{s_2} \hat{\xi}_{s_2} \mathcal{D}_{\mathbf{c}}^{(0)} \cdots \rho_{s_k} \hat{\xi}_{s_k} \mathcal{D}_{\mathbf{c}}^{(0)} \\ &= \prod_{i=1}^{k} \rho_{s_i} \prod_{i=1}^{k} \hat{\xi}_{s_i} \left(\mathcal{D}_{\mathbf{c}}^{(0)}\right)^k, \end{aligned}$$

which yields (9.45). Applied to l arbitrary components of $\mathcal{D}_{\mathbf{c}}$ we get (9.46) and (9.47) in view of the general formula (5.25), valid for (eventually singular) divisors on an irreducible Abelian variety. Let $\rho := \prod_{i=0}^{l} \rho_i$, where we recall that $\rho_0 = 1$, and use (9.43) to compute

$$\begin{aligned} g_a\left(\sum_{i=0}^{l} r_i \mathcal{D}_{\mathbf{c}}^{(i)}\right) - l + 1 &= \dim L\left(\sum_{i=0}^{l} r_i \mathcal{D}_{\mathbf{c}}^{(i)}\right) \\ &= \frac{1}{\rho^l} \frac{\left(\rho \sum_{i=0}^{l} r_i \mathcal{D}_{\mathbf{c}}^{(i)}\right)^l}{l!} \end{aligned}$$

$$
\begin{aligned}
&= \frac{1}{l!\rho^l}\left(\sum_{i=0}^{l}\left(\frac{\rho r_i}{\rho_i}\right)\rho_i\mathcal{D}_{\mathbf{c}}^{(i)}\right)^l \\
&= \frac{1}{l!\rho^l}\left(\sum_{i=0}^{l}\left(\frac{\rho r_i}{\rho_i}\right)\rho_i\hat{\xi}_i\mathcal{D}_{\mathbf{c}}^{(0)}\right)^l \\
&= \left(\sum_{i=0}^{l} r_i\hat{\xi}_i\right)^l \frac{\left(\mathcal{D}_{\mathbf{c}}^{(0)}\right)^l}{l!} \\
&= \left(\sum_{i=0}^{l} r_i\hat{\xi}_i\right)^l \dim L(\mathcal{D}_{\mathbf{c}}^{(0)}) \\
&= \left(\sum_{i=0}^{l} r_i\hat{\xi}_i\right)^l \left(g_a(\mathcal{D}_{\mathbf{c}}^{(0)}) - l + 1\right).
\end{aligned}
$$

as was to be shown. □

The above proposition specializes to the following formula that relates the arithmetic genus of $\mathcal{D}_{\mathbf{c}}^{(\alpha)}$ to the one of $\mathcal{D}_{\mathbf{c}}^{(0)}$:

$$
g_a(\mathcal{D}_{\mathbf{c}}^{(\alpha)}) - l + 1 = \hat{\xi}_\alpha^l \left(g_a(\mathcal{D}_{\mathbf{c}}^{(0)}) - l + 1\right). \tag{9.48}
$$

9.5.4 Behavior of the Principal Balances Near the Lower Ones

In this paragraph we wish to point out in terms of the root system (Dynkin diagram) of $L(\mathfrak{g},\nu)$ how the principal balances (depending on $2l$ free parameters) degenerate into lower balances depending on $2l-1$ free parameters. This will imply a few geometric facts on the intersection of the divisors and their position with respect to the Toda vector field, that we will explore in the subsequent paragraphs.

We fix two simple roots $\alpha \neq \beta$ in Π and we consider the lower balance that corresponds to $\{\alpha,\beta\}$. Specializing (9.41) we can write it as

$$
\begin{aligned}
x_\alpha(t;\{\alpha,\beta\}) &= \frac{n'_\alpha}{t^2}(1+O(t^2)), \\
x_\beta(t;\{\alpha,\beta\}) &= \frac{n'_\beta}{t^2}(1+O(t^2)), \\
x_\gamma(t;\{\alpha,\beta\}) &= \frac{b'_\gamma}{t^{a_{\gamma\alpha}n'_\alpha + a_{\gamma\beta}n'_\beta}}(1+c'_\gamma t + O(t^2)) \text{ for } \gamma \notin \{\alpha,\beta\}, \\
y_\alpha(t;\{\alpha,\beta\}) &= -\frac{2}{t}(1+O(t)), \\
y_\beta(t;\{\alpha,\beta\}) &= -\frac{2}{t}(1+O(t)), \\
y_\gamma(t;\{\alpha,\beta\}) &= -\frac{a_{\gamma\alpha}n'_\alpha + a_{\gamma\beta}n'_\beta}{t} + c'_\gamma + O(t), \text{ for } \gamma \notin \{\alpha,\beta\}.
\end{aligned}
$$

We also consider the principal balance that goes with α, see (9.40),

$$\begin{aligned}
x_\alpha(t;\alpha) &= \frac{1}{t^2}(1 + b_\alpha t^2 + O(t^3)), \\
x_\gamma(t;\alpha) &= \frac{b_\gamma}{t^{a_{\gamma\alpha}}}(1 + c_\gamma t + O(t^2)) \text{ for } \gamma \in \Pi \setminus \{\alpha\}, \\
y_\alpha(t;\alpha) &= -\frac{2}{t}(1 - b_\alpha t^2 + O(t^2)), \\
y_\gamma(t;\alpha) &= -\frac{a_{\gamma\alpha}}{t} + c_\gamma + O(t) \text{ for } \gamma \in \Pi \setminus \{\alpha\}.
\end{aligned}$$

We will distinguish six[5] cases, according to the values of $a_{\alpha\beta}$ and $a_{\beta\alpha}$; said differently, according to the relative position of α and β in the Dynkin diagram of $L(\mathfrak{g},\nu)$. We display in the six tables that follow for each of these cases how the series $x(t;\{\alpha,\beta\}), y(t;\{\alpha,\beta\})$ can be obtained from the series $x(t;\alpha), y(t;\alpha)$: for each of the series $x_i(t;\alpha)$, or for their inverses, as given in (9.49), one substitutes for the parameters the values given in the tables and one lets $\varsigma \to 0$ (for the justification, see below). This leads to the series $x_i(t;\{\alpha,\beta\})$, as given in (9.49). The fact that the series $y_i(t;\alpha)$ tend to the series $y_i(t;\{\alpha,\beta\})$ then follows from it, since $\dot{x}_i = x_i y_i$, for $i = 0,\ldots,l$.

Table 9.3. Case $a_{\alpha\beta} = 0$.

γ	b_γ	c_γ
$\gamma = \alpha$	$O(\varsigma^0)$	
$\gamma = \beta$	$\varsigma^{-2}/4 + O(\varsigma^{-1})$	ς^{-1}
$a_{\beta\gamma} < 0$	$O(\varsigma)$	$O(\varsigma^{-1})$
$a_{\beta\gamma} = 0,\ \gamma \neq \alpha$	b'_γ	c'_γ

We give the proof in the first case, which correspond to Table 9.3. Here, α and β are two roots that are orthogonal, $a_{\alpha\beta} = 0$, which means that they are not connected to each other in the Dynkin diagram of $L(\mathfrak{g},\nu)$. For simplicity, let us assume that, in the Dynkin diagram, β has precisely two neighbors, which we denote by β' and β'', and that $\xi_{\beta'} = \xi_\beta = \xi_{\beta''}$; these assumptions actually represent the typical case. Since $a_{\alpha\beta} = 0$, (9.49) specializes to

$$\begin{aligned}
x_\alpha(t;\alpha) &= \frac{1}{t^2}(1 + b_\alpha t^2 + O(t^3)), \\
x_\beta(t;\alpha) &= b_\beta(1 + c_\beta t + O(t^2)), \\
x_\gamma(t;\alpha) &= \frac{b_\gamma}{t^{a_{\gamma\alpha}}}(1 + c_\gamma t + O(t^2)) \text{ for } \gamma \in \Pi \setminus \{\alpha\},
\end{aligned}$$

[5] There is a seventh case, namely when $a_{\alpha\beta} = a_{\beta\alpha} = -2$, but then the root system is $\mathfrak{a}_1^{(1)}$, which has rank 1, so there are no balances corresponding to $\{\alpha,\beta\}$.

Table 9.4. Case $a_{\alpha\beta} = a_{\beta\alpha} = -1$.

γ	b_γ	c_γ
$\gamma = \alpha$	$\varsigma^{-2}/9 + O(\varsigma^{-1})$	
$\gamma = \beta$	$-8\varsigma^{-3}/27 + O(\varsigma^{-2})$	ς^{-1}
$a_{\alpha\gamma} < 0,\ \gamma \neq \beta$	$O(\varsigma)$	$O(\varsigma^{-1})$
$a_{\beta\gamma} < 0,\ \gamma \neq \alpha$	$O(\varsigma^2)$	$O(\varsigma^{-1})$
$a_{\alpha\gamma} = a_{\beta\gamma} = 0$	b'_γ	c'_γ

Table 9.5. Case $a_{\alpha\beta} = -2$ and $a_{\beta\alpha} = -1$.

γ	b_γ	c_γ
$\gamma = \alpha$	$\varsigma^{-2}/3 + O(\varsigma^{-1})$	
$\gamma = \beta$	$-512\varsigma^{-3} + O(\varsigma^{-2})$	ς^{-1}
$a_{\beta\gamma} < 0,\ \gamma \neq \alpha$	$O(\varsigma^3)$	$-\varsigma^{-1} + O(\varsigma^0)$
$a_{\beta\gamma} = 0$	b'_γ	c'_γ
$\gamma = \alpha$	$O(\varsigma^0)$	
$\gamma = \beta$	$O(\varsigma^{-1})$	ς
$a_{\beta\gamma} < 0,\ \gamma \neq \alpha$	$O(\varsigma)$	$O(\varsigma)$
$a_{\beta\gamma} = 0$	b'_γ	c'_γ

Table 9.6. Case $a_{\alpha\beta} = -1$ and $a_{\beta\alpha} = -2$.

γ	b_γ	c_γ
$\gamma = \alpha$	$\varsigma^{-2}/12 + O(\varsigma^{-1})$	
$\gamma = \beta$	$\varsigma^{-4}/4 + O(\varsigma^{-3})$	ς^{-1}
$a_{\alpha\gamma} < 0,\ \gamma \neq \beta$	$O(\varsigma^2)$	$-\varsigma^{-1} + O(\varsigma^0)$
$a_{\alpha\gamma} = 0$	b'_γ	c'_γ

so that

$$\frac{1}{x_\alpha}(t;\alpha) = t^2 - b_\alpha t^4 + O(t^5)),$$
$$\frac{1}{\sqrt{x_\alpha x_\beta}}(t;\alpha) = \pm\frac{1}{\sqrt{b_\beta}}t(1 - c_\beta t/2 + O(t^2)).$$

Similarly, the principal balance that goes with β is given by

Table 9.7. Case $a_{\alpha\beta} = -3$ and $a_{\beta\alpha} = -1$.

γ	b_γ	c_γ
$\gamma = \alpha$	$\varsigma^{-2} + O(\varsigma^{-1})$	
$\gamma = \beta$	$(8 \pm 16/\sqrt{3})\varsigma^{-3} + O(\varsigma^{-2})$	ς^{-1}
$\gamma \notin \{\alpha, \beta\}$	$O(\varsigma^{9})$	$-3\varsigma^{-1}$

Table 9.8. Case $a_{\alpha\beta} = -1$ and $a_{\beta\alpha} = -3$.

γ	b_γ	c_γ
$\gamma = \alpha$	$\varsigma^{-2}/9 + O(\varsigma^{-1})$	
$\gamma = \beta$	$8\varsigma^{-5}/27 + O(\varsigma^{-4})$	ς^{-1}
$\gamma \notin \{\alpha, \beta\}$	$O(\varsigma^{10})$	$-2\varsigma^{-1}$
$\gamma = \alpha$	$O(\varsigma^{0})$	
$\gamma = \beta$	$O(\varsigma^{-1})$	ς
$\gamma \notin \{\alpha, \beta\}$	$O(\varsigma^{2})$	-2ς

$$
\begin{aligned}
x_\alpha(t;\beta) &= \tilde{b}_\alpha(1 + \tilde{c}_\alpha t + O(t^2)),\\
x_\beta(t;\beta) &= \frac{1}{t^2}(1 + \tilde{b}_\beta t^2 + O(t^3)),\\
x_\gamma(t;\alpha) &= \frac{\tilde{b}_\gamma}{t^{a_{\gamma\alpha}}}(1 + \tilde{c}_\gamma t + O(t^2)) \text{ for } \gamma \in \Pi \setminus \{\alpha\},\\
\frac{1}{x_\alpha}(t;\alpha) &= \frac{1}{\tilde{b}_\alpha}(1 - \tilde{c}_\alpha t + +O(t^2)),\\
\frac{1}{\sqrt{x_\alpha x_\beta}}(t;\alpha) &= \pm\frac{1}{\sqrt{\tilde{h}_\beta}} t(1 - \tilde{c}_\alpha t/2 + O(t^2)).
\end{aligned}
$$

For the lower balance that goes with $\Pi' = \{\alpha, \beta\}$ (9.49) gives

$$
\begin{aligned}
x_\alpha(t;\{\alpha,\beta\}) &= \frac{1}{t^2}(1 + b'_\alpha t^2 + O(t^3)),\\
x_\beta(t;\{\alpha,\beta\}) &= \frac{1}{t^2}(1 + b'_\beta t^2 + O(t^3)),\\
x_\gamma(t;\{\alpha,\beta\}) &= \frac{b'_\gamma}{t^{a_{\gamma\alpha}+a_{\gamma\beta}}}(1 + c'_\gamma t + O(t^2)) \text{ for } \gamma \notin \{\alpha,\beta\}\,,.
\end{aligned}
$$

In a neighborhood of $\mathcal{D}_{\mathbf{c}}^{(\alpha)} \cap \mathcal{D}_{\mathbf{c}}^{(\beta)}$, the functions x_α^{-1}, $1/\sqrt{x_\alpha x_\beta}$ and x_γ ($\gamma \neq \alpha, \beta$) are finite since they are finite in the affine, and since no other $\mathcal{D}_{\mathbf{c}}^{(\gamma)}$ contains a component of the intersection $\mathcal{D}_{\mathbf{c}}^{(\alpha)} \cap \mathcal{D}_{\mathbf{c}}^{(\beta)}$ (Paragraph 9.5.2).

Therefore, by Hartog's Theorem, they are holomorphic about $\mathcal{D}_{\mathbf{c}}^{(\alpha)} \cap \mathcal{D}_{\mathbf{c}}^{(\beta)}$, and so, the functions x_α^{-1}, $1/\sqrt{x_\alpha x_\beta}$ and x_γ, evaluated along the integral curves that start at points of $\mathcal{D}_{\mathbf{c}}^{(\alpha)} \cap \mathcal{D}_{\mathbf{c}}^{(\beta)}$ are bona fide limits of the series $x_\alpha^{-1}(t;\alpha)$, $1/\sqrt{x_\alpha x_\beta}(t;\alpha)$ and $x_\gamma(t;\alpha)$. This leads at once to the term by term estimates for the local behavior of the free parameters

$$\begin{array}{lll} b_\alpha \to b'_\alpha & & \\ 1/\sqrt{b_\beta} \to 0 & c_\beta/\sqrt{b_\beta} \to -2 & \\ b_\gamma \to 0 & b_\gamma c_\gamma \to b'_\gamma & \gamma \in \{\beta', \beta''\} \\ b_\gamma \to b'_\gamma & c_\gamma \to c'_\gamma & \gamma \notin \{\alpha, \beta, \beta', \beta''\}\,; \end{array} \tag{9.49}$$

setting $c_\beta = 1/\varsigma$, we get the two first and fourth lines of Table 9.3. Using the Casimir $C_1 = \prod_{i=0}^{l} x_i^{\xi_i}$ and our assumption that $\xi_{\beta'} = \xi_\beta = \xi_{\beta''}$ one concludes

$$b_\beta b_{\beta'} b_{\beta''} = O(\varsigma^0).$$

Since $b_\beta = 1/(4\varsigma^2) + O(\varsigma^{-1})$ it follows that $b_{\beta'} b_{\beta''} = O(\varsigma^2)$ and from the third line of (9.49) conclude

$$c_{\beta'} c_{\beta''} = O(\varsigma^{-2}). \tag{9.50}$$

Substituting the series (9.49) and the estimates above (lines 1, 2 and 4 of Table 9.3) into the Hamiltonian $\frac{1}{2}\langle X \,|\, X\rangle = \frac{1}{2}\,\mathrm{Trace}\, X^2$, where X is given by (9.23), yields

$$\begin{aligned} &\frac{1}{2}\left\langle \hat\lambda_{\beta'} \,|\, \hat\lambda_{\beta'} \right\rangle c_{\beta'}^2 + \frac{1}{2}\left\langle \hat\lambda_{\beta''} \,|\, \hat\lambda_{\beta''} \right\rangle c_{\beta''}^2 + \left\langle \hat\lambda_{\beta'} \,|\, \hat\lambda_{\beta''} \right\rangle c_{\beta'} c_{\beta''} + \\ &\quad + \frac{1}{\varsigma}\left(\left\langle \hat\lambda_{\beta} \,|\, \hat\lambda_{\beta'} \right\rangle c_{\beta'} + \left\langle \hat\lambda_{\beta} \,|\, \hat\lambda_{\beta''} \right\rangle c_{\beta''} \right) \\ &\quad \frac{1}{2\varsigma^2}\left(\frac{1}{\langle \beta \,|\, \beta \rangle} - \left\langle \hat\lambda_{\beta} \,|\, \hat\lambda_{\beta} \right\rangle \right) + O(\varsigma^{-1}), \end{aligned}$$

which together with (9.50) yields, upon rescaling variables, the relations

$$\varsigma c_{\beta'} = O(\varsigma^0), \qquad \varsigma c_{\beta''} = O(\varsigma^0).$$

Thus, from the third line of (9.49) deduce

$$b_{\beta'} = O(\varsigma), \qquad b_{\beta''} = O(\varsigma),$$

and the third line of Table 9.3, completing the proof of the first case. The other cases are similar, but one uses other functions, for example for the case that corresponds to Table 9.4 one uses $1/(x_\alpha x_\beta)$ and $1/(x_\alpha^2 x_\beta)$.

9.5.5 Tangency of the Toda Flows to the Painlevé Divisors

In this paragraph we analyze the tangency of the Toda flow along the Painlevé divisors. As we will show, the tangency locus of the Toda vector field to the Painlevé divisor $\mathcal{D}_{\mathbf{c}}^{(\alpha)}$ is contained in the lower-dimenional loci $\mathcal{D}_{\mathbf{c}}^{(\alpha)} \cap \mathcal{D}_{\mathbf{c}}^{(\beta)}$, where β is a simple root that is not orthogonal to α, i.e., a root that is connected to α in the Dynkin diagram of $L(\mathfrak{g}, \nu)$, and the degree of tangency of $\mathcal{X}_H$ to $\mathcal{D}_{\mathbf{c}}^{(\alpha)}$ along $\mathcal{D}_{\mathbf{c}}^{(\alpha)} \cap \mathcal{D}_{\mathbf{c}}^{(\beta)}$ is expressible in terms of the Cartan integer $a_{\alpha\beta}$, as indicated in Table 9.9.

Let us first show how the order of tangency of a linear vector field on an Abelian variety to a divisor can be described analytically.

Lemma 9.19. *Let $\mathcal{D}$ be an analytic hypersurface on an Abelian variety $\mathbf{T}^l$, let $m \in \mathcal{D}$ and choose a local defining function f for $\mathcal{D}$ at m. Also, let $\partial/\partial t_1, \dots, \partial/\partial t_l$ be the l commuting vector fields, corresponding to a system of linear coordinates $(t_1, \dots, t_l)$ on $\mathbf{T}^l$. The order of tangency to $\mathcal{D}$ of $\partial/\partial t_i$ at m is the order of vanishing of ϕ_{ij} at m, where j is arbitrary and*

$$\phi_{ij} := \left.\frac{\partial f/\partial t_i}{\partial f/\partial t_j}\right|_{\mathcal{D}} = \left.\frac{\partial f^n/\partial t_i}{\partial f^n/\partial t_j}\right|_{\mathcal{D}} . \tag{9.51}$$

Proof. Introduce, as in (7.45), the holomorphic differentials on $\mathcal{D}$,

$$\omega_i := \mathrm{d}t_1 \wedge \ldots \wedge \widehat{\mathrm{d}t_i} \wedge \ldots \wedge \mathrm{d}t_l \Big|_{\mathcal{D}} . \tag{9.52}$$

Since $\mathrm{d}f_{|\mathcal{D}} = 0$ we have that $\sum_{j=1}^{l} \frac{\partial f}{\partial t_j} \mathrm{d}t_{j|\mathcal{D}} = 0$, which implies that

$$\frac{\partial f}{\partial t_j}_{|\mathcal{D}} \omega_i = \pm \frac{\partial f}{\partial t_i}_{|\mathcal{D}} \omega_j,$$

so that $\phi_{ij} = \omega_i/\omega_j$. As we already pointed out, the points of tangency of $\partial/\partial t_i$ are the zeros of ω_i, with multiplicity, hence also of any ϕ_{ij}, as the generic point of tangency of $\partial/\partial t_i$ is not a point of tangency of $\partial/\partial t_j$. $\square$

In order to compute the tangency we first show that the Toda Hamiltonians can be localized in terms of molecules that have to do with simple roots. See Paragraph 9.5.1 for the notation $\partial\Pi'$ and Π'', for a given $\Pi' \subset \Pi$. Also, the Toda system that is associated with $\Pi' \subseteq \Pi$ will be referred to as $X_{\Pi'}$; if $\Pi' \neq \Pi$ then $X_{\Pi'}$ is a non-periodic Toda lattice, see Remark 9.13.

We fix $\Pi' \subset \Pi$ and we denote a basis of generators of the (polynomial) invariants of the Toda system, associated to Π', by I'. Similarly for the exterior Π'', where the basis will be denoted by I''. Also, we denote by I^δ the $\#(\partial\Pi') - 1$ dimensional vector space, defined by

$$I^\delta := \left\{ \sum_{\alpha \in \Pi} v_\alpha y_\alpha \text{ such that } \forall \beta \in \Delta' \cup \Delta'' \ : \ \beta \perp \sum_{\alpha \in \Pi} v_\alpha \alpha \right\} . \tag{9.53}$$

Proposition 9.20. *With these notations,*

(1) All polynomial invariants of the Toda system, associated to Π, are of the form

$$P(I',I'',I^\delta)+\sum_{\alpha\in\partial\Pi'}x_\alpha G_\alpha(x,y), \tag{9.54}$$

where P and G_α are polynomials in their arguments;

(2) Under the a.c.i. assumption, the polynomials I', I'' and I^δ, restricted to $\mathbf{T}^l_{\mathbf{c}}$, are holomorphic in a neighborhood of the divisors $\mathcal{D}^{(\alpha)}_{\mathbf{c}}$, if $\alpha\in\Pi'\cup\Pi''$. In particular, they are holomorphic in a neighborhood of the intersection $\bigcap_{\alpha\in\Pi'\cup\Pi''}\mathcal{D}^{(\alpha)}_{\mathbf{c}}$.

Proof. For any $\Pi'\subset\Pi$ the subspace of M, defined by

$$M':=\bigcup_{\alpha\in\partial\Pi'}\left\{(x,y)\in M\mid x_\alpha=0\right\},$$

is preserved by the flow of the Toda system X_Π, in view of the equations $\dot x_\alpha=x_\alpha y_\alpha$, for $\alpha\in\Pi'$. Moreover, since Π' and Π'' are two totally unconnected Dynkin subdiagrams of Π the restriction of X_Π to M' yields the Toda systems $X_{\Pi'}$ and $X_{\Pi''}$ on the one hand, and the following *linear* differential equation for the y_α, with $\alpha\in\partial\Pi'$,

$$\dot y_\alpha=\sum_{\beta\in\Pi'\cup\Pi''}a_{\alpha\beta}x_\beta. \tag{9.55}$$

It follows that the invariants of the Toda system X_Π, restricted to M', are generated by those of $X_{\Pi'}$, $X_{\Pi''}$ and the invariants of the linear system.

In order to compute the invariant of the latter, let us first look for linear invariants $\sum_{\alpha\in\Pi}v_\alpha y_\alpha$. Such a function will be an invariant for (9.55) if and only if

$$0=\sum_{\alpha\in\Pi}v_\alpha\dot y_\alpha=\sum_{\alpha\in\Pi}\sum_{\beta\in\Pi'\cup\Pi''}v_\alpha a_{\alpha\beta}x_\beta=\sum_{\beta\in\Pi'\cup\Pi''}\left\langle\sum_{\alpha\in\Pi}v_\alpha\alpha,H_\beta\right\rangle x_\beta,$$

for arbitrary x_β, with $\beta\in\Pi'\cup\Pi''$. This means that the invariants of (9.55) of the form $\sum_{\alpha\in\Pi}v_\alpha y_\alpha$ are precisely the elements of I^δ (see (9.53). Since (9.55) is a vector field on a $\#(\partial\Pi')-1$ dimensional space (recall that $C_0=0$), the linear invariants that we have found generate all polynomial invariants of (9.55). This shows that every invariant of X_Π, restricted to M', must be a polynomial of I', I'' and I^δ. The same is true for the unrestricted Toda system X_Π, up to terms of the form $x_\alpha G_\alpha$, with $\alpha\in\partial\Pi'$. This shows (1).

Under the a.c.i. assumption, the pole order of a meromorphic function f on $\mathbf{T}^l_{\mathbf{c}}$ along the divisor $\mathcal{D}^{(\alpha)}_{\mathbf{c}}$ is given by the pole order of the series $f(t;\mathcal{D}^{(\alpha)}_{\mathbf{c}})$ (Proposition 6.14). Recall that this series is obtained by substituting the principal balance $x(t;\mathcal{D}^{(\alpha)}_{\mathbf{c}})$, $y(t;\mathcal{D}^{(\alpha)}_{\mathbf{c}})$ in f (which in our case is a polynomial).

Since $x_\beta(t;\mathcal{D}_{\mathbf{c}}^{(\alpha)})$ or $y_\beta(t;\mathcal{D}_{\mathbf{c}}^{(\alpha)})$ can only have a pole when α and β are joined by edge in the Dynkin diagram of Π (see Theorem 9.16), each element of I', which clearly involves only the x_β and y_β with $\beta \in I'$, will be holomorphic along $\mathcal{D}_{\mathbf{c}}^{(\alpha)}$, with $\alpha \in \Pi''$. To show that each element of I' is holomorphic along $\mathcal{D}_{\mathbf{c}}^{(\alpha)}$, with $\alpha \in \Pi'$ one use the Laurent solutions. By symmetry, the elements of $\mathcal{I}''$ do not have a pole along the divisors $\mathcal{D}_{\mathbf{c}}^{(\alpha)}$, with $\alpha \in \Pi' \cup \Pi''$.

We now turn to the linear invariants, i.e., the elements of I^δ. Since for any $\alpha,\ \beta \in \Pi$ one has that $y_\alpha(t;\mathcal{D}_{\mathbf{c}}^{(\beta)}) = -a_{\alpha\beta}/t+O(t^0)$ we find for $\sum_\alpha v_\alpha y_\alpha \in I^\delta$, i.e., with the v_α satisfying $\sum_{\alpha\in\Pi} v_\alpha a_{\alpha\beta} = 0$ for all $\beta \in \Pi' \cup \Pi''$, we find that

$$\sum_{\alpha\in\Pi} v_\alpha y_\alpha(t;\mathcal{D}_{\mathbf{c}}^{(\beta)}) = -\sum_{\alpha\in\Pi} v_\alpha a_{\alpha\beta} t^{-1} + O(t^0) = O(t^0),$$

for all $\beta \in \Pi' \cup \Pi''$. Summarizing, the functions in I', I'' and I^δ do not have a pole along any of the divisors $\mathcal{D}_{\mathbf{c}}^{(\beta)}$, with $\beta \in \Pi' \cup \Pi''$. This means that they are holomorphic in a neighborhood of these divisors and, a fortiori, in a neighborhood of the intersection of these divisors. □

Theorem 9.21. *The Toda vector field*

$$\dot{x} = x \cdot y, \qquad \dot{y} = Ax,$$

on the hyperplane $\sum \xi_i y_i = 0$ is transversal or tangent to the Painlevé divisors $\mathcal{D}_{\mathbf{c}}^{(\alpha)}$ and $\mathcal{D}_{\mathbf{c}}^{(\beta)}$ along the generic points of $\mathcal{D}_{\mathbf{c}}^{(\alpha)} \cap \mathcal{D}_{\mathbf{c}}^{(\beta)}$, as dictated by the Dynkin subdiagram of Π, spanned by α and β, and this in a way, indicated in Table 9.9.

Proof. We apply Lemma 9.19 to the divisor $\mathcal{D}_{\mathbf{c}}^{(\alpha)}$, with local defining function $x_\alpha^{-1/2}$. We assume that the first two linear coordinates t_1 and t_2 correspond to the Hamiltonians H_1 and H_2, where H_2 is the weight homogeneous invariant of weight m_2+1. According to the lemma the order of tangency of $\partial/\partial t_1$ to $\mathcal{D}_{\mathbf{c}}^{(\alpha)}$ at m is given by the order of vanishing at m of the function ϕ_{12}, defined by

$$\phi_{12} := \left.\frac{\partial x_\alpha/\partial t_1}{\partial x_\alpha/\partial t_2}\right|_{\mathcal{D}_{\mathbf{c}}^{(\alpha)}} = \left.\frac{x_\alpha y_\alpha}{x_\alpha \langle \mu_\alpha \,|\, \nabla_y H_2\rangle}\right|_{\mathcal{D}_{\mathbf{c}}^{(\alpha)}}, \tag{9.56}$$

where $\langle\cdot\,|\,\cdot\rangle$ denotes the standard inner product on $\mathbf{C}^l$ and $\nabla_y H_2 = \frac{\partial H_2}{\partial y}$. Also, $\partial x_\alpha/\partial t_2$ was computed using the Poisson brackets (9.6), so that μ_α is explicitly given by

$$\mu_\alpha = \frac{1}{2} a_{\alpha\beta} \langle e_\beta \,|\, e_\beta\rangle\,. \tag{9.57}$$

The computation of this vanishing may be done by using the principal balances that go with the divisor $\mathcal{D}_{\mathbf{c}}^{(\alpha)}$ (Proposition 6.14).

Table 9.9. The Toda vector field on $\mathbf{T}^l_{\mathbf{c}}$ is tangent to the irreducible components $\mathcal{D}^{(\alpha)}_{\mathbf{c}}$ of the Painlevé divisor $\mathcal{D}_{\mathbf{c}}$, along some of the subvarieties $\mathcal{D}^{(\alpha)}_{\mathbf{c}} \cup \mathcal{D}^{(\beta)}_{\mathbf{c}}$, as given in the present table. The precise degree of tangency depends on the Cartan integers $a_{\alpha\beta}$; in two of the cases the divisor has two branches and we give the order of tangency along each of the branches.

$a_{\alpha\beta}$	$a_{\beta\alpha}$	Dynkin	$\mathcal{D}^{(\alpha)}_{\mathbf{c}}$	$\mathcal{D}^{(\beta)}_{\mathbf{c}}$
0	0	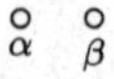	transversal	transversal
−1	−1	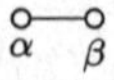	simply tangent	simply tangent
−2	−1	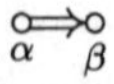	doubly tangent + transversal	doubly tangent
−3	−1	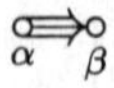	fourfold tangent	fourfold tangent + transversal

Since $y_\alpha(t; \mathcal{D}^{(\alpha)}_{\mathbf{c}}) = -2/t + O(t^0)$ this means that for a generic $m \in \mathcal{D}^{(\alpha)}_{\mathbf{c}}$

$$\phi_{12}^{-1}(m) = \lim_{t\to 0} y_\alpha^{-1} \langle \mu_\alpha \,|\, \nabla_y H_2 \rangle_{|_{\mathcal{D}^{(\alpha)}_{\mathbf{c}}}} = -\frac{1}{2}\,\mathrm{Res}_t \left\langle \mu_\alpha \,|\, \nabla_y H_2(t; \mathcal{D}^{(\alpha)}_{\mathbf{c}}) \right\rangle. \quad (9.58)$$

We use now Proposition 9.20, with $\Pi' = \{\alpha, \beta\}$, i.e., we write H_2 as

$$H_2 = P(I', I'', I^\delta) + \sum_{\gamma \in \partial \Pi'} x_\gamma G_\gamma(x, y).$$

By direct substitution in (9.58) we get

$$\phi_{12}^{-1}(m) = -\frac{1}{2}\,\mathrm{Res}_t \left(\sum \frac{\partial P}{\partial I'_j}(t; \mathcal{D}^{(\alpha)}_{\mathbf{c}}) \left\langle \mu_\alpha \,|\, \nabla_y I'_j(t; \mathcal{D}^{(\alpha)}_{\mathbf{c}}) \right\rangle + \sum_{\gamma \in \partial \Pi'} x_\gamma \left\langle \mu_\alpha \,|\, \nabla_y G_\gamma(t; \mathcal{D}^{(\alpha)}_{\mathbf{c}}) \right\rangle \right),$$

up to two terms, which we show to vanish. For the first one, notice that since the elements of I^δ are linear in y, the vector $\nabla_y I^\delta_j$ is constant, for any I^δ_j in I^δ. Since, according to Proposition 9.20 P (and hence also $\partial P/\partial I^\delta$) is a polynomial in I', I'' and I^δ, the latter being holomorphic at $\mathcal{D}^{(\alpha)}_{\mathbf{c}}$,

$$\sum \frac{\partial P}{\partial I^\delta_j}(t; \mathcal{D}^{(\alpha)}_{\mathbf{c}}) \langle \mu_\alpha \,|\, \nabla_y I^\delta_j \rangle$$

is also holomorphic, hence has zero residue.

For the second one one proceeds in the same way, using that

$$\langle \mu_\alpha \,|\, \nabla_y I'' \rangle = \sum_{\beta \in \Pi''} \frac{1}{2} a_{\alpha\beta} \langle e_\beta \,|\, e_\beta \rangle \frac{\partial I''}{\partial y_\beta} = 0,$$

as $a_{\alpha\beta} = 0$ for all $\beta \in \Pi''$ (recall that $\alpha \in \Pi'$).

The upshot is that we only need to estimate the two terms in $\phi_{12}^{-1}(m)$. We will do that for each of them separately. We choose a system of local parameters $\varsigma = (\varsigma_1, \dots, \varsigma_{l-1})$ for $\mathcal{D}_{\mathbf{c}}^{(\alpha)}$ in a neighborhood of m.

We first prove the following estimate.

$$\operatorname{Res}_t \left(\sum_{\gamma \in \partial \Pi'} x_\gamma \left\langle \mu_\alpha \,|\, \nabla_y G_\gamma(t; \mathcal{D}_{\mathbf{c}}^{(\alpha)}) \right\rangle \right) = O(\varsigma^0) \text{ near } m. \tag{9.59}$$

Since H_2 has weight $m_2 + 1$, all elements of the vector $\nabla_y G_\gamma$ have weight $m_2 - 2$, where the weighted degree $m_2 + 1$ of H_2 is three or four for the classical Lie algebras, while it is five for $\mathfrak{e}_6$, six for $\mathfrak{e}_7$, $\mathfrak{f}_4$, $\mathfrak{g}_2$ and eight for $\mathfrak{e}_8$, as is read off at once from Table 2.2. By weight homogeneity,

$$\left\langle \mu_\alpha \,|\, \nabla_y G_\gamma(t; \mathcal{D}_{\mathbf{c}}^{(\alpha)}) \right\rangle = t^{2-m_2} (R_0 + R_1(b,c) + R_2(b,c) t^2 + \cdots),$$

with $R_i(b,c)$ homogeneous of weigth i. Observe for $\gamma \in \partial \Pi'$

$$x_\gamma(t; \mathcal{D}_{\mathbf{c}}^{(\alpha)}) = b_\gamma t + O(t^2), \text{ or } x_\gamma(t; \mathcal{D}_{\mathbf{c}}^{(\alpha)}) = b_\gamma O(t). \tag{9.60}$$

In the first three cases of Table 9.9 we have that $b_\gamma \leqslant O(\varsigma^0)$ and that $m_2 \in \{2, 3\}$, so that

$$x_\gamma \left\langle \mu_\alpha \,|\, \nabla_y G_\gamma(t; \mathcal{D}_{\mathbf{c}}^{(\alpha)}) \right\rangle \leqslant \frac{O(\varsigma^0)}{t},$$

accounting for the classical Lie algebras. In the fourth case, $b_\gamma \leqslant O(\varsigma^2)$; if $m_2 = 5$, so that $2 - m_2 = -3$ then we have

$$\begin{aligned} \operatorname{Res}_t x_\gamma \left\langle \mu_\alpha \,|\, \nabla_y G_\gamma(t; \mathcal{D}_{\mathbf{c}}^{(\alpha)}) \right\rangle &= b_\gamma \tilde{R}_1(b,c), \text{ when } x_\gamma = b_\gamma t(1 + c_\gamma t + \cdots) \\ &= b_\gamma \tilde{R}_2(b,c), \text{ when } x_\gamma = b_\gamma (1 + c_\gamma t + \cdots). \end{aligned}$$

According to Tables 9.7 and 9.8 we always have that

$$b_\gamma \leqslant O(\varsigma^2) \text{ and } R_i(b,c) \leqslant O(\varsigma^{-i})$$

which implies that

$$\operatorname{Res}_t x_\gamma \left\langle \mu_\alpha \,|\, \nabla_y G_\gamma(t; \mathcal{D}_{\mathbf{c}}^{(\alpha)}) \right\rangle \leqslant O(\varsigma^0).$$

Since we have supposed that $m_2 = 5$ this covers the Lie algebras $\mathfrak{g}_2^{(1)}$ and $\mathfrak{d}_4^{(3)}$ only, but similar arguments hold for the remaining cases. This yields the first estimate (9.59).

For the second estimate we will use the following terminology. In Tables 9.5 and 9.8 we had two possibilities for obtaining lower balances from the principal balances; in both cases the second possibility requires a separate argument, and yields in fact a different order of vanishing. We will refer in the rest of the proof to these two cases as the *alternative case*, referring to the other cases in Tables 9.3 till 9.8 as the *normal case.* Using this terminology, the estimate that we will derive near $m \in \mathcal{D}_{\mathbf{c}}^{(\alpha)} \cap \mathcal{D}_{\mathbf{c}}^{(\beta)}$ is the following:

$$\mathrm{Res}_t \sum_{j=1}^{2} \tfrac{\partial P}{\partial I_j'}(t;\mathcal{D}_{\mathbf{c}}^{(\alpha)}) \left\langle \mu_\alpha \,|\, \nabla_y I_j'(t;\mathcal{D}_{\mathbf{c}}^{(\alpha)})\right\rangle \tag{9.61}$$

$$\begin{cases} = O(\varsigma^{1-m_2'}) & \text{in the normal case} \\ \leqslant O(\varsigma^{1}) & \text{in the alternative case}\end{cases}$$

Since $\Pi' = \{\alpha,\beta\}$, the sum in the expression above ranges over the two invariants I_1' and I_2', listed in Table 9.1; they are extensions of the Weyl invariants going with Π' and have weigthed degrees m_1' and m_2'. Moreover, by (2) in Proposition 9.20, for any element I of I', I'' and I^δ, the limit $\lim_{p\to p_0} I(t;\mathcal{D}_{\mathbf{c}}^{(\alpha)})$ is finite ($p_0 \in \mathcal{D}_{\mathbf{c}}^{(\alpha)} \cap \mathcal{D}_{\mathbf{c}}^{(\beta)}$) and so is this limit for $I = \partial P/\partial I_i'$; we will call the latter limit $c_i(p_0)$. In both the normal and the alternative case one checks that $c_2(p_0) \neq 0$, as $\partial P/\partial I_i'$ is either constant or a linear function of the $y_i's$; moreover, by adding a power of H_1 to H_2 (H_2 is not canonical) we can always assume that $c_1(p_0) \neq 0$. Thus we have:

$$\begin{aligned}
&\mathrm{Res}_t \sum_{j=1}^{2} \frac{\partial P}{\partial I_j'}(t;\mathcal{D}_{\mathbf{c}}^{(\alpha)}) \left\langle \mu_\alpha \,|\, \nabla_y I_j'(t;\mathcal{D}_{\mathbf{c}}^{(\alpha)})\right\rangle \\
&= \frac{1}{2}\,\mathrm{Res}_t \sum_{j=1}^{2} \frac{\partial P}{\partial I_j'}(t;\mathcal{D}_{\mathbf{c}}^{(\alpha)}) \sum_{\gamma=\alpha,\beta} a_{\beta\alpha} \left\langle e_\alpha \,|\, e_\alpha\right\rangle \frac{\partial I_j'}{\partial y_\gamma}(t;\mathcal{D}_{\mathbf{c}}^{(\alpha)}) \\
&\sim \frac{1}{2}\left\langle e_\alpha \,|\, e_\alpha\right\rangle \sum_{i=1}^{2} c_i(p_0)\,\mathrm{Res}_t \left(2\frac{\partial I_j'}{\partial y_\alpha}(t;\mathcal{D}_{\mathbf{c}}^{(\alpha)}) + a_{\beta\alpha}\frac{\partial I_j'}{\partial y_\alpha}(t;\mathcal{D}_{\mathbf{c}}^{(\beta)})\right) \\
&\sim \frac{1}{2}\left\langle e_\alpha \,|\, e_\alpha\right\rangle \left(c_1(p_0) R^{m_1'-1}(b,c) + c_2(p_0) R^{m_2'-1}(b,c)\right),
\end{aligned}$$

where $R^{(m_1'-1)}$ and $R^{(m_2'-1)}$ are weigth homogeneous polynomials in b and in c, of weight $m_1'-1$ and $m_2'-1$. Of course, $m_1' = 1$ in all cases and $m_2'-1 = 0, 1, 2$ or 4 according to whether α and β are linked in their Dynkin subdiagram by 0, 1, 2 or 3 edges (in that order). It follows that *at worst*

$$R^{(m_1'-1)}(b,c) = R^{(0)} = \text{constant } \neq 0$$

and, near p_0,

$$R^{(m_2'-1)}(b,c)\begin{cases} = O(\varsigma^{1-m_2'}) & \text{in the normal case} \\ \leqslant O(\varsigma^{1}) & \text{in the alternative case.}\end{cases}$$

In order to show that equality actually holds one uses the explicit invariant $I_2' = H_2$ from Table 9.1, which yields the result upon using that the parameters $b_\alpha^{(2)}$ and $c_\beta^{(1)}$ blow up (in ς) according to their weights, while the other $b_\gamma^{(2)}$ and $c_\gamma^{(1)}$ blow up less forcefully than their weights. This leads to the second estimate.

To conclude, the second estimate determines the leading behaviour (in ς) of ϕ_{12}^{-1}, which blows up as $\varsigma^{-m_2'-1}$ near $\mathcal{D}_{\mathbf{c}}^{(\alpha)} \cap \mathcal{D}_{\mathbf{c}}^{(\beta)}$, in the normal case. This means that ϕ_{12} behaves like $\varsigma^{m_2'-1}$ near $\mathcal{D}_{\mathbf{c}}^{(\alpha)} \cap \mathcal{D}_{\mathbf{c}}^{(\beta)}$ in the normal case, while it does not vanish at all in the alternative case. This yield Table 9.9. □

9.5.6 Intersection Multiplicity of Two Painlevé Divisors

We have seen that the tangency locus of the Toda vector field $\mathcal{X}_H$ to the Painlevé divisors is contained in the intersection with the neighboring divisors, as dictated by the Dynkin diagram of $L(\mathfrak{g}, \nu)$; moreover, the order of tangency is encoded in the Cartan integers (see Table 9.9). This suggests that the intersection multiplicity of two (or many, see Paragraph 9.5.8) is also determined by the Dynkin diagram. We will now show that this is indeed the case.

Theorem 9.22. *Under the a.c.i. assumption, the following formula for the intersection multiplicity of two irreducible components of the Painlevé divisors holds: if $\alpha, \beta \in \Pi$, with $\alpha \neq \beta$ then*

$$\operatorname{mult}(\mathcal{D}_{\mathbf{c}}^{(\alpha)} \cdot \mathcal{D}_{\mathbf{c}}^{(\beta)}) = \frac{\#\mathbf{W}^{(\alpha\beta)}}{\det A^{(\alpha\beta)}}, \tag{9.62}$$

where $\mathbf{W}^{(\alpha\beta)}$ and $A^{(\alpha\beta)}$ denote the Weyl group and the Cartan matrix, associated with the roots $\{\alpha, \beta\}$. In particular, the above multiplicity is independent of the irreducible component of $\mathcal{D}_{\mathbf{c}}^{(\alpha)} \cap \mathcal{D}_{\mathbf{c}}^{(\beta)}$.

Table 9.10 contains the three numbers that appear in this equation, besides some other related information. The proof of Theorem 9.22 consists in computing for all possible values of $a_{\alpha\beta}$ local equations for the two divisors $\mathcal{D}_{\mathbf{c}}^{(\alpha)}$ and $\mathcal{D}_{\mathbf{c}}^{(\beta)}$, in the neighborhood of their intersection, which amounts to establishing Table 9.10. We will give the proof in the two cases $a_{\alpha\beta} = 0$ and $a_{\alpha\beta} = a_{\beta\alpha} = -1$ only.

We first consider the case $a_{\alpha\beta} = 0$ (see Table 9.3), i.e., $A^{(\alpha\beta)} = \begin{pmatrix} 2 & 0 \\ 0 & 2 \end{pmatrix}$. It follows from (9.41) that, for generic $\mathbf{c}$, the function $\psi_{\alpha\beta} := (x_\alpha x_\beta)^{-1/2}$ vanishes simply along $\mathcal{D}_{\mathbf{c}}^{(\alpha)}$ and $\mathcal{D}_{\mathbf{c}}^{(\beta)}$, it is finite and non-zero in the affine, since $C_1 = \prod_{i=0}^{l} x_i^{\xi_i}$ is a Casimir. Therefore in a neighborhood $U_{\alpha\beta}$ in $\mathbf{T}_{\mathbf{c}}^l$ of the $(l-2)$-dimensional variety $\mathcal{D}_{\mathbf{c}}^{(\alpha)} \cap \mathcal{D}_{\mathbf{c}}^{(\beta)}$ the function $\psi_{\alpha\beta}$ is holomorphic and vanishes along $\mathcal{D}_{\mathbf{c}}^{(\alpha)} \cup \mathcal{D}_{\mathbf{c}}^{(\beta)}$ only, i.e., $\psi_{\alpha\beta}(t; \alpha)$ is a Taylor series in t,

Table 9.10. The intersection multiplicity for the case of two divisors. The table is organized according to the number of edges between the two roots α and β in the Dynkin diagram. For the corresponding Lie algebra we give its Cartan matrix $A^{(\alpha\beta)}$ with its determinant and the order of its Weyl group $\mathbf{W}^{(\alpha\beta)}$, as read off from Table 2.1. From the local equations of $\mathcal{D}_c^{(\alpha)}$ and $\mathcal{D}_c^{(\beta)}$, as computed in the text, we make a picture of the singularity at their intersection and we compute the intersection multiplicity, which clearly matches (9.62).

Dynkin	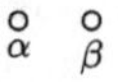	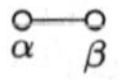	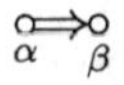	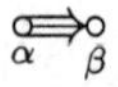
Lie algebra	$\mathfrak{a}_1 \oplus \mathfrak{a}_1$	$\mathfrak{a}_2$	$\mathfrak{b}_2$	$\mathfrak{g}_2$
$A^{(\alpha\beta)}$	$\begin{pmatrix} 2 & 0 \\ 0 & 2 \end{pmatrix}$	$\begin{pmatrix} 2 & -1 \\ -1 & 2 \end{pmatrix}$	$\begin{pmatrix} 2 & -2 \\ -1 & 2 \end{pmatrix}$	$\begin{pmatrix} 2 & -1 \\ -3 & 2 \end{pmatrix}$
$\det A^{(\alpha\beta)}$	4	3	2	1
$\#\mathbf{W}^{(\alpha\beta)}$	4	3!	8	12
$\mathcal{D}_c^{(\alpha)}$	$y=0$	$y=x^2$	$x(y-x^3)=0$	$(y-ax^5)(y-bx^5)=0$
$\mathcal{D}_c^{(\beta)}$	$x=0$	$y=ax^2$	$y=ax^3$	$x(y-cx^5)=0$
Singularity	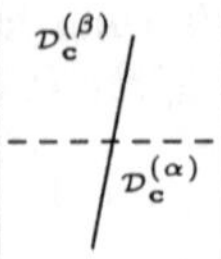	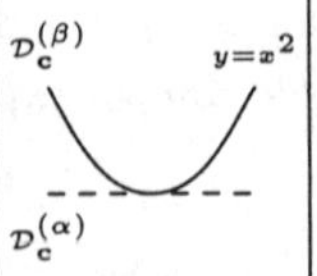	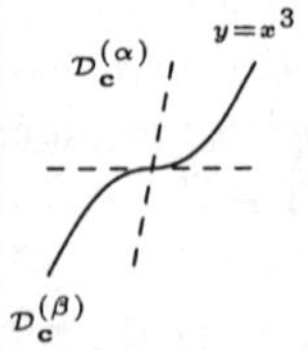	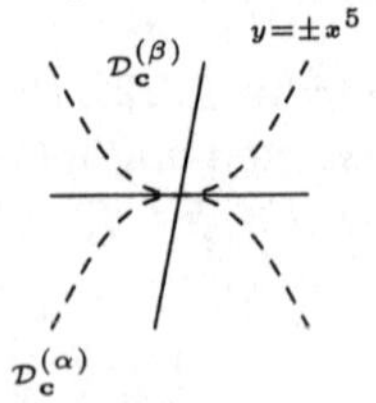
Multiplicity	1	2	4	12

with holomorphic coefficients in the free parameters $(\varsigma, b_\gamma, c_\gamma)$ of the principal balance going with $\mathcal{D}^{(\alpha)}$; their behavior can be deduced from Table 9.3. Upon confining this principal balance to a fixed generic torus $\mathbf{T}_\mathbf{c}^l$, some of the (b_γ, c_γ) account for the values of the constants of motion and others are running parameters describing $\mathcal{D}_\mathbf{c}^{(\alpha)}$. Since by Table 9.9 the Toda flow is in this case transversal to $\mathcal{D}_\mathbf{c}^{(\alpha)}$ (and to $\mathcal{D}_\mathbf{c}^{(\beta)}$) near $\mathcal{D}_\mathbf{c}^{(\alpha)} \cap \mathcal{D}_\mathbf{c}^{(\beta)}$, the divisors $\mathcal{D}_\mathbf{c}^{(\alpha)}$ and $\mathcal{D}_\mathbf{c}^{(\beta)} \subset \mathbf{T}_\mathbf{c}^l$ may behave near $\mathcal{D}_\mathbf{c}^{(\alpha)} \cap \mathcal{D}_\mathbf{c}^{(\beta)}$ in many different ways, i.e., several sheets of $\mathcal{D}_\mathbf{c}^{(\alpha)}$ and $\mathcal{D}_\mathbf{c}^{(\beta)}$ may meet at the intersection $\mathcal{D}_\mathbf{c}^{(\alpha)} \cap \mathcal{D}_\mathbf{c}^{(\beta)}$, and they may or may not be tangent to one another. Now using the estimates in Table 9.3 compute

$$\begin{aligned} \psi_{\alpha\beta}(t; \mathcal{D}_\mathbf{c}^{(\alpha)}) &= b_\beta^{-1/2} t(1 - c_\beta t/2 + \cdots) \\ &= 2\varsigma t + t^2 + tO_2(t, \varsigma) \\ &= (2\varsigma + t)t \cdot \text{unit} \quad \text{near } \mathcal{D}_\mathbf{c}^{(\alpha)} \cap \mathcal{D}_\mathbf{c}^{(\beta)}, \end{aligned}$$

where $O_2(t,\varsigma)$ stands for a holomorphic series, quadratic or more in t and ς, whose coefficients depend on the (b_γ, c_γ). Since near $\mathcal{D}_{\mathbf{c}}^{(\alpha)} \cap \mathcal{D}_{\mathbf{c}}^{(\beta)}$ we have $\psi_{\alpha\beta}(t;\mathcal{D}_{\mathbf{c}}^{(\alpha)}) = 0$ precisely on $\mathcal{D}_{\mathbf{c}}^{(\alpha)} \cup \mathcal{D}_{\mathbf{c}}^{(\beta)}$, exactly one sheet of $\mathcal{D}_{\mathbf{c}}^{(\alpha)}$ must cross one sheet of $\mathcal{D}_{\mathbf{c}}^{(\beta)}$ transversally at $\mathcal{D}_{\mathbf{c}}^{(\alpha)} \cap \mathcal{D}_{\mathbf{c}}^{(\beta)}$, leading to the singularity picture advertised in the second column of Table 9.10.

We now consider the case $a_{\alpha\beta} = a_{\beta\alpha} = -1$, i.e., $A^{(\alpha\beta)} = \begin{pmatrix} 2 & -1 \\ -1 & 2 \end{pmatrix}$, corresponding to the third column in Table 9.10. The proof proceeds along similar lines as the previous case, choosing now $\psi_{\alpha\beta} := (x_\alpha x_\beta)^{-1}$. Using the estimates in Table 9.4, compute

$$\begin{aligned} \psi_{\alpha\beta}(t;\mathcal{D}_{\mathbf{c}}^{(\alpha)}) &= b_\beta^{-1} t(1 - c_\beta t + (c_\beta^2 - b_\alpha + b_\beta)t^2/2 + f_\beta t^3 + \cdots) \qquad (9.63)\\ &= -\left(\frac{3\varsigma}{2}\right)^3 t + \left(\frac{3}{2}\right)^3 \varsigma^2 t^2 - \frac{3}{2}\varsigma t^2 + \frac{t^4}{4} + tO_4(\varsigma, t) \end{aligned}$$

near $\mathcal{D}_{\mathbf{c}}^{(\alpha)} \cap \mathcal{D}_{\mathbf{c}}^{(\beta)}$. In this case a new complication arises: by Table 9.9 the Toda vector field $\mathcal{X}_H$ is now simply tangent to both $\mathcal{D}_{\mathbf{c}}^{(\alpha)}$ and $\mathcal{D}_{\mathbf{c}}^{(\beta)} \subset \mathbf{T}_{\mathbf{c}}^{l}$; thus, all sheets of $\mathcal{D}_{\mathbf{c}}^{(\alpha)}$ (say m_α) and of $\mathcal{D}_{\mathbf{c}}^{(\beta)}$ (say m_β) must be simply tangent to the u-axis (direction of $\mathcal{X}_H$); hence $\mathcal{D}_{\mathbf{c}}^{(\alpha)}$ and $\mathcal{D}_{\mathbf{c}}^{(\beta)}$ are at least simply tangent to each other at $\mathcal{D}_{\mathbf{c}}^{(\alpha)}$. Let v be a transversal direction to u, making the uv-plane into a transversal slice of $\mathcal{D}_{\mathbf{c}}^{(\alpha)}$ and $\mathcal{D}_{\mathbf{c}}^{(\beta)}$ near $\mathcal{D}_{\mathbf{c}}^{(\alpha)} \cap \mathcal{D}_{\mathbf{c}}^{(\beta)}$; let w account for the (b_γ, c_γ) axes (see Figure 9.2).

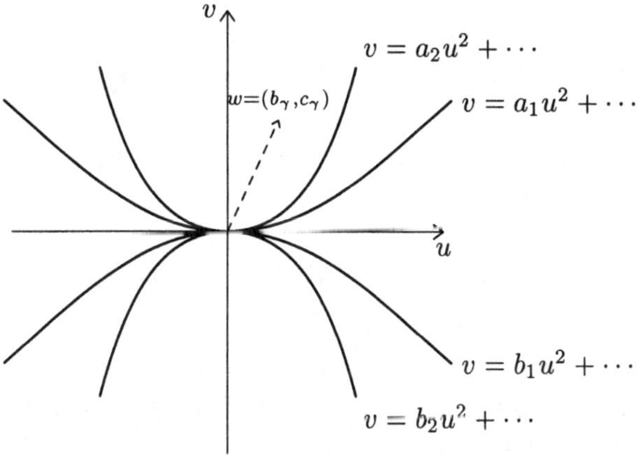

Fig. 9.2. Here we have depicted two sheets (locally) of $\mathcal{D}_\alpha$ and two sheets of $\mathcal{D}_\beta$, simply tangent along the UV-plane. The U-axis is in the direction of the vector field $\mathcal{V}_1$.

Since the function $\psi_{\alpha\beta}$ vanishes precisely on $\mathcal{D}_{\mathbf{c}}^{(\alpha)} \cup \mathcal{D}_{\mathbf{c}}^{(\beta)}$ the Weierstrass Preparation Theorem (Theorem 5.7) implies

$$\psi_{\alpha\beta} = \prod_{i=1}^{m_\alpha}(v - a_i u^2) \prod_{j=1}^{m_\beta}(v - a'_j u^2) \cdot \text{unit} \qquad a_i, a'_j \neq 0.$$

The Toda flow moves an arbitrary point (u_0, v_0) on $\mathcal{D}_{\mathbf{c}}^{(\alpha)}$ (assume it is situated on the branch $v_0 = a_1 u_0^2 + \cdots$) to $(u, v) = (a_0 + t, a_1 u_0^2 + \cdots)$. Therefore $\psi_{\alpha\beta}$ evaluated along a trajectory near that branch takes on the form

$$\begin{aligned}
&\psi_{\alpha\beta}(t; \mathcal{D}_{\mathbf{c}}^{(\alpha)}) \\
&\quad = (t(t + 2u_0) + O_3(t, u_0)) \times \\
&\qquad \prod_{i=2}^{m_\alpha}\left(t(t + 2u_0) + \left(1 - \frac{a_1}{a_i}\right) u_0^2 + O_3(t, u_0)\right) \times \\
&\qquad \prod_{j=1}^{m_\beta}\left(t(t + 2u_0) + \left(1 - \frac{a_1}{a'_j}\right) u_0^2 + O_3(t, u_0)\right) \times \text{unit} \\
&\quad = \left(2\prod_{i=2}^{m_\alpha}\left(1 - \frac{a_1}{a_i}\right)\prod_{j=1}^{m_\beta}\left(1 - \frac{a_1}{a'_j}\right) u_0^{2m_\alpha + 2m_\beta - 1} t + \cdots + t^{2m_\alpha + 2m_\beta}\right) \times \\
&\qquad \text{unit} \times \text{Weierstrass polynomial in } t \text{ and } u_0.
\end{aligned} \tag{9.64}$$

Comparing the two expressions (9.63) and (9.64) one finds $m_\alpha + m_\beta = 2$ and thus, $m_\alpha = m_\beta = 1$, since $m \geqslant 1$ and $n \geqslant 1$; one also finds that the coefficient of $u_0^{2m+2n-1} t$ in (9.64) does not vanish, and that $u_0 = \varsigma$. It implies moreover that the $\mathcal{D}_{\mathbf{c}}^{(\alpha)}$ and $\mathcal{D}_{\mathbf{c}}^{(\beta)}$ sheets are simply tangent to each other near $\mathcal{D}_{\mathbf{c}}^{(\alpha)} \cap \mathcal{D}_{\mathbf{c}}^{(\beta)}$.

In the last two cases one instead takes $\psi_{\alpha\beta}(t) = x_\alpha^{-3/2} x_\beta^{-2},\ x_\alpha^{-3} x_\beta^{-5}$ respectively, using Table 9.9.

9.5.7 Toda Lattices Leading to Abelian Surfaces

Under the a.c.i. assumption, there are six cases of periodic Toda lattices that lead to Abelian surfaces. In this paragraph we use the results of the previous paragraphs to fully describe the geometry of these Abelian surfaces. As in these cases the Abelian varieties $\mathbf{T}_{\mathbf{c}}^l$ are of dimension $l = 2$ they correspond to the twisted affine Lie algebras $L(\mathfrak{g}, \nu)$ for which the rank of $\mathfrak{g}_0$ is two. From Tables (2.4) and (2.5) we read off that $L(\mathfrak{g}, \nu)$ must correspond to one of the following root systems: $\mathfrak{a}_2^{(1)}$, $\mathfrak{c}_2^{(1)}$, $\mathfrak{g}_2^{(1)}$, $\mathfrak{a}_4^{(2)}$, $\mathfrak{d}_3^{(2)}$ and $\mathfrak{d}_4^{(3)}$. Notice also that since the divisor $\mathcal{D}_{\mathbf{c}}$ always has $l + 1$ irreducible components, $\mathcal{D}_{\mathbf{c}}$ will in this case always have three components, which are (possibly singular) curves.

In this paragraph we will compute the number of intersection points of any of these curves, their genus, the polarization type they induce on the tori and the dimension of their linear systems. We first specialize Proposition 9.18 and Theorem 9.22 to the case of $l = 2$.

Theorem 9.23. *Under the a.c.i. assumption, in the case of surfaces ($l = 2$),*

$$g_a(\mathcal{D}_{\mathbf{c}}^{(0)}) - 1 = \frac{\mathcal{D}_{\mathbf{c}}^{(0)} \cdot \mathcal{D}_{\mathbf{c}}^{(0)}}{2} = \frac{\mathcal{D}_{\mathbf{c}}^{(\alpha)} \cdot \mathcal{D}_{\mathbf{c}}^{(\beta)}}{2\hat{\xi}_\alpha \hat{\xi}_\beta} = \frac{g_a(\mathcal{D}_{\mathbf{c}}^{(\alpha)}) - 1}{\hat{\xi}_\alpha^2}, \tag{9.65}$$

where $\alpha \neq \beta$ are any two of the three simple roots of Π. Also, every intersection point of $\mathcal{D}_{\mathbf{c}}^{(\alpha)}$ and $\mathcal{D}_{\mathbf{c}}^{(\beta)}$ has multiplicity

$$\operatorname{mult}(\mathcal{D}_{\mathbf{c}}^{(\alpha)} \cdot \mathcal{D}_{\mathbf{c}}^{(\beta)}) = \frac{\#\mathbf{W}^{(\alpha\beta)}}{\det A^{(\alpha\beta)}}, \tag{9.66}$$

where $\mathbf{W}^{(\alpha\beta)}$ and $A^{(\alpha\beta)}$ denote the Weyl group and the Cartan matrix of the semi-simple Lie algebra generated by the simple roots α and β.

We derive from this Theorem some useful formulas for the three curves that constitute the Painlevé divisor.

Proposition 9.24. *Under the a.c.i. assumption, the number of distinct intersection points of $\mathcal{D}_{\mathbf{c}}^{(\alpha)}$ and $\mathcal{D}_{\mathbf{c}}^{(\beta)}$ is given, in the case of surfaces ($l = 2$), by*

$$\#(\mathcal{D}_{\mathbf{c}}^{(\alpha)} \cap \mathcal{D}_{\mathbf{c}}^{(\beta)}) = \xi_\gamma \frac{\#\mathbf{W}^{(\alpha_1\alpha_2)}}{\#\mathbf{W}^{(\alpha\beta)}}, \tag{9.67}$$

and

$$\mathcal{D}_{\mathbf{c}}^{(\alpha)} \cdot \mathcal{D}_{\mathbf{c}}^{(\beta)} = \frac{1}{\hat{\xi}_\gamma} \frac{\#\mathbf{W}^{(\alpha_1\alpha_2)}}{\det A^{(\alpha_1\alpha_2)}}, \tag{9.68}$$

where $\{\alpha, \beta, \gamma\} = \Pi = \{\alpha_0, \alpha_1, \alpha_2\}$, the vector ξ (resp. $\hat{\xi}$) is the normalized[6] *null vector of $A^\top$ (resp. of A) and η is defined by (9.35). As a consequence*

$$g_a(\mathcal{D}_{\mathbf{c}}^{(i)}) - 1 = \hat{\xi}_i^2 \frac{\eta_1 \eta_2}{\hat{\xi}_1 \hat{\xi}_2}, \tag{9.69}$$

$$\mathcal{D}_{\mathbf{c}}^{(i)} \cdot \mathcal{D}_{\mathbf{c}}^{(j)} = 2\hat{\xi}_i \hat{\xi}_j \frac{\eta_1 \eta_2}{\hat{\xi}_1 \hat{\xi}_2}, \tag{9.70}$$

where $0 \leqslant i, j \leqslant 2$, and

$$\delta_1 \delta_2 = g_a(\mathcal{D}_{\mathbf{c}}^{(0)}) - 1 = \frac{\eta_1 \eta_2}{\hat{\xi}_1 \hat{\xi}_2},$$

where (δ_1, δ_2) denotes the type of the polarization that $\mathcal{D}_{\mathbf{c}}^{(0)}$ induces on $\mathbf{T}_{\mathbf{c}}^2$.

[6] $\hat{\xi}_0 = 1 = \xi_0$.

Table 9.11. The Painlevé divisors in the case of Toda lattices leading to Abelian surfaces. Besides the integers $\hat{\xi}_i$ and η_i we give in each case the genera of the irreducible components of the Painlevé divisor and the intersection numbers of each pair of components.

Lie algebra	$\mathfrak{a}_2^{(1)}$	$\mathfrak{c}_2^{(1)}$	$\mathfrak{g}_2^{(1)}$	$\mathfrak{a}_4^{(2)}$	$\mathfrak{d}_3^{(2)}$	$\mathfrak{d}_4^{(3)}$
$(\hat{\xi}_1, \hat{\xi}_2)$	$(1,1)$	$(1,1)$	$(1,2)$	$(1,\frac{1}{2})$	$(2,1)$	$(2,3)$
(η_1, η_2)	$(1,1)$	$(2,1)$	$(3,2)$	$(1,2)$	$(1,2)$	$(3,2)$
(δ_1, δ_2)	$(1,1)$	$(1,2)$	$(1,3)$	$(2,2)$	$(1,1)$	$(1,1)$
Π_2	$\alpha_0 - \alpha_1$	$\alpha_0 \Rightarrow \alpha_1$	$\alpha_0 \quad \alpha_1$	$\alpha_0 \Rightarrow \alpha_1$	$\alpha_1 \Rightarrow \alpha_0$	$\alpha_0 \Rightarrow \alpha_1$
Π_0	$\alpha_1 - \alpha_2$	$\alpha_2 \Rightarrow \alpha_1$	$\alpha_2 \Rrightarrow \alpha_1$	$\alpha_1 \Rightarrow \alpha_2$	$\alpha_1 \Rightarrow \alpha_2$	$\alpha_2 \Rrightarrow \alpha_1$
Π_1	$\alpha_0 - \alpha_2$	$\alpha_0 \quad \alpha_2$	$\alpha_0 - \alpha_2$	$\alpha_0 \quad \alpha_2$	$\alpha_0 \quad \alpha_2$	$\alpha_0 \quad \alpha_2$
$g_a(\mathcal{D}_{\mathbf{c}}^{(0)})$	2	3	4	5	2	2
$g_a(\mathcal{D}_{\mathbf{c}}^{(1)})$	2	3	4	5	5	5
$g_a(\mathcal{D}_{\mathbf{c}}^{(2)})$	2	3	13	2	2	10
$\mathcal{D}_{\mathbf{c}}^{(0)} \cdot \mathcal{D}_{\mathbf{c}}^{(1)}$	2	4	6	8	4	4
$\mathcal{D}_{\mathbf{c}}^{(1)} \cdot \mathcal{D}_{\mathbf{c}}^{(2)}$	2	4	12	4	4	12
$\mathcal{D}_{\mathbf{c}}^{(0)} \cdot \mathcal{D}_{\mathbf{c}}^{(2)}$	2	4	12	4	2	6
$\mathcal{D}_{\mathbf{c}}^{(0)} \cap \mathcal{D}_{\mathbf{c}}^{(1)}$	1	1	6	2	1	2
$\mathcal{D}_{\mathbf{c}}^{(1)} \cap \mathcal{D}_{\mathbf{c}}^{(2)}$	1	1	1	1	1	1
$\mathcal{D}_{\mathbf{c}}^{(0)} \cap \mathcal{D}_{\mathbf{c}}^{(2)}$	1	4	6	4	2	6

Proof. Since all intersection points of $\mathcal{D}_{\mathbf{c}}^{(\alpha)}$ and $\mathcal{D}_{\mathbf{c}}^{(\beta)}$ have the same multiplicity, Theorem 9.23 implies that

$$\#(\mathcal{D}_{\mathbf{c}}^{(\alpha)} \cap \mathcal{D}_{\mathbf{c}}^{(\beta)}) = \frac{\mathcal{D}_{\mathbf{c}}^{(\alpha)} \cdot \mathcal{D}_{\mathbf{c}}^{(\beta)}}{\operatorname{mult}(\mathcal{D}_{\mathbf{c}}^{(\alpha)} \cdot \mathcal{D}_{\mathbf{c}}^{(\beta)})} = \hat{\xi}_\alpha \hat{\xi}_\beta \left(\mathcal{D}_{\mathbf{c}}^{(0)}\right)^2 \frac{\det A^{(\alpha\beta)}}{\#\mathbf{W}^{(\alpha\beta)}}.$$

It follows that

$$\frac{\#(\mathcal{D}_{\mathbf{c}}^{(\alpha)} \cap \mathcal{D}_{\mathbf{c}}^{(\beta)})}{\#(\mathcal{D}_{\mathbf{c}}^{(1)} \cap \mathcal{D}_{\mathbf{c}}^{(2)})} = \frac{1}{\hat{\xi}_\gamma} \frac{\det A^{(\alpha\beta)}}{\det A^{(\alpha_1\alpha_2)}} \frac{\#\mathbf{W}^{(\alpha_1\alpha_2)}}{\#\mathbf{W}^{(\alpha\beta)}} = \xi_\gamma \frac{\#\mathbf{W}^{(\alpha_1\alpha_2)}}{\#\mathbf{W}^{(\alpha\beta)}},$$

where we have used (9.37) in the last step. Thus, in order to show (9.67) it suffices to show that $\mathcal{D}_{\mathbf{c}}^{(\alpha)}$ and $\mathcal{D}_{\mathbf{c}}^{(\beta)}$ intersect in one point. This is done by investigating the lower balance that goes with $\{\alpha_1, \alpha_2\}$. As follows from Proposition 9.14 this balance will have free parameters at steps $1, 1+m_1, 1+m_2$ and $2 - \langle \alpha_0, h' \rangle = 2(\xi_0 + \xi_1 + \xi_2)$, where m_1 and m_2 are the exponents of $\mathfrak{g}_0$.

Table 9.12. The zoo: curves completing the Toda invariant surfaces into Abelian surfaces. Both genera (g) of these curves $\mathcal{D}_i := \mathcal{D}_{\mathbf{c}}^{(i)}$ are given: for singular curves the integer appearing after the comma corresponds to the genus of the smooth version. We also give the vectors ξ and $\hat{\xi}$, the linear equivalence relation between the curves, and the polarization type $\delta = (\delta_1, \delta_2)$. The singularities are those of Table 9.10.

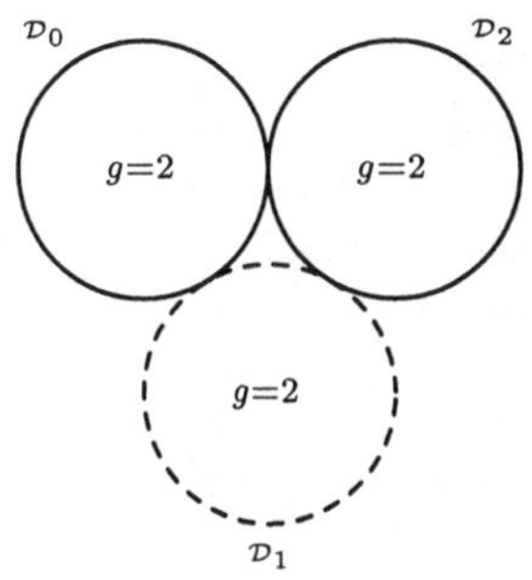

$\mathfrak{a}_2^{(1)}$	$\mathfrak{a}_4^{(2)}$
ξ=(1,1,1)	ξ=(1,2,2)
$\hat{\xi}$=(1,1,1)	$\hat{\xi}$=(1,1,1/2)
$3\mathcal{D}^{(0)} \sim 3\mathcal{D}^{(1)}$	$\mathcal{D}^{(1)} \sim \mathcal{D}^{(0)}$
$3\mathcal{D}^{(0)} \sim 3\mathcal{D}^{(2)}$	$2\mathcal{D}^{(2)} \sim \mathcal{D}^{(0)}$
δ=(1,1)	δ=(2,2)

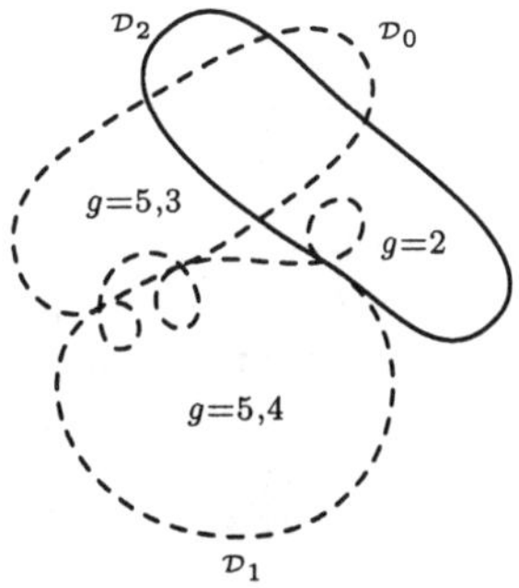

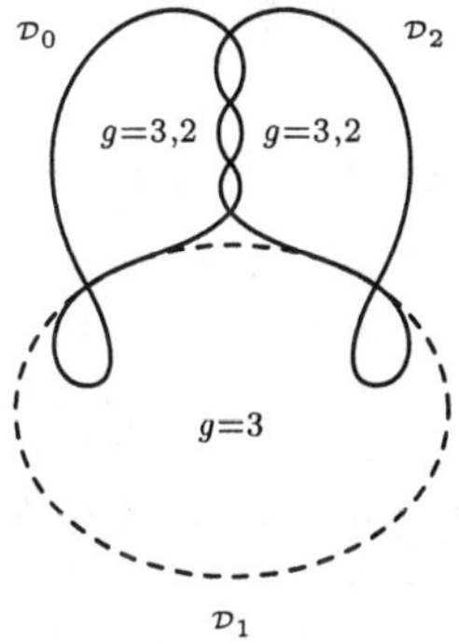

$\mathfrak{c}_2^{(1)}$	$\mathfrak{d}_3^{(2)}$
ξ=(1,2,1)	ξ=(1,1,1)
$\hat{\xi}$=(1,1,1)	$\hat{\xi}$=(1,2,1)
$2\mathcal{D}^{(1)} \sim 2\mathcal{D}^{(0)}$	$\mathcal{D}^{(1)} \sim 2\mathcal{D}^{(0)}$
$\mathcal{D}^{(2)} \sim \mathcal{D}^{(0)}$	$2\mathcal{D}^{(2)} \sim 2\mathcal{D}^{(0)}$
δ=(1,2)	δ=(1,1)

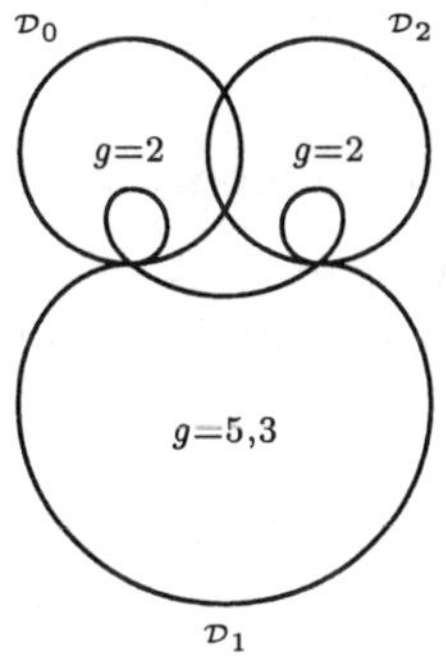

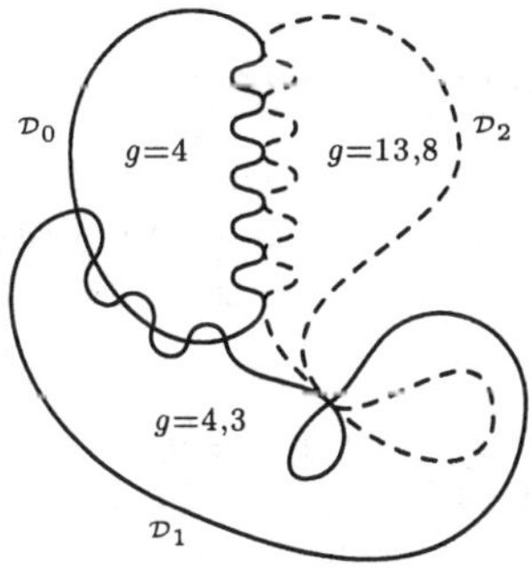

$\mathfrak{g}_2^{(1)}$	$\mathfrak{d}_4^{(3)}$
ξ=(1,3,2)	ξ=(1,2,1)
$\hat{\xi}$=(1,1,2)	$\hat{\xi}$=(1,2,3)
$\mathcal{D}^{(1)} \sim \mathcal{D}^{(0)}$	$\mathcal{D}^{(1)} \sim 2\mathcal{D}^{(0)}$
$\mathcal{D}^{(2)} \sim 2\mathcal{D}^{(0)}$	$\mathcal{D}^{(2)} \sim 3\mathcal{D}^{(0)}$
δ=(1,3)	δ=(1,1)

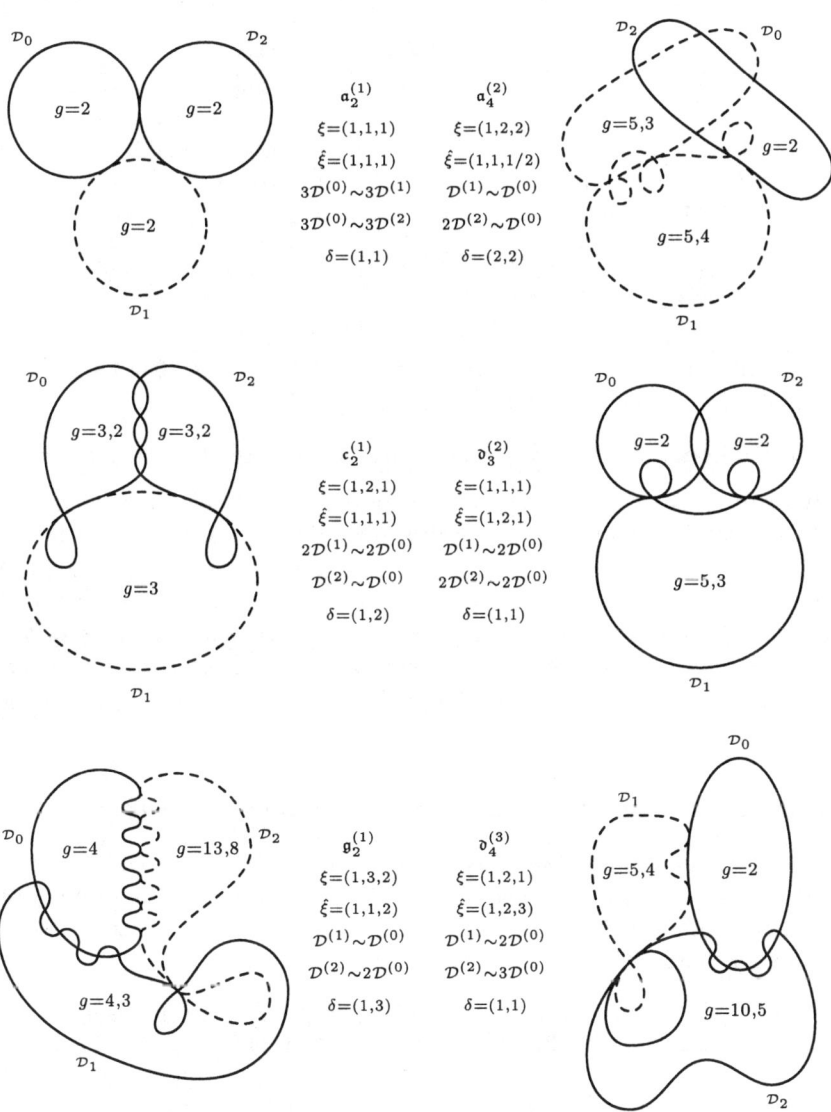

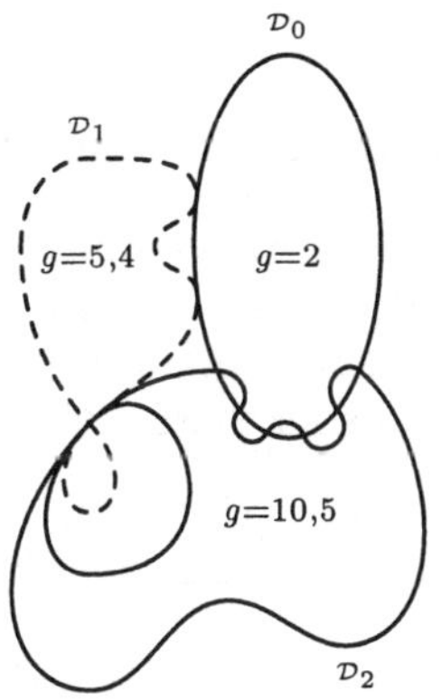

Now 1 and $2(\xi_0+\xi_1+\xi_2)$ are the weighted degrees of the Casimirs C_0 and C_1 (see (9.7)) and the commuting Hamiltonians, which are extensions of the Weyl polynomials of $\mathfrak{g}_0$, have weighted degrees $1+m_1$ and $1+m_2$. As we know that this balance corresponds to points, the fact that the degrees match the Kowalevski exponents implies, in view of Remark 7.35 that all four parameters are trivial parameters and that the balance corresponds to precisely one point. This yields (9.67). Multiplying (9.67) with (9.66) we get

$$\mathcal{D}_{\mathbf{c}}^{(\alpha)}\cdot\mathcal{D}_{\mathbf{c}}^{(\beta)}=\xi_\gamma\frac{\#\mathbf{W}^{(\alpha_1\alpha_2)}}{\det A^{(\alpha\beta)}}=\frac{1}{\hat{\xi}_\gamma}\frac{\#\mathbf{W}^{(\alpha_1\alpha_2)}}{\det A^{(\alpha_1\alpha_2)}},$$

where we used again (9.37), yielding (9.68). In view of (9.48) it suffices to show (9.69) for $i=0$ in order to have it for all i. To do this, take $\alpha=1$ and $\beta=2$ in (9.68), so that $\hat{\xi}_\gamma=\hat{\xi}_0=1$ and substitute it in (9.65) to find

$$g_a(\mathcal{D}_{\mathbf{c}}^{(0)})-1=\frac{\mathcal{D}_{\mathbf{c}}^{(1)}\cdot\mathcal{D}_{\mathbf{c}}^{(2)}}{2\hat{\xi}_\alpha\hat{\xi}_\beta}=\frac{1}{2\hat{\xi}_1\hat{\xi}_2}\frac{\#\mathbf{W}^{(\alpha_1\alpha_2)}}{\det A^{(\alpha_1\alpha_2)}}=\frac{\eta_1\eta_2}{\hat{\xi}_1\hat{\xi}_2},$$

where we have used (2.24) in the last step. This shows (9.69) for $i=0$. Formula (9.70) then follows from it by using (9.45). □

The information that we have obtained leads to the following Theorem.

Theorem 9.25. *Under the a.c.i. assumption, the level manifolds of the Toda lattices that correspond to the twisted affine Lie algebras $L(\mathfrak{g},\nu)$ with root systems $\mathfrak{a}_2^{(1)}$, $\mathfrak{c}_2^{(1)}$, $\mathfrak{g}_2^{(1)}$, $\mathfrak{a}_4^{(2)}$, $\mathfrak{d}_3^{(2)}$ and $\mathfrak{d}_4^{(3)}$ compactify into Abelian surfaces by adjoining a divisor which has precisely three irreducible components. These components, which are (often singular) curves, are depicted in Table 9.12, where we also give their genera (arithmetic and geometric) and the type of the polarization that is induced on $\mathbf{T}_{\mathbf{c}}^2$ by $\mathcal{D}_{\mathbf{c}}^{(0)}$. These curves intersect according to the four patterns listed in Table 9.10.*

9.5.8 Intersection Multiplicity of Many Painlevé Divisors

In order to describe the geometry of the higher dimensional ($l>2$) periodic Toda lattices we need the generalization of Theorem 9.22 to more than two divisors (l, in fact). Based on the formula for two divisors and some specific results for more divisors we conjecture the following.

Conjecture 9.26. Let $\emptyset\neq\Pi'\subset\Pi$ be arbitrary. Along any component of $\bigcap_{\alpha\in\Pi'}\mathcal{D}_{\mathbf{c}}^{(\alpha)}$ the following formula for the intersection multiplicity holds:

$$\operatorname{mult}(\mathcal{D}_{\mathbf{c}}^{(\alpha_1)}\cdot\mathcal{D}_{\mathbf{c}}^{(\alpha_2)}\cdots\mathcal{D}_{\mathbf{c}}^{(\alpha_k)},\ \alpha_i\in\Pi')=\frac{\#\mathbf{W}'}{\det A'},\tag{9.71}$$

where $\mathbf{W}'$ and A' denote the Weyl group and the Cartan matrix, associated with the roots $\Pi'\subset\Pi$.

Assuming the conjecture we can prove the analog of Proposition 9.24.

Proposition 9.27. *Consider the Toda lattice, associated to $L(\mathfrak{g}, \nu)$, which we assume to be a.c.i. and for whose Painlevé divisors Conjecture 9.26 is assumed to hold. For any $i \in \{0, \dots, l\}$ the number of distinct intersection points of all divisors $\mathcal{D}_{\mathbf{c}}^{(j)}$, except $\mathcal{D}_{\mathbf{c}}^{(i)}$, is given by*

$$\#\left(\bigcap_{j\neq i} \mathcal{D}_{\mathbf{c}}^{(j)}\right) = \xi_i \frac{\#\mathbf{W}^{(0)}}{\#\mathbf{W}^{(i)}}, \tag{9.72}$$

where $\mathbf{W}^{(j)}$ is the Weyl group of the root system $\Pi \setminus \{\alpha_j\}$, for $j \in \{0, \dots, l\}$. The arithmetic genus of $\mathcal{D}_{\mathbf{c}}^{(0)}$ is given by

$$g_a(\mathcal{D}_{\mathbf{c}}^{(0)}) - l + 1 = \frac{1}{l!}\left(\mathcal{D}_{\mathbf{c}}^{(0)}\right)^l = \frac{\mathcal{D}_{\mathbf{c}}^{(0)} \cdot \mathcal{D}_{\mathbf{c}}^{(1)} \cdots \widehat{\mathcal{D}_{\mathbf{c}}^{(i)}} \cdots \mathcal{D}_{\mathbf{c}}^{(l)}}{l! \prod_{j\neq i} \hat{\xi}_j}, \tag{9.73}$$

where the global intersection number is given by

$$\mathcal{D}_{\mathbf{c}}^{(0)} \cdot \mathcal{D}_{\mathbf{c}}^{(1)} \cdots \widehat{\mathcal{D}_{\mathbf{c}}^{(i)}} \cdots \mathcal{D}_{\mathbf{c}}^{(l)} = \frac{1}{\hat{\xi}_i} \frac{\#\mathbf{W}^{(0)}}{\det A^{(0)}} = \frac{l!}{\hat{\xi}_i} \prod_{j=1}^{l} \eta_j. \tag{9.74}$$

As a consequence, the arithmetic genus of $\mathcal{D}_{\mathbf{c}}^{(i)}$, for $i \in \{0, \dots, l\}$, is given by

$$g_a(\mathcal{D}_{\mathbf{c}}^{(i)}) - l + 1 = \hat{\xi}_i^l \prod_{j=1}^{l} \frac{\eta_j}{\hat{\xi}_j}. \tag{9.75}$$

The divisor $\mathcal{D}_{\mathbf{c}}^{(0)}$ defines a principal polarization on the generic torus $\mathbf{T}_{\mathbf{c}}^l$ when ν is different from the identity, while it defines a polarization of type[7]

$$(\delta_1, \dots, \delta_l) = \left(\frac{\xi_1}{\hat{\xi}_1}, \dots, \frac{\xi_l}{\hat{\xi}_l}\right). \tag{9.76}$$

otherwise. In particular all $\mathfrak{a}_l^{(1)}$, $\mathfrak{d}_l^{(1)}$ and $\mathfrak{e}_l^{(1)}$ also lead to a principal polarization.

Proof. Formula (9.73) is an immediate consequence of Proposition 9.18. By the same proposition we have as in the proof of Proposition 9.24 that

$$\frac{\#\left(\cap_{j\neq i}\mathcal{D}_{\mathbf{c}}^{(j)}\right)}{\#\left(\cap_{j=1}^{l}\mathcal{D}_{\mathbf{c}}^{(j)}\right)} = \xi_i \frac{\#\mathbf{W}^{(0)}}{\#\mathbf{W}^{(i)}},$$

[7] In the case of $\mathfrak{f}_4^{(1)}$ there is another candidate which we cannot rule out, namely $(\delta_1, \dots, \delta_4) = (1, 1, 1, 4)$, instead of $(1, 1, 2, 2)$.

and (9.72) follows by showing, as in the proof of Proposition 9.24, that the l divisors $\mathcal{D}_{\mathbf{c}}^{(1)},\dots,\mathcal{D}_{\mathbf{c}}^{(l)}$ meet in a single point, using Remark 7.35. Using (9.71) and (9.72) we find that

$$\mathcal{D}_{\mathbf{c}}^{(0)}\cdot\mathcal{D}_{\mathbf{c}}^{(1)}\cdots\widehat{\mathcal{D}_{\mathbf{c}}^{(i)}}\cdots\mathcal{D}_{\mathbf{c}}^{(l)}=\xi_i\frac{\mathbf{W}^{(0)}}{\det A^{(i)}},$$

so that, in view of (9.37) and (2.24),

$$\mathcal{D}_{\mathbf{c}}^{(0)}\cdot\mathcal{D}_{\mathbf{c}}^{(1)}\cdots\widehat{\mathcal{D}_{\mathbf{c}}^{(i)}}\cdots\mathcal{D}_{\mathbf{c}}^{(l)}=\frac{1}{\hat\xi_i}\frac{\mathbf{W}^{(0)}}{\det A^{(0)}}=\frac{l!}{\hat\xi_i}\prod_{i=1}^{l}\eta_i,$$

which proves (9.74). The formula (9.75) for $g_a(\mathcal{D}_{\mathbf{c}}^{(i)})$ follows from it, upon using (5.25) and (9.43). In order compute the polarization type of the tori $\mathbf{T}_{\mathbf{c}}^l$ we use (9.75) and the general formula (5.25) for Abelian varieties, which gives, for $\mathcal{D}_{\mathbf{c}}^{(0)}$, two different expressions for

$$g_a(\mathcal{D}_{\mathbf{c}}^{(0)})-l+1=\prod_{i=1}^{l}\frac{\eta_i}{\hat\xi_i}=\prod_{i=1}^{l}\delta_i. \tag{9.77}$$

We distinguish three cases. **(1)** If ν has order two or three, but the root system of $L(\mathfrak{g},\nu)$ is not $\mathfrak{a}_{2l}^{(2)}$ then $\eta_i=\hat\xi_i$, for all i, and thus, (9.77) implies that $\prod_{i=1}^{l}\delta_i=1$, which means that $\delta_1=\cdots=\delta_l=1$, i.e., $\mathbf{T}_{\mathbf{c}}^l$ is principally polarized. **(2)** in the case of $\mathfrak{a}_{2l}^{(2)}$ we use $\mathcal{D}_{\mathbf{c}}^{(l)}$ instead of $\mathcal{D}_{\mathbf{c}}^{(0)}$ to find that $\prod_{i=1}^{l}\delta_i=1$, leading to principally polarized tori $\mathbf{T}_{\mathbf{c}}^l$ as well. **(3)** Suppose now that ν is the identity, so that $\eta=\xi$. If A is symmetric then $\xi=\hat\xi$, so that $\delta_1=\cdots=\delta_l=1$, which yields a principal polarization for $\mathfrak{a}_l^{(1)}$, $\mathfrak{d}_l^{(1)}$ and $\mathfrak{e}_l^{(1)}$. In the case of $\mathfrak{g}_2^{(1)}$ we have already seen that $(\delta_1,\delta_2)=(1,3)$. For $\mathfrak{f}_4^{(1)}$ we have $\xi=(1,2,3,4,2)^\top$ and $\hat\xi=(1,2,3,2,1)^\top$ which leaves $(\delta_1,\dots,\delta_4)=(1,1,2,2)$ and $(\delta_1,\dots,\delta_4)=(1,1,1,4)$ as the only possibilities. For $\mathfrak{b}_l^{(1)}$ we get $\xi=(1,1,2,\dots,2)^\top$ and $\hat\xi=(1,1,2,\dots,2,1)^\top$, so that $(\delta_1,\dots,\delta_l)=(1,\dots,1,2)$. Finally, for $\mathfrak{c}_l^{(1)}$ we get $\xi=(1,2,2,\dots,2,1)^\top$ and $\hat\xi=(1,1,\dots,1)^\top$, leaving several possibilities, but it is shown in [15, ¶ 4] that $(\delta_1,\dots,\delta_l)=(1,\dots,1,2,\dots,2)$. Thus, except maybe in the case of $\mathfrak{f}_4^{(1)}$ we have that (9.76) holds. □

Conjecture 9.28. The following identity holds, as an identity in integral homology

$$\mathcal{D}_{\mathbf{c}}^{(\beta_1)}\cdot\mathcal{D}_{\mathbf{c}}^{(\beta_2)}\cdots\mathcal{D}_{\mathbf{c}}^{(\beta_k)}\sim\left(\prod_{i=1}^{k}\xi_{\beta_i}\right)\prod_{j=1}^{l}\left(\frac{\eta_j}{\hat\xi_j}\right)k!H_0^{(k)},$$

where

$$H_0^{(k)}=\frac{\left(\mathcal{D}_{\mathbf{c}}^{(0)}\right)^k}{k!\prod_{j=1}^{l}(\eta_j/\hat\xi_j)}.$$

10 Integrable Spinning Tops

10.1 Spinning Tops

10.1.1 Equations of Motion and Poisson Structure

A *spinning top* is by definition a rigid body with a fixed point that rotates in a constant gravitational field. The equations of motion of a spinning top derive from the rotational version of Newton's law, which states that

torque = instantaneous change in angular momentum.

We will derive these equations in two different — but equivalent — forms; the equivalence will be based on the dictionary between $\mathbf{R}^3$ (viewed as column vectors) and $\mathfrak{so}(3)$ (the Lie algebra of skew-symmetric 3×3 matrices). Namely, a vector space isomorphism between $\mathbf{R}^3$ and $\mathfrak{so}(3)$ is obtained by

$$\begin{array}{rccc} \phi: & \mathbf{R}^3 & \to & \mathfrak{so}(3) \\ & x = \begin{pmatrix} x_1 \\ x_2 \\ x_3 \end{pmatrix} & \mapsto X = & \begin{pmatrix} 0 & -x_3 & x_2 \\ x_3 & 0 & -x_1 \\ -x_2 & x_1 & 0 \end{pmatrix}. \end{array} \tag{10.1}$$

Since $\phi(x)y = x \times y$ we have that ϕ is naturally described, in terms of the standard basis (e_1, e_2, e_3) of $\mathbf{R}^3$, by

$$\phi(e_i)e_j = \epsilon_{ijk}e_k,$$

where $1 \leqslant i, j, k \leqslant 3$, and where ϵ_{ijk} is the skew-symmetric tensor for which $\epsilon_{123} = 1$. Using ϕ the standard operations on vectors in $\mathbf{R}^3$ are related to some standard operations on skew-symmetric 3×3 matrices, as given in Table 10.1.

We will derive the equations of motion that describe the spinning top in terms of an orthogonal coordinate system that is attached to the top, and whose origin coincides with the fixed point. The coordinates of any point of space, of any point of the top, or of any physical quantity, with respect to this coordinate system will be denoted by a lower case letter when it is represented by a vector, and as the corresponding upper case letter when it is represented

Table 10.1. The properties of the basic correspondence ϕ between $\mathbf{R}^3$ and $\mathfrak{so}(3)$, as defined in (10.1). For $x \in \mathbf{R}^3$ we denote its image $\phi(x)$ by X, and similarly $Y := \phi(y)$.

	$\mathbf{R}^3$	$\mathfrak{so}(3)$
standard inner product	$\langle x \mid y \rangle$	$-\frac{1}{2}\,\mathrm{Trace}(XY)$
vector product	$x \times y$	$[X, Y]$
vector product	$x \times y$	$\phi(Xy)$
action by $R \in \mathbf{SO(3)}$	Rx	$RXR^{\top}$

by an element of $\mathfrak{so}(3)$, using the above isomorphism ϕ. We also consider an absolute orthogonal coordinate system, whose origin is again the fixed point of the top, but which is inertial in the sense that Newton's laws hold in them in their usual form (think of this coordinate system as being attached to the earth). The coordinates of the physical quantities with respect to this coordinate system will again be denoted by upper- and lowercase letters, according to whether we represent them by elements of $\mathfrak{so}(3)$ or by vectors in $\mathbf{R}^3$, but, in addition, these letters will be underlined. Table 10.2 gives a list of all quantities that will be used to derive the equations of motion of the spinning top. As the top spins, most of these coordinates vary, but some may be time-independent, such as the center of mass in body coordinates.

Table 10.2. The notation that we use to represent the different quantities that are associated with a spinning top. The first two columns are the coordinates with respect to an absolute (inertial) coordinate system, with the fixed point as its origin; the last two columns are the coordinates with respect to a moving coordinate system, attached to the top, also with the fixed point as its origin.

	vector abs.	matrix abs.	vector mov.	matrix mov.
body point	$\underline{q}(t)$	$\underline{Q}(t)$	q	Q
center of mass	$\underline{m}(t)$	$\underline{M}(t)$	m	M
gravitational field	$\underline{\gamma}$	$\underline{\Gamma}$	$\gamma(t)$	$\Gamma(t)$
angular velocity	$\underline{\omega}(t)$	$\underline{\Omega}(t)$	$\omega(t)$	$\Omega(t)$
angular momentum	$\underline{a}(t)$	$\underline{A}(t)$	$a(t)$	$A(t)$

We denote by $R(t)$ the element of $\mathbf{SO(3)}$ that relates the moving frame to the fixed frame, so that

$$\underline{q}(t) = R(t)q,$$

and similarly for the other quantities, represented in vector form. It follows from Table 10.1 that if these quantities are represented in matrix form then they are related by conjugation, e.g.,

$$\underline{Q}(t) = R(t)QR(t)^\top.$$

Thus, we know explicitly how to go from any column of Table 10.2 to any other column of that table.

The physical quantities *angular velocity* and *angular momentum* may also require some explanation. The linear velocity $\dot{\underline{q}}$ of any point of a body that is rotating can be written as $\dot{\underline{q}} = \underline{\omega} \times \underline{q}$, where $\underline{\omega} = \underline{\omega}(t)$ is a vector in the direction of the axis of rotation, and is called the angular velocity. Defining $R(t)$ as above we have that

$$\dot{\underline{q}} = \underline{\omega} \times \underline{q} = \underline{\Omega}\underline{q} = \underline{\Omega} Rq,$$

which yields, compared to $\dot{\underline{q}} = \dot{R}q$ (which follows from $\underline{q}(t) = R(t)q$), that

$$\underline{\Omega}(t) = \dot{R}(t)R(t)^\top. \tag{10.2}$$

The angular momentum a derives from $\underline{\Omega}$ via the inertia matrix I of the body with respect to the fixed point: in body coordinates we have that $a = I\omega$. The inertia matrix I is the symmetric 3×3 matrix, defined by

$$I_{ij} := \int_{\text{body}} x_i x_j d\mu,$$

where $\mu : \mathbf{R}^3 \to \mathbf{R}$ is the *mass distribution* or *mass density.* Notice that, with this definition, the formula $a = I\omega$ for the angular momentum is the obvious generalization of the corresponding formula for the angular momentum of a point mass, with respect to a given point.

We are now ready to derive the equations of motion of a spinning top. Let us first express in vector notation that the gravitational field is constant, $\dot{\underline{\gamma}} = 0$. Since $\underline{\gamma} = R(t)\gamma(t)$ we have by using (10.2) and by Table 10.1,

$$\dot{\gamma} = -R^\top \dot{R} R^\top \underline{\gamma} = -R^\top \underline{\Omega}\underline{\gamma} = R^\top(\underline{\gamma} \times \underline{\omega}) = R^\top \underline{\gamma} \times R^\top \underline{\omega} = \gamma \times \omega.$$

The corresponding equation, in matrix form, follows immediately from it by using Table 10.1. Still, we cannot resist to give a direct matrix computation, because in this form it is clearer why it leads to a Lax equation. In this case we consider the derivative of $\Gamma(t) = R(t)^\top \underline{\Gamma} R(t)$, and we use now that $\underline{\Gamma}$ is constant to find

$$\dot{\Gamma} = \dot{R}^\top \underline{\Gamma} R + R^\top \underline{\Gamma} \dot{R} = -R^\top \underline{\Omega}\underline{\Gamma} R + R^\top \underline{\Gamma}\underline{\Omega} R = \Gamma\Omega - \Omega\Gamma = [\Gamma, \Omega].$$

There is a second set of equations, that comes from the rotational version of Newton's law. For computing the torque which is exerted on the body by gravity one may think of all mass as being concentrated in its center of gravity, so that the torque is given by $\underline{\gamma} \times \underline{m}$, and Newton's equation becomes

$$\underline{\gamma} \times \underline{m} = \dot{\underline{a}}, \qquad \text{or} \qquad [\underline{\Gamma}, \underline{M}] = \dot{\underline{A}}. \tag{10.3}$$

Thus, in vector form, we have that

$$(Ra)^{\cdot} = R\gamma \times Rm = R(\gamma \times m),$$

so that

$$\dot{a} = \gamma \times m - R^\top \dot{R} a = \gamma \times m - R^\top \underline{\Omega} R a = \gamma \times m - \Omega a = \gamma \times m + a \times \omega.$$

In matrix form we have that

$$[\underline{\Gamma}, \underline{M}] = \dot{\underline{A}} = \dot{R} A R^\top + R \dot{A} R^\top + RA\left(R^\top\right)^{\cdot} = \underline{\Omega} R A R^\top + R \dot{A} R^\top - RAR^\top \underline{\Omega},$$

so that

$$\begin{aligned} \dot{A} &= R^\top \left[\underline{\Gamma}, \underline{M}\right] R - R^\top \underline{\Omega} R A + A R^\top \underline{\Omega} R \\ &= [\Gamma, M] - \Omega A + A \Omega \\ &= [\Gamma, M] + [A, \Omega]. \end{aligned}$$

Summarizing, when a top spins in a constant gravitational field, the angular velocity and the gravitational field, as seen by the top, satisfy the following system of coupled equations,

$$\begin{aligned} \dot{\Gamma} &= [\Gamma, \Omega], \\ \dot{A} &= [\Gamma, M] + [A, \Omega]. \end{aligned} \tag{10.4}$$

In these equations, M is a given constant matrix, and the matrices Ω and A are related via the inertia matrix I, as we make more precise now. As we said $a = I\omega$ and I is symmetric. Therefore, we can choose the body coordinates such that I is a diagonal matrix, $I = \operatorname{diag}(I_1, I_2, I_3)$, where the diagonal entries I_i are called the *principal moments of inertia.* Assuming that the body is not planar, $\det I \neq 0$, we introduce $\lambda_i := I_i^{-1}$, for $i = 1, 2, 3$. It follows that if we write $a = (a_1, a_2, a_3)^\top$ then

$$A = \begin{pmatrix} 0 & -a_3 & a_2 \\ a_3 & 0 & -a_1 \\ -a_2 & a_1 & 0 \end{pmatrix} \qquad \text{and} \qquad \Omega = \begin{pmatrix} 0 & -\lambda_3 a_3 & \lambda_2 a_2 \\ \lambda_3 a_3 & 0 & -\lambda_1 a_1 \\ -\lambda_2 a_2 & \lambda_1 a_1 & 0 \end{pmatrix}.$$

If we denote, in addition, $m = (m_1, m_2, m_3)^\top$ and $\gamma = (\gamma_1, \gamma_2, \gamma_3)^\top$ then the equations (10.4) are easily written out as

$$\begin{aligned}
\dot\gamma_1 &= \lambda_3 a_3 \gamma_2 - \lambda_2 a_2 \gamma_3, \\
\dot\gamma_2 &= \lambda_1 a_1 \gamma_3 - \lambda_3 a_3 \gamma_1, \\
\dot\gamma_3 &= \lambda_2 a_2 \gamma_1 - \lambda_1 a_1 \gamma_2, \\
\dot a_1 &= m_3 \gamma_2 - m_2 \gamma_3 + (\lambda_3 - \lambda_2) a_2 a_3, \\
\dot a_2 &= m_1 \gamma_3 - m_3 \gamma_1 + (\lambda_1 - \lambda_3) a_3 a_1, \\
\dot a_3 &= m_2 \gamma_1 - m_1 \gamma_2 + (\lambda_2 - \lambda_1) a_1 a_2.
\end{aligned} \tag{10.5}$$

This vector field on $\mathbf{R}^6$ is Hamiltonian with respect to the linear Poisson structure $\{\cdot\,,\cdot\}$, whose Poisson matrix with respect to $\gamma_1, \dots, \gamma_3, a_1, \dots, a_3$ (in that order) is given by

$$\begin{pmatrix}
0 & 0 & 0 & 0 & -\gamma_3 & \gamma_2 \\
0 & 0 & 0 & \gamma_3 & 0 & -\gamma_1 \\
0 & 0 & 0 & -\gamma_2 & \gamma_1 & 0 \\
0 & -\gamma_3 & \gamma_2 & 0 & -a_3 & a_2 \\
\gamma_3 & 0 & -\gamma_1 & a_3 & 0 & -a_1 \\
-\gamma_2 & \gamma_1 & 0 & -a_2 & a_1 & 0
\end{pmatrix}. \tag{10.6}$$

One easily recognizes it as the Lie-Poisson structure on $\mathfrak{e}(3)$ (see (8.113)). The Hamiltonian is precisely the energy of the top,

$$H = \frac{1}{2}(\lambda_1 a_1^2 + \lambda_2 a_2^2 + \lambda_3 a_3^2) + m_1\gamma_1 + m_2\gamma_2 + m_3\gamma_3.$$

Two physically obvious constants of motion are the length of the gravity vector and the component of the angular momentum in the direction of gravity. They turn out to be Casimirs for the above Poisson structure, namely the functions

$$\begin{aligned}
K_1 &:= \gamma_1^2 + \gamma_2^2 + \gamma_3^2, \\
K_2 &:= a_1\gamma_1 + a_2\gamma_2 + a_3\gamma_3,
\end{aligned} \tag{10.7}$$

generate the algebra of Casimirs of $\{\cdot\,,\cdot\}$. Since the rank of $\{\cdot\,,\cdot\}$ is 4 we need another, independent, constant of motion in order for the vector field to be integrable. Such an extra constant of motion will only exist when the moments of inertia of the top and its center of gravity bear very special relationships, as we will see. For future use, notice that the rank of the matrix (10.6) is 4 precisely when one of γ_1, γ_2 or γ_3 is different from zero. It follows that the locus where the rank of $\{\cdot\,,\cdot\}$ is smaller than 4 is the three-plane $\gamma_1 = \gamma_2 = \gamma_3 = 0$. Since the Casimir K_1 is constant on this three-plane, Proposition 7.56 implies that the commuting vector fields of any integrable system on this Poisson manifold are independent on the generic fiber of the momentum map of this integrable system.

Remark 10.1. We have obtained the equations of motion as a polynomial vector field on $\mathbf{R}^6$, which is Hamiltonian with respect to a linear Poisson structure on $\mathbf{R}^6$. In the sequel we will consider its complexification, and we will still refer to it as describing a spinning top.

Remark 10.2. The equations (10.4) do not fully describe the motion of the top. In order to give a full description one must supplement the equations with the linear equation (10.2), which can also be written as $\dot{R} = R\Omega$. Then, one first solves (10.4), yielding $A(t)$, hence also $\Omega(t)$ and one next solves the non-autonomous linear equation $\dot{R} = R\Omega$. This is similar to what we said in Paragraph 8.1.

10.1.2 A.c.i. Tops

Three are three known integrable cases of (10.5), namely

(1) $m_1 = m_2 = m_3 = 0$ (*Euler-Poinsot top*);
(2) $m_1 = m_2 = 0$ and $I_1 = I_2$ (*Lagrange top*);
(3) $m_2 = m_3 = 0$ and $I_1 = I_2 = 2I_3$ (*Kowalevski top*).

Let us first show that the three above cases are the only ones that may lead to a weight homogeneous, irreducible a.c.i. system; they will then be studied in detail in the sections that follow. Since (10.5) becomes weight homogeneous when the variables are given the following weights : $\varpi(\gamma_i) = 2\varpi(a_i) = 2$, where $i = 1, 2, 3$, one has that the indicial locus $\mathcal{I}$ of (10.5)is explicitly given by

$$\begin{aligned}
0 &= 2\gamma_1^{(0)} + \lambda_3 a_3^{(0)} \gamma_2^{(0)} - \lambda_2 a_2^{(0)} \gamma_3^{(0)}, \\
0 &= 2\gamma_2^{(0)} + \lambda_1 a_1^{(0)} \gamma_3^{(0)} - \lambda_3 a_3^{(0)} \gamma_1^{(0)}, \\
0 &= 2\gamma_3^{(0)} + \lambda_2 a_2^{(0)} \gamma_1^{(0)} - \lambda_1 a_1^{(0)} \gamma_2^{(0)}, \\
0 &= a_1^{(0)} + m_3 \gamma_2^{(0)} - m_2 \gamma_3^{(0)} + (\lambda_3 - \lambda_2) a_2^{(0)} a_3^{(0)}, \\
0 &= a_2^{(0)} + m_1 \gamma_3^{(0)} - m_3 \gamma_1^{(0)} + (\lambda_1 - \lambda_3) a_3^{(0)} a_1^{(0)}, \\
0 &= a_3^{(0)} + m_2 \gamma_1^{(0)} - m_1 \gamma_2^{(0)} + (\lambda_2 - \lambda_1) a_1^{(0)} a_2^{(0)}.
\end{aligned} \tag{10.8}$$

We first show, using Proposition 7.32, that if all the moments of inertia are different then this can only be the vector field of an a.c.i. system if $m_1 = m_2 = m_3 = 0$, which corresponds to the Euler top. Indeed, let us assume that $I_1 < I_2 < I_3$, so that $\lambda_1 > \lambda_2 > \lambda_3$ and consider any of the four non-zero solutions to (10.8) for which $\gamma_1^{(0)} = \gamma_2^{(0)} = \gamma_3^{(0)} = 0$, namely

$$\left(a_i^{(0)}\right)^2 = \frac{1}{(\lambda_k - \lambda_i)(\lambda_i - \lambda_j)}, \tag{10.9}$$

where (i, j, k) is any cyclic permutation of $(1, 2, 3)$ and with the following compatibility condition

$$a_1^{(0)}a_2^{(0)}a_3^{(0)} = \frac{1}{(\lambda_1-\lambda_2)(\lambda_2-\lambda_3)(\lambda_3-\lambda_1)}. \tag{10.10}$$

Notice that since $\lambda_1 > \lambda_2 > \lambda_3$ we have that $a_1^{(0)}$ and $a_3^{(0)}$ are purely imaginary, while $a_2^{(0)}$ is real. The Kowalevski matrix at such a solution is given by

$$\mathcal{K} = \begin{pmatrix} 2 & \lambda_3 a_3^{(0)} & -\lambda_2 a_2^{(0)} & 0 & 0 & 0 \\ -\lambda_3 a_3^{(0)} & 2 & \lambda_1 a_1^{(0)} & 0 & 0 & 0 \\ \lambda_2 a_2^{(0)} & -\lambda_1 a_1^{(0)} & 2 & 0 & 0 & 0 \\ 0 & m_3 & -m_2 & 1 & (\lambda_3-\lambda_2)a_3^{(0)} & (\lambda_3-\lambda_2)a_2^{(0)} \\ -m_3 & 0 & m_1 & (\lambda_1-\lambda_3)a_3^{(0)} & 1 & (\lambda_1-\lambda_3)a_1^{(0)} \\ m_2 & -m_1 & 0 & (\lambda_2-\lambda_1)a_2^{(0)} & (\lambda_2-\lambda_1)a_1^{(0)} & 1 \end{pmatrix}$$

Using (10.9) and (10.10) we find by direct computation that its characteristic polynomial is given by

$$|\mu \operatorname{Id}_6 - \mathcal{K}| = (\mu+1)(\mu-1)(\mu-2)^3(\mu-3).$$

Proposition 7.32 implies that the eigenspace of $\mathcal{K}$ that corresponds to the eigenvalue 2 is three-dimensional, which means that all 4×4 minors of $2\operatorname{Id}_6 - \mathcal{K}$ are zero. Computing the determinant of the 4×4 matrix, which is obtained by removing the last two columns, as well as the first and last rows from the matrix $2\operatorname{Id}_6 - \mathcal{K}$, we find that

$$0 = m_1 + m_2(\lambda_3-\lambda_1)a_3^{(0)} + m_3(\lambda_1-\lambda_2)a_2^{(0)}.$$

Since $a_2^{(0)}$ is real, while $a_3^{(0)}$ is purely imaginary we must have that

$$m_2 = 0, \qquad \text{and} \qquad 0 = m_1 + m_3(\lambda_1-\lambda_2)a_2^{(0)}.$$

In particular we have that $m_1 = 0$ if and only if $m_3 = 0$. If $m_1 = m_2 = m_3 = 0$ then we are in the case of the Euler top. Suppose therefore that $m_2 = 0$, but that m_1 and m_3 are both different from zero (still assuming that all moments of inertia are different). Then we can consider the following solution to the indicial equation (10.8):

$$a^{(0)} = \left(0, \frac{-2\sqrt{-1}}{\lambda_2}, 0\right), \quad \gamma^{(0)} = \frac{1}{\lambda_2(m_3 - \sqrt{-1}m_1)}\left(-2\sqrt{-1}, 0, 2\right).$$

The characteristic polynomial of the corresponding Kowalevski matrix is given by

$$(\mu+1)(\mu-2)(\mu-3)(\mu-4)(\mu(\mu-1)-2K),$$

where

$$K = \frac{(2\lambda_3-\lambda_2)(\lambda_1-\lambda_2)}{\lambda_2^2} + m_1 \frac{\lambda_3-\lambda_1}{\lambda_2}\frac{m_1-\sqrt{-1}m_3}{m_1^2+m_3^2}.$$

Since all roots of the characteristic polynomial must be integer, $-2K$, which is the product of two of the roots, must be an integer. But K cannot be real since $m_1 m_3 \neq 0$. It follows that when all moments of inertia are different then, under the a.c.i. assumption, we must be in the case of the Euler top.

We next consider the case in which at least two of the moments of inertia are equal. When all three are equal then we may choose the axes that are attached to the body such that $m_1 = m_2 = 0$, leading to a special case of the Lagrange top. Therefore we will actually assume that precisely two of them are equal, say $I_1 = I_2 \neq I_3$, so that $\lambda_1 = \lambda_2 \neq \lambda_3$. Then we may assume, in addition, that $m_2 = 0$ by a simple rotation of the body axes, perpendicular to the axis that corresponds to λ_3. Following Kowalevski's idea we only look for weight homogeneous principal balances to (10.5), for $\lambda_1 = \lambda_2 \neq \lambda_3$ and $m_2 = 0$, and we investigate the indicial locus $\mathcal{I}$, as given by (10.8), with the latter relations between the parameters that define the top. We have that

$$
\begin{aligned}
a_1^{(0)}\gamma_1^{(0)} + a_2^{(0)}\gamma_2^{(0)} + a_3^{(0)}\gamma_3^{(0)} &= 0,\\
\lambda_1(a_1^{(0)}\gamma_1^{(0)} + a_2^{(0)}\gamma_2^{(0)}) + \lambda_3 a_3^{(0)}\gamma_3^{(0)} &= 0
\end{aligned}
$$

where the first equation follows from the fact that $a_1\gamma_1 + a_2\gamma_2 + a_3\gamma_3$ is a constant of motion (in fact a Casimir), while the second equation follows from taking a simple linear combination of the first three equations in (10.8), using $\lambda_1 = \lambda_2$ (multiply, for $i = 1, 2, 3$, the i-th equation by $\lambda_i a_i^{(0)}$ and add up the resulting equations). Since $\lambda_1 \neq \lambda_3$ it follows that $a_3^{(0)}\gamma_3^{(0)} = 0$. We apply now the Kowalevski-Painlevé Criterion (Theorem 6.13): if (10.5) is one of the vector fields of an irreducible weight homogeneous a.c.i. system then either a_3 is a constant of motion or there must exist a Laurent solution for which $a_3^{(0)} \neq 0$. In the first case the last equation in (10.5) leads to $m_1 = 0$, yielding the Lagrange top. We pursue the second case, in which there must exist such a solution with $\gamma_3^{(0)} = 0$. The indicial equation takes then the following simple form.

$$
\begin{aligned}
0 &= 2\gamma_1^{(0)} + \lambda_3 a_3^{(0)}\gamma_2^{(0)},\\
0 &= 2\gamma_2^{(0)} - \lambda_3 a_3^{(0)}\gamma_1^{(0)},\\
0 &= a_2^{(0)}\gamma_1^{(0)} - a_1^{(0)}\gamma_2^{(0)},\\
0 &= a_1^{(0)} + m_3\gamma_2^{(0)} + (\lambda_3 - \lambda_1)a_2^{(0)}a_3^{(0)},\\
0 &= a_2^{(0)} - m_3\gamma_1^{(0)} + (\lambda_1 - \lambda_3)a_3^{(0)}a_1^{(0)},\\
0 &= a_3^{(0)} - m_1\gamma_2^{(0)}.
\end{aligned}
\tag{10.11}
$$

We find from the last equation that $a_3^{(0)} = m_1\gamma_2^{(0)}$, in particular $m_1 \neq 0$ and $\gamma_2^{(0)} \neq 0$, and then from the second equation that $\gamma_1^{(0)} = 2/(\lambda_3 m_1)$. Substituting this in the first equation in (10.11) we find $\gamma_2^{(0)} = \pm 2\sqrt{-1}/(\lambda_3 m_1)$, while the third equation yields $a_2^{(0)} = \pm\sqrt{-1}a_1^{(0)}$. Then the fourth and fifth equation in (10.11) both reduce to

$$2\sqrt{-1}m_3 \pm a_1^{(0)} m_1(2\lambda_1 - \lambda_3) = 0. \tag{10.12}$$

Now there are two possibilities. Either $2\lambda_1 - \lambda_3 = 0$ so that $m_3 = 0$. Then $a_1^{(0)}$ is free and we find that

$$\left(\gamma^{(0)}, a^{(0)}\right) = \left(\frac{1}{\lambda_1 m_1}, \pm\frac{\sqrt{-1}}{\lambda_1 m_1}, 0, \alpha, \pm\sqrt{-1}\alpha, \pm\frac{\sqrt{-1}}{\lambda_1}\right),$$

where α is the free parameter. This is Kowalevski's case: the moments of inertia satisfy $I_1 = I_2 = 2I_3$ and the center of mass belongs to the equatorial plane (corresponding to the moments I_1 and I_2) through the fixed point. This case will be studied in more detail in the next section. Let us investigate now the other possibility, i.e., $2\lambda_1 - \lambda_3 \neq 0$. Then we may solve (10.12) for $a_1^{(0)}$, leading to the following two solutions to the indicial equation

$$(\gamma, a) = \left(\frac{2}{\lambda_3 m_1}, \pm\frac{2\sqrt{-1}}{\lambda_3 m_1}, 0, \mp\frac{2\sqrt{-1}m_3}{(2\lambda_1 - \lambda_3)m_1}, \frac{2m_3}{(2\lambda_1 - \lambda_3)m_1}, \pm\frac{2\sqrt{-1}}{\lambda_3}\right). \tag{10.13}$$

Consider now the Kowalevski matrix of (10.5), which is for $\lambda_2 = \lambda_1$ and $m_2 = 0$ explicitly given by

$$\begin{pmatrix} 2 & \lambda_3 a_3^{(0)} & -\lambda_1 a_2^{(0)} & 0 & -\lambda_1\gamma_3^{(0)} & \lambda_3\gamma_2^{(0)} \\ -\lambda_3 a_3^{(0)} & 2 & \lambda_1 a_1^{(0)} & \lambda_1\gamma_3^{(0)} & 0 & -\lambda_3\gamma_1^{(0)} \\ \lambda_1 a_2^{(0)} & -\lambda_1 a_1^{(0)} & 2 & -\lambda_1\gamma_2^{(0)} & \lambda_1\gamma_1^{(0)} & 0 \\ 0 & m_3 & 0 & 1 & (\lambda_3 - \lambda_1)a_3^{(0)} & (\lambda_3 - \lambda_1)a_2^{(0)} \\ -m_3 & 0 & m_1 & (\lambda_1 - \lambda_3)a_3^{(0)} & 1 & (\lambda_1 - \lambda_3)a_1^{(0)} \\ 0 & -m_1 & 0 & 0 & 0 & 1 \end{pmatrix}. \tag{10.14}$$

Substituting (10.13) in it it has the following characteristic polynomial,

$$(\mu + 1)(\mu - 2)(\mu - 3)(\mu - 4)\left(\mu^2 - \mu - 2\frac{(\lambda_1 - \lambda_3)(2\lambda_1 - \lambda_3)}{\lambda_3^2}\right). \tag{10.15}$$

Since we want that the latter solution of the indicial equation leads to a principal balance we must have that the two roots of the quadratic polynomial in (10.15) are non-negative integers. But the form of the polynomial shows that their sum must be 1, leaving 0 and 1 as the only possibility. However, $\lambda_1 \neq \lambda_3$ and $2\lambda_1 \neq \lambda_3$, so 0 is certainly not a root (hence neither is 1). Therefore, this solution to the indicial equation cannot lead to a principal balance.

10.2 The Euler-Poinsot and Lagrange Tops

The Euler-Poinsot and Lagrange tops are relatively simple in the sense that they do not linearize on Abelian surfaces, but on elliptic curves. However this does not make these integrable tops entirely trivial, as we will see.

10.2.1 The Euler-Poinsot Top

When $m_1 = m_2 = m_3 = 0$ the corresponding top is called an Euler-Poinsot top. Then the last three equations in (10.5) reduce to the Euler equations (6.7), which we studied in Example 6.4. Physically this case means that the center of mass coincides with the fixed point, which is equivalent to the absence of gravity. Once the Euler equations have been integrated (in terms of elliptic functions) the remaining three equations in (10.5) are linear (time-dependent) equations in the remaining variables γ_i. Thus we will concentrate here on the Euler equations on $\mathbf{C}^3$, which take, with $x_i := a_i$ and $\lambda_{ij} := \lambda_i - \lambda_j$ the simple form

$$\begin{aligned} \dot{x}_1 &= \lambda_{32} x_2 x_3, \\ \dot{x}_2 &= \lambda_{13} x_3 x_1, \\ \dot{x}_3 &= \lambda_{21} x_1 x_2, \end{aligned} \tag{10.16}$$

which has the constants of motion

$$\begin{aligned} K &= \frac{1}{2}(x_1^2 + x_2^2 + x_3^2), \\ H &= \frac{1}{2}(\lambda_1 x_1^2 + \lambda_2 x_2^2 + \lambda_3 x_3^2), \end{aligned}$$

which define the momentum map $\mathbf{F} = (K, H)$ of the Euler top. The generic fiber $\mathbf{F_c}$ of the momentum map, with $\mathbf{c} = (c_1, c_2)$, is the affine curve $\mathcal{E}_\mathbf{c} \subset \mathbf{C}^3$, defined by

$$\begin{aligned} c_1 &= \frac{1}{2}(x_1^2 + x_2^2 + x_3^2), \\ c_2 &= \frac{1}{2}(\lambda_1 x_1^2 + \lambda_2 x_2^2 + \lambda_3 x_3^2). \end{aligned}$$

$\mathcal{E}_\mathbf{c}$ is an elliptic curve, as it is a double cover of the conic

$$C_\mathbf{c} : \lambda_{12} x_1^2 - \lambda_{23} x_3^2 = 2(c_2 - \lambda_2 c_1),$$

with four ramification points, namely the following points on $C_\mathbf{c}$,

$$P^{\epsilon_1 \epsilon_2} := \left(\epsilon_1 \sqrt{\frac{2(-\lambda_3 c_1 + c_2)}{\lambda_{13}}}, 0, \epsilon_2 \sqrt{\frac{2(\lambda_1 c_1 - c_2)}{\lambda_{13}}}, \right), \tag{10.17}$$

where $\epsilon_1^2 = \epsilon_2^2 = 1$.

The Euler equations define a (weight) homogeneous vector field (all variables have weight 1). Let us show that all its balances $(x_1(t), x_2(t), x_3(t))$ are weight homogeneous. We denote the pole order of $x_i(t)$ by n_i, for $i = 1, 2, 3$, and notice that if one of the $x_i(t)$ has a pole then all series $(x_1(t), x_2(t), x_3(t))$ have a pole. Therefore, these pole orders are positive, $n_i \geqslant 0$, and they satisfy, in view of (10.16),

$$\begin{aligned} n_1 + 1 &= n_2 + n_3, \\ n_2 + 1 &= n_3 + n_1, \\ n_3 + 1 &= n_1 + n_2, \end{aligned}$$

with $n_1 = n_2 = n_3 = 1$ as the unique solution. This shows that all balances $(x_1(t), x_2(t), x_3(t))$ are weight homogeneous. The indicial equation, which is a special case of (10.8), is given by

$$\begin{aligned} 0 &= x_1^{(0)} + \lambda_{32} x_2^{(0)} x_3^{(0)}, \\ 0 &= x_2^{(0)} + \lambda_{13} x_3^{(0)} x_1^{(0)}, \\ 0 &= x_3^{(0)} + \lambda_{21} x_1^{(0)} x_2^{(0)}. \end{aligned} \tag{10.18}$$

and it has four non-zero solutions, to wit, $\left(x_i^{(0)}\right)^2 = \lambda_{ki}^{-1}\lambda_{ij}^{-1}$, where (i, j, k) is any cyclic permutation of $(1, 2, 3)$ and with the following compatibility condition

$$x_1^{(0)} x_2^{(0)} x_3^{(0)} = \frac{1}{\lambda_{12}\lambda_{23}\lambda_{31}}.$$

For any of these four solutions the Kowalevski matrix takes the form

$$\mathcal{K} = \begin{pmatrix} 1 & \lambda_{32} x_3^{(0)} & \lambda_{32} x_2^{(0)} \\ \lambda_{13} x_3^{(0)} & 1 & \lambda_{13} x_1^{(0)} \\ \lambda_{21} x_2^{(0)} & \lambda_{21} x_1^{(0)} & 1 \end{pmatrix},$$

with characteristic polynomial (at these points)

$$\lambda^3 - 3\lambda^2 + 4 = (\lambda + 1)(\lambda + 2)^2.$$

This was to be expected: weight homogeneity accounts for the eigenvalue -1 (see Proposition 7.11), while the two invariants have as differentials, evaluated at these points,

$$\begin{aligned} \mathrm{d}K(x^{(0)}) &= x_1^{(0)} \mathrm{d}x_1 + x_2^{(0)} \mathrm{d}x_2 + x_3^{(0)} \mathrm{d}x_3, \\ \mathrm{d}H(x^{(0)}) &= \lambda_1 x_1^{(0)} \mathrm{d}x_1 + \lambda_2 x_2^{(0)} \mathrm{d}x_2 + \lambda_3 x_3^{(0)} \mathrm{d}x_3, \end{aligned}$$

which are obviously independent (since the λ_i are all different), which accounts for the double eigenvalue 2 (Theorem 7.30).

It is then clear that at step 1 one finds the zero solution, $x^{(1)} = (0,0,0)$ and the equations at step 2 are given by the linear equations

$$
\begin{aligned}
x_1^{(2)} &= \lambda_{32}(x_2^{(0)}x_3^{(2)} + x_3^{(0)}x_2^{(2)}),\\
x_2^{(2)} &= \lambda_{13}(x_3^{(0)}x_1^{(2)} + x_1^{(0)}x_3^{(2)}),\\
x_3^{(2)} &= \lambda_{21}(x_1^{(0)}x_2^{(2)} + x_2^{(0)}x_1^{(2)}).
\end{aligned}
$$

One easily finds the following two independent solutions,

$$
(x_1^{(2)}, x_2^{(2)}, x_3^{(2)}) = \begin{cases} (1, 0, \lambda_{21}x_2^{(0)}),\\ (0, 1, \lambda_{21}x_1^{(0)}),\end{cases}
$$

so that the first few terms of the four balances are given by

$$
\begin{aligned}
x_1(t) &= \frac{x_1^{(0)}}{t} + at + O(t^2),\\
x_2(t) &= \frac{x_2^{(0)}}{t} + bt + O(t^2),\\
x_3(t) &= \frac{x_3^{(0)}}{t} + \lambda_{21}(x_2^{(0)}a + x_1^{(0)}b)t + O(t^2).
\end{aligned}
$$

The free parameters a and b are trivial parameters, as follows again from Theorem 7.30: substituting the above first terms in $K = c_1$ and $H = c_2$ yields

$$
\begin{aligned}
c_1 &= (x_1^{(0)} + \lambda_{21}x_2^{(0)}x_3^{(0)})a + (x_2^{(0)} + \lambda_{21}x_1^{(0)}x_3^{(0)})b,\\
c_2 &= (\lambda_1 x_1^{(0)} + \lambda_3\lambda_{21}x_2^{(0)}x_3^{(0)})a + (\lambda_2 x_2^{(0)} + \lambda_3\lambda_{21}x_1^{(0)}x_3^{(0)})b,
\end{aligned}
$$

which is easily solved for a and b. We claim that the embedding of $\mathcal{E}_{\mathbf{c}}$ in $\mathbf{P}^3$, defined by $\varphi_{\mathbf{c}} : (x_1, x_2, x_3) \mapsto (1 : x_1 : x_2 : x_3)$ is the Kodaira map, associated to the points (divisor) in $\overline{\mathcal{E}_{\mathbf{c}}} \setminus \mathcal{E}_{\mathbf{c}}$, where $\overline{\mathcal{E}_{\mathbf{c}}}$ is the smooth completion of $\mathcal{E}_{\mathbf{c}}$. To show this, we first observe that the 4 points of the indicial locus (10.18) all belong to the following intersection of quadrics in $\mathbf{P}^2$:

$$
\begin{aligned}
0 &= \left(x_1^{(0)}\right)^2 + \left(x_2^{(0)}\right)^2 + \left(x_3^{(0)}\right)^2,\\
0 &= \lambda_1\left(x_1^{(0)}\right)^2 + \lambda_2\left(x_2^{(0)}\right)^2 + \lambda_3\left(x_3^{(0)}\right)^2.
\end{aligned}
$$

This can be shown by direct computation, or by expressing that the Laurent series $K(x(t))$ and of $H(x(t))$, which of course must be constant, have no pole of order 2. Since two conics in $\mathbf{P}^2$ intersect in four distinct points at most the four points that come from the indicial locus account for all intersection points.

The above Laurent solution of (10.16) yields the following local parametrization of these points:

$$x_1 = x_1^{(0)}\varsigma^{-1}, \quad x_2 = x_2^{(0)}\varsigma^{-1} + O(1), \quad x_3 = x_3^{(0)}\varsigma^{-1} + O(1),$$

which shows that these points are actually smooth points in the simplistic compactification of $\mathcal{E}_{\mathbf{c}}$ in $\mathbf{P}^3$, which is the closure of $\varphi_{\mathbf{c}}(\mathcal{E}_{\mathbf{c}})$. Our map $\varphi_{\mathbf{c}}$ is the Kodaira map that goes with the divisor that consists of the four points at infinity, since on the one hand the embedding functions have only a simple pole at these points and no other poles, and since on the other hand $\dim L(4 \text{ points}) = 4$ (see (5.25) or Theorem (5.8)), so we have a full basis of these functions.

Let us also write down the quadratic differential equations for the Euler top. We will do this in the chart $Z_1 \neq 0$, as it is trivial in the chart $Z_0 \neq 0$. We need to compute the Wronskians $W(x_2, x_1)$ and $W(x_3, x_1)$. We have for example that

$$\begin{aligned} W(x_2, x_1) &= \lambda_{13}x_1^2x_3 + \lambda_{23}x_2^2x_3 \\ &= \left(\lambda_1 x_1^2 + \lambda_2 x_2^2 + \lambda_3 x_3^2\right) x_3 - \lambda_3 \left(x_1^2 + x_2^2 + x_3^2\right) x_3 \\ &= 2(c_2 - \lambda_3 c_1)x_3, \end{aligned}$$

and similarly $W(x_3, x_1) = 2(\lambda_2 c_1 - c_2)x_2$. It follows that if we define $y_i := x_i/x_1$ then, in the chart $Z_1 \neq 0$ the quadratic differential equations are given explicitly by

$$\begin{aligned} \dot{y}_0 &= \lambda_{23}y_2y_3, \\ \dot{y}_1 &= 0, \\ \dot{y}_2 &= 2(c_2 - \lambda_3 c_1)y_1y_3, \\ \dot{y}_3 &= 2(\lambda_2 c_1 - c_2)y_1y_2. \end{aligned}$$

To finish the example of the Euler top we discuss its Lax equation (with spectral parameter), which is obtained by viewing the motion of the Euler top as a geodesic flow on $\mathbf{SO(3)}$ for the Manakov metric (see Paragraph 8.2.1). Namely, for fixed distinct parameters λ_1, λ_2 and λ_3 defining the Euler top, as before, we let

$$A := \operatorname{diag}(\lambda_1, \lambda_2, \lambda_3) \qquad \text{and} \qquad B := \operatorname{diag}(\lambda_2\lambda_3, \lambda_1\lambda_3, \lambda_1\lambda_2),$$

we define $\Lambda_{ij} := -\lambda_k$, where $\{i, j, k\} = \{1, 2, 3\}$; these Λ_{ij} are viewed as the entries of a symmetric matrix Λ (with, say, zeros on the diagonal). Let

$$X := \begin{pmatrix} 0 & x_3 & -x_2 \\ -x_3 & 0 & x_1 \\ x_2 & -x_1 & 0 \end{pmatrix}.$$

Then it is easily verified that the Euler equations are equivalent to the Lax equation (with spectral parameter $\mathfrak{h}$)

$$\begin{aligned}(A\mathfrak{h}+X)^{\cdot} &= [A\mathfrak{h}+X, B\mathfrak{h}+\Lambda\cdot X] \\ &= \left[A\mathfrak{h}+X, \left(f'\left(\frac{A\mathfrak{h}+X}{\mathfrak{h}}\right)\mathfrak{h}\right)_{+}\right],\end{aligned}$$

where $f(z)=\lambda_1\lambda_2\lambda_3 \ln z$. The spectral curve $|\mathfrak{z}\,\mathrm{Id}_3 - A\mathfrak{h} - X| = 0$ is easily computed to be given explicitly by

$$\mathcal{E}'_{\mathbf{c}} : \prod_{i=1}^{3}(\mathfrak{z}-\lambda_i\mathfrak{h}) + 2(c_1\mathfrak{z}-c_2\mathfrak{h}) = 0.$$

We claim that $\mathcal{E}'_{\mathbf{c}}$ and $\mathcal{E}_{\mathbf{c}}$ are isomorphic elliptic curves (precisely, their smooth compactifications are isomorphic). In order to do this we first rewrite $\mathcal{E}'_{\mathbf{c}}$ in the form

$$2v^2(c_1u-c_2) + \prod_{i=1}^{3}(u-\lambda_i) = 0, \tag{10.19}$$

where we have defined $(u,v) := (\mathfrak{z}/\mathfrak{h}, 1/\mathfrak{h})$. Formula (10.19) exhibits $\mathcal{E}'_{\mathbf{c}}$ as a double cover of $\mathbf{P}^1$, ramified at the points $u=\lambda_1$, λ_2, λ_3 and $u=c_2/c_1$, in particular $\mathcal{E}'_{\mathbf{c}}$ is an elliptic curve, just like $\mathcal{E}_{\mathbf{c}}$. In order to show that both curves are isomorphic it suffices to show that the cross-ratio of their ramification points are the same. The cross-ratio of the latter ramification points is easily computed to be given by

$$\begin{aligned}(\lambda_1,\lambda_2,\lambda_3,c_2/c_1) &= \frac{\lambda_1-\lambda_3}{\lambda_1-c_2/c_1} : \frac{\lambda_2-\lambda_3}{\lambda_2-c_2/c_1} \\ &= \frac{\lambda_{13}}{\lambda_{23}}\frac{c_1\lambda_2-c_2}{c_1\lambda_1-c_2}.\end{aligned}$$

We need to compare this to the cross-ratio of the points P^{++}, P^{+-}, P^{-+}, and P^{--}, which lie on the conic $\lambda_{12}x_1^2 - \lambda_{23}x_3^2 = 2(c_2-\lambda_2c_1)$. Therefore we need to parametrize the conic, which is done by using the slope of the lines that pass through a fixed point of the conic. Thus we let

$$x_3 = x_1\varsigma + c_3, \qquad x_1 = -\frac{2c_3\lambda_{32}}{\lambda_{32}\varsigma+\lambda_{12}\varsigma^{-1}},$$

with

$$c_3 := \sqrt{\frac{2(c_2-\lambda_2c_1)}{\lambda_{32}}}.$$

This yields the following parameters for the ramification points $P^{\epsilon_1\epsilon_2}$ of $\mathcal{E}_{\mathbf{c}}$ (see (10.17)):

$$\varsigma(P^{\epsilon_1\epsilon_2}) = \frac{\epsilon_2\sqrt{\frac{\lambda_1 c_1 - c_2}{\lambda_{13}}} - \sqrt{\frac{c_2 - \lambda_2 c_1}{\lambda_{32}}}}{\epsilon_1\sqrt{\frac{-\lambda_3 c_1 + c_2}{\lambda_{13}}}}.$$

It is now easily computed that

$$\begin{aligned}(P^{++}, P^{-+}, P^{--}, P^{+-}) &= \frac{\varsigma(P^{++}) - \varsigma(P^{--})}{\varsigma(P^{++}) - \varsigma(P^{+-})} : \frac{\varsigma(P^{-+}) - \varsigma(P^{--})}{\varsigma(P^{-+}) - \varsigma(P^{+-})} \\ &= \frac{\lambda_{13}}{\lambda_{23}} \frac{c_1\lambda_2 - c_2}{c_1\lambda_1 - c_2}.\end{aligned}$$

This shows that the cross-ratios agree, and hence that the elliptic curves are isomorphic. Notice that the pencil $\bar{H} - \lambda\bar{K}$ of rank four quadrics, where

$$\begin{aligned}2\bar{K} &:= x_1^2 + x_2^2 + x_3^2 - 2c_1x_0^2, \\ 2\bar{H} &:= \lambda_1 x_1^2 + \lambda_2 x_2^2 + \lambda_3 x_3^2 - 2c_2 x_0^2,\end{aligned}$$

contains four quadrics of rank three, namely when $\lambda = \lambda_1$, λ_2, λ_3 or c_2/c_1. These points are precisely the ramification points of the elliptic curve $\mathcal{E}'_{\mathbf{c}}$, a property reminiscent of what we have seen in each of the integrable geodesic flows on $\mathbf{SO}(4)$ (Chapter 8). The Lax equation is easily seen to satisfy the Linearization Criterion of Corollary 6.43, as only the pole condition needs to be checked.

10.2.2 The Lagrange Top

We now turn to the Lagrange top, which is characterized by $I_1 = I_2$ (two moments of inertia with respect to the fixed point are equal) and $m_1 = m_2 = 0$ (the center of gravity lies on the axis, passing through the fixed point, and corresponding to the equal moments of inertia, the *symmetry* axis). The symmetry axis explains the occurrence of the extra integral a_3 (the component of the moment of angular momentum in the direction of the symmetry axis). Indeed, the Hamiltonian group action corresponding to the S^1 action, simultaneous by an angle θ in the (a_1, a_2) and γ_1, γ_2 planes has co-momentum map a_3. Thus, by Noether's Theorem, since the Hamiltonian H of $\mathcal{V}_1$ is invariant under the above rotation when $\lambda_1 = \lambda_2$ and $m_1 = m_2 = 0$, the co-momentum a_3 provides a constant of motion and hence the corresponding Hamiltonian vector field $\mathcal{V}_2 := \mathcal{X}_{a_3}$ commutes with $\mathcal{V}_1$. One also speaks of the *symmetric top*.[1]

[1] This is the type of top most of us used to play with.

By rescaling the variables we may assume that $I_1 = 1$, so that $\lambda_1 = \lambda_2 = 1$. Substituting this and $m_1 = m_2 = 0$ in (10.5), yields to following vector field $\mathcal{V}_1$, displayed with the commuting vector field $\mathcal{V}_2 := \mathcal{X}_{a_3}$.

$$\begin{aligned}
\dot{\gamma}_1 &= \lambda_3 a_3 \gamma_2 - a_2 \gamma_3, & \gamma_1' &= \gamma_2, \\
\dot{\gamma}_2 &= a_1 \gamma_3 - a_3 \gamma_1, & \gamma_2' &= -\gamma_1, \\
\dot{\gamma}_3 &= a_2 \gamma_1 - \lambda_3 a_1 \gamma_2, & \gamma_3' &= 0, \\
\dot{a}_1 &= m_3 \gamma_2 + (\lambda_3 - 1) a_2 a_3, & a_1' &= a_2, \\
\dot{a}_2 &= -m_3 \gamma_1 - (\lambda_3 - 1) a_1 a_3, & a_2' &= -a_1, \\
\dot{a}_3 &= 0, & a_3' &= 0.
\end{aligned} \tag{10.20}$$

In these coordinates, the momentum map of the Lagrange top is $\mathbf{F} := (F_1, F_2, F_3, F_4)$, where the constants of motion F_i are given by

$$\begin{aligned}
F_1 &= \gamma_1^2 + \gamma_2^2 + \gamma_3^2, \\
F_2 &= \gamma_1 a_1 + \gamma_2 a_2 + \gamma_3 a_3, \\
F_3 &= \frac{1}{2}\left(a_1^2 + a_2^2 + \lambda_3 a_3^2\right) + m_3 \gamma_3, \\
F_4 &= a_3,
\end{aligned}$$

F_4 being a Hamiltonian for $\mathcal{V}_2$ (see (10.6)). Clearly the complexified rotation group $\mathbf{C}^*$ acts on $\mathbf{C}^6$, leaving the level manifolds $\mathbf{F_c}$ of the momentum map invariant; $\mathcal{V}_2$ is the fundamental vector field that corresponds to this action. This fact is classically used to reduce the Lagrange top to a system with one degree of freedom, which is then integrated in terms of elliptic functions.

We will give here a Lax equation for the Lagrange top and discuss its algebraic integrability. Since in the case of the Lagrange top $[M, \Omega - A] = 0$, as we have taken $\lambda_1 = 1 (= \lambda_2)$, the equations (10.4) are equivalent to the Lax equation

$$(M\mathfrak{h}^2 + A\mathfrak{h} + \Gamma)^{\cdot} = \left[M\mathfrak{h}^2 + A\mathfrak{h} + \Gamma, M\mathfrak{h} + \Omega\right]. \tag{10.21}$$

This means that, explicitly, the Lax operator is given in this case by the following skew-symmetric matrix,

$$X(\mathfrak{h}) = \begin{pmatrix} 0 & -m_3\mathfrak{h}^2 - a_3\mathfrak{h} - \gamma_3 & a_2\mathfrak{h} + \gamma_2 \\ m_3\mathfrak{h}^2 + a_3\mathfrak{h} + \gamma_3 & 0 & -a_1\mathfrak{h} - \gamma_1 \\ -a_2\mathfrak{h} - \gamma_2 & a_1\mathfrak{h} + \gamma_1 & 0 \end{pmatrix}, \tag{10.22}$$

whose spectral curve, corresponding to the values of the constants of motion $\mathbf{c} = (c_1, \ldots, c_4)$, is given by

$$\mathcal{C}_{\mathbf{c}} : \mathfrak{z}(\mathfrak{z}^2 + m_3^2\mathfrak{h}^4 + 2m_3c_4\mathfrak{h}^3 + 2c_3\mathfrak{h}^2 + 2c_2\mathfrak{h} + c_1) = 0,$$

which consists of a line and an elliptic curve $\mathcal{E}_{\mathbf{c}}$, whose points at infinity will be denoted by ∞^+ and ∞^-. Thus, for generic $\mathbf{c}$ the spectral curve $\mathcal{C}_{\mathbf{c}}$ is singular and reducible, and we cannot use the Linearization Criterion directly to show that the Lagrange top linearizes on the elliptic curves $\mathcal{E}_{\mathbf{c}}$. An elegant solution to this problem was proposed in [63], who notice that under the Lie algebra isomorphism of $\mathfrak{so}(3)$ and $\mathfrak{sl}(2)$, given by

$$\begin{pmatrix} 0 & -z & y \\ z & 0 & -x \\ -y & x & 0 \end{pmatrix} \longleftrightarrow \frac{1}{\sqrt{2}}\begin{pmatrix} \epsilon x & \epsilon z + \bar{\epsilon} y \\ \epsilon z - \bar{\epsilon} y & -\epsilon x \end{pmatrix}, \qquad \epsilon = \exp\frac{\sqrt{-1}\pi}{4},$$

the above Lax operator $X(\mathfrak{h})$ becomes a Lax operator $\tilde{X}(\mathfrak{h})$ for which the spectral curve is $\mathcal{E}_{\mathbf{c}}$, in particular it is reducible and the Linearization Criterion applies to it. Moreover, using the Lax equation they show the following theorem.

Theorem 10.3. *The Lagrange top is a generalized a.c.i. system: the generic fiber* $\mathbf{F}_{\mathbf{c}}$ *of its momentum map is isomorphic to an affine part of the generalized Jacobian* $\mathrm{Jac}(\mathcal{C}_{\mathbf{c}};\infty^{\pm})$ *and the flow of the vector fields* $\mathcal{V}_1$ *and* $\mathcal{V}_2$ *is linear on it.*

The *generalized Jacobian* $\mathrm{Jac}(\mathcal{E}_{\mathbf{c}};\infty^{\pm})$ is the non-compact Abelian group, which is defined as

$$\mathrm{Jac}(\mathcal{E}_{\mathbf{c}};\infty^{\pm}) := \mathrm{Div}^0(\mathcal{E}_{\mathbf{c}})/\sim,$$

where $\sim$ is defined, for divisors $\mathcal{D}$ and $\mathcal{D}'$ of degree zero on $\mathcal{E}_{\mathbf{c}}$, by $\mathcal{D}\sim\mathcal{D}'$ if and only if there exists a meromorphic function f on $\overline{\mathcal{E}_{\mathbf{c}}}$, such that

$$(f) = \mathcal{D} - \mathcal{D}', \qquad \text{and} \qquad f(\infty^+) = f(\infty^-) = 1.$$

It is an extension of the usual Jacobian $\mathrm{Jac}(\mathcal{E}_{\mathbf{c}})$ of $\mathcal{E}_{\mathbf{c}}$, namely there is an exact sequence

$$0 \longrightarrow \mathbf{C}^* \longrightarrow \mathrm{Jac}(\mathcal{E}_{\mathbf{c}},\infty^{\pm}) \longrightarrow \mathrm{Jac}(\mathcal{E}_{\mathbf{c}}) \longrightarrow 0.$$

For more information of generalized Jacobians, and further generalizations, see [155].

10.3 The Kowalevski Top

As we have said in the previous section the Kowalevski top is a spinning top whose principal moments of inertia (I_1, I_2, I_3) (with regard to the fixed point) satisfy the relation $I_1 = I_2 = 2I_3$ and whose center of mass belongs to the equatorial plane (corresponding to the moments I_1 and I_2) through the fixed point. By properly picking the axes of inertia in the equatorial plane and by rescaling, we can achieve $m = (2, 0, 0)$ and $\lambda = (1, 1, 2)$. Then the equations of motion (10.5) take the following simple form.

$$\begin{aligned} \dot{\gamma}_1 &= 2a_3\gamma_2 - a_2\gamma_3, & \dot{a}_1 &= a_2 a_3, \\ \dot{\gamma}_2 &= a_1\gamma_3 - 2a_3\gamma_1, & \dot{a}_2 &= 2\gamma_3 - a_1 a_3, \\ \dot{\gamma}_3 &= a_2\gamma_1 - a_1\gamma_2, & \dot{a}_3 &= -2\gamma_2. \end{aligned} \tag{10.23}$$

10.3.1 Liouville Integrability and Lax Equation

In the above form (10.23) the Kowalevski top has the following independent constants of motion.

$$\begin{aligned} Q_1 &:= \frac{a_1^2}{2} + \frac{a_2^2}{2} + a_3^2 + 2\gamma_1, \\ Q_2 &:= a_1\gamma_1 + a_2\gamma_2 + a_3\gamma_3, \\ Q_3 &:= \gamma_1^2 + \gamma_2^2 + \gamma_3^2, \\ Q_4 &:= K_+ K_-, \end{aligned} \tag{10.24}$$

where

$$K_\pm := \left(\frac{a_1 \pm \sqrt{-1}a_2}{2}\right)^2 - (\gamma_1 \pm \sqrt{-1}\gamma_2). \tag{10.25}$$

Q_2 is the angular momentum in the direction of gravity, Q_3 is the length of the gravity vector, and Q_1 is the energy. The extra constant of motion Q_4 was found by Kowalevski.

As we pointed out in the previous section the vector field (10.23) is Hamiltonian with respect to the linear Poisson structure $\{\cdot\,,\cdot\}$, whose Poisson matrix with respect to $\gamma_1, \ldots, \gamma_3, a_1, \ldots, a_3$ (in that order) is given by (10.6), the Hamiltonian being given by Q_1, while Q_2 and Q_3 are two independent Casimirs; another Hamiltonian vector field, that commutes with (10.23) is given by $\{\cdot\,, Q_4\}$. The fact that Q_3 is a Casimir justifies that we can fix its value to 1; for the computations that we will do there is no advantage of fixing the value of Q_3, hence we will not do this. With this convention, the Kowalevski top becomes a Liouville integrable system on $(\mathbf{C}^6, \{\cdot\,,\cdot\})$, with momentum map (Q_1, Q_2, Q_3, Q_4), and where the Q_i are defined in (10.24).

Different Lax equations (with spectral parameter) were proposed for the Kowalevski top. We will give here the most natural one, which is due to Reiman and Semenov-Tian-Shanskii, see [149] and [32]; for the other ones, see [13] and [76]. Let

$$X(\mathfrak{h}) := \begin{pmatrix} 0 & \sqrt{-1}\gamma_2 - \gamma_1 & \frac{\sqrt{-1}a_2 - a_1}{2}\mathfrak{h} & -\gamma_3 \\ \gamma_1 + \sqrt{-1}\gamma_2 & 0 & \gamma_3 & \frac{a_1 + \sqrt{-1}a_2}{2}\mathfrak{h} \\ \frac{-a_1 - \sqrt{-1}a_2}{2}\mathfrak{h} & -\gamma_3 & -a_3\mathfrak{h} & \gamma_1 + \sqrt{-1}\gamma_2 - \mathfrak{h}^2 \\ \gamma_3 & \frac{a_1 - \sqrt{-1}a_2}{2}\mathfrak{h} & -\gamma_1 + \sqrt{-1}\gamma_2 + \mathfrak{h}^2 & a_3\mathfrak{h} \end{pmatrix}$$

and

$$Y(\mathfrak{h}) := \begin{pmatrix} -a_3 & 0 & \frac{a_1 - \sqrt{-1}a_2}{2} & 0 \\ 0 & a_3 & 0 & -\frac{a_1 + \sqrt{-1}a_2}{2} \\ \frac{a_1 + \sqrt{-1}a_2}{2} & 0 & a_3 & \mathfrak{h} \\ 0 & \frac{\sqrt{-1}a_2 - a_1}{2} & -\mathfrak{h} & -a_3 \end{pmatrix}.$$

Note that if one thinks of $\mathfrak{h}$ as a formal, but purely imaginary parameter, and the variables a_i, γ_i as being real then $X(\mathfrak{h})$ is a skew-hermitian matrix.

It is easily checked by direct computation that (10.23) can be written as $\dot{X}(\mathfrak{h}) = \sqrt{-1}[X(\mathfrak{h}), Y(\mathfrak{h})]$. For an explanation of how this Lax equation was obtained, and for generalizations, see [32].

We investigate the geometry of the spectral curve $\det(\mathfrak{z}\,\mathrm{Id}_4 - X(\mathfrak{h})) = 0$, that corresponds to the Lax operator $X(\mathfrak{h})$ (see Figure 10.1 for an overview of the relations between the maps, the curves and the special points on them). Explicitly, this curve is given by

$$\mathcal{C}_{\mathbf{c}} : \mathfrak{z}^4 + \mathfrak{z}^2(\mathfrak{h}^4 - \mathfrak{h}^2 c_1 + 2c_3) + \mathfrak{h}^4 c_4 + \mathfrak{h}^2(c_2^2 - c_1 c_3) + c_3^2 = 0, \tag{10.26}$$

where the values of the constants of motion Q_i have been denoted by $\mathbf{c} = (c_1, \ldots, c_4)$; in what follows these values will always be assumed generic. Then $\mathcal{C}_{\mathbf{c}}$ is smooth and is a double cover of the curve $\mathcal{C}'_{\mathbf{c}}$, given by

$$\mathcal{C}'_{\mathbf{c}} : z^2 = \left((\mathfrak{h}^2 - c_1)^2 + 4(c_3 - c_4)\right)\mathfrak{h}^2 - 4c_2^2,$$

where the double cover that links the two curves is explicitly given by

$$\begin{aligned} \psi' : \ \mathcal{C}_{\mathbf{c}} &\to \mathcal{C}'_{\mathbf{c}} \\ (\mathfrak{z}, \mathfrak{h}) &\mapsto (z, \mathfrak{h}) = \left(\mathfrak{h}^{-1}(2\mathfrak{z}^2 + \mathfrak{h}^4 - c_1\mathfrak{h}^2 + 2c_3), \mathfrak{h}\right). \end{aligned} \tag{10.27}$$

It is clear that $\mathcal{C}'_{\mathbf{c}}$ is a hyperelliptic curve; the projection map $\pi' : \mathcal{C}'_{\mathbf{c}} \to \mathbf{C}$ that is defined by $\pi'(z, \mathfrak{h}) = \mathfrak{h}$ realizes $\mathcal{C}'_{\mathbf{c}}$ as a double cover of $\mathbf{C}$ that is ramified at six points, hence $\mathcal{C}'_{\mathbf{c}}$ compactifies into a hyperelliptic Riemann surface $\overline{\mathcal{C}'_{\mathbf{c}}}$ of genus two. The cover ψ' has 4 branch points: they are the points $(z, \mathfrak{h})$ on $\mathcal{C}'_{\mathbf{c}}$ for which $\mathfrak{z} = 0$, i.e.,

$$z\mathfrak{h} = \mathfrak{h}^4 - c_1\mathfrak{h}^2 + 2c_3.$$

Squaring this equation and using the equation of $C'_{\mathbf{c}}$ yields that the $\mathfrak{h}$ coordinates of these points must be roots to the quartic polynomial

$$\mathfrak{h}^4 c_4 + \mathfrak{h}^2 (c_2^2 - c_1 c_3) + c_3^2 = 0,$$

leading to 4 points $p'_1 \dots, p'_4$ on $C'_{\mathbf{c}}$, which are the branch points of ψ'; the corresponding ramification points on $\mathcal{C}_{\mathbf{c}}$ are denoted by $p_1^+, \dots, p_4^+$. When ψ' is extended to a double cover $\overline{\psi'} : \overline{\mathcal{C}_{\mathbf{c}}} \to \overline{C'_{\mathbf{c}}}$ there will be no new branch points: the points ∞'_1 and ∞'_2 in $\overline{C'_{\mathbf{c}}} \setminus C'_{\mathbf{c}}$ are given in terms of a local parameter ς by

$$(\mathfrak{h}, z) = \left(\varsigma^{-1}, \pm\varsigma^{-3}(1 - c_1\varsigma^2 + 2(c_3 - c_4)\varsigma^4 + O(\varsigma^6))\right),$$

which, substituted in

$$\mathfrak{z} = \sqrt{\frac{1}{2}\left(z\mathfrak{h} - \mathfrak{h}^4 + c_1\mathfrak{h}^2 - 2c_3\right)}$$

yields $\mathfrak{z} = \pm\sqrt{-\varsigma^{-4}(1 + O(\varsigma^2))} = \pm\sqrt{-1}\varsigma^{-2}(1 + O(\varsigma^2))$ for one sign of z and $\mathfrak{z} = \pm\sqrt{-c_4 + O(\varsigma^2)} = \pm\sqrt{-c_4}(1 + O(\varsigma^2))$ for the other sign. It follows that over both points the cover $\overline{\psi'}$ is unramified, leading to 4 points $\infty_1^\pm$ and $\infty_2^\pm$ on $\mathcal{C}_{\mathbf{c}}$. We conclude that $\overline{\psi'} : \overline{\mathcal{C}_{\mathbf{c}}} \to \overline{C'_{\mathbf{c}}}$ is a double cover of a genus 2 Riemann surface, branched over 4 points, so that the genus of $\overline{\mathcal{C}_{\mathbf{c}}}$ is 5, by Riemann-Hurwitz.

The curve $C'_{\mathbf{c}}$ is also a double cover of the elliptic curve

$$\mathcal{E}_{\mathbf{c}} : v^2 = \left((u - c_1)^2 + 4(c_3 - c_4)\right) u^2 - 4c_2^2 u,$$

where the covering map is given by

$$\begin{aligned} \chi' : \ & C'_{\mathbf{c}} \to \mathcal{E}_{\mathbf{c}} \\ & (z, \mathfrak{h}) \mapsto (v, u) = (z\mathfrak{h}, \mathfrak{h}^2). \end{aligned} \tag{10.28}$$

This cover is unramified, but its extension to a cover $\overline{\chi'} : \overline{C'_{\mathbf{c}}} \to \overline{\mathcal{E}_{\mathbf{c}}}$ must be ramified at two points, according to Riemann-Hurwitz. This can also be seen directly from the fact that $C'_{\mathbf{c}}$ and $\mathcal{E}_{\mathbf{c}}$ both have two points at infinity. Notice that the points p'_i are identified in pairs by χ', say $\chi'(p'_1) = \chi'(p'_3)$ and $\chi'(p'_2) = \chi'(p'_4)$; we will denote these two image points, in that order by p_1 and p_2, while we also denote $\overline{\chi'}(\infty'_i) = \infty_i$, for $i = 1, 2$.

The resulting $4 : 1$ cover $\mathcal{C}_{\mathbf{c}} \to \mathcal{E}_{\mathbf{c}}$ also factorizes over another curve, namely consider the curve $C''_{\mathbf{c}}$, defined by

$$C''_{\mathbf{c}} : \mathfrak{z}^4 + \mathfrak{z}^2(u^2 - uc_1 + 2c_3) + u^2 c_4 + u(c_2^2 - c_1 c_3) + c_3^2 = 0,$$

which is obtained by putting $u = \mathfrak{h}^2$ in the equation for $\mathcal{C}_{\mathbf{c}}$.

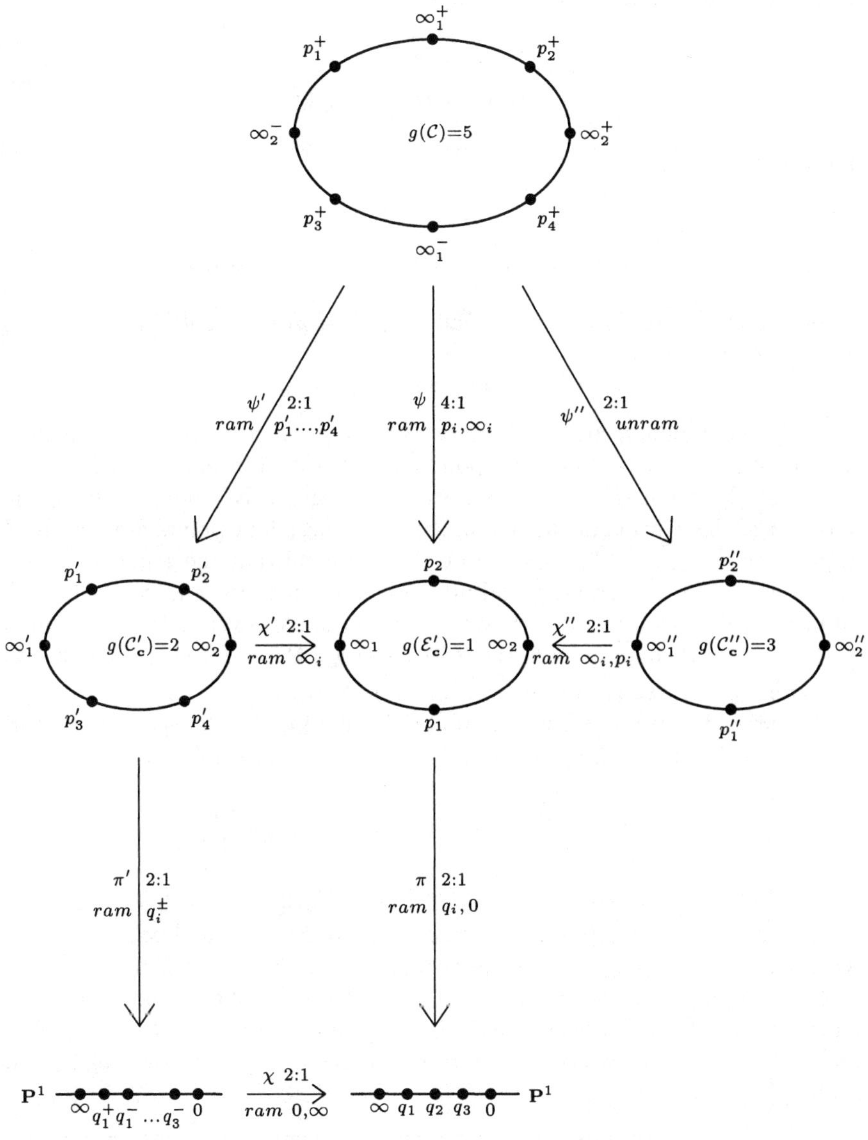

Fig. 10.1. The spectral curve, associated with the above Lax equation for the Kowalevski top, and some of its quotients. The points at infinity are identified in pairs by $\overline{\psi'}$ and $\overline{\psi''}$, but they are ramification points for $\overline{\chi'}$ and $\overline{\chi''}$. The points labeled p_i are identified in pairs by $\overline{\chi'}$ and $\overline{\psi''}$, but they are ramification points for $\overline{\psi'}$ and $\overline{\chi''}$. Also, $\overline{\chi}$ maps the six branch points of $\overline{\pi'}$ to three of the branch points of $\overline{\pi}$ (0 is the other branch point of $\overline{\pi}$).

Then we have a double cover $\psi'' : \mathcal{C}_{\mathbf{c}} \to \mathcal{C}''_{\mathbf{c}}$, given by

$$\begin{aligned} \psi'' : \ \mathcal{C}_{\mathbf{c}} &\to \mathcal{C}''_{\mathbf{c}} \\ (\mathfrak{z}, \mathfrak{h}) &\mapsto (\mathfrak{z}, u) = (\mathfrak{z}, \mathfrak{h}^2) \end{aligned} \tag{10.29}$$

and a double cover

$$\begin{aligned} \chi'' : \ \mathcal{C}''_{\mathbf{c}} &\to \mathcal{E}_{\mathbf{c}} \\ (\mathfrak{z}, u) &\mapsto (v, u) = (2\mathfrak{z}^2 + u^2 - c_1 u + 2c_3, u). \end{aligned} \tag{10.30}$$

It follows easily from the above that p_1 and p_2 are the only branch points of χ'': the map

$$\chi' \circ \psi' = \chi'' \circ \psi'' : \mathcal{C}_{\mathbf{c}} \to \mathcal{E}_{\mathbf{c}}$$

has only the points p_1 and p_2 as branch points, with corresponding ramification points $p_1^+, \dots, p_4^+$, but these points are identified in pairs by ψ''. Notice that this also implies that ψ'' is unramified. Similarly, one sees using the above local description of the points $\infty_i^{\pm}$ that these four points are identified in pairs by $\overline{\psi''} : \overline{\mathcal{C}_{\mathbf{c}}} \to \overline{\mathcal{C}''_{\mathbf{c}}}$, implying on the one hand that the genus of $\overline{\mathcal{C}''_{\mathbf{c}}}$ is 3 and on the other hand that $\overline{\chi''}$ admits also the points ∞_1 and ∞_2 as branch points. Finally, notice that if we view $\mathcal{E}_{\mathbf{c}}$ as a double cover of $\mathbf{P}^1$ by the map $\pi : (v, u) \mapsto u$ then $\overline{\chi'}$ induces the $2:1$ cover $\chi : \mathbf{P}^1 \to \mathbf{P}^1$ that maps $\mathfrak{h} \in \mathbf{C}$ to $u = \mathfrak{h}^2$. Of course, χ is ramified at 0 and at ∞ (only).

Let us now verify the Linearization Criterion for the above Lax operator. We will do this by checking the conditions in Corollary 6.43. First notice that

$$Y(\mathfrak{h}) = -\frac{X(\mathfrak{h})}{\mathfrak{h}} + \operatorname{diag}(-a_3, a_3, 0, 0) + \frac{C}{\mathfrak{h}},$$

where C is independent of $\mathfrak{h}$, so that $Y(\mathfrak{h})$ satisfies (6.54), with $\Psi = 1$. Thus, it suffices to show that $\xi/\mathfrak{h}$ has no pole at the points $\infty_1^{\pm}$ and $\infty_2^{\pm}$, where we recall that $\xi = (1, \xi_2, \xi_3, \xi_4)^{\top}$ is the normalized null-vector of $\mathfrak{z}\operatorname{Id}_4 - X(\mathfrak{h})$, for $(\mathfrak{z}, \mathfrak{h})$ belonging to the spectral curve $\mathcal{C}_{\mathbf{c}}$. In terms of the cofactors A_{ij} of the matrix $\mathfrak{z}\operatorname{Id}_4 - X(\mathfrak{h})$ the components ξ_i are explicitly given by $\xi_i = A_{1i}/A_{11}$, for $i = 2, 3, 4$. Also, substituting $\mathfrak{h} = 1/\varsigma$ in the equation (10.26) of $\mathcal{C}_{\mathbf{c}}$ and solving for $\mathfrak{z}$ we find

$$\mathfrak{z} = \pm\frac{\sqrt{-1}}{\sqrt{2}\varsigma^2}\sqrt{1 - c_1\varsigma^2 + 2c_3\varsigma^4 \pm \sqrt{1 - 2c_1\varsigma^2 + (c_1^2 + 4(c_3 - c_4))\varsigma^4 - 4c_2^2\varsigma^6}}, \tag{10.31}$$

giving as local parameterization

$$\mathfrak{h} = \frac{1}{\varsigma}, \qquad \mathfrak{z} = \pm\sqrt{-c_4}\left(1 + \frac{c_2^2 - c_1(c_3 - c_4)}{2c_4}\varsigma^2 + O(\varsigma^4)\right), \tag{10.32}$$

for two of the points and

$$\mathfrak{h} = \frac{1}{\varsigma}, \qquad \mathfrak{z} = \pm\frac{\sqrt{-1}}{\varsigma^2}\left(1 - \frac{1}{2}c_1\varsigma^2 - \frac{1}{8}(c_1^2 + 4c_4 - 8c_3)\varsigma^4 + O(\varsigma^6)\right), \tag{10.33}$$

for the other two points. If we substitute (10.32) in $\xi_i = A_{1i}/A_{11}$ then we find

$$\begin{aligned}
\xi_2 &= \mp\frac{1}{\sqrt{-c_4}}\left(\left(\frac{a_1 + \sqrt{-1}a_2}{2}\right)^2 - \gamma_1 - \sqrt{-1}\gamma_2\right) + O(\varsigma) = \mp\frac{c_4^+}{\sqrt{-c_4}} + O(\varsigma),\\
\xi_3 &= \pm\frac{1}{2}(a_1 + \sqrt{-1}a_2)\frac{c_4^+}{\sqrt{-c_4}}\varsigma + O(\varsigma^2),\\
\xi_4 &= -\frac{1}{2}(a_1 + \sqrt{-1}a_2)\varsigma + O(\varsigma^2),
\end{aligned}$$

where c_4^+ is the value of the factor K_+ of the Kowalevski integral (see (10.25)). When substituting (10.33) in ξ_i then we find very similar results, namely

$$\begin{aligned}
\xi_2 &= \pm\sqrt{-1}c_4^+ + O(\varsigma),\\
\xi_3 &= \pm\frac{\sqrt{-1}}{2}(a_1 + \sqrt{-1}a_2)c_4^+\varsigma + O(\varsigma^2),\\
\xi_4 &= -\frac{1}{2}(a_1 + \sqrt{-1}a_2)\varsigma + O(\varsigma^2),
\end{aligned}$$

so that the vector ξ, and hence also $\xi/\mathfrak{h}$ is finite at the four points $\infty_1^\pm$ and $\infty_2^\pm$. This completes the check of the Linearization Criterion and we may conclude that the Kowalevski vector field linearizes on the Jacobians of the spectral curves $\mathcal{C}_{\mathbf{c}}$.

As suggested by Kowalevski's last invariant Q_4, it is more convenient to consider the new variables

$$\begin{aligned}
x_1 &= \tfrac{1}{2}(a_1 + \sqrt{-1}a_2), & x_4 &= \tfrac{1}{4}\left(a_1 + \sqrt{-1}a_2\right)^2 - (\gamma_1 + \sqrt{-1}\gamma_2),\\
x_2 &= \tfrac{1}{2}(a_1 - \sqrt{-1}a_2), & x_5 &= \tfrac{1}{4}\left(a_1 - \sqrt{-1}a_2\right)^2 - (\gamma_1 - \sqrt{-1}\gamma_2),\\
x_3 &= a_3, & x_6 &= \gamma_3.
\end{aligned} \tag{10.34}$$

In these new coordinates the constants of motion (10.24) take the following form.

$$\begin{aligned}
F_1 &= (x_1 + x_2)^2 + x_3^2 - x_4 - x_5,\\
F_2 &= x_1x_2(x_1 + x_2) - x_4x_2 - x_5x_1 + x_3x_6,\\
F_3 &= (x_1^2 - x_4)(x_2^2 - x_5) + x_6^2,\\
F_4 &= x_4x_5.
\end{aligned} \tag{10.35}$$

In the sequel we will take the corresponding map $\mathbf{F} := (F_1, F_2, F_3, F_4)$ as the momentum map of the Kowalevski top.

The Poisson structure is (up to a constant factor $\sqrt{-1}$) given by the following matrix,

$$\begin{pmatrix} 0 & -x_3/2 & x_1 & 0 & x_6 - x_2x_3 & (x_1^2 - x_4)/2 \\ x_3/2 & 0 & -x_2 & x_1x_3 - x_6 & 0 & (x_5 - x_2^2)/2 \\ -x_1 & x_2 & 0 & -x_1^2 - x_4 & x_2^2 + x_5 & 0 \\ 0 & x_6 - x_1x_3 & x_1^2 + x_4 & 0 & 2\star & x_1(x_1^2 - x_4) \\ x_2x_3 - x_6 & 0 & -x_2^2 - x_5 & -2\star & 0 & x_2(x_5 - x_2^2) \\ (x_4 - x_1^2)/2 & (x_2^2 - x_5)/2 & 0 & x_1(x_4 - x_1^2) & x_2(x_2^2 - x_5) & 0 \end{pmatrix} \tag{10.36}$$

where $\star = (x_1 + x_2)x_6 - x_1x_2x_3$. With respect to this Poisson structure the vector field $\mathcal{V}_1 := \{\cdot, F_1\}$ and the commuting vector field $\mathcal{V}_2 := -\{\cdot, F_4\}$ take the following form.

$$\begin{aligned} \dot{x}_1 &= x_1x_3 - x_6, & x_1' &= x_4(x_2x_3 - x_6), \\ \dot{x}_2 &= x_6 - x_2x_3, & x_2' &= x_5(x_6 - x_1x_3), \\ \dot{x}_3 &= x_2^2 - x_1^2 + x_4 - x_5, & x_3' &= x_1^2x_5 - x_2^2x_4, \\ \dot{x}_4 &= 2x_3x_4, & x_4' &= -2x_4(x_6(x_1 + x_2) - x_1x_2x_3), \\ \dot{x}_5 &= -2x_3x_5, & x_5' &= 2x_5(x_6(x_1 + x_2) - x_1x_2x_3), \\ \dot{x}_6 &= x_1(x_2^2 - x_5) - x_2(x_1^2 - x_4), & x_6' &= x_1x_5(x_1^2 - x_4) - x_2x_4(x_2^2 - x_5). \end{aligned} \tag{10.37}$$

Observe that the vector field $\mathcal{V}_1$ is weight homogeneous when we assign the following weights to the variables:

$$\varpi(x_1, x_2, x_3, x_4, x_5, x_6) := (1, 1, 1, 2, 2, 2).$$

Then the invariants have the following weights:

$$\varpi(F_1, F_2, F_3, F_4) = (2, 3, 4, 4).$$

On $\mathbf{C}^6$ there are two natural involutions σ and τ which preserve the constants of motion, hence leave the fibers of the momentum map invariant. They are given by

$$\begin{aligned} \sigma &: (x_1, x_2, x_3, x_4, x_5, x_6) \mapsto (x_2, x_1, -x_3, x_5, x_4, -x_6), \\ \tau &: (x_1, x_2, x_3, x_4, x_5, x_6) \mapsto (x_1, x_2, -x_3, x_4, x_5, -x_6). \end{aligned} \tag{10.38}$$

Since σ is a Poisson morphism it preserves the vector fields $\mathcal{V}_1$ and $\mathcal{V}_2$; τ is however an anti-Poisson map, so it reverses $\mathcal{V}_1$ as well as $\mathcal{V}_2$,

$$\sigma^*\mathcal{V}_1 = \mathcal{V}_i, \text{ and } \tau^*\mathcal{V}_2 = -\mathcal{V}_2.$$

10.3.2 Algebraic Complete Integrability

We look for weight homogeneous Laurent solutions to $\mathcal{V}_1$. Since $\varpi(x_{i+3}) = 2\varpi(x_i) = 2$ for $i = 1, \dots, 3$, the indicial equation is given by

$$\begin{aligned}
0 &= x_1^{(0)} + x_1^{(0)} x_3^{(0)} - x_6^{(0)},\\
0 &= x_2^{(0)} + x_6^{(0)} - x_2^{(0)} x_3^{(0)},\\
0 &= x_3^{(0)} + (x_2^{(0)})^2 - (x_1^{(0)})^2 + x_4^{(0)} - x_5^{(0)},\\
0 &= x_4^{(0)}(1 + x_3^{(0)}),\\
0 &= x_5^{(0)}(1 - x_3^{(0)}),\\
0 &= 2x_6^{(0)} + x_1^{(0)}\left((x_2^{(0)})^2 - x_5^{(0)}\right) - x_2^{(0)}\left((x_1^{(0)})^2 - x_4^{(0)}\right).
\end{aligned}$$

These equations are easily solved, yielding the following (non-zero) solutions.

$$(x_1^{(0)}, x_2^{(0)}, x_3^{(0)}, x_4^{(0)}, x_5^{(0)}, x_6^{(0)}) = \begin{cases} (0, a, 1, 0, 1 + a^2, 0),\\ (a, 0, -1, 1 + a^2, 0, 0),\\ \mp\sqrt{-1}(1, -1, 0, 0, 0, 1), \end{cases} \tag{10.39}$$

where a is a free parameter. Thus, the indicial locus $\mathcal{I}$ consists of two disjoint curves $\mathcal{I}'$ and $\mathcal{I}''$, besides two points that do not belong to any of these curves. The involution σ induces an involution on $\mathcal{I}$ which permutes the two points and the two curves; it will also be denoted by σ. The Kowalevski matrix at an arbitrary $x^{(0)} \in \mathcal{I}$ is given by

$$\mathcal{K}\left(x^{(0)}\right) = \begin{pmatrix} 1 + x_3^{(0)} & 0 & x_1^{(0)} & 0 & 0 & -1\\ 0 & 1 - x_3^{(0)} & -x_2^{(0)} & 0 & 0 & 1\\ -2x_1^{(0)} & 2x_2^{(0)} & 1 & 1 & -1 & 0\\ 0 & 0 & 2x_4^{(0)} & 2(1 + x_3^{(0)}) & 0 & 0\\ 0 & 0 & -2x_5^{(0)} & 0 & 2(1 - x_3^{(0)}) & 0\\ \star & -\sigma(\star) & 0 & x_2^{(0)} & -x_1^{(0)} & 2 \end{pmatrix},$$

where $\star = x_2^{(0)}(x_2^{(0)} - 2x_1^{(0)}) - x_5^{(0)}$. If we substitute each of the solutions (10.39) in the characteristic polynomial $|\mu \operatorname{Id}_6 - \mathcal{K}(x^{(0)})|$ then we find (in that order)

$$|\mu \operatorname{Id}_6 - \mathcal{K}(x^{(0)})| = \begin{cases} (\mu + 1)\mu(\mu - 1)(\mu - 2)(\mu - 3)(\mu - 4),\\ (\mu + 1)\mu(\mu - 1)(\mu - 2)(\mu - 3)(\mu - 4),\\ (\mu + 1)^2(\mu - 2)^2(\mu - 3)(\mu - 4), \end{cases}$$

so that the last one certainly does not lead to a principal balance.

Note that Propositions 7.11 and Theorem 7.30 a priori tell us that -1, 2, 3 and 4 must be in the spectrum, and in the first two cases also 0. Since $\mathcal{V}_1$ is divergence free, Proposition 7.12 implies that $\operatorname{Trace}\mathcal{K}(x^{(0)}) = 9$, which in the first two cases determines completely the spectrum of $\mathcal{K}(x^{(0)})$, but not quite in the third case.

In order to show that the two curves $\mathcal{I}'$ and $\mathcal{I}''$ in the indicial locus lead to principal balances, which we will denote by $x(t;\mathcal{I}')$ and $x(t;\mathcal{I}'')$, we need to exhibit the first five terms (coming from steps 0 – 4) of them, with the free parameters that appear. Since $\mathcal{I}'$ and $\mathcal{I}''$ correspond under the involution σ, so that $x(t;\mathcal{I}'')$ is obtained from $x(t;\mathcal{I}')$ by applying σ, it suffices to exhibit the first five terms of $x(t;\mathcal{I}')$. We actually only give the first four terms, except for $x_4(t)$ where we give the first five (sic!) terms: the fifth term of the other $x_i(t)$ is uniquely specified, given the fifth term of $x_4(t)$.

$$\begin{aligned}
x_1(t) &= b - tab^2 + \frac{t^2}{4}Z'(b) + O(t^3),\\
x_2(t) &= \frac{a}{t} + b(a^2+1) + \frac{ta}{6}(4a^2b^2 + 4b^2 - c_1)\\
&\qquad + \frac{t^2}{12}(a^2(4a^2b^3 + 8b^3 - 2c_1b - c_2) - Z'(b)) + O(t^3),\\
x_3(t) &= \frac{1}{t} - ab - \frac{t}{3}((a^2+4)b^2 - c_1) + \frac{t^2a}{4}(4b^3 + c_2) + O(t^3),\\
x_4(t) &= t^2(Z(b) - a^2b^4) + O(t^3), \qquad\qquad (10.40)\\
x_5(t) &= (a^2+1)\left(\frac{1}{t^2} + \frac{2ab}{t} + \frac{7a^2b^2 + 4b^2 - c_1}{3}\right.\\
&\qquad \left. + \frac{ta}{6}(12(a^2+1)b^3 - 4c_1b - c_2) + O(t^2)\right),\\
x_6(t) &= \frac{b}{t} - ab^2 + \frac{t}{6}(4a^2b^3 - 2b^3 - c_1b + 3c_2) + O(t^2),
\end{aligned}$$

where $Z(b) := -b^4 + c_1b^2 - 2c_2b + c_3$, and $Z'(b)$ is its derivative. We have already replaced the three trivial parameters by the values of the first three constants of motion: we have that $F_i(x(t)) = c_i$ for $i = 1, \ldots, 3$. An equation for the abstract Painlevé divisor $\Gamma'_{\mathbf{c}}$, where $\mathbf{c} = (c_1, \ldots, c_4)$ is then found by substituting $x(t;\mathcal{I}')$ in the equation $x_4x_5 = c_4$. This yields the affine curve $\Gamma'_{\mathbf{c}}$ in $\mathbf{C}^2$, defined by

$$\Gamma_{\mathbf{c}} : (1+a^2)^2b^4 - (1+a^2)(c_1b^2 - 2c_2b + c_3) + c_4 = 0. \qquad (10.41)$$

The same equation is found for the abstract Painlevé divisor $\Gamma''_{\mathbf{c}}$.

In the sequel we will often have no reason to distinguish the two isomorphic curves $\Gamma'_{\mathbf{c}}$ and $\Gamma''_{\mathbf{c}}$; in this case we will write simply $\Gamma_{\mathbf{c}}$ for either of these curves.

The curve $\Gamma_{\mathbf{c}}$ is a double ramified cover of the elliptic curve $C_{\mathbf{c}}$ in $\mathbf{C}^2$, defined by

$$C_{\mathbf{c}} : c_4u^2 - u(c_1v^2 - 2c_2v + c_3) + v^4 = 0, \tag{10.42}$$

the covering map being given by

$$(u, v) = \psi(a, b) = (1/(a^2 + 1), b).$$

Rewriting the equation of $\Gamma_{\mathbf{c}}$ as

$$a^4b^4 + a^2(b^4 - Z(b)) + c_4 - Z(b) = 0,$$

we see that the map has 4 ramification points, which are the 4 points $(0, b_i)$, where b_i is any root of $c_4 - Z(b)$. When ψ is extended to a covering map $\overline{\psi} : \overline{\Gamma_{\mathbf{c}}} \to \overline{C_{\mathbf{c}}}$ between the smooth compactifications of $\Gamma_{\mathbf{c}}$ and of $C_{\mathbf{c}}$ the map has no other ramification points, so that the genus of $\overline{\Gamma_{\mathbf{c}}}$ is 3, by Riemann-Hurwitz. In order to check that the $\overline{\psi}$ is unramified at infinity, use the data given in Table 10.3: the six points in the table are obviously identified in pairs by $\overline{\psi}$.

Table 10.3. The six points ∞_{ϵ_1} and $\infty_{\epsilon_2\epsilon_3}$ (where $\epsilon_1^2 = \epsilon_2^2 = \epsilon_3^2 = 1$) in $\overline{\Gamma_{\mathbf{c}}} \setminus \Gamma_{\mathbf{c}}$ in terms of a local parameter ς. The value of δ is one fixed square root of $c_1^2 - 4c_4$.

point	a	b
∞_{ϵ_1}	$\epsilon_1 \frac{\sqrt{c_3}}{\varsigma^2}\left(1 - \frac{c_2}{c_3}\varsigma + O(\varsigma^2)\right)$	ς
$\infty_{\epsilon_2\epsilon_3}$	$\epsilon_2\sqrt{-1}\left(1 + \frac{\epsilon_3\delta - c_1}{4}\varsigma^2 + c_2\frac{\delta - \epsilon_3 c_1}{2\delta}\varsigma^3 + O(\varsigma^4)\right)$	ς^{-1}

We now look for weight homogeneous polynomials which have a simple pole in t at most, when any of the two principal balances is substituted in it. In the notation of Paragraph 7.7.1 this means that we wish to construct a basis of $\mathcal{Z}_\rho$ as a $\mathcal{H}$-module, where the pole vector ρ is chosen as $\rho := (1, 1)$. The results, displayed in Table 10.4, suggest that no more independent functions will be found beyond weight 4.

Let us verify the adjunction formula, to give more support to this belief. To do this we need the residue terms of the Laurent solutions $z_0(t; \Gamma'_{\mathbf{c}}), \dots, z_7(t; \Gamma'_{\mathbf{c}})$. They are given in Table 10.5.

We investigate the map $\varphi'_{\mathbf{c}} : \Gamma'_{\mathbf{c}} \to \mathbf{P}^7$, which is given by these residues. It follows from Table 10.5 that $\varphi'_{\mathbf{c}}$ is given by

$$\varphi'_{\mathbf{c}}(a, b) = (0 : 0 : a : 1 : b : ab : (1 + a^2)b^2 : 2a^2b^3 - Z'(b)/2), \tag{10.43}$$

where we recall that $Z'(b) = -4b^3 + 2c_1b - 2c_2$. This map is clearly an embedding of the affine curve $\Gamma'_{\mathbf{c}}$.

Table 10.4. The polynomials of weight at most 5 which have a simple pole at most when any of the principal balances is substituted in them. In order to verify that $\dim \mathcal{Z}_\rho^5 = 13$, as is asserted in the last line, one extra term in the principal balances needs to be computed.

k	$\dim \mathcal{F}^k$	$\dim \mathcal{H}^k$	$\dim \mathcal{Z}_\rho^k$	# dep	ζ_k	indep. functions
0	1	1	1	0	1	$z_0 = 1$
1	3	0	3	0	3	$z_1 := x_1, \dots, z_3 := x_3$
2	9	1	3	1	2	$z_4 := x_6,\ z_5 := x_1x_2$
3	19	1	5	4	1	$z_6 := (x_1+x_2)x_6 - x_1x_2x_3$
4	39	3	9	8	1	$z_7 := x_3\dot{x}_6 - \dot{x}_3x_6$
5	69	1	13	13	0	—

Table 10.5. The residues of the functions $z_i(t; \Gamma_{\mathbf{c}}')$ and $z_i(t; \Gamma_{\mathbf{c}}'')$ define two embeddings $\varphi_{\mathbf{c}}'$ and $\varphi_{\mathbf{c}}''$ of the curves $\Gamma_{\mathbf{c}}' = \Gamma_{\mathbf{c}}''$ into $\mathbf{P}^7$.

	z_0	z_1	z_2	z_3	z_4	z_5	z_6	z_7
$\varphi_{\mathbf{c}}'$	0	0	a	1	b	ab	$(1+a^2)b^2$	$2a^2b^3 - Z'(b)/2$
$\varphi_{\mathbf{c}}''$	0	a	0	-1	$-b$	ab	$-(1+a^2)b^2$	$2a^2b^3 - Z'(b)/2$

In order to see that it extends to an embedding of $\overline{\Gamma_{\mathbf{c}}'}$, substitute the first and second line of Table 10.3 in (10.43) to find the following first two terms:

$$\varphi_{\mathbf{c}}'(\infty_{\epsilon_1}) \sim \left(0:0:\epsilon_1\sqrt{c_3} - \varsigma\frac{\epsilon_1 c_2}{\sqrt{c_3}} : 0 : 0 : \varsigma\epsilon_1\sqrt{c_3} : c_3 - 2\varsigma c_2 : 2\varsigma c_3\right),$$

$$\varphi_{\mathbf{c}}'(\infty_{\epsilon_2\epsilon_3}) \sim \left(0:0:\varsigma\sqrt{-1}\epsilon_2 : \varsigma : 1 : \sqrt{-1}\epsilon_2 : \frac{\varsigma}{2}(c_1 - \epsilon_3\delta) : \right.$$
$$\left. : -\epsilon_3\delta + \varsigma c_2\frac{2\epsilon_3 c_1 - \delta}{\delta}\right).$$

Letting $\varsigma \to 0$ we find the following image points:

$$P_{\epsilon_1}' := \lim_{p\to\infty_{\epsilon_1}} \varphi_{\mathbf{c}}'(p) = (0:0:1:0:0:0:\epsilon_1\sqrt{c_3}:0),$$

$$Q_{\epsilon_2\epsilon_3}' := \lim_{p\to\infty_{\epsilon_2\epsilon_3}} \varphi_{\mathbf{c}}'(p) = \left(0:0:0:0:1:\epsilon_2\sqrt{-1}:0:-\epsilon_3\delta\right).$$

We see at once that these six points are different, so that $\varphi_{\mathbf{c}}'$ is injective, and since the linear terms in ς are non-vanishing we conclude that the image curve is non-singular and isomorphic to $\Gamma_{\mathbf{c}}'$.

Applying the involution σ, which acts, according to (10.38), on the z_i in the following way

$$\sigma(z_0, \dots, z_7) = (z_0, z_2, z_1, -z_3, -z_4, z_5, -z_6, z_7),$$

we find the image of these six points under the map $\varphi''_{\mathbf{c}} : \Gamma''_{\mathbf{c}} \to \mathbf{P}^7$, which is given by the residues of the functions $z_0(t; \Gamma''_{\mathbf{c}}), \dots, z_7(t; \Gamma'_{\mathbf{c}})$. Namely, we find the following image points in $\mathbf{P}^7$:

$$P''_{\epsilon_1} := \lim_{p \to \infty_{\epsilon_1}} \varphi''_{\mathbf{c}}(p) = (0 : 1 : 0 : 0 : 0 : 0 : -\epsilon_1\sqrt{c_3} : 0),$$
$$Q''_{\epsilon_2\epsilon_3} := \lim_{p \to \infty_{\epsilon_2\epsilon_3}} \varphi''_{\mathbf{c}}(p) = \left(0 : 0 : 0 : 0 : 1 : -\epsilon_2\sqrt{-1} : 0 : \epsilon_3\delta\right).$$

Thus, we get 2 new points P''_{ϵ_1}, while $Q'_{\epsilon_2\epsilon_3} = Q''_{-\epsilon_2,-\epsilon_3}$ which leads to 4 intersection points of $\mathcal{D}'_{\mathbf{c}} := \overline{\varphi'_{\mathbf{c}}(\Gamma'_{\mathbf{c}})}$ and $\mathcal{D}''_{\mathbf{c}} := \overline{\varphi''_{\mathbf{c}}(\Gamma''_{\mathbf{c}})}$. Comparing the term in ς of $\varphi'_{\mathbf{c}}(P)$, for P close to $\infty_{\epsilon_2\epsilon_3}$ with the term in ς of $\varphi''_{\mathbf{c}}(P)$, for P close to $\infty_{-\epsilon_2,-\epsilon_3}$, we conclude that the two images curves intersect transversally in these four points, as indicated in Figure 10.2. Using the fact that the Euler characteristic of an ordinary double point is 1, we find by using Proposition 5.42 that the genus of the image divisor is given by $2 \times 3 + 1 - 2 + 4 \times 1 = 9$. Since $9 = 7 + 2$ the adjunction formula is verified for $\varphi_{\mathbf{c}}$.

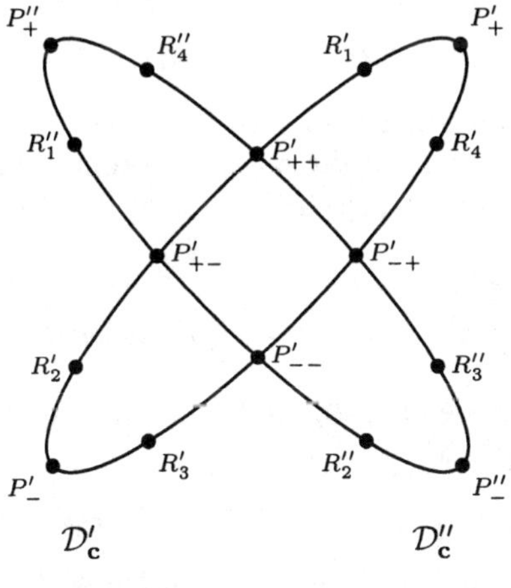

Fig. 10.2. The Painlevé divisor $\mathcal{D}_{\mathbf{c}}$ of the Kowalevski top consists of two non-singular genus 3 curves $\mathcal{D}'_{\mathbf{c}}$ and $\mathcal{D}''_{\mathbf{c}}$ which intersect transversally in the 4 points $P'_{\epsilon_2\epsilon_3}$. They correspond under an involution σ which will be shown to be a translation over a half period on the Abelian surface $\mathbf{T}^2_{\mathbf{c}}$, obtained by adjoining $\mathcal{D}_{\mathbf{c}}$ to the fiber $\mathbf{F}_{\mathbf{c}}$ of the momentum map. There are 12 half periods on $\mathcal{D}_{\mathbf{c}}$, among which the 8 points R'_i and R''_i where $\mathcal{D}'_{\mathbf{c}}$ and $\mathcal{D}''_{\mathbf{c}}$ are branched over the elliptic curve $C_{\mathbf{c}}$ and 4 other smooth points P'_{ϵ_1} and P''_{ϵ_1} of $\mathcal{D}_{\mathbf{c}}$. The latter 4 points are the (double) tangency points of the vector field $\overline{\mathcal{V}}_1$.

We are now in a very similar situation as in the case of geodesic flow for metric II (see Paragraph 8.3.2). Namely we have an isomorphic embedding $\varphi_{\mathbf{c}}$ of $\mathbf{F_c}$, satisfying the adjunction formula, and it will turn out later that these functions provide an embedding of the compactification of $\mathbf{F_c}$, which will turn out to be an Abelian surface of type $(2,4)$. But not all the Wronskians $W(z_i, z_j)$ can be written as a quadratic polynomial in $z_0, \dots, z_7$, with coefficients in $\mathcal{H}$. The reason is again that the Painlevé divisor is very ample but is not normally generated. The remedy is in this case slightly simpler than in the case of geodesic flow for metric II, because the divisor at infinity is now reducible, having two components, so instead of looking for a basis (over $\mathcal{H}$) for the space of polynomials which have a double pole at most when any of the two principal balance is substituted in them, we look for a basis (over $\mathcal{H}$) of the smaller space of polynomials which have a double pole at most when the principal balance $x(t;\Gamma'_{\mathbf{c}})$ is substituted in it, but have a simple pole at most when the principal balance $x(t;\Gamma''_{\mathbf{c}})$ is substituted in them. This means that we choose as pole vector $\rho := (2,1)$. Already the functions $z_0, \dots, z_7$ defined in Table 10.4 have this property; we find the following additional functions.

$$\begin{aligned} z_8 &= x_2^2, & z_{13} &= x_1^2 x_5,\\ z_9 &= x_2 x_3, & z_{14} &= x_1^3 x_5,\\ z_{10} &= x_2 x_6, & z_{15} &= x_1 x_2^2,\\ z_{11} &= x_5 & z_{16} &= x_2 z_6,\\ z_{12} &= x_1 x_5, & z_{17} &= x_2 z_7. \end{aligned}$$

We will denote the embedding of $\mathbf{F_c}$ into $\mathbf{P}^{17}$, given by these functions by $\psi_{\mathbf{c}}$. It leads to an embedding $\psi'_{\mathbf{c}}$ of the abstract Painlevé divisor $\Gamma'_{\mathbf{c}}$ into $\mathbf{P}^{17}$ by taking the coefficients in t^{-2} of $z_0(t;\Gamma'_{\mathbf{c}}), \dots, z_{17}(t;\Gamma'_{\mathbf{c}})$, to wit

$$\begin{aligned} (a,b) \mapsto (0:\dots:0:a^2:a:ab:1+a^2:(1+a^2)b:(1+a^2)b^2:\\ (1+a^2)b^3:a^2b:a(1+a^2)b^2:a(2a^2b^3-Z'(b)/2)). \end{aligned} \tag{10.44}$$

Similarly, we get an embedding $\psi''_{\mathbf{c}}$ of $\Gamma''_{\mathbf{c}}$ into $\mathbf{P}^{17}$ by taking the residues of $z_0(t;\Gamma''_{\mathbf{c}}), \dots, z_{17}(t;\Gamma''_{\mathbf{c}})$, to wit

$$\begin{aligned} (a,b) \mapsto (0:a:0:-1:-b:ab:-(1+a^2)b^2:2a^2b^3-Z'(b)/2:0:-b:-b^2\\ :0:0:0:-a^3(a^2b^4-Z(b)):ab^2:-(1+a^2)b^3:b(2a^2b^3-Z'(b)/2)). \end{aligned}$$

It is checked as before that the closure of each of these embeddings yields two non-singular curves, which we will denote again by $\mathcal{D}'_{\mathbf{c}}$ and $\mathcal{D}''_{\mathbf{c}}$. Again, they intersect transversally in 4 distinct points, namely in the points

$$\lim_{p\to\infty_{\epsilon_2\epsilon_3}} \psi'_{\mathbf{c}}(p) = \lim_{p\to\infty_{-\epsilon_2,-\epsilon_3}} \psi''_{\mathbf{c}}(p).$$

This corresponds to the following points[2] in projective space:

$$Q'_{\epsilon_2\epsilon_3} := \left(0:\cdots:0:1:0:0:0:\epsilon_2\sqrt{-1}\,\frac{\epsilon_3\delta - c_1}{2}:\epsilon_2\sqrt{-1}:0:-\epsilon_3\delta\right). \tag{10.45}$$

The other points are all distinct and are given[3] by

$$P'_{\epsilon_1} := \lim_{p\to\infty_{\epsilon_1}} \psi'_{\mathbf{c}}(p) = (0:\cdots:0:1:0:0:1:0:0:0:0:\epsilon_1\sqrt{c_3}:0),$$
$$P''_{\epsilon_1} := \lim_{p\to\infty_{\epsilon_1}} \psi''_{\mathbf{c}}(p) = (0:1:0:0:0:0:-\epsilon_1\sqrt{c_3}:0:\cdots:0:\star:0:0:0),$$

We now show that $(\psi_{\mathbf{c}})_*\mathcal{V}_1$ extends to a holomorphic vector field $\overline{\mathcal{V}}_1$ on $\mathbf{P}^{17}$ by writing down quadratic differential equations for this vector field in the charts $Z_0 \neq 0$ and $Z_1 \neq 0$. For the first chart, we have

$$\begin{aligned}
\dot z_1 &= z_1z_3 - z_4, & \dot z_9 &= (z_1+z_2)(z_8-z_{11}) + z_3(2z_4-z_9) - c_2,\\
\dot z_2 &= z_4 - z_2z_3, & \dot z_{10} &= z_4^2 - z_5z_{11} - z_2(c_2 - 2z_{15} + z_{12}),\\
\dot z_3 &= z_3^2 + 2(z_5+z_8-z_{11}) - c_1, & \dot z_{11} &= -2z_3z_{11},\\
\dot z_4 &= z_3z_4 + 2(z_{15}-z_{12}) - c_2, & \dot z_{12} &= z_7 - 2z_3z_{12},\\
\dot z_5 &= z_4(z_1-z_2), & \dot z_{13} &= -2z_4z_{12},\\
\dot z_6 &= z_4^2 + z_5^2 - 2z_{13} - c_3 + c_4, & \dot z_{14} &= z_3z_{14} - 3z_4z_{13},\\
\dot z_7 &= 2z_5z_{12} - (c_1z_1 - c_2)z_{11}, & \dot z_{15} &= (2z_1 - z_2)z_{10} - z_3z_{15},\\
\dot z_8 &= 2(z_2z_4 - z_3z_8), & \dot z_{16} &= z_4z_6 - c_2(z_5+z_8) + z_5(c_1z_2 - 2z_{12}),\\
& & \dot z_{17} &= z_7(z_4 - 2z_9) - z_2z_{14} + z_{12}z_{15} + c_4(z_5 - z_8).
\end{aligned}$$

In the chart $Z_1 \neq 0$ we define $y_i := z_i/z_1$ for $i = 0,\dots,17$. Then the vector field $(\psi_{\mathbf{c}})_*\mathcal{V}_1$ can be written as follows.

$$\begin{aligned}
\dot y_0 &= y_0y_4 - y_3,\\
\dot y_1 &= 0,\\
\dot y_2 &= y_0y_6 - y_9,\\
\dot y_3 &= -2y_{11} + y_3y_4 + 2y_5(y_2+1) - c_1y_0,\\
\dot y_4 &= y_4^2 - 2(y_0y_{13} - y_{15}) - c_2y_0,\\
\dot y_5 &= y_6 - y_{10},\\
\dot y_6 &= y_4y_6 - 2y_{14}y_0 + c_1y_5 - c_2(y_0y_5 + 1),\\
\dot y_7 &= y_{14}(1+4y_2) - 2c_1y_0y_{13} + (c_2 + c_3y_0 - c_4y_0)y_{11} - c_4(1 - y_2^2) + 2y_4y_7,\\
\dot y_8 &= 3y_0y_{16} - y_4(y_5 + 2y_8),\\
\dot y_9 &= 2y_{15}(1+y_2) + y_3y_6 - y_5(c_1y_0 + 2y_{11}),\\
\dot y_{10} &= y_5y_{15} - y_0(y_{14} + c_3 - c_4) - y_2y_{13} - y_4(y_{10} - 2y_6),\\
\dot y_{11} &= -2y_3y_{12} - y_0y_7,\\
\dot y_{12} &= -2(y_4y_{12} + y_7),
\end{aligned}$$

---

[2] We use the same notation for these image point in $\mathbf{P}^{17}$ as their images in $\mathbf{P}^7$, but this should not cause any confusion.

[3] The (complicated) value of $\star$ is irrelevant.

$$\begin{aligned}
\dot{y}_{13} &= -y_3y_{14} - y_4y_{13}, \\
\dot{y}_{14} &= -2y_4y_{14}, \\
\dot{y}_{15} &= 2(y_{16} - y_4y_{15}), \\
\dot{y}_{16} &= y_6^2 + c_1y_5^2 - 2y_2y_{14} - c_2(y_5 + y_8), \\
\dot{y}_{17} &= y_{14}y_5 - 3y_7(y_{10} - y_6) - c_2(y_8(y_8 - y_{11} + y_5 - c_1y_0) + y_9(y_9 - y_4) \\
&\qquad +y_0(c_2y_2 + 2y_{13})) - (c_3 - c_4)y_{11}(y_2 - 2) - c_4(y_2y_8 - 2y_2y_5 + y_5).
\end{aligned}$$

We now show that the flow of $\overline{\mathcal{V}}_1$, starting from the 8 points P'_{ϵ_1}, P''_{ϵ_1} and $Q'_{\epsilon_2\epsilon_3}$ goes immediately into the affine. For the four intersection points $Q'_{\epsilon_2\epsilon_3}$ of $\mathcal{D}'_{\mathbf{c}}$ and $\mathcal{D}''_{\mathbf{c}}$ we read off from (10.44) and Table 10.3 that the leading coefficient of $z_{10}(t;\Gamma'_{\mathbf{c}})$ has a pole for $\varsigma = 0$ that is maximal among the leading coefficients of $z_8(t;\Gamma'_{\mathbf{c}}),\dots,z_{17}(t;\Gamma'_{\mathbf{c}})$, and thus, z_{10} defines the chart about these points. With some effort, one computes that

$$\lim_{P\to\infty_{\epsilon_2\epsilon_3}} \frac{1}{z_{10}(t;\Gamma'_{\mathbf{c}})} = t^3 + O(t^4),$$

which shows that the flow of $\overline{\mathcal{V}}_1$, that starts from the points $Q'_{\epsilon_2\epsilon_3}$, goes into the affine immediately. For the other four points one arrives at the same conclusion by checking that the following limits are different from zero.

$$\begin{aligned}
&\lim_{P\to\infty_{\epsilon_1}} \frac{1}{z_8(t;\Gamma'_{\mathbf{c}})} = \frac{c_3}{9}t^6 + O(t^7), \\
&\lim_{P\to\infty_{\epsilon_1}} \frac{1}{z_1(t;\Gamma''_{\mathbf{c}})} = \lim_{P\to\infty_{\epsilon_1}} \frac{1}{z_2(t;\Gamma'_{\mathbf{c}})} = \frac{\epsilon_1\sqrt{c_3}}{3}t^3 + O(t^4),
\end{aligned}$$

Finally, we need to check that $\overline{\psi_{\mathbf{c}}(\mathbf{F_c})} \setminus \psi_{\mathbf{c}}(\mathbf{F_c})$ contains no other irreducible components than $\mathcal{D}'_{\mathbf{c}}$ and $\mathcal{D}''_{\mathbf{c}}$. For the intersection points (10.45) we have in terms of a local parameter ς in a neighborhood of $\infty_{\epsilon_2,\epsilon_3} \in \overline{\Gamma'_{\mathbf{c}}}$ that

$$\frac{1}{z_{10}(t;\Gamma'_{\mathbf{c}})} = \sqrt{-1}\epsilon_2\varsigma t^2 + O(t^3),$$

so that $\overline{\psi_{\mathbf{c}}(\mathbf{F_c})} \setminus \psi(\mathbf{F_c})$ has degree three at each of their four image points. Since z_{10} has a double pole on $\mathcal{D}'_{\mathbf{c}}$ and a simple pole on $\mathcal{D}''_{\mathbf{c}}$ there are no other divisors passing through these points, i.e., condition (7.57) is fulfilled. At the points P'_{ϵ_1} and P''_{ϵ_1} one uses the vector field $\mathcal{V}_2$, since $\mathcal{V}_1$ is tangent to the divisor at those points, using the function z_7 to define the chart at the points $P'_{\epsilon_1} \in \mathcal{D}'_{\mathbf{c}}$ and z_1 for the points $P''_{\epsilon_1} \in \mathcal{D}''_{\mathbf{c}}$. For a full proof one checks first that $\mathcal{V}_2$ also extends to a holomorphic vector field on $\mathbf{P}^{17}$ which can be done, as above, by writing down explicitly the quadratic differential equations for this vector field (in two affine charts). This leads to the following theorem that states the algebraic complete integrability of the Kowalevski top, and describes some of its algebraic geometric features.

Theorem 10.4. *Let* $(\mathbf{C}^6, \{\cdot,\cdot\}, \mathbf{F})$ *denote the integrable system that describes the Kowalevski top, where* $\mathbf{F}$ *and* $\{\cdot,\cdot\}$ *are given by (10.35) and (10.36), with commuting vector fields (10.37). The weights of the phase variables are given by* $\varpi(x_1, x_2, x_3, x_4, x_5, x_6) = (1, 1, 1, 2, 2, 2)$.

(1) $(\mathbf{C}^6, \{\cdot,\cdot\}, \mathbf{F})$ *is a weight homogeneous a.c.i. system;*
(2) For generic $\mathbf{c}$ *the fiber* $\mathbf{F_c}$ *of its momentum map completes into an Abelian surface* $\mathbf{T}^2_{\mathbf{c}}$ *by adding a singular divisor* $\mathcal{D}_{\mathbf{c}}$*;*
(3) $\mathcal{D}_{\mathbf{c}}$ *consists of two irreducible components* $\mathcal{D}'_{\mathbf{c}}$ *and* $\mathcal{D}''_{\mathbf{c}}$*, which are both non-singular curves of genus three;*
(4) $\mathcal{D}'_{\mathbf{c}}$ *and* $\mathcal{D}''_{\mathbf{c}}$ *are translates of each other over a half period in* $\mathbf{T}^2_{\mathbf{c}}$ *and they intersect each other transversally in 4 points;*
(5) The line bundle $[\mathcal{D}_{\mathbf{c}}]$ *defines a polarization of type* $(2,4)$ *on* $\mathbf{T}^2_{\mathbf{c}}$ *and leads to an embedding of* $\mathbf{T}^2_{\mathbf{c}}$ *in* $\mathbf{P}^7$*; it is not normally generated, but the line bundle* $[2\mathcal{D}'_{\mathbf{c}} + \mathcal{D}''_{\mathbf{c}}]$ *is;*
(6) $\mathbf{T}^2_{\mathbf{c}}$ *is dual to* $\operatorname{Prym}(\mathcal{D}'_{\mathbf{c}}/C_{\mathbf{c}})$*, where* $C_{\mathbf{c}}$ *is the elliptic curve, given by (10.42).*

Proof. We have verified the conditions of the complex Liouville Theorem, which allows us to conclude that we have a weight homogeneous a.c.i. system, and that the tori $\mathbf{T}^2_{\mathbf{c}}$ that compactify the generic fibers $\mathbf{F_c}$ are obtained by adding the images

$$\mathcal{D}'_{\mathbf{c}} = \overline{\varphi'_{\mathbf{c}}(\Gamma'_{\mathbf{c}})} \qquad \text{and} \qquad \mathcal{D}''_{\mathbf{c}} = \overline{\varphi''_{\mathbf{c}}(\Gamma''_{\mathbf{c}})},$$

for which we have already obtained the intersection pattern. Since each of these curves have genus 3 they induce on $\mathbf{T}^2_{\mathbf{c}}$ a polarization of type $(1,2)$, and together they provide a very ample line bundle, as follows from Ramanan's Theorem (Theorem 5.18). Also Example 5.32 implies that $\mathbf{T}^2_{\mathbf{c}}$ is dual to $\operatorname{Prym}(\mathcal{D}'_{\mathbf{c}}/C'_{\mathbf{c}})$, where $C'_{\mathbf{c}}$ will be shown later to be equal to $C_{\mathbf{c}}$. We can identify σ as a translation over a half period on $\mathbf{T}^2_{\mathbf{c}}$, because it is an involution of $\mathbf{T}^2_{\mathbf{c}}$ that leaves the vector fields $\overline{\mathcal{V}}_1$ and $\overline{\mathcal{V}}_2$ invariant. On the other hand, τ yields the -1 involution on $\mathbf{T}^2_{\mathbf{c}}$ because it is an involution that flips the sign of both vector fields. Since τ acts on the embedding variables as follows,

$$\tau(z_0 : \cdots : z_7) = (z_0 : z_1 : z_2 : -z_3 : -z_4 : z_5 : -z_6 : -z_7)$$

it acts on the parameters that appear in the principal balances according to

$$\tau(t, a, b) = (-t, -a, b),$$

so that the half periods of $\mathbf{T}^2_{\mathbf{c}}$ that lie on $\mathcal{D}_{\mathbf{c}}$ are given by the 12 points of $\mathcal{D}_{\mathbf{c}}$ corresponding to $a = 0$ and $a = \infty$; the former 8 points are the points on $\mathcal{D}_{\mathbf{c}}$ where $\mathcal{D}'_{\mathbf{c}}$ and $\mathcal{D}''_{\mathbf{c}}$ are branched over $\mathcal{E}_{\mathbf{c}}$, the points $(0, b_i)$ where b_i is any root of $c_4 - Z(b)$; the points corresponding to $a = \infty$ are the points P'_{ϵ_1} and P''_{ϵ_1}, where $\epsilon_1^2 = 1$. □

We now turn to the holomorphic differentials on the divisor $\mathcal{D}_{\mathbf{c}}$. In the notation of Paragraph 7.6.7 (see especially Example 7.54) we choose $y_0 := x_3$ and $y := x_2$ so that

$$\begin{aligned} y_0^{(0)} &= 1, & y_0^{(1)} &= -ab, \\ y^{(0)} &= a & y^{(1)} &= b(a^2+1) \end{aligned}$$

and, using (10.37) and (10.40) compute

$$\mathcal{V}_2\left[\frac{1}{y_0}\right]_{|\mathcal{D}'_{\mathbf{c}}} = -b^2(a^2+1), \qquad \mathcal{V}_2\left[\frac{y}{y_0}\right]_{|\mathcal{D}'_{\mathbf{c}}} = -(1+a^2)(b^3+Z'(b)/2).$$

It follows that

$$\begin{aligned} \delta &= \begin{vmatrix} 1 & -b^2(a^2+1) \\ b(2a^2+1) & -(1+a^2)(b^3+Z'(b)/2) \end{vmatrix} \\ &= (a^2+1)\left(2a^2b^3 - Z'(b)/2\right). \end{aligned}$$

The holomorphic differentials $\mathrm{d}t_1$ and $\mathrm{d}t_2$, restricted to $\mathcal{D}'_{\mathbf{c}}$, are therefore given by

$$\begin{aligned} \omega_1 &= \frac{\mathrm{d}a}{(a^2+1)(2a^2b^3 - Z'(b)/2)}, \\ \omega_2 &= \frac{b^2\mathrm{d}a}{2a^2b^3 - Z'(b)/2}. \end{aligned}$$

If we write ω_1 in terms of a local parameter ς on a neighborhood of the two points ∞_{ϵ_1} by using Table 10.3 then we find

$$\omega_1 = (-\epsilon_1 c_3^{-3/2}\varsigma^2 + O(\varsigma^3))\mathrm{d}\varsigma,$$

which shows that ω_1 has a double zero at the two half periods P'_{ϵ_1}, so that $\mathcal{V}_1$ is doubly tangent to $\mathcal{D}'_{\mathbf{c}}$ at these two points. Since the degree of the canonical bundle of a Riemann surface of genus 3 equals $2(3-1)=4$ this accounts for all points of tangency of $\mathcal{V}_1$ along the smooth curve $\mathcal{D}'_{\mathbf{c}}$. Using the involution σ, which preserves $\mathcal{V}_1$, it follows that $\mathcal{V}_1$ is also doubly tangent to $\mathcal{D}''_{\mathbf{c}}$ at the points P''_{ϵ_1}. Finally, notice also that both ω_1 and ω_2 are odd with respect to the involution τ, as $\tau(a,b) = (-a,b)$. Since $\mathcal{D}'_{\mathbf{c}}/\tau = C_{\mathbf{c}}$ this shows that $\mathbf{T}^2_{\mathbf{c}}$ is dual to $\mathrm{Prym}(\mathcal{D}'_{\mathbf{c}}/C_{\mathbf{c}})$, as announced in the proof of Theorem 10.4.

10.4 The Goryachev-Chaplygin Top

10.4.1 Liouville Integrability and Lax Equation

In this paragraph we discuss another integrable top, that was first introduced by Goryachev in 1900 (see [66]) and that was first integrated by Chaplygin (see [41]). The physical characteristics of the top are that the principal moments of inertia satisfy $I_1 = I_2 = 4I_4$ and that the center of mass belongs to the equatorial plane (corresponding to the moments I_1 and I_2) through the fixed point. This top is special in (at least) two different ways. First, it is integrable on the Poisson submanifold $K_2 = 0$ only; recall from (10.7) that K_2 is the component of the angular momentum in the direction of gravity. Second, it is not a.c.i., although it is closely related to an a.c.i. system; this will be explained in Paragraph 10.4.3.

By a simple rescaling we may assume that $(\lambda_1, \lambda_2, \lambda_3) = (1, 1, 4)$ and that $(m_1, m_2, m_3) = (-4, 0, 0)$, so that the equations of motion (10.5) take the simple form

$$\begin{aligned} \dot{\gamma}_1 &= 4a_3\gamma_2 - a_2\gamma_3, & \dot{a}_1 &= 3a_2a_3, \\ \dot{\gamma}_2 &= a_1\gamma_3 - 4a_3\gamma_1, & \dot{a}_2 &= -4\gamma_3 - 3a_3a_1, \\ \dot{\gamma}_3 &= a_2\gamma_1 - a_1\gamma_2, & \dot{a}_3 &= 4\gamma_2. \end{aligned} \tag{10.46}$$

With these scalings the energy of the top is given by

$$H = \frac{1}{2}(a_1^2 + a_2^2 + 4a_3^2) - 4\gamma_1.$$

As we said we consider the Poisson subspace M of $(\mathbf{R}^6, \{\cdot\,,\cdot\})$, defined by

$$K_2 := a_1\gamma_1 + a_2\gamma_2 + a_3\gamma_3 = 0, \tag{10.47}$$

which is smooth away from the origin ($\{\cdot\,,\cdot\}$ still denotes the $\mathfrak{e}(3)$-Lie-Poisson structure, whose Poisson matrix is given by (10.6)). On M the algebra of Casimirs is generated by

$$K_1 := \gamma_1^2 + \gamma_2^2 + \gamma_3^2.$$

The extra invariant, which was found by Goryachev, is given by

$$K_3 := (a_1^2 + a_2^2)a_3 + 4a_1\gamma_3.$$

Notice that it K_3 is only in involution with H upon using $K_2 = 0$, since $\{K_3, H\} = 4a_2K_2$. The Casimir K_1 and the integrals K_3 and H are independent on a dense open subset of M so that $(M, \{\cdot\,,\cdot\}, (K_1, K_3, H))$ is a Liouville integrable system.

A natural Lax equation, with spectral parameter, for the Goryachev-Chaplygin top was obtained by Bobenko and Kuznetsov (see [31]). Their recipe, which still remains a mystery is this: remove from the Lax pair of the Kowalevski top the first row and first column and you get a Lax pair for the Goryachev-Chaplygin top, except let $t \to 2t$ and add $\sqrt{-1}a_3$ to the $(1,1)$ entry of the generator of the flow, which we call $M(\mathfrak{h})$ instead of $\sqrt{-1}Y(\mathfrak{h})$. Precisely, by the above procedure we get

$$L(\mathfrak{h}) := \begin{pmatrix} 0 & \gamma_3 & \frac{a_1+\sqrt{-1}a_2}{2}\mathfrak{h} \\ -\gamma_3 & -a_3\mathfrak{h} & \gamma_1+\sqrt{-1}\gamma_2-\mathfrak{h}^2 \\ \frac{a_1-\sqrt{-1}a_2}{2}\mathfrak{h} & -\gamma_1+\sqrt{-1}\gamma_2+\mathfrak{h}^2 & a_3\mathfrak{h} \end{pmatrix}$$

and

$$M(\mathfrak{h}) := \begin{pmatrix} 3\sqrt{-1}a_3 & 0 & a_2-\sqrt{-1}a_1 \\ 0 & 2\sqrt{-1}a_3 & 2\sqrt{-1}\mathfrak{h} \\ -(a_2+\sqrt{-1}a_1) & -2\sqrt{-1}\mathfrak{h} & -2\sqrt{-1}a_3 \end{pmatrix}.$$

The Lax equation $\dot{L}(\mathfrak{h}) = [L(\mathfrak{h}), M(\mathfrak{h})]$ leads then to the following vector field (still using that $K_2 = 0$),

$$\begin{aligned} \dot\gamma_1 &= 4a_3\gamma_2 - a_2\gamma_3, & \dot a_1 &= 3a_2a_3, \\ \dot\gamma_2 &= a_1\gamma_3 - 4a_3\gamma_1, & \dot a_2 &= 4\gamma_3 - 3a_3a_1, \\ \dot\gamma_3 &= a_2\gamma_1 - a_1\gamma_2, & \dot a_3 &= -4\gamma_2. \end{aligned} \tag{10.48}$$

which is easily seen to be equivalent to (10.46), upon rescaling the variables (multiply all variables, including time, by -1). The Linearization Criterion is verified in the same way as in the case of the Kowalevski top. Namely, the spectral curve is given by

$$\mathcal{C}_{\mathbf{c}} : \mu^3 + \mu(\mathfrak{h}^4 - \frac{c_2}{2}\mathfrak{h}^2 + c_1) + \frac{c_3}{4}\mathfrak{h}^3 = 0,$$

where (c_1, c_2, c_3) are the values of (K_1, H, K_3). We have that

$$M(\mathfrak{h}) = -2\sqrt{-1}\frac{L(\mathfrak{h})}{\mathfrak{h}} + \operatorname{diag}(3\sqrt{-1}a_3, 0, 0) + \frac{C}{\mathfrak{h}},$$

so that it suffices to verify that $\xi/\mathfrak{h}$ has no pole at the three points at infinity of the spectral curve, where $\xi = (1, \xi_2, \xi_3)^\top$ is the normalized null-vector of $\mathfrak{z}\,\mathrm{Id}_3 - L(\mathfrak{h})$, for $(\mathfrak{z}, \mathfrak{h})$ belonging to the spectral curve $\mathcal{C}_{\mathbf{c}}$,

$$\xi_2 = \frac{(-a_1+\sqrt{-1}a_2)\mathfrak{h}^3 + (\sqrt{-1}a_1\gamma_2 - \sqrt{-1}a_2\gamma_1 + a_3\gamma_3)\mathfrak{h} - 2\gamma_3\mu}{2\left(\mu^2 + \mathfrak{h}^4 - (a_3^2 + 2\gamma_1)\mathfrak{h}^2 + \gamma_1^2 + \gamma_2^2\right)},$$

$$\xi_3 = \frac{((a_1 - \sqrt{-1}a_2)a_3 - 2\gamma_3)\mathfrak{h}^2 + (a_1 - \sqrt{-1}a_2)\mu\mathfrak{h} + 2\gamma_3(\gamma_1 - \sqrt{-1}\gamma_2)}{2\left(\mu^2 + \mathfrak{h}^4 - (a_3^2 + 2\gamma_1)\mathfrak{h}^2 + \gamma_1^2 + \gamma_2^2\right)}.$$

For the first two points at infinity we have that

$$\mathfrak{h} = \frac{1}{\varsigma}, \qquad \mu = \pm\frac{\sqrt{-1}}{\varsigma^2}(1 - \frac{c_2}{4}\varsigma^2 + O(\varsigma^3)).$$

Direct substitution in ξ_2 and ξ_3 yields

$$\xi_2 = \frac{a_1 - \sqrt{-1}a_2}{4\gamma_1 + 2a_3^2 - c_2}\frac{1}{\varsigma}, \qquad \xi_3 = \mp\frac{a_2 + \sqrt{-1}a_1}{4\gamma_1 + 2a_3^2 - c_2}\frac{1}{\varsigma},$$

so that $\xi_2/\mathfrak{h}$ and $\xi_3/\mathfrak{h}$ are finite at these points. Similarly, the third point at infinity is given by

$$\mathfrak{h} = \frac{1}{\varsigma}, \qquad \mu = -\frac{c_3}{4}\varsigma - \frac{1}{8}c_2c_3\varsigma^3 + O(\varsigma^5),$$

which yields, by direct substitution in ξ_2 and ξ_3,

$$\xi_2 = -\frac{a_1 - \sqrt{-1}a_2}{2}\varsigma + O(\varsigma^3), \quad \xi_3 = \frac{1}{2}\left(a_3(a_1 - \sqrt{-1}a_2) - 2\gamma_3\right)\varsigma^2 + O(\varsigma^4),$$

so that $\xi_2/\mathfrak{h}$ and $\xi_3/\mathfrak{h}$ are also finite at this point. Therefore, we have checked the conditions in Corollary 6.43 and we may conclude that the Lax operator $L(\mathfrak{h})$ satisfies the Linearization Criterion and hence that the Goryachev-Chaplygin vector field linearizes on the Jacobians of the spectral curves $\mathcal{C}_{\mathbf{c}}$. We will see in Paragraph 10.4.3 that, despite this fact, the Goryachev-Chaplygin top is not a.c.i.

10.4.2 The Bechlivanidis-van Moerbeke System

In this paragraph we consider an integrable system on $\mathbf{C}^7$ which was constructed by Bechlivanidis and van Moerbeke (see [30]) in order to understand the geometry of the Goryachev-Chaplygin top. We call this system the *Bechlivanidis-van Moerbeke system.* We consider on $\mathbf{C}^7$, with coordinates $x_1 \dots, x_7$ the Poisson structure defined by the following matrix,

$$\begin{pmatrix} 0 & 0 & -16x_5 & 0 & -8 & 0 & 16x_2 \\ 0 & 0 & 0 & 0 & 0 & 0 & 4 \\ 16x_5 & 0 & 0 & 2x_5 & 2x_4 & -4x_2x_5 & 4x_2x_4 \\ 0 & 0 & -2x_5 & 0 & -1 & 0 & -2x_2 \\ 8 & 0 & -2x_4 & 1 & 0 & -2x_2 & 0 \\ 0 & 0 & 4x_2x_5 & 0 & 2x_2 & 0 & -4x_2^2 - x_1 \\ -16x_2 & -4 & -4x_2x_4 & 2x_2 & 0 & 4x_2^2 + x_1 & 0 \end{pmatrix}. \tag{10.49}$$

Also, consider the following 5 functions,

$$\begin{aligned} F_1 &= x_1 - 4x_2^2 - 8x_4, \\ F_2 &= x_1x_2 + 4x_6, \\ F_3 &= x_3 + x_4^2 - x_5^2, \\ F_4 &= x_2x_3 + x_4x_6 + x_5x_7, \\ F_5 &= -x_1x_3 - x_6^2 + x_7^2. \end{aligned} \tag{10.50}$$

It is easy to see that F_1, F_2 and F_3 are Casimirs for this Poisson structure, which has rank 4 generically, and that F_4 and F_5 are in involution; they lead to the vector fields (up to a constant)

$$\begin{aligned} \dot{x}_1 &= -8x_7, & x_1' &= 8(x_1x_5 + 2x_2x_7), \\ \dot{x}_2 &= 4x_5, & x_2' &= 4x_7, \\ \dot{x}_3 &= 2(x_4x_7 + x_5x_6), & x_3' &= 4(x_2x_5x_6 + x_2x_4x_7 - 2x_3x_5), \\ \dot{x}_4 &= -4x_2x_5 - x_7, & x_4' &= x_1x_5 - 2x_2x_7, \\ \dot{x}_5 &= x_6 - 4x_2x_4, & x_5' &= x_1x_4 + 2x_2x_6 - 4x_3, \\ \dot{x}_6 &= -x_1x_5 + 2x_2x_7, & x_6' &= -x_1x_7 - 2x_1x_2x_5 - 4x_2^2x_7, \\ \dot{x}_7 &= x_1x_4 + 2x_2x_6 - 4x_3, & x_7' &= 8x_2x_3 + 2x_1x_2x_4 - 4x_2^2x_6 - x_1x_6. \end{aligned} \tag{10.51}$$

We denote these vector fields by $\mathcal{V}_1$ and $\mathcal{V}_2$. Consider also the bivector field $\{\cdot,\cdot\}'$ on $\mathbf{C}^7$, defined by the following matrix,

$$\begin{pmatrix} 0 & 0 & -16x_7 & 0 & 0 & 0 & -8x_1 \\ 0 & 0 & 0 & 0 & -4 & 0 & 0 \\ 16x_7 & 0 & 0 & 2x_7 & -2x_6 & -4x_2x_7 & 8x_3 - 4x_2x_6 \\ 0 & 0 & -2x_7 & 0 & 4x_2 & 0 & -x_1 \\ 0 & 4 & 2x_6 & -4x_2 & 0 & -x_1 & 0 \\ 0 & 0 & 4x_2x_7 & 0 & x_1 & 0 & 2x_1x_2 \\ 8x_1 & 0 & 4x_2x_6 - 8x_3 & x_1 & 0 & -2x_1x_2 & 0 \end{pmatrix}. \tag{10.52}$$

It follows by direct computation that $\{\cdot,\cdot\}'$ satisfies the Jacobi identity, as well as $\{\cdot,\cdot\} + \{\cdot,\cdot\}'$. Moreover, it is easy to see that F_1, F_2 and F_5 are Casimirs of $\{\cdot,\cdot\}'$ and that F_3 and F_4 generate (up to a constant) the vector fields $\mathcal{V}_1$ and $\mathcal{V}_2$. Thus, $\mathcal{V}_1$ and $\mathcal{V}_2$ are bi-Hamiltonian vector fields.

We now show that the Bechlivanidis-van Moerbeke system is a.c.i. We see that $\mathcal{V}_1$ becomes weight homogeneous, by assigning to the variables x_i the following weights,

$$\nu = \varpi(x_1, \ldots, x_7) = (2, 1, 4, 2, 2, 3, 3).$$

With respect to these weights the constants of motion F_i are also weight homogeneous, with weights

$$\varpi(F_1,\dots,F_5)=(2,3,4,5,6),$$

and the time involution is given by

$$\tau(x_1,\dots,x_7):=(x_1,x_2,x_3,x_4,-x_5,x_6,-x_7). \tag{10.53}$$

The indicial locus $\mathcal{I}$ is given by

$$\begin{aligned}
&2x_1^{(0)}-8x_7^{(0)}=0,\\
&x_2^{(0)}+4x_5^{(0)}=0,\\
&4x_3^{(0)}+2(x_4^{(0)}x_7^{(0)}+x_5^{(0)}x_6^{(0)})=0,\\
&2x_4^{(0)}-4x_2^{(0)}x_5^{(0)}-x_7^{(0)}=0,\\
&2x_5^{(0)}+x_6^{(0)}-4x_2^{(0)}x_4^{(0)}=0,\\
&3x_6^{(0)}-x_1^{(0)}x_5^{(0)}+2x_2^{(0)}x_7^{(0)}=0,\\
&3x_7^{(0)}+x_1^{(0)}x_4^{(0)}+2x_2^{(0)}x_6^{(0)}-4x_3^{(0)}=0,
\end{aligned}$$

and it is easy to check that it consists of the following five points,

$$\begin{aligned}
\mathcal{I}^\epsilon&:=\left(0,-\frac{\epsilon}{2},0,-\frac{1}{8},\frac{\epsilon}{8},0,0\right),\\
\mathcal{I}_1^\epsilon&:=\left(2,-\epsilon,0,-\frac{1}{4},\frac{\epsilon}{4},\frac{\epsilon}{2},\frac{1}{2}\right),\\
\mathcal{I}_2&:=\left(-4,0,-\frac{1}{4},-\frac{1}{2},0,0,-1\right),
\end{aligned}$$

where $\epsilon=\pm1$. The Kowalevski matrix at an arbitrary $\mathcal{K}(w)$ is given by

$$\mathcal{K}(w)=\begin{pmatrix}
2 & 0 & 0 & 0 & 0 & 0 & -8\\
0 & 1 & 0 & 0 & 4 & 0 & 0\\
0 & 0 & 4 & 2x_7^{(0)} & 2x_6^{(0)} & 2x_5^{(0)} & 2x_4^{(0)}\\
0 & -4x_5^{(0)} & 0 & 2 & -4x_2^{(0)} & 0 & -1\\
0 & -4x_4^{(0)} & 0 & -4x_2^{(0)} & 2 & 1 & 0\\
-x_5^{(0)} & 2x_7^{(0)} & 0 & 0 & -x_1^{(0)} & 3 & 2x_2^{(0)}\\
x_4^{(0)} & 2x_6^{(0)} & -4 & x_1^{(0)} & 0 & 2x_2^{(0)} & 3
\end{pmatrix},$$

so that we find the following expressions for the characteristic polynomial $\chi(\mu;w):=|\mu\,\mathrm{Id}_7-\mathcal{K}(w)|$ at these five points:

$$\chi(\mu;w)=\begin{cases}
(\mu+1)(\mu-1)(\mu-2)(\mu-3)^2(\mu-4)(\mu-5), & w=\mathcal{I}^\epsilon,\\
(\mu+2)(\mu+1)(\mu-2)(\mu-3)(\mu-4)(\mu-5)(\mu-6), & w=\mathcal{I}_1^\epsilon,\\
(\mu+2)(\mu+1)(\mu-2)(\mu-3)(\mu-4)(\mu-5)(\mu-6), & w=\mathcal{I}_2.
\end{cases}$$

In fact, one easily computes that

$$
\begin{aligned}
\mathrm{d}F_1(\mathcal{I}^\epsilon) &= \mathrm{d}x_1 + 4\epsilon \mathrm{d}x_2 - 8\mathrm{d}x_4,\\
\mathrm{d}F_2(\mathcal{I}^\epsilon) &= -\frac{\epsilon}{2}\mathrm{d}x_1 + 4\mathrm{d}x_6,\\
\mathrm{d}F_3(\mathcal{I}^\epsilon) &= -\frac{1}{4}\mathrm{d}x_4 - \frac{\epsilon}{4}\mathrm{d}x_5 + \mathrm{d}x_3,\\
\mathrm{d}F_4(\mathcal{I}^\epsilon) &= -\frac{1}{8}\mathrm{d}x_6 + \frac{\epsilon}{8}\mathrm{d}x_7 - \frac{\epsilon}{2}\mathrm{d}x_3,\\
\mathrm{d}F_5(\mathcal{I}^\epsilon) &= 0,
\end{aligned}
$$

which explains, in view of Theorem 7.30, the occurrence of the eigenvalues $2, 3, 4$ and 5 of $\mathcal{K}(\mathcal{I}^\epsilon)$ (recall that $\varpi(F_i) = i+1$), while one similarly computes that all $\mathrm{d}F_i(\mathcal{I}_1^\epsilon)$ and $\mathrm{d}F_i(\mathcal{I}_2)$ are different from zero ($i = 1, \dots, 5$), explaining by the same Theorem the occurrence of the eigenvalues $i+1$, with $i = 1, \dots, 5$, in $\mathcal{K}(\mathcal{I}_1^\epsilon)$ and $\mathcal{K}(\mathcal{I}_2)$. Moreover, Proposition 7.11 implies in the three cases the occurrence of the eigenvalue -1, and the sum of all eigenvalues is $\mathrm{Trace}(\mathcal{K}) = \sum \varpi(x_i) = 17$, in view of Proposition 7.12. This yields for the points $\mathcal{I}_1^\epsilon$ and $\mathcal{I}_2$ the characteristic polynomial of $\mathcal{K}$, as given above, while for the points $\mathcal{I}^\epsilon$ we could predict, using these arguments, only that the eigenvalues of $\mathcal{K}(\mathcal{I}^\epsilon)$ are $\{-1, 2, 3, 4, 5, \lambda, 4-\lambda\}$, with λ unknown.

It follows that only the points $\mathcal{I}^\epsilon$ can lead to a principal balance and that, if $\mathcal{I}^\epsilon$ leads indeed to a principal balance, then trivial free parameters b, c, e and f will appear at steps $2, 3, 4$ and 5, while effective parameters a and d will appear at steps 1 and 3. We give the first 4 terms of these balances, thereby determining the effective parameters; two other trivial parameters (e and f) will appear at steps 4 and 5.

$$
\begin{aligned}
x_1(t;\mathcal{I}^\epsilon) &= \frac{2\epsilon a}{t} - 2a^2 - \left(8d + 2\epsilon a(a^2 + c_1/3)\right)t + O(t^2),\\
x_2(t;\mathcal{I}^\epsilon) &= -\frac{\epsilon}{2t} - \frac{a}{2} + \frac{3a^2 + c_1}{6}\epsilon t + \frac{\epsilon c}{4}t^2 + O(t^3),\\
x_3(t;\mathcal{I}^\epsilon) &= \frac{1}{16t}(c - 4d) + O(1),\\
x_4(t;\mathcal{I}^\epsilon) &= -\frac{1}{8t^2}\left(1 + (a^2 + c_1/3)t^2 + t^3(8d - c) + O(t^4)\right), \qquad (10.54)\\
x_5(t;\mathcal{I}^\epsilon) &= \frac{\epsilon}{8t^2}(1 + (a^2 + c_1/3)t^2 + ct^3 + O(t^4)),\\
x_6(t;\mathcal{I}^\epsilon) &= \frac{1}{4t^2}(a + \epsilon t^2(8d - c + a\epsilon(a^2 + c_1/3)) + O(t^3)),\\
x_7(t;\mathcal{I}^\epsilon) &= \frac{1}{4t^2}\left(a\epsilon + t^2(4d + a\epsilon(a^2 + c_1/3)) + O(t^3)\right).
\end{aligned}
$$

One of the trivial parameters, b, was already replaced by the value c_1 of the constant of motion F_1: we have that $F_1(x(t;\mathcal{I}^\epsilon)) = c_1$.

If we put these principal balances into the remaining equations $F_i = c_i$, $i = 2, \dots, 5$, where $\mathbf{c} = (c_1, \dots, c_5)$ is generic, and if we eliminate[4] the remaining three trivial free parameters c, e and f, then we get the equation of an affine curve $\Gamma_{\mathbf{c}}^{\epsilon}$, to wit,

$$64d^2 = (4a^3 + ac_1 - c_2)^2 - 16(4a^2c_3 + 4ac_4 - c_5). \tag{10.55}$$

Notice that the equation of this curve is independent of ϵ, so that the two Painlevé divisors $\Gamma_{\mathbf{c}}^{+}$ and $\Gamma_{\mathbf{c}}^{-}$ are isomorphic. It is seen at once from (10.55) that $\Gamma_{\mathbf{c}}^{\epsilon}$ is hyperelliptic and has genus two. It has two points at infinity, denoted $\infty_{\mathbf{c}}^{\eta}$, where $\eta^2 = 1$. A neighborhood of these points is described in terms of a local parameter ς by

$$a = \varsigma^{-1}, \qquad d = \frac{\eta}{(2\varsigma)^3}\left(4 + c_1\varsigma^2 - c_2\varsigma^3 - 8c_3\varsigma^4 - 8c_4\varsigma^5 + 0(\varsigma^6)\right). \tag{10.56}$$

We now look for weight homogeneous polynomials which have a double pole in t at most, when any of the two principal balance is substituted in it. In the notation of Paragraph 7.7.1 this means that we wish to construct a basis of $\mathcal{Z}_\rho$ as a $\mathcal{H}$-module, where the pole vector ρ is chosen as $\rho := (2, 2)$. The following observation is useful in this case. Assume that the Bechlivanidis-van Moerbeke system defines indeed an irreducible weight homogeneous a.c.i. system and denote by $\mathcal{D}_{\mathbf{c}}^{\epsilon}$ the image of the curves $\overline{\Gamma_{\mathbf{c}}^{\epsilon}}$ by the corresponding embedding. Then each of these divisors will define a principal polarization on the smooth compactification of the generic fiber $\mathbf{F_c}$ of the momentum map, which is an Abelian surface $\mathbf{T}_{\mathbf{c}}^2$. This means that $\mathbf{T}_{\mathbf{c}}^2$ is (isomorphic to) the Jacobi surface $\operatorname{Jac}(\mathcal{D}_{\mathbf{c}}^{\epsilon})$, and, by (5.38) that a basis of $\mathcal{Z}_\rho$ consists of

$$\dim L(2(\mathcal{D}_{\mathbf{c}}^{+} + \mathcal{D}_{\mathbf{c}}^{-})) = \frac{(2\mathcal{D}_{\mathbf{c}}^{+} + 2\mathcal{D}_{\mathbf{c}}^{-})^2}{2} = \frac{(4\mathcal{D}_{\mathbf{c}}^{+})^2}{2} = 16(g(\mathcal{D}_{\mathbf{c}}) - 1) = 16$$

elements, assuming that $2\mathcal{D}_{\mathbf{c}}^{+}$ and $2\mathcal{D}_{\mathbf{c}}^{-}$ are linearly equivalent. In view of Piovan's Theorem (Theorem 5.37) we can construct such a basis by using products and Wronskians (with respect to both vector fields) of the functions in $\mathcal{Z}_{(1,1)}$; a basis for the latter space consists of $(1, x_1, x_2, x_3)$ as is read off at once from (10.54). Besides the functions $z_0 := 1$ and $z_i := x_i$, where $i = 1, \dots, 7$, we arrive at the following functions, which are listed together with their leading behavior.

$$\begin{aligned} z_8 &:= z_2 z_3 \sim \frac{\epsilon(4d - c)}{32t^2}, \\ z_9 &:= z_1 z_3 \sim -\frac{a(4d - c)\epsilon}{8t^2}, \\ z_{10} &:= z_1 z_4 - 2z_2 z_6 \sim \frac{a^2}{2t^2}, \end{aligned}$$

[4] In the sequel we will only need the value of c, which is given by $c = 12d + a\epsilon(c_1 + 4a^2) - c_2\epsilon$, as follows from $F_2(x(t; \mathcal{I}^{\epsilon})) = c_2$.

$$\begin{aligned}
z_{11} &:= z_1 z_5 + 2 z_2 z_7 \sim -\frac{a^2 \epsilon}{2t^2}, \\
z_{12} &:= z_4 z_7 + z_5 z_6 \sim \frac{4d - c}{32t^2}, \\
z_{13} &:= z_2 z_{12} - 2 z_3 z_5 \sim \frac{a(4d - c)}{32t^2}, \\
z_{14} &:= z_3^2 \sim \frac{(4d - c)^2}{256t^2}, \\
z_{15} &:= z_1 z_{12} + 4 z_3 z_7 \sim -\frac{a^2(4d - c)}{4t^2}.
\end{aligned}$$

The resulting embedding of $\mathbf{F}_c$ into $\mathbf{P}^{15}$ will be denoted by $\varphi_{\mathbf{c}}$; notice that it is an isomorphic embedding since each of the phase variables $x_1, \dots, x_7$ is taken as one of the embedding variables. Taking the coefficients of t^{-2} of the balances $z_0(t; \Gamma_{\mathbf{c}}^{\epsilon}), \dots, z_{15}(t; \Gamma_{\mathbf{c}}^{\epsilon})$, we get an embedding $\varphi_{\mathbf{c}}^{\epsilon}$ of each of the Painlevé divisors $\Gamma_{\mathbf{c}}^{\epsilon}$ into $\mathbf{P}^{15}$. Explicitly,

$$\begin{aligned}
\varphi_{\mathbf{c}}^{\epsilon}: \ \Gamma_{\mathbf{c}}^{\epsilon} &\to \mathbf{P}^{15} \\
(a, d) &\mapsto (0:0:0:0:1:-\epsilon:-2a:-2a\epsilon:16\epsilon\delta:-64a\epsilon\delta:-4a^2:4a^2\epsilon: \\
&\qquad 16\delta:16a\delta:-128\delta^2:-128a^2\delta),
\end{aligned} \tag{10.57}$$

where $\delta := (c - 4d)/64$. Notice that this is indeed an embedding of the affine curves since a and d appear linearly in the map; also the two affine curves are embedded disjointly because ϵ also appears by itself in the embedding. In order to see what happens at the points at infinity of the embedded curves, write a neighborhood of the points ∞^{η} by using (10.56) and let $t \to 0$. For doing this computation the reader will find the following local expression of δ near ∞^{η} useful:

$$\delta = \frac{1}{64\varsigma^3}\left(4(\eta + \epsilon) + c_1(\eta + \epsilon)\varsigma^2 - c_3(\eta + \epsilon)\varsigma^3 - 8c_3\eta\varsigma^4 - 8c_4\eta\varsigma^5 + O(\varsigma^6)\right).$$

Notice that $\delta \sim \pm(2\varsigma)^{-3}$ when $\eta = \epsilon$, while $\delta \sim -c_3\varsigma/8$ otherwise. Taking limits we find

$$\begin{aligned}
\lim_{p \to \infty^+} \varphi_{\mathbf{c}}^{+}(p) &= \lim_{p \to \infty^-} \varphi_{\mathbf{c}}^{-}(p) = (0:\cdots:0:1:0) =: O, \\
\lim_{p \to \infty^-} \varphi_{\mathbf{c}}^{+}(p) &= (0:\cdots:1:-1:0:0:0:0) =: P^+, \\
\lim_{p \to \infty^+} \varphi_{\mathbf{c}}^{-}(p) &= (0:\cdots:1:1:0:0:0:0) =: P^-,
\end{aligned}$$

so that the two curves $\mathcal{D}_{\mathbf{c}}^{\epsilon} := \overline{\varphi_{\mathbf{c}}^{\epsilon}(\Gamma^{\epsilon})}$ have one point in common (the point O). It is actually a tangency point, as follows by computing the linear term in ς in the above limits.

Notice that this is equivalent to saying that the adjunction formula (7.50) is satisfied. For by Remark 5.43 and Table 5.1, the adjunction formula requires that

$$\begin{aligned}\dim(L(2\mathcal{D}_{\mathbf{c}}^{+}+2\mathcal{D}_{\mathbf{c}}^{-})) &= g(2(\mathcal{D}_{\mathbf{c}}^{+}+\mathcal{D}_{\mathbf{c}}^{-}))-1\\ &= 4\left(\frac{(\mathcal{D}_{\mathbf{c}}^{+})^2}{2}+\frac{(\mathcal{D}_{\mathbf{c}}^{-})^2}{2}\right)+4\mathcal{D}_{\mathbf{c}}^{+}\cdot\mathcal{D}_{\mathbf{c}}^{-}\\ &= 4(g(\mathcal{D}_{\mathbf{c}}^{+})-1+g(\mathcal{D}_{\mathbf{c}}^{-})-1)+8\\ &= 4\times 2+8=16.\end{aligned}$$

We represent the geometric configuration of these two divisors in $\mathbf{P}^{15}$ in Figure 10.3.

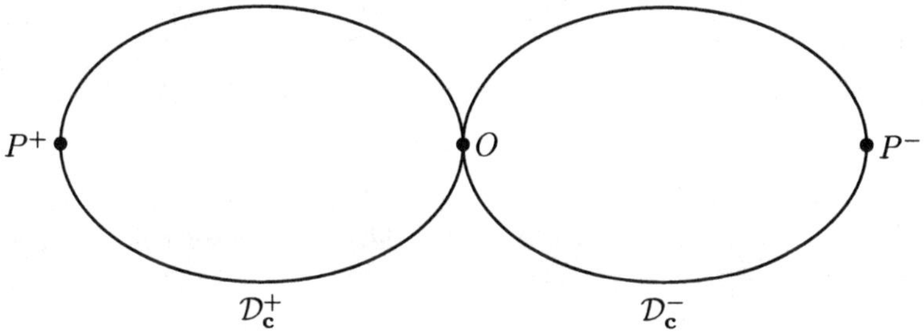

Fig. 10.3. For generic $\mathbf{c}$ the Bechlivanidis-van Moerbeke divisor $\mathcal{D}_{\mathbf{c}}$ consists of 2 isomorphic non-singular hyperelliptic curves $\mathcal{D}_{\mathbf{c}}^{+}$ and $\mathcal{D}_{\mathbf{c}}^{-}$ of genus 2. They intersect in the point O, which is a point of tangency; moreover, the vector field $\overline{\mathcal{V}}_1$ is tangent to $\mathcal{D}_{\mathbf{c}}$ at O and at the points P^+ and P^-.

We now check that the two vector fields $(\varphi_{\mathbf{c}})_*\mathcal{V}_1$ and $(\varphi_{\mathbf{c}})_*\mathcal{V}_2$ extend to holomorphic vector fields on $\mathbf{P}^{15}$, that we will denote by $\overline{\mathcal{V}}_1$ and $\overline{\mathcal{V}}_2$. To show this we establish that these vector fields can be written as quadratic vector field in the two charts $Z_0 \neq 0$ and $Z_2 \neq 0$ (see Lemma 7.58). For the vector field $(\varphi_{\mathbf{c}})_*\mathcal{V}_1$ this is almost obvious in the chart $Z_0 \neq 0$ because $\mathcal{V}_1$ is itself quadratic (at most), and because the functions $z_8, \dots, z_{15}$ were constructed by Piovan's Theorem. We only give the three quadratic equations that are a bit harder to find:

$$\begin{aligned}\dot{z}_{12} &= (z_{10}-4z_3)z_4 - z_5 z_{11} - z_9 - c_5,\\ \dot{z}_{13} &= -2z_3 z_6 - \frac{3}{2}z_1 z_8 - \frac{1}{2}(c_1 z_3 - 2c_3 z_1 + 4c_4 z_2 + 2c_5)z_2,\\ \dot{z}_{15} &= -2z_1 z_9 - 16 z_3^2 + 2c_2 z_8 + (c_3 z_1 - 2c_4 z_2 - c_5)z_1.\end{aligned}$$

In the same chart we can write $(\varphi_{\mathbf{c}})_*\mathcal{V}_2$ as

$$
\begin{aligned}
z_1' &= 8z_{11}, & z_6' &= -z_1z_7 - 2z_2z_{11},\\
z_2' &= 4z_7, & z_7' &= 8z_8 - z_1z_6 + 2z_2z_{10},\\
z_3' &= 4z_{13}, & z_8' &= 4(z_3z_7 + z_2z_{13}),\\
z_4' &= z_{11} - 4z_2z_7, & z_9' &= 4z_2z_{15},\\
z_5' &= 4(z_2z_6 - z_3) + z_{10}, & z_{10}' &= z_{11}(2z_1 - c_1) - 2c_2z_7,
\end{aligned} \tag{10.58}
$$

$$
\begin{aligned}
z_{11}' &= z_1z_{10} - 4z_9 + 8z_7^2 + 8z_5z_{11} + 2z_2(8z_8 + c_1z_6 + 2c_2z_4),\\
z_{12}' &= -2z_1z_8 - c_1z_2z_3 - c_2z_3 + 2c_3z_1z_2 - 4c_4z_2^2 - 2c_5z_2,\\
z_{13}' &= 4(z_7z_{12} - z_6z_8 + 6z_{14}) + (c_1 - z_1)(c_1z_3/4 + c_4z_2) + c_2z_2(2c_3 - 3z_3)\\
&\qquad -16c_3z_3 - 2c_5z_2^2,\\
z_{14}' &= 8z_3z_{13},\\
z_{15}' &= 16(z_{11}z_{12} + z_8z_{10} + 4z_2z_{14}) - c_2z_3(c_1 + 3z_1) + 2c_2c_3z_1\\
&\qquad -4c_4(2z_{10} + 8z_3 + c_2z_2) - 2c_5z_1z_2,
\end{aligned}
$$

and so in the chart $Z_0 \neq 0$ both vector fields are quadratic. For the chart $Z_2 \neq 0$, define $y_i := z_i/z_2$ for $i = 0, \ldots, 15$. Then the vector field $(\varphi_{\mathbf{c}})_*\mathcal{V}_1$ can in this chart be written as follows,

$$
\begin{aligned}
\dot y_0 &= -4y_0y_5, & \dot y_5 &= 8y_4^2 - 4y_5^2 + y_6 - (y_1 - c_1y_0)y_4,\\
\dot y_1 &= -4y_0y_{11}, & \dot y_6 &= y_{11} + 4y_5y_6 - 2c_2y_0y_5,\\
\dot y_2 &= 0, & \dot y_7 &= -y_{10} - 4c_4y_0^2 + 2(c_2y_0 - 2y_6)y_4,\\
\dot y_3 &= 2y_0y_{13}, & \dot y_8 &= 2(y_{13} + 2y_5y_8),\\
\dot y_4 &= 4y_4y_5 + y_7 - (y_{11} - c_1y_5)y_0, & \dot y_9 &= 2(y_1y_{13} - 4y_7y_8),
\end{aligned}
$$

$$
\begin{aligned}
\dot y_{10} &= (4y_6 - c_2y_0)y_7 + \tfrac{y_{11}}{2}(c_1y_0 - y_1)\\
\dot y_{11} &= 2c_2(y_0y_6 - y_4) - (c_1 + 8y_6)y_6 - 8y_8 - 4y_5y_{11},\\
\dot y_{12} &= -\tfrac{1}{2}y_6y_{10} - c_2y_5^2 + \tfrac{y_4}{2}(2c_2y_4 - 8y_8 + c_1y_6) + \tfrac{y_7}{2}(c_1y_5 - y_{11}),\\
\dot y_{13} &= 4(y_6y_8 - y_5y_{13}) - \tfrac{1}{2}c_1y_8 - \tfrac{3}{2}c_2y_3 - c_3(4y_6 - c_2y_0)\\
&\qquad -\tfrac{c_4}{2}(y_1 - c_1y_0 - 8y_4) - c_5,\\
\dot y_{14} &= 4(y_8y_{12} - y_5y_{14}),\\
\dot y_{15} &= -8y_8y_{10} + 4c_1y_3y_6 + c_2((12y_3 + y_{10})y_4 - y_1y_3 - y_5y_{11})\\
&\qquad -4c_4(y_{10} + 4y_3)y_0 - c_5y_1,
\end{aligned}
$$

while the vector field $(\varphi_{\mathbf{c}})_*\mathcal{V}_2$ can in this chart be written as follows,

$$
\begin{aligned}
y_0' &= -4y_0y_7,\\
y_1' &= 4(2y_{11} - y_1y_7),\\
y_2' &= 0,\\
y_3' &= 4(y_{13} - y_3y_7),\\
y_4' &= -y_{11} - 4(y_0y_{12} + y_5y_6) + 2c_2y_0y_5,\\
y_5' &= -y_{10} - 4(y_4y_6 + c_4y_0^2) + 2c_2y_0y_4,\\
y_6' &= \tfrac{1}{2}(8y_4 - y_1 + c_1y_0)y_{11} - c_2y_0y_7,\\
y_7' &= 8(y_8 + y_6^2) - 4y_7^2 + c_1y_6 + 2c_2(y_4 - y_0y_6),\\
y_8' &= y_{13}(y_1 - 8y_4 - c_1y_0),\\
y_9' &= -16(2y_3y_{12} + y_6y_{13}) - 4(c_1y_3y_7 - c_2y_{12}),\\
y_{10}' &= -4y_7y_{10} - (c_1 + 8y_6)y_{11} + 2c_2(y_0y_{11} - y_7),\\
y_{11}' &= 8y_1y_8 - (4y_6 + c_1)y_{10} + c_2(y_1y_4 + y_0y_{10}) - 8c_3y_1 - 4c_4(8y_4 + c_1y_0)y_0,\\
y_{12}' &= 16y_3^2 + 8y_5y_{13} - \tfrac{1}{4}y_1y_9 + c_1(c_4 + \tfrac{1}{4}c_1y_3)y_0 + c_2(2c_3y_0 - c_4y_0^2 - 2y_3)\\
&\quad -16c_3y_0y_3 - 2c_5(1 + 2y_0y_4),\\
y_{13}' &= 4(2y_{14} + y_7y_{13}) + \tfrac{c_1}{4}(4y_6 + c_1)y_3 + \tfrac{c_2}{4}(y_4y_{10} - y_5y_{11} - 10y_8 - 8y_3y_4)\\
&\quad -16c_3y_3 + \tfrac{c_2c_3}{4}(8 - y_0y_1) + c_4(16y_0y_3 + c_1) + \tfrac{1}{2}(c_1c_5 - c_2c_4)y_0 - \tfrac{c_5}{2}y_1,\\
y_{14}' &= 4(2y_8y_{13} - y_7y_{14}),\\
y_{15}' &= -4y_7y_{15} + 16y_{11}y_{13} + 4y_9y_{10} + 32y_1y_{14} - 4c_1(c_1y_3y_4 + 2y_6y_8 - 8y_3^2\\
&\quad -4(y_5y_{13} + c_3y_6 - c_4y_4)) + c_2(2c_3y_1 - 3y_1y_8 - 2c_5y_0 - 4c_4 - c_1y_3)\\
&\quad -32c_3y_3(3y_1 - 16y_4 - c_1) + 32y_8(8c_3 + c_4y_0) - 8(c_4y_{10} - c_5y_6),
\end{aligned}
$$

and thus, in the chart $Z_2 \neq 0$ both vector fields are quadratic.

We now show that the flow of $\overline{\mathcal{V}}_1$ that starts from the points O and $P^{\pm}$ goes into the affine immediately. Since the only non-zero coordinate of O is the one that corresponds to z_{14} we need to show that the series $1/z_{14}(t;\Gamma_{\mathbf{c}}^{+})$ has a non-zero limit at O; to do this, write the coefficients of the latter series in terms of a local parameter ς around O (by using (10.56), with $\eta = +1$) to find that

$$\lim_{\varsigma\to 0} \frac{1}{z_{14}}(t;\Gamma_{\mathbf{c}}^{+}) = 3812t^8 + O(t^9).$$

For the points $P^{\pm}$ one considers the series $z_{10}(t;\Gamma_{\mathbf{c}}^{\pm})$, writing the coefficients of the series in terms of a local parameter around $P^{\mp}$ to find, with slightly less effort that

$$\lim_{\varsigma\to 0} \frac{1}{z_{10}}(t;\Gamma_{\mathbf{c}}^{\pm}) = 2t^4 + O(t^6).$$

To conclude, we need to check that no other divisors pass through the points O and $P^{\pm}$. Since the vector field $\overline{\mathcal{V}}_1$ is tangent to each irreducible component of the Painlevé divisor at these points, as we will see, we need to check this by using the vector field $\overline{\mathcal{V}}_2$, which we checked to be holomorphic.

Notice that since z_{14} and z_{10} have a double pole along $\mathcal{D}_{\mathbf{c}}^{+}$ and define the local chart at O and $P^{\pm}$ respectively, the expected degree of the divisor at 0 is 4, while it is 2 at the points P^{+} and P^{-}. We first compute at $P^{\pm}$ the principal balances of $1/z_{10}$ with respect to $\mathcal{V}_2 = {}'$, which can be done by computing the first two terms in the holomorphic series only. Indeed, using (7.39) and (10.58) we compute the first few terms of the holomorphic series $z_{10}^{-1}(t_2;\Gamma_{\mathbf{c}}^{\pm})$ (it is holomorphic since $\mathcal{V}_2$ is holomorphic and $z_{10} \neq 0$ defines the chart):

$$\begin{aligned}\frac{1}{z_{10}}(t_2;\Gamma_{\mathbf{c}}^{\pm}) &= \frac{1}{z_{10}}(t;\Gamma_{\mathbf{c}}^{\pm})_{|t=0} + \left(\frac{1}{z_{10}}\right)'(t;\Gamma_{\mathbf{c}}^{\pm})_{|t=0} t_2 + O(t_2^2),\\ &= 8\varsigma t_2 + O(t_2^2),\end{aligned}$$

so that the degree of the divisor at infinity at $P^{\pm}$ is indeed 2. The check at the point O is simplified by the fact that the function z_{14} which defines the chart at O is a square, $z_{14} = z_3^2$. With some effort one shows that the *holomorphic* series:

$$\begin{aligned}&\frac{1}{z_3}(t_2;\Gamma_{\mathbf{c}}^{\pm})\\ &= \frac{1}{z_3}(t;\Gamma_{\mathbf{c}}^{\pm})_{|t=0} + \left(\frac{1}{z_3}\right)'(t;\Gamma_{\mathbf{c}}^{\pm})_{|t=0} t_2 + \left(\frac{1}{z_3}\right)''(t;\Gamma_{\mathbf{c}}^{\pm})_{|t=0} \frac{t_2^2}{2} + O(t_2^3),\\ &= 4\varsigma^2 t_2 + 16 t_2^2 + O(\varsigma t_2^2, t_2^3),\end{aligned}$$

so that $\frac{1}{z_3}(t_2;\Gamma_{\mathbf{c}}^{\pm})$ has multiplicity 2 at O and hence $1/z_{14}^2 = 1/z_3^2$ has multiplicity 4, as desired. Strictly speaking, about O, where the chart is defined by z_{14}, we only know that the series $z_i/z_{14}(t_2;\Gamma_{\mathbf{c}}^{\pm})$, with $0 \leqslant i \leqslant 15$ is holomorphic, but $z_3/z_{14} = 1/z_3$, so $1/z_3(t_2;\Gamma_{\mathbf{c}}^{\pm})$ is holomorphic and so we must have as a holomorphic series identity

$$\frac{1}{z_{14}}(t_2;\Gamma_{\mathbf{c}}^{\pm}) = \left(\frac{1}{z_3}(t_2;\Gamma_{\mathbf{c}}^{\pm})\right)^2,$$

and this is why it is honest to just compute $1/z_3(t_2;\Gamma_{\mathbf{c}}^{\pm})$ and square it. This shows that the divisor at infinity has no other components than $\mathcal{D}_{\mathbf{c}}^{+}$ and $\mathcal{D}_{\mathbf{c}}^{-}$. We conclude that the generic fiber $\mathbf{F_c}$ of the momentum map of the Bechlivanidis-van Moerbeke system completes into the Jacobian of the curve $\overline{\Gamma_{\mathbf{c}}}$ by adjoining two copies $\mathcal{D}_{\mathbf{c}}^{\pm}$ of $\overline{\Gamma_{\mathbf{c}}}$ that are tangent at their common intersection point 0. Let us show that the vector field $\mathcal{V}_1$ is (simply) tangent to $\mathcal{D}_{\mathbf{c}}^{\pm}$ at O and at the points $P^{\pm}$. To do this we follow the method that was given in Section 7.6.7.

In the notations of Example 7.54 we choose $y_0 := x_1$ and $y := x_2$ so that, using (10.54)

$$\begin{aligned}y_0^{(0)} &= 2\epsilon a, & y_0^{(1)} &= -2a^2,\\ y^{(0)} &= -\epsilon/2 & y^{(1)} &= -a/2\end{aligned}$$

and using (10.51) and (10.54),

$$\mathcal{V}_2 \left[\frac{1}{y_0}\right]_{|\mathcal{D}'_c} = \epsilon, \qquad \mathcal{V}_2 \left[\frac{y}{y_0}\right]_{|\mathcal{D}'_c} = -\frac{a^3\epsilon + 2d}{a^2}.$$

It follows that

$$\delta = \frac{1}{4a^2} \begin{vmatrix} 2\epsilon a & \epsilon \\ -2\epsilon a^2 & -\frac{a^3\epsilon+2d}{a^2} \end{vmatrix} = -\frac{\epsilon d}{a^3}.$$

The holomorphic one-form on $\mathcal{D}_c$ is therefore given by $\omega_1 = \mathsf{d}a/(8d)$. In a neighborhood of the points 0 and $P^{\pm}$ we have that $\omega_1 = -\varsigma \mathsf{d}\varsigma/(4\eta)$, where $\eta = \pm 1$, as follows by substituting (10.56) in ω_1. This shows that $\mathcal{V}_1$ is indeed simply tangent to the divisors $\mathcal{D}_c^{\pm}$ at the points O and $P^{\pm}$; see Figure 10.3.

10.4.3 Almost Algebraic Complete Integrability

Following [30] we now relate the Goryachev-Chaplygin top to the Bechlivanidis-van Moerbeke system. Recalling that $M = \left\{(\gamma, a) \in C^6 \mid K_2(\gamma, a) = 0\right\}$, consider the regular map $\pi : M \to \mathbf{C}^7$, defined by

$$(x_1, \dots, x_7) = \left(a_1^2 + a_2^2, \sqrt{-1}a_3, \gamma_3^2, \gamma_1, \sqrt{-1}\gamma_2, \sqrt{-1}a_1\gamma_3, a_2\gamma_3\right). \qquad (10.59)$$

Using (10.47) one easily checks that it preserves the constants of motion, in the sense that

$$(F_1, \dots, F_5) \circ \pi = (2H, \sqrt{-1}K_3, K_1, 0, 0)$$

and that the Goryachev-Chaplygin vector field (10.46) is π-related to $\mathcal{V}_1$, the first Bechlivanidis-van Moerbeke vector field (10.51). Although π is not a Poisson map with respect to the Poisson structures defined by (10.49) or (10.52) there is another Poisson structure on $\mathbf{C}^7$ with respect to which π is Poisson. Its Poisson matrix is given as follows.

$$\begin{pmatrix} 0 & 0 & -4(x_5x_6 + x_4x_7) & 2x_7 & -2x_6 & 4x_2x_7 + 2x_1x_5 & 4x_2x_6 - 2x_1x_4 \\ 0 & 0 & 0 & -x_5 & -x_4 & x_7 & x_6 \\ 4(x_5x_6 + x_4x_7) & 0 & 0 & 0 & 0 & -2x_3x_5 & 2x_3x_4 \\ -2x_7 & x_5 & 0 & 0 & 0 & 0 & x_3 \\ 2x_6 & x_4 & 0 & 0 & 0 & x_3 & 0 \\ -4x_2x_7 - 2x_1x_5 & -x_7 & 2x_3x_5 & 0 & x_3 & 0 & -2x_2x_3 \\ 2x_1x_4 - 4x_2x_6 & -x_6 & -2x_3x_4 & x_3 & 0 & 2x_2x_3 & 0 \end{pmatrix}$$

This Poisson structure has F_3, F_4 and F_5 as Casimirs, while F_1 and F_2 generate the two commuting vector fields (10.51) of the Bechlivanidis-van Moerbeke system.

It is clear that the Goryachev-Chaplygin top is not a.c.i.: if we solve (10.59) for $(a_1, \dots, \gamma_3)$ and we substitute the principal balance (10.54) in it then we find a solution which is not single-valued (see Paragraph 6.2.2): one finds for example that

$$a_2 = \pm x_7 \sqrt{\frac{x_1}{x_7^2 - x_6^2}},$$

so that $a_2(t; \mathcal{I}') \sim t^{-3/2}$, which is not single-valued.

The map π, restricted to the generic fiber of the momentum map, is $2:1$ and unramified and maps the integrable vector fields to the linear vector fields on a torus. By the above the map must be ramified at infinity, so the generic fiber of the momentum map of the Goryachev-Chaplygin top naturally compactifies into a double cover of a genus two hyperelliptic Jacobian, where the cover is only ramified at infinity; the affine part supports two holomorphic commuting vector fields that are independent at each point. The Goryachev-Chaplygin top is an example of an almost a.c.i. system, defined as follows.

Definition 10.5. Let $(M, \{\cdot\,,\cdot\}, \mathbf{F})$ be a complex integrable system, where M is a non-singular affine variety and where $\mathbf{F} = (F_1, \dots, F_s)$. We say that $(M, \{\cdot\,,\cdot\}, \mathbf{F})$ is an *almost algebraic completely integrable system* or an *almost a.c.i. system* if there exists for generic $\mathbf{c} \in \mathbf{C}^s$ an r-dimensional Abelian variety $\mathbf{T}^r_{\mathbf{c}}$ and a regular map $\pi_{\mathbf{c}} : \mathbf{F}'_{\mathbf{c}} \to \mathbf{T}^r_{\mathbf{c}}$, such that

(1) $\pi_{\mathbf{c}}$ is finite and unramified;
(2) $\pi_{\mathbf{c}}(\mathbf{F}_{\mathbf{c}})$ is an affine part of $\mathbf{T}^r_{\mathbf{c}}$;
(3) Each of the integrable vector fields $\mathcal{X}_{F_i}$ is $\pi_{\mathbf{c}}$-related to a holomorphic (hence linear) vector field on $\mathbf{T}^r_{\mathbf{c}}$.

Clearly, every a.c.i. system is an almost a.c.i. system. We have shown that the Goryachev-Chaplygin top is an example of an almost a.c.i. system that is not a.c.i. For another example, see [169], where one also finds a natural family of Liouville integrable systems that are neither a.c.i. nor almost a.c.i.

10.4.4 The Relation Between the Toda and the Bechlivanidis-van Moerbeke System

It has been shown by Bechlivanidis and van Moerbeke in [30] that the Bechlivanidis-van Moerbeke system and the periodic $\mathfrak{sl}(3)$ Toda lattice are intimately related. Namely, considering the geometry of the fibers of the momentum map of both systems, and comparing in particular the divisors to be glued at infinity they first construct a basis for the functions that have a simple pole at most at only two of the three curves in the Toda divisor. In order to present these functions explicitly and compute from it the map from this Toda lattice to the Bechlivanidis-van Moerbeke system, we first recall from Section 9.4 that the periodic $\mathfrak{sl}(3)$ Toda lattice (i.e., the $\mathfrak{a}_2^{(1)}$ Toda lattice) is given by[5]

$$\begin{aligned} \dot y_0 &= y_0(y_5 - y_4), & \dot y_3 &= y_1 - y_2,\\ \dot y_1 &= y_1(y_3 - y_5), & \dot y_4 &= y_2 - y_0,\\ \dot y_2 &= y_2(y_4 - y_3), & \dot y_5 &= y_0 - y_1, \end{aligned} \tag{10.60}$$

[5] We did a simple relabeling $x_i \to y_i$ and $F_i \to G_i$ because in this section the coordinates x_i and the constants of motion F_i are reserved for the Bechlivanidis-van Moerbeke system.

with the following constants of motion

$$\begin{aligned} G_1 &:= y_0y_1y_2, \\ G_2 &:= y_3 + y_4 + y_5 = 0, \\ G_3 &:= \tfrac{1}{2}(y_3^2 + y_4^2 + y_5^2) - y_0 - y_1 - y_2, \\ G_4 &:= y_3y_4y_5 + y_0y_3 + y_1y_4 + y_2y_5. \end{aligned} \tag{10.61}$$

Notice that we are considering here the Toda lattice on the hyperplane $y_3 + y_4 + y_5 = 0$ (see 9.4). Using (9.30) it is easy to verify that $(1, y_3, y_1y_2, y_0+y_4y_5)$ is a basis for the functions that are finite along one of the Toda curves (i.e., for each of these functions f the series $f(t; w_0)$ has no pole in t), while having a simple pole at most along the two other Toda curves (i.e., for each of these functions f the series $f(t; w_1)$ and $f(t; w_2)$ have at most a simple pole in t). If one looks for a polynomial map that respects the weights of the variables then the fact that $(1, x_1, x_2, x_3)$ is a basis for the functions that have a simple pole along the Bechlivanidis-van Moerbeke divisor suggest to search for a map for which $x_1 \sim y_0 + y_4y_5$ and $x_2 \sim y_3$ and $x_3 \sim y_1y_2$. Comparing the invariants (10.61) with (10.50) and the vector field (10.60) with (10.51) leads to the map

$$\begin{aligned} \psi: \ M_1 \subset \mathbf{C}^6 &\to M_2 \subset \mathbf{C}^7 \\ (y_1, \ldots, y_6) &\mapsto (x_1, \ldots, x_7) \end{aligned} \tag{10.62}$$

given by

$$\begin{aligned} x_1 &= y_0 + y_4y_5, & x_5 &= (y_2 - y_1)/8, \\ x_2 &= -y_3/2, & x_6 &= -(y_2y_5 + y_1y_4)/8, \\ x_3 &= -y_1y_2/16, & x_7 &= -(y_2y_5 - y_1y_4)/8, \\ x_4 &= -(y_1 + y_2)/8, \end{aligned}$$

where

$$\begin{aligned} M_1 &= \left\{y = (y_1, \ldots, y_6) \in \mathbf{C}^6 \mid G_2(y) = 0\right\}, \\ M_2 &= \left\{x = (x_1, \ldots, x_7) \in \mathbf{C}^7 \mid F_3(x) = F_4(x) = 0, \ x_3 \neq 0, \ x_4 \neq x_5\right\}. \end{aligned}$$

It is easy to verify that this injective map ψ maps the Toda vector field (10.60) to (10.51); to check this one uses a few times that $G_2 = y_3 + y_4 + y_5 = 0$. Similarly, one easily checks that

$$(F_1, \ldots, F_5) \circ \psi = (G_2(y_4 + y_5 - y_3)/2 - G_3, \ -G_4/2, \ 0, \ y_1y_2G_2/32, \ 16G_1),$$

which implies, since $G_2 = 0$, that the map ψ maps the constants of motion of the Bechlivanidis-van Moerbeke system to the constants of motion of the Toda lattice.

The inverse map which is only defined on the subspace $F_3 = F_4 = 0$, and which is rational, is given by

$$\begin{aligned} y_0 &= (x_1x_3 + x_6^2 - x_7^2)/x_3, & y_3 &= -2x_2, \\ y_1 &= -4(x_4 + x_5), & y_4 &= (x_6 - x_7)/(x_4 - x_5), \\ y_2 &= 4(x_5 - x_4), & y_5 &= (x_6 + x_7)/(x_4 - x_5). \end{aligned}$$

Notice that the restriction of ψ to the generic fiber of the momentum map will be a regular map with rational inverse, which accounts for the fact that the divisors to be added to these fibers consist of three translates of the theta divisor in the Toda case, and only two in the Bechlivanidis-van Moerbeke case.

The Bechlivanidis-van Moerbeke system contains an affine part of *any* two-dimensional (hyperelliptic) Jacobian as a fiber of its momentum map. It is isomorphic to the two-dimensional even Mumford system, which has, for any g, a natural g-dimensional generalization; the latter contains an affine part of any g-dimensional hyperelliptic Jacobian as a fiber of its momentum map. For this, we refer the interested reader to [169, Chapter VI].

References

1. R. Abraham and J. E. Marsden. *Foundations of mechanics.* Benjamin/Cummings Publishing Co. Inc. Advanced Book Program, Reading, Mass., 1978. Second edition, revised and enlarged, With the assistance of T. Ratiu and R. Cushman.
2. M. Adams, J. Harnad, and J. Hurtubise. Isospectral Hamiltonian flows in finite and infinite dimensions. II. Integration of flows. *Comm. Math. Phys.*, 134(3):555–585, 1990.
3. M. Adams, J. Harnad, and J. Hurtubise. Darboux coordinates and Liouville-Arnol′d integration in loop algebras. *Comm. Math. Phys.*, 155(2):385–413, 1993.
4. M. Adams, J. Harnad, and E. Previato. Isospectral Hamiltonian flows in finite and infinite dimensions. I. Generalized Moser systems and moment maps into loop algebras. *Comm. Math. Phys.*, 117(3):451–500, 1988.
5. M. Adler. On a trace functional for formal pseudo differential operators and the symplectic structure of the Korteweg-de Vries type equations. *Invent. Math.*, 50(3):219–248, 1978/79.
6. M. Adler, L. Haine, and P. van Moerbeke. Limit matrices for the Toda flow and periodic flags for loop groups. *Math. Ann.*, 296(1):1–33, 1993.
7. M. Adler and P. van Moerbeke. Completely integrable systems, Euclidean Lie algebras, and curves. *Adv. in Math.*, 38(3):267–317, 1980.
8. M. Adler and P. van Moerbeke. Linearization of Hamiltonian systems, Jacobi varieties and representation theory. *Adv. in Math.*, 38(3):318–379, 1980.
9. M. Adler and P. van Moerbeke. The algebraic integrability of geodesic flow on SO(4). *Invent. Math.*, 67(2):297–331, 1982.
10. M. Adler and P. van Moerbeke. Kowalewski's asymptotic method, Kac-Moody Lie algebras and regularization. *Comm. Math. Phys.*, 83(1):83–106, 1982.
11. M. Adler and P. van Moerbeke. Geodesic flow on SO(4) and the intersection of quadrics. *Proc. Nat. Acad. Sci. U.S.A.*, 81(14, Phys. Sci.):4613–4616, 1984.
12. M. Adler and P. van Moerbeke. The intersection of four quadrics in $\mathbf{P}^6$, abelian surfaces and their moduli. *Math. Ann.*, 279(1):25–85, 1987.
13. M. Adler and P. van Moerbeke. The Kowalewski and Hénon-Heiles motions as Manakov geodesic flows on SO(4)—a two-dimensional family of Lax pairs. *Comm. Math. Phys.*, 113(4):659–700, 1988.
14. M. Adler and P. van Moerbeke. The complex geometry of the Kowalewski-Painlevé analysis. *Invent. Math.*, 97(1):3–51, 1989.
15. M. Adler and P. van Moerbeke. The Toda lattice, Dynkin diagrams, singularities and abelian varieties. *Invent. Math.*, 103(2):223–278, 1991.
16. A. Alekseev, A. Malkin, and E. Meinrenken. Lie group valued moment maps. *J. Differential Geom.*, 48(3):445–495, 1998.

17. E. Arbarello, M. Cornalba, P. A. Griffiths, and J. Harris. *Geometry of algebraic curves. Vol. I*, volume 267 of *Grundlehren der Mathematischen Wissenschaften [Fundamental Principles of Mathematical Sciences]*. Springer-Verlag, New York, 1985.
18. V. I. Arnold. *Mathematical methods of classical mechanics.* Springer-Verlag, New York, 1978. Translated from the Russian by K. Vogtmann and A. Weinstein, Graduate Texts in Mathematics, 60.
19. M. Audin. *The topology of torus actions on symplectic manifolds*, volume 93 of *Progress in Mathematics.* Birkhäuser Verlag, Basel, 1991. Translated from the French by the author.
20. M. Audin. *Spinning tops*, volume 51 of *Cambridge Studies in Advanced Mathematics.* Cambridge University Press, Cambridge, 1996. A course on integrable systems.
21. M. Audin. Hamiltonian monodromy via Picard-Lefschetz theory. *Comm. Math. Phys.*, 229(3):459–489, 2002.
22. M. Audin and R. Silhol. Variétés abéliennes réelles et toupie de Kowalevski. *Compositio Math.*, 87(2):153–229, 1993.
23. O. Babelon, D. Bernard, and M. Talon. *Introduction to classical integrable systems.* Cambridge Monographs on Mathematical Physics. Cambridge University Press, Cambridge, 2003.
24. O. Babelon, P. Cartier, and Y. Kosmann-Schwarzbach, editors. *Lectures on integrable systems*, River Edge, NJ, 1994. World Scientific Publishing Co. Inc. In memory of J.-L. Verdier.
25. O. Babelon and C.-M. Viallet. Hamiltonian structures and Lax equations. *Phys. Lett. B*, 237(3-4):411–416, 1990.
26. W. Barth. Abelian surfaces with $(1,2)$-polarization. In *Algebraic geometry, Sendai, 1985*, volume 10 of *Adv. Stud. Pure Math.*, pages 41–84. North-Holland, Amsterdam, 1987.
27. W. Barth. Affine parts of abelian surfaces as complete intersections of four quadrics. *Math. Ann.*, 278(1-4):117–131, 1987.
28. W. Barth, C. Peters, and A. Van de Ven. *Compact complex surfaces*, volume 4 of *Ergebnisse der Mathematik und ihrer Grenzgebiete (3) [Results in Mathematics and Related Areas (3)]*. Springer-Verlag, Berlin, 1984.
29. L. Bates and R. Cushman. What is a completely integrable nonholonomic dynamical system? In *Proceedings of the XXX Symposium on Mathematical Physics (Toruń, 1998)*, volume 44, pages 29–35, 1999.
30. C. Bechlivanidis and P. van Moerbeke. The Goryachev-Chaplygin top and the Toda lattice. *Comm. Math. Phys.*, 110(2):317–324, 1987.
31. A. I. Bobenko and V. B. Kuznetsov. Lax representation and new formulae for the Goryachev-Chaplygin top. *J. Phys. A*, 21(9):1999–2006, 1988.
32. A. I. Bobenko, A. G. Reyman, and M. A. Semenov-Tian-Shansky. The Kowalewski top 99 years later: a Lax pair, generalizations and explicit solutions. *Comm. Math. Phys.*, 122(2):321–354, 1989.
33. O. I. Bogoyavlensky. On perturbations of the periodic Toda lattice. *Comm. Math. Phys.*, 51(3):201–209, 1976.
34. O. I. Bogoyavlensky. Integrable Euler equations on SO(4) and their physical applications. *Comm. Math. Phys.*, 93(3):417–436, 1984.
35. F. Bottacin. Poisson structures on Hilbert schemes of points of a surface and integrable systems. *Manuscripta Math.*, 97(4):517–527, 1998.

36. N. Bourbaki. *Éléments de mathématique.* Hermann, Paris, 1975. Fasc. XXXVIII: Groupes et algèbres de Lie. Chapitre VII: Sous-algèbres de Cartan, éléments réguliers. Chapitre VIII: Algèbres de Lie semi-simples déployées, Actualités Scientifiques et Industrielles, No. 1364.
37. N. Bourbaki. *Lie groups and Lie algebras. Chapters 4–6.* Elements of Mathematics. Springer-Verlag, Berlin, 2002. Translated from the 1968 French original by A. Pressley.
38. P. Bueken and P. Vanhaecke. The moduli problem for integrable systems: the example of a geodesic flow on SO(4). *J. London Math. Soc. (2)*, 62(2):357–369, 2000.
39. A. Cannas da Silva and A. Weinstein. *Geometric models for noncommutative algebras*, volume 10 of *Berkeley Mathematics Lecture Notes.* American Mathematical Society, Providence, RI, 1999.
40. R. W. Carter. *Simple groups of Lie type.* John Wiley & Sons, London-New York-Sydney, 1972. Pure and Applied Mathematics, Vol. 28.
41. S. A. Chaplygin. A new case of rotation of a rigid body, supported at one point. In *Collected Works*, pages 118–124. 1948.
42. S. S. Chern and J. G. Wolfson. A simple proof of Frobenius theorem. In *Manifolds and Lie groups (Notre Dame, Ind., 1980)*, volume 14 of *Progr. Math.*, pages 67–69. Birkhäuser Boston, Mass., 1981.
43. R. H. Cushman and L. M. Bates. *Global aspects of classical integrable systems.* Birkhäuser Verlag, Basel, 1997.
44. P. A. Damianou. Master symmetries and R-matrices for the Toda lattice. *Lett. Math. Phys.*, 20(2):101–112, 1990.
45. P. Deift, L. C. Li, T. Nanda, and C. Tomei. The Toda flow on a generic orbit is integrable. *Comm. Pure Appl. Math.*, 39(2):183–232, 1986.
46. R. Donagi and E. Markman. Spectral covers, algebraically completely integrable, Hamiltonian systems, and moduli of bundles. In *Integrable systems and quantum groups (Montecatini Terme, 1993)*, volume 1620 of *Lecture Notes in Math.*, pages 1–119. Springer, Berlin, 1996.
47. B. A. Dubrovin. Theta-functions and nonlinear equations. *Uspekhi Mat. Nauk*, 36(2(218)):11–80, 1981. With an appendix by I. M. Krichever.
48. J. J. Duistermaat. On global action-angle coordinates. *Comm. Pure Appl. Math.*, 33(6):687–706, 1980.
49. H. R. Dullin, P. H. Richter, and A. P. Veselov. Action variables of the Kovalevskaya top. *Regul. Chaotic Dyn.*, 3(3):18–31, 1998. J. Moser at 70 (Russian).
50. N. Ercolani and E. D. Siggia. Painlevé property and geometry. *Phys. D*, 34(3):303–346, 1989.
51. G. Falqui, F. Magri, and M. Pedroni. Bihamiltonian geometry and separation of variables for Toda lattices. *J. Nonlinear Math. Phys.*, 8(suppl.):118–127, 2001. Nonlinear evolution equations and dynamical systems (Kolimbary, 1999).
52. J. D. Fay. *Theta functions on Riemann surfaces.* Springer-Verlag, Berlin, 1973. Lecture Notes in Mathematics, Vol. 352.
53. R. L. Fernandes and P. Vanhaecke. Hyperelliptic Prym varieties and integrable systems. *Comm. Math. Phys.*, 221(1):169–196, 2001.
54. H. Flaschka. The Toda lattice. I. Existence of integrals. *Phys. Rev. B (3)*, 9:1924–1925, 1974.

55. H. Flaschka. Integrable systems and torus actions. In *Lectures on integrable systems (Sophia-Antipolis, 1991)*, pages 43–101. World Sci. Publishing, River Edge, NJ, 1994.
56. H. Flaschka and L. Haine. Torus orbits in G/P. *Pacific J. Math.*, 149(2):251–292, 1991.
57. H. Flaschka and L. Haine. Variétés de drapeaux et réseaux de Toda. *Math. Z.*, 208(4):545–556, 1991.
58. A. T. Fomenko. *Integrability and nonintegrability in geometry and mechanics*, volume 31 of *Mathematics and its Applications (Soviet Series)*. Kluwer Academic Publishers Group, Dordrecht, 1988. Translated from the Russian by M. V. Tsaplina.
59. A. T. Fomenko. *Symplectic geometry*, volume 5 of *Advanced Studies in Contemporary Mathematics*. Gordon and Breach Publishers, Luxembourg, second edition, 1995. Translated from the 1988 Russian original by R. S. Wadhwa.
60. A. T. Fomenko and V. V. Trofimov. *Integrable systems on Lie algebras and symmetric spaces*, volume 2 of *Advanced Studies in Contemporary Mathematics*. Gordon and Breach Science Publishers, New York, 1988. Translated from the Russian by A. Karaulov, P. D. Rayfield and A. Weisman.
61. J.-P. Françoise. Integrability of quasi-homogeneous vector fields. *Unpublished preprint.*
62. L. Gavrilov. Generalized Jacobians of spectral curves and completely integrable systems. *Math. Z.*, 230(3):487–508, 1999.
63. L. Gavrilov and A. Zhivkov. The complex geometry of the Lagrange top. *Enseign. Math. (2)*, 44(1-2):133–170, 1998.
64. V. L. Ginzburg, V. Guillemin, and Y. Karshon. Assignments and abstract moment maps. *J. Differential Geom.*, 52(2):259–301, 1999.
65. H. Goldstein. *Classical Mechanics.* Addison-Wesley Press, Inc., Cambridge, Mass., 1951.
66. D. Goryachev. On the motion of a rigid material body about a fixed point in the case $A = B = 4C$. *Mat. Sb.*, 21(3), 1900.
67. G. Grelaud, C. Quitté, and P. Tauvel. Bases de Chevalley et $\mathfrak{sl}(()2)$-triples des algèbres de Lie simples exceptionnelles. *Université de Poitiers preprint series*, 53:1–47, 1990.
68. P. Griffiths. Linearizing flows and a cohomology interpretation of Lax equations. In *Seminar on nonlinear partial differential equations (Berkeley, Calif., 1983)*, volume 2 of *Math. Sci. Res. Inst. Publ.*, pages 37–46. Springer, New York, 1984.
69. P. Griffiths and J. Harris. *Principles of algebraic geometry.* Wiley Classics Library. John Wiley & Sons Inc., New York, 1994. Reprint of the 1978 original.
70. A. Grothendieck. Technique de descente et théorèmes d'existence en géométrie algébrique. V. Les schémas de Picard: théorèmes d'existence. In *Séminaire Bourbaki, Vol. 7*, pages Exp. No. 232, 143–161. Soc. Math. France, Paris, 1995.
71. A. Grothendieck. Technique de descente et théorèmes d'existence en géométrie algébrique. VI. Les schémas de Picard: propriétés générales. In *Séminaire Bourbaki, Vol. 7*, pages Exp. No. 236, 221–243. Soc. Math. France, Paris, 1995.
72. M. A. Guest. *Harmonic maps, loop groups, and integrable systems*, volume 38 of *London Mathematical Society Student Texts.* Cambridge University Press, Cambridge, 1997.

73. V. Guillemin and S. Sternberg. *Symplectic techniques in physics.* Cambridge University Press, Cambridge, second edition, 1990.
74. L. Haine. Geodesic flow on SO(4) and abelian surfaces. *Math. Ann.*, 263(4):435–472, 1983.
75. L. Haine. The algebraic complete integrability of geodesic flow on SO(N). *Comm. Math. Phys.*, 94(2):271–287, 1984.
76. L. Haine and E. Horozov. A Lax pair for Kowalevski's top. *Phys. D*, 29(1-2):173–180, 1987.
77. J. Harnad. Isospectral flow and Liouville-Arnol'd integration in loop algebras. In *Geometric and quantum aspects of integrable systems (Scheveningen, 1992)*, volume 424 of *Lecture Notes in Phys.*, pages 1–42. Springer, Berlin, 1993.
78. R. Hartshorne. *Algebraic geometry.* Springer-Verlag, New York, 1977. Graduate Texts in Mathematics, No. 52.
79. S. Helgason. *Differential geometry, Lie groups, and symmetric spaces*, volume 34 of *Graduate Studies in Mathematics.* American Mathematical Society, Providence, RI, 2001. Corrected reprint of the 1978 original.
80. M. W. Hirsch. *Differential topology*, volume 33 of *Graduate Texts in Mathematics.* Springer-Verlag, New York, 1994. Corrected reprint of the 1976 original.
81. N. Hitchin. *Monopoles, minimal surfaces and algebraic curves*, volume 105 of *Séminaire de Mathématiques Supérieures [Seminar on Higher Mathematics].* Presses de l'Université de Montréal, Montreal, QC, 1987.
82. N. Hitchin. Stable bundles and integrable systems. *Duke Math. J.*, 54(1):91–114, 1987.
83. N. Hitchin. Riemann surfaces and integrable systems. In *Integrable systems (Oxford, 1997)*, volume 4 of *Oxf. Grad. Texts Math.*, pages 11–52. Oxford Univ. Press, New York, 1999. Notes by J. Sawon.
84. N. J. Hitchin, G. B. Segal, and R. S. Ward. *Integrable systems*, volume 4 of *Oxford Graduate Texts in Mathematics.* The Clarendon Press Oxford University Press, New York, 1999. Twistors, loop groups, and Riemann surfaces, Lectures from the Instructional Conference held at the University of Oxford, Oxford, September 1997.
85. J. Hoppe. *Lectures on integrable systems*, volume 10 of *Lecture Notes in Physics. New Series m: Monographs.* Springer-Verlag, Berlin, 1992.
86. E. Horozov and P. van Moerbeke. The full geometry of Kowalewski's top and (1, 2)-abelian surfaces. *Comm. Pure Appl. Math.*, 42(4):357–407, 1989.
87. J. E. Humphreys. *Introduction to Lie algebras and representation theory*, volume 9 of *Graduate Texts in Mathematics.* Springer-Verlag, New York, 1978. Second printing, revised.
88. J. C. Hurtubise. Integrable systems and algebraic surfaces. *Duke Math. J.*, 83(1):19–50, 1996.
89. S. Iitaka. *Algebraic geometry*, volume 76 of *Graduate Texts in Mathematics.* Springer-Verlag, New York, 1982. An introduction to birational geometry of algebraic varieties, North-Holland Mathematical Library, 24.
90. C. Jacobi. Sur le mouvement d'un point et sur un cas particulier du problème des trois corps. *Compt. Rend.*, 3:59–61, 1836.
91. C. Jacobi. Note von der geodätischen linie auf einem ellipsoid und den verschiedenen anwendungen einer merkwürdigen analytischen substitution. *J. Reine Angew. Math.*, 19:309–313, 1839.

92. M. Kac and P. van Moerbeke. On an explicitly soluble system of nonlinear differential equations related to certain Toda lattices. *Advances in Math.*, 16:160–169, 1975.
93. V. G. Kac. *Infinite-dimensional Lie algebras.* Cambridge University Press, Cambridge, second edition, 1985.
94. V. Kanev. Spectral curves and Prym-Tjurin varieties. I. In *Abelian varieties (Egloffstein, 1993)*, pages 151–198. de Gruyter, Berlin, 1995.
95. D. Kazhdan, B. Kostant, and S. Sternberg. Hamiltonian group actions and dynamical systems of Calogero type. *Comm. Pure Appl. Math.*, 31(4):481–507, 1978.
96. H. Knörrer. Geodesics on the ellipsoid. *Invent. Math.*, 59(2):119–143, 1980.
97. H. Knörrer. Geodesics on quadrics and a mechanical problem of C. Neumann. *J. Reine Angew. Math.*, 334:69–78, 1982.
98. H. Knörrer. Integrable Hamiltonsche Systeme und algebraische Geometrie. *Jahresber. Deutsch. Math.-Verein.*, 88(2):82–103, 1986.
99. Y. Kosmann-Schwarzbach. Lie bialgebras, Poisson Lie groups and dressing transformations. In *Integrability of nonlinear systems (Pondicherry, 1996)*, volume 495 of *Lecture Notes in Phys.*, pages 104–170. Springer, Berlin, 1997.
100. Y. Kosmann-Schwarzbach and F. Magri. Poisson-Nijenhuis structures. *Ann. Inst. H. Poincaré Phys. Théor.*, 53(1):35–81, 1990.
101. B. Kostant. The solution to a generalized Toda lattice and representation theory. *Adv. in Math.*, 34(3):195–338, 1979.
102. S. Kowalevski. Sur le problème de la rotation d'un corps solide autour d'un point fixe. In *The Kowalevski property (Leeds, 2000)*, volume 32 of *CRM Proc. Lecture Notes*, pages 315–372. Amer. Math. Soc., Providence, RI, 2002. Reprinted from Acta Math. **12** (1889), 177–232.
103. V. B. Kuznetsov, editor. *The Kowalevski property*, volume 32 of *CRM Proceedings & Lecture Notes*, Providence, RI, 2002. American Mathematical Society.
104. H. Lamb. *Hydrodynamics.* Cambridge Mathematical Library. Cambridge University Press, Cambridge, sixth edition, 1993. With a foreword by R. A. Caflisch.
105. H. Lange and C. Birkenhake. *Complex abelian varieties*, volume 302 of *Grundlehren der Mathematischen Wissenschaften [Fundamental Principles of Mathematical Sciences]*. Springer-Verlag, Berlin, 1992.
106. A. Lesfari. Abelian surfaces and Kowalewski's top. *Ann. Sci. École Norm. Sup. (4)*, 21(2):193–223, 1988.
107. P. Libermann and C.-M. Marle. *Symplectic geometry and analytical mechanics*, volume 35 of *Mathematics and its Applications.* D. Reidel Publishing Co., Dordrecht, 1987. Translated from the French by B. E. Schwarzbach.
108. A. T. Lundell. A short proof of the Frobenius theorem. *Proc. Amer. Math. Soc.*, 116(4):1131–1133, 1992.
109. A. M. Lyapunov. *Comm. Kharkov Math. Soc*, (2) 4(25):123–140, 1894.
110. S. V. Manakov. A remark on the integration of the Eulerian equations of the dynamics of an n-dimensional rigid body. *Funkcional. Anal. i Priložen.*, 10(4):93–94, 1976.
111. A. Marshakov. *Seiberg-Witten theory and integrable systems.* World Scientific Publishing Co. Inc., River Edge, NJ, 1999.
112. A. McDaniel. Representations of sl$(n, \mathbf{C})$ and the Toda lattice. *Duke Math. J.*, 56(1):47–99, 1988.

113. A. McDaniel and L. Smolinsky. A Lie-theoretic Galois theory for the spectral curves of an integrable system. I. *Comm. Math. Phys.*, 149(1):127–148, 1992.
114. A. McDaniel and L. Smolinsky. The flow of the G_2 periodic Toda lattice. *J. Math. Phys.*, 38(2):926–945, 1997.
115. A. McDaniel and L. Smolinsky. A Lie-theoretic Galois theory for the spectral curves of an integrable system. II. *Trans. Amer. Math. Soc.*, 349(2):713–746, 1997.
116. A. McDaniel and L. Smolinsky. Lax equations, weight lattices, and Prym-Tjurin varieties. *Acta Math.*, 181(2):283–305, 1998.
117. H. P. McKean. Integrable systems and algebraic curves. In *Global analysis (Proc. Biennial Sem. Canad. Math. Congr., Univ. Calgary, Calgary, Alta., 1978)*, volume 755 of *Lecture Notes in Math.*, pages 83–200. Springer, Berlin, 1979.
118. J. J. Morales Ruiz. *Differential Galois theory and non-integrability of Hamiltonian systems*, volume 179 of *Progress in Mathematics.* Birkhäuser Verlag, Basel, 1999.
119. J. Moser. Finitely many mass points on the line under the influence of an exponential potential–an integrable system. In *Dynamical systems, theory and applications (Rencontres, BattelleRes. Inst., Seattle, Wash., 1974)*, pages 467–497. Lecture Notes in Phys., Vol. 38. Springer, Berlin, 1975.
120. J. Moser. Three integrable Hamiltonian systems connected with isospectral deformations. *Advances in Math.*, 16:197–220, 1975.
121. J. Moser. Geometry of quadrics and spectral theory. In *The Chern Symposium 1979 (Proc. Internat. Sympos., Berkeley, Calif., 1979)*, pages 147–188. Springer, New York, 1980.
122. J. Moser. Various aspects of integrable Hamiltonian systems. In *Dynamical systems (Bressanone, 1978)*, pages 137–195. Liguori, Naples, 1980.
123. J. Moser. *Integrable Hamiltonian systems and spectral theory.* Lezioni Fermiane. [Fermi Lectures]. Scuola Normale Superiore, Pisa, 1983.
124. M. Mulase. Algebraic geometry of soliton equations. *Proc. Japan Acad. Ser. A Math. Sci.*, 59(6):285–288, 1983.
125. M. Mulase. Cohomological structure in soliton equations and Jacobian varieties. *J. Differential Geom.*, 19(2):403–430, 1984.
126. D. Mumford. On the equations defining abelian varieties. I. *Invent. Math.*, 1:287–354, 1966.
127. D. Mumford. On the equations defining abelian varieties. II. *Invent. Math.*, 3:75–135, 1967.
128. D. Mumford. On the equations defining abelian varieties. III. *Invent. Math.*, 3:215–244, 1967.
129. D. Mumford. *Abelian varieties.* Tata Institute of Fundamental Research Studies in Mathematics, No. 5. Published for the Tata Institute of Fundamental Research, Bombay, 1970.
130. D. Mumford. Varieties defined by quadratic equations. In *Questions on Algebraic Varieties (C.I.M.E., III Ciclo, Varenna, 1969)*, pages 29–100. Edizioni Cremonese, Rome, 1970.
131. D. Mumford. Prym varieties. I. In *Contributions to analysis (a collection of papers dedicated to Lipman Bers)*, pages 325–350. Academic Press, New York, 1974.

132. D. Mumford. An algebro-geometric construction of commuting operators and of solutions to the Toda lattice equation, Korteweg deVries equation and related nonlinear equation. In *Proceedings of the International Symposium on Algebraic Geometry (Kyoto Univ., Kyoto, 1977)*, pages 115–153, Tokyo, 1978. Kinokuniya Book Store.
133. D. Mumford. *Tata lectures on theta. II*, volume 43 of *Progress in Mathematics*. Birkhäuser Boston Inc., Boston, MA, 1984. Jacobian theta functions and differential equations, With the collaboration of C. Musili, M. Nori, E. Previato, M. Stillman and H. Umemura.
134. D. Mumford. *Algebraic geometry. I*. Classics in Mathematics. Springer-Verlag, Berlin, 1995. Complex projective varieties, Reprint of the 1976 edition.
135. D. Mumford. *The red book of varieties and schemes*, volume 1358 of *Lecture Notes in Mathematics*. Springer-Verlag, Berlin, expanded edition, 1999. Includes the Michigan lectures (1974) on curves and their Jacobians, With contributions by E. Arbarello.
136. W. Nahm. Self-dual monopoles and calorons. In *Group theoretical methods in physics (Trieste, 1983)*, volume 201 of *Lecture Notes in Phys.*, pages 189–200. Springer, Berlin, 1984.
137. J. M. Nunes da Costa and P. A. Damianou. Toda systems and exponents of simple Lie groups. *Bull. Sci. Math.*, 125(1):49–69, 2001.
138. M. A. Olshanetsky and A. M. Perelomov. Completely integrable Hamiltonian systems connected with semisimple Lie algebras. *Invent. Math.*, 37(2):93–108, 1976.
139. S. Pantazis. Prym varieties and the geodesic flow on SO(n). *Math. Ann.*, 273(2):297–315, 1986.
140. A. M. Perelomov. *Integrable systems of classical mechanics and Lie algebras. Vol. I*. Birkhäuser Verlag, Basel, 1990. Translated from the Russian by A. G. Reyman [A. G. Reĭman].
141. L. A. Piovan. Algebraically completely integrable systems and Kummer varieties. *Math. Ann.*, 290(2):349–403, 1991.
142. L. A. Piovan. Cyclic coverings of abelian varieties and the Goryachev-Chaplygin top. *Math. Ann.*, 294(4):755–764, 1992.
143. L. A. Piovan and P. Vanhaecke. Integrable systems and projective images of Kummer surfaces. *Ann. Scuola Norm. Sup. Pisa Cl. Sci. (4)*, 29(2):351–392, 2000.
144. S. Poisson. Mémoire sur la variation des constantes arbitraires dans les questions de mécanique. *J. Ecole Polytec.*, 8:266–344, 1809.
145. S. Ramanan. Ample divisors on abelian surfaces. *Proc. London Math. Soc. (3)*, 51(2):231–245, 1985.
146. T. Ratiu. Euler-Poisson equations on Lie algebras and the N-dimensional heavy rigid body. *Amer. J. Math.*, 104(2):409–448, 1982.
147. T. Ratiu and P. van Moerbeke. The Lagrange rigid body motion. *Ann. Inst. Fourier (Grenoble)*, 32(1):viii, 211–234, 1982.
148. A. G. Reyman and M. A. Semenov-Tian-Shansky. A new integrable case of the motion of the 4-dimensional rigid body. *Comm. Math. Phys.*, 105(3):461–472, 1986.
149. A. G. Reyman and M. A. Semenov-Tian-Shansky. Lax representation with a spectral parameter for the Kowalewski top and its generalizations. In *Plasma theory and nonlinear and turbulent processes in physics, Vol. 1, 2 (Kiev, 1987)*, pages 135–152. World Sci. Publishing, Singapore, 1988.

150. A. G. Reyman and M.A. Semenov-Tian-Shansky. Reduction of Hamiltonian systems, affine Lie algebras and Lax equations. *Invent. Math.*, 54(1):81–100, 1979.
151. A. G. Reyman and M.A. Semenov-Tian-Shansky. Reduction of Hamiltonian systems, affine Lie algebras and Lax equations. II. *Invent. Math.*, 63(3):423–432, 1981.
152. M. A. Semenov-Tyan-Shanskiĭ. What a classical r-matrix is. *Funktsional. Anal. i Prilozhen.*, 17(4):17–33, 1983.
153. J.-P. Serre. Géométrie algébrique et géométrie analytique. *Ann. Inst. Fourier, Grenoble*, 6:1–42, 1955–1956.
154. J.-P. Serre. Morphismes universels et variété d'albanese. In *Seminaire C. Chevalley E.N.S.* 1958/59.
155. J.-P. Serre. *Algebraic groups and class fields*, volume 117 of *Graduate Texts in Mathematics*. Springer-Verlag, New York, 1988. Translated from the French.
156. J.-P. Serre. *Complex semisimple Lie algebras.* Springer Monographs in Mathematics. Springer-Verlag, Berlin, 2001. Translated from the French by G. A. Jones, Reprint of the 1987 edition.
157. T. Shiota. Characterization of Jacobian varieties in terms of soliton equations. *Invent. Math.*, 83(2):333–382, 1986.
158. M. Spivak. *A comprehensive introduction to differential geometry. Vol. I.* Publish or Perish Inc., Wilmington, Del., second edition, 1979.
159. W. W. Symes. Systems of Toda type, inverse spectral problems, and representation theory. *Invent. Math.*, 59(1):13–51, 1980.
160. W. W. Symes. The QR algorithm and scattering for the finite nonperiodic Toda lattice. *Phys. D*, 4(2):275–280, 1981/82.
161. C. L. Terng and K. Uhlenbeck, editors. *Surveys in differential geometry: integral systems [integrable systems].* Surveys in Differential Geometry, IV. International Press, Boston, MA, 1998. Lectures on geometry and topology, sponsored by Lehigh University's Journal of Differential Geometry, A supplement to the Journal of Differential Geometry.
162. A. Thimm. Integrable geodesic flows on homogeneous spaces. *Ergodic Theory Dynamical Systems*, 1(4):495–517 (1982), 1981.
163. A. Treibich and J.-L. Verdier. Solitons elliptiques. In *The Grothendieck Festschrift, Vol. III*, volume 88 of *Progr. Math.*, pages 437–480. Birkhäuser Boston, Boston, MA, 1990. With an appendix by J. Oesterlé.
164. I. Vaisman. *Lectures on the geometry of Poisson manifolds*, volume 118 of *Progress in Mathematics.* Birkhäuser Verlag, Basel, 1994.
165. P. van Moerbeke and D. Mumford. The spectrum of difference operators and algebraic curves. *Acta Math.*, 143(1-2):93–154, 1979.
166. P. Vanhaecke. Linearising two-dimensional integrable systems and the construction of action-angle variables. *Math. Z.*, 211(2):265–313, 1992.
167. P. Vanhaecke. A special case of the Garnier system, $(1,4)$-polarized abelian surfaces and their moduli. *Compositio Math.*, 92(2):157–203, 1994.
168. P. Vanhaecke. Integrable systems and symmetric products of curves. *Math. Z.*, 227(1):93–127, 1998.
169. P. Vanhaecke. *Integrable systems in the realm of algebraic geometry*, volume 1638 of *Lecture Notes in Mathematics.* Springer-Verlag, Berlin, second edition, 2001.

170. J.-L. Verdier. Algèbres de Lie, systèmes hamiltoniens, courbes algébriques (d'après M. Adler et P. van Moerbeke). In *Bourbaki Seminar, Vol. 1980/81*, volume 901 of *Lecture Notes in Math.*, pages 85–94. Springer, Berlin, 1981.
171. A. P. Veselov. Conditions for the integrability of Euler equations on so(4). *Dokl. Akad. Nauk SSSR*, 270(6):1298–1300, 1983.
172. A. Weil. *Variétés abéliennes et courbes algébriques.* Actualités Sci. Ind., no. 1064 = Publ. Inst. Math. Univ. Strasbourg 8 (1946). Hermann & Cie., Paris, 1948.
173. A. Weinstein. The local structure of Poisson manifolds. *J. Differential Geom.*, 18(3):523–557, 1983.
174. E. T. Whittaker. *A treatise on the analytical dynamics of particles and rigid bodies.* Cambridge Mathematical Library. Cambridge University Press, Cambridge, 1988. With an introduction to the problem of three bodies, Reprint of the 1937 edition, With a foreword by W. McCrea.
175. H. Yoshida. Integrability of generalized Toda lattice systems and singularities in the complex T-plane. In *Nonlinear integrable systems—classical theory and quantum theory (Kyoto, 1981)*, pages 273–289. World Sci. Publishing, Singapore, 1983.
176. H. Yoshida. Necessary condition for the existence of algebraic first integrals. I. Kowalevski's exponents. *Celestial Mech.*, 31(4):363–379, 1983.
177. S. L. Ziglin. Bifurcation of solutions and the nonexistence of first integrals in Hamiltonian mechanics. I. *Funktsional. Anal. i Prilozhen.*, 16(3):30–41, 96, 1982.
178. S. L. Ziglin. Bifurcation of solutions and the nonexistence of first integrals in Hamiltonian mechanics. II. *Funktsional. Anal. i Prilozhen.*, 17(1):8–23, 1983.
179. N. T. Zung. Symplectic topology of integrable Hamiltonian systems. I. Arnold-Liouville with singularities. *Compositio Math.*, 101(2):179–215, 1996.

Index

Printing: Strauss GmbH, Mörlenbach
Binding: Schäffer, Grünstadt